W9-CMI-619

From Topology to Computation: Proceedings of the Smalefest

M.W. Hirsch J.E. Marsden M. Shub
Editors

From Topology to Computation: Proceedings of the Smalefest

With 76 Illustrations

Springer-Verlag
New York Berlin Heidelberg London Paris
Tokyo Hong Kong Barcelona Budapest

Morris W. Hirsch
Department of Mathematics
University of California
Berkeley, CA 94720 USA

Michael Shub
IBM Research
T.J. Watson Research Center
Yorktown Heights, NY 10598-0218 USA

Jerrold E. Marsden
Department of Mathematics
University of California
Berkeley, CA 94720 USA

Library of Congress Cataloging-in-Publication Data
Smalefest (1990: Berkeley, Calif.)
From topology to computation: proceedings of the Smalefest / M.W. Hirsch, J.E. Marsden & M. Shub, editors.
p. cm.
Papers presented at an international research conference in honor of Stephen Smale's 60th birthday, held Aug. 5–9, 1990 in Berkeley.
Includes bibliographical references.
ISBN 0-387-97932-8
1. Differential topology—Congresses. 2. Numerical calculations—Congresses. 3. Differentiable dynamical systems—Congresses. I. Hirsch, Morris W., 1933– . II. Marsden, Jerrold E. III. Shub, Michael, 1943– . IV. Smale, Stephen, 1930– . V. Title.
QA613.6.S63 1990
514′.72—dc20 92-31991

Printed on acid-free paper.

Production managed by Ellen Seham; manufacturing supervised by Vincent Scelta.
Typeset by Asco Trade Typesetting Ltd., Hong Kong.
Printed and bound by Edwards Brothers Inc., Ann Arbor, MI.
Printed in the United States of America.

9 8 7 6 5 4 3 2 1

ISBN 0-387-97932-8 Springer-Verlag New York Berlin Heidelberg
ISBN 3-540-97932-8 Springer-Verlag Berlin Heidelberg New York

G. Paul Bishop, Jr., Berkeley, CA, photographer

Preface

An extraordinary mathematical conference was held 5–9 August 1990 at the University of California at Berkeley:

From Topology to Computation:
Unity and Diversity in the Mathematical Sciences
An International Research Conference
in Honor of Stephen Smale's 60th Birthday

The topics of the conference were some of the fields in which Smale has worked:

- Differential Topology
- Mathematical Economics
- Dynamical Systems
- Theory of Computation
- Nonlinear Functional Analysis
- Physical and Biological Applications

This book comprises the proceedings of that conference.

The goal of the conference was to gather in a single meeting mathematicians working in the many fields to which Smale has made lasting contributions. The theme "Unity and Diversity" is enlarged upon in the section entitled "Research Themes and Conference Schedule." The organizers hoped that illuminating connections between seemingly separate mathematical subjects would emerge from the conference. Since such connections are not easily made in formal mathematical papers, the conference included discussions after each of the historical reviews of Smale's work in different fields. In addition, there was a final panel discussion at the end of the conference.

These discussions, with many contributions from the audience, turned out to be extremely useful in bringing out insights and patterns in Smale's work, and in bringing to light informative incidents and facts—some from decades ago—that are too often lost to the history of mathematics. We considered these discussions to be so valuable that we persuaded Springer-Verlag to have them transcribed from the rather crude videotapes of the conference. The transcription was done with great patience and precision by Esther Zack. Some of the speakers had the opportunity to edit their transcriptions or submit manuscripts of their remarks; for mistakes in the other transcriptions, the editors are responsible. Whereas these discussions are not all of the utmost clarity, they present remarkable commentaries on important parts of contemporary mathematics by some of the people who developed it.

The main part of this volume consists of research papers contributed by speakers, colleagues, and students of Smale. All of these were refereed under the supervision of the editors. Also included are informal talks about Smale's life and career which were given at social events, and autobiographical writings by Smale, some of which are published here for the first time.

No discussion of Smale's work would be complete without reference to his political activities against oppression, some of which are discussed in his autobiographical articles. An interesting evening panel on Smale's politics was chaired by Serge Lang, but unfortunately we were not able to make a transcript.

The choice of topics and speakers inevitably reflects the interests of the organizing committee. Whereas we had intended to cover all aspects of Smale's mathematics, as it turned out some very interesting parts of his work were neglected: There were unfortunately no talks or contributed papers about mathematical biology, game theory, linear programming, or smooth solutions to partial differential equations. But to do justice to the full breadth of Smale's research would have required more than five days.

The organizing committee consisted of Gerard Debreu, Morris Hirsch, Nancy Kopell, Jerrold Marsden, Jacob Palis, Michael Shub, Anthony Tromba, and Alan Weinstein—all colleagues or former students of Smale.

The conference was generously supported by

IBM Research
Mathematical Sciences Research Institute
National Science Foundation
Springer-Verlag
University of California at Berkeley

Contents

CONTRIBUTED PAPERS

Part 6: Theory of Computation

DISCUSSIONS

CONTRIBUTED PAPERS

Part 7: Nonlinear Functional Analysis

Part 8: Applications

Part 9: Final Panel

Contributors

R. Abraham
Department of Mathematics
University of California
Santa Cruz, CA

M. Barge
Department of Mathematics
Montana State University
Bozeman, MT

S. Batterson
Department of Mathematics
Emory University
Atlanta, GA

L. Blum
International Computer Science Institute
Berkeley, CA

R. Bott
Department of Mathematics
Harvard University
Cambridge, MA

G. Chichilnisky
Department of Economics
Columbia University
New York, NY

G. Debreu
Economics Department
University of California
Berkeley, CA

J. Demmel
Mathematics Department and Computer Science Division
University of California
Berkeley, CA

A.T. Fomenko
Department of Geometry and Topology
Moscow State University
Moscow, Russia

J. Franks
Department of Mathematics
Northwestern University
Evanston, IL

F. Gao
Department of Computer Science
University of British Columbia
Vancouver, Canada

J. Guckenheimer
Department of Mathematics
Cornell University
Ithaca, NY

M.D. Hirsch
Department of Mathematics and Computer Science

Emory University
Atlanta, GA

M.W. Hirsch
Department of Mathematics
University of California
Berkeley, CA

P. Holmes
Department of Theoretical and Applied Mechanics and Mathematics
Cornell University
Ithaca, NY

M. Jakobson
Department of Mathematics
University of Maryland
College Park, MD

S. Johnson
Department of Mathematics
Williams College
Williamstown, MA

N. Kopell
Department of Mathematics
Boston University
Boston, MA

E. Kostlan
Department of Mathematics
Kapi Olani Community College
Honolulu, HI

I.A.K. Kupka
Department of Mathematics
University of Toronto
Ontario, Canada

J.E. Marsden
Department of Mathematics
University of California
Berkeley, CA

N. Megiddo
Almaden, IBM Research Center
San Jose, CA

J. Palis
Instituto de Matematica Pura e Aplicada
Rio de Janeiro, Brazil

J. Palmore
Department of Mathematics
University of Illinois at Urbana Champaign
Urbana, IL

M.M. Peixoto
Instituto de Matematica Pura e Aplicada
Rio de Janeiro, Brazil

T. Ratiu
Department of Mathematics
University of California
Santa Cruz, CA

J. Renegar
School of Operations Research
Cornell University
Ithaca, NY

M. Shub
IBM T.J. Watson Research Center
Yorktown Heights, NY

S. Smale
Department of Mathematics
University of California
Berkeley, CA

D. Spring
Department of Mathematics
Glendon College
York University
Toronto
Ontario, Canada

R. Thom
Institut des Hautes
Etudes Scientifiques
Bures-sur-Yvette, France

P. Tomter
Matematisk Institutt
Universitetet i Oslo
Oslo, Norway

J.F. Traub
Department of Computer Science
Columbia University
New York, NY

A.J. Tromba
Mathematisches Institut
Munich, Germany

V.A. Vassil'ev
Institute of Systems Studies
Academy of Science
Moscow, Russia

X. Wang
Department of Mathematics
Hangzhou University
Hangzhou, China

R.F. Williams
Department of Mathematics
University of Texas
Austin, TX

X. Xuan
Research and Development
Hewlett Packard Company
Santa Rosa, CA

L.-S. Young
Department of Mathematics
University of Arizana
Tucson, AZ

E.C. Zeeman
Hertford College
Oxford University
Oxford, England

Research Themes

Introduction

Many mathematicians have contributed powerful theorems in several fields. Smale is one of the very few whose work has opened up to research vast areas that were formerly inaccessible. From his early papers in differential topology to his current work in theory of computation, he has inspired and led the development of several fields of research: topology of nonlinear function spaces; structure of manifolds; structural stability and chaos in dynamical systems; applications of dynamical systems to mathematical biology, economics, electrical circuits; Hamiltonian mechanics; nonlinear functional analysis; complexity of real-variable computations. This rich and diverse body of work is outlined in the following subsection.

There are deep connections between Smale's work in apparently disparate fields, stemming from his unusual ability to use creatively ideas from one subject in other, seemingly distant areas. Thus, he used the homotopy theory of fibrations to study immersions of manifolds, and also the classification of differentiable structures. In another area, he applied handle body decompositions of manifolds to structural stability of dynamical systems. Smale applied differential geometry and topology to the analysis of electrical circuits, and to several areas of classical mechanics. He showed how qualitative dynamical systems theory provides a natural framework for investigating complex phenomena in biology. A recent example is his application of algebraic topology to complexity of computation. In each case his innovative approach quickly became a standard research method. His ideas have been further developed by his more than 30 doctoral students, many of whom are now leading researchers in the fields he has pioneered.

This conference brought together mathematicians who are currently making important contributions to these fields. It had two purposes: First, to present recent developments in these fields; and second, to explore the connections between them. This was best done by examining the several areas of Smale's research in a single conference which crosses the traditional boundary lines between mathematical subjects. In this way, a stimulating environment encouraged a fruitful exchange of ideas between mathematicians working in topics that are formally separate, but which, as Smale's work demonstrates, have strong intellectual connections.

Through this conference proceedings we hope that important new insights may be achieved into the extraordinary diversity and unity of mathematics.

Topics

Differential Topology

Smale's first work in differential topology, on the classification of immersions of spheres, led to the general classification of immersions of manifolds. But it also presented, for the first time, the use of fibrations of function spaces in what is now called geometric topology. Through fibrations, the powerful tools of algebraic topology were applied in new ways to a host of geometrical problems. This became, in the hands of Smale and many others, a standard approach to many areas: embeddings, diffeomorphisms, differential structures, piecewise linear theory, submersions, and other fields. The classification of differential structures on topological manifolds due to R. Kirby and L. Siebernmann, and many of the profound geometrical theories of M. Gromov, are based on Smale's technique of function space fibrations.

In 1960, Smale startled the mathematical world with his proofs of the Generalized Poincaré Conjecture and what is now called the H-Cobordism Theorem. Up to that time, the topological classification of manifolds was stuck at dimension three. John Milnor's exciting discovery in 1956 of exotic differential structures on the 7-sphere had pointed to the need for a theory of differential structures, but beyond his examples nothing was known about sufficient conditions for diffeomorphism. Smale had the audacity to attack the problem in dimensions five and above. His results opened the floodgates of research in geometric topology. His techniques of handle cancellation and his constructive use of Morse theory proved enormously fruitful in a host of problems and have become standard approaches to the structural analysis of manifolds. Michael Freedman's recent topological classification of 4-dimensional manifolds is a far-reaching generalization of Smale's handle-canceling methods. It is closely related to exciting developments in Yang–Mills theory by Donaldson, Uhlenbeck, Taubes, and others. This work involves other areas of nonlinear functional analysis and mechanics that will be discussed below.

Dynamical Systems

In the early 1960s, Smale embarked on the study of dynamical systems. Like topology, this subject was founded by Poincaré, who called it the qualitative theory of differential equations. Intensively developed by G.D. Birkhoff, by 1960 it seemed played out as a source of new ideas. At this point, Smale introduced a new approach, based on geometrical assumptions about the dynamical process, rather than the standard method of examining specific equations coming from physics and engineering. The key notion was a hyperbolic structure for the nonwandering set, a far-reaching generalization of the standard notion of hyperbolic fixed point. Under this hypothesis, Smale proved that the nonwandering set (of points that are recurrent in a certain sense) breaks up into a finite number of compact invariant sets in a unique way; these he called basic sets. Each basic set was either a single periodic orbit or contained infinitely many periodic orbits that were tangled in a way that today would be called "chaotic." Moreover, he proved the dynamics in a basic set to be structurally stable.

These new ideas led to a host of conjectures, proofs, examples, and counterexamples by Smale, his many students, and collaborators. Above all, they led to new ways of looking at dynamical systems. These led to precise constructions and rigorous proofs for phenomena that, until then, were only vaguely describable, or only known in very special cases.

For example, Smale's famous Horseshoe is an easily described transformation of the two-dimensional sphere that he proved to be both chaotic (in a precise sense) and structurally stable, and completely describable in combinatorial terms. Moreover, this construction and analysis generalized to all manifolds of all dimensions. But it was more than merely an artificial class of examples, for Smale showed that any system satisfying a simple hypothesis going back to Poincaré (existence of a transverse homoclinic orbit) must have a horseshoe system embedded in it. Such a system is, therefore, not only chaotic, but the chaos is stable in the sense that it cannot be eliminated by arbitrarily small perturbations. In this way, many standard models of natural dynamical processes have been proved to be chaotic.

Smale's new dynamical ideas were quickly applied, by himself and many others, to a variety of dynamical systems in many branches of science.

Nonlinear Functional Analysis

Smale has made fundamental contributions to nonlinear analysis. His application (with R.S. Palais) of Morse's critical point theory to infinite-dimensional Hilbert space has been extensively used for nonlinear problems in both ordinary and partial differential equations. The "Palais–Smale" condition, proving the existence of a critical point for many variational problems, has been used to prove the existence of many periodic solutions for nonlinear Hamiltonian systems. Another application has been to prove the existence of minimal spheres and other surfaces in Riemannian manifolds.

Smale also was a pioneer in the development of the theory of manifolds of maps. The well-known notes of his lectures by Abraham and the related work of Eells has undergone active development ever since. For example, manifolds of maps were used by Arnol'd, Ebin, Marsden, and others in their work on the Lagrangian representation of ideal incompressible fluids, in which the basic configuration space is the group of volume-preserving diffeomorphisms, and for which the Poisson reduced equations are the standard Euler equations of fluid mechanics.

In 1965, Smale proved a generalization of the famous Morse–Sard theorem on the existence of regular values to a wide class of nonlinear mappings in infinite-dimensional Banach spaces. This permitted the use of transversality methods, so useful in finite-dimensional dynamics and topology, for many questions in infinite-dimensional dynamics. An important example is A. Tromba's proof that, generically (in a precise sense), a given simple closed curve in space bounds only a finite number of minimal surfaces of the topological type of the disk. A similar result was proved by Foias and Temam for stationary solutions to the Navier–Stokes equations.

Physical and Biological Applications

Smale's first papers in mechanics are the famous ones on "Topology and Mechanics." These papers appeared in 1970 around the beginning of the geometric formulation of mechanics and its applications, when Mackey's book on the foundations of quantum mechanics and Abraham's book on the foundations of mechanics had just come out. Smale's work centered on the use of topological ideas, principally on the use of Morse theory and bifurcation theory to obtain new results in mechanics. Probably the best-known result in this work concerns relative equilibria in the planar n-body problem, which he obtained by exploiting the topological structure of the level sets of conserved quantities and the reduced phase space, so that Morse theory gave interesting results. For example, he showed that a result of Moulton in 1910, that there are $\frac{1}{2}n!$ collinear relative equilibria, is a consequence of critical point theory. Smale went on to determine the global topology and the bifurcation of the level sets of the conserved angular momentum and the energy for the problem. These papers were a great influence: for example, they led to further work of his former student Palmore on relative equilibria in the planar n-body problem and in vortex dynamics, as well as a number of studies by others on the topology of simple mechanical systems such as the rigid body. This work also was the beginning of the rich symplectic theory of reduction of Hamiltonian systems with symmetry. Smale investigated the case of the tangent bundle with a metric invariant under a group action, which was later generalized and exploited by Marsden, Weinstein, Guillemin, Sternberg, and others for a variety of purposes, ranging from fluids and plasmas to representation theory. The international influence of these papers on a worldwide generation of young workers in the now burgeoning area of geometric mechanics was tremendous.

Smale's work on dynamical systems also had a great influence in mechanics. In particular, the Poincaré–Birkhoff–Smale horseshoe construction has led to studies by many authors with great benefit. For example, it was used by Holmes and Marsden to prove that the PDE for a forced beam has chaotic solutions, by Kopell and Howard to find chaotic solutions in reaction diffusion equations, by Kopell, Varaiya, Marsden, and others in circuit theory, by Levi in forced oscillations, and by Wiggins and Leonard to establish connections between dynamical chaos and Lagrangian or particle mixing rates in fluid mechanics. This construction is regarded as a fundamental one in dynamical systems, and it is also one that is finding the most applications.

In 1972, Smale published his paper on the foundations of electric circuit theory. This paper, highly influenced by an interaction between Smale, Desoer, and Oster, examines the dynamical system defined by the equations for an electric circuit, and gives a study of the invariant sets defined by Kirchhoff's laws and the dynamical systems on these sets. Smale was the first to deal with the implications of the topological complexities that this invariant set might have. In particular, he raised the question of how to deal with the hysteresis or jump phenomena due to singularities in the constraint sets of the form $f(x, dx/dt) = 0$, and he discussed various regularizing devices. This had an influence on the electrical engineering community, such as the 1981 paper of Sastry and Desoer, "The Jump Behavior of Circuits and Systems," which provided the answers to some of the questions raised by Smale's work. (This paper actually originated with Sastry's Masters thesis written in the Department of Mathematics at Berkeley.) This work also motivated the studies of Takens on constrained differential equations.

The best known of Smale's several papers in mathematical biology is the first, in which he constructed an explicit nonlinear example to illustrate the idea of Turing that biological cells can interact via diffusion to create new spatial and/or temporal structure. His deep influence on mathematical biology came less from the papers that he wrote in this field than from the impetus that his pure mathematical work gave to the study of qualitative dynamical systems. Because of the difficulty in measuring all of the relevant variables and the need for clarifying simplification, qualitative dynamical systems provides a natural framework for investigating dynamically complex phenomena in biology. These include "dynamical diseases" (Glass and Mackey), oscillatory phenomena such as neural "central pattern generators" (Ermentrout and Kopell), complexity in ecological equations (May) and immune systems (Perelson), and problems involving spontaneous pattern formation (Howard and Kopell, Murray, Oster).

Another important paper constructed a class of systems of classical competing species equations in $\mathbb{R}^n$ with the property that the simplex Δ^{n-1} spanned by the n coordinate unit vectors is invariant and the trajectories of the large system asymptotic to those of *any* dynamical system in Δ^{n-1}; this demonstrated the possible complexity in systems of competing species. Hirsch has shown that arbitrary systems of competing species decompose into pieces which are virtually identical with Smale's construction. Thus, Smale's example, seemingly a very special case, turns out to be the basic building block for the general case.

Economics

Smale's geometrical approach to dynamics proved fruitful not only in the physical and biological sciences, but also in economic theory. In 1973, he began a series of papers investigating the approach to equilibria in various economic models. In place of the then standard linear methods relying heavily on convexity, he used nonlinear differentiable dynamics with an emphasis on generic behavior. In a sense, this represented a return to an older tradition in mathematical economics, one that relied on calculus rather than algebra, but with intuitive arguments replaced by the rigorous and powerful methods of modern topology and dynamics.

Smale showed that under reasonable assumptions the number of equilibria in a large market economy is generically finite, generalizing work of Debreu. He gave a rigorous treatment of Pareto optimality. His interest in economic processes inspired his work on global Newton algorithms (see the section below). This led to an important paper on price-adjustment processes.

His interest in theoretical economics led Smale to work in the theory of games. In an original approach to the "Prisoner's Dilemma," which is closely related to economic competition, he showed that two players employing certain kinds of reasonable strategies will, in the long run, achieve optimal gains.

Smale's work in economics led to this research in the average stopping time for the simplex algorithm in linear programming, discussed below.

Theory of Computation

Smale's work in economic equlibrium theory led him to consider questions about convergence of algorithms. The economists H. Scarf and C. Eaves had turned Sperner's classical existence proof for the Brouwer fixed-point theorem into a practical computational procedure for approximating a fixed point. Their methods were combinatorial; Smale transformed them into the realm of differentiable dynamics. His "global Newton" method was a simple-looking variant of the classical Newton–Raphson algorithm for solving $f(x) = 0$, where f is a nonlinear transformation of n-space. Unlike the classical Newton's method, which guarantees convergence of the algorithm only if it is started near a solution, Smale proved that, under reasonable assumptions, the global Newton algorithm will converge to a solution for almost every starting point which is sufficiently far from the origin. Because it is easy to find such starting points, this led to algorithms that are guaranteed to converge to a solution with probability one. The global Newton algorithm was the basis for an influential paper on the theory of price adjustment by Smale in mathematical economics.

This geometrical approach to computation was developed further in a series of papers on Newton's method for polynomials, several of which were joint work with M. Shub. These broke new ground by applying to numerical analysis ideas from dynamical systems, differential topology and probability, together with mathematical techniques from many fields: algebraic geometry, geometric measure theory, complex function theory, and differential geometry.

Smale then turned to an algorithm that is of great practical importance, the simplex method for linear programming. It was known that this algorithm usually converged quickly, but that there are pathological examples requiring a number of steps that grows exponentially with the number of variables. Smale asked: What is the average number of steps required for m inequalities in n variables if the coefficients are bounded by 1 in absolute value? He translated this into a geometric problem which he then solved. The surprising answer is that the average number of steps is sublinear in n (or m) if m (or n) is kept fixed.

These new problems, methods, and results led to a great variety of papers by Smale and many others, attacking many questions of computational complexity. Smale always emphasized that he looks at algorithms as mathematicians do, in terms of real numbers, and not as computer scientists do, in terms of a finite number of bits of information. Most recently this led Smale, L. Blum, and M. Shub to a new algebraic approach to the general theory of computability and some surprizing connections with Gödel's theorem.

Smalefest, August 5–9, 1990

Unity and Diversity in the Mathematical Sciences

Sunday, August 5, **DIFFERENTIAL TOPOLOGY AND ECONOMICS**

9:00–9:15	*Dean G. Chew*, Opening Remarks
9:15–10:00	*M. Hirsch*, A Survey of Smale's Work in Differential Topology
10:00–10:45	Discussion, *J. Stallings and A. Haefliger*
11:15–12:15	*M. Freedman*, On Related Examples due to R.H. Bing and Ya. Zeldovich

Afternoon Session Chaired by Y-H. Wan

2:00–3:00	*G. Debreu*, Stephen Smale and the Economic Theory of General Equilibrium
3:00–3:30	Discussion
4:00–5:00	*J. Geanakoplos*, Solving Systems of Simultaneous Equations in Economics
5:00–6:00	*A. MasCollel*, Smale and First Order Conditions in Economics

Monday, August 6, **DYNAMICAL SYSTEMS**

Morning Sesstion Chaired by Jenny Harrison

9:00–9:45	*J. Palis*, A Survey of Smale's Work in Dynamical Systems
9:45–10:30	Discussion, *S. Newhouse and R.F. Williams*
11:00–12:00	*L. Young*, Ergodic Theory of Chaotic Dynamical Systems

Afternoon Session Chaired by Charles Pugh

2:00–3:00	*F. Takens*, Homoclinic Bifurcations
3:30–4:30	*D. Sullivan*, Universal Geometric Structure of Quasi-Periodic Orbits
4:30–5:30	*J. Franks*, Rotation Numbers for Surface Homeomorphisms
6:15–7:30	Reception
7:30	Banquet, *J. Kelley, R.F. Williams, M. Peixoto, C. Zeeman*

Tuesday, August 7, **THEORY OF COMPUTATION**

9:00–9:45 *M. Shub*, A Survey of Smale's Work in Computational Theory
9:45–10:30 Discussion, *J. Traub, S. Batterson, J. Demmel*

11:00–12:00 *L. Blum*, Godel's Incompleteness Theorem and Decidability over a Ring
12:15 Film: Turning the Sphere Inside Out

2:00–3:00 *J. Renegar*, Approximating Solutions for Algebraic Formulae

3:30–4:30 *V.A. Vasil'ev*, Cohomology of Braid Groups and Complexity
4:30–5:30 *C. McMullen*, Braids, Algorithms and Rigidity of Rational Maps

7:00–8:00 Evening Lecture, *S. Lang*, Smale's Political Work

Wednesday, August 8, **NONLINEAR FUNCTIONAL ANALYSIS**

Session Chair, Murray Protter
9:00–9:45 *A.J. Tromba*, A Survey of Smale's Work in Functional Analysis
9:45–10:30 Discussion, *R. Abraham and F. Browder*

11:00–12:00 *K. Uhlenbeck*, The Topology of Moduli Spaces: A 90's Tribute to Smale's 60's

12:30–2:30 Banquet Lunch, *R. Bott and R. Thom*
2:30–3:30 *R.S. Palais*, Critical Point Theory when Condition C is Hiding
3:30–4:30 *P. Rabinowitz*, Variational Methods and Hamiltonian Systems

5:00–6:00 *N. Smale*, Complete Conformally Flat Metrics with Constant Scalar Curvature

Thursday, August 9, **PHYSICAL AND BIOLOGICAL APPLICATIONS**

Morning Session Chaired by R. Abraham
9:00–9:45 *J. Marsden*, Steve Smale and Geometric Mechanics
9:45–10:30 Discussion, *A. Weinstein and T. Ratiu*

11:00–12:00 *N. Kopell*, Dynamical Systems and the Geometry of Singularly Perturbed Equations

Afternoon Session Chaired by C. Desoer
2:00–3:00 *P. Holmes*, From Nonlinear Oscillations to Horseshoes and on to Turbulence—Perhaps

3:30–4:30 *S. Sastry*, Smale's Work in Electrical Circuit Theory
4:30–6:00 *M. Shub and R. Williams* (*Organizers*), Overview and Discussion, *D. Karp, D. Brown, O. Lanford, R. Bott, R. Thom, E.C. Zeeman, M. Peixoto*
6:00 Close of Conference

From Topology to Computation: Proceedings of the Smalefest

Part 1
Autobiographical Material

1
Some Autobiographical Notes*

STEVE SMALE

Part I. Origins and Youth

THE MAELSTROM, October 1, 1924, Vol. I, Number 1

The Fortnightly Publication of the Maelstrom Society of Albion College

Lawrence Albert Smale,† Editor

The Maelstrom is the official organ of the Maelstrom Society. Unfortunately the present membership is limited to one; but it is hoped before the next issue of the Maelstrom appears, such a youthful host of poets, philosophers, artists, disciples of free-love, Bohemians, budding anarchists, communists, nihilists, atheists, agnostics, pacifists, citizens-of-the-world, and any others of the anti-Philistine persuasion attending Albion College, if such there be, will have clamored for admittance, as to constitute a group that will pale the heterogeneity of Mendoza's band of outlaws in Barnard Shaw's celebrated play. Before petitions for membership will be considered however, it will be necessary for conscientious members of the following organizations to fully apostatize:

All Greek-letter fraternities
Ku Klux Klan
All Other 100-Percenters
Kiwanis
Rotary
American Legion
Republican Party
Democratic Party
National League of Women Voters
W.C.T.U.
...

* For the proceedings of the "Smalefest," Berkeley, August 1990.

† Lawrence Albert Smale died in March 1991.

The Maelstrom at various times will attempt to expose the follies that these organizations are guilty of. The editor laments that he has not the pen of a Nietzche for invective cogent enough to hurl against their absurd doctrines, mawkish traditions, and stupid customs.

October 3, 1924 Flint Daily Herald page 1 Headline

FLINT STUDENT EDITOR OF ALBION COLLEGE PAPER
ATTACKS PRESIDENT AND IS PROMPTLY EXPELLED

Within one half hour after the first sale an Albion college campus this forenoon of the "Maelstrom", a self-confessed free-lance sheet, the lone editor, Lawrence Albert Smale, of Flint, a freshman, had been expelled from the college.

The "Maelstrom's" 12 pages were devoted to attacks on President, religion, fraternities of the college, college spirit and the usually accepted conventions of life.

October 6, 1924 Detroit Free Press page 1 Headline

OUSTED ABELION COLLEGE 'FREE THOUGHT' EDITOR JAILED

Like the fabled Frankenstein, which destroyed its man-creator, the "Maelstrom", exotic publication which shocked, horrified and interested Albion College, arose today from its ashes of oblivion to cause the arrest of its editor and sponsor, Lawrence Smale. Smale is charged with selling and distributing obscene literature, in a warrant issued in Marshall.

Pop had a dual life. He worked in a ceramic laboratory at AC Spark Plug, a General Motors plant in Flint, Michigan. It was a modest white-collar job, and he didn't enjoy it. He was also a Marxist. His party, the Proletarian Party, had diverged from the Communist Party in the twenties, considering the latter to be too reformist. Perhaps even more than a Marxist, he was an atheist. He had been expelled from college for publishing his own magazine in which he had asked if God could make a stone so large that he couldn't lift it. Pop also didn't like partiotism. To him, the Boy Scouts of America represented God and Country, the worst aspects of American life.

I remember that one of the few early traumas in my life occurred when I decided to join the Boy Scouts. Pop was gentle and easy-going, but never would allow that to happen. Even though he bought me presents to compensate, I cried for days.

My father certainly has been a big influence in my life. I was 20 years old when I first set foot in a church. That was Notre Dame in Paris. I have kept to this day a healthy skepticism about our country's institutions.

On the other hand, I tended to be more concerned with current problems. The Proletarian Party believed in emphasizing theory and analysis; and Pop followed it in considering some of my activities as reforming the capitalist system. That would only postpone the day of socialism.

My father, mother, younger sister Judy, and I lived in the country 10 miles from Flint. At first we had neither electricity nor plumbing. Judy and I walked a mile to a one-room schoolhouse which we attended until high school. I still marvel at that little school. One teacher, a woman with a year or two of college, taught nine grades of students; each grade had classes of reading, math, history, and so on. In addition, she was librarian, did the janitorial work, the cooking of school lunches, and everything else. Yet, we did obtain a good education.

I never completely adjusted to high school. Coming from a rural area tended to make me feel socially awkward. Many of my interests were outside of school. I became intensely involved in chess, participating in three national chess tournaments. I also studied organic chemistry and built a laboratory of sorts in the loft of a former chicken house.

In high school, my first political conflict took place. The story of the Scopes "Monkey" trial had made a strong impression on me, and I was upset that our high school biology teacher, Joe Jewett, wouldn't teach evolution. He simply skipped that chapter in the text (at least it was better than my grade school, where such books were screened from the library). I circulated a petition among my fellow students to get evolution into the course. But only one classmate signed!

Some years later, my parents sent me a clipping from the "Flint Journal", Nov. 15, 1959, headed:

MATH SHARK IMPRESSED TEACHERS

It quoted Jewett:

... Joseph L. Jewett, who had Smale in one of his biology classes, described him as a "very attentive boy, very much interested in the subject and always asking questions".

Jewett said it wasn't a 'one-way' proposition for Smale because he 'contributed a lot' to the class.

"It wasn't a case of all learning from him," Jewett said. "He gave out some ideas also."

Jewett described Smale as quiet and courteous. He was respected by his classmates, Jewett said.

A little different perspective is given by the characterization of me in the high school yearbook, on my graduation in 1948:

"I agree with no man's opinions—I have some of my own"

Going to college at Ann Arbor opened up a new world. I found many students who were similar to me, made friends (male friends, mostly; it was still several years more before I could relate to women comfortably).

I took part in campus life, and, for example, helped organize a chess club. More important for our story, I gradually became involved in the left-wing political life at the University of Michigan.

At that time the liberal-left political party was the Progressive Party; it ran

Henry Wallace for president in 1948 and Vincent Hallinan in 1952. There was a campus branch called the Young Progressives, or YP. Even though it wasn't very visible, I think that it is fair to say that the Communist Party was the main force on the left. Its youth group, the Labor Youth League, or LYL, held secret meetings, secret membership, but had certain "open" spokesmen.

I went to a communist lead "Youth Festival" in East Berlin in the summer of 1951 and then returned to organize the "Society for Peaceful Alternatives" on campus. I was active in the YP, became a campus leader of the LYL, and, for a short time, was even a member of the Communist Party.

There were two related phenomena going on at the same time in this political movement. On the one hand, there was the vocal activism on such issues as civil rights, the Korean War, nuclear weapons, McCarthyism, and the Rosenberg case (reformism, my father would say, although he did contribute to the Rosenbergs). The left's position and activities on these issues weren't compromised (compared to the liberals') by McCarthyist pressures. That is why I (and the others) was drawn in; I have no regrets for my participation in these programs.

On the other hand, we were involved in studying and teaching a Soviet version of Marxist studies, and eventually recruiting for the LYL and the CP. This aspect of the movement, especially, had much in common with fundamentalist religion. Stalin, before his death, was God, and Soviet policy was gospel. We studied the works of Marx, Lenin, and, of course, Stalin and his followers. A main document we read was the *History of the Communist Party of the U.S.S.R.*, written in Moscow. As Balza Baxter, the LYL state chairman put it, as quoted by the *Michigan Daily*, around the end of 1952:

> The LYL is a youth organization in which Communists participate as equals with non-Communists, and LYL uses a guide to the explanation of things what we call the scientific study of Marxism–Leninism.
>
> I might say that the League has a fraternal relationship with the Communist Party. We do study the Stalin versions of Marx and Lenin. After all, the Soviet Communist Party has had much more experience with Communism than anyone else.

There was also a system of rewards and discipline. The rewards included admission to membership in the LYL and the CP, or even leadership in those organizations. Social ostracism and excpulsion were the ultimate discipline. Discipline was threatened for such things as personal political independence and fraternizing with Trotskyites.

I remember once on a trip to New York City visiting briefly the headquarters of the Socialist Workers Party (followers of Trotsky) with a friend. On my return, I mentioned this visit casually to fellow members of the LYL. They did not take the matter lightly and eventually brought it to the State Chairman, Balza Baxter. He was a relatively easy-going fellow, who I liked a lot. Baxter smoothed it over. "Just youthful intellectual curiosity," he said.

Although there were some strains and some signs of my independence, I had accepted the basic doctrines of the Communist Party and defended them.

Today, when I ask myself how I could have acted so foolishly, it is not difficult to find an answer.

There are strong tendencies for just about everyone to want something basic to believe in without questioning. Most often this takes the form of traditional religion. But even atheists and revolutionaries must rest their beliefs on some foundation. They are not immune to the problem of reasoning in circles.

But still, to accept the Communist Party?

Consider my frame of reference at that time. I was sufficiently skeptical of the country's institutions to the point that I couldn't accept the negative newspaper reports about the Soviet Union. I so believed in the goal of a utopian society that brutal means to achieve it could be justified. I was unsure of myself on social grounds, and the developing social network of leftists around me gave me security. Then, these were the times of McCarthyism, the Rosenberg executions, the Korean War hysteria; the CP was the main group giving unqualified resistance to these forces.

In the spring of 1952, there was an extended campus conflict at the U. of M. over the free speech question, anticipating in a minor way the FSM struggle at Berkeley 12 years later. There were big differences. McCarthyism was strong, HUAC was in its heyday, and old left tactics dominated the protest. It is useful to go into greater depth on this story, as it illuminates the tenor of those times; it also illustrates my early political style.

On March 4, 1952 the front page headline story of the *Michigan Daily* read:

TWO SPEAKERS HALTED BY "U"

Committee Fears Talks 'Subversive'

Two men associated with allegedly subversive organizations were temporarily banned from speaking on campus yesterday in an apparently unprecedented move by the University Lecture Committee.

The speakers, proposed by two student organizations, were denied permission to appear "until sufficient evidence is produced" to satisfy the committee that the speeches would not be subversive.

Permission was withheld from:

1) Abner Greene ...

2) Arthur McPhaul, excecutive secretary of the Michigan Chapter of the Civil Rights Congress, also branded subversive.

Some students weren't accepting the speakers' ban and a mild form of defiance took place. On May 7, 1952, the *Michigan Daily* ran the headline story:

MCPHAUL ADDRESSES 'PRIVATE MEETING' AT UNION

Blasting the House Un-American Activities Committee as "an arm of the millionaire forces of Wall Street", Arthur McPhaul, banned Monday from speaking on campus by the University, put in a colorful appearance at a private Union dinner last night under mysterious sponsorship ...

Next in the *Michigan Daily*:

PROBE BEGUN OF MCPHAUL APPEARANCE

A University investigation was launched yesterday into mysterious circumstances surrounding the appearance of banned speaker Arthur McPhaul at a private Union dinner Thursday night.

and then:

CHARGE STUDENTS VIOLATED 'U' RULE IN MCPHAUL TALK

Charges of violation of a University student conduct by-law have been made before the Joint Judiciary Council against all of the students known to have attended the McPhaul dinner March 6 at the Union, it was learned yesterday ...

The By-law allegedly violated states: No permission for the use of University property for meetings or lectures shall be granted to any student organization not recognized by University authorities, nor shall such permission be granted to any individual student.

But as we defendents were to say in our full page ad:

We could not possibly have violated the regulation ...

As it is now written, this regulation is a restriction upon the discretion of University authorities to grant the use of University property. Since no student is in any position to grant permission for the use of University property, he cannot possibly violate this regulation.

My reaction to some of these events is contained in the letter I wrote to my parents at that time, reproduced here without change:

Dear Mom and Pop,

Just a word to let you know more details of the disciplinary aciton against us. There is some question of expulsion as indicated in personal feelings of some people in the newspapers. I, however, feel that the "officials" date not take such a step, although they would like to. The Joint Studies Jucic met Sat. morn., Sat. afternoon, Sun. morning and Sun. afternoon (today). So far it has been impossible (no time yet) to get the defense of the accused much publicity. Nevertheless the support of the student body is remarkable. For example, we wanted to put 1/2 page ad in the daily with a long statement of our case ... Cost $80. Well Fred and a friend of his and mine went around to the J.C. Students here and collected $80. Maybe more later. In just a few hours we have collected $70. It is really wonderful to find the students standing by us. All the details of the case you can find in the clippings.

As yet I am the only student that refused to tell whether I was at the dinner.

I had a most pleasant time today. We went petitioning to put the Progressive Party on the ballot in Willow Run. 3000 homes worse than tenements without a single tree. A Negro girl from Detroit and I went throuigh the Negro district there. There was a wonderful response. The residents never had a good word for the Democrats or Republicans. 80% signed that were registered. I found much more courtesy than in the corresponding white district.

Sat. night we had a large party—3 cars from Detroit.

Well all for now,
love-Steve

The investigation, as well as the number of news stories and editorials kept espanding. Then, finally, May 4, headline story of the *Michigan Daily*:

FIVE STUDENTS ON PROBATION

The smoke-screen lifted over the controversial McPhaul dinner investigation yesterday as the Joint Judiciary Council and the Sub-committee on Discipline cleared fifteen students of breaking a Regents' by-law, but put five of them on probation for their conduct before the council.

Eight confusing weeks of deliberations involving four University bodies and hundreds of pages of testimony reached a climax as students received official notification of their acquital or punishment from University officials yesterday morning.

Five of them were put on probation for 'failure to give the Judiciary the cooperation students should reasonably be expected to give a student disciplinary body'. All will be forced to drop out of extra-curricular activities where regular eligibility is required. This consists of elective offices or other positions where the student represents his group or the University, according to an Office of Student Affairs ruling. The penalty will last until Jan. 31, 1953. Those disciplined are:
VALERIE M. COWEN ....
STEVEN SMALE, Grad., 21 years old who will be required to drop out as secretary-treasurer of the Chess Club and treasurer of Society for Peaceful Alternatives.
Shaffer, Sharpe, Smale and Luce all graduated with honors and Luce is a Phi Beta Kappa.

We appealed without success.

In June, 1953, the department chairman, T. Hildebrandt, called me into his office. He told me that unless my mathematics grades improved, I wouldn't be allowed to continue in graduate school. This had some effect on me as I already was moving away from the political arena. A couple of years earlier I had failed a physics couse, and then, as a senior, changed my major from physics to mathematics. It was mainly out of momentum that I continued in mathematics into graduate school at the University of Michigan. But now I was 23 years old and beginning to wonder where I was headed.

In the fall of 1953, the combination of concern about my future, Hildebrandt's warning, and a great mathematics teacher, Raoul Bott, converted me into a very serious student of mathematics. It turned out that this mathematics phase was to last until the FSM in the fall of 1964.

I was lucky to meet Clara Davis in the fall of 1954, and we were married at the beginning of 1955. This relationship was important for me in many ways. In particular, it was helpful in my moving away from a rigid political ideology and solidifying my work in mathematics.

I received a Teaching Assistantship and started teaching in the mathematics department in the fall of 1954. After five class meetings, Hildebrandt called me in again and told me that I was fired because of my previous leftist activities. He put the responsibility on the University administration. On the other hand, Hildebrandt had found a research contract that would continue my pay. I didn't protest publicly since the money was coming anyway, and I was more interested in doing mathematics. However, the next fall, the pay stopped too. Fortunately, Clara obtained a job as a librarian in Dearborn, and her wages supported us during the last year of my graduate work.

It was not surprising that my political past continued to haunt me as I sought a job. For example, I recall a friendly professor (Ray Wilder) telling me that I should discontinue letters of reference from Chairman Hildebrandt who was warning potential employers that I was a leftist.

In the fall of 1956, we moved to Chicago where I had accepted my first teaching job. It was not in a regular mathematics department, but at the "College" of the University of Chicago. I just remember a couple of political notes from that time. One was that Soviet military intervention in Hungary helped reinforce my evolving alienation from Communism. Also, one morning I left my Chicago apartment to find two FBI men waiting for me. They wished to discuss the political events at Ann Arbor; but they accepted my quick refusal to do so.

Nat was born in 1957, Laura in 1959. The mathematics went well. After two years in Chicago, I held successive positions in the Institute for Advanced Study (Princeton, N.J.), Instituto de Matematica, Pura ed Aplicada (Rio de Janeiro), the University of California (Berkeley), and Columbia University (New York).

Clara, Nat, Laura, and I were living in New York in October of 1962 when we heard the first news that the Soviets were putting nuclear missiles in Cuba. I reacted sharply to the growing threat that atomic war could start any day. I became intensely angry at Kennedy, being aware that the United States had already missiles located on the Soviet border in Turkey. When I became convinced that Soviet missiles were en route to Cuba, I became angry at Kruschev as well. It didn't make sense to die in a nuclear war due to the insane militarism of the two countries.

So, Clara and I with Nat and Laura packed a few of our belongings and started driving to Mexico!

There were a few faculty at Columbia who knew what we were doing and they were helpful. I still have a debt of gratitude to Ralph Abraham and Serge Lang who saw that my class kept running smoothly. My father and mother were visiting us in New York at the time and were also sympathetic and cooperative. The long drive to Mexico helped to ease my tension; also, the tension of the superpower confrontation eased too, as we passed into Mexico.

We were in touch by phone with my friends at Columbia, and as the missile crisis was fading, I learned that my departure from Columbia was by no means irrevocable. Thus, I flew back to New York, taking up my class again. Clara drove our car back with the kids, arriving a few days later. Few people were aware of what had happened.

Were we crazy or were we acting quite intelligently? Even today it is hard for me to analyze our decision to leave the country as atomic war seemed imminent. Clara, at the time, put the question on our minds this way: At what time should a Jew have decided to leave Germany?

We moved to Berkeley from New York in the summer of 1964. The events of the Free Speech Movement, or FSM, erupted soon after the start of classes in the fall. The huge scale, audacity, drama and sincerity of that movement captivated me. The New Left" politics were refreshing.

Friends of ours, Mike Shub, Kathy, and David Frank were students at Columbia University who moved out to Berkeley at the same time we did; they drove our car for us when we came by plane. All three, together with Mike's wife Beth Pessen, were arrested in the famous FSM sit-in. I had visited them at Sproul Hall just before their arrest, and afterwards Clara and I helped get them out of jail. Eventually, Mike was to write a thesis with me and he remains one of my closest friends.

I didn't become part of the FSM itself, but supported the movement through activity in faculty groups. I was among that very small minority of faculty which supported the FSM, both goals and tactics, without qualifications and with enthusiasm.

After the big arrest and further struggle, the faculty senate voted to support the demands of the FSM: The FSM victory became clear. The whole fall experience was one of great exhilaration for me.

During that activity I came to know some of the FSM leaders, including Mario Savio and Steve Weissman. I admired both Savio and Weissman. They were brilliant strategists and articulate speakers. Savio's moving speeches on the steps of Sproul Hall were crucial to the FSM success. Weissman played a central role in obtaining faculty support. I could never have guessed that within a year Weissman and I would be on opposite sides for the crucial decision.

Part II. Vietnam Day

May 23, 1965 The New York Times

33-HOUR TEACH-IN ATTRACTS 10,000

MANY CAMP OUT FOR NIGHT AT BERKELEY VIETNAM DEBATE

At 12:55 this morning a bleary-eyed, bearded young man whose gray sweatshirt bore the inscription, "Let's Make Love, Not War", stretched out on the grass outside the Student Union and went to sleep in the darkness.

Nearaby, a girl with straight black hair and bare feet wrapped herself in an Army blanket to ward off the night-time chill and plaintively asked her esort, "Don't they ever run out of things to say?"

Orators addressing the 33-hour teach-in demonstration on the University of California's Berkeley campus this weekend never did run out of things to say, or of students to say them to.

The demonstration, billed as a protest against Government policies in Vietnam, attracted a peak crowd of nearly 10,000 to hear such speakers as Senator Ernest Gruening, Democrat of Alaska, and Dr. Benjamin Spock, the specialist in child care.

Even as the speeches droned on past midnight into the chill morning hours, as many as 4,000 students camped out on the field in sleeping bags and blankets to listen. Scores of stray dogs raced yelping in and around the sprawled bodies hunting for scraps of food in the darkness.

... Show-business trappings were in evidence at Berkeley this weekend. A score of folk singers strummed out tunes about civil rights and the draft. A satiric group called The Committee performed a short play about an automated sit-in.

One group of militants collected pints of blood for the Dominican rebels while another collected signatures from students promising that they would refuse to answer a draft call.

Tables were set up around the field selling everything from "The Young Socialist" to "The Soul Book"; one leaflet urged students to 'join the revolutionary organization of your choice', arguing that any revolution would be better than the status quo.

If the Berkeley teach-in had its denigrators on the faculty, it also found its champions. In opening the teach-in, David Krech, professor of psychology, said the demonstration represented 'the finest hour of this great university'.

Among those who accepted invitations were Dick Gregory, the comedian; Kenneth Rexroth, the poet; Norman Mailer, the novelist; Norman Thomas, and Ruben Brache of the Dominican rebels.

A majority of the audience clearly agreed with the critics of the Administration. Senator Gruening drew a standing ovation when he demanded that the President seek an immediate negotiated settlement in Vietnam on any terms available.

"The white man cannot settle the internal problems of Asia", he said.

Johnson extended the war against Vietnam in February 1965 by bombing the North. I followed the stories from Vietnam with growing apprehension, only too conscious of the danger of atomic war. For more than a decade, life had gone especially well. I felt a great satisfaction with my work, my family, my job. My main insecurity came from the knowledge that all this would likely be destroyed if the U.S. and U.S.S.R. went to war.

This knowledge affected my actions.

Three years earlier during the Cuban missile crisis, I abandoned (temporarily, it turned out) an enviable position on the Columbia University faculty to take my family to Mexico. Now I reacted differently. I became increasingly committed to political action to try to stop Johnson.

Why was I so presumptuous as to think that anything I could do would have a significant effect? There were several reasons. First, I had an activist background. In my student days, for example, I had organized a "Society for Peaceful Alternatives." Secondly, I had just seen and been part of the unfolding of the Free Speech Movement. The power generated by this force was already enormous. Finally, I had the time, security, and prestige that came from a tenured professorship.

My involvement in the Vietnam protest came slowly at first. A teach-in on the Berkeley campus was organized by some faculty to support the first teach-in at Ann Arbor in March of 1965. I said a few words there. Then I represented the local faculty AFT (The American Federation of Teachers) union as chairman of "The Political Affairs Committee" in the organization of an anti-war march in San Francisco in April 1965.

Although the march was called to support an SDS (Students for Democratic Society) march in Washington, D.C., it was pretty much dominated by the old left. Every poster or slogan had to be approved in advance by the organizing committee. The march turned out to be moderately large, but lacked drama and media coverage. The more radical groups had split off to have their own march and it was quite small.

It must have been at one of those organizational meetings that Jerry Rubin had noticed me.

The war kept escalating. So, when Barebara Gullahorn and her boyfriend, Jerry Rubin, came by my office toward the end of April 1965 to ask if I would help organize a town meeting on Vietnam, I was ready. That was before Jerry became famous, and I didn't recognize either him or Barbara. But I liked their spirit and approach. I supported them in bringing together one or two dozen anti-war activists of varied backgrounds. Our first meetings were in Jerry's tiny apartment on Telegraph Avenue near the campus. We decided early on the form of the protest. It was to be a combination "teach-in," community meeting, and educational protest, to be held at the University of May 21. It would be called Vietnam Day.

Leaflets were distributed to get help and funds. Speakers were invited and some accepted, the organizational meetings became larger, and the major media picked up the story. Support for our effort grew.

I obtained the early endorsement of the faculty AFT. This was important to legitimize our activity and to obtain campus space.

I tried, with some success, to enlist the help of people closest to me.

Clara was involved in "Women for Peace" then and we began to work together on aspects of Vietnam Day. She designed some of the leaflets and I would take them to the printer.

Moe Hirsch and I had been close friends for almost 10 years. We worked in the same field of mathematics, attended conferences together, and were now colleagues at Berkeley. Moreover, Moe and I had a lot in common when it came to politics. We had participated in the FSM together. Now he joined the organization of the big teach-in with enthusiasm. Moe was a good speaker and writer, and he spoke for us on Sproul Hall steps and helped draft some of the key documents we would need, both before and after Vietnam Day.

Another mathematics teacher in my department, John Lewis, joined our efforts. Mike Shub was a graduate student in mathematics. Recall that I had known him from Columbia and that he had been arrested in the FSM. Mike worked with us especially on questions of liaison with the campus community. Among many others, I remember especiany an older Socialist Workers Party Member, Paul Montauk, who worked as a chef in San Francisco.

Paul was especially constructive and his many connections were helpful to us.

And, of course, there was Jerry. I liked him a lot, and we became good friends. Jerry was there all the time with his great enthusiasm, sharp insights about dealing with the media, and, especially, his "do it" philosophy.

We didn't spend much time on analysis and theory. He knew what we wanted and had some good instincts about how to obtain it. Therefore, our mode was one of continually doing things, all kinds of things, which would make Vietnam Day into a bigger and sharper anti-war protest. Jerry and I did, however, talk for many hours about how to deal with various problems that came up.

At some point the de facto organization that was preparing for the teach-in was formalized. It was called the Vietnam Day Committee or VDC, and Jerry and I were chosen co-chairmen.

It was a lot of fun to work in the Vietnam Day Committee. We didn't think in terms of duty. It was more like an exiting creative challenge: How to make something that would be the greatest teach-in, the biggest Vietnam war protest. How to make Johnson cringe.

For example, we would sit around in Jerry's apartment making suggestions for speakers. It was like a competition to see who could propose the biggest, most provocative names. There was practically no limit to our ambitions; Bertrand Russell, Fidel Castro, the ex-President of the Dominican Republic, Norman Mailer, U.S. Senators, Jean-Paul Sartre, and so on. Then we would call them up then and there, or send telegrams. And sometimes it worked and the invitations were accepted.

But there was another important aspect to sending out these invitations. It helped gain the attention of the media, as the following example shows.

At that time, the U.S. invasion of the Dominican Republic was a big news story. It was only natural that the *Berkeley Gazette* would put at the top of its front page, May 10:

EX-COMINICAN CHIEF BOSCH INVITED HERE

Juan Bosch, deposed president of the Dominican Republic, has been invited by the sponsors of the May 21 University of California 'teach-in' on Viet Nam to give his view of the island crisis.

A cablegram, signed by Stephen Smale, professor of mathematics, and Jerry Rubin, co-chairmen of the Viet Nam Day being planned for 30 continuous hours May 21 and 22, was sent to Bosch who now lives in Puerto Rico.

Then on May 15 the Gazette carried the story:

REBEL LEADER DUE

Juan Bosch, former political leader of the Dominican Republic, has informed the organizers of the Viet Nam teach-in scheduled for May 21 on the University of California campus, that he is very interested in speaking at the event ...

The appearance of Rubin Brache has been confirmed. France and the Soviet Union have been attempting to seat Brache as the United Nations representative for the rebel government of the Dominican Republic.

This bold style helped build the big teach-in.

Another important example was our invitation to Dean Rusk, the Secretary of State: The *San Francisco Chronicle* wrote on May 2:

UC FACULTY CHALLENGE TO RUSK

Three University of California faculty groups challenged Secretary of State Dean Rusk yesterday to debate this Nation's "disastrous policy in Vietnam" at a "teach-in" demonstration on campus May 21.

The challenge was issued in a telegram sent to Rusk by mathematics professors Morris W. Hirsch and Stephen Smale ...

Then a little later in the *Daily Cal*:

Letters and telegrams continue to pour in from people who are accepting invitations to speak at a massive Community Meeting on Vietnam scheduled May 21 and 22.

The State Department has replied to a telegram sent by Professors Stephen Smale and Morris Hirsch of mathematics. In a phone call to Hirsch a spokesman for the State Department said department officials were thinking of sending a high level official such as Averell Harriman or William Bundy.

Finally, on the front page of the *Berkeley Gazette*:

STATE DEPT. OUT OF TEACH-IN GATHERING HERE

The Viet Nam Day 'Teach-In' on the University of California campus Friday will be carried on without the two representatives of the State Department, leaders of the 'teach-in' announced today.

The U.S. State Department has withdrawn its offer to send two representatives to the event, one of whom was to have been William Bundy.

This marks the second State Department withdrawal from debate on U.S. policies in Viet Nam in the last four days. McGeorge Bundy withdrew from the national radio debate at the last minute on Saturday, just before the event ...

The committee also confirmed the appearance of comedian Dick Gregory at the 'teach-in' ...

Of course, we were busy writing press releases, on the phone to the newspapers, and holding press conferences during this time.

There were other audacious steps we took that kept us on the front pages as Vietnam Day approached. An example from the front page of the *Daily Cal*, May 20, speaks for itself:

REQUEST CLASSES BE STOPPED FOR VIETNAM DAY TOMORROW

The much-heralded Vietnam Day—a solid 30 hours of intense discussion of the war in Vietnam—gets off the ground at noon tomorrow.

Its organizers are requesting that all classes be cancelled beginning at noon tomorrow so that all students may attend.

Acknowledging that 'this is a very unusual request', a telegram sent last night to the Regents and to Acting Chancellor Martin Meyerson declared, "We have become bold enough to make this request because of the extraordinary circumstances that developed to make Vietnam Day an event of international prominence and importance for the University of California."

Besides support from foreign individuals and peace groups, explained Jerry Rubin of the Vietnam Day Committee, the 'extraordinary circumstances' included a sympathy strike scheduled for tomorrow by a national Japanese student union.

The telegram, signed by Rubin and Stephen Smale, professor of mathematics, said 'Professors are already cancelling classes and office hours and many students have already accepted the fact that they will not attend classes during this event'.

The University local of the teaching assistants' union, said the telegram, has asked all TA's not to hold their classes.

"The University encourages maximum discussion of all pressing social issues," replied Chancellor's Special Assistant Neil Smelser when informed of the telegram, "but such discussion should not interfere with regular activities."

"Class schedules and office hours will proceed as normal ..."

The list of speakers, performers, debaters, comedians and singers is staggering—and so is the marathon schedule.

It was crucial for the success of our venture to have a good campus location. Luckily, the specter of the FSM was still haunting the university administration. A year earlier, a Vietnam Day would have been impossible. Now we were not willing to compromise significantly on this issue and the Chancellor's office was accommodating. The story in the *Berkeley Gazette* May 14 describes the cooperation well:

UC CURBS RULES FOR TEACH-IN

University of California rules on campus political activity are being relaxed for the May 21–22 Viet Nam Day teach-in, according to sponsors of the event and Prof. Neil Smelser of the Chancellor's office.

The University will make available the lower Student Union Plaza and the adjacent baseball field and will construct a passageway to the field to accommodate the crowds expected at the event. The speakers' platform also will be raised to make it more visible from all sides.

The 72-hour notification rule applying to off-campus speakers will be reduced to notification on the day of the event ...

Co-chairmen for the event, Prof. Stephen Smale and Jerry Rubin, stated "The University has been very cooperative in making available the areas needed for the teach-in."

Vietnam Day was obtaining support now from a wide spectrum. It was almost unprecedented to see so many different liberal and radical groups who had so often fought each other come together to help organize, endorse, and appear on the program together.

We also had some international help, for example, from the eminent mathematicians Laurent Schwartz, Paris, S. Iyanaga, Tokyo, and Bertrand Russell, England.

We also had opposition.

As the teach-in drew near, two defenders of the Vietnam War, whom we had invited, made a public attack. They were Robert Scalopino and Eugene Burdick, both professors at the University in Berkeley.

The *San Francisco Chronicle*, May 20, wrote:

BURDICK BLAST AT TEACH-IN 'CIRCUS'

This wasn't so bad; we had worked to create a carnival aspect to Vietnam Day. We wanted it to be fun to go to, as well as a big protest. The *Berkeley Gazette* on the eve of the event (May 20) carried a big red front page headline:

UC'S SCALAPINO RIPS TEACH-IN MEETING ON CAMPUS TOMORROW

A further element of tension as the teach-in hour arrived was indicated in the *San Francisco Examiner*, May 21:

Berkeley Police Capt. William N. Beall Jr., head of the patrol division, said he will have all his men on standby alert for possible duty.

This will be the first time such an alert has been ordered since the Sproul Hall sit-ins, Dec. 2–3.

Vietnam Day, May 21, 1965, arrived and we waited anxiously; Jerry, me, Barbara, Moe, and the others who had contributed so much effort. The success of the big teach-in became an end in itself and would help justify our own existence. This success hinged on the number of those who would come. And whether they would enjoy it enough to stay. It was supposed to last 33 hours. Of course, real success couldn't be measured just by the number in the audiences, but that number was the most tangible sign.

I had been quoted in the *San Francisco Examiner* that morning (May 21, p. 18):

Professor Stephen Smale, one of the Vietnam Day Committee members, said hopefully yesterday he expects 'about 10,000 people to attend at the peak hours' of the thing.

Well, a few hours later, we saw it happen. The first report confirmed it. The *Berkeley Gazette* afternoon, May 21 in a red front-page headline:

10,000 ASSEMBLE FOR INAUGURAL SESSION OF 'TEACH-IN' ON CAMPUS

The Vietnam "teach-in" started on the University of California campus at noon today as some 10,000 filled the lower Student Union plaza.

The carnival air prevailed and the scene was gaudy with banners. Any number of food concessions were open. Earlier charges that the program had been turned into an outright protest of U.S. policy instead of an academic discussion appeared to have no effect on the spirit of the crowd.

We also had succeeded in making Vietnam Day an enjoyable event. One of the stories on Vietnam Day in the May 22 *San Francisco Examiner* headlined:

VIETNAM DAY AT U.C.! ALL FUN OF A CARNIVAL

Indeed, the crowds had filled up the available space and they stayed. The newspapers gave widely varying estimates of peak crowds, but 10,000–12,000 seemed to be typical numbers. The *Oakland Tribune* was the newspaper most biased against us. There were three different headings about Vietnam Day in their May 23rd paper:

WOMAN HIT IN EYE NEAR TEACH-IN (front page)

TEACH-IN LIMPS TO FINALE (front page of city news section)

and

VIET POLICY PROTESTS FAR SHORT OF GOALS (same)

Yet, their news story stated:

... Finally, about 1 p.m. an estimated 9,000 persons were mustered to hear and cheer [Norman] Thomas' blustery harangue against United States involvement in Vietnam ...

The peak crowds were one thing, and the total attendance something else, since it was to last a day and a half. As the *San Francisco Chronicle* wrote May 22:

... There was a constant ebb and flow of spectators, but their total never waned below the 5,000 mark despite gusty winds and mostly sun-less sky ...

One could only guess at the total number of participants, and Jerry and I did that in a subsequent VDC letter. We said 50,000 came.

During the first evening of the event, there was a fund raising dinner where Benjamin Spock, Senator Gruening, Norman Thomas, I.F. Stone, and others were present. This generated a substantial amount of money. Other funds were solicited directly from the crowds. I believe, altogether, we obtained quite enough to pay for all the expenses, including travel expenses for most of the speakers. Clara was at a VDC member apartment to help count the cash and recalls how it was piled high on the floor. We had a system of monitors which, besides collecting money, used a system of walkie-talkies for meeting troubles that might have arisen. However, Vietnam Day turned out to be rather peaceful.

The media coverage was heavy. Besides the national press, Bay Area newspapers carried several main front-page stories, and there was some rather sympathetic reporting. A local radio station, KPFA, ran most of the program with a live broadcast.

Vietnam Day lasted the full programmed 33 hours and actually went beyond that for two or three hours more. Activity never stopped, even during the early morning hours, although the crowds dwindled. I myself did go home to get a couple of hours sleep.

The main message of the speakers was "Get out of Vietnam" and this was greeted enthusiastically.

The May 23rd *San Francisco Examiner* wrote on its front page:

> Cal's teach-in a-go-go hit full stride yesterday with a full roster of a-go-go-out-of-Vietnam advocates...
>
> Sitting on the grass, munching hot dogs and swizzling soft drinks, they listened raptly and cheered lustily such speakers as Socialist Norman Thomas, Assemblyman John Burton, comedian Dick Gregory, novelist Norman Mailer and dozens of others.
>
> The marathon protest continued until 11:45 p.m. last night, or 3 hours and 15 minutes past its scheduled end, caused in part by the reappearance of several of the guest speakers, Gregory, for instance, made three separate addresses to the students.

I still remember vividly the powerful oration of Isaac Deutscher. After his socialist-oriented attack on the United States in the chilly morning hour of 1:30 A.M., 10,000 or so gave him a prolonged standing ovation. He is reported to have said afterward: "This is the most exciting speaking engagement I have had since I spoke to the Polish workers thirty years ago. It is extraordinary—simply extraordinary" (*Teach-ins*: *USA*, eds. L. Menashe and R. Radosh, Praeger, New York, 1967, p. 33).

And I also remember well Norman Mailer's description of Lyndon Johnson. Ralph Gleason in a review of Vietnam Day in the *San Francisco Chronicle* said:

> ... a veteran journalist remarked, as Mailer stood on the platform, the applause thundering in the air, that his was the most impressive political speech in a generation.

Here is more from Gleason:

> The revolution of creativity marked the two-day affair at the University of California, too. It throbbed day and night with the vitality of improvisation. At times over 8000 people packed the lawns and the plaza in a atmosphere of morality that, despite the criticisms of what Paul Krassner called the 'teach-in drop-outs' was as peaceful as a religious revival.
>
> ... Norman Mailer spoke of the reality of death and life and how the society's very existence depends upon a moral revolution, a return from formality to reality, from symbol to the thing itself.

Before Vietnam Day, I was quite undecided as to whether I would continue my involvement, especially at such a level. In fact, I had made plans to go to Hawaii at the end of June for a vacation with my family. With the success of Vietnam Day, I recommitted myself with such fervor that I would become close to abandoning my career as a mathematician. In fact, I kept up an ever-increasing pace of organizing demonstrations until October 16. But that is getting a little ahead.

What could out-do Vietnam Day? A clue came from the teach-in talk of Straughton Lynd. He said:

... is the creation of civil disobedience so massive and so persistent that the Tuesday Lunch Club that is running this country—Johnson, McNamara, Bundy, and Rusk—will forthwith resign ...

Jerry and I agreed to call a new teach-in coupled with massive civil disobedience. We chose the dates October 15–16, to maximize the momentum, taking into account that classes started in September. Our friends in the VDC concurred.

Here is a summary that Jerry and I wrote at the time (from old notes of mine):

Vietnam Day, Berkeley
by J. Rubin & S. Smale

On May 21–22 a student faculty group sponsored a community protest meeting on the Berkeley campus which lasted for 35 hours, bringing about 50,000 different people to the event over the period and with broadcasts reaching an estimated half a million people in the Bay Area.

The speeches centered around U.S. participation in Vietnam with most of the talks attacking various aspects of U.S. policy, with liberal and radical points of view both well represented.

Initially the sponsor of the protest meeting ws the Berkeley Faculty AFT. This was later joined by the Faculty Peace Committee and the employed graduate student AFT. However, the people from these groups and others who were actually doing the preparation formed an organization called the Vietnam Day Committee which came to be an important organization in its own right, and by the time Vietnam Day had arrived, was the de facto sponsoring organization. The Vietnam Day Committee still flourishes as an active peace group in Berkeley.

The working principle of the Vietnam Day Committee was to put forth an interesting and exiting program with wide publicity. Emphasis was on peaceful alternatives to Johnson's policy, but strong attempts, with partial success, were made to put defenders of Administration policy on the platform at Vietnam Day. The meeting was called an educational protest.

It was an important policy of the Vietnam Day Committee to make no concessions to respectability, or to weaken the protest. For example, we resisted much pressure to de-emphasize the student and campus element of the program by removing the event from the campus. (Berkeley students are not "respectable"!). Also, we opposed pressure to remove the most radical speakers. One of the major successes of the project was the fact that speaking on the same platform were such diverse men as Norman Thomas, Senator Gruening, Dr. Benjamin Spock, Isaac Deutscher, Dick Gregory, Norman Mailer, the Editor of the National Guardian, and speakers from the Dubois Club, YSA, Progressive Labor, and pacifist groups.

Publicity, besides dramatic use of the press, included the distribution of thousands of posters and hundreds of thousands of fliers.

Following similar principles, the Vietnam Day Committee in future activity is implementing its educational protest with direct action.

In addition to a picket of Johnson when he comes to San Francisco for the 20th Anniversary of the U.N. June 25–26, the Vietnam Day Committee is plAnning another protest meeting on October 15 with massive civil disobedience on October 16.

All peace and political groups, nationally and even internationally are being asked to support these days of protest ...

[Smale ends his chapter here. He is writing his autobiography in chapters like these, several of which are reproduced in this volume. Future chapters will continue the story. (*Editor's note*)]

2
On How I Got Started in Dynamical Systems,* 1959–1962

STEVE SMALE

1. Let me first give a little mathematical background. This is conveniently divided into two parts. The first is the theory of ordinary differential equations having a finite number of periodic solutions; and the second has to do with the case of infinitely many solutions, or, roughly speaking, with "homoclinic behavior."

For an ordinary differential equation in the plane (or a 2nd-order equation in one variable), generally there are a finite number of periodic solutions. The Poincaré–Bendixson theory yields substantial information. If a solution is bounded and has no equilibria in its limit set, its asymptotic behavior is periodic. The Van der Pol equation without forcing is an outstanding example which has played a large role historically in the qualitative theory of ordinary differential equations.

There was some systemization of this theory by Andronov and Pontryagin in the 1930s when these scientists introduced the notion of structural stability. There is a school now in the Soviet Union, the "Gorki school," which reflects this tradition. Andronov's wife, Andronova–Leontevich, was a member of this school, and I met her in Kiev in 1961 (Andronov had died earlier). A fine book (translated into English) by Andronov, Chaiken, and Witt gives an account of this mathematics.

The work of Andronov–Pontryagin was picked up by Lefschetz after World War II. Lefschetz's book and influcnce helped establish the study of structural stability in America.

The second part of the background mathematics has to do with the phenomenon of an infinite number of periodic solutions (persistent under perturbation) or the closely related notion of homoclinic solutions. Again. Poincaré wrote on these problems; Birkhoff found a deeper connection between the two concepts, homoclinic solutions and an infinite number of periodic points for transformations of the plane. From a different direction, Cartwright and Littlewood, in their extensive studies of the Van der Pol equation with forcing

---

* Partly based on a talk given at a Berkeley seminar circa 1976. Reprinted with permission from *The Mathematics of Time: Essays on Dynamical Systems, Economic Processes and Related Topics*, S. Smale, Springer-Verlag, New York, 1980.

term, came across the same phenomena. Then Levinson simplifed some of this work.

2. Next, I would like to give a little personal background. I finished my thesis in topology with Raoul Bott in 1956 at the University of Michigan; and that summer I attended my first mathematics conference, in Mexico City, with my wife, Clara. It was an international conference in topology and my introduction to the international mathematics community. There I met René Thom and two graduate students from the University of Chicago, Moe Hirsch and Elon Lima. Thom was to visit the University of Chicago that fall, and I was starting my first teaching job at Chicago then (not in the mathematics department, but in the College). In the fall, I became good friends with all three. I attended Thom's lectures on transversality theory and was happy that Thom and Hirsch became interested in my work on immersion theory.

My main interests were in topology throughout these years, but already my thesis contained a section on ordinary differential equations. Also at Chicago, I used an ordinary differential equation argument to find the homotopy structure of the space of diffeomorphisms of the 2-sphere. Having both Dick Palais and Shlomo Sternberg at Chicago was very helpful in getting some understanding of dynamical systems.

3. It was around 1958 that I first met Mauricio Peixoto. We were introduced by Lima who was finishing his Ph.D. at that time with Ed Spanier. Through Lefschetz, Peixoto had become interested in structural stability and he showed me his own results on structural stability on the disk D^2 (in a paper which was to appear in the *Annals of Mathematics*, 1959). I was immediately enthusiastic, not only about what he was doing, but with the possibility that, using my topology background. I could extend his work to n dimensions. I was extremely naive about ordinary differential equations at that time and was also extremely presumptuous. Peixoto told me that he had met Pontryagin, who said that he didn't believe in structural stability in dimensions greater than two, but that only increased the challenge.

In fact, I did make one contribution at that time. Peixoto had used the condition on D^2 of Andronov and Pontryagin that "no solution joins saddles." This was a necessary condition for structural stability. Having learned about transversality from Thom, I suggested the generalization for higher dimensions: the stable and unstable manifolds of the equilibria (and now also of the nontrivial periodic solutions) intersect transversally. In fact. this was a useful condition, and I wrote a paper on Morse inequalities for a class of dynamical systems incorporating it.

However, my overenthusiasm led me to suggest in the paper that these systems were almost all (an open dense set) of ordinary differential equations! If I had been at all familiar with the literature (Poincaré, Birkhoff, Cartwright–Littlewood), I would have seen how crazy this idea was.

On the other hand, these systems, though sharply limited, would find a place in the literature, and were christened Morse–Smale dynamical systems by Thom. This work gave me entry into the mathematical world of ordinary differential equations. In this way, I met Lefschetz and gave a lecture at a conference on this subject in Mexico City in the summer of 1959.

We had moved from Chicago in the summer of 1958 to the Institute for Advanced Study in Princeton; Peixoto and Lima invited me to Rio to finish the second year of my NSF postdoctoral fellowship. Thus, Clara and I and our two kids, Nat and Laura, left Princeton in December, 1959, for Rio.

4. With kids aged $\frac{1}{2}$ and $2\frac{1}{2}$, and most of our luggage consisting of diapers, Clara and I stopped to see the Panamanian jungles (from a taxi). I had heard about and always wanted to see the famous railway from Quito in the high Andes to the jungle port of Guayaquil in Ecuador. So, around Christmas of 1959, the four of us were on that train. Then we spent a few days in Lima, Peru, during which we were all quite sick. We finally took the plane to Rio. I still remember well arriving at night and going out several times trying to get milk for the crying kids, always returning with cream or yogurt, etc. We learned later that milk was sold only in the morning in Rio.

But with the help of the Limas, the Peixotos, and the Nachbins, life got straightened out for us. In fact, it happened that just before we arrived, the leader of an abortive coup, an Air Force officer, had escaped to Argentina. We got his luxurious Copacabana apartment and maids as well.

Shortly after arriving in Rio, my paper on dynamical systems appeared, and Levinson wrote me that one couldn't expect my systems to occur so generally. His own paper (which, in turn, had been inspired by work of Cartwright and Littlewood) already contained a counter-example. There were an infinite number of periodic solutions and they could not be perturbed away.

Still partly with disbelief, I spent a lot of time studying his paper, eventually becoming convinced. In fact, this led to my second result in dynamical systems, the horseshoe, which was an abstract geometrization of what Levinson and Cartwright–Littlewood had found more analytically before. Moreover, the horseshoe could be analysed completely qualitatively and shown to be structurally stable.

For the record, the picture I abstracted from Levinson looked like this:

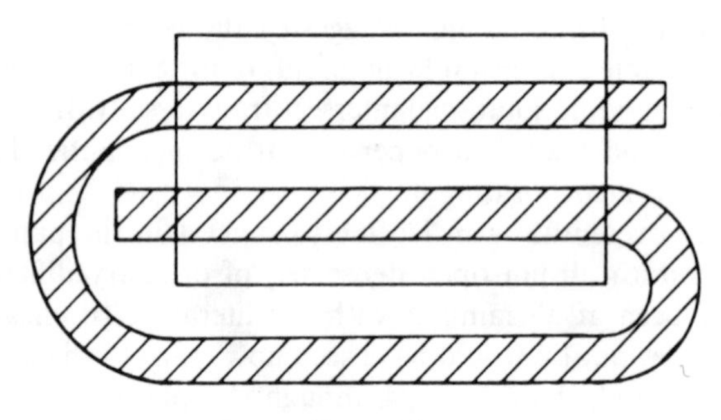

FIGURE 1

When I spoke on the subject that summer (1960) at Berkeley, Lee Neuwirth said: "Why don't you make it look like this?"

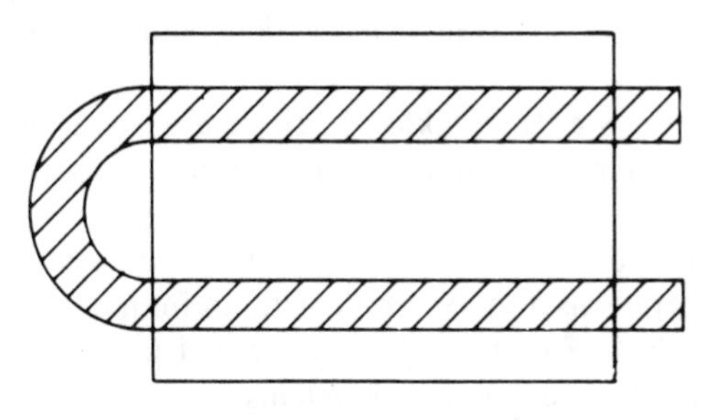

FIGURE 2

I said "fine" and called it the horseshoe.

I still considered myself mainly a topologist, and when considering some questions of gradient dynamical systems, I could see possibilities in topology. This developed into the "higher-dimensional Poincaré conjecture" and was the genesis of my being quoted later as saying I did my best-known work on the beaches of Rio. In fact, I often spent the mornings on those beaches with a pad of paper and a pen. Sometimes Elon Lima was with me. In June, I flew to Bonn and Zurich to speak of my results in topology. This turned out to be a rather traumatic trip, but that is another story.

I had accepted a job at Berkeley (at about the same time as Chern, Hirsch, and Spanier, all from Chicago) and arrived there from Rio in July, 1960. Except for a few lectures on "the horseshoe," I was preoccuped the next year with topology. But, in the summer of 1961, I announced to my friends that I had become so enthusiastic about dynamical systems that I was giving up topology. The explicit reason I gave was that no problem in topology was as important and exciting as the topological conjugacy problem for diffeomorphisms, already on the 2-sphere. This conjugacy problem represented the essence of dynamical systems, I felt.

5. During this year, I had an irresistible offer from Columbia University, so we sold a house we had just bought and moved to New York in the summer of 1961. But before taking up teaching duties at Columbia, I spoke at a conference on ordinary differential equations in Colorado Springs and then, in September, 1961, went to the Soviet Union. At a meeting on nonlinear oscillations in Kiev, I gave a lecture on the horseshoe example, "the first structurally stable dynamical system with an infinite number of periodic solutions." I had a distinguished translator, the topologist, Postnikov, whom I had just met in Moscow. Postnikov agreed to come to Kiev and translate my talk in return for my playing go with him. He said he was the only go player in the Soviet Union. My roommate in Kiev was Larry Markus.

I met and saw much of Anosov in Kiev. Anosov had followed the Gorki school, but he was based in Moscow. After Kiev, I went back to Moscow where Anosov introduced me to Arnold, Novikov, and Sinai. I must say I was extraordinarily impressed to meet such a powerful group of four young mathematicians. In the following years, I often said there was nothing like that in the West.

I gave some lectures at the Steklov Institute and made some conjectures on the structural stability of certain toral diffeomorphisms and geodesic flows of negative curvature.

After I had worked out the horseshoe. Thom brought to my attention the toral diffeomorphisms as an example with an infinite number of periodic points which couldn't be perturbed away. Then I had examined the stable manifold structure of these dynamical systems.

6. After teaching dynamical systems the fall semester at Columbia. I was off again, with Clara and the kids, this time to visit André Haefliger in Lausanne for the spring quarter. Besides lecturing in Lausanne, I gave lectures at the Collège de France, Urbino, Copenhagen, and, finally, Stockholm, all on dynamical systems and emphasizing the global stable manifolds, which I found more and more to lie close to the heart of the subject. In Stockholm, at the International Congress, I saw Sinai again and he told me that Anosov had proved all the conjectures I had made the preceding year in the Soviet Union.

In Lausanne I had begun to start thinking about the calculus of variations and infinite-dimensional manifolds, and this preoccupation took me away from dynamical systems for the next three years.

3
The Story of the Higher Dimensional Poincaré Conjecture (What Actually Happened on the Beaches of Rio)*

STEVE SMALE

This paper is dedicated to the memory of Allen Shields.

Although these pages tell mainly a personal story, let us start with a description of the "n-dimensional Poincaré Conjecture." It asserts:

A compact n-dimensional manifold M^n that has the homotopy type of the n-dimensional sphere

$$S^n = \{x \in \mathbf{R}^{n+1} \mid \|x\| = 1\}$$

is homeomorphic to S^n.

A "compact n-dimensional manifold" could be taken as a closed and bounded n-dimensional surface (differentiable and nonsingular) in some Euclidean space.

The homotopy condition could be alternatively defined by saying there is a continuous map $f: M^n \to S^n$ inducing an isomorphism on the homotopy groups; or that every continuous map $g: S^k \to M^n$, $k < n$ (or just $k \leq n/2$) can be deformed to a point. One could equivalently demand that M^n be simply connected and have the homology groups of S^n.

Henri Poincaré studied this problem in his pioneering papers in topology. In [13], 1900, he announced a proof of the general n-dimensional case. A counter-example to his method is exhibited in a subsequent paper [14] 1904, where he limits himself this time to 3 dimensions. In this paper he states his famous problem, but not as a conjecture. The traditional description of the problem as "Poincaré's Conjecture" is inaccurate in this respect.

Many other mathematicians after Poincaré have claimed proofs of the 3-dimensional case. See, for example, [18] for a popular account of some of these attempts. On the other hand, there has been a solidly developing body of theorems and techniques of topology since Poincaré.

* This article is an expanded version of a talk given at the 1989 annual joint meeting of the American Mathematical Society, and the Mathematical Association of America. Reprinted with permission from *Mathematical Intelligencer*, Vol. 12, No. 2 (1990), pp. 44–51.

In 1960 I showed that the assertion is true for all $n > 4$, and this is an account of that discovery. The story here is complemented by two articles [15] and [16], but the overlap is minimal.

I first heard of the Poincaré conjecture in 1955 in Ann Arbor at the time I was writing a thesis on a problem of topology. Just a short time later, I felt that I had found a proof (3 dimensions). Hans Samelson was in his office, and very excitedly I sketched my ideas to him. First triangulate the 3-manifold and remove one 3-dimensional simplex. It is sufficient to show the remaining manifold is homeomorphic to a 3-simplex. Then remove one 3-simplex at a time. This process doesn't change the homeomorphism type and finally one is left with a single 3-simplex. Q.E.D. Samelson didn't say much. After leaving his office, I realized that my "proof" hadn't used any hypothesis on the 3-manifold.

Less than 5 years later in Rio de Janeiro, I found a counterexample to the 3-dimensional Poincaré Conjecture. The "proof" used an invariant of the Leningrad mathematician Rohlin, and I wrote out the mathematics in detail. It would complement nicely the proof I had just found that for dimensions larger than 4; the result was true. Luckily, in reviewing the counterexample, I noticed a fatal mistake.

Let us go back in time. I was bom into the "Golden Age of Topology." Today it is easy to forget how much topology in that era dominated the frontiers of mathematics. It has been said that half of all the Sloan Postdoctoral Fellowships awarded in those years went to topologists. Today it would be hard to conceive of such a lopsided distribution. Topology of that time, in fact, had revolutionizing effects in algebra (K-theory, algebraic geometry) and analysis (dynamical systems, the global study of partial differential equations).

In 1954 Thom's cobordism paper was published. That theory was used by Hirzebruch to prove his "signature theorem" (as part of his development of Riemann-Roch). In turn, already by 1956, Milnor used the signature theorem to prove the existence of exotic 7-dimensional spheres. I followed these results closely. Also as a student I learned from Raoul Bott about Serre's use of spectral sequences and Morse theory to find information on the homotopy groups of spheres. A little later, Bott himself was proving his periodicity theorems, also with the use of Morse Theory.

I received my doctorate in Ann Arbor in 1956 with Bott. My first encounter with the mathematical world at large occurred that summer with the Mexico City meeting in Algebraic Topology. I had never been to any conference before. And this conference was a historic event in mathematics by any standard. I have not seen that concentration of creative mathematics matched. My wife Clara and I took a bus from Ann Arbor to Mexico City, originally knowing only Bott and Samelson. Before leaving Mexico, I had met most of the stars of topology. There I also met two graduate students from the University of Chicago, Moe Hirsch and Elon Lima, who were to become part of our story.

FIGURE 1. Steve Smale with his son Nat, Chicago, 1958.

That fall I took up my first regular position as an instructor in the college at the University of Chicago, primarily teaching set theory to humanities students. I had good relations with the mathematics department and went to the lectures of visiting professor René Thom (whom I had met in Mexico City) on transversality theory. I also pursued work in topology showing that one could "turn a sphere inside out."

Chicago was a leading mathematics center at that time, before Weil, Chern, and a number of other important mathematicians left. An important part of the environment was created by the younger mathematicians, especially Moe Hirsch, Elon Lima, Dick Lashof, Dick Palais, and Shlomo Sternberg. I was lucky to be there during that period.

A two-year National Science Foundation (NSF) post-doctoral fellowship enabled me to go to the Institute for Advanced Study in the fall of 1958. Topology was very active in Princeton then. I shared an office with Moe Hirsch and we attended Milnor's crowded lectures on characteristic classes and Borel's seminar on transformation groups. I frequently encountered Deane Montgomery, Marston Morse, and Hassler Whitney, played go (with handicap) with Ralph Fox, and met his students Lee Neuwirth and John Stallings.

Moreover, in the summer of 1958, Lima introduced me to Mauricio Peixoto, who sparked my interest in structural stability. That interest led to an invitation to spend the last six months of my NSF in Rio de Janeiro at I.M.P.A. (Instituto de Matematica, Pura e Aplicada).

Thus, at the beginning of January 1960, with Clara and our children, Laura and Nat, I arrived in Rio to meet our Brazilian friends. We arrived in Brazil just after a coup had been attempted by an Air Force colonel. He fled the country to take refuge in Argentina, and we were able to rent his apartment! It was an 11-room luxurious place, and we also hired his two maids. The U.S. dollar went a long way in those days.

Most of our immediate neighbors were in the U.S. or Brazilian military. We could sit in our highly elevated gardened patio and look across to the hill of the *favela* (Babylonia) where *Black Orpheus* was filmed. In the hot, humid

Dynamics and Manifold Decomposition

Consider a manifold M^n, with some Riemannian metric, and some function $f\colon M^n \to \mathbf{R}$. Construct the dynamical system defined by the differential equation $dx/dt = -\text{grad } f$ on M.

If p is a non-degenerate critical point of f, then the set $W^s(p)$ of all points tending to p under the dynamics as $t \to \infty$ is an imbedded cell; the same applies to the set $W^u(p)$ of points tending to p as $t \to -\infty$. In the 2-dimensional picture, both $W^s(p)$ and $W^u(p)$ are 1-cells (or arcs).

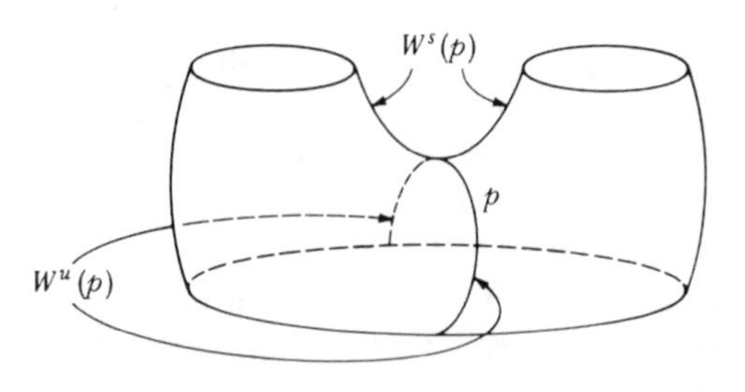

FIGURE 2

Suppose now that f has only non-degenerate critical points. We may assume that $W^s(p) \pitchfork W^u(q)$ for all critical points p, q of f; or, one could say that "the stable and unstable manifolds of grad f meet transversally." Under this hypothesis, the $W^s(p)$ give a nice decomposition of M. The boundary of a cell is a union of lower dimensional cells. René Thom had earlier considered such a decomposition, but without our transversality hypothesis so that the boundary of a cell could be in higher dimensional cells.

Next, from this dynamical system I constructed what I called handlebodies.

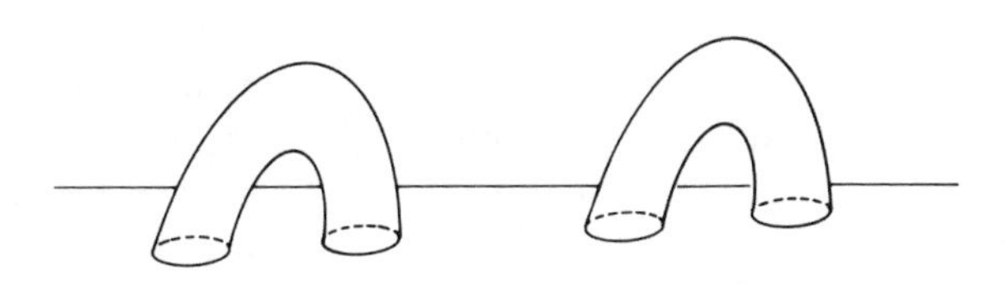

FIGURE 3

This is a diagram of a 1-handlebody, which consists of thickened 1-cells attached to a 3-disk. Then one can add all the 2-handles at one time, etc., until the manifold is given a filtrated structure by these handles.

This gives a starting point for their elimination, a pair at a time, using finally the homotopy and dimension hypotheses on M.

Eventually, one obtains the following description of our original manifold, which was to be proved.

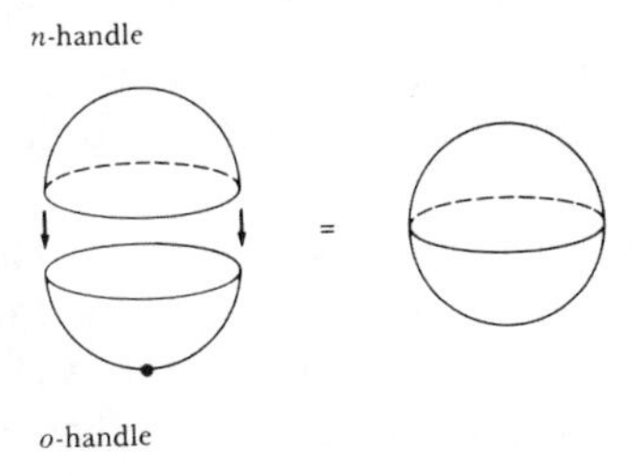

FIGURE 4

evenings preceding *carnaval*, we would watch hundreds of the favela dwellers winding down their path to dance the samba in the streets. Sometimes I would join their wild dancing, which paraded for many miles.

Just one block in the opposite direction from the hill lay the famous beaches of Copacabana (the Leme end). I would spend the mornings on that wide, beautiful, sandy beach sometimes swimming, or, depending on the height of the waves, body surfing. Also I took a pen and pad of paper and would work on mathematics.

Afternoons I would spend at I.M.P.A. discussing differential equations with Peixoto and topology with Lima. At that time I.M.P.A. was located in a small old building on a busy street. The next time I was to visit Rio, I.M.P.A. was in a much bigger building in a much busier street. I.M.P.A. has now moved to an enormous modern palace surrounded by jungle in the suburbs of Rio.

Returning to the story, my mathematical attention was at first directed towards dynamical systems and I constructed the "horseshoe" [15]. As I continued working in gradient dynamical systems, I noticed how the dynamics led to a new way of decomposing a manifold simply into cells. The possibilities of using this decomposition to attack the Poincaré conjecture soon developed, and before long all my work focused on that problem.

With apparent success in dimensions greater than 4, I reviewed my proof carefully; then I went through the details with Lima. Gaining confidence, I wrote Hirsch in Princeton and sent off a research announcement to Sammy Eilenberg. The box titled "Dynamics and Manifold Decomposition" contains a mathematical description of what was happening.

I was already planning to leave Rio for three weeks in Europe during June of that year, 1960. There was the famous *Arbeitstagung*, an annual mathematics event organized by Hirzebruch. This was to be followed by a topology conference in Zürich to which I had been invited. The two meetings provided a good opportunity to present my results. For a change of pace and with his consent, I have put two recent letters of Stallings to Zeeman in a box later in this article. These letters describe well the events that took place in Europe. While my memory in general is consistent with Stallings', I don't believe Hirsch helped me as Stallings conjectured. I do recall spending some relaxed days in St. Moritz with Moe Hirsch and Raoul Bott, after a more dramatic and traumatic week in Bonn.

Sometimes I have become upset at what I feel are inaccuracies in historical accounts of the discovery of the higher-dimensional Poincaré conjecture. For example, Andy Gleason wrote in 1964 [1]: "... It was a great surprise, therefore, when Stallings in 1960 (4) proved that the generalized Poincaré conjecture is true for dimensions 7 and up. His result was extended by Zeeman (10) shortly thereafter to cover dimensions 5 and 6." (I wasn't mentioned in his article.)

Paul Halmos in his autobiography [8, p. 398] writes of my anger with him. I am sorry that I was angry with Paul, and I wish I could be more relaxed about this subject in general. My recent correspondence with Jack Milnor illustrates the issue.

Dear Jack,

I am writing about your article on the work of M. Freedman in connection with his winning the Fields Medal (*Proc. of the Int. Congress* 1986). There you say in discussing the "n-dimensional Poincaré Hypothesis"

> The cases $n = 1, 2$ were known in the nineteenth century, while the cases $n \geq 5$ were proved by Smale, and independently by Stallings and Zeeman and by Wallace, in 1960–61.

The word independently seems inconsistent with the history of that discovery.

Stallings in his preprint "The topology of high-dimensional piecewise-linear manifolds" writes

> When I heard that Smale had proved the Generalized Poincaré Conjecture for manifolds of dimension 5 or more, I began to look for a proof of this fact myself.

I also recall Zeeman having written accurately of these events and giving a similar picture.

After my announcement appeared, Wallace wrote me (Sept. 29, 1960) asking me for details of my proof and telling me where he was blocked in his attempts. I sent him my preprints of the proof which he acknowledged in October (I still have his letters).

I made a mistake in the first draft of my proof, which I easily repaired; but that doesn't affect these issues.

Certainly Stallings and Zeeman, and Wallace have done fine work on this subject. Yet I do wish that mathematicians were more aware of the facts that I have just described. I am sending copies of this to Stallings, Zeeman and Wallace, and to a few other mathematicians.

Best regards,
Steve Smale

Milnor wrote back (as mildly modified and expanded by him):

February 27, 1988

Dear Steve,

I am very sorry that my attempt at sketching the history of the Poincaré conjecture was inaccurate. This was partly a result of not doing my homework properly, but more a matter of expressing myself very badly. What I should have said of course is that the Stallings proof, completed by Zeeman for dimensions 5 and 6, was *logically* independent of your proof; but it is certainly true that yours came first. Similarly, I should have said that Wallace's proof, for dimensions strictly greater than five, definitely came later, and was not really different from your proof.

It is worth noting that the Smale (or Wallace) proof, at the cost of the stronger hypothesis of differentiability, gives a much stronger conclusion. Namely, it shows that the smooth n-manifold contains a smoothly embedded standard $(n - 1)$-sphere which cuts it up into two smoothly embedded standard closed n-balls. The Stallings–Zeeman proof starts with the weaker hypothesis that M is a combinatorial manifold, and obtains the weaker conclusion that M with a single point removed is combinatorially homeomorphic to the standard Euclidean space.

Sincerely,
John Milnor

I wrote him, thanking him, saying that I appreciated his letter.

John Stallings to Christopher Zeeman

March 10, 1988

Dear Chris,

I got your letter to Milnor, which must be about something he said or wrote about Smale, you, and me in the 1960 era. I think I got some comment from Smale about this too, but I cleaned up my office between then and now, and I have no idea what Smale said; and I don't know what Milnor said. And probably I should just go back to sleep.

Your memory of this is much more specific than mine. I remember spring of 1960. I remember the Smale rumor, which I disbelieved until Papakyriakopoulos visited; I think Papa said something like, Eilenberg had looked over Smale's proof of the high-dimensional PC and thought it was OK. I vaguely remember your giving a talk; it was about a paper by Penrose, Whitehead, and Zeeman; and it had some sort of engulfing thing in it; and it mainly strikes me that you or somebody said that it was the only joint paper JHC Whitehead had been involved in, in which his name didn't come last in alphabetical order. I remember sitting up in my little office in the olden Maths. Institute at Oxford and staring at the blackboard, trying to think about how to cancel handles; for some reason this was how I thought Smale would go about it, but I don't think anybody told me that. I couldn't figure out how to get rid of the pesky 1-handles, because I know some awful presentations of the trivial group. Then it occurred to me that I could, so to speak, push all the trouble out to infinity, using the engulfing stuff plus what I think of as an idea due to Barry Mazur in the Morton Brown formulation.

Then I remember being driven to Bonn by Ioan James; there was some other American along too, maybe Bob Hunter or Dick Swan. Somewhere in the middle of Holland, Ioan got out of the car to piss at a farm; we Americans thought this was very British and were too scared to do that ourselves. In Bonn, both Smale and I talked. I remember that A. Borel complained in my talk that I was waving my hands too much on matters in the foundations of PL topology that hadn't been properly proved; and this complaint obsessed me for the next six or seven years. Of Smale's talk, I remember thinking that he hadn't really considered the possibility that there could be awful presentations of the trivial group; in other words, he was just going to cancel the 1-handles with some 2-handles without thinking hard about it; I thought this was a fatal error in Smale's method, and this made me secretly very gleeful. However, the next week, at Zürich, Smale had fixed that up; in fact, if I had memorized Whitehead's "Simplicial Spaces, Nuclei and *m*-Groups," I would have seen how to do this myself. My impression, for which I have no evidence at all, is that Moe Hirsch, who is probably a much better scholar than Smale (in other words, Smale is the Mad Genius and Hirsch is the Hard Worker), told Smale how to fix it up. The point was that you could add trivial pairs of 2- and 3-cells, which on the 2-skeleton is equivalent to adding trivial relators; this is the extra, delicate type of "Tietze transformation" which you need in order to manipulate the 2-cells algebraically to get them to cancel the 1-cells. If the dimension is high enough, this manipulation can be done geometrically. I guess it is in the delicate part of this geometry that the question of whether Smale did it in dimension 5 or not comes up.

After that, I remember enjoying your hospitality, changing at Bletchley, and doing a lot of nice geometrical mathematics during the summer of 1960.

I do not really feel the reason why people are so interested in what exactly happened then. After all, it is not a big nationalistic thing like Newton versus Leibniz. We have all done more mathematics since then; and what I am really interested in is what I am doing and learning about now. Smale is a colleague who really means a lot to Berkeley and to me.

There is a funny thing about time. Not only does existence bifurcate infinitely towards the future, but also towards the past. In other words, if I manage to cross the street and not get flattened by a truck, I think, well, in an alternate universe I was flattened by the truck; I'm just not on that path towards the future, but another me was there. Now, I had the opportunity to receive a visit a few months ago from a mathematician I had last seen in 1977, namely Passi. I remember his appearance in 1977 very dearly. Now he looks 10 or 11 years older, but other than that very similar to how he used to look; *except* that nowadays he is wearing a *different nose*! So I think there were several Passis ten years ago; my existence passed by one of them then, but somehow a slightly different one came out of a different past to visit Berkeley recently.—And so I take these reminiscences of what happened 28 years ago with a little sense of amusement, and I presume that we are all coming together out of slightly different pasts.

Yours,
John R. Stallings

March 14, 1988

Dear Chris,

On March 10, I sent you a description of my memories of PL days in 1960. Now, instead of trying to be purely factual, I want to add some things about my impressions and my philosophy. I plan to send a copy of this to Smale and to Milnor.

I was a graduate student at Princeton, 1956–59; not as well-prepared as some of the other students, and with an idea that doing math was much more important than learning it. I learned something of the background of these high-dimensional topology results that I have been discussing. The methods of JHC Whitehead, MHA Newman, V Guggenheim, etc., were taught in Fox's classes. Milnor had had his construction of exotic structures on S^7 and was giving classes on differential topology. So the time was ripe for something to happen.

Most of the graduate students seemed to be trying to show off to each other about their vast knowledge and intellectual sharpness;

sheaves and spectral sequences were big topics of conversation. Then came Mazur's sphere-embedding theorem. Suddenly (in winter or spring of 57–58), Barry Mazur reappeared, after a long stretch of god-knows-what in Paris, with his theorem. It was a very easy and immediately understandable proof of something that had seemed so complex that no one dared conjecture it. The result was completed and polished up by others later on, but the Genius was Mazur's; this Genius consisted in the audacity and simplicity of Callow Youth, mathematical mode.

In mathematics there are plenty of people who are quick and quite enough people who have assimilated detailed volumes of knowledge. I consider the Mazur-like phenomenon to be far more attractive and important. After someone like Mazur makes his incursion into new territory, the mathematical masses follow, licking up drops of the leader's sweat, computing lists and tables and obstructions and sheaves and spectral sequences. This gives employment to many.

I did my thesis on Grushko's Theorem, on a level that seemed at the time mundane (in my old age, I have learned to appreciate this thesis much more). But what I admired and envied, and who I hoped to imitate—Barry Mazur.

Here is the story I make up about Smale's work on the high-dimensional PC. Smale had already been working in high-dimensional differential topology; he was just a few steps behind Milnor in the incursion into this subject. Then the PC idea came to him "on the beaches of Rio."

I imagine that if I had been on the beaches of Rio, the idea wouldn't have come to me so easily, because, as I pointed out before, I perceived a big problem in getting rid of the 1-handles. I imagine Smale did not know enough to be worried by this, and so he plowed into this big result. I do think there was a little bug in his proof; but this was a non-fatal bug. The 1-handle problem had been, in a somewhat different context, dealt with by JHC Whitehead 20 years earlier. I really think that Smale's Big Theorem was a matter of putting together a new audacity in high dimensions with techniques from olden times due to Whitehead and Whitney, and with his own particular developments. Perhaps the finishing-up of this theorem, and the pushing onward (to computations and obstructions and tables) were done by others. In particular, I think that Milnor had a great deal to do with creating the fundamental techniques of differential topology and expounding them clearly; and that the biggest fruit of this was to make the proof of Smale's Theorem quite solid.

Then there was the engulfing-theory proof of the high-dimensional PC. I'll call this "my theorem." In fact, my theorem and Smale's theorem, although they have a similar sound, are logically distinct. Smale's theorem was that if a differentiable manifold is a homotopy-sphere, then its underlying PL-structure is that of a sphere. My theorem was that if a PL-manifold is a homotopy-sphere, then its underlying topological structure is that of a sphere.

There were several things which paved the way for my theorem. I was familiar with Mazur's argument (I think I was the first person he convinced; and I had come up with a funny fact, when I was an undergraduate, that a group in which you could define an infinite product was the trivial group, a trivial fact). And I was familiar with PL topology from Fox's courses. The mental block against proving something about high-dimensional homotopy-spheres had been reduced by hearing about the fact that Smale's alleged proof was being favorably received. And then, as you pointed out, I had the Penrose-Whitehead-Zeeman trick at my fingertips.

Once I had my theorem, it seems to me that its proof was much simpler and more obvious than Smale's proof of his theorem. Although you and I and others wrote up a lot about the foundations of PL topology, the truth is that there is nothing deep and interesting in this (comparable, say, to Sard's Theorem in differential topology); all the stuff about general position is somehow not too impressive or hard to believe, in spite of the fact that the details are painful. What was interesting about the books on PL topology were the end results, such as the unknotting theorems and h- and s-cobordism theorems; most of these results have both engulfing versions and handle versions. And they have, of course, later on given birth to new generations of mathematical theories.

Now, both my theorem and Smale's are valid from dimension 5 on up. They are easier to prove in dimensions 7 and higher. The intuition that they work in dimension 5 was there from the start. It was mainly a question of formulating the right lemmas and making plausible arguments. With reference to my theorem, your idea of "piping away the highest-dimensional singularities" was a good formulation.

(It occurs to me that I am saying something here that would warm the cockles of the hearts of Thurston and Gromov. The crimes of Thurston and Gromov, which consist of asserting things that are true if interpreted correctly, without giving really good proofs, thus claiming for themselves whole regions of mathematics and all the theorems therein, depriving the hard workers of well-earned credit—something like these crimes was indulged in by both Smale and me in 1960, because we knew some of these theorems were true, and said so, at least in private, without really knowing how to prove them, or perhaps without wanting to indulge in the labor this would involve.)

As I see it, Smale did his work on his own mostly, with perhaps some very minor collaboration with Hirsch. You and I did a considerable amount of collaboration in the summer of 1960, both in polishing the arguments down to dimension 5 and in using similar arguments for various unknotting results. Eventually, we published our own papers separately, although there were little pieces of joint work here and there. If we were really concerned about who thought of what first, we should have had a secretary taking down all our conversations. Neither of us was interested in producing joint papers. But that collaboration was stimulating and helped us produce enough good theorems for both of us.

Finally, one last word: After working on this subject for a few years, I lost interest in it. There were these big, easy theorems. They were done. Full stop. If someone else wants to lecture about them, they are welcome to do so. If someone wants to use these theorems to compute and classify, they have my minor blessing. But it all sounds tiring and boring to me.

Cheers,
John R. Stallings

After the Zürich conference, I returned to join my family in Rio and shortly thereafter took up my position in Berkeley.

During the next year I wrote several papers extending the results, culminating with a paper proving the "*h*-cobordism theorem" in June 1961. A mathematical survey of all these matters, with references, is given in [17].

Because of Serge Lang and an irresistible offer from Columbia, we sold our house and left Berkeley in the summer of 1961. After three years of studying various aspects of global analysis, we returned from New York to Berkeley because of the prospect of better working conditions.

That was the fall of 1964, and the Free Speech Movement (FSM) caught my attention. After the big sit-in, I helped obtain the release from jail of mathematics graduate students David Frank and Mike Shub.

The Vietnam war was drastically escalated early in 1965. I felt that the U.S. heavy bombing of Vietnam was indefensible and threatened world peace. My involvement in the protest increased, and included organizing teach-ins and militant troop train demonstrations. I became cochairman with Jerry Rubin of the VDC, or Vietnam Day Committee. (Our headquarters near campus was later destroyed by a bomb.)

Already during the fall of 1965, I was becoming disillusioned with the VDC and returned to proving theorems. The subsequent part of this story is described in [16]. In Moscow, August 1966, I criticized the U.S. in Vietnam (and Russia as well) to return home under a storm of criticism. The University of California stopped payment of my NSF summer salary.

Dan Greenberg [2] gives a full account of what happened next.

Upon being informed of the agitation surrounding him, and the withholding of his check, Smale sent to Connick an account of his summer researches—an account which will probably be a classic document in the literature of science and government.

He quickly established that he had satisfied the requirement of 2 months of research for 2 months of salary.

Connick was Vice-Chancellor of Academic Affairs at Berkeley. Greenberg's "classic document" surprised me, but my letter did contain my most famous quote. As recounted by Greenberg, I wrote to Connick:

> However, during the remainder of this time I was also doing mathematics, e.g., in campgrounds, hotel rooms, or on a steamship. On the *S.S. France*, for example, I discussed problems with top mathematicians and worked on mathematics in the lounge of the boat. (My best-known work was done on the beaches of Rio de Janeiro, 1960!)
>
> I would like to repeat that I resent your stopping of my NSF support money for superficial technicalities. The reason goes back to my being issued a subpoena by the House Un-American Activities Committee and the subsequent congressional and newspaper attacks on me.

Before long, I did receive the NSF funds and things quieted down. However, within a year, a much bigger explosion took place when the NSF returned to me my new proposal amidst Congressional pressures.

The articles [3–7] in *Science* by Dan Greenberg chronicle these events. See also [11] and [12].

Eventually as the situation was settling down once more, my statement about the beaches of Rio surfaced. This time it was the science adviser to President Johnson, Donald Hornig [9], who wrote in *Science*:

> This blithe spirit leads mathematicians to seriously propose that the common man who pays the taxes ought to feel that mathematical creation should be supported with public funds on the beaches of Rio de Janeiro or in the Aegean Islands.

(I also had visited the Greek islands in August 1966 but, of course, not with NSF money.)

I was very happy at the response of the Council and President C. Morrey of the American Mathematical Society. Morrey's letter [10] in *Science* started:

> The Council of the American Mathematical Society at its meeting on 28 August asked me to forward the following comments to *Science*:
>
> Many mathematicians were dismayed and shocked by the excerpts of the speech by Donald Hornig, the Presidential Science Adviser (19 July, p. 248). His ... comments about mathematics and mathematicians are ... uncalled for. Implicit in Hornig's remarks about vacations on the beaches of Rio or the Aegean Islands was a thinly veiled attack on Stephen Smale. The allegations against Smale were adequately disproved by Daniel S. Greenberg in his articles in *Science* on the Smale-NSF controversy.

At the same time a letter to the *Notices* wound up:

> ... the policy of creating a major scandal involving the implied application of political criteria in grant administration, apparently for the purpose of placating and warding off the demagogic attack of a single Congressman, is not a policy that will cause

anyone (and least of all Congress) to have any great respect for the principles and integrity of those who adopt it. The conspicuous public silence of the whole class of Federal science administrators with regard to the future effects of current draft policy and the Vietnam war upon American science has been positively deafening. In this context, Hornig's apparent attempt to turn the discussion of the current crisis in Federal support for basic science into a hunt for scapegoats in the form of "mathematicians on the beaches" would be ludicrous if it were not so destructive.

Hyman Bass	E.R. Kolchin
F.E. Browder	S. Lang
William Browder	M. Loève
S.S. Chern	R.S. Palais
Robert A. Herrmann	M.H. Protter
I.N. Herstein	G. Washnitzer

It was especially gratifying to see such support from my friends.

Let me end, as I did in Phoenix: "Thanks very much for listening to my story."

References

1. Gleason, A., Evolution of an active mathematical theory, *Science* 145 (July 31, 1964) 451–457.
2. Greenberg, Dan, The Smale case: NSF and Berkeley pass through a case of jitters, *Science* 154 (Oct. 7, 1966) 130–133.
3. ———, Smale and NSF: A new dispute erupts, *Science* 157 (Sept. 15, 1967) 1285.
4. ———, Handler statements on Smale case, *Science* 157 (Sept. 22, 1967) 1411.
5. ———, The Smale case: Tracing the path that led to NSF's decision, *Science* 157 (Sept. 29, 1967) 1536–1539.
6. ———, Smale: NSF shifts position, *Science* 158 (Oct. 6, 1967) 98.
7. ———, Smale: NSF's records do not support the charges, *Science* 158 (Nov. 3, 1967) 618–619.
8. Halmos, P., *I Want to be a Mathematician, an Automathography.* New York: Springer-Verlag (1985).
9. Hornig, D., A point of view, *Science* 161 (July 19, 1968) 248.
10. Morrey, C., Letter to the editor, *Science* 162 (Nov. 1, 1968) 514–515.
11. ———, The case of Stephen Smale, *Notices of the American Math. Soc.* 14 (Oct. 1967) 778–782
12. ———, The case of Stephen Smale: Conclusion, *Notices of the American Math. Soc.* 15 (Jan. 1968) 49–52 and 16 (Feb. 1968) 297 (by Serge Lang).
13. Poincaré, H., *Oeuvres*, VI, Gauthier-Villars, Paris 1953, Deuxième Complément à L'Analysis Situs, 338–370.
14. ———, Cinquième Complément à L'Analysis Situs, 435–498.
15. Smale, S., On How I Got Started in Dynamical Systems, *The Mathematics of Time.* New York: Springer-Verlag (1980).
16. ———, On the Steps of Moscow University, *Math. Intelligencer* 6 no. 2 (1984) 21–27.
17. ———, A survey of some recent developments in differential topology, *Bull. Amer. Math. Soc.* 69 (1963) 131–145.
18. Taubes, G., What happens when Hubris meets Nemesis, *Discover*, July 1987.

4
On the Steps of Moscow University*

STEVE SMALE

From the front page of the *New York Times*, Saturday, August 27, 1966:

Moscow Silences a Critical American

By RAYMOND H. ANDERSON
Special to The New York Times

MOSCOW, Aug. 26—A University of California mathematics professor was taken for a fast and unscheduled automobile ride through the streets of Moscow, questioned and then released today after he had criticized both the Soviet Union and the United States at an informal news conference.

Speaking on the steps of the University of Moscow, the professor, Dr. Stephen Smale . . .

August 15, 1966, was a typical hot day in Athens and I was alone at the airport. My wife and children had just left by car for the northern beaches. My plane to Moscow was to depart shortly.

I had arrived in Paris from Berkeley at the end of May. It was a great pleasure to see Laurent Schwartz again. Though tall and imposing, Schwartz was soft in voice and demeanor. He was a man of many achievements. As a noted collector of butterflies, he had visited jungles throughout the world; Schwartz was an outstanding mathematician; he was also a leader of the French left. With Jean-Paul Sartre, he had been in the forefront of intellectuals opposing the French war in Algeria. He was accused of disloyalty, and his apartment bombed. More recently, Schwartz and I had been in frequent correspondence about internationalizing the protest against the Vietnam War. He had asked me to speak at "Six Hours for Vietnam" which he and other French radicals had organized.

The Salle de la Mutualité—I remembered that great old hall on the left bank for the political rallies I had attended there fifteen years earlier. Now

* The following article is a portion of a full length autobiography the author is currently writing. Reprinted with permission from *Mathematical Intelligencer*, Vol. 6, No. 2 (1984), pp. 21–27.

there were several thousand exhuberant young people in the audience, and I was at the microphone. My French was poor and since my talk could be translated I decided to speak in English. I still wasn't at ease giving non-mathematical talks; even though I had scribbled out my brief talk on scratch paper, I was nervous.

There was a creative tension in that atmosphere that inspired me to communicate my feelings about the United States in Vietnam. As I was interrupted with applause my emotional state barely permitted me to give the closing lines: "... As an American, I feel very ashamed of my country now, and I appreciate very much your organizing and attending meetings like this. Thank you." Then I was led across the stage to M. Vanh Bo, the North Vietnamese representative in Paris; we embraced and the applause reached its peak.

The next day I eagerly read the news story in *Le Monde*; and Joseph Barry, "datelined Paris", in the *Village Voice* June 2, 1966, gave an account of the meeting:

> ... The dirty war was one's own country's and several thousand came to the Mutualité to condemn it. "Six Hours for Vietnam" it was titled. Familiar names, such as Sartre, Ricoeur, Schwartz, Vidal Naquet, were appended to the call. And it was organized by the keepers of the French conscience: French teachers and students.
>
> There was some comfort in the presence of American professors—from Princeton, Boston, and Berkeley. Indeed Professor Smale of the University of California got

FIGURE 1. Steve Smale, 1966. Photo by Caroline Abraham.

almost as big a hand as Monsieur Vanh Bo of North Vietnam. Such is the sentimentality of the left. Even the chant that followed Smale's speech—"U.S. go home! U.S. go home!"—was obviously a left-handed tribute ...

After that I very quickly returned to the world of mathematics. I remember spending the next evening at dinner with René and Suzanne Thom. René Thom was a deeply original mathematician who was to become known to the public for his "catastrophe theory." A few days later René and I drove together to Geneva to a mathematics conference, and I to a rendezvous with my wife, Clara, and young children, Laura and Nat.

Both Schwartz and Thom had won the Fields medal. Two—sometimes four—of these medals are awarded at each International Congress of Mathematics (held every four years). It is the most esteemed prize in mathematics, and I had been greatly disappointed in not getting the Fields Medal at the previous International Congress in Stockholm in 1962. That disappointment led me to deemphasize the importance of the Fields Medal; I was (and I still am) conscious of the considerations of mathematical politics in its choice. Thus I showed little concern about getting the medal at the Congress that summer in Moscow. Nevertheless, when René Thom told me during the ride to Geneva that I was to be a Fields medalist at the Moscow Congress, it was a great thrill. Thom had been on the awarding committee and was giving me informal advance notice. Georges de Rham notified me officially in Geneva a few days later.

The time in Geneva was beautiful: it included intense mathematical activity, seeing old friends, and Alpine hikes with my family.

Medal or no medal, I had been planning to go to the Moscow conference where I was to give an hour-long address. Clara and I had ordered a Volkswagen camper to be delivered in Europe, and we drove it from Geneva to Greece via Yugoslavia, camping with our kids on the route. The plan was for Clara and the kids to stay in Greece while I was in Moscow. They would meet me at the airport on my return. We would go by auto-ferry to Istanbul, where we had reservations at the Hilton Hotel (a change from camping!). The return was to feature an overnight roadside stop in the region of Dracula in Transylvania.

It was our first visit to Greece and we enjoyed the beaches, visited monasteries at Meteora and traveled to the isle of Mykonos. I was especially moved by the home of the oracle at Delphi, I knew it well from stories my father had read to me when I was a child.

Now at the Athens airport, I reviewed those beautiful memories; tomorrow I would be receiving the Fields medal in front of thousands of mathematicians gathered in Moscow from all over the world.

At first I was only slightly annoyed when the customs official stopped me, objecting to something about my passport. I knew my passport was ok; what was this little hassle about? Slowly I began to understand. When we had come across the Greek border, customs had marked in my passport that we were bringing in a car. (The government was concerned that we would sell the car without paying taxes.) Now the Greek officials were not letting me leave the country without that car. The customs officials were adamant. Unfortunately, the car was out of reach for the next ten days. The plane left the ground—it was the only one to Moscow until the next day. My heart sank. There was no possibility of getting to Moscow in time to get my medal.

The American embassy was closed, but after some frantic efforts, I managed to find a helpful consular official, Mr. Paul Sadler. He became convinced of my story, waived the protocol of the embassy, and wrote to the Director of Customs:

Dear Sir:

As explained in our telephone conversation with Mr. Psilopoulos ...

The Embassy is aware of the customs regulations which require an automobile imported by a tourist to be sealed in bond if the tourist leaves the country without exporting the automobile. Unfortunately, Mr. Smale's wife has departed Athens for Thessaloniki with the car and Mr. Smale is unable to deliver the car to the customs.

Since it is of such urgency that Mr. Smale be in Moscow no later than the afternoon of August 16, the Embassy would greatly appreciate your approval to permit Mr. Smale to leave Greece while his car remains in the Country. The Embassy will guarantee to the Director of Customs any customs duties which may be due on the automobile in the event Mr. Smale does not return to Greece and export the car by the end of August, 1966 ...

I was able to take the plane the next day. At a stop in Budapest, Paul Erdos, a Hungarian mathematician I knew, boarded the plane. Erdos gave me the startling news that the House Unamerican Activities Committee had issued a subpoena for me to testify in Congressional hearings. I guessed immediately it was because of my activity on the Vietnam Day Committee in Berkeley; I had been co-chairman with Jerry Rubin. We had organized Vietnam Day and made an attempt to stop troop trains. Most of all, there were the days of protest—the "International Days of Protest."

I arrived late in Moscow and rushed from the airport to the Kremlin where I was to receive the Fields Medal at the opening ceremonies of the International Congress. Without a registration badge the guards at the gate refused me admission into the palace. Finally, through the efforts of a Soviet mathematician who knew me, I obtained entrance and found a rear seat. René Thom was speaking about me and my work:

... si les oeuvres de Smale ne possedent peut-etre pas la perfection formelle du travail definitif, c'est que Smale est un pionnier qui prend ses risques avec un courage tranquille; ...

After Thom's talk I found a letter from my friend, Serge Lang, waiting for me. He wrote that the *San Francisco Examiner* (August 5) had written a slanderous article about me and added ... "I hope you sue them for everything they are worth ... I promise you 2,000 dollars as help in a court fight ..." The article started as follows:

UC Prof Dodges Subpena, Skips U.S. for Moscow

Stephen Smale
Supporter of VDC-FSM

By ED MONTGOMERY
Examiner Staff Writer

Dr. Stephen Smale, University of California professor and backer of the Vietnam Day Committee and old Free Speech Movement, is either on his way or is in Moscow, The Examiner learned today.

In leaving the country, he has dodged a subpoena directing him to appear before the House Committee on Un-American Activities in Washington.

One of a number of subpoenaed Berkeley anti-war activist, Dr. Smale took a leave of absence from UC and leased his home there before his trip abroad.

And it went on to quote a University of California spokesman:

According to press reports the committee has subpoenaed several persons. No Berkeley *student* [my emphasis] was included ...

Not only was I under fire from HUAC, but my University, it seemed, wouldn't even acknowledge me. On the other hand, my friends and the many other mathematicians came to my defense. Serge, who was in Berkeley at the time, and my department chairman, Leon Henkin, acted quickly to clear my name. Supportive petitions were circulated in Berkeley and Moscow. An article in the *San Francisco Chronicle* (Aug. 6) which followed the *Examiner* was terrific. (See box.)

UC Prof's Subpoena May Boomerang on Probers

By Jack Smith

The attempt to subpoena a brilliant University of California mathematician for anti-war activities may boomerang on the House Un-American Activities Committee. The Chronicle was told yesterday.

"American [sic] is going to be the laughing stock of the world for treating such a way in that way," his colleague Dr. Serge Lang declared.

The highly respected mathematician, Professor Stephen Smale, was named along with seven others in the Bay Area to appear before the Congressional committee on August 16 because of anti-Vietnam war activities.

Subpoenas were served on a number of those involved Thursday, but Professor Smale, a one-time Vietnam Day Committee leader, was out of the country.

REPORT

Refuting a somewhat exaggerated published report that the 36-year-old professor had "dodged" the subpoena, colleagues noted that he had applied for a leave of absence at least six months ago. He left here in June and spent two months at the University of Geneva in Switzerland.

At present, according to Dr. Lang, a visiting professor of mathematics at UC from Columbia University, Dr. Smale is driving with his wife and two children from Geneva to Moscow for the International Congress on Mathematics, scheduled August 16 through 26.

During the congress, Dr. Smale will be given the Field Medal, the highest honor in mathematics and comparable to the Nobel Prize.

Smale was one of the organizing members of Berkeley's Vietnam Day Committee, which began as the sponsor of a 36-hour "teach-in" in May, 1965. Its anti-war activities continued afterward, and Smale was one of the leaders of demonstrations seeking to halt troop trains last fall and took part in campus rallies.

Next, I received a letter from Beverly Axelrod, which further lifted my spirit. (See box.) I took special pleasure at her addressing me as "Provocateur Extraordinaire"; and that preceded my most provocative act by a week!

The HUAC hearings in Washington were the most tumultuous ever. Jerry Rubin appeared in costume as an American Revolutionary soldier; members of the Maoist Progressive Labor party proudly proclaimed themselves communists. In the *New York Times*, August 18, (1966) Beverly Axekod was quoted as saying that she felt in physical danger staying after lawyer Arthur Kinoy was dragged out of the hearings by court police.

Another *Times* article *that day* was captioned:

AWARD IN MOSCOW BALKS HOUSE UNIT
COAST MATHEMATICIAN UNABLE TO
TESTIFY AT HEARINGS

A year earlier I would have welcomed the chance to testify in Washington. Now, politics was playing a secondary role to mathematics, and so it was just as well to have missed the subpoena. I had the best of both worlds, the "prestige" of having been called, but not the problem of going.

In Moscow the scientific aspects of the Congress were quite exciting. I renewed friendships with the Russians, Anosov, Arnold and Sinai, whom I had met in Moscow in 1961. The four of us were working in the fast developing branch of mathematics called dynamical systems and had much to discuss in a short time.

I became caught up in the social life of a group of non-establishment Muscovites, visiting their homes and attending their parties. These Russians were scornful of communism, sarcastic about their government, and, in the privacy of their own company, free and uninhibited. This was quite a change

BEVERLY AXELROD
ATTORNEY AT LAW

345 FRANKLIN STREET
SAN FRANCISCO 2, CALIFORNIA
TELEPHONE
MArket 1-3968

August 14, 1966

Dr. Steven Smale
Provocateur Extraordinaire
Moscow, U.S.S.R.

Dear Steve,

THIS lawsuit is not just another legal exercise; it offers the possibility of really dealing a death blow to HUAC. Most, if not all, the others who have been subpoenaed will be parties to this action, and we intend to get "prestige" names to be added afterwards. Yours would be the most important name, for obvious reasons. We hope you will agree and authorize the use of your name as plaintiff.

Also, as I explained to Leon Henkin, we would like to be able to use the names of as many of your colleagues as are willing to participate. The theory is that everyone suffers an injury from the activities of the Committee, and particularly those who, because of their profession, feel a kinship with you.

I'm working on this in association with a number of New York lawyers, including Kuntsler & Kinoy, and national A.C.L.U., which is footing the bill.

Please read the enclosed complaint, and, as soon as possible, at any hour, phone me, or Arthur Kinoy, or William Kunstler, at the Congressional Hotel, Wash. D.C. phone 202 LI 6-6611. (collect)

We can take telephone authorization for use of names, to be followed up by a letter similar to the one enclosed.

Hope your stay in Moscow is a great experience- I envy you.

Best regards,

Beverly

FIGURE 2

from the serious, cautious, apolitical manner of Soviet mathematicians to which I had become accustomed. Considering that dissent was being punished at that time by prison (recall Daniel and Sinavsky), it was surprising to find how open they were in my presence.

I loved the Russians in that crowd. Nevertheless, we argued heatedly about Vietnam. Some of those dissidents were so anti-communist that they wished for an American victory in Vietnam. Chinese communism was akin to

Stalinism and they hoped that the Americans could stop both Soviet and Chinese expansion.

There was political agitation among the 5,000 mathematicians in Moscow on two fronts. One aim was to obtain condemnation of the United States intervention in Vietnam, and I was quite active in that movement. The other aim was to pressure the International Congress to condemn HUAC's attempt to subpoena me. A *Times* story, datelined Moscow, Aug. 21, led off with:

SCIENTISTS URGED TO CONDEMN U.S.
WAR ROLE AND HOUSE INQUIRY
TARGETS AT WORLD PARLEY

The *New York Times* story caught the attention of the State Department which became concerned enough to initiate an investigation. The National Academy of Sciences, Office of the Foreign Secretary, cooperated by contacting the official delegates at the Congress in Moscow:

... we would appreciate any comments you may be able to make to us about the political subjects raised in the context of the Congress which we in turn could pass on to the Department of State ...

Indeed, the writing and circulation of a petition was in progress:

... We the undersigned mathematicians from all parts of the world express our support for the Vietnamese people and their right to self-determination, and our solidarity with those American colleagues who oppose the war which is dishonoring their country.

This activity brought me together with Laurent Schwartz and Chandler Davis once more. Davis had served a prison term stemming from his HUAC appearance back in Michigan. Unable to find a regular job in the United States, he had become professor of mathematics at the University of Toronto. I had been friends with Chandler as a student at Ann Arbor and had been present at his HUAC hearings.

The three of us were invited by four Vietnamese to a banquet on Tuesday, August 23, in appreciation for our opposition to the war. Without any assistance from the Russians, these Hanoi mathematicians put a great amount of energy into preparing a fine Vietnamese feast under difficult conditions; the banquet took place in a dormitory. I gained a convincing picture of the struggle that these particular Vietnamese were going through, continuing to do their mathematics while American bombs fell on their capital city. And for the first time I could see directly the meaning of the American anti-war movement for the Vietnamese. Even though our language and culture were different, much warmth flowed between us.

At that banquet, I was asked to give an interview to a reporter named Hoang Thinh from Hanoi. I didn't know what to say and struggled with the problem for the next day. I felt a great debt and obligation to the Vietnamese—after all, it was my country that was causing them so much pain. It was

FIGURE 3. Steve Smale in Moscow, 1967.

my tax money that was supporting the U.S. Air Force, paying for the napalm and cluster bombs. On the other hand, I was a mathematician, with compelling geometrical ideas to be translated into theorems. There was a limit to my ability to survive as a scientist and weather further political storms. I was conscious of the problems that could develop for me from a widely publicized anti-U.S. interview given to a Hanoi reporter in Moscow. In particular, I knew that what I said might come out quite differently in the North Vietnamese newspaper, and even more so when translated back into the U.S. press.

This was the background for my rather unusual course of action. On the one hand, I would give the interview; on the other hand, I would ask the American reporters in Moscow to be present so that my statements could be reported more directly. I would give a press conference on Friday morning, August 26.

The mechanics for holding such a press conference in Moscow were not simple. I surely wasn't going to become dependent on the Russians by asking for a room. Since Moscow University was the location for the Congress, it was natural to hold the interview on the steps of this university. Out of respect for my hosts, I invited the Soviet press. There was obviously a provocative aspect to what I was doing, so for some protection, I asked a number of friends to be present, including Chandler Davis, Laurent Schwartz, Leon Henkin, and a close friend and colleague, Moe Hirsch. I was already scheduled to leave for Greece early the following morning.

The interview with Hoang Thinh was now set up under my conditions. Then late Thursday I received word that Thinh wouldn't be present! His questions were given me in writing and I was to return my answers in writing. I began to draft a statement.

As I wrote down my words attacking the United States from Moscow, I felt that I had to censure the Soviet Union as well. This would increase my jeopardy, but, having just received the Fields Medal and being the center of much additional attention because of HUAC, I was as secure as anybody. If *I* couldn't make a sharp anti-war statement in Moscow and criticize the Soviets, who could?

The Congress was coming to a close and my rendezvous with the journalists was close at hand. Amidst my friends and all the mathematicians I felt a certain loneliness. No one was giving me encouragement for what I was doing now. Chandler showed some sympathy, and Thursday night the two of us walked together locked in discussion and debate.

That morning, Friday, August 26, as the hour arrived, the main question on my mind was: how would the Soviets react to my press conference? Everything so far seemed to be going smoothly as the participants assembled on those broad steps with the enormous main building of the University behind. A woman from a Soviet news agency asked if she could interview me personally, alone. I said yes, but afterwards. Then I read the following statement:

> This meeting was prompted by an invitation to an interview by the North Vietnamese Press. After much thought, I accepted, never having refused an interview before. At the same time, I invited representatives from Tass and the American Press, as well as a few friends.
>
> I would like to say a few words first. Afterwards I will answer questions.
>
> I believe the American Military Intervention in Vietnam is horrible and becomes more horrible every day. I have great sympathy for the victims of this intervention, the Vietnamese people. However, in Moscow today, one cannot help but remember that it was only 10 years ago that Russian troops were brutally intervening in Hungary and that many courageous Hungarians died fighting for their independence. Never could I see justification for Military Intervention, 10 years ago in Hungary or now in the much more dangerous and brutal American Intervention in Vietnam.
>
> There is a real danger of a new McCarthyism in America, as evidenced in the actions of the House Un-American Activities Committee. These actions are a serious threat to the right of protest, both in the hearings and in the legislation they are proposing. Again saying this in the Soviet Union, I feel I must add that what I have seen here in the discontent of the intellectuals on the Sinavsky-Daniel trial and their lack of means of expressing this discontent, shows indeed a sad state of affairs. Even the most basic means of protest are lacking here. In all countries it is important to defend and expand the freedoms of speech and the press.

After I had read my prepared remarks, a woman, whom I didn't recognize, approached me to say that I was wanted immediately by Karmonov, the organizing secretary of the Congress. Keeping cool, I replied, "O.K., but wait till I finish here." There were a few more questions from the American correspondents, which I answered. Now there were the two remaining pieces of business, which seemed to be converging. The woman from the Congress

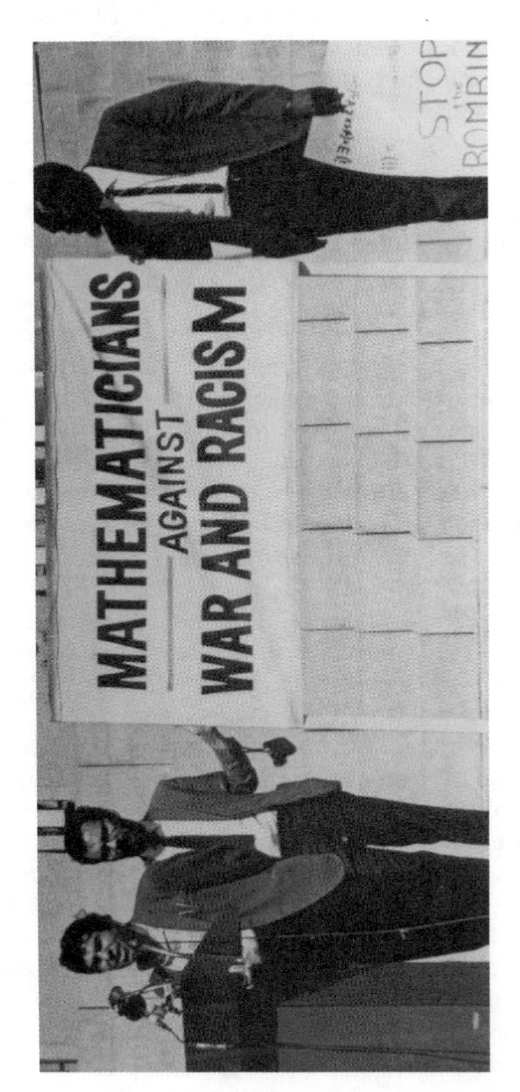

FIGURE 4. Steve Smale speaks at the Chicago National Convention, 1968.

Committee reappeared to lead me to Karmonov, and the Soviet reporter was content to follow along. Sensing a story, the American press followed, as did Moe Hirsch and some other friends.

Karmanov was very friendly; no mention of politics or the press conference came up. It was an unhurried conversation about my impressions of Moscow, inconsistent with the urgency with which I had been called. Karmonov was quite solicitious, and gave me a big picture book about the treasures of the Kremlin (in German!). He next wished to accommodate me in seeing the museums and other touristic aspects of Moscow. A car and guide were put at my disposal. At that point I had no interest whatsoever in sight-seeing. However, I had promised the woman from the Soviet press a personal interview and reluctantly acceded to go with her by car, not at all clear who was going, where we were going or why. I felt pressured and a little scared. But all the while I was treated not just politely, but like a dignitary. It was hard to resist.

As several of us left Karmonov's office, the American reporters were waiting, and asked me what was happening. I really didn't know; in any case, I had no chance to answer as I was rushed out into a waiting car. The newsmen were pushed aside by Russians on each side of me. Moe yelled out, "Are you all right, Steve?" I only had time for a hurried "I think so," before the car was driven off at high speed with the American press in hot pursuit. I was to read later in the newspapers how most, but not all, of the American reporters were eluded. The car stopped at the headquarters of the Soviet news agency, Novosti, and I was taken inside. The same kind of red carpet treatment continued. The top administrators showed me around, but there was no interview. My notes were copied, but otherwise my visit to Novosti seemed to have no purpose. We were just passing the time of day, and, finally, I insisted that I be returned to the Congress activities.

I felt an enormous release of tension as I rejoined my colleagues at the closing reception and ceremonies of the International meeting. There had been a lot of concern for me, and friends told me: "Stay in crowds." With trepidation I returned alone after midnight to my room at the Hotel Ukraine. The phone was ringing. I. Petrovskii, the President of the Congress, wanted to see me the first thing next morning. That was impossible since I was leaving by air at 7:00 A.M. The phone rang again. This time it was the American Embassy. Was I all right? Did I need any assistance? I replied that I probably didn't need any help.

Indeed, I was on that 7:00 A.M. plane. Clara, Nat and Laura had been enjoying the sea and sun during this time and were unaware of the mix-up with customs and the events in Moscow. It certainly was a welcome sight to see them at the Athens airport.

5
Professional Biography, Bibliography, and Graduate Students

STEVE SMALE

Biography

1930	Born July 15 at Flint, Michigan
1952	B.S., University of Michigan
1953	M.S., University of Michigan
1957	Ph.D., University of Michigan
1956–68	Instructor, University of Chicago
1958–60	Member, Institute for Advanced Study, Princeton
1962	Visiting Professor, College de France, Paris (Spring)
1960–61	Associate Professor of Mathematics, University of California, Berkeley
1961–64	Professor of Mathematics, Columbia University
1964–	Professor of Mathematics, University of California, Berkeley
1960–62	Alfred P. Sloan Research Fellow
1966	Member, Institute for Advanced Study, Princeton (Fall)
1967–68	Research Professor, Miller Institute for Basic Research in Science, Berkeley
1969–70	Visiting Member, Institut des Hautes Études Scientifiques (Fall)
1972–73	Visiting Member, Institut des Hautes Études Scientifiques (Fall)
1972–73	Visiting Professor, University of Paris, Orsay (Fall)
1974	Visiting Professor, Yale University (Fall)
1976	Visiting Professor, Instituto de Matematica Pura e Aplicada, Rio de Janeiro (April)
1976	Visiting Professor, Institut des Hautes Études Scientifiques (May, June)
1976–	Professor without stipend, Department of Economics, University of California, Berkeley
1979–80	Research Professor, Miller Institute for Basic Research in Science, Berkeley

1987	Visiting Scientist, IBM Corporation, Yorktown Heights (Fall)
1987	Visiting Professor, Columbia University (Fall)

Honors

1960–62	Alfred P. Sloan Research Fellowship
1964–	Foreign Membership, Brazilian Academy of Science
1965	Veblen Prize for Geometry, American Mathematical Society
1966	Fields Medal, International Mathematical Union
1967	University of Michigan Sesquicentennial Award
1967–	Membership, American Academy of Arts and Sciences
1967–68	Research Professorship, Miller Institute for Basic Research in Science, Berkeley
1970–	Membership, National Academy of Sciences
1972	Colloquium Lecturership, American Mathematical Society
1974	Honorary D.Sc., University of Warwick, England
1979–80	Research Professorship, Miller Institute for Basic Research in Science, Berkeley
1983–	Election as Fellow, The Econometric Society
1983	Faculty Research Lecturership, University of California, Berkeley
1987	Honorary D.Sc., Queens University, Kingston, Ontario
1988	Chauvenet Prize, Mathematical Association of America
1989	Von Neumann Award, Society for Industrial and Applied Mathematics
1990	Honorary Membership in Instituto de Mathematica, Pura e Aplicada (IMPA), Rio de Janeiro
1990	Research Professorship, Miller Institute for Basic Research in Science, Berkeley

Bibliography

1. A note on open maps, *Proceedings of the AMS* **8** (1957), 391–393.
2. A Vietoris mapping theorem for homotopy, *Proceedings of the AMS* **8** (1957), 604–610.
3. Regular curves on Riemannian manifolds, *Transactions of the AMS* **87** (1958), 492–512.
4. On the immersions of manifolds in Euclidean space (with R.K. Lashof), *Annals of Mathematics* **68** (1958), 562–583.
5. Self-intersections of immersed manifolds (with R.K. Lashof), *Journal of Mathematics and Mechanics* **8** (1959), 143–157.
6. A classification of immersions of the two-sphere, *Transactions of the AMS* **90** (1959), 281–290.
7. The classification of immersions of spheres in Euclidean space, *Annals of Mathematics* **69** (1959), 327–344.

8. Diffeomorphisms of the two-sphere, *Proceedings of the AMS* **10** (1959), 621–626.
9. On involutions of the 3-sphere (with Morris Hirsch), *American Journal of Mathematics* **81** (1959), 893–900.
10. Morse inequalities for a dynamical system, *Bulletin of the AMS* **66** (1960), 43–49.
11. The generalized Poincaré conjecture in higher dimensions, *Bulletin of the AMS* **66** (1960), 373–375.
12. On dynamical systems, *Boletin de la Sociedad Mathematica Mexicana* (1960), 195–198.
13. On gradient dynamical systems, *Annals of Mathematics* **74** (1961), 199–206.
14. Generalized Poincaré conjecture in dimensions greater than 4, *Annals of Mathematics* **74** (1961), 391–406.
15. Differentiable and combinatorial structures on manifolds, *Annals of Mathematics* **74** (1961), 498–502.
16. On the structure of 5-manifolds, *Annals of Mathematics* **75** (1962), 38–46.
17. On the structure of manifolds, *American Journal of Mathematics* **84** (1962), 387–399.
18. Dynamical systems and the topological conjugacy problem for diffeomorphisms, *Proceedings of the International Congress of Mathematicians*, Stockholm, 1962. Almquist and Wiksells Uppsala 1963.
19. A survey of some recent developments in differential topology, *Bulletin of the AMS* **69** (1963), 131–146.
20. Stable manifolds of diffeomorphisms and differential equations, *Annali della Scuola Normali Superioro di Pisa, Serie III* **XVII** (1963), 97–116.
21. A generalized Morse theory (with R. Palais), *Bulletin of the AMS* **70** (1964), 165–172.
22. Morse theory and non-linear generalization of the Dirichlet problem, *Annals of Mathematics* **80** (1964), 382–396.
23. Diffeomorphisms with many periodic points, *Differential and Combinatorial Topology (A symposium in honor of Marston Morse)*, Princeton University Press, Princeton, NJ, 1965, pp. 63–80.
24. An infinite dimensional version of Sard's theorem, *American Journal of Mathematics* **87** (1965), 861–866.
25. On the Morse index theorem, *Journal of Mathematics and Mechanics* **14** (1965), 1049–1056.
26. A structurally stable differentiable homeomorphism with an infinite number of periodic points, *Report on the Symposium on Non-Linear Oscillations*, Kiev Mathematics Institute, Kiev, 1963, pp. 365–366.
27. On the calculus of variations, *Differential Analysis*, Bombay, Tata Institute 1964, pp. 187–189.
28. Structurally stable systems are not dense, *American Journal of Mathematics* **88** (1966), 491–496.
29. Dynamical systems on n-dimensional manifolds, *Differential Equations and Dynamical Systems* (1967), pp. 483–486. Academic Press N.Y. 1967.
30. Differentiable dynamical systems, *Bulletin of the AMS* **73** (1967), 747–817.
31. What is global analysis? *American Mathematics Monthly* Vol. 76, (1969), 4–9.
32. Nongenericity of Ω-stability (with R. Abraham), *Global Analysis. Proc. of Symposia in Pure Mathematics*, Vol. 14, American Mathematical Society, Providence, RI, 1970, pp. 5–8.

33. Structural stability theorems (with J. Palis), *Global Analysis. Proc. of Symposia in Pure Mathematics*, Vol. 14, American Mathematical Society, Providence, RI, 1970, pp. 223–231.
34. Notes on differential dynamical systems, *Global Analysis. Proc. of Symposia in Pure Mathematics*, Vol. 14, American Mathematical Society, Providence, RI, 1970, pp. 277–287.
35. The Ω-stability theorem, *Global Analysis. Proc. of Symposia in Pure Mathematics*, Vol. 14, American Mathematical Society, Providence, RI, 1970, pp. 289–297.
36. Topology and mechanics, *I. Inventiones Mathematicae* **10** (1970), 305–331.
37. Topology and mechanics, *II. Inventiones Mathematicae* **11** (1970), 45–64.
38. Stability and genericity in dynamical systems, Seminaire Bourbaki (1969–70) pp. 177–186, Springer, Berlin.
39. Problems on the nature of relative equilibria in celestial mechanics, *Proc of Conference on Manifolds, Amsterdam*, Springer Verlag, Berlin, 1970, pp. 194–198.
40. On the mathematical foundations of electrical circuit theory, *Journal of Differential Geometry* **7** (1972), 193–210.
41. Beyond hyperbolicity (with M. Shub), *Annals of Mathematics* **96** (1972), 597–591.
42. Personal perspectives on mathematics and mechanics, *Statistical Mechanics: New Concepts, New Problems, New Applications*, edited by Stuart A. Rice, Karl F. Freed, and John C. Light, University of Chicago Press, Chicago, 1972, pp. 3–12.
43. Stability and isotopy in discrete dynamical systems, *Dynamical Systems*, edited by M.M. Peixoto, Academic Press, New York, 1973.
44. Global analysis and economics I, *Pareto Optimum and a Generalization of Morse Theory, Dynamical Systems*, edited by M.M. Peixoto, Academic Press, New York, 1973.
45. Global analysis and economics IIA, *Journal of Mathematical Economics*, April **1** 1974, pp. 1–14.
46. Global analysis and economics III, Pareto optima and price equilibria, *Journal of Mathematical Economics* **I** (1974), 107–117.
47. Global analysis and economics IV, Finiteness and stability of equilibria with general consumption sets and production, *Journal of Mathematical Economics* **I** (1974), 119–127.
48. Global analysis and economics V, Pareto theory with constraints, *Journal of Mathematical Economics* **I** (1974), 213–221.
49. Optimizing several functions, *Manifolds—Tokyo 1973, Proc. of the International Conference on Manifolds and Related Topics in Topology, Tokyo, 1973*, University of Tokyo Press, Tokyo 1975, pp. 69–75.
50. A mathematical model of two cells via Turing's equation, *Lectures on Mathematics in the Life Sciences* **6** (1974), 17–26.
51. Sufficient conditions for an optimum, *Warwick Dynamical Systems 1974*, Lecture Notes in Mathematics, Springer-Verlag, Berlin, 1975, pp. 287–292.
52. Differential equations (with I.M. Singer), *Encyclopedia Britannica Fifteenth Edition*, 1974, pp. 736–767. Helen Hemingway Berton Publisher.
53. Dynamics in general equilibrium the theory, *American Economic Review* **66** (1976), 288–294.
54. Global analysis and economics VI: Geometric analysis of Parieto Optima and price equilibria under classical hypotheses, *Journal of Mathematical Economics* **3** (1976), 1–14.

55. The qualitative analysis of a difference equation of population growth (with R.F. Williams), *Journal of Mathematical Biology* **3** (1976), 1–4.
56. On the differential equations of species in competition, *Journal of Mathematical Biology* **3** (1976), 5–7.
57. A convergent process of price adjustment, *Journal of Mathematical Economics* **3** (1976), 107–120.
58. Exchange processes with price adjustment, *Journal of Mathematical Economics* **3** (1976) 211–226.
59. Global stability questions in dynamical systems, *Proceedings of Conference on Qualitative Analysis, Washington D.C.*
60. Dynamical systems and turbulence, *Turbulence Seminar: Berkeley 1976/77*, Lecture Notes in Mathematics edited by P. Bernard and T. Ratiu, No. 615, Springer-Verlag, Berlin, 1977, pp. 48–71.
61. Some dynamical questions in mathematical economics, *Colloques Internationaux du Centre National de la Recherche Scientifique*, No. 259: Systemes Dynamiques et Modeles Economiques, (1977).
62. An approach to the analysis of dynamic processes in economic systems, *Equilibrium and disequilibrium in Economic Theory*, Reidel, Dordrecht, 1977, pp. 363–367.
63. On comparative statics and bifurcation in economic equilibrium theory, *Annals of the New York Academy of Sciences* **316** (1979), 545–548.
64. On algorithms for solving $f(x) = 0$ (with Morris W. Hirsch), *Communications on Pure and Applied Mathematics* **55** (1980), 1–12.
65. Smooth solutions of the heat and wave equations, *Commentarii Mathematici Helvetici* **55** (1980), 1–12.
66. The fundamental theorem of algebra and complexity theory, *Bulletin of the American Mathematical Society* **4** (1981), 1–36.
67. The prisoner's dilemma and dynamical systems associated to non-cooperative games, *Econometrica* **48** (1980), 1617–1634.
68. Global analysis and economics, *Handbook of Mathematical Economics*, Vol. 1, edited by K.J. Arrow and M.D. Intrilligator, North-Holland, Amsterdam, 1981, 331–370.
69. *The Mathematics of Time: Essays on Dynamical Systems, Economic Processes and Related Topics*, Springer-Verlag, Berlin, 1980.
70. The problem of the average speed of the simplex method, *Proceedings of the XIth International Symposium on Mathematical Programming*, edited by A. Bachem, M. Grotschel, and B. Korte, Springer-Verlag, Berlin, 1983, pp. 530–539.
71. On the average number of steps of the simplex method of linear programming, *Mathematical Programming* **27** (1983), 241–262.
72. On the steps of Moscow University, *Mathematical Intelligencer*, **6**, No. 2 (1984), 21–27.
73. Gerard Debreu wins the Nobel Prize, *Mathematical Intelligencer* **6**, No. 2 (1984), 61–62.
74. Scientists and the arms race (in German translation), *Natur-Wissenschafter Gegen Atomrustung*, edited by Hans-Peter Durr, Hans-Peter Harjes, Matthias Krech, and Peter Starlinge, Spiegel-Buch, Hamburg, 1983, pp. 327–334.
75. Computational complexity: On the geometry of polynomials and a theory of cost Part I (with Mike Shub), *Annales Scientifiques de l'Ecole Normale Superieure* **18** (1985), 107–142.
76. On the efficiency of algorithms of analysis, *Bulletin of the American Mathematical Society* **13** (1985), 87–121.

77. Computational complexity: On the geometry of polynomials and a theory of cost: Part II (with M. Shub), *SIAM Journal of Computing* **15** (1986), 145–161.
78. On the existence of generally convergent algorithms (with M. Shub), *Journal of Complexity* **2** (1986), 2–11.
79. Newton's method estimates from data at one point, *The Merging of Disciplines: New Directions in Pure, Applied, and Computational Mathematics*, edited by Richard E. Ewing, Kenneth I. Gross, and Clyde F. Martin, Springer-Verlag, New York, 1986, pp. 185–196.
80. On the topology of algorithms, I, *Journal of Complexity* **3** (1987), 81–89.
81. Algorithms for solving equations, *Proceedings of the International Congress of Mathematicians, August 3–11, 1986, Berkeley, California USA*, American Mathematical Society, Providence, RI, 1987, pp. 172–195.
82. Global analysis in economic theory, *The New Palgrave: A Dictionary of Economics*, Vol. 2, edited by John Eatwell, Murray Milgate, and Peter Newman, Macmillan, London, 1987, pp. 532–534.
83. The Newtonian contribution to our understanding of the computer, *Queen's Quarterly* **95** (1988), 90–95.
84. On a theory of computation and complexity over the real numbers: NP-completeness, recursive functions and universal machines (with Lenore Blum and Mike Shub), *Bulletin of the American Mathematical Society* (*New Series*) **21** (1989), 1–46.
85. Some remarks on the foundations of numerical analysis, *SIAM Review* **32**(2) (1990), 221–270.
86. The story of the higher dimensional Poincaré conjecture, *Mathematical Intelligencer* **12**, No. 2 (1990), 40–51.
87. Dynamics retrospective: Great problems, attempts that failed, *Physica D* **51** (1991), 267–273.

Work in Progress

The Gödel Incompleteness Theorem and Decidability over a Ring (with L. Blum), this volume.

Submitted

Complexity of Bezout's Theorem I: Geometric Aspects (with M. Shub).

Other Work

Book reviews

R1. Review of "Global Variational Analysis: Weierstrass Integrals on a Riemannian Manifold" by Marston Morse, *Bulletin of the American Mathematical Society* **83** (1977), 683–693.
R2. Review of "Catastrophe Theory: Selected Papers, 1972–1977" by E.C. Zeeman, *Bulletin of the American Mathematical Society* **84** (1978), 1360–1368.

Obituaries

O1. Marston Morse, *Mathematical Intelligencer* **1** No. 1 (1978), 33–34.

Calendar

C1. Crystals 1981 (with Clara Smale). Pacific Rotoprinting, 1980.
C2. Beautiful Crystals 1992 and 1993, Golden Turtle, San Francisco.

Other Articles

A1. The annual meeting of the National Academy of Sciences, *New York Review of Books*, July 1, 1981, 38–39.
A2. The major in mathematics, *Notices of the American Mathematical Society* (1977).

Graduate Students

Name and Address	**Date and Thesis Title**
Edward Bishop	Fall 1991 Machines and Complexity over the p-adic numbers
Robert E. Bowen	June 1970 Topological entropy and axiom A
Alberto Buffo Department of Mathematics Dartmouth College Hanover, NH 03755	May 1985 Inverse problem of potential scattering in three dimensions
William Paul Byers Department of Mathematics Concordia University Quebec, Canada H4B 1R6	September 1969 Anosov flows
Hildeberto Cabral Departmento de Matematica Universidad do Recife Recife, Brazil	December 1972 On the integral manifolds of the N-body problem
Cesar Camacho IMPA Estrada Dona Castorina 110 Jardin Botanico CEP 22460 Rio de Janeiro, Brazil	June 1971 $R^n \times S^1$-actions
Robert Devaney Mathematics Department Boston University Boston, MA 02215	June 1973 Reversible diffeomorphisms and flows

FIGURE 1. Smale and some of his students at the Smalefest.

James B. Dowling	September 1968 Finsler geometry on Sobolev manifolds
John Franks Mathematics Department Northwestern University Chicago, IL 60201	September, 1968 Anosov diffeomorphisms
David Fried Department of Mathematics Boston University Boston, MA 02215	June 1976 Cross-section flows
Joel Friedman Department of Computer Science Princeton University Princeton, NJ 08544	
Dan Friedman Economics Department University of California Santa Cruz, CA 95064	June 1977
Feng Gao Department of Computer Science University of British Columbia Vancouver CANADA V6T 1W5	

John Guckenheimer
Department of Mathematics
Cornell University
Ithaca, NY 14853-7901
June 1970
Endomorphisms of the Riemann sphere

Michael D. Hirsch
Computer Science Department
Princeton, NJ
May 1990
Applications of topology to lower bound estimates in computer science

Morris W. Hirsch
Mathematics Department
University of California
Berkeley, CA 94720
Spring 1958
Immersions of manifolds

Angelyn Jewell
National Security Agency
Arlington VA
December 1979
Rational maps of the Riemann sphere

Nancy Kopell
Department of Mathematics
Boston University
Boston, MA 02215
September 1967
Commuting diffeomorphisms

Eric Kostlan
Department of Mathematics
University of Hawaii
Honolulu, HI
May 1985
Statistical complexity of numerical linear algebra

Ernesto Lacomba-Zamora
Departmento de Matematica
Universidad Autonoma Metropolitana
Mexico (13 DF) Mexico
September 1972
Relative equilibria and bifurcation sets for geometries on homogeneous spaces

Jean-Pierre Langlois
Department of Mathematics
San Francisco State
San Francisco, CA 94132
December 1982
Repeated play in non-cooperative game theory

Alejandro Romino Lopez-Yanez
Departamento de Matematicas
Universidad Nacional de Mexico
Mexico, D.F. Mexico
December 1974
Integral manifolds of the planar N-body problem

Sheldon Newhouse
Department of Mathematics
University of North Carolina
Chapel Hill, NC 27599-3250
June 1969
On generic properties of differentiable automorphisms of S^2

Zbigniew Nitecki
Mathematics Department
Tufts University
Medford, MA 02144

September 1968
The orbit structure of endomorphisms

Chong Pin Ong

December 1973
Curvature and mechanics

J Palis
Instituto de Matematica Pura e Applicada
Estrada Dona Castorina
110-Jardim Botanico
Caixa Postal 34021, Rio de Janeiro, Brazil

March, 1968
On Morse–Smale diffeomorphisms

Michael Noel Payne
Department of Mathematics
College of Alameda
Alameda, CA 94501

June 1973
Structural stability and quadratic dynamical systems

Themistocles Rassias
Department of Mathematics
University of La Verne
Athens, Greece

June 1976
Global analysis: Morse theory and the problem of plateau

George Rassias
4, Kasson St, Halandri
Athens, Greece

December 1977
Differential topology—geometry of manifolds

Jim Renegar
Operations Research
Cornell University
Ithaca, NY 14853

June 1983
On the computational complexity of simplicial algorithms in approximating zeroes of complex polynomials

Gregg Saunders
1665 Collingwood Ave.
San Jose, CA 95125

June 1984
Iteration of rational functions of one complex variable and basis of attractive fixed points

Stephen Schechter
Department of Mathematics
North Carolina State University
Raleigh, NC 27695

June 1975
Smooth Pareto economic systems

Chester Ray Schneider

September 1968
Differentiable $SL(n, R)$ actions

Siavash Mirshams Shashahani
Department of Mathematics
Institute of Theoretical Physics and Mathematics
Tehran, Iran

Steptember 1969
Global theory of second order ordinary differential equations on manifolds

M. Shub
T.J. Watson IBM Research Center
Yorktown Heights, NY 10598-0218

September, 1967
Endomorphisms of compact differentiable manifolds

Per Tomter
Institute of Mathematical Sciences
University of Oslo
Oslo, Norway

September 1968
Anosov flows on infrahomogeneous space

Tony Tromba
Department of Mathematics
University of California
Santa Cruz, CA 95064

June, 1968
A Brouwer Degree for Mappings of Infinite Dimensional Oriented Manifolds and its Applications to Non-linear Analysis

Yieh-Hei Wan
Mathematics Department
SUNY at Buffalo
Buffalo, NY 14214-3093

June 1973
Morse theory for two functions

Xiaohua Xuan
Research Development
Hewlett Packard Co.
Santa Rosa, CA 95401

May 1990
Non-linear boundary value problems: Computations, convergence and complexity

Part 2
Informal Talks

6
Luncheon Talk and Nomination for Stephen Smale

R. Bott

Thank you, Moe. It is a great pleasure to be here, but I must say that I feel a bit like Karen who just remarked that she comes from a different world. I come from the beach! In fact, I'm hardly dry. On my beach, you don't usually wear clothes, but as you see, I put some on for this great occasion.

Now, I'm quite amazed to find what deep philosophical things we have becomes involved in here at lunch. But let me warn you about becoming too philosophical—it is a sure sign of age! In fact, this whole discussion reminds me of a happening at the Institute in Princeton circa 1950. Niels Bohr had come to lecture. He was grand, beautiful, but also very philosophical. After a while, von Neumann, who was sitting next to me, leaned over and whispered in my ear, "It's calcium, it's the calcium in the brain." Keep that in mind, youngsters!

In a sense, thinking back to those days at the Institute, the man who inspired me most was Carl Ludwig Siegel, and I seem to have inherited his feelings that it is not so much the theory that constitutes mathematics, but the theorem and, above all, the crucial example. He felt that mathematics had started to go down the road of abstraction with Riemann! He was skeptical of Hilbert's proofs; they were too abstract.

But I'm *not* going to become philosophical! In fact, tomorrow I'm heading straight back to my beach. You see, it is very easy for you youngsters to enjoy Steve's birthday, but after all, he was my student, and if he is 60, you can well imagine what that means about my age.

So instead of philosophy, let me reminisce, and tell you a little bit about how it all began. I came to Michigan in 1951 from the Institute having learned some topology there. In fact, I had a stellar cast: personal instruction from Specker and Reidemeister and Steenrod! Morse was there, of course, and I was great friends with Morse, but he was not really my mathematical mentor. He was interested in other things at that time—Hilbert spaces, etc. Anyway, I came to Michigan as a young instructor, a rather lowly position, and I cannot help telling you the anecdote that greeted me, so to speak, on arrival.

In the summer, there had been an International Congress at Harvard where I had met my Chairman-to-be, Professor Hildebrand. I had grown a full beard for the occasion as a lark and, indeed, the only other people whom I recall having beards at that time were the Italians. When I presented myself to Hildebrand, I had already shaved off my beard long ago. He looked at me and said, "You mean you don't have a beard?" And I said, "No, I don't usually wear one." Thereupon he said in his customary frank manner, "Well, young man. That beard at the Congress cost you $500. When I came home, I took $500 off your salary; I don't approve of beards." But there, I could entertain you all evening with Hildebrand stories.

Michigan had a great tradition in topology at that time. Samelson was there. Also Moise, Wilder, and Young; Steenrod and Sammy Eilenberg had been there. Samelson and I hit it off from the very start and we both started a very business-like seminar on Serre's thesis. This was algebraic topology with a capital A. In fact, that whole generation had lost faith in proving purely geometric theorems. That is why it was so amazing to my generation when later on Steve, and then progressively more and more other people, were able to go back to geometry and with their bare hands do things which at that time we thought were undoable.

When I offered my first course in topology, I had three enrolled students, but there were quite a few other people who attended. One was Steve, the other one was Jim Munkres, and the third, as I like to say, was a really bright guy. He was a universalist and just did mathematics on the side. I expect that those of you who know Steve well can imagine what he was like as a student. He sat in the back and it wasn't clear whether he was paying attention, but he always looked benign. I think it is fair to say that he wasn't really considered a star at Michigan. In fact, there was some question whether he should be allowed to stay! Maybe that's why he picked me. But I think I gave you, Steve, a good question which in some ways still has unsolved parts to it today. My problem was this. If we take the space of regular curves, starting from a point and project each curve onto its end tangent, does this projection have the fiber homotopy property? This turned out to be very difficult. There was no clean way of doing it. But that didn't stop Steve! Actually I had only a small concrete application in mind related to a theorem of Whitney's about the winding numbers of curves in the plane. So that when Steve later elaborated his technique to immersion theory I was completely amazed. And in case you have not heard this story, when Steve wrote me that he could turn the sphere inside out, I immediately sent him a counterexample! That's true. I wrote him back, "Don't be foolish. It can't be done." So, you see, this shows how good it is to have a stupid thesis advisor. Once you overcome him, the rest will be easy.

Well, so many good things have been said about Steve that I do not want to add to them. I mean, he will become quite insufferable. Let me rather tell you about some of his other gifts. These include many techniques of nearly getting rid of people. Several times I've gone on excursions with Steve, and on returning I've often ended up on my knees: "Back home again and still *alive*!" There was a wonderful excursion organized by Steve on the beaches of the Olympic Peninsula. Maybe he did his best job there. This beach had many promontories which one could only traverse when the tide was low enough. This is what intrigued Steve and he had researched it carefully. Only he had done the trip the wrong way around, so that when we came along we were walking, so to speak, against the grain of his experience. On the last day, he told us that the end was in sight and that we could surely take a rather daring shortcut along the beach. The tide also did look very low. And so we were all walking along very nicely, relaxed, doing mathematics all over the place. My wife and one enterprising daughter were along. And then suddenly I remember turning a corner only to see a quite unexpected bay which we still had to traverse. And, indeed, that already my wife, who was quite a bit ahead of me, was being ushered through the ice-cold waters that were rising along the cliffs by one of our wonderful God-sent mountaineers who had joined our expedition. By the time I got to the spot I did not trust myself to carry my little daughter through the water, but rather put her on the back of another wonderful mountaineer, "Pham." Pham is about half my height, but he took her over to safety, and then ran back to alert all the people behind us, such as Steve, his son, the Shubs, etc. Then I got into the water with all my packs. By this time the waves were crashing in and between their onslaught I managed to get to the place where I was supposed to climb out again. I tried but couldn't quite make it. I tried once again, but of course was weaker. It never occurred to me to remove my pack! Instead, what went through my mind was, "Aha! This is how one drowns!" But at my third attempt that wonderful girl, Megan, another of our mountaineers, came back just in time to get me on my feet.

All right. This is enough. I think I have now lowered the level of discourse sufficiently! But in closing, I cannot resist one serious word. It was a great pleasure to have you as a student, Steve—even though you tried to drown me. I've always been very proud of you, and I am sure you will not cease to do great mathematics for many years to come. Happy birthday—you have now finally come of age!

Nomination for Stephen Smale*

It is a pleasure and an honour for me to hereby place Stephen Smale's name in nomination for the Presidency of the American Mathematical Society.

Smale is one of the leading mathematicians of his generation, whose work has been foundational in differential topology, dynamical systems, and many

* Reprinted from *Notices of the American Mathematical Society*, **38**, 7, 758–760, 1991.

aspects of nonlinear analysis and geometry. It has been his genius also to bring these subjects to bear in a significant way on mechanics, economics, the theory of computation, and other brands of applied mathematics.

Smale's characteristic quality is courage, combined with great geometric insight, patience, and power. He is single-minded in his pursuit of understanding a subject on his own terms and will follow it wherever it leads him. These qualities also emerge in many episodes from his personal life. Once his sails are set, it is literally impossible to stop him. Thus in a few short years he moved from Sunday sailor to skippering his own boat across the Pacific with a mathematical crew of hardy souls; and so his interest in minerals has, over the years, not only taken him to obscure and dangerous places all over the globe, but has culminated in one of the finest mineral collections anywhere.

Smale characteristically takes on one project at a time, thinks deeply about it and then turns to the next. He likes to share his thoughts with others, keeps his office door open, never seems in a hurry, and inspires his students with his own confidence. His willingness to run for this position therefore assures me that if he were elected he would grace our Society not only with his great mathematical distinction, ecumenical interests and quiet—almost shy—manner, but that he would also do his homework thoroughly and give the serious problems that face our subject and our institutions his "prime time."

It was my good fortune to have Smale as one of three enrolled students in the first course on topology which I taught at Michigan (1952–1953). (Munkres was another one, but the third, whose name now escapes me, was—as I liked to say—"the really smart one." Indeed, he could play blindfolded chess, compose operas, etc.) Smale's manner in class was the same then as it is now. He preferably sits in the back; says little, and seems to let the mathematical waves wash over him, rather than confront them. However, Smale's courage surfaced soon at Michigan when he chose me, the greenhorn of topology—actually of mathematics altogether—as his thesis advisor. I proposed a problem concerning regular curves (i.e., curves with nowhere zero tangents) on manifolds; namely, that the projection of the space of such curves on its final tangent-direction satisfied the "covering homotopy property." This notion had just been invented in the late forties, and I had learned it from Steenrod in Princeton just the year before. The combination of analysis and topology in this question appealed to Smale and he went to work. I was pleased and impressed by the geometric insight and technical power of his eventual solution, but completely amazed when he in the next few years extended these techniques to produce his famous "inside-out turning of the sphere" in $\mathbb{R}^3$ through regular deformations. In fact, when he wrote me about this theorem I replied curtly with a *false* argument which purported to prove the impossibility of such a construction!

More precisely, what Smale had proved was that the regular immersion classes of a k-sphere S^k in $\mathbb{R}^n$ correspond bijectively to $\pi_k(V_{k,n})$, the kth homotopy group of the Stiefel-manifold of k-frames in $\mathbb{R}^n$. He had thus managed to reduce a difficult differential topology question to one in pure homotopy theory and so had set the stage for an obstruction theory of immersions. The final beautiful development of this train of thought came in the thesis of Morris Hirsch, directed by Spanier, but also with great interest and encouragement from Smale.

In the 1960s, Smale produced his "Generalized Poincaré Conjecture" and—what is still to this day the most basic tool in differential topology—the "H-cobordism theorem." All these were corollaries of his deep rethinking of Morse theory, which he perfected to a powerful tool in all aspects of differential topology. Above all, Steve had the courage to look for concrete geometric results, where my generation was by and large taught to be content with algebraic ones.

Smale's rethinking of the Morse theory involved fitting it into the broader framework of dynamical systems. This enabled him not only to extend the Morse inequalities to certain dynamical systems, but also to use concepts from dynamical systems to understand the Morse theory more profoundly. By clearly formulating and using the transversality condition on the "descending" and "ascending" cells furnished by the gradient flow, he was able to control these cell-subdivisions much more accurately than Thom had been able to do ten years earlier. I distinctly remember when he retaught me Morse theory in this new and exciting guise during a Conference in 1960 in Switzerland.

By that time he had actually also constructed his famous "horseshoe" map of the 2-sphere, and so was well on his way to laying the foundations of a subject we now call "chaos." Indeed, he showed that under the assumption of a hyperbolic structure on a "non-wandering" set, this set breaks up into a finite number of compact invariant sets in a unique way. Each of these was either a single periodic orbit or else was an infinite union of such orbits so inextricably tangled that we would call them "chaotic" today. Moreover, he showed that these basic sets, as he called them, were structurally stable (the chaos cannot be removed by a small perturbation) as well as ubiquitous!

In the years since these spectacular results—which earned him the Field Medal in 1966—Smale has not ceased to find new and exciting quests for his geometric and dynamical imagination.

In the later sixties, he and his students studied the Morse theory in infinite dimensions as a tool in non-linear differential equations. He proved a Sard type theorem in this framework and the Palais-Smale Axioms are now the foundation on which the modern school build their Morse theory beyond "Palais-Smale."

In mechanics, Smale was one of the initiators of the "geometric reduction" theory which occurs so prominently in the work of the symplectic school. In economics, Smale reintroduced differential techniques in the search for equilibrium with great success, and in computation theory his "probabilistic growth theory" applied to algorithms—in particular his estimates for a modified form of Newton's algorithm is an exciting new development in that subject. And the last time I heard Smale talk, he was explaining Gödel's theorem in dynamical system terms!

In fact, with an oeuvre of this magnitude and with more than thirty distinguished doctoral students dispersed all over the world, one would have to invoke truly legendary names to best Smale's impact on today's mathematical world.

He is clearly a candidate of the first order whom we must not pass up.*

---

* The election was won by Ron Graham.

7
Some Recollections of the Early Work of Steve Smale

M.M. PEIXOTO

Ladies and gentlemen:

It is for me a great honor and a great pleasure to speak at this Smalefest.

I will focus here on the personal and mathematical contacts I had with Steve during the period September 1958–June 1960. From September 1958 to June 1959, I was living in Baltimore and Steve was living in Princeton at the Institute for Advanced Study. I visited him several times there. In June 1959, I went back to IMPA (Instituto de Matemática Pura e Aplicada) in Rio de Janeiro. Steve spent the first six months of 1960 at IMPA and during that period I had daily contacts with him, mathematical and otherwise. This period of Steve's career is sometimes associated with the "Beaches of Rio" episode. But this episode—originated by an out-of-context reference to a letter of Steve's—is something that happened in the United States and several years later. It will be mentioned here only indirectly.

I was introduced to Steve on the initiative of Elon Lima in the late summer of 1958 in Princeton. After spending 1 year there, I was about to leave in a few weeks and Steve was coming for a 2-year stay at the Institute. He was already a substantial mathematician having, among other things, turned the sphere S^2 inside out.

As Lima had predicted Steve showed immediately great interest in my work on structural stability on the 2-disk. I was delighted to see this interest; at that time, hardly anybody besides Lefschetz cared about structural stability.

In those days, one burning question that I had much discussed with Lefschetz was: How to express in n-dimensions the "no saddle connection" condition of Andronov–Pontryagin? Before the end of 1958, Steve had the right answer to this question: The stable and unstable manifolds of singularities and closed orbits should meet transversally.

Simple-minded and rather trivial as this may look to the specialist today, the transversality condition applied to stable and unstable manifolds was a very crucial step in Steve's work in both topology and dynamical systems.

In fact, in 1949, René Thom in a seminal short paper considered a Riemannian manifold M together with a Morse function on it $f: M \to R$ and to these objects he associated the decomposition $\Delta^s[\Delta^u]$ of M into cells given by the stable [unstable] manifolds of grad f. Thom used this decomposition Δ^s to derive, among other things, the Morse inequalities.

The important remark that the transversality condition applied to this case implies that the boundary of one of the cells of Δ^s is the union of lower dimentional cells led Steve to his celebrated work on the high dimensional Poincaré conjecture. Whereas Thom considered only Δ^s, Steve considered both Δ^s and Δ^u and added the transversality condition.

The transversality condition in a more general setting than the one of grad f also played an important role in Steve's subsequent—and no less important—work on dynamical systems.

So these two major developments in topology and dynamical systems can be traced back to the way Steve handled the "no saddle connection" condition of Andronov–Pontryagin. Very humble beginnings indeed!

Steve, his wife Clara and their kids, Nathan aged 2 and Laura aged 1, arrived in Rio in early January 1960. He was then working hard on the Poincaré conjecture but kept an eye on dynamical systems. We used to talk almost every day, in the afternoons, about my two-dimensional structural stability problem.

During his 6-month stay in Rio, Steve made another important contribution to structural stability, the horseshoe diffeomorphism on S^2. This was a landmark in dynamical systems theory, showing that structural stability is consistent with the existence of infinitely many periodic orbits.

In mid-June 1960 at the end of his stay, he made a quick trip to Bonn and Zurich to speak about his work on the high dimensional Poincaré conjecture that he was then finishing.

During Steve's absence I discovered a major flaw in my own work, due to the fact that I had bumped into what nowadays is called the closing lemma.

So I became very anxious to discuss this with him.

On his arrival back in Rio, I invited Clara and Nat to join me and go to the airport to meet Steve at dawn.

Even at a distance it was clear that something was wrong with Steve. Haggard, tense and tired he had on his face what to me looked like a black eye.

In contrast with his appearance, his speech soon became smooth and even. Yes, he was in trouble, there were objections to his proof, some were serious. He hoped to be able to fix everything. He mentioned names but indicated no hardfeelings toward anyone.

Later, he wrote about this trip and called it "rather traumatic" and "dramatic and traumatic."

These words of Steve fit very well the overall impression that he left on me in that singular meeting at Rio's airport late June 1960 as the day was just breaking in: cold, measured ferocity of purpose.

When I was told about the "Beaches of Rio" episode, with the implication associated to it of a frolicsome trip South of the Rio Grande, I found it a funny counterpoint to what I had actually witnessed.

So I was there and saw from a privileged position the birth of two of Stephen Smale's major contributions to mathematics.

As it is the case with many great mathematical achievments, one detects there the presence of that magic triad—simplicity, beauty, depth—that we mathematicians strive for. But simplicity is perhaps the more easily detectable of these qualities and simplicity is also a striking feature of Steve's work mentioned above.

In this connection, it is perhaps appropriate to mention here some poetic reflexions of A. Grothendieck about invention and discovery. Translated from the French this is part of Grothendieck's poetical outburst about his craft:

> A truly new idea does not burst out all made up of diamonds, like a flash of sparkling light, also it does not come out from any machine tool no matter how sophisticated and powerful it might be. It does not announce itself with great noise proclaiming its pedigree: I am this and I am that ... It is something humble and fragile, something delicate and alive ... ."

I will stop here, well aware of the saying according to which "dinner speakers are like the wheel of a vehicle: The longer the spoke, the greater the tire."

Thank you for your attention.

8 Luncheon Talk

R. Thom

Dear Steve,

Probably more than 35 years have elapsed since the first time I heard about you. It was in Princeton in 1955 when Raoul Bott spoke to me about a very bright student, who had recently found a wonderful theorem about immersions. I was startled by the result, and by the straightforward geometric aspect of the proof (of which I gave an account at the Bourbaki Seminar a little later). Moe Hirsch gave us, in his opening lecture, a beautiful review of all the consequences of your theorem. If you will excuse me for some extra technicalities, I would like to add to these a further one, in another—not mentioned —direction; namely, the connection with the theory of singularities of smooth maps, which started to attract my interest after the founding theorems of H. Whitney on the cusp and the cuspidal point. The problem of classifying immersions can be generalized as follows: Given two smooth manifolds X^n and Y^p, and some "singularity type" (s) of local smooth maps: g; $\mathbb{R}^n \to \mathbb{R}^p$, find conditions for a homotopy class $h: X \to Y$ such that no smooth map f of the class h exhibits the singularity (s). Following Ehresmann's theory of jets, one may associate to any singularity of type (s) a set of orbits for the algebraic action of the group $L^k(n) \times L^k(p)$ of k-jets of local isomorphisms of source and target spaces in the space $J^k(n, p)$ of local jets from $\mathbb{R}^n$ to $\mathbb{R}^p$. (See, for instance, H. Levine's Notes on my Lectures as Gastprofessor in Bonn.) If it happens that this set of orbits is an algebraic cycle in homology theory, then it was proved by A. Haefliger that the dual cohomology class to this singular cycle $s(f)$ for a given map f is some specific characteristic class (mod Z or Z_2) of the quotient bundle $(T(X/f^*(T(Y))$ [this class is a polynomial characterizing the singularity (s)]. This may be seen as a faraway consequence of your immersion theorem.

Then came, a few years later, your astonishing proof of the Poincaré conjecture, which, with the preceding results, led to your winning the Fields Medal at the Moscow Congress of 1966. This was also the time of your engagement in the Berkeley Movement, and against the Vietnam War. Around that time also, I had more opportunities to get to know you. I remember happily the two trips we made in your VW Camping bus from

Paris to Eastern France, first to some Mineral Fair in the Vosges mountains, later on to Geneva. I was deeply impressed by the straightforward generosity of your political action. We, as old Europeans, are far more cautious about general political conditions; we know that things are not going to change so easily. May I seize this opportunity to express a personal comment on the present political situation? As a citizen of France, a nation which, in little more than 25 years, was twice rescued from disaster and annihilation thanks to the American Army, I look forward with apprehension to the day when the last G.I. will move out from European Soil ...

In those years, this youthful enthusiasm of yours made a strong impression on me, and—excuse again this flowery expression—I had the feeling that you wanted to make of your life a "hymn to freedom."

Scientifically also, those were the years when, with your "horseshoe" example, you started a wonderful revival of Qualitative Dynamics, a revival which, for its global impact on general science, could be compared to the glorious period of the years 1935–45 for Algebraic Topology. Not quite, nevertheless, for the revival of Dynamics you started affected only the Western part of the World's Mathematics, whereas in the Soviet Union the Dynamical School with names such as Kolmogorov, Sinai, Arnol'd was still in full bloom.

In the years 1975–80, I cannot hide the fact that our relations went through a delicate passage: After the mediatic explosion which popularized Catastrophe Theory in 1975 came a backlash reaction in which you took part. I must say that of all the criticism that Christopher Zeeman and myself had to hear around that time, that which came from you was the most difficult to bear; for being criticized by people you admire is one of the most painful experiences there is. Now, after 15 years have passed, we may look at the matter more serenely. Had we chosen for "catastrophe theory" a less fragrant (or should I say flagrant?) name, had we called it "the use of Qualitative Dynamics methods in the interpretation of natural phenomena," probably nothing of the sort would have happened. But even now, I do not feel too guilty about the mediatic aspect of the catastrophe terminology; I chose it because of its deep philosophical bearing: namely, any phenomenon entails an element of discontinuity—something must be generated by the local phenomenon, and fall upon your eye (or your measuring apparatus). In this light, the validity of Catastrophe Theory as an interpretative methodology of phenomena can hardly be denied.

Let me end with a general comment on your mathematical style. It could be said that there are basically two styles of mathematical writing: the constipated style, and the easy-going—or free air—style. Needless to say, the constipated style originated with our old master Bourbaki (and it found very early, as I discovered in Princeton in 1951, very strong supporters on this side of the Atlantic Ocean). Formalists of the Hilbertian School believed that mathematics can be done—or has to be done—without meaning. But when we have to choose between rigor and meaning, the Free Air people will decidedly choose meaning. In this we concur, and we shall leave the Old

Guard to their recriminations about laxist writing. Rigor comes always sooner or later; as General de Gaulle said: *L'intendance, ça suit toujours*, meaning that the ammunition always follows the fighters.

You always considered Mathematics as a game (this may justify some laxity in style), and we enjoyed your playing. It is now your 60th birthday. For an Academic, this is just an end of youth. We sincerely wish you to continue in the same way, to gratify us with the wonderful findings of your ever creative mind, and this for many years still to come.

9
Banquet Address at the Smalefest

E.C. Zeeman

It is a pleasure to wish Steve a happy birthday and to acknowledge the profound influence that he has had upon British mathematics. Prior to that influence, I remember the first lecture that I ever heard Henry Whitehead give way back in 1950: Henry was saying rather pessimistically that topologists had to give up the homeomorphism problem because it was too difficult and should content themselves with algebraic topology. Marshall Stone said much the same thing in a popular article called "The Revolution in Mathematics," that topology had been nearly completely swallowed up by algebra.

Then, during the next decade, two spectacular results came whizzing across the Atlantic completely dispelling that pessimism. The first was Barry Mazur's proof of the Schönflies Conjecture (later beautifully refined by Mort Brown), and the second was Steve's proof of the Poincaré Conjecture. The secret in each case was the audacity to bypass the main obstruction that had blocked research for half a century. In Barry's case, the blockage was the Alexander horned sphere which he bypassed by the hypothesis of local flatness, and in Steve's case it was dimension 3 which he bypassed by going up to dimensions 5 to infinity. Steve's result, in particular, gave a tremendous boost to the resurgent British interest in geometric topology.

But even more profound has been his influence in dynamical systems. British mathematics had suffered since the war from an artificial apartheid between pure and applied, causing research in differential equations in the UK to fall between two stools and almost disappear. But thanks to the influence of Steve and René Thom, there is now a flourishing school of dynamical systems in the UK, which is having the beneficial side effect of bringing pure and applied together again. In particular, Steve and many of his students came to the year-long Warwick symposia in 1968–69 and 1973–74 which had the effect of drawing widespread attention to the field, and in 1974 we were very proud to be able to give him an honorary degree at Warwick.

I would like to focus on three characteristics of Steve's work. Firstly, his perception is very geometric, and this has always enabled him to see through to the heart of the matter. It has given a unifying thread to all his work. His lectures are beautifully simple, and yet profound, because he always chooses exactly the right little sketch that will enable his audience to visualise the essence and remember it.

The second characteristic is his excellent mathematical taste. Or to put it another way, I happen to like the same kind of mathematics as he does. We both started in topology, and then both moved into dynamical systems. I remember at one point we both independently became interested in game theory, and when we turned up at the next conference—I think it was at Northwestern—we discovered we had both inadvertently advertised talks with the same title, dynamics in game theory, but luckily the talks were quite different.

Of course, the real evidence for the excellence of Steve's mathematical judgement is the number of mathematicians worldwide who now follow his taste. In fact, I have only known him to make one serious error of judgement, and that was his opinion of catastrophe theory.

The third characteristic of Steve's work to which I would like to draw attention is his audacity and courage. These two qualities are complementary. His audacity is his desire to make a splash, to shock people, to get under their skins, and to make them confront themselves. But one can forgive his audacity because of his courage, his courage to stand by his beliefs even when swimming against the tide. These two personality traits pervade all his activities, his mathematics, his sport and his politics.

In mathematics, he has the audacity to let his intuition leap ahead of proof, and the courage to publish that intuition as bold conjectures. In sport, he has the audacity to tackle mountains and set sail across the oceans, and the courage to carry through with these achievements. In politics, he had the audacity to rebuke the Soviet Union on the steps of Moscow University for invading Hungary, and the courage to face the KGB afterwards if necessary; and in the same breath the audacity to rebuke his own country for invading Vietnam, and the courage to face the consequences afterwards of having to defend his funding against attacks by politicians. This week he had the audacity to refuse to have his broken ankle set in plaster, and the courage to endure the resulting pain so that he would not disappoint us at the Smalefest.

Of course, much of Steve's courage springs from Clara, who has always stood by him through thick and thin, giving him a secure harbour within which to anchor, and from which he could then sail out to conquer the world. It was she who introduced him to the collecting of minerals, to which he has devoted so much of his enthusiasm over the years. I would like to say specially to her tonight that we all include her in the celebrations. So I call upon you to drink a toast to Clara and Steve.

Part 3
Differential Topology

10
The Work of Stephen Smale in Differential Topology

Morris W. Hirsch

Background

The theme of this conference is "Unity and Diversity in Mathematics." The diversity is evident in the many topics covered. Reviewing Smale's work in differential topology will reveal important themes that pervade much of his work in other topics, and thus exhibit an unexpected unity in seemingly diverse subjects.

Before discussing his work, it is interesting to review the status of differential topology in the middle 1950s, when Smale began his graduate study.

The full history of topology has yet to be written (see, however, Pont [52], Dieudonné [8]). Whereas differentiable manifolds can be traced back to the smooth curves and surfaces studied in ancient Greece, the modern theory of both manifolds and algebraic topology begins with Betti's 1871 paper [4]. Betti defines "spaces" as subsets of Euclidean spaces define by equalities and inequalities on smooth functions.[1] Important improvements in Betti's treatment were made by Poincaré in 1895. His definition of manifold describes what we call a real analytic submanifold of Euclidean space; but it is clear from his examples, such as manifolds obtained by identifying faces of polyhedra, that he had in mind abstract manifolds.[2] Curiously, Poincaré's "homéomorphisme" means a C^1 diffeomorphism. Abstract smooth manifolds in the modern sense—described in terms of coordinate systems—were defined (for the two-dimensional case) by Weyl [73] in his 1913 book on Riemann surfaces.

Despite these well-known works, at mid-century there were few studies of the global geometrical structure of smooth manifolds. The subject had

[1] This is the paper defining, rather imprecisely, what are now called Betti numbers. Pont [52] points out that the same definition is given in unpublished notes of Riemann, who had visited Betti.

[2] It is not obvious that such manifolds imbed in Euclidean space!

not yet been named.[3] Most topologists were not at all interested in smooth maps. The "topology of manifolds" was a central topic, and the name of an important book by Ray Wilder, but it dealt only with algebraic and point-set topology. Steenrod's important book *The Topology of Fibre Bundles* was published in 1956. The de Rham theorems were of more interest to differential geometers than to topologists, and Morse theory was considered part of analysis.

A great deal was known about algebraic topology. Many useful tools had been invented for studying homotopy invariants of CW complexes and their mappings (Eilenberg–MacLane spaces, Serre's spectral sequences, Postnikov invariants, Steenrod's algebras of cohomology operations, etc.) Moreover, there was considerable knowledge of nonsmooth manifolds—or more accurately, manifolds that were not assumed to be smooth, such as combinatorial manifolds, homology manifolds, and so forth. Important results include Moise's theory of triangulations of 3-manifolds, Reidemeister's torsion classification of lens spaces, Bing's work on wild and tame embeddings and decompositions, and "pathology" such as the Alexander horned sphere and Antoine's Necklace (a Cantor set in $\mathbf{R}^3$ whose complement is not simply connected). The deeper significance of many of these theories emerged later, in the light of the h-cobordism theorem and its implications.

A great deal was known about 3-dimensional manifolds, beginning with Poincaré's examples and Heegard's decomposition theory of 1898. The latter is especially important for understanding Smale's work, because it is the origin of the theory of handlebody decompositions.

The deepest results known about manifolds were the duality theorems of Poincaré, Alexander, and Lefschetz; H. Hopf's theorem that the indices of singularities of a vector field on a manifold add up to the Euler characteristic; de Rham's isomorphism between singular real cohomology and the cohomology of exterior differential forms; Chern's generalized Gauss–Bonnet formula; the foliation theories of Reeb and Haefliger; theories of fiber bundles and characteristic classes due to Pontryagin, Stiefel, Whitney, and Chern, with further developments by Steenrod, Weil, Spanier, Hirzebruch, Wu, Thom, and others; Rohlin's index theorem for 4-dimensional manifolds; Henry Whitehead's little-known theory of simple homotopy types; Wilder's work on generalized manifolds; P.A. Smith's theory of fixed points of cyclic group actions. Most relevant to Smale's work was M. Morse's calculus of variations in the large, Thom's theory of cobordism and transversality, and Whitney's studies of immersions, embeddings, and other kinds of smooth maps.

[3] The term "differential topology" seems to have been coined by John Milnor in the late 1950s, but did not become current for some years. The word "diffeomorphism" did not yet exist—it may be due to W. Ambrose. While Smale was at the Institute for Advanced Study, he showed me a letter from an editor objecting to "diffeomorphism," claiming that "differomorphism" was etymologically better!

No one yet knew of any examples of homeomorphic manifolds that were not diffeomorphic, or of topological manifolds not admitting a differentiable structure—Milnor's invention of an exotic 7-sphere was published in 1956. Work on the classification of manifolds, and many other problems, was stuck in dimension 2 by Poincaré's conjecture in dimension 3 (still unsolved).

The transversality methods developed by Pontryagin and Thom were not widely known. The use of manifolds and dynamical systems in mechanics, electrical circuit theory, economics, biology, and other applications is now common[4]; but in the fifties it was quite rare.

Conversely, few topologists had any interest in applications. The spirit of Bourbaki dominated pure mathematics. Applications were rarely taught or even mentioned; computation was despised; classification of structure was the be-all and end-all. Hardly anyone, pure or applied, used computers (of which there were very few). The term "fractals" had not yet been coined by Mandelbrot; "chaos" was a biblical rather than a mathematical term.

In this milieu, Smale began his graduate studies at Michigan in 1952.[5] The great man in topology at Michigan being Ray Wilder, most topology students chose to work with him. Smale, however, for some reason became the first doctoral student of a young topologist named Raoul Bott. In view of Smale's later work in applications, it is interesting that Bott had a degree in electrical engineering; and the "Bott–Duffin Theorem" in circuit theory is still important.

Immersions

Smale's work in differential topology was preceded by two short papers on the topology of maps [56, 57]. His theorems are still interesting, but not closely related to his later work. Nevertheless, the theme of much subsequent work by Smale, in many fields, is found in these papers: *fibrations*, and more generally, *the topology of spaces of paths.*

Given a (continuous) map $p: E \to B$, a *lift* of a map $f: X \to B$ is a map $g: X \to E$ such that the following diagram commutes:

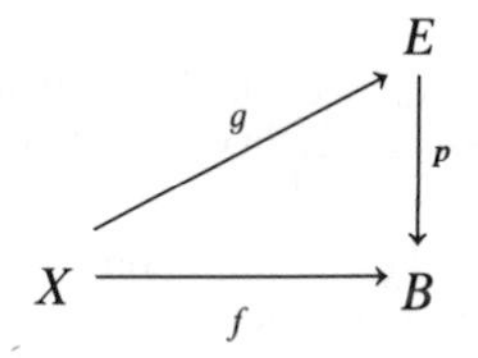

That is, $p \circ g = f$. We say (p, E, B) is a *fibration* if every path $f{:}\to B$ can be

[4] Thanks largely to Smale's pioneering efforts in these fields.

[5] Smale's autobiographical memoir in this volume recounts some of his experiences in Michigan.

lifted, the lift depending continuously on specified initial values in E.[6] Fibrations, the subject of intense research in the fifties, are the maps for which the tools of algebraic topology are best suited.

In his doctoral thesis [65], Smale introduced the use of fibrations of spaces of differentiable maps as a tool for classifying immersions. This novel technique proved to be of great importance in many fields of geometric topology, as will be discussed below.

Smale's first work in differential topology was about immersions. An *immersion* $f: M \to N$ is a smooth map between manifolds M, N such that at every $x \in M$, the tangent map $T_x f: T_x M \to T_{f(x)} N$ is injective. Here TM denotes the tangent vector bundle of M, with fiber Tf_x over $x \in M$. A *regular homotopy* is a homotopy f_t, $0 \leq t \leq 1$ of immersions such that Tf_t is a homotopy of bundle maps.[7] An immersion is an *embedding* if it is a homeomorphism onto its image, which is necessarily a locally closed smooth submanifold. A regular homotopy of embeddings is an *isotopy*.

Here is virtually everything known about immersions in the early fifties: In a *tour de force* of differential and algebraic topology and geometric intuition in 1944, Hassler Whitney [77, 78] had proved that every (smooth) n-dimensional manifold could be embedded in $\mathbf{R}^{2n}$ for $n \geq 1$, and immersed in $\mathbf{R}^{2n-1}$ for $n \geq 2$. On the other hand, it was known that the projective plane and other nonorientable surfaces could not be embedded in $\mathbf{R}^3$, and Whitney had proved other impossibility results using characteristic classes. Steenrod had a typewritten proof that the Klein bottle does not embed in real projective 3-space. The Whitney–Graustein theorem [76] showed that immersions of the circle in the plane are classified by their winding numbers. As a student working with Ed Spanier I proved the complex projective plane, which could be embedded in $\mathbf{R}^7$, could not be immersed in $\mathbf{R}^6$.[8]

Immersions of Circles

The problem Smale solved in his thesis is that of classifying regular homotopy classes of immersions of the circle into an arbitrary manifold N. More generally, he classified immersions $f: I \to N$ of the closed unit interval $I =$

[6] Precisely: Given a compact polyhedron P and maps $F: P \times I \to B$, $g: P \times 0 \to E$ such that $p \circ g = F|P \times 0 \to E$, there is an extension of g to a map $G: P \times I \to E$ such that $p \circ G = F$.

[7] What is important and subtle here is joint continuity in (t, x) of $\partial f_t(x)/\partial x$. Without it, "regular homotopy" would be the same as "homotopy of immersions." In the plane, for example, the identity immersion of the unit circle is not regularly homotopic to its reflection in a line, but these two immersions *are* homotopic through immersions, as can be seen by deforming a figure-eight immersion into each of them.

[8] The proof consisted of computing the secondary obstruction to a normal vector field on an embedding in $\mathbf{R}^7$, using a formula of S.D. Liao [38], another student of Spanier. This calculation was immediately made trivial by a general result of W.S. Massey [40].

[0, 1] having fixed boundary data, i.e., fixed initial and terminal tangent vectors $f'(0)$ and $f'(1)$.

Smale's approach was to study the map $p: E \to B$, where

- E is the space of immersions[9] $f: I \to N$ having fixed initial value $f(0)$ and fixed initial tangent $f'(0)$;
- B is the space of nonzero tangent vectors to N;
- p assigns to f the terminal tangent vector $f'(1)$.

The classification problem is equivalent to enumerating the path components of the fibers because it can be seen that such a path component is a regular homotopy class for fixed boundary data.

Bott asked Smale an extraordinarily fruitful question: *Is (p, E, B) a fibration?* This amounts to asking for a *Regular Homotopy Extension Theorem.* In his thesis [65], Smale proved the following:

Theorem. *Let $\{u_t, 0 \leq t \leq 1\}$ be a deformation of $f'(1)$ in B, i.e., a path of nonzero tangent vectors beginning with $f'(1)$. Then there is a regular homotopy $F: S^1 \times I \to N$, $F(x, t) = f_t(x)$ such that $f_0 = f$, all f_t have the same initial tangent, and the terminal tangent of f_t is u_t. Moreover, F can be chosen to depend continuously on the data f and the deformation $\{u_t\}$.*

This result is nontrivial, as can be seen by observing that it is false if N is 1-dimensional (exercise!).

It is not hard to see that the total space E is contractible. Therefore the homotopy theory of fibrations implies that the kth homotopy group of any fiber F is naturally isomorphic to the $(k + 1)$st homotopy group of the base space B. Now B has a deformation retraction onto the space $T_1 N$ of unit tangent vectors. By unwinding the homotopies involved, Smale proved the following result theorem for Riemannian manifolds N of dimension $B \geq 2$:

Theorem. *Assume N is a manifold of dimension $n \geq 2$. Fix a base point x_0 in the circle, and a nonzero "base vector" v_0 of length 1 tangent to N. Let F denote the space of immersions $f: S^1 \to N$ having tangent v_0 at x_0. To f assign the loop $f_\#: S^1 \to T_1 N$, where $f_\#$ sends $x \in S^1$ to the normalized tangent vector to f at x, namely $f'(x)/\|f'(x)\|$. Then $f_\#$ induces a bijection between the set of path components of F and the fundamental group $\pi_1(T_1 N, v_0)$.*

For the special case where N is the plane, this result specializes to the Whitney–Graustein theorem [76] stated above.

---

[9] A space of immersions is given the C^1 topology. This means that two immersions are close if at each point their values are close and their tangents are close. It turns out that the homotopy type of a space of immersions is the same for all C^r topologies, $1 \leq r \leq \infty$.

Immersions of Spheres in Euclidean Spaces

Smale soon generalized the classification to immersions of the k-sphere S^k in Euclidean n-space $\mathbf{R}^n$. Again the key was a fibration theorem. Let E now denote the space of immersions of the closed unit k-disk D^k into $\mathbf{R}^n$, and B the space of immersions of S^{k-1} into $\mathbf{R}^n$. The map $p: E \to B$ assigns to an immersion $f: D^k \to \mathbf{R}^n$ its restriction to the boundary. Smale proved that (p, E, B) is a fibration *provided* $k < n$. Geometrically, this says regular homotopies of $f|S^{k-1}$ can be extended over D^k to get a regular homotopy of f, and similarly for k-parameter families of immersions.

Using this and similar fibration theorems, Smale obtained the following result [58, 59]:

Theorem. *The set of regular homotopy classes of immersions $S^k \to \mathbf{R}^n$ corresponds bijectively to $\pi_k(V_{n,k})$, the kth homotopy group of the Stiefel manifold of k-frames in $\mathbf{R}^n$, provided $k < n$.*

To an immersion $f: S^k \to \mathbf{R}^n$ Smale assigned the homotopy class of a map $\theta: S^k \to V_{n,k}$ as follows. By a regular homotopy, we can assume f coincides with the standard inclusion $S^k \to \mathbf{R}^n$ on a small open k-disk in S^k, whose complement is a closed k-disk B. Let $e(x)$ denote a field of k-frames tangent to B. Form a k-sphere Σ by gluing two copies B_0 and B_1 of B along the boundary. Define a map $\sigma(f): \Sigma \to V_{n,k}$ by mapping x to $f_* e(x)$ if $x \in B_0$, and to $e(x)$ if $x \in B_1$. Here f_* denotes the map of frames induced by Tf. The homotopy class of $\sigma(f)$ is called the *Smale invariant* of the immersion f.

The calculation of homotopy groups is a standard task for algebraic topology. While it is by no means trivial, in any particular case a lot can usually be calculated. The Stiefel manifold $V_{n,k}$ has the homotopy type of the homogeneous space $O(n)/O(n-k)$, where $O(m)$ denotes the Lie group of real orthogonal $m \times m$ matrices. Therefore explicit classifications of immersions were possible for particular values of k and n, thanks to Smale's theorem.[10]

A surprising application of Smale's classification is his theorem that *all immersions of the 2-sphere in 3-sphere are regularly homotopic*, the reason being that $\pi_2(O(3)) = 0$.[11] In particular *the identity map of S^2, considered as an immersion into $\mathbf{R}^3$, is regularly homotopic to the antipodal map.* The analogous statement is false for immersions of the circle in the plane.

When Smale submitted his paper on immersions of spheres for publication, one reviewer claimed it could not be correct, since the identity and

---

[10] Even where the homotopy group $\pi_k(V_{n,k})$ has been calculated, there still remains the largely unsolved geometric problem of finding an explicit immersion $f: S^k \to \mathbf{R}^n$ representing a given homotopy class. Some results for $k = 3$, $n = 4$ were obtained by J. Hass and J. Hughes [18].

[11] Always remember: π_2 of *any* Lie group is 0.

antipodal maps of S^2 have Gauss maps of different degrees![12] Exercise: Find the reviewer's mistake!

It is not easy to visualize such a regular homotopy, now called an *eversion* of the 2-sphere. After Smale announced his result, verbal descriptions of the eversion were made by Arnold Shapiro (whom I could not understand), and later by Bernard Morin (whom I could).[13]

One way to construct an eversion is to first regularly homotop the identity map of the sphere into the composition of the double covering of the projective plane followed by *Boy's surface*, an immersion of the projective plane into 3-space pictured in *Geometry and the Imagination* [24]. Since this identifies antipodal points, the antipodal can also be regularly homotoped to this same composition.

Tony Phillips' *Scientific American* article [49] presents pictures of an eversion. Charles Pugh made prizewinning wire models of the eversion through Boy's surface, unfortunately stolen from Evans Hall on the Berkeley campus. There is also an interesting film by Nelson Max giving many visualizations of eversions. Even with such visual aids, it is a challenging task to understand the deformation of the identity map of S^2 to the antipodal map through immersions.

Smale's proof of the Regular Homotopy Extension Theorem (for spheres and disks of all dimensions) is based on integration of certain vector fields, foreshadowing his later work in dynamics.

There is no problem in extending a regular homotopy of the boundary restriction of f to a smooth homotopy of f; the difficulty is to make the extension a regular homotopy. Smale proceeded as follows.

Since D^k is contractible, the normal bundle to an immersion $f\colon D^k \to \mathbf{R}^n$ is trivial. Therefore, to each $x \in D^k$, we can continuously assign a nonzero vector $w(x)$ normal to the tangent plane to $f(D^k)$ at $f(x)$.[14] (Note the use of the hypothesis $k < n$.) Now $f(D^k)$ is not an embedded submanifold, and w is not a well-defined vector field on $f(D^k)$, but f is locally an embedding, and w extends locally to a vector field in $\mathbf{R}^n$. This is good enough to use integral curves of w to push most of $f(D^k)$ along these integral curves, out of the way of the given deformation of f along S^{k-1}. Because of the extra dimension, Smale was able to use this device to achieve regularity of the extension. Of course, the details, containing the heart of the proof, are formidable. But the concept is basically simple.

In his theory of immersions of spheres in Euclidean spaces, Smale introduced two powerful new methods for attacking geometrical problems:

---

[12] The *Gauss map*, of an embedding f of a closed surface S into 3-space, maps the surface to the unit 2-sphere by sending each point $x \in S$ to the unit vector outwardly normal to $f(S)$ at $f(x)$.

[13] Morin is blind.

[14] More precisely, $w(x)$ is normal to the image of $Df(x)$, the derivative of f at x.

Dynamical systems theory (i.e., integration of vector fields) was used to construct deformations in order to prove that certain restriction maps on function spaces are fibrations; and then algebraic topology was used to obtain isomorphisms between homotopy groups. These techniques were soon used in successful attacks on a variety of problems.

Further Development of Immersion Theory

I first learned of Smale's thesis at the 1956 Symposium on Algebraic Topology in Mexico City. I was a rather ignorant graduate student at the University of Chicago; Smale was a new Ph.D. from Michigan.[15] While I understood very little of the talks on Pontryagin classes, Postnikov invariants and other arcane subjects, I thought I could understand the deceptively simple geometric problem Smale addressed: *Classify immersed curves in a Riemannian manifold.*

In the fall of 1956, Smale was appointed Instructor at the Universtiy of Chicago. Having learned of Smale's work in Mexico City, I began talking with him about it, and reading his immersion papers. I soon found much simpler proofs of his results. Every day I would present them to Smale, who would patiently explain to me why my proofs were so simple as to be wrong. By this process I gradually learned the real difficulties, and eventually I understood Smale's proofs.

In my own thesis [25] directed by Ed Spanier, I extended Smale's theory to the classification of immersions $f: M \to N$ between arbitrary manifolds, provided $\dim N > \dim M$. In this I received a great deal of help from both Smale and Spanier. The main tool was again a fibration theorem: the restriction map, going from immersions of M to germs of immersions of neighborhoods of a subcomplex of a smooth triangulation of M, is a fibration.

The proof of this fibration theorem used Smale's fibration theorem for disks as a local result; the globalization was accomplished by means of a smooth triangulation of M, the simplices of which are approximately disks.

The classification took the following form. Consider the assignment to f of its tangent map $TF: TM \to TN$ between tangent vector bundles. This defines a map Φ going from the space of immersions of M in N to the space of (linear) bundle maps from TM to TN that are injective on each fiber. The homotopy class of this map (among such bundle maps) generalizes the Smale invariant. Using the fibration theorem and Smale's theorems, I showed Φ induces isomorphisms on homotopy groups. By results of Milnor and J.H.C. Whitehead, this implies Φ is a homotopy equivalence. The proof of the classification is a bootstrapping induction on the dimension of M; the inductive step uses the fibration theorem.

---

[15] For an account of the atmosphere in Chicago in the fifties, see my memoir [28].

Thus immersions are classified by certain kinds of bundle maps, whose classification is a standard task for algebraic topology. A striking corollary of the classification is that every parallelizable manifold is immersible in Euclidean space of one dimension higher.[16]

Several topologists[17] reformulated the classification of immersions of an m-dimensional manifold M into a Euclidean space $\mathbf{R}^{n+k}$ as follows. Let Ψ_M: $M \to BO$ be the classifying map (unique up to homotopy) for the stable normal bundle of M. An immersion f: $M \to \mathbf{R}^{n+k}$ determines a lift of Ψ_M over the natural map $BO(k) \to BO$. Using homotopy theory, it can be deduced from the classification theorem that regular homotopy classes of immersions correspond bijectively in this way to homotopy classes of lifts of Ψ_M.

Subsequently many other classification problems were solved by showing them to be equivalent to the homotopy classification of certain lifts, or what is the same thing, crosssections of a certain fibration. The starting point for this approach to geometric topology was the extraordinarily illuminating talk of R. Thom at the International Congress of 1958 [70], in which he stated that smoothings of a piecewise linear manifold correspond to sections of a certain fibration.[18]

Other proofs of the general immersion classification theorem were obtained by R. Thom [69], A. Phillips [48], V. Poenaru [50], and M. Gromov and Ja. Eliasberg [14, 15] (see also A. Haefliger [16]). Each of these different approaches gave new insights into the geometry of immersions.

Many geometrically minded topologists were struck by the power of the fibration theorem and attacked a variety of mapping and structure problems with fibration methods.

The method of fibrations of function spaces was applied to *submersions* (smooth maps f: $M \to N$ of rank equal to dim N) by A. Phillips [48]. Again the key was a fibration theorem, and the classification was by induced maps between tangent bundles. This was generalized to *k-mersions* (maps of rank $k >$ dim N) by S. Feit [10]. General immersion theory was made applicable to immersions between manifolds of the *same* dimension, provided the domain manifold has no closed component, by V. Poenaru [50] and myself [26].

Fibration methods were used to classify piecewise linear immersions by Haefliger and Poenaru [17]. Topological immersions were classified by J.A. Lees [37] and R. Lashof [36].

M. Gromov [11–13] made a profound study of mapping problems amenable to fibration methods, and successfully attacked many geometric problems of the most diverse types. The article by D. Spring in this volume discusses

[16] Exercise (unsolved): Describe an explicit immersion of real projective 7-space in $\mathbf{R}^8$!

[17] The first may have been M. Atiyah [2].

[18] This may have been only a conjecture—Thom was not guilty of excessive clarity.

Gromov's far-reaching extensions of immersion theory to other mapping problems such as immersions which are symplectic, holomorphic, or isometric; see Chapters 2 and 3 of Gromov's book [13].

R. Thom [69] gave a new, more conceptual proof of Smale's theorem. R.S. Palais [47] proved an isotopy extension theorem, showing that the restriction map for *embeddings* is not merely a fibration, it is a locally trivial fiber bundle (see also E. Lima [39]). R. Edwards and R. Kirby [9] proved an isotopy extension theorem for topological manifolds.

The 1977 book [35] by R. Kirby and L. Siebenmann contains a unified treatment of many classification theories for structures on topological, piecewise linear, and smooth manifolds. Besides many new ideas, it presents developments and analogues of Smale's fibration theories, Gromov's ideas, and Smale's later theory of handlebodies. See in particular, Siebenmann's articles [54] and [55], a reprinting of [53].

Diffeomorphisms of Spheres

In 1956 Milnor astounded topologists with his construction of an exotic differentiable structure on the 7-sphere, that is, *a smooth manifold homeomorphic but not diffeomorphic to* S^7. This wholly unexpected phenomenon triggered intense research into the classification of differentiable structures, and the relation between smooth, piecewise linear, and topological manifolds.

Milnor's construction was based on a diffeomorphism of the 6-sphere which, he proved, could not be extended to a diffeomorphism of the 7-ball; it was, therefore, not isotopic to any element of the orthogonal group $O(7)$ considered as acting on the 6-sphere. His exotic 7-sphere was constructed by gluing together two 7-balls by this diffeomorphism of their boundaries. These ideas stimulated investigation into diffeomorphism groups.

Two-spheres

In 1958 Smale [59] published the following result:

Theorem. *The space* $\mathrm{Diff}(S^2)$ *of diffeomorphism of the 2-sphere admits the orthogonal group* $O(3)$ *as a deformation retract.*[19]

Again a key role in the proof was played by dynamical systems. I recall Smale discussing his proof of this at Chicago. At one stage he did not see

[19] Around this time an outline of a proof attributed to Kneser was circulating by word of mouth; it was based on an alleged version of the Riemann mapping theorem which gives smoothness at the boundary of smooth Jordan domains, and smooth dependence on parameters. I do not know if such a proof was ever published.

why the proof did not go through for spheres of all dimensions, except that he knew that the analogous result for the 6-sphere would contradict Milnor's constructions of an exotic differentiable structure on the 7-sphere! It turned out that the Poincaré–Bendixson Theorem, which is valid only in dimension 2, played a key role in his proof.

In 1958 Smale went to the Institute for Advanced Study. This was a very fertile period for topology, and a remarkable group of geometers and topologists were assembled in Princeton. These included Shiing–Shen Chern, Ed Spanier, Armand Borel, Ed Floyd, Dean Montgomery, Lester Dubins, Andy Gleason, John Moore, Ralph Fox, Glenn Bredon, John Milnor, Richard Palais, Jim Munkres, André Weil, Henry Whitehead, Norman Steenrod, Bob Williams, Frank Raymond, S. Kinoshita, Lee Neuwirth, Stewart Cairns, John Stallings, Barry Mazur, Papakyriakopoulos, and many others.

I shared an office with Smale and benefited by discussing many of his ideas at an early stage in their development. Among the many questions that interested him was a famous problem of P.A. Smith: Can an involution (a map of period 2) of S^3 have a knotted circle of fixed points? He did not solve it, but we published a joint paper [29] on smooth involutions having only two fixed points. Unfortunately it contains an elementary blunder, and is totally wrong.[20]

Three-spheres

Smale worked on showing that the space $\mathrm{Diff}(S^3)$ of diffeomorphism of the 3-sphere admits the orthogonal group $O(4)$ as a deformation retract. Using several fibrations, such as the restriction map going from diffeomorphisms of S^3 to embeddigs of D^3 in S^3, and from the latter to embeddings of S^2 in S^3, and so forth, he reduced this to the same problem for the space of embeddings of S^2 in $\mathbf{R}^3$. Although it failed, his approach was important, and stimulated much further research. Hatcher [20] proved Smale's conjecture in 1975.

In a manuscript for this work Smale analyzed an embedding in Euclidean space by considering a *height function*, i.e., the composition of the embedding with a nonzero linear function.[21]

Smale tried to find a height function which, for a given compact set of embeddings of S^2 in $\mathbf{R}^3$, would look like a Morse function for each embedding, exhibiting it as obtained from the unit sphere by extruding pseudopods in a manageable way. These could then all be pushed back, following the height function, until they all became diffeomorphisms of S^2.

---

[20] I am glad to report that other people have also made mistakes in this problem.

[21] This idea goes back to Möbius [45], who used it in an attempt to classify surfaces; see Hirsch [27] for a discussion. J. Alexander [1] had used a similar method to study piecewise linear embeddings of surfaces.

At that point, appeal to his theorem on diffeomorphisms of S^2 would finish the proof.

He had a complicated inductive proof; but Robert Williams, Henry Whitehead and I (all at the Institute then) found that the induction failed at the first step!

Nevertheless the idea was fruitful. J. Cerf [6] succeeded in proving that Diff(S^3) has just two path components. Cerf used a more subtle development of Smale's height function: He showed that for a one-parameter family of embeddings, there is a function having at worst cubic singularities, but behaving topologically like a Morse function for each embedding in the family. Cerf's ideas were to prove useful in other deformation problems in topology and dynamics, and surprisingly, in algebraic K-theory. See Hatcher [19], Hatcher and Wagoner [21], and Cerf [5, 7].

Smale would return to the use of height functions as tools for dissecting manifolds in his spectacular attack on the generalized Poincaré conjecture.

In using height functions to analyze embedded 2-spheres, Smale was grappling with a basic problem peculiar to the topology of manifolds: *There is no easy way to decompose a manifold.* Unlike a simplicial complex, which come equipped with a decomposition into the simplest spaces, a smooth manifold —without any additional structure such as a Riemannian metric—is a homogeneous global object. If it is "closed"—compact, connected and without boundary—it contains no proper closed submanifold of the same dimension, is not a union of a countable family of closed submanifolds of lower dimension. This is a serious problem if we need to analyze a closed manifold because it means *we cannot decompose it into simpler objects of the same kind.*

Before 1960 the traditional tool for studying the geometric topology of manifolds was a smooth triangulation. Cairns and Whitehead had shown such triangulations exist and are unique up to isomorphic subdivisions. Thus to every smooth manifold there is associated a combinatorial manifold. In this way simplicial complexes, for which combinatorial techniques and induction on dimension are convenient tools, are introduced into differential topology. But useful as they are for algebraic purposes, they are not well-suited for studying differentiable maps.[22]

Smale would shortly return to the use of Morse functions to analyze manifolds. His theory of theory of handlebodies was soon to supply topologists with a highly succussful technique for decomposing smooth manifolds.

---

[22] Simplicial complexes were introduced, as were so many other topological ideas, by Poincaré. Using them he gave a new and much more satisfactory definition of Betti numbers, which had originally been defined in terms of boundaries of smooth submanifolds. It is interesting that while the old definition was obviously invariant under Poincaré's equivalence relation of "homéomorphisme," which meant what we call "C^1 diffeomorphism," invariance of simplicially defined Betti numbers is not at all obvious. (It was later proved by J. Alexander.) Thus the gap between simplicial and differentiable techniques has plagued topology from its beginnings.

The Generalized Poincaré Conjecture and the *h*-Cobordism Theorem

In January of 1960 Smale arrived in Rio de Janeiro to spend six months at the Instituto de Matematica Pura e Aplicada (IMPA). Early in 1960, he submitted a research announcement: *The generalized Poincaré conjecture in higher dimensions* [60], along with a handwritten manuscript outlining the proof. The editors of the *Bulletin of the American Mathematical Society* asked topologists in Princeton to look over the manuscript. I remember Henry Whitehead, who had once published his own (incorrect) proof, struggling with Smale's new techniques.[23]

The theorem Smale announced in his 1960 *Bulletin* paper is, verbatim:

Theorem (Theorem A). *If M^n is a closed differentiable (C^∞) manifold which is a homotopy sphere, and if $n \neq 3, 4$, then M^n is homeomorphic to S^n.*

The notation implies M^n has dimension n. "Closed" means compact without boundary. Such a manifold is a *homotopy sphere* if it is simply connected and has the same homology groups as the n-sphere (which implies it has the same homotopy type as the n-sphere).

Poincaré [51] had raised the question of whether a simply connected 3-manifold having the homology of the 3-sphere is homeomorphic to the 3-sphere S^3.[24] Some form of the generalized conjecture (i.e., the result proved by Smale without any dimension restriction) had been known for many years; it may be have been due originally to Henry Whitehead.

Very little progress had been made since Poincaré on his conjecture.[25] Because natural approaches to the generalized conjecture seemed to require knowledge of manifolds of lower dimension, Smale's announcement was

---

[23] Whitehead was very good about what he called "doing his homework," that is, reading other people's papers. "I would no more use someone's theorem without reading the proof," he once remarked, "than I would use his wallet without permission." He once published a paper relying on an announcent by Pontryagin, without proof, of the formula $\pi_4(S^2) = 0$, which was later shown (also by Pontryagin) to have order 2. Whitehead was quite proud of his footnote stating that he had not seen the proof. Smale, on the other hand, told me that if he respected the author, he would take a theorem on trust.

[24] As Smale points out in his *Mathematical Intelligencer* article [67], Poincaré does not hazard a guess as the the answer. He had earlier mistakenly announced that a 3-manifold is a 3-sphere provided it has the same homology. In correcting his mistake, by constructing the dodecahedral counterexample, he invented the fundamental group. Thus we should really call it Poincaré's *question*, not conjecture.

[25] There is still no good reason to believe in it, except a lack of counterexamples; and some topologists think the opposite conjecture is more likely. *Maybe* it is undecidable!

astonishing. Up to then, no one had dreamed of proving things only for manifolds of higher dimension, three dimensions already being too many to handle.

Nice Functions, Handles, and Cell Structures

Smale's approach is intimately tied to his work, both later and earlier, on dynamical systems. At the beginning of his stay in Princeton, he had been introduced to Mauricio Peixoto, who got Smale interested in dynamical systems.[26]

Smale's proof of Theorem A begins by decomposing the manifold M (dropping the superscript) by a special kind of Morse function $f: M \to \mathbf{R}$, which he called by the rather dull name of "nice function." He wrote:

> The first step in the proof is the construction of a nice cellular type structure on any closed C^∞ manifold M. More precisely, define a real-valued f on M to be a *nice* function if it possesses only nondegenerate critical points and for each critical point β, $f(\beta) = \lambda(\beta)$, the index of β.

It had long been known (due to M. Morse) that any Morse function gives a homotopical reconstruction of M as a union of cells, with one s-cell for each critical point of index s.

Smale observed that the s-cell can be "thickened" in M to a set which is diffeomorphic to $D^s \times D^{n-s}$. Such a set he calls a *handle of type s*; the type of a handle is the dimension of its *core* $D^s \times 0$. Thus from a Morse function he derived a description of M as a union of handles with disjoint interiors.

But Smale wanted the handles to be successively adjoined in the order of their types: First 0-handles (n-disks), then 1-handles, and so on. For this, he needed a "nice" Morse function: The value of the function at a critical point equals the index of the critical point. A little experimentation shows that most Morse functions are not nice. Smale stated:

Theorem (Theorem B). *On every closed C^∞ manifold there exist nice functions.*

To get a nice function, Smale had to rearrange the k-cell handle cores, and to do this he first needed to make the stable and unstable manifolds of all the critical points to meet each other transversely.[27]

---

[26] See Peixoto's article on Smale's early work, in this volume; and also Smale's autobiographical article [66].

[27] If p is a singular point of a vector field, its *stable manifold* is the set of points whose trajectories approach p as $t \to \infty$. The *unstable manifold* is the stable manifold for $-f$, comprising trajectories going to p in negative time.

Smale referred to his article "Morse Inequalities for a Dynamical System" [61] for the proof that a gradient vector field on a Riemannian manifold can be C^1 approximated by a gradient vector field for which the stable and unstable manifolds of singular points meet each other transversely. From this he was able to construct a nice function. The usefulness of this will be seen shortly.

In his *Bulletin* announcement [60] Smale then made a prescient observation:

> The stable manifolds of the critical points of a nice function can be thought of as the cells of a complex while the unstable manifolds are the dual cells. This structure has the advantage over previous structures that both the cells and the duals are differentiably imbedded in M. We believe that nice functions will replace much of the use of C^1 triangulations and combinatorial methods in differential topology.

With nice functions at his disposal, Smale could decompose any closed manifold into a union of handles, successively adjoined in the same order as their type. This is a far-reaching generalization of the work of Möbius [45], who used what we call Morse functions in a similar way to decomposed surfaces.

Eliminating Superfluous Handles

The results about nice functions stated so far apply to all manifolds. To prove Theorem A required use of the hypothesis that M is a homotopy sphere of dimension at least five. What Smale proved was that in this case there is a Morse function with *exactly* two critical points—necessarily a maximum and a minimum. It then follows easily, using the grid of level surfaces and gradient lines, that M is the union of two smooth n-dimensional submanifolds with boundary, meeting along their common boundary, such that each is diffeomorphic to D^n.

From this it is simple to show that M is homeomorphic to S^n.[28] Actually more is true. In the first place, it is not hard to show from the decomposition of M into two n-balls that *the complement of point in M is diffeomorphic to $\mathbf{R}^n$*. Second, it follows from the theory of smooth triangulations that *the piecewise linear (PL) manifold*[29] *which smoothly triangulates M is PL isomorphic to the standard PL n-sphere.*

How did Smale get a Morse function with only two critical points? He used the homotopical hypothesis to eliminate all other critical points. To

---

[28] In fact, it takes some thought to see why one cannot immediately deduce that M is *diffeomorphic* to S^n; but recall Milnor's celebrated 7-dimensional counterexample [42].

[29] A *piecewise linear manifold* has a triangulation in which the closed star of every vertex is isomorphic to a rectilinear subdivision of a simplex.

visualize the idea behind his proof, imagine a sphere embedded in 3-space in the form of a U-shaped surface. Letting the height be the nice Morse function, we see that there are two maxima, one minimum, and one saddle. The stable manifold of the saddle is a curve, the two ends of which limit at the two maxima. Now change the embedding by pushing down on one of the maxima until the part of the U capped by that maximum has been mashed down to just below the level of the saddle. This can be done so that on the final surface the saddle point has become noncritical, and no new saddle has been introduced. Thus on the new surface, which is diffeomorphic to the original, there is a Morse function with only two critical points. *If we had not known that the original surface is diffeomorphic to the 2-sphere, we would realize it now.*

The point to observe in this process is that *we canceled the extra maximum against the saddle point*; both disappeared at the same time.

Smale's task was to do this in a general way. Because M is connected, there is no topological reason for the existence of more than one maximum and one minimum. If there are two maxima, the Morse inequalities, plus some topology, require the existence of a saddle whose stable manifold is one-dimensional and limits at two maxima, just as in the U-shaped example earlier. Smale redefined the Morse function on the level surfaces above this saddle, and just below it, to obtain a new nice function having one fewer saddle and one fewer maximum. In this way, he proved [62] there exists, on any connected manifold, a nice function with only one maximum and one minimum (and possibly other critical points).

The foregoing had already been proved by M. Morse [46]. Smale went further. Assuming that M is simply connected and of dimension at least five, he used a similar cancellation of critical points to eliminate all critical points of index 1 and $n - 1$.

Each handle corresponds both to a critical point and to a generator in a certain relative singular chain group. Under the assumptions that homology groups vanish, it follows that these generators must cancel algebraically in a certain sense. The essence of Smale's proof of Poincaré's conjecture was to show how to imitate this algebraic calculation vith a geometric one: By isotopically rearranging the handles, he showed that a pair of handles of successive dimensions fit together to form an n-disk. By absorbing this disk into previously added handles, he produced a new handle decompostion with two fewer handles, together with a new Morse function having two fewer critical points. To make the algebra work and to perform the isotopies, Smale had to assume the manifold is simply connected and of dimension at least five.

In this way he proved the following important result. Recall that M is r-connected if every map of an i-sphere into M is contractible to a point for $0 \leq i \leq r$. The ith *type number* μ_i of a Morse function is the number of critical points of index i.

Theorem (Theorem D). *Let M^n be a closed $(m-1)$-connected C^∞ manifold, with $n \geq 2m$ and $(n,m) \neq (4,2)$. Then there is a nice function on M whose type numbers satisfy $\mu_0 = \mu_n = 1$ and $\mu_i = 0$ for $0 < i < m$, $n - m < i < n$.*

This is a kind of converse to the theorem on Morse inequalities.

Smale applied Theorem D to obtain structure theorems for certain simply connected manifolds manifold having no homology except in the bottom, top, and middle dimensions. To state them, we need Smale's definition of a *handlebody of type* (n,k,s): This is an n-dimensional manifold H obtained "by attaching s-disks, k in number, to the n-disk and 'thickening' them."[30] The class of such handlebodies Smale denoted by $\mathscr{H}(n,k,s)$. Notice that a manifold in $\mathscr{H}(n,0,s)$ is a homotopy n-sphere that is the union of two n-disks glued along their boundaries.

For odd-dimensional manifolds, Smale generalized the classical Heegard decomposition of a closed 3-manifold [22, 23]:

Theorem (Theorem E). *Let M be a closed $(m-1)$-connected C^∞ manifold of dimension $2m+1$, $m \neq 2$. Then M is the union of two handlebodies H, $H' \in \mathscr{H}(2m+1,k,m)$.*

For highly connected even-dimensional manifolds, Smale proved the following result which generalizes the classification of closed surfaces:

Theorem (Theorem F). *Let M be a closed $(m-1)$-connected C^∞ $2m$-manifold, $m \neq 2$. Then there is a nice function on M whose type numbers equal the Betti numbers of M.*

For a surface $m = 1$, and one should additionally assume M is orientable (otherwise the projective plane is a counterexample). Suppose M is a connected compact orientable surface of genus g. If $g = 0$ then the first Betti number is zero, and Theorem F says there is a Morse function with only two critical points, which implies M is a sphere. For higher genus, one can derive from Theorem F the usual picture of sphere with g hollow handles.[31]

---

[30] "Handlebody" is from the German "henkelkörper," a term common in the fifties (but J. Eells always said "Besselhagen"). Although it sounds innocuous today, at the time "handlebody" struck many people as a clumsy neologism—which only made Smale use it more.

[31] Both Jordan [31] and Möbius [45] published proofs of the classification of compact surfaces in the 1860s. From a modern standpoint these are failures. The authors lacked even the language to define what we mean by homeomorphic spaces. It is striking that the following "definition" was used by both of them: Two surfaces are equivalent if each can be decomposed into infinitely small pieces so that contiguous pieces of one correspond to contiguous pieces of the other. While we find it hard to make sense of this, apparently none of their readers was disturbed by it!

Theorem (Theorem H). *There exists a triangulated manifold with no differentiable structure.*

In fact he proved a significantly stronger result: There is a closed PL manifold which does not have the homotopy type of any smooth manifold.

Smale started with a certain 12-dimensional handlebody $H \in \mathscr{H}(12, 8, 6)$ previously constructed by Milnor in 1959 [43]. Milnor had shown that the boundary is a homotopy sphere which could not be diffeomorphic to a standard sphere because H has the wrong index. Smale's results showed that the boundary homeomorphic to S^{11} and a smooth triangulation makes the boundary PL homeomorphic to S^{11}. By gluing a 12-disk to H along the boundary, Smale constructed a closed PL 12-manifold M. Milnor's index argument implied that M did not have the homotopy type of any smooth closed manifold.

An entirely different example of this kind was independently constructed by M. Kervaire [33].[32]

The h-Cobordism Theorem

In his address to the Mexico City symposium in 1956 [71], Thom introduced a new equivalence relation between manifolds, which he called "J-equivalence." This was renamed "h-cobordism" by Kervaire and Milnor [34]. Two closed smooth n-manifolds M_0, M_1 are *h-cobordant* if there is a smooth compact manifold W of dimension $n + 1$ whose boundary is diffeomorphic to the disjoint union of submanifolds $V_i, i = 0, 1$, such that M_i and N_i are diffeomorphic, and each N_i is a deformation retract of W. Such a W is an *h-cobordism* between M_0 and $M - 1$.

This is a very convenient relation, linking differential and algebraic topology. It defines an equivalence relation between manifolds in terms of another manifold, just as a homotopy between maps is defined as another map, thus allowing knowledge about manifolds to be used in studying the equivalence relation. Whereas the geometric implications of two manifolds being h-cobordant is not clear, nevertheless it is often an easy task to verify that a given manifold W is a cobordism: It suffices to prove that all the relative homotopy groups of (W, N_i) vanish, and for this the machinery of algebraic topology is available. In contrast, there are very few methods available for proving that two manifolds are diffeomorphic; and a diffeomorphism is a very different object from a manifold.

---

[32] Kervaire's example is constructed by a similar strategy from a 10-dimensional handlebody in $\mathscr{H}(10, 2, 5)$. It has an elegant description: Take two copies of the unit disk bundle of S^5 and "plumb" them together, interchanging fiber disks and base disks in a product representation over the upper hemisphere. In place of the index, Kervaire used an Arf invariant.

For these reasons there was great excitement when, shortly after the announcement of the generalized Poincaré conjecture, Smale proved the following result [64]:

Theorem (The h-Cobordism Theorem). *Let W be an h-cobordism between M_0 and M_1. If W is simply connected and has dimension at least 6, then W is diffeomorphic to $M_0 \times I$. Therefore, M_1 and M_0 are diffeomorphic.*

So important is the h-cobordism theorem that it deserves to be called The Fundamental Theorem of Differential Topology.

Kervaire and Milnor studied oriented homotopy n-spheres under the relation of h-cobordism. Using the operation of connected sum, they made the set of h-cobordism classes of homotopy n-spheres into an Abelian group Θ_n [34]. They proved these groups to be finite for all $n \neq 3$ (the case $n = 3$ is still open), and computed their orders for $1 \leq n \leq 17$, $n \neq 3$. For example, the order is 1 for $n = 1, 2, 4, 5, 612$; it is 2 for $n = 8, 14, 16$; and it is 992 for $n = 11$. In this work, they did not use the h-cobordism theorem. Use of that theorem, however, sharpens their results, as they remark: "For $n \neq 3$, 4, Θ_n can be described as the set of all diffeomorphism classes of differentiable structures on the topological n-sphere," where it should be understood that the diffeomorphisms preserve orientation.

From Milnor and Kervaire's work Smale proved, as a corollary to the h-cobordism theorem, that *every smooth homotopy 6-sphere is diffeomorphic to S^6.*

The Structure of Manifolds

In this paper "On the Structure of 5-Manifolds" [63], Smale puts handle theory to work in classifying certain manifolds more complicated than homotopy spheres, namely, boundaries of handlebodies of type $(2m, k, m)$.

Using Milnor's surgery methods he is able to show, for example, that a smooth, closed 2-connected 5-manifold, whose second Stiefel–Whitney class vanishes, is the boundary of a handlebody of type $(6, k, 3)$. He then shows that such a 5-manifold is completely determined up to diffeomorphism by its second homology group, and he constructs examples in every diffeomorphism class.

Another result Smale states in this paper is that *every smooth, closed 2-connected 6-manifold is homeomorphic either to S^6 or to a connected sum of $S^3 \times S^3$ with copies of itself.*

In the same issue of the *Annals*, C.T.C. Wall has a paper [72] called "Classification of $(n - 1)$-connected $2n$-manifolds" containing a detailed study of the smooth, combinatorial and homotopical structure of such manifolds. The h-cobordism theorem is the main tool (in addition to results of Milnor and Kervaire, plus a lot of algebra). Wall proves:

Theorem. *Let $n \geq 3$ be congruent modulo 8 to 3, 5, 6 or 7.*[33] *Let M, N be smooth, closed $(n-1)$-connected $2n$-manifolds of the same homotopy type. Then M is diffeomorphic to the connected sum of N with a homotopy $2n$-sphere. If $n = 3$ or 6 then they are diffeomorphic.*

Milnor had shown in 1956 that there are smooth manifolds homeomorphic but not diffeomorphic to S^7. Kervaire and Milnor's work [34], plus the h-cobordism theorem, showed that up to orientation-preserving diffeomorphism there are exactly 28 such manifolds. Wall [72] proved a surprising result about the product of such manifolds:

Theorem. *The product of two smooth manifolds, each homeomorphic to S^7, is diffeomorphic to $S^7 \times S^7$.*

The s-Cobordism Theorem

There is no room to chronicle the all consequences, generalizations, and applications of the h-cobordism theorem and its underlying idea of handle cancellation. But one—the *s-cobordism theorem*—is worth citing here for its remarkable blend of homotopy theory, algebra and differential topology.

As with much of topology, this story starts with J.H.C. Whitehead. In 1939, he published a paper with the mysterious title "Simplicial Spaces, Nuclei and m-Groups" [74], followed a year later by "Simple Homotopy Types" [75]. In these works, he introduced the notion of a *simple* homotopy equivalence between simplicial (or CW) complexes. Very roughly, this means a homotopy equivalence which does not overly distort the natural bases for the cellular chain groups. He answered the question of when a given homotopy equivalence is homotopic to a simple one, by inventing an obstruction, lying in what is now called the Whitehead group of the fundamental group, whose vanishing is necessary and sufficient for the existence of such a homotopy. Whitehead proved that his invariant vanishes—because the Whitehead group is trivial—whenever the fundamental group is cyclic of order 1, 2, 3, 4, 5 or ∞. See Milnor's excellent exposition [44].

Several people independently realized that Whitehead's invariant was the key to extending Smale's h-cobordism theorem to manifolds whose fundamental groups are nontrivial: D. Barden [3], B. Mazur [41], and J. Stallings [68]. The result is this:

Theorem (The s-Cobordism Theorem). *Let W be an h-cobordism between M_0 and M_1. If W has dimension at least 6, and the inclusion of M_0 (or equivalently, of M_1) into W is a simple homotopy equivalence, then W is diffeomorphic to $M_0 \times I$. Therefore, M_1 and M_0 are diffeomorphic.*

---

[33] These are the dimensions for which $\pi_{n-1}(SO) = 0$.

Using Whitehead's calculation we immediately obtain:

Corollary. *The conclusion of the h-cobordism theorem is true even if W is not simply connected, provided its fundamental group is infinite cyclic or cyclic of order* ≤ 5.

The *s*-cobordism theorem has been expounded by J. Hudson [30] and M. Kervaire [32].

References

1. J.W. Alexander, On the subdivision of 3-space by a polyhedron, *Proc. Nat. Acad. Sci. U.S.A.* **9** (1924), 406–407.
2. M.F. Atiyah, Immersions and embeddings of manifolds, *Topology* **1** (1962), 125–132.
3. D. Barden, The structure of manifolds, Ph.D. thesis, Cambridge University, 1963.
4. E. Betti, Sopra gli spazi di un numero qualunque di dimensioni, *Ann. Mathemat. Pura Appl.* **4** (1871).
5. J. Cerf, Isotopie et pseudo-isotopie, Proceedings of International Congress of Mathematicians (Moscow), pp. 429–437 (1966).
6. ———, *Sur les difféomorphismes de la sphère de dimension trois* ($\Gamma_4 = 0$), Springer Lecture Notes in Mathematics Vol. 53, Springer-Verlag, Berlin, 1968.
7. ———, Les stratification naturelles des espaces de fonctions différentiables reelles et le théorème de la pseudo-isotopie, *Publ. Math. Inst. Hautes Etudes Scient.* **39** (1970).
8. J.A. Dieudonné, A History of Algebraic and Differential Topology, 1900–1960, Birkhauser, Boston, 1989.
9. R.D. Edwards and R.C. Kirby, Deformations of spaces of embeddings, *Ann. Math.* **93** (1971), 63–88.
10. S.D. Feit, *k*-mersions of manifolds, *Acta Math.* **122** (1969), 173–195.
11. M.L. Gromov, Transversal maps of foliations, *Dokl. Akad. Nauk. S.S.S.R.* **182** (1968), 225–258 (Russian). English Translation: *Soviet Math. Dokl.* **9** (1968), 1126–1129. *Math. Reviews* **38**, 6628.
12. ———, Stable mappings of foliations into manifolds, *Izv. Akad. Nauk. S.S.S.R. Mat.* **33** (1969), 707–734 (Russian). English translation: *Trans. Math. U.S.S.R. (Izvestia)* **3** (1969), 671–693.
13. ———, *Partial Differential Relations*, Springer-Verlag, New York, 1986.
14. M.L. Gromov and Ja.M. Eliasberg, Construction of non-singular isoperimetric films, *Proceedings of Steklov Institute* **116** (1971), 13–28.
15. ———, Removal of singularities of smooth mappings, *Math. USSR (Izvestia)* **3** (1971), 615–639.
16. A. Haefliger, *Lectures on the Theorem of Gromov, Liverpool Symposium II*, Lecture Notes in Mathematics Vol. 209, (C.T.C. Wall, ed.), Springer-Verlag, Berlin, pp. 118–142.
17. A. Haefliger and V. Poenaru, Classification des immersions combinatoires, *Publ. Math. Inst. Hautes Etudes Scient.* **23** (1964), 75–91.
18. J. Hass and J. Hughes, Immersions of surfaces in 3-manifolds, *Topology* **24** (1985), 97–112.

19. A. Hatcher, Concordance spaces, higher simple homotopy theory, and applications, *Proceedings of the Symposium on Pure Mathematics*, Vol. 32, American Mathematical Society, Providence, RI.
20. ———, A proof of the Smale Conjecture, Diff(S^3) = $O(4)$, *Ann. Math.* **117** (1983), 553–607.
21. A. Hatcher and J. Wagoner, Pseudo-isotopies of compact manifolds, *Astérisque* **6** (1973).
22. P. Heegard, Forstudier til en topolgisk teori för de algebraiske sammenhang, Ph.D. thesis, University of Copenhagen, 1898.
23. ———, Sur l'analysis situs, *Bull. Soc. Math. France* **44** (1916), 161–242.
24. D. Hilbert and S. Cohn-Vossen, *Geometry and the Imagination*, Chelsea, New York, 1956.
25. M.W. Hirsch, Immersions of manifolds, *Trans. Amer. Math. Soc.* **93** (1959), 242–276.
26. ———, On imbedding differentiable manifolds in euclidean space, *Ann. Math.* **73** (1961), 566–571.
27. ———, *Differential Topology*, Springer-Verlag, New York, 1976.
28. ———, *Reminiscences of Chicago in the Fifties*, *The Halmos Birthday* Volume (J. Ewing, ed.), Springer-Verlag, New York, 1991.
29. M.W. Hirsch and S. Smale, On involutions of the 3-sphere, *Amer. J. Math.* **81** (1959), 893–900.
30. J.F.P. Hudson, *Piecewise Linear Topology*, Benjamin, New York, 1969.
31. C. Jordan, Sur les déformations des surfaces, *J. Math. Pures Appl.* (2) **11** (1866), 105–109.
32. M. Kervaire, A manifold which does not admit any differentiable structure, *Comment. Math. Helv.* **34** (1960), 257–270.
33. ———, Le théorème de Barden-Mazur-Stallings, *Comment. Math. Helv.* **40** (1965), 31–42.
34. M. Kervaire and J. Milnor, Groups of homotopy spheres, I, *Ann. Math.* **77** (1963), 504–537.
35. R.C. Kirby and L. Siebenmann, *Foundational Essays on Topological Manifolds, Smoothings, and Triangulations*, Annals of Mathematics Studies Vol. 88, Princeton University Press and Tokyo University Press, Princeton, NJ, 1977.
36. R. Lashof, Lees' immersion theorem and the triangulation of manifolds, *Bull. Amer. Math. Soc.* **75** (1969), 535–538.
37. J. Lees, Immersions and surgeries on manifolds, *Bull. Amer. Math. Soc.* **75** (1969), 529–534.
38. S.D. Liao, On the theory of obstructions of fiber bundles, *Ann. Math.* **60** (1954), 146–191.
39. E. Lima, On the local triviality of the restriction map for embeddings, *Comment. Math. Helv.* **38** (1964), 163–164.
40. W.S. Massey, On the cohomology ring of a sphere bundle, *J. Math. Mechanics* **7** (1958), 265–290.
41. B. Mazur, Relative neighborhoods and the theorems of Smale, *Ann. Math.* **77** (1936), 232–249.
42. J. Milnor, On manifolds homeomorphic to the 7-sphere, *Ann. Math.* **64** (1956), 395–405.
43. ———, Differentiable manifolds which are homotopy spheres, Mimeographed, Princeton University, 1958 or 1959.

44. ———, Whitehead torsion, *Bull. Amer. Math. Soc.* **72** (1966), 358–426.
45. A.F. Möbius, Theorie der elementaren Verwandtschaft, *Leipziger Sitzungsberichte math.-phys.* Classe **15** (1869), also in *Werke*, Bd. 2.
46. M. Morse, The existence of polar nondegenerate functions on differentiable manifolds, *Ann. Math.* **71** (1960), 352–383.
47. R.S. Palais, Local triviality of the restriction map for embeddings, *Comment. Math. Helv.* **34** (1960), 305–312.
48. A. Phillips, Submersions of open manifolds, *Topology* **6** (1966), 171–206.
49. ———, Turning a sphere inside out, *Sci. Amer.* (1966), **223**, May, 112–120.
50. V. Poenaru, Sur la théorie des immersions, *Topology* **1** (1962), 81–100.
51. H. Poincaré, Cinquième complément à l'Analysis Situs, *Rend. Circ. Mat. Palermo* **18** (1904), 45–110.
52. Jean-Claude Pont, *La topologie algébrique: des origines à Poincaré*, Presses Universitaires de France, Paris, 1974.
53. L. Siebenmann, Topological manifolds, *Proceedings of the International Congress of Mathematicians in Nice, September 1970 (Paris 6ᵉ)*, Vol. 2, Gauthiers-Villars, Paris (1971) pp. 133–163.
54. ———, Classification of sliced manifold structures, Appendix A, *Foundational Essays on Topological Manifolds, Smoothings, and Triangulations*, Princeton University Press and Tokyo University Press, Princeton, NJ, 1977; *Ann. Math.* Study 88, pp. 256–263.
55. ———, Topological manifolds, *Foundational Essays on Topological Manifolds, Smoothings, and Triangulations*, Vol. 88, Princeton University Press and Tokyo University Press, Princeton, NJ, 1977; *Ann. Math.* Study 88, pp. 307–337.
56. S. Smale, A note on open maps, *Proc. Amer. Math. Soc.* **8** (1957), 391–393.
57. ———, A Vietoris mapping theorem for homotopy, *Proc. Amer. Math. Soc.* **8** (1957), 604–610.
58. ———, The classification of immersions of spheres in euclidean spaces, *Ann. Math.* **69** (1959), 327–344.
59. ———, Diffeomorphisms of the 2-sphere, *Proc. Amer. Math. Soc.* **10** (1959), 621–626.
60. ———, The generalized Poincaré conjecture in higher dimensions, *Bull. Amer. Math. Soc.* **66** (1960), 373–375.
61. ———, Morse inequalities for a dynamical system, *Bull. Amer. Math. Soc.* **66** (1960), 43–49.
62. ———, Generalized Poincaré conjecture in dimensions greater than four, *Ann. Math.* **74** (1961), 391–406.
63. ———, On the structure of 5-manifolds, *Ann. Math.* **75** (1962), 38–46.
64. ———, On the structure of manifolds, *Amer. J. Math.* **84** (1962), 387–399.
65. ———, Regular curves on Riemannian manifolds, *Trans. Amer. Math. Soc.* **87** (1968), 492–512.
66. ———, On how I got started in dynamical systems, *The Mathematics of Time*, Springer-Verlag, New York, 1980, pp. 147–151.
67. ———, The story of the higher dimensional Poincaré conjecture (what actually happened on the beaches of Rio), *Math. Intelligencer* **12**, No. 2 (1990), 44–51.
68. J. Stallings, Lectures on polyhedral topology, Technical report, Tata Institute of Fundamental Research, Bombay, 1967, Notes by G. Ananada Swarup.
69. R. Thom, La classifications des immersions, *Seminaire Bourbaki, Exposé* 157, 1957–58 (mineographed).

70. ———, Des variétés triangulées aux variétés différentiables, *Proceedings of the International Congress of Mathematicians* 1962 (J.A. Todd, ed.), Cambridge University Press, Cambridge, 1963, pp. 248–255.
71. ———, Les classes charactèristiques de Pontryagin des variétés triangulés, Symposium Internacional de Topologia Algebraica (Mexico City), Universidad Nacional Autonomia, pp. 54–67.
72. C.T.C. Wall, Classification of $(n-1)$-connected $2n$-manifolds, *Ann. Math.* **75** (1962), 163–189.
73. H. Weyl, *Über die Idee der Riemannschen Flächen*, B.G. Teubner Verlagsgesellschaft, Stuttgart 1973. Translated as *The Concept of a Riemann Surface*, Addison-Wesley, New York, 1955.
74. J.H.C. Whitehead, Simplicial spaces, nuclei and m-groups, *Proc. London Math. Soc.* **45** (1939), 243–327.
75. ———, Simple homotopy types, *Amer. J. Math.* **72** (1940), 1–57.
76. H. Whitney, On regular closed curves in the plane, *Compositio Math.* **4** (1937), 276–284.
77. ———, The self-intersections of a smooth n-manifold in $(2n-1)$-space, *Ann. Math.* **45** (1944), 220–246.
78. ———, The singularities of a smooth n-manifold in $(2n-1)$-space, *Ann. Math.* **45** (1944), 247–293.

11
Discussion

J. STALLINGS, A. HAEFLIGER, AND AUDIENCE MEMBERS

John Stallings: I'm not exactly sure what to tell you. If only I were a better stand-up comic, I could put something together. Maybe I will just say a word about the PL versus the differential stuff. I think it's always been a kind of a two-sided thing here that even very long ago when Poincaré was committing his duality theorem, for instance, he first of all defined Betti numbers using ... it's not really clear what he was using, but I think he was using smooth things. And then he decided to make a more precise definition and used his triangulations and dual cells and stuff like that, so that at that point you might say he saw that there was some kind of logical difficulty in the smooth case. Anyhow, maybe not a real logical difficulty, but somehow a difficulty, and he got this PL version of things. And then at the same time he also more or less invented differential forms, which means no triangulations at all, and brought it back to the differential world. So I think it's a matter of the interaction of these things, it's a matter of taste. For example, my idea is that the reason I've never been really happy with the real numbers is that I must have had a very bad experience as a child and not that good a time in freshman calculus. I personally have a preference even for just incredibly horrible objects, or finite things. But this is obviously just a matter of preference. Anyway, that's all for my stand-up comedy.

I suppose since this is a discussion section, maybe somebody should discuss

Dennis Sullivan: I was once a graduate student at Princeton and I remember hearing this story that Smale proved Poincare's conjecture and the h-cobordism theorem, and Stallings could be seen walking down the corridor saying, "I wonder how he did it?" And so he came back with a whole new proof of the PL case. Is that true?

John Stallings: Well, I think we read this paper of gossip in the *Mathematical Intelligencer*. Smale had this paper, and I have these couple of letters I wrote to Zeeman that were reproduced there word for word. I think, as I recall it, I was on post-doc in Oxford and I heard these rumors about Smale's stuff. I didn't believe them at all at first. And then Papakyriakopoulos had to somehow get out of the United States for two weeks so that he could go back home, or some stupid thing like that, and so he came to visit Oxford and told me this rumor, and it seemed more reasonable hearing it from him. So then I tried to figure it out, and it turned out that what I figured out was somewhat different from what Smale had done. That was in Oxford rather than Princeton.

André Haefliger: I was reading the survey article that Steve wrote in 1963, and as you know he likes a strong statement. One of them is related to the fact that "Today, one can substantially develop differential topology most simply without any reference to the combinatorial manifold." At that time I would have completely agreed with him. But maybe a few years later it became clear, for instance, that some very strong embedding theorems were easy to prove, like the one John proved—a homotopy embedding—which shows much more clarity, like Zeeman's unknotting theorem. And a way of proving or studying differentiable embeddings would be through PL embeddings and then using smoothing theory as Moe explained. So my feeling is that maybe this is not quite right, and I would like to ask the specialists to comment about that.

Christopher Zeeman: Regarding what you said about the techniques about proof: I happen to think they're also statements about theories, too. For example, if you look at the group of knots of some sphere inside another from the smooth point of view, you often get a direct sum of something to do with differential structures, and something to do with the topological knots themselves. And you can use the PL structure to get the real topological group of knots and siphon off the rather artificial bit due to the differential structures. So regarding the PL, the real value is separating what's deep topology from what's artificial smoothing. But given that, I would agree wholeheartedly with Steve in the sense that a smooth rounded thing is aesthetically much more beautiful than this kinky thing.

Rob Kirby: I have a question, an historical question. I only became mathematically aware around 1962 of all of this history that Hirsch gave us; not all of it, but most of it was before then, and my curiosity concerns the work on the bordism theory by Thom, the stuff that was presented in the paper of Milnor on surgery and handlebodies, and so on. I'm puzzled as to how much of that was in the air. In particular, you seemed to imply that the notion of handlebodies was not really there until Smale got it the way you described. I mean, Thom's work on cobordism sure was there, and where did Milnor fit in? Tell us some more history, someone.

René Thom: About Poincaré's conjecture: Papakyriakopoulos had known for some time an algebraic formulation saying that if he could solve the corresponding algebraic problem, then the Poincaré conjecture was proved. Then it turned out that the statement of Papakyriakopoulos was wrong, so that now we are still part of the same situation. But if we could really find a strictly algebraic equivalent of the Poincaré conjecture, even if we could not later prove that a counterexample to Poincaré's conjecture exists, then in my view we could say Poincaré's conjecture is solved. But, it's a question of taste.

John Stallings: Well, there is such an algebraic formulation of Poincaré's conjecture. But it's about direct products of free groups and things like that.

Moe Hirsch: I'd like to maybe respond to Prof. Kirby's question. Though it's true in cobordism theory, which was done in 1950—the early fifties—it goes back earlier to Pontryagin, the idea that when you go past the critical point of the function, you're attaching a handle. But I don't think it was known if you could add the handles in strict order of dimension. It's true, surgery was also done separately and independently by Milnor, and Milnor and Kervaire studied the groups of homotopy spheres under h-cobordism, but it wasn't until Poincaré's conjecture that we knew that h-cobordism was the same as diffeomorphism, that these really were spheres. Certainly all these different algebraic theories were influential on each other and interacted in a very rich and wonderful way. I myself don't know the history of all that subject. No one person does, really. That's one of the reasons we're here.

André Haefliger: The idea of surgery was suggested to Milnor by Thom in '58, just two years before the solution of Poincaré's conjecture by Steve. And it was then used extensively by Milnor and Kervaire. And Kervaire, back in 1960, was able to construct the first topological manifold without a differential structure. They computed a lot of groups of homotopy spheres, but the topological equivalence relation was always h-cobordism. They did not even know if they were spheres, except for the first example of 7-spheres which Milnor knew were really homeomorphic to usual spheres. So after Smale's discovery that these things go together, it was clear that they were the same. Really, Milnor and Kervaire studied differential structures on spheres up to diffeomorphism, but they didn't know that before.

John Franks: Moe spoke of the unity in particular in the use of dynamical techniques in proving this theorem of topology, and at the risk of engaging in revisionism, I'd like to go a little farther and actually claim this is a theorem in dynamical systems, and not topology at all. It's actually a much nicer theorem when viewed from that point of view. It answers the question of what kind of gradient dynamical systems can exist on a manifold of given homotopy type. Necessary conditions, the Morse inequalities, which really just say the homology is compatible with the dynamical system, were known for a long time, and the remarkable fact about his theorem is that it says if the manifold is sufficiently high dimensional and it's simply connected, those conditions are actually sufficient as well as necessary—a strikingly remarkable theorem, I think. And the Poincaré conjecture then just falls out as an interesting corollary of dynamical systems.

Moe Hirsch: I'd like to make one historical comment. Haefliger mentioned Kervaire's theorem, "Construction of a Manifold with No Differentiable Structure." But that also was in Steve's original announcement of Poincaré's conjecture. He also constructed a manifold with no differentiable structure. He used the Poincaré conjecture to know that the boundary of a certain manifold was a sphere and that this sphere could not be an ordinary, standard sphere using Milnor's theory, so that you could glue on a cell.

Steve Smale: Even today there is confusion between surgery and handlebodies. They are conceptually quite different things.

Moe Hirsch: Steve, were you aware of this, say, in '59? Were you thinking along those lines?

Steve Smale: Which ones?

Moe Hirsch: Well, the difference between surgery and handlebodies.

Steve Smale: Yeah, I guess I had some ideas then, probably.

Moe Hirsch: That's progress.

Jack Wagoner: I want to mention, in defense of smooth topology, that Steve won the Fields medal studying the difference between smooth and PL. You all know that, right? I wanted to talk about the idea that Moe mentioned before, that one of the ideas in Steve's work was to take a Morse function on the manifold and simplify it so that it only had two critical points. And that was the basic idea behind the proof of Poincaré's conjecture. But then with the work of Cerf in studying one-parameter families of functions, using smooth topology, the idea is that by using smoothness you could really carefully analyze what happened in the path of functions as you eliminated the critical points. And so by consideration of the space of all functions on the manifold, you could put things in transverse or nice position, and you could read off, in a God-given way, the singularities that occur. Now that was a very important difference between PL topology because those transitions in the PL category were not as obvious. And that whole theme of studying spaces of functions and so on turned out to be very important in algebraic K-theory. For example, in the definition of the algebraic K-group K_2, Milnor had defined K_2—back before there was K_0 or K_1—he defined K_2 using some ideas of Steinberg. But then in studying the one-parameter Cerf theory in the nonsimply connected case, the Milnor relations were right there when you studied the singularities of the path of functions. So these ideas were carried on. I think this is a very important concept, to take the space of functions, or the space of all structures, and try to analyze the internal strata. This has had many ramifications for K-theory and L-theory, and in symbolic dynamics in more recent years.

Alan Weinstein: I'd like to think, too, that in studying differentiable manifolds, we can do everything with a decomposition given by a Morse function, but I just wanted to mention a recent example that seems to suggest that triangulations aren't dead as a tool. This is the work of Eliasberg in classifying contact structures on manifolds, structures which are quite flexible and almost as flexible as differentiable structures. He proved a theorem about the equivalence of contact structures, and the way it's done involves actually triangulating a three-dimensional manifold, then approximating its three-dimensional simplices by smoothing and working with a triangle or simplex at a time. And, at present, it seems that that's the only known proof of this classification theorem. It'll be interesting to see whether, in fact, this problem can also be attacked by using Morse functions. I'm not that familiar with it, but it's my understanding that some of Thurston's work on foliations also, at least at some point along the way, had this character of working with the triangulations.

Moe Hirsch: Well, I think one of the unsettled issues is the exact role of these different structures. There's no doubt that a triangulation has beautiful combinatorial and algebraic properties. For example, it's very easy to prove that there are triangulations with arbitrarily small simplices. And from that you can prove, there are handle decompositions with arbitrarily small handles, but I don't know an easy direct proof that you have handle decompositions with small handles. And, certainly, certain kinds of inductive arguments are easier with simplices. On the other hand, you usually have many—too many—simplices.

Another direction which I wanted to mention, but didn't, is the non-simply-connected version of the h-cobordism theorem. It's called the s-cobordism theorem, where s stands for simple. This goes back to simple homotopy type, and it's very hard to even define simple homotopy type and define the invariants that Henry Whitehead invented without reference to CW structures and cell decomposition of that kind. So it's still not clear to me in what problems it's appropriate to use a triangulation? There could be several. In what problems is it appropriate to use a CW structure or some more general kind of structure? These are issues that are still going on, and to some extent they are questions of style, what people know. So, I don't think anyone takes seriously the idea that there's a rivalry between them. On the other hand, you should mention that Dennis Sullivan showed the only difference between the differential and the PL case is really a mod 2 invariant, right? So differentiable stuff is really only Z_2, so we can ignore it if we want. So maybe as the conference goes on we can discuss more of these issues.

Karen Uhlenbeck: Nowadays the Witten–Jones invariants for 3-manifolds are basically constructed from combinatorial methods; I won't say triangulations because the 3-manifold is essentially chopped up in combinatorial fashion. However, at the moment I can't say that this is all proved, but the only known proof I know where you actually get an invariant is in Morse theory. The only one I know. So I just present the fact that I think the two methods are still sort of in tension with each other, even in work that isn't complete.

Steve Smale: Yeah, with all that's been said, I'd like to retract my 1963 statement!

Moe Hirsch: I wanted to mention an historical note. When Poincaré first defined differentiable manifolds, he defined "homéomorphisme" to be what we would call a C^1 diffeomorphism; I think it was Steve who pointed this out to me. Stallings said he didn't know exactly how Poincaré defined Betti numbers. Well, he used Betti's definition, which really wasn't very good, but he used basically smooth submanifolds. And this really wasn't adequate, so he invented triangulations and then immediately lost the invariance of Betti numbers because it's obvious that diffeomorphic manifolds have isomorphic Betti numbers if you define them in terms of differentiable submanifolds. But if you define them in terms of simplicies, and triangulation, he immediately lost that. So even in the early days there was what Karen Uhlenbeck described as a tension between the two approaches, and this is still going on today.

Audience Member: I am not really a specialist of this subject, but I remember as a graduate student hearing of a beautiful result of Dennis Sullivan's concerning Lipschitz manifolds. I was wondering if we could explain where this fits in the line of history?

Dennis Sullivan: Well, I think Terry Wall may be the first one to have asked the question "Do manifolds have Lipschitz structures?" I remember hearing that as a graduate student. I guess another point is that Milnor's discovery that there are various differentiable structures on a manifold is, in a sense, kind of disturbing. When you start classifying using the Hirsch–Mazur theory, you get all this complicated structure stuff involving homotopy groups. So the Lipschitz category is something which avoids the problems of the Hauptvermutung which is sort of almost true, except for that mod 2 invariant; but it has the advantage that you get some approximately differentiable manifold which you can do global analysis on. And then the final state of affairs is that for all manifolds, except in dimension 4, every topological manifold has a unique Lipschitz structure. In dimension 4, the Lipschitz structure behaves like a smooth structure. There are lots of them, and two smooth manifolds can be homeomorphic without being lipeomorphic. If you replace the word "Lipschitz" with "quasiconformal," it's the same story. Anyway, it's sort of a long story.

Moe Hirsch: Well, maybe we should have private discussion. And in theory, there are refreshments on the balcony, one floor up on the west side of the building.

12
Note on the History of Immersion Theory

David Spring

In his survey of Smale's work in Differential Topology, Moe Hirsch observed that there seemed to be no significant results in immersion theory published in the period between the work of H. Whitney in the forties and Smale's classification of immersions of spheres (1958, 1959). This observation overlooks the C^1-isometric immersion results of J. Nash (1954) and N. Kuiper (1955). At the time, the Nash isometric immersion theory was viewed as a separate result in Riemannian geometry; it had no relevance to Smale's immersion theory or to the flowering of immersion-theoretic topology that took place in the sixties following Smale's ground-breaking results. However, the Nash theory had an unexpected role to play as a precursor to M. Gromov's remarkable refinement of immersion theory, known as Convex Integration Theory (1973). The history of immersion theory is rich and complex. It is perhaps appropriate on the occasion of the Smalefest, at which there were interesting informal discussions on the history of Differential Topology, to tie together some historical loose ends on the relation of the Nash C^1-isometric immersion theory to these much later developments in immersion-theoretic topology due to M. Gromov.

During the 1950s J. Nash published two papers in Riemannian geometry on the realization problem of inducing an abstract Riemannian metric on a smooth manifold by an immersion or embedding of the manifold into Euclidean space. His first fundamental paper, Nash (1954) solved this realization problem in the case of an arbitrary continuous metric on a C^1-manifold, provided the codimension is ≥ 2 (the realization map is of class C^1). N. Kuiper (1955) refined the Nash construction so that only codimension ≥ 1 is required; this theorem is commonly referred to as the Nash–Kuiper C^1-isometric immersion theorem. Nash's subsequent work on realizing smooth metrics involves the so-called Nash implicit function theorem in functional analysis and is not relevant to the discussion here.

The Nash–Kuiper process begins with a C^1-immersion into some $\mathbf{R}^q$ which, after a dilatation (in the case of compact manifolds) may be assumed to be *short*: the difference between the given metric and the induced metric is positive definite. The main part of the Nash–Kuiper process consists of an infinite sequence of successive corrections to a C^1-short immersion, each correction taking place in a chart of the manifold. A correction consists of a spiraling in the normal directions to the immersion whose effect is to stretch lengths locally to produce a C^1-immersion which is less short; in the limit, one obtains a C^1-immersion which realizes the given Riemannian metric. The main interest lies in the controlled normal spiraling process, expressed by Nash in analytic terms. Gromov (1973) abstracted this spiraling process to an ingenious analytic approximation result which applies to a wide variety of geometric situations. Gromov (1973, 1986) showed that that this approximation result can be fed back into immersion-theoretic topology to provide a different sort of proof (i.e., which does not require a covering homotopy step) of homotopy classification theorems for both open and closed differential relations in jet spaces.

We briefly describe a local version of Gromov's analytic approximation theorem which lies at the center of Convex Integration Theory [this term was coined in Gromov (1973) to describe his proof of this result] and which provides an interesting analysis of the operator $\partial^r/\partial t^r$, $r \geq 1$, acting on C^r-functions $f\colon [a,b] \times K \to \mathbf{R}^q$, K an n-rectangle in the auxiliary space $\mathbf{R}^n$ (coordinates $(t,x) \in [a,b] \times \mathbf{R}^n$):

Suppose that the values of the "pure" derivative $\partial^r f/\partial t^r$ lie in the interior of the convex hull of an arcwise-connected set $A \subset \mathbf{R}^q$. Fix $\varepsilon > 0$. Then f may be perturbed to a C^r-function g such that the values of $\partial^r g/\partial t^r$ lie in a preassigned neighborhood of A and such that the following approximations obtain ($\| \ \|$ is a sup-norm):

$$\|D^\alpha(f-g)\| < \varepsilon,$$

where D^α runs over all partial derivatives in the variables (t,x) such that $|\alpha| \leq r$ and $D^\alpha \neq \partial^r/\partial t^r$. In particular, f, g are C^{r-1}-close. Employing Nash's terminology, which is still current in Convex Integration Theory, the map f is short because its derivative $\partial^r f/\partial t^r$ lies in the interior of the convex hull of A. The approximation result consists of "stretching" $\partial^r f/\partial t^r$ in a controlled manner to $\partial^r g/\partial t^r$ which lies in a neighborhood of A. This approximation result was first proved for the case $r = 1$ in Gromov (1973); the case $r \geq 1$ appeared in the work of Spring (1983) and the most general version is by Gromov (1986).

Homotopy classification results in Convex Integration Theory are obtained by following the usual scheme of working locally in charts, with one important difference: The approximation result above replaces the homotopy lifting property used in the Smale–Hirsch approach to immersion theory. Furthermore, the sphere parameters in the homotopy groups may be absorbed into the auxiliary $\mathbf{R}^n$-parameters in the formulation above. In this way, for example, Gromov (1973, 1986) reproved the Smale–Hirsch classification of immersions theorem and he gave a particularly elegant and illuminating proof of the Nash–Kuiper C^1-isometric immersion theorem. We mention here that Convex Integration applies also to obtain homotopy classification theorems for open relations which may not be invariant under diffeomorphisms of the base space, i.e., for open relations in jet spaces which lie beyond the scope of the various generalizations to the Smale–Hirsch theory that were developed during the 1960s.

One of the more interesting applications of Convex Integration Theory is an existence (and classification) theory for certain families of closed differential relations in spaces of r-jets. These closed relations correspond essentially to nonlinear rth-order systems of underdetermined partial differential equations which satisfy some general convexity conditions which ensure that Gromov's approximation theorem applies. The model example is the closed condition in the space of 1-jets which corresponds to the C^1-isometric immersion problem studied by Nash and Kuiper. Until Gromov (1973), the immersion-theoretic topology developed during the 1960s had no general techniques for solving closed conditions in jet spaces. We mention here an amusing example in Convex Integration Theory [Gromov (1986), §1.2.1] which ties together the Nash–Kuiper result with Smale's famous inversion of the 2-sphere in $\mathbf{R}^3$: The inversion of the 2-sphere can be carried out by a regular homotopy of isometric C^1-immersions.

To summarize, the Nash–Kuiper C^1-isometric immersion theorem was an important early result which, some 20 years later, was generalized in a remarkable way by Gromov in his Convex Integration Theory. This theory provided a new analytic approach in immersion-theoretic topology for solving both open and closed relations in spaces of r-jets.

References

M. Gromov, Convex integration of differential relations, *Izv. Akad. Nauk S.S.S.R.* **37**, (1973), 329–343.

M. Gromov, *Partial Differential Relations*, Springer-Verlag, New York, 1986.

M. Hirsch, Immersions of manifolds, *Trans. Amer. Math. Soc.* **93** (1959), 242–276.

N. Kuiper, On C^1-isometric imbeddings. I, II. *Indag. Math.* **17** (1955), 545–556, 683–689.

J. Nash, C^1-isometric imbeddings, *Ann. Math.* **63** (1954), 383–396.

S. Smale, A classification of immersions of the two-sphere, *Trans. Amer. Math. Soc.* **90** (1958), 281–290.

S. Smale, The classification of immersions of spheres in Euclidean spaces, *Ann. Math.* **69** (1959), 327–344.

D. Spring, Convex integration of non-linear systems of partial differential equations, *Ann. Inst. Fourier* **33** (1983), 121–177.

13
The Smale–Hirsch Principle in Catastrophe Theory

V.A. VASSIL'EV

The Smale–Hirsch principle is an assertion that the space of smooth mappings $M \to N$ without singularities of a certain kind is similar to the space of admissible sections of the jet bundle $J(M, N) \to M$ (i.e., of the sections which do not meet the singular set in the jet space). The initial examples, in which this assertion holds, were found in [Smale, Hirsch 1, 2].

The catastrophists calls the space of mappings with nongeneric singularities a discriminant. The discriminants and their complements were studied in many works; see, for example, [Ar1, 3, 7, AVGL, Br1, 2, Cerf1, 2, Cohen1, 2, Fuchs, Igusa1, 2, Looijenga, Segal, Vain].

I present here some cases in which these complements satisfy (some variants of) the Smale–Hirsch principle, and hence the study of their topology reduces to a purely homotopical problem. In this way, we obtain simple interpretations of the May–Segal theorem on the homology of braid groups [Segal] and of the Arnol'd theorem on the homology of spaces of real polynomials without roots of a high multiplicity [Ar7]. We also obtain a strengthening (and many generalizations) of K. Igusa's theorem on the space of quasi-Morse functions [Igusa1, Cerf2] and the complete solution of V.I. Arnol'd's problem on the stable cohomology ring of the complements to the discriminants of complex hypersurface singularities [Ar5, 6]. The general method of the proof has also many other applications (see Section 3.2): a new series of knot invariants, an effective method for calculation of the homology of spaces of mappings of m-dimensional spaces into m-connected spaces; a simple proof of the Goresky–MacPherson theorem for the homology of the complements of affine plane arrangements [GM].

Many results of this article were obtained by solving the problems of V.I. Arnol'd; I thank him for stating these problems. I thank also V.I. Arnol'd D.B. Fuchs, and A.A. Beilinson for many useful discussions.

1. One-Dimensional Case

1.1. Functions on $\mathbf{R}^1$ and S^1 without Multiple Zeros

Notations. Let $\mathscr{F}$ be the space of smooth functions $\mathbf{R}^1 \to \mathbf{R}^1$ equal to 1 outside a compact set; let Φ be the space of functions $S^1 \to \mathbf{R}^1$, and P_d the space of real polynomials

$$X^d + a_1 X^{d-1} + \cdots + a_d. \tag{1}$$

Let Σ_k be the subset in $\mathscr{F}$, Φ, or P_d consisting of functions having a root of multiplicity k or more. $j^k(\cdot)$ is the k-jet extension operation: For any function $f: M^1 \to \mathbf{R}^1$, $j^k(f)$ is a map $M^1 \to \mathbf{R}^{k+1}$; $j^k(f)(X) = (f(X), f'(X), \ldots, f^{(k)}(X))$.

Theorem 1. *The space $\mathscr{F} - \Sigma_k$ is homotopy equivalent to the loop space $\Omega S^{k-1} \sim \Omega(\mathbf{R}^k - 0)$; the space $\Phi - \Sigma_k$ to the free loop space $\Omega_f S^{k-1} \equiv (S^{k-1})^{S^1}$; these homotopy equivalences are realized by the $(k-1)$-jet extensions of functions.*

Theorem 2. *There exists an embedding of the space $P_d - \Sigma_k$, $d > k > 2$, into ΩS^{k-1}, which induces an isomorphism of the homotopy groups π_i, $i < [d/k]\cdot(k-2)$, an epimorphism of π_i, $i = [d/k]\cdot(k-2)$, and an isomorphism of all the cohomology groups H^i, $i \leq [d/k]\cdot(k-2)$. For $i > [d/k]\cdot(k-2)$, $H^i(P_d - \Sigma_k) = 0$.*

Example. The space $\mathscr{F} - \Sigma_3$ is homotopically equivalent to ΩS^2; in particular, $\pi_1(\mathscr{F} - \Sigma_3) \cong \pi_2(\mathscr{F} - \Sigma_3) \cong Z$. The generators of these groups are shown in the Figs. 1 and 2. Namely, consider i-spheroids in $\mathscr{F}$ as i-parametrical families $F(X, \lambda)$, $X \in \mathbf{R}^1$, $\lambda \in \mathbf{R}^i$, such that $F(X, \lambda) \equiv 1$ for large $|X| + |\lambda|$. The generator of the group $\pi_1(\mathscr{F} - \Sigma_3)$ is realized by any family $F(X, \lambda)$ whose zero locus is shown in Fig. 1; see [Ar7]. Similarly, any generic 2-spheroid in $\mathscr{F}$ defines a smooth compact surface $\Gamma \subset \mathbf{R}^3 = \mathbf{R}^1_X + \mathbf{R}^2_\lambda$: the

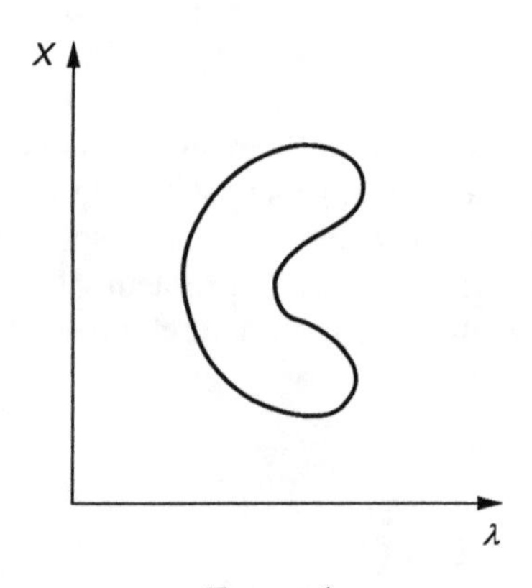

Figure 1

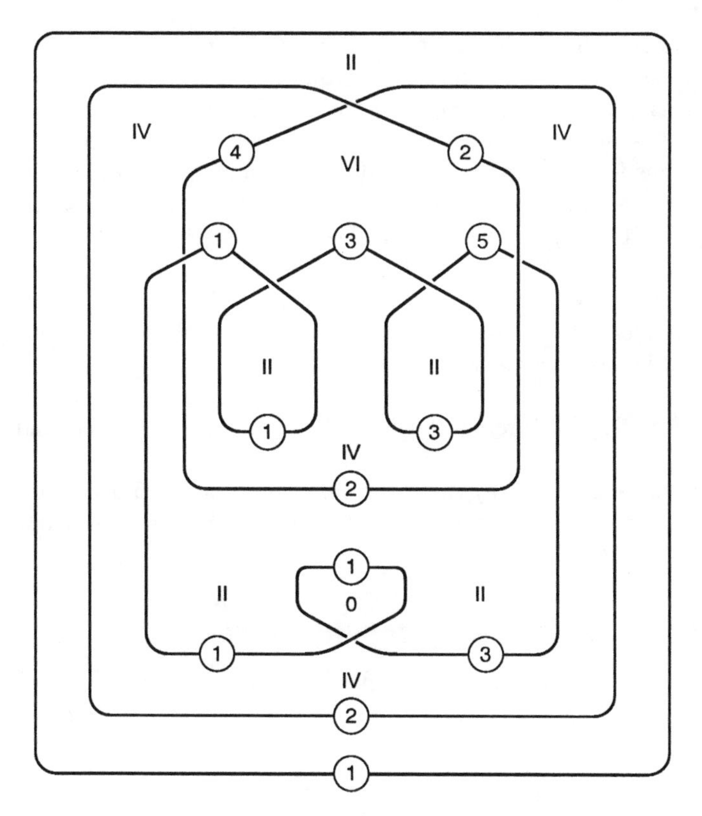

FIGURE 2

zero locus of the function F. If our spheroid is in $\mathscr{F} - \Sigma_3$, then the projection $\Gamma \to \mathbf{R}^2$ has only fold singularities (but no cusps or nonstable singularities). The curve of critical values in $\mathbf{R}^2_1$ of the projection of a basic 2-spheroid in $\mathscr{F} - \Sigma_3$ is shown in Fig. 2. The Roman numerals in the complement to the curve denote the number of zeros of the functions $F(\cdot, \lambda)$ for λ in the corresponding component of the complement; the number ⓘ on a smooth arc of the curve means that when λ goes from the region with a larger Roman numeral into the region with a smaller one crossing this arc, the ith and the $(i + 1)$th zeros of $F(\cdot, \lambda)$ (taken in their order on $\mathbf{R}^1$) coalesce and vanish.

Note. The groups $H_i(P_d - \Sigma_k)$, $\pi_1(\mathscr{F} - \Sigma_3)$, $\pi_1(\Phi - \Sigma_3)$ were calculated in [Ar7]; all the other results of Theorems 1 and 2 were obtained in [V2].

1.2. The May–Segal Theorem

This theorem (see [May1, Segal]) affirms that the cohomology ring of the stable braid group is isomorphic to that of any connected component of $\Omega^2 S^2$; by the Hopf fibration, $S^3 \to S^2$, the later is isomorphic to $H^*(\Omega^2 S^3)$.

This theorem has the following elementary interpretation.

The classifying space of the braid group of d strings. $K(\mathrm{Br}(d), 1)$, can be realized as the space of complex polynomials (1) without multiple roots; denote this space by $\mathbf{C}^d - \Sigma$. The space $\Omega^2 S^3$ can be considered as the space of mappings $\mathbf{R}^2 \to S^3$ with some conditions at infinity. Any point $A \in \mathbf{C}^d - \Sigma$ defines such a mapping: To any $X \in \mathbf{C}^1 = \mathbf{R}^2$, it assigns the direction of the vector $\langle A(X), A'(X) \rangle \in (\mathbf{C}^2 - 0) = (\mathbf{R}^4 - 0)$.

Theorem 3 (see [V3]). *The above embedding $(\mathbf{C}^d - \Sigma) \to \Omega^2 S^3$ induces a cohomology isomorphism in dimensions not exceeding $[d/2] + 1$, and an epimorphism in all other dimensions.*

The May–Segal theorem is the limit variant of this one when d tends to the infinity.

Of course, the homology isomorphism from Theorem 3 does not reflect a homotopy one: the group $\pi_1(\Omega^2 S^3)$ is $\mathbf{Z} \neq$ the braid group. This trouble does not occur in the following situation.

1.3. The Complements of the Resultants

Denote by $\mathbf{C}^{c+d}$ the space of polynomial systems

$$\begin{gathered} X^c + a_1 X^{c-1} + \cdots + a_c, \\ X^d + b_1 X^{d-1} + \cdots + b_d; \end{gathered} \tag{2}$$

let $\Sigma(c, d)$ be the resultant in $\mathbf{C}^{c+d}$, i.e., the set of systems (2) having common roots. It is easy to check that if $c > d$, then $\mathbf{C}^{c+d} - \Sigma(c, d)$ is homotopically equivalent to $\mathbf{C}^{d+d} - \Sigma(d, d)$.

Theorem 4. *There exists a natural inclusion $\mathbf{C}^{d+d} - \Sigma(d, d) \to \Omega^2 S^3$ which induces isomorphisms of groups π_i, $i \leq d - 1$, and H^i, $i \leq d + 1$, and epimorphisms of π_d and of all the groups H^i.*

2. Multidimensional Theorems

2.1. Real Maps without Complicated Singularities

Notations. Let $\mathfrak{A}$ be a singularity class of mappings of m-dimensional manifolds into $\mathbf{R}^n$, i.e., a closed semialgebraic subset of the jet space $J^k(\mathbf{R}^m, \mathbf{R}^n)$, invariant under the natural action of the group $\mathrm{Diff}(\mathbf{R}^m)$. Then, for any m-fold M, the corresponding singular set $\mathfrak{A}(M) \subset J^k(M, \mathbf{R}^n)$ is well-defined.

Denote by $A(M, \mathfrak{A})$ the set of smooth maps $M \to \mathbf{R}^n$ having no singularities of the class $\mathfrak{A}$. Let φ be a map, $\varphi \in A(M, \mathfrak{A})$. Denote by $A(M, \mathfrak{A}, \varphi)$ the subspace of maps in $A(M, \mathfrak{A})$ which coincide with φ in a neighborhood of

∂M. Let $B(M, \mathfrak{A}, \varphi)$ be the space of all smooth sections of the jet bundle $J^k(M, \mathbf{R}^n) \to M$ which do not meet $\mathfrak{A}(M)$ and which coincide with the k-jet extension of φ in a neighborhood of ∂M. Note that k-jet extensions define a natural embedding

$$A(M, \mathfrak{A}, \varphi) \to B(M, \mathfrak{A}, \varphi). \tag{3}$$

Theorem 5. *Let M be a compact smooth manifold, and assume the codimension of $\mathfrak{A}$ in $J^k(\mathbf{R}^m, \mathbf{R}^n)$ is greater than $m + 1$. Then the map* (3) *induces a homology isomorphism in all dimensions. Moreover, this map is a weak homotopy equivalence if one of three additional conditions holds*:

(1) codim $\mathfrak{A} > m + 2$;
(2) $n = 1$ *and* $\mathfrak{A}$ *is the closure of the set of singularities* A_3 (*see* [Ar4], [AVGL]), *so that* $A(M, \mathfrak{A})$ *is the space of quasi-Morse functions*;
(3) $m = n = 1$.

Note. Case (2) was considered by K. Igusa (see [Igusa1, Cerf2]); he proved that in this case the map (3) is m-connected. The "homotopical" part of Theorem 5 follows from the "homological" one by the Whitehead theorem (in all three cases), the Igusa's theorem [in case (2)], and the considerations of Section 1 [in case (3)].

2.2. The Stable Cohomology of the Complements of the Discriminants and the Caustics of Singularities of Holomorphic Functions

Let f be a holomorphic function $(\mathbf{C}^n, 0) \to (\mathbf{C}, 0)$ and F its deformation, i.e., a function $(\mathbf{C}^n \times \mathbf{C}^t, 0) \to (\mathbf{C}, 0)$ such that $f = F(\cdot, 0)$.

The discriminant (respectively, the caustic) of the deformation F is the set of all values $\lambda \in \mathbf{C}^t$ such that the function $F(\cdot, \lambda)$ has a critical point with the critical value 0 near the origin in $\mathbf{C}^n$ (respectively, a non-Morse critical point).

Let $I^*(F)$, $\tilde{I}^*(F)$ be the rings of local cohomologies of complements of the discriminant and of the caustic, i.e., the cohomology rings of these complements in any sufficiently small disk centered at $0 \in \mathbf{C}^t$.

Example. Let $n = 1$, $f = X^d$, and $F = F(X, a)$ be the set of all complex polynomials of the form (1). Then $I^*(F) = H^*(\mathrm{Br}(d))$, $\tilde{I}^*(F) = H^*(\mathrm{Br}(d - 1))$.

The hierarchy of singularities allows us to define homomorphisms of such rings. Indeed, any deformation of a singularity may be considered as a subfamily of any sufficiently large deformation of any more complicated singularity. [See Fig. 3, where (the real version of) a natural embedding of the parameter spaces of standard deformations of singularities A_2, A_3 is shown; for the exact definitions, see [V1, AVGL, §II.6].]

This embedding maps the discriminant into the discriminant and the caus-

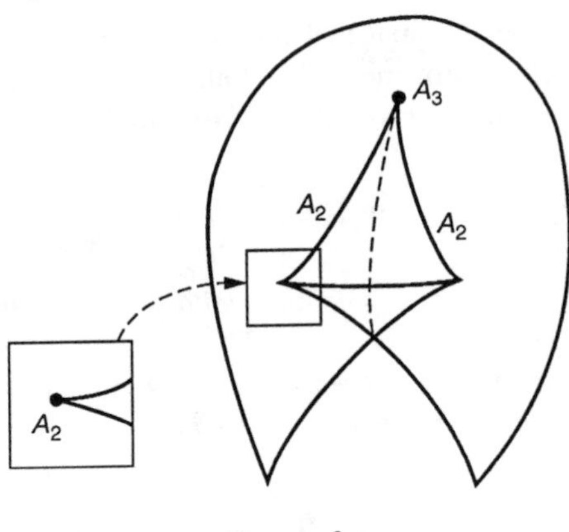

FIGURE 3

tic into the caustic and, hence, defines homomorphisms of the rings $I^*(\cdot)$, $\tilde{I}^*(\cdot)$.

These homomorphisms allow us to define stable objects: A stable cohomology class of complements of the discriminants (the caustics) is a set of elements in $I^*(F)$ [or $\tilde{I}^*(F)$] (one for each deformation) consistent with respect to the natural embeddings of the parameter spaces.

Theorem 6. *The stable cohomology ring of complements to the discriminants (the caustics) of deformations of isolated singularities of n complex variables is naturally isomorphic to the ring* $H^*(\Omega^{2n}S^{2n+1})$ *(respectively, to the ring* $H^*(\Omega^{2n}\Sigma^{2n}\Lambda(n))$, *where* $\Lambda(n) = U(n)/O(n)$ *is the nth Lagrange Grassmannian).*

Both these isomorphisms are examples of the (homological) Smale–Hirsch principle; note that $\Sigma^{2n}\Lambda(n)$ is the homotopy type of the space of holomorphic germs $(\mathbf{C}^n, 0) \to \mathbf{C}$ for which 0 is a regular or a Morse singular point. The May–Segal theorem (see Section 1.2) is a special case of this theorem which corresponds to $n - 1$.

2.3. Complements of Resultants

Let $f: (\mathbf{C}^m, 0) \to (\mathbf{C}^n, 0)$, $m < n$, be a holomorphic map such that $0 \in \mathbf{C}^m$ is an isolated point of $f^{-1}(0)$. Let $F: (\mathbf{C}^m \times \mathbf{C}^t, 0) \to (\mathbf{C}^n, 0)$ be a deformation of $f = F(\cdot, 0)$.

The resultant of the deformation F is the set of all $\lambda \in \mathbf{C}^t$ such that the equation $F(\cdot, \lambda) = 0$ has a solution near $0 \in \mathbf{C}^m$. As in Section 2.2, we can define the ring $\mathscr{H}^*(m, n)$ of stable local cohomologies of complements of such resultant varieties.

Theorem 7. *For any $m < n$, the ring $\mathscr{H}^*(m,n)$ is naturally isomorphic to the ring $H^*(\Omega^{2m}S^{2n-1})$; if $m < n - 1$, then all the stable homotopy groups of complements of the resultants are also isomorphic to the corresponding groups $\pi_i(\Omega^{2m}S^{2n-1}) \equiv \pi_{i+2m}(S^{2n-1})$.*

Perhaps the restriction $m < n - 1$ in the last statement can be omitted.

3. On the Method

3.1. The Main Spectral Sequence

The proofs of Theorems 1–7 are based on a spectral sequence, different versions of which calculate the cohomology of all the considered spaces. All the statements on the cohomology isomorphisms in our theorems are the comparison theorems of the corresponding spectral sequences; the homotopy statements are deduced by using the Whitehead theorem and comparing the π_1 and π_2 groups of the considered spaces.

To describe our spectral sequence for the complement to the usual resultant $\Sigma(c,d)$, $c \geq d$, see Section 1.3.

Theorem 8. *There exists a spectral sequence $E_r^{p,q} \to H^{p+q}(\mathbb{C}^{c+d} - \Sigma(c,d))$, whose term$E_1^{p,q}$ is trivial if $p > 0$ and is equal to $H^{q+2p}(\mathbf{C}^1(-p), \pm\mathbb{Z})$ for $p \leq 0$. Here $\mathbf{C}^1(t)$ is the configuration space consisting of all t-element subsets in $\mathbf{C}^1$ and $\pm\mathbb{Z}$ is a local system, locally isomorphic to $\mathbb{Z}$ but changing the orientation over a loop defining odd permutation of t points.*

Proof. By the Alexander duality theorem, $H^i(\mathbf{C}^{c+d} - \Sigma(c,d)) = \bar{H}_{2(c+d)-i-1}(\Sigma(c,d))$, where $\bar{H}_*$ denotes the homology of the one-point compactification.

Let us fix an embedding $I: \mathbf{C}^1 \to \mathbf{R}^N$ (N sufficiently large) such that for every d points in $\mathbf{C}^1$, the convex hull of their images in $\mathbf{R}^N$ is a $(d-1)$-simplex. For any $t = 1, 2, \ldots, d$, consider the set $\sigma_t \in \mathbf{C}^{c+d} \times \mathbf{R}^N$ formed by pairs {(two polynomials $(A, B) \in \mathbf{C}^{c+d}$ having common zeros in some t points $X_1, \ldots, X_t \in \mathbf{C}^1$); a point of the simplex in $\mathbf{R}^N$ spanned by the points $I(X_1), \ldots, I(X_t)$. The union $\bigcup_{t\leq d} \sigma_t$ is denoted by σ.

Lemma. $\bar{H}_x(\sigma) = \bar{H}_x(\Sigma(c,d))$.

The space σ has a natural filtration: Its term $F_p\sigma$ is the union of spaces σ_t, $t \leq p$. Consider the homological spectral sequence $E^r_{p,q} \to \bar{H}(\sigma)$ generated by this filtration; see Fig. 4. By definition, $E^1_{p,q} = H_{p+q}(\sigma_p - \sigma_{p-1})$. Reorganize this spectral sequence into a cohomological one using the change of indices:

$$E_r^{p,q} \equiv E^r_{-p,2(c+d)-q-1}; \tag{4}$$

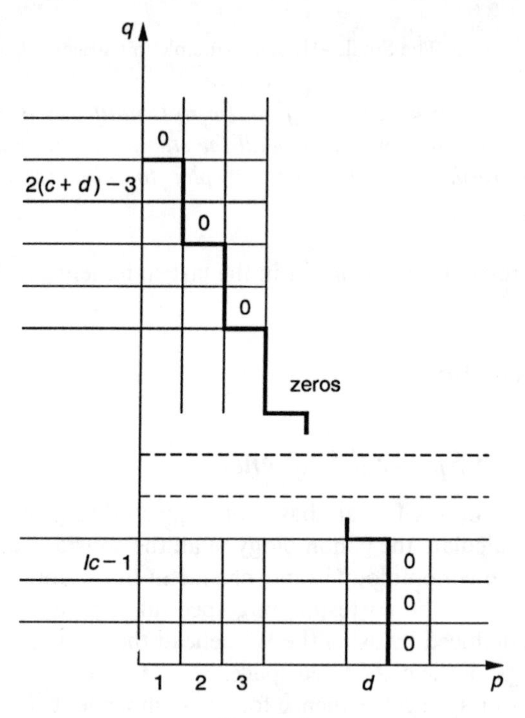

FIGURE 4

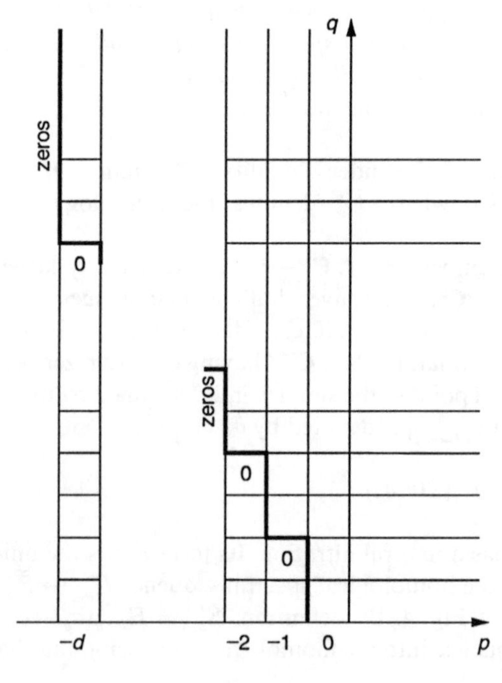

FIGURE 5

see Fig. 5. By the Alexander duality, this sequence converges to the group $H^*(\mathbb{C}^{c+d} - \Sigma(c,d))$ and is the desired one.

Note. This spectral sequence degenerates in the term E_1.

Similar sequences calculate the cohomology of spaces $P_d - \Sigma_k$ (Theorem 2), $\mathbf{C}^d - \Sigma$ (Theorem 3), and of complements to the discriminants and resultants in Theorems 6 and 7. In the other cases (when the examined space is infinite-dimensional), we proceeds as follows:

1. Choose an exhausting increasing system of finited-dimensional approximating subspaces.
2. Construct, as above, spectral sequences which converge to the cohomologies of the complements of the discriminants in these subspaces.
3. Prove the stabilisation theorem for these spectral sequences: For any natural T, all our spectral sequences (except for an initial set) are naturally isomorphic to each other in the domain $\{(p,q)|p+q<T\}$; see [V1–3].

3.2. Some Other Applications of the Main Spectral Sequence

3.2.1. Complements to the Arrangements

Consider a finite family of affine subspaces $V_1, \ldots, V_d$ in $\mathbf{R}^n$. The topology of the set $\mathbf{R}^n - \bigcup V_i$ in some special cases was studied in [Ar2, Br2, OS] and many other works; Goresky and MacPherson [GM] have obtained a formula expressing the cohomology group $H^*(\mathbf{R}^n - \bigcup V_i)$ in the terms of the dimensions of all spaces $\bigcap_{i\in L} V_i$, $L \subset \{1,\ldots,d\}$. In our terms, the formula of [GM] has the following interpretation.

Let us embed all the spaces V_i generically into a space $\mathbf{R}^N$, N very large. For any point $x \in V_{i_1} \cap \cdots \cap V_{i_s}$, consider the $(s-1)$-simplex in $\mathbf{R}^N$ spanned by the corresponding points in the images of $V_{i_1}, \ldots, V_{i_s}$. The union of all such simplices has the same closed homology as $\bigcup V_i$ and a natural filtration by the dimensions of the simplices.

The Goresky–MacPherson formula is nothing but the E_2 term, the corresponding spectral sequence (which degenerates at this term by trivial reasons).

3.2.2. Invariants of Knots

Consider the space of all smooth maps $S^1 \to \mathbf{R}^3$ and define its discriminant as the set of mappings having singularities or self-intersections. Tautologically, a numerical invariant of knots is a zero-dimensional cohomology class of the complement to this invariant. In [V4], a spectral sequence is constructed, which presents such classes; the construction is similar to that of Section 3.1. The support of this sequence belongs to the sector $\{(p,q)|p<0,\ p+q\geq 0\}$, the invariants comes from the line $p+q=0$.

3.2.3. The Cohomology of Spaces of Maps of m-Dimensional Simplicial Polyhedra into m-Connected Ones

A variant of the spectral sequence of Section 3.1 calculates these cohomologies; see [V2, 3]; in the most popular case of $\Omega^m S^n$, $m < n$, its term $E_1^{p,q}$ is isomorphic to $H^{q+p(n+1-m)}(\mathbb{R}^m(-p), (+Z)^{(n-m)})$ if $p < 0$ and is trivial if $p > 0$; here $\mathbf{R}^m(-p)$ is the configuration space of m-point subsets in $\mathbf{R}^m$, $\pm\mathbb{Z}$ as in Theorem 8. If n is odd, this spectral sequence degenerates in the term E_1; in particular, we obtain a new proof of the known splitting formula $H^i(\Omega^\infty S^{\infty+j}) = \bigoplus_{t=0}^{\infty} H^{i-tj}(S(t), (\pm\mathbb{Z})^j)$; see [CMT, May2, Snaith].

References

[Adams] J.F. Adams, *Infinite Loop Spaces*, Princeton University Press and University of Tokyo Press, Princeton, NJ, 1978.

[Ar1] V.I. Arnol'd, On some topological invariants of the algebraic functions, *Proc. Moscow Math. Soc.* **21** (1970), 27–46 (in Russian).

[Ar2] — Cohomology of the colored braids group, *Mat. Notices (Zametki)* **5**(2) (1969).

[Ar3] — Topological invariants of algebraic functions. II, *Funct. Anal. Appl.* **4**(2) (1970), 1–9.

[Ar4] — Critical points of smooth functions, Proc. Intern. Math. Congr in Vancouver, 1974. pp. 19–39. Publisher: Canadian Mathematical Company. Editor: Ralph D. James. Date of publish: 1975.

[Ar5] — On some problems in singularity theory, Geometry and Analysis, Bombay, 1981, pp. 1–10. Publisher: Tata Institute Fundamental Research.

[Ar6] — Some unsolved problems of the singularity theory, Proc. Intern. Math. Congr.

[Ar7] — The spaces of functions with mild singularities, *Funct. Anal. Appl.* **23**(3) (1989), pp. 1–10.

[AVG] V.I. Arnol'd, A.N. Varchenko and S.M. Gussein-Zade, *Singularities of Differentiable Mappings. I*, Nauka, Moscow, 1982.

[AVGL] V.I. Arnol'd, V.A. Vasil'ev, V.V. Gorjunov, and O.V. Ljashko, *Singularities. I. (Dynamical systems-6)*, VINITI, Moscow, 1988. English transl.: *Encyclopedia of Mathematical Science*, Vol. 6, Springer-Verlag, Berlin, in prep.

[Br1] E. Brieskorn, Singular elements of semi-simple algebraic groups, *Actes Congr. Intern. Math. Nice*; Paris, 1971, v.II, pp. 279–284.

[Br2] — *Sur les groupes de tresses (d'après V.I. Arnold)*, Lecture Notes in Mathematics No. 317, Springer-Verlag, Berlin, 1973, pp. 21–44.

[Cerf1] J. Cerf, La stratification naturelle des espaces de fonctions et le theoreme de la pseudo-isotopie, *Publ. Math. IHES* **39** (1970) 6–173.

[Cerf2] — Suppression des singularités de codimension plus grande que 1 dans les familles de fonctions différentiables réeles (d'après K. Igusa), *Seminaire Bourbaki*, 1983/84, No. 627.

[Cohen1] F.R. Cohen, The homology C_{n+1}-spaces, $n \geq 0$. In [CLM], pp. 207–353.

[Cohen2] — Artin braid groups, classical homotopy theory and sundry other curiosities, Preprint, 1986, 40pp.

[CLM] —, T.J. Lada, and J.P. May, *The Homology of Iterated Loop Spaces*, Lecture Notes in Mathematics No. 533, Springer-Verlag, New York, 1976.

[CMT] —, J.P. May, and L.R. Taylor, Splitting of certain spaces CX, *Math. Proc. Cambridge Philos. Soc.* **84**(3) (1978), 465–496.

[DL] E. Dyer and R.K. Lashof, Homology of iterated loop spaces, *Amer. J. Math.* **84** (1962), 35–88.

[Ep] S.I. Epstein, Fundamental groups of the spaces of sets of polynomials without common roots, *Funct. Anal. Appl.* **7**(1) (1973), 90–91.

[Fuchs] D.B. Fuchs, Cohomology of the braid groups mod 2, *Funct. Anal. Appl.* **4**(2) (1970), 62–73.

[GM] M. Goresky and R. MacPherson, *Stratified Morse Theory*, Springer-Verlag, Berlin, 1986.

[Gromov] M. Gromov, *Partial Differential Relations*, Springer-Verlag, Berlin, 1986.

[Hirsch1] M. Hirsch, Immersions of manifolds, *Trans. Amer. Math. Soc.* **93**(2) (1959), 242–276.

[Hirsch2] — On embedding differentiable manifolds in Euclidean space, *Ann. Math* **73** (1961), 566–571.

[Igusa1] K. Igusa, Higher singularities of smooth functions are unnecessary, *Ann. Math.* **119** (1984), 1–58.

[Igusa2] — On the homotopy type of the space of generalised Morse functions, *Topology* **23**(2) (1984), 245–256.

[Looijenga] E. Looijenga, The complement of the bifurcation variety of a simple singularity, *Invent. Math.* **23** (1974), 105–116.

[May1] J.P. May, *The Geometry of Iterated Loop Spaces*, Lecture Notes in Mathematics, No. 268, Springer-Verlag, Berlin, 1972.

[May2] — Infinite loop space theory, *Bull. Amer. Math. Soc.* **83**(4) (1977), 456–494.

[Mil] R.J. Milgram, Iterated loop spaces, *Ann. Math.* **84** (1966), 386–403.

[Milnor] J. Milnor, *Singular Points of Complex Hypersurfaces*, Princeton University Press and Tokyo University Press, Princeton, NJ, 1968.

[OS] P. Orlik and L. Solomon, Combinatorics and topology of complements of hyperplanes, *Invent. Math.* **56** (1980), 167–189.

[Segal] G.B. Segal, Configuration-spaces and iterated loop-spaces, *Invent. Math.* **21**(3) (1973), 213–221.

[Smale] S. Smale, The classification of immersions of spheres in euclidean space, *Ann. Math.* **69**(2) (1959), 327–344.

[Snaith] V.P. Snaith, A stable decomposition of $\Omega^n S^n X$, *J. London Math. Soc.* **2** (1974), 577–583.

[Van] F.V. Vainstein, Cohomology of the braid groups, *Funct. Anal. Appl.* **12**(2) (1978), 72–73.

[V1] V.A. Vassil'ev, The stable cohomology of complements of the discriminants of the singularities of smooth functions, *Contemporary*

Problems of Mathematics, VINITI, Vol. 33, Moscow, 1988 (to be translated in *Sov. Math. J.*).

[V2] — Topology of spaces of functions without complicated singularities, *Funct. Anal. Appl.* **23**(4) (1989), 24–36.

[V3] — Topology of complements of the discriminants and loop-spaces, *Advances in Soviet Mathematics: Theory of Singularities and its Applications*, American Mathematical Society. Providence, RI, 1990.

[V4] — Cohomology of knot spaces. *Advances in Soviet Mathematics: Theory of Singularities and its Applications*, American Mathematical Society, Providence, RI, 1990.

[V5] V.A. Vassil'ev, Complements of discriminants of smooth maps: Topology and applications (in preparation).

Part 4
Economics

14
Stephen Smale and the Economic Theory of General Equilibrium*

GERARD DEBREU

During a 9-year time interval, Stephen Smale's dominant research interest was the economic theory of general equilibrium. His bibliography for the period 1973–1981 contains 15 articles that addressed problems at the core of that theory or, in the case of two of them, that were directly motivated by those problems. The 11 other papers authored or co-authored by Smale during the same time interval dealt with topics ranging from a model of two cells to complexity theory. One of them was devoted to the game theoretical problem of the prisoner's dilemma.

Nearly half of Smale's 15 articles on the theory of general equilibrium appeared in the *Journal of Mathematical Economics*, and one was included in the May 1976 issue of the *American Economic Review*, after having been presented by invitation at the 1975 annual meeting of the American Economic Association. The latter recognition by the economics profession came at the same time as his appointment in 1976 as Professor of Economics at the University of California at Berkeley, concurrent with his appointment as Professor of Mathematics. Both accolades by economists were later followed by his election as a Fellow of the Econometric Society in 1983.

By 1981 Smale had left a mark on the economic theory of general equilibrium that will be the subject of this lecture. After that, he published one more article on economics [1987] returning to that theory, a survey valuable to the reader of Smale's work on economics for the broad views, and the insights into his research that it gives.

The four closely interrelated areas of the economic theory of general equilibrium to which Smale contributed will now be briefly described. A more detailed discussion of each area and of Smale's contributions will then follow.

* I thank all those who commented on an earlier version of this paper, or on its presentation on August 5, 1990, and I am especially grateful to Donald Brown, Graciela Chichilnisky, John Geanakoplos, Andreu Mas-Collel, Carl Simon, Stephen Smale, Hal Varian, and Karl Vind for their remarks.

0. General Equilibrium

The economic theory of general equilibrium describes the observed state of an economy as the result of the interaction through markets of two types of agents, consumers and producers. Each one of those agents decides on the basis of prices what quantity of each one of the various commodities to produce, exchange, and consume. The large number of commodities, the equally large number of their prices, and the large number of interacting agents make the use of a mathematical model imperative for that description. In 1874–77, Léon Walras provided the first explanation of general equilibrium in the framework of such a model. Walrasian theory has been developed by several generations of economists; its present formulation (as in the Arrow–Debreu model [1954]) makes it possible to convey some of its main ideas concisely to mathematicians.

The number of commodities is assumed to be a given integer ℓ. Once a measurement unit has been chosen for each commodity, and a sign convention has been made to distinguish inputs from outputs, a commodity-vector $\mathbf{z}_a$ in the commodity-space R^ℓ describes the quantity of each one of the ℓ commodities that agent a decides to consume or to produce.

The ith consumer ($i = 1, \ldots, m$) is characterized by a consumption-set X_i, a non-empty subset of R^ℓ, and a total preference preorder $\precsim_i$ on X_i. Those concepts have the following economic interpretation.

For the ith consumer, inputs are positive, outputs are negative, and his consumption is denoted by $x_i \in R^\ell$. The hth coordinate of x_i is the quantity of the hth commodity that he consumes if $x_i^h > 0$, or the negative of the quantity that he produces if $x_i^h < 0$. The consumption-set X_i is the set of possible consumption-vectors. As an example, an x_i having, over the next month, a small food input and a large (in absolute value) labor output may be impossible.

Given two consumptions x and x' both in X_i: if the ith consumer restricted to consume x or x' chooses x', we write $x \precsim_i x'$ and read "x' is at least as desired by the ith consumer as x." The preference relation $\precsim_i$ is assumed to be a total preorder (i.e., reflexive and transitive).

The jth producer ($j = 1, \ldots, n$) is characterized by a production-set Y_j, a nonempty subset of R^ℓ. The economic interpretation is now the following.

For the jth producer inputs are negative, outputs are positive, and his production is denoted by $y_j \in R^\ell$. The hth coordinate of y_j is the quantity of the hth commodity that he produces if $y_j^h > 0$, or the negative of the quantity that he consumes if $y_j^h < 0$. The production-set Y_j is the set of possible production-vectors. As an example, an input–output vector y_j may be impossible given the technological knowledge available to the jth producer.

The economy $\mathscr{E}_0$ is then described by the m preordered consumption-sets X_i, by the n production-sets Y_j, and by the total endowment-vector $\mathbf{e} \in \mathscr{R}^\ell$ characterizing the given quantity of each commodity available to the economy as a whole. Thus, $\mathscr{E}_0 = ((X_i, \precsim_i)_{i=1,\ldots,m}, (Y_j)_{j=1,\ldots,n}, \mathbf{e})$.

In this description of the economy, the ownership of its resources is not specified. Note also that the particular case of a *pure exchange economy* corresponds to the situation in which every Y_j degenerates to $\{0\}$. The only possible production-vector then is the origin.

An attainable state s_0 of the economy $\mathscr{E}_0$ is a list of the decisions made by the $m + n$ agents, compatible with the endowment-vector $\mathbf{e}$. In mathematical terms, an attainable state of $\mathscr{E}_0$ is a list $s_0 = ((x_i)_{i=1,\dots,m}, (y_j)_{j=1,\dots,n})$ of m consumptions x_i and n productions y_j satisfying the conditions

(1) for every i, $x_i \in X_i$;
(2) for every j, $y_j \in Y_j$;
(3) $\sum_{i=1}^{m} x_i - \sum_{j=1}^{n} y_j = e$.

1. A natural preorder on the set of attainable states was introduced by Pareto who defined s_0' as superior (or indifferent) to s_0 if, for every consumer, his consumption in s_0' is preferred (or indifferent) to his consumption in s_0. In symbols:

$$s_0 \precsim s_0' \qquad \text{if for every } i,\ x_i \precsim_i x_i',$$

Except in trivial cases, the Pareto preorder is not total, and attempting to compare two attainable states s_0 and s_0', one may find that one consumer prefers s_0, whereas another prefers s_0'. This gives special importance to the concept of a Pareto optimum defined as an attainable state of $\mathscr{E}_0$ that is maximal for the Pareto preorder. If s_0 is a Pareto optimum, it is then impossible to find an attainable state s_0' that is superior to s_0 according to the unanimity principle formalized by the Pareto criterion. Except in trivial cases again, there are infinitely many Pareto optima which cannot be compared to each other according to the Pareto preorder.

The characterization of Pareto optima by means of prices is one of the fundamental insights provided by welfare economics. Let $\mathbf{p}$ be a price-vector associating the price p^h with one unit of the hth commodity. Thus, the value of a vector $\mathbf{z}$ in the commodity space R^ℓ relative to a vector $\mathbf{p}$ in the (dual) price space R^ℓ is the bilinear form $\mathbf{p}\cdot\mathbf{z} = \sum_{h=1}^{\ell} p^h z^h$. A consumption x_i in X_i is in equilibrium relative to $\mathbf{p}$ for the ith consumer if x_i is best according to $\precsim_i$ in the set $\{\mathbf{z} \in X_i | \mathbf{p}\cdot\mathbf{z} \leqq \mathbf{p}\cdot x_i\}$. It is then impossible for the ith consumer to find in his consumption-set X_i a commodity-vector preferred to x_i unless he spends more than he does on x_i. Similarly, y_j in Y_j is in equilibrium relative to $\mathbf{p}$ for the jth producer if $[\mathbf{z} \in Y_j]$ implies $[\mathbf{p}\cdot\mathbf{z} \leqq \mathbf{p}\cdot y_j]$. It is then impossible for the jth producer to find in his production-set Y_j a commodity-vector yielding a greater profit than y_j does. An attainable state s_0 of $\mathscr{E}_0$ is in equilibrium relative to $\mathbf{p}$ if every agent of the economy $\mathscr{E}_0$ is in equilibrium relative to $\mathbf{p}$.

One of the main goals of welfare economics is to give conditions on $\mathscr{E}_0$ ensuring that with every Pareto optimum s_0 is associated a price-vector $\mathbf{p}$ relative to which s_0 is in equilibrium.

2. If the resources of the economy are privately owned, its description also

requires a specification of the endowment e_i in R^ℓ of the ith consumer, as well as of the fraction θ_{ij} of the profit of the jth producer allocated to the ith consumer. Thus, $\sum_{i=1}^m e_i = \mathbf{e}$; for every pair (i, j), $\theta_{ij} \geqq 0$; and for every j, $\sum_{i=1}^m \theta_{ij} = 1$.

The mathematical description of the private ownership economy $\mathscr{E}$ now is

$$\mathscr{E} = ((X_i, \precsim_i, e_i)_{i=1,\ldots,m}, (Y_j)_{j=1,\ldots,n}, (\theta_{ij})_{\substack{i=1,\ldots,m \\ j=1,\ldots,n}}).$$

In this new context, an attainable state s of $\mathscr{E}$ is a list $s = ((x_i)_{i=1,\ldots,m}, (y_j)_{j=1,\ldots,n}, \mathbf{p})$ of $\mathbf{m}$ consumptions x_i, n productions y_j, and a price-vector $\mathbf{p}$ satisfying the conditions:

(i) for every i, $x_i \in X_i$;
(ii) for every j, $y_j \in Y_j$;
(iii) $\sum_{i=1}^m x_i - \sum_{j=1}^n y_j = \sum_{i=1}^m e_i$;
(iv) $\mathbf{p}$ belongs to the price-space.

Such an attainable state is an equilibrium of $\mathscr{E}$ if (a) for every j, $\mathbf{p} \cdot y_j$ is the maximum profit relative to $\mathbf{p}$ in Y_j and (b) for every i, the consumption x_i is best according to $\precsim_i$ in the budget set

$$\left\{\mathbf{z} \in X_i | \mathbf{p} \cdot \mathbf{z} \leqq \mathbf{p} \cdot e_i + \sum_{j=1}^{n} \theta_{ij} \mathbf{p} \cdot y_j\right\}.$$

The inequality expresses that the value of the ith consumer's consumption is at most equal to the sum of the value of his initial endowment and his shares of the profits of the n producers.

A fundamental question arises here; namely, under what conditions on the private ownership economy $\mathscr{E}$ can one assert that there is an equilibrium state?

3. Beyond existence theorems the question of uniqueness of general equilibrium presents itself for investigation. The requirement of global uniqueness, however, turns out to be too demanding, and, instead, conditions on $\mathscr{E}$ that guarantee generic local uniqueness of general equilibrium are searched for.

4. Define now a state σ of the private ownership economy as a list $\sigma = (x_i)_{i=1,\ldots,m}, (y_j)_{j=1,\ldots,n}, \mathbf{p})$ of m consumptions x_i, n productions y_j, and a price-vector $\mathbf{p}$ satisfying the conditions

(i) for every i, $x_i \in X_i$;
(ii) for every j, $y_j \in Y_j$;
(iv) p belongs to the price-space.

This definition differs from the definition of an *attainable* state of $\mathscr{E}$ given earlier in that condition (iii) $\sum_{i=1}^m x_i - \sum_{j=1}^n y_j = \sum_{i=1}^m e_i$ is no longer imposed.

From the foundation for a theory of dynamic stability being laid by the study of local uniqueness of general equilibrium, one can begin to examine in the space of states of $\mathscr{E}$ processes that converge to the set of equilibria of $\mathscr{E}$.

Alternatively, processes of that type can be investigated in the search for efficient algorithms for the computation of equilibria of $\mathscr{E}$.

1. The Characterization of Pareto Optima

If, with every Pareto optimum s_0 of the economy $\mathscr{E}_0$, one can associate a price-vector $\mathbf{p}$ such that s_0 is an equilibrium relative to $\mathbf{p}$, then prices acquire a new significance as a means of achieving Pareto optimality. This insight due to Pareto [1909] was studied mathematically by him and by several of his successors at first in a differential calculus framework. This phase ended in the 1940s with solutions, notably by Lange [1942] and Allais [1943], that gave conditions confirming (and qualifying) Pareto's insight.

In the early 1950s, Arrow [1951] and Debreu [1951] pointed out that an alternative approach via convex analysis, resting on Minkowski's supporting plane theorem, provided a proof that was more rigorous, more general, and simpler than the calculus proofs offered up to that time. Convexity properties of preference relations and of production sets played an essential role in that approach. When Aumann [1964] introduced into economics the measure theoretical concept of a continuum of economic agents, those convexity properties became consequences of Lyapunov's theorem on the convexity and compactness of the range of a finite-dimensional atomless vector measure (Vind [1964]) in sectors of the economy all of whose agents are small. The case of an industry with large producers having nonconvex production-sets was not covered however. In the case of nonconvexities, another difficulty arises as well with the loss of an elementary property of convex sets for which a local maximizer of a linear function is also a global maximizer.

The move of economic theory away from the differential calculus approach in the 1950s and 1960s was reversed when the methods of global analysis became basic to the study of generic local uniqueness of general equilibrium that will be the subject of Section 3. It is in this new context that Smale reconsidered the question of characterization of Pareto optima in a differential setup, in a series of seven articles ([1973], [1974b], [1974d], [1975a], [1975b], [1976b], [1981]). Smale's reconsideration differs from the earlier characterization of Pareto optima by differential methods by its rigor, its generality, its new results, and its powerful techniques. Two main themes run through those seven publications. One is an abstract theory of optimization for several functions, the other is the application to economics of that abstract theory. Some of the main features of both themes will now be outlined.

Smale consistently uses for each consumer, say the ith, a utility function $u_i\colon X_i \to R$ representing his preference relation $\precsim_i$ on X_i in the sense that $[x \precsim_i x']$ is equivalent to $[u_i(x) \leqq u_i(x')]$. The preference relation is observable, and, therefore, a better primitive concept than one of its representations. But the function u_i can be used instead of the underlying relation $\precsim_i$ if the conditions of one of the available representation theorems prevail.

The abstract theory of optimization for several functions is studied by Smale in the following terms. W is an open subset of R^n and $v_1, \ldots, v_m$ are C^2 functions from W to R. A point x' in W is said to be (Pareto) superior to a point x in W if for every i, $v_i(x') \geqq v_i(x)$ and, for some j, $v_j(x') > v_j(x)$. A (Pareto) optimum is a point x of W such that there is no point of W superior to x. A strict optimum is a point $x \in W$ such that $[v_i(x') \geqq v_i(x)$ for every $i]$ implies $[x' = x]$. A general theorem of central importance is given by Smale ([1975b] as the end product of a development started in [1973] and pursued in subsequent papers [1974b], [1974d], [1975a], and by Wan [1975].

In its statement, $Dv_i(x)$ denotes the derivative of v_i at x, a real-valued linear function on R^n, and $D^2v_i(x)$, the second derivative of v_i at x, a quadratic form on R^n.

If $x \in W$ is a local optimum, then there exist $\lambda_1, \ldots, \lambda_m$, non-negative, not-all zero, such that $\sum_i \lambda_i Dv_i(x) = 0$.

Moreover, if $\sum_i \lambda_i D^2v_i(x)$ is negative definite on the space $\{\mathbf{z} \in R^n | \forall i = 1, \ldots, m, \lambda_i Dv_i(x)\mathbf{z} = 0\}$, then x is a local strict optimum.

In the case of a pure exchange economy $\mathscr{E}_0$ with ℓ commodities and m consumers, for every i the consumption set X_i of the ith consumer is taken to be P, the interior of R^ℓ_+ the closed positive orthant of R^ℓ. The utility function of that consumer is a C^2 function u_i: $P \to R$ with no critical point, satisfying the conditions (i) differentiable monotonicity, (ii) differentiable convexity, and (iii) a boundary condition, for the statement of which the following notation and terminology are required.

The set of points of P indifferent to x for the ith consumer (the indifference surface of x) is $u_i^{-1}(c)$, where $c = u_i(x)$. We denote by $g_i(x)$ the oriented unit normal vector to that surface at x,

$$g_i(x) = \frac{\operatorname{grad} u_i(x)}{\|\operatorname{grad} u_i(x)\|}.$$

(i) The differentiable monotonicity condition then requires that all the coordinates of $g_i(x)$ be strictly positive.

(ii) The differentiable convexity condition says that for every $x \in P$, the second derivative $D^2u_i(x)$ is negative definite when restricted to the tangent plane to the indifference surface through x at x.

(iii) The boundary condition says that every indifference surface is closed in R^ℓ.

For the pure exchange economy $\mathscr{E}_0$, the set W of attainable states is the set of m-lists $x = (x_1, \ldots, x_m)$ such that for every i, $x_i \in P$, and the sum of the consumptions x_i equals the endowment-vector $\mathbf{e}$ of the economy $\mathscr{E}_0$:

$$W = \left\{ x \in P^m \middle| \sum_i x_i = \mathbf{e} \right\}.$$

For every i, the utility function u_i on P induces the function v_i on W according to the equality

$$v_i(x) = u_i(x_i) \qquad \text{for every } x \in W.$$

The main result of Smale [1981, Section 4] is obtained in part as an application of his abstract theory of optimization for several functions. Under the assumptions just specifled on the pure exchange economy $\mathscr{E}_0$:

The following three conditions on a point x of W are equivalent

(a) *x is a local Pareto optimum,*
(b) *x is a strict Pareto optimum,*
(c) *$g_i(x_i)$ is independent of i.*

Let θ be the set of $x \in W$ satisfying one of these conditions. Then θ is a C^1 submanifold of W of dimension $m - 1$.

The common value $\mathbf{p}$ of the vectors $g_i(x_i)$ is the price-vector relative to which x is an equilibrium. Of the corollaries of this result given by Smale, one will be singled out as an incisive version of the two fundamental theorems of welfare economics.

The pair $(x, \mathbf{p}) \in W \times S^{\ell-1}$ of an attainable state x of $\mathscr{E}_0$ and of a unit price-vector $\mathbf{p}$ is called a Pareto equilibrium of $\mathscr{E}_0$ if x is an equilibrium relative to $\mathbf{p}$. The set of those Pareto equilibria is denoted by Λ. Then (Smale [1976b] and [1981]):

The map from Λ to W defined by $(x, \mathbf{p}) \mapsto X$ is a C^1 diffeomorphism of Λ onto θ.

The main result and its corollary that have just been quoted, as well as the general theorem on optimization for several functions from which they are derived, go from well-known (and easy to prove) first-order conditions for a Pareto optimum to new delicate second-order conditions, and to the original characterization of the set θ of Pareto optima as a submanifold of W diffeomorphic to the set Λ of Pareto equilibria. They are representative of the approach to Pareto optimality taken by Smale. But they account for only a small part of the many, sometimes complex, contributions that he made to that subject in the seven articles that he wrote in that area. For instance, the results mentioned earlier call for generalizations several of which were provided by Smale. In particular, closed consumption-sets X_i rather than open sets of the form $P = \operatorname{Int} R_+^\ell$ are covered in his work. So are economies with production. So are nonconvexities for consumers' preferences and for producers' production-sets. Smale's writings on Pareto optimality amount to an extensive mathematical study of a topic of central importance in economic theory.

2. The Existence of a General Equilibrium

The theory of general equilibrium must provide conditions on the private ownership economy $\mathscr{E}$ ensuring that the equilibrium state on which it is centered exists.

The existence problem can be approached along the following line. First, exclude by economic considerations the case of a zero price-vector, and note that two positively collinear price-vectors $\mathbf{p}_1$ and $\mathbf{p}_2$ are equivalent in the sense that all the economic agents make the same decisions on the basis of $\mathbf{p}_1$ as they do on the basis of $\mathbf{p}_2$. Having chosen a norm in the price-space, one can then restrict the price-vector to be in the set Γ of vectors with norm 1. Consider a price-vector $\mathbf{p}$ in Γ and the reactions to $\mathbf{p}$ of the agents of $\mathscr{E}$. The jth producer ($j = 1, \ldots, n$) tries to maximize his profit relative to $\mathbf{p}$ by choosing an element y_j of Y_j. The ith consumer ($i = 1, \ldots, m$) tries to satisfy his preferences $\precsim_i$ by choosing an element x_i of X_i constrained by the inequality $\mathbf{p} \cdot x_i \leqq \mathbf{p} \cdot \mathbf{e}_i + \sum_{j=1}^{n} \theta_{ij} \mathbf{p} \cdot y_j$. The set of $\mathbf{p}$ for which these operations can be carried out for every agent of $\mathscr{E}$ is a subset Π of Γ. Given $\mathbf{p}$ in Π, denote by y_j a production chosen by the jth producer, and by x_i a consumption chosen by the ith consumer, the corresponding excess demand is the commodity-vector

$$\sum_{i=1}^{m} x_i - \sum_{j=1}^{n} y_j - \sum_{i=1}^{m} e_i.$$

That vector need not be unique, however, and one is, therefore, led to associate with $\mathbf{p}$ in Π, the set $\zeta(\mathbf{p})$ of all the excess demand vectors in the commodity space R^ℓ to which $\mathbf{p}$ can give rise. The vector $\mathbf{p}^*$ is an equilibrium price-vector if and only if $0 \in \zeta(\mathbf{p}^*)$. Thus, the problem of existence of a general equilibrium can be formulated in the following manner. A correspondence ζ associating with every element $\mathbf{p}$ of Π a nonempty subset $\zeta(\mathbf{p})$ of R^ℓ is given. Under what conditions on ζ can one assert that there is a $\mathbf{p}^*$ in Π such that $0 \in \zeta(\mathbf{p}^*)$? Several ways of dealing with this existence question have been proposed. Notably, starting in the 1950s (Arrow and Debreu [1954], McKenzie [1954], Gale [1955], Nikaido [1956], Debreu [1956]), many existence proofs have been based on fixed point arguments.

Smale ([1974a], [1976c], [1981], [1987]) takes a different approach and studies first the simple case where for each $\mathbf{p}$ in the set

$$\Delta_1 = \left\{ \mathbf{p} \in R_+^\ell \,\middle|\, \sum_{h=1}^{\ell} p^h = 1 \right\},$$

the excess demand vector is unique. Consequently, an excess demand function z from Δ_1 to R^ℓ is defined, and one seeks an element $\mathbf{p}^*$ in Δ_1 for which $z(\mathbf{p}^*) = 0$.

The function z is assumed to be of class C^2 for the application of Sard's theorem that will be the basis of the argument.

The function z is also assumed to satisfy Walras' Law: for every $p \in \Delta_1$ one has $\mathbf{p} \cdot \mathbf{z}(\mathbf{p}) = 0$. This law is directly implied by the insatiability of consumers.

Finally, a boundary condition is imposed: If $p^h = 0$, then $z^h(\mathbf{p}) \geqq 0$. If the hth commodity is free, then its excess demand is non-negative.

Smale states [1976c], [1981], [1987]:

If the C^2 function z satisfies Walras' Law and the boundary condition, then there is an equilibrium price-vector.

Smale's proof has several important new features that are now sketched in an extremely brief summary of the complex arguments he develops in the first section of [1981], and especially as he proves Theorems 1.2 and 1.3.

Define $\Delta_0 = \{q \in R^\ell | \sum_{h=1}^{\ell} q^h = 0\}$, and let $\varphi(\mathbf{p}) = z(\mathbf{p}) - [\sum_h z^h(\mathbf{p})]\mathbf{p}$. The map φ is a C^2 function from Δ_1 to Δ_0, and $z(\mathbf{p}) = 0$ is equivalent to $\varphi(\mathbf{p}) = 0$. The last equality, therefore, expresses that $\mathbf{p}$ is an element of $E = \varphi^{-1}(0)$, the set of equilibrium price-vectors that must be proved to be nonempty. Next, define $\hat{\varphi}(\mathbf{p}) = \varphi(\mathbf{p})/\|\varphi(\mathbf{p})\|$, a function from $\Delta_1 \backslash E$ to $S^{\ell-2}$, the unit sphere in Δ_0.

Assume that there is a point p_0 in the boundary of Δ_1 such that $Dz(p_0)$ is nondegenerate (a weak assumption that can be dispensed with). Sard's theorem implies that there is a regular value y of $\hat{\varphi}$ in $S^{\ell-2}$ near $\hat{\varphi}(p_0)$ and such that $\hat{\varphi}^{-1}(y)$ is nonempty. According to the inverse function theorem, the set $\gamma = \hat{\varphi}^{-1}(y)$ is a 1-dimensional C^2 submanifold, i.e., a C^2 curve in Δ_1. The curve γ is closed in $\Delta_1 \backslash E$. A component of γ leaves $\partial\Delta_1$ at a point near p_0. It cannot leave Δ_1. Because it has no endpoint, it must tend to E. Therefore, E is not empty.

Smale's proof considers the existence problem that he studies from the viewpoint of global analysis of which he uses several basic results. It is constructive in the following sense. The curve γ is a solution of his Global Newton differential equation

$$D\varphi(\mathbf{p})\frac{dp}{dt} = -\lambda\varphi(\mathbf{p}),$$

that will be discussed in Section 4, where λ is $+1$ or -1 and sign(λ) is determined by sign (determinant of $D\varphi(\mathbf{p})$). A discrete algorithm approximating the solution of that differential equation, therefore, will yield price-vectors approximately satisfying the equilibrium conditions.

Smale then extends the preceding existence theorem and weakens the differentiability assumption to a continuity assumption, approximating a given continuous function by a sequence of C^∞ functions. He also shows how correspondences rather than functions can be included in his study by means of another approximation process. Thus, existence theorems having the full generality of results previously obtained directly by fixed-point methods are derived from the global analysis foundation that he laid in a solution that emphasizes constructive aspects.

3. The Generic Local Uniqueness of General Equilibria

The theory of general equilibrium would be entirely determinate if the equilibrium price-vector were unique. But the conditions on the private ownership economy $\mathscr{E}$ guaranteeing that uniqueness are severely restrictive; and weakening the requirement of global uniqueness to that of local uniqueness is not by itself sufficient to overcome the severity of those restrictions. Under differentiability assumptions, however, the property of local uniqueness holds for all economies outside a negligible set. The conjecture of a result of that type (Debreu [1970]) was the occasion, in the summer of 1968, for my first meeting with Smale who introduced me to global analysis and who extended that result in several directions in five articles ([1974al, [1974c], [1976b], [1979], [1981]).

The theorems that Smale states in [1981, Section 5] are indicative of several of the genericity results obtained in various contexts over the past two decades. His proofs are centered on a study of the equilibrium manifold which appears with slightly different definitions in the preceding five papers. In [1981], he adopts the following definition in the pure exchange case that he considers. As in Section 1, there are ℓ commodities and m consumers in the economy. The ith consumer ($i = 1,\ldots,m$) has a *fixed* consumption set $P = \operatorname{Int} R^{\ell}_{+}$, and a *fixed* utility function u_i: $P \to R$ which is assumed to be of class C^2, to have no critical point, and to satisfy the differentiable monotonicity condition (i), the differentiable convexity condition (ii), and the boundary condition (iii). The ith consumer also has a *variable* endowment-vector $\mathbf{e}_i$ in P. Thus, the parameter defining the economy is the m-list $\mathbf{e} = (\mathbf{e}_1,\ldots,\mathbf{e}_m)$ in P^m, whereas an allocation of the total resources $\sum_i \mathbf{e}_i$ to the consumers is an m-list $x = (x_1,\ldots,x_m)$ in P^m such that $\sum_i x_i = \sum_i \mathbf{e}_i$. The equilibrium manifold then is the set Σ of pairs $(\mathbf{e}, x)$ satisfying the equilibrium conditions:

$$\Sigma = \left\{(\mathbf{e}, x) \in P^m \times P^m \middle| \forall i, g_i(x_i) = \mathbf{p} \text{ and } \mathbf{p}\cdot x_i = \mathbf{p}\cdot \mathbf{e}_i; \sum_i x_i = \sum_i \mathbf{e}_i \right\}.$$

For every $\mathbf{e} \in P^m$, denote also by $E(\mathbf{e})$ the set $\{x \in P^m | (\mathbf{e}, x) \in \Sigma\}$ of equilibrium allocations associated with $\mathbf{e}$. The main results of [1981, Section 5] state that

(1) *Σ is a C^1 submanifold of $P^m \times P^m$ of dimension ℓm.*
(2) *There is a closed subset F of P^m of measure zero such that if $\mathbf{e} \notin F$, then the set $E(\mathbf{e})$ is finite. Moreover, $E(\mathbf{e})$ varies continuously with $\mathbf{e}$ in $P^m \backslash F$.*

The proof of (2) can be briefly outlined. Let π: $P^m \times P^m \to P^m$ be the projection defined by $\pi(\mathbf{e}, x) = \mathbf{e}$; let π_0 be the restriction of π to Σ; and let $C \subset \Sigma$ be the closed set of critical points of π_0. The set $F = \pi_0(C)$ of critical values of π_0 has measure zero by Sard's theorem; it is closed because π_0 is a closed map. The remaining assertions in (2) follow from the inverse function theorem.

Thus, every regular economy $\mathbf{e}$ outside the negligible critical set F has a finite number of equilibria which vary continuously in a neighborhood of $\mathbf{e}$.

The argument that establishes the result is focused on properties of the equilibrium manifold Σ, and of the projection π_0 (as is the approach independently taken by Balasko [1975]).

Smale generalizes the previous genericity theorems to economies with a fixed total endowment, with variable utility functions, with general consumption sets, and with production. He dispenses with convexity and monotonicity assumptions by introducing the concept of an *extended* price-equilibrium in which the attainable state of the economy is only required to satisfy the first-order conditions for an equilibrium. The five articles in which Smale addressed the question of generic local uniqueness of general equilibrium are given a remarkable unity of method by their global analysis approach.

4. The Global Newton Method

Two major problems in the theory of general equilibrium have led to the investigation of processes converging to the zeros of the excess demand function.

The first is the description of the dynamic behavior of a private ownership economy out of equilibrium. Assume that by reacting to a price-vector $\mathbf{p}$ in $S_+^{\ell-1} = R_+^\ell \cap S^{\ell-1}$, the non-negative part of the unit sphere in the price space, the agents of that economy generate an excess demand function $z: S_+^{\ell-1} \to R^\ell$. The point $\mathbf{p}^*$ is an equilibrium price-vector if and only if $z(\mathbf{p}^*) = 0$, and a simple example of a dynamic process of economic disequilibrium has been formalized as the differential equation

$$\frac{dp}{dt} = z(\mathbf{p}). \tag{1}$$

Processes of this type were introduced by Samuelson [1941], and extensively studied by Arrow and Hurwicz [1958], and Arrow, Block, and Hurwicz [1959]. Examples of Scarf [1960], however, showed that (1) may not be globally stable. This assertion was confirmed by the results of Sonnenschein [1972], [1973], Mantel [1974], and Debreu [1974] who proved, with increasing generality, that the excess demand function z generated by a private ownership economy is arbitrary (except for continuity and Walras' Law).

The second problem is the development of efficient algorithms for the computation of general equilibria. Scarf, who played a leading role in that area, proposed solutions based on combinatorial methods [1973].

In an article that stands out by its originality and by its importance among the contributions he made to the theory of general equilibrium, Smale [1976c] provides, in an abstract setting, a process that can be applied in the solution of the first problem and that yields a differential analogue to the combinatorial algorithms of Scarf, and Eaves and Scarf [1976]. An antecedent for that analogue appeared in an article by Hirsch [1963]. Simultaneously, Kellogg, Li, and Yorke [1977] gave an algorithm converging to the set of fixed points of a function on a closed disk and starting from almost any point in the boundary.

Smale considers a function f defined on a closed set $\overline{M}$, where M is a nonempty, bounded, connected, open subset of R^n, whose boundary ∂M is a smooth submanifold; the function f takes its values in R^n; it is of class C^2. The goal of the analysis is to give a process starting from a point in the boundary of M and converging to the set E of zeros of f, and thereby to obtain a constructive solution of the equation $f(x) = 0$.

The process proposed by Smale is a solution of the "Global Newton" differential equation

$$Df(x)\frac{dx}{dt} = -\lambda(x)f(x), \tag{2}$$

where $Df(x)$ is the derivative of f at x, and λ is a real-valued function defined on $\overline{M}$ whose sign is determined by the sign of the determinant of $Df(x)$ in the following manner.

Assume that for every x in ∂M, one has $\operatorname{Det} Df(x) \neq 0$, and that $Df(x)^{-1}f(x)$ is not tangent to ∂M. For every x in ∂M, choose now $\operatorname{sign}(\lambda(x))$ so that the vector $-\lambda(x)\ Df(x)^{-1}f(x)$ points into M. It is assumed that (a) for every x in ∂M, $\operatorname{sign}(\lambda(x)) = \operatorname{sign}(\operatorname{Det} Df(x))$ or (b) for every x in ∂M, $\operatorname{sign}(\lambda(x)) = -\operatorname{sign}(\operatorname{Det} Df(x))$.

At any x in M for which $\operatorname{Det} Df(x) \neq 0$, the value $\lambda(x)$ is chosen so that in case (a), $\operatorname{sign}(\lambda(x)) = \operatorname{sign}(\operatorname{Det} Df(x))$; in case (b), $\operatorname{sign}(\lambda(x)) = -\operatorname{sign}(\operatorname{Det} Df(x))$.

Smale's main result asserts that in these conditions, there is a subset Σ of measure zero in ∂M, depending only on f, such that starting from x_0 in ∂M but not in Σ, there exists a unique C^1 solution $\varphi\colon [t_0, t_1) \to \overline{M}$ of (2) with $\|d\varphi/dt\| = 1$ and t_1 maximal. Moreover, this solution converges to the set E of zeros of f as t tends to t_1.

If, in addition, $Df(x)$ is nonsingular for every x in E (a property that almost all C^1 functions $f\colon \overline{M} \to R^n$ have), then E is finite, the exceptional set Σ is closed, and the solution φ converges to a single zero x^* of f.

The first of the two questions mentioned at the beginning of this section, the description of the dynamic behavior of a private ownership economy out of equilibrium, was considered by Smale as "the fundamental problem of equilibrium theory" [1977]. In the process formalized by differential equation (1), no trading takes place in the economy as it seeks an equilibrium. Three articles ([1976d], [1977], [1978]) of Smale studied an alternative process in which pure exchange occurs at every instant at the current disequilibrium prices. This is a "non-tâtonnement" process of the type surveyed by Arrow and Hahn [1971, Ch. 13] and by Hahn [1982, Ch. 16]. But Smale's new approach focuses on cone fields of derivatives, rather than on fields of derivatives. Thus, for every feasible allocation $x = (x_1, \ldots, x_m)$ of the fixed total resources e in the interior of R^ℓ_+ among the m-consumers of the economy, a cone C_x (of vectors tangent at x to the set W of feasible allocations) is given. A curve ξ from the time-interval $[a, b)$ to the set W is an exchange curve if and only if for every t, $d\xi/dt \in C_{\xi(t)}$. In [1976d], giving a central role to the concept of a cone field of derivatives, Smale studied conditions under which an exchange curve converges to an equilibrium allocation.

5. Smale and the Theory of General Equilibrium

In his recent article on global analysis in economic theory [1987], Smale lists several of his motivations for the systematic use of differential methods in the treatment of general equilibrium.

(1) The proofs of existence of equilibrium are simpler. Kakutani's fixed point theorem is not used, the main tool being the calculus of several variables.
(2) Comparative statics is integrated into the model in a natural way, the first derivatives playing a fundamental role.
(3) The calculus approach is closer to the older traditions of the subject.
(4) Insofar as possible the proofs of equilibrium are constructive. These proofs may be implemented by a speedy algorithm, which is Newton's method modified to give global convergence. On the other hand, the existence proofs are sufficiently powerful to yield the generality of the Arrow-Debreu theory.

This short survey has outlined some of the principal contributions that Smale made in response to those motivations. Several of them have become important, sometimes essential, parts of the reformulation of the theory of general equilibrium by differential methods that took place over the past 20 years (extensive references for which are the books by Mas–Colell [1985] and by Balasko [1988]).

But Smale's economics period has also influenced the orientation of his mathematical research during the past decade. At the end of that period, his attention was attracted by the problem of the average speed of the simplex method of linear programming, to the solution of which he contributed two major articles ([1983a], [1983b]). Earlier, his interest in the Global Newton method had led to his collaboration with Hirsch on algorithms for finding the zeros of a function from R^n to R^n [1979b]. Both lines of work later merged in the approach to the theory of computation that has become the main subject of his research.

Bibliography

[1943] Allais, M., *A la Recherche d'une Discipline Economique*, Imprimerie Nationale, Paris.

[1951] Arrow, K.J., "An Extension of the Basic Theorems of Classical Welfare Economics," *Proceedings of the Second Berkeley Symposium on Mathematical Statistics and Probability*, J. Neyman, ed., University of California Press, Berkeley, pp. 507–532.

[1954] Arrow, K.J. and G. Debreu, "Existence of an Equilibrium for a Competitive Economy," *Econometrica*, **22**, 265–290.

[1958] Arrow, K.J. and L. Hurwicz, "On the Stability of the Competitive Equilibrium, I," *Econometrica*, **26**, 522–552.

[1959] Arrow, K.J., H.D. Block, and L. Hurwicz, "On the Stability of the Competitive Equilibrium, II," *Econometrica*, **27**, 82–109.

[1971] Arrow, K.J. and F.H. Hahn, *General Competitive Analysis*, Holden-Day, San Francisco.

[1964] Aumann, R.J., "Market with a Continuum of Traders," *Econometrica*, **32**, 39–50.

[1975a] Balasko, Y., "Some Results on Uniqueness and on Stability of Equilibrium in General Equilibrium Theory," *Journal of Mathematical Economics*, **2**, 95–118.

[1975b] Balasko, Y., "The Graph of the Walras Correspondence" *Econometrica*, **43**, 907–912.

[1988] Balasko, Y., *Foundations of the Theory of General Equilibrium*, Academic Press, New York.

[1951] Debreu, G., "The Coefficient of Resource Utilization," *Econometrica*, **19**, 273–292.

[1956] Debreu, G., "Market Equilibrium," *Proceedings of the National Academy of Sciences*, **42**, 876–878.

[1970] Debreu, G., "Economies with a Finite Set of Equilibria," *Econometrica*, **38**, 387–392.

[1974] Debreu, G., "Excess Demand Functions," *Journal of Mathematical Economics*, **1**, 15–21.

[1974] Dierker, E. *Topological Methods in Walrasian Economics*, Springer-Verlag, New York.

[1982] Dierker, E., "Regular Economies," *Handbook of Mathematical Economics II*, K.J. Arrow and M.D. Intriligator. eds., North-Holland, Amsterdam, pp. 795–830.

[1976] Eaves, B.C. and H. Scarf, "The Solution of Systems of Piecewise Linear Equations," *Mathematics of Operations Research*, **1**, 1–27.

[1955] Gale, D., "The Law of Supply and Demand," *Mathematica Scandinavica*, **3**, 155–169.

[1982] Hahn, F.H., "Stability," *Handbook of Mathematical Economics. II*, K.J. Arrow and M.D. Intriligator, eds., North-Holland, Amsterdam, 745–793.

[1963] Hirsch, M., "A Proof of the Non-retractibility of a Cell onto its Boundary," *Proceedings of the American Mathematical Society*, **14**, 364–365.

[1979] Hirsch, M., "On Algorithms for Solving $f(x) = 0$" (with S. Smale), *Communications on Pure and Applied Mathematics*, **32**, 281–312.

[1977] Kellogg, R.B., T.Y. Li, and J. Yorke, "A Method of Continuation for Calculating a Brouwer Fixed Point," *Fixed Points. Algorithms and Applications*, S. Karamardian. ed., Academic Press, New York.

[1942] Lange, O., "The Foundations of Welfare Economics," *Econometrica*, **10**, 215–228.

[1974] Mantel, R., "On the Characterization of Aggregate Excess Demand," *Journal of Economic Theory*, **7**, 348–353.

[1985] Mas-Colell, A., *The Theory of General Economic Equilibrium. A Differential Approach*, Cambridge University Press, Cambridge.

[1954] McKenzie, L.W., "On Equilibrium in Graham's Model of World Trade and Other Competitive Systems," *Econometrica*, **22**, 147–161.

[1956] Nikaido, H., "On the Classical Multilateral Exchange Problem," *Metroeconomica*, **8**, 135–145.

[1909] Pareto, V., *Manuel d'Economie Politique*, Giard, Paris.

[1941] Samuelson, P.A., "The Stability of Eauilibrium: Comparative Statics and Dynamics," *Econometrica*, **9**, 97–120.

[1960] Scarf, H., "Some Examples of Global Instability of Competitive Equilibrium," *International Economic Review*, **1**, 157–172.

[1973] Scarf, H., (with the collaboration of T. Hansen), *The Computation of Economic Equilibria*, Yale University Press, New Haven, CT.

[1973] Smale, S., "Global Analysis and Economics I, Pareto Optimum and a Generalization of Morse Theory," *Dynamical Systems*, M.M. Peixoto, ed., Academic Press New York.

[1974a] Smale, S., "Global Analysis and Economics IIA, Extension of a Theorem of Debreu," *Journal of Mathematical Economics*, **1**, 1–14.

[1974b] Smale, S., "Global Analysis and Economics III, Pareto Optima and Price Equilibria," *Journal of Mathematical Economics*, **1**, 107–117.

[1974c] Smale, S., "Global Analysis and Economics IV, Finiteness and Stability of Equilibria with General Consumption Sets and Production," *Journal of Mathematical Economics*, **1**, 119–127.

[1974d] Smale, S., "Global Economics and Economics V, Pareto Theory with Constraints," *Journal of Mathematical Economics*, **1**, 213–221.

[1975a] Smale, S., "Optimizing Several Functions," *Manifolds Tokyo 1973*, University of Tokyo Press, Tokyo, pp. 69–75.

[1975b] Smale, S., "Sufficient Conditions for an Optimum," *Warwick Dynamical Systems 1974*, Springer-Verlag, Berlin, pp. 287–292.

[1976a] Smale, S., "Dynamics in General Equilibrium Theory," *American Economic Review*, **66**, 288–294.

[1976b] Smale, S., "Global Analysis and Economics VI, Geometric Analysis of Pareto Optima and Price Equilibria under Classical Hypotheses," *Journal of Mathematical Economics*, **3**, 1–14.

[1976c] Smale, S., "A Convergent Process of Price Adjustment and Global Newton Methods," *Journal of Mathematical Economics*, **3**, 107–120.

[1976d] Smale, S., "Exchange Processes with Price Adjustment," *Journal of Mathematical Economics*, **3**, 211–226.

[1977] Smale, S., "Some Dynamical Questions in Mathematical Economics," *Colloques Internationaux du Centre National de la Recherche Scientifique, No. 259: Systèmes Dynamiques et Modèles Economiaues.* Centre National de la Recherche Scientifique, Paris.

[1978] Smale, S., "An Approach to the Analysis of Dynamic Processes in Economic Systems," *Equilibrium and Disequilibrium in Economic Theory*, G. Schwödiauer, ed., D. Reidel Publishing Co., Boston, pp. 363–367.

[1979] Smale, S., "On Comparative Statics and Bifurcation in Economic Equilibrium Theory," *Annals of the New York Academy of Sciences*, **316**, 545–548.

[1980a] Smale, S., "The Prisoner's Dilemma and Dynamical Systems Associated to Non-cooperative Games," *Econometrica*, **48**, 1617–1634.

[1980b] Smale, S., *The Mathematics of Time: Essays on Dynamical Systems, Economic Processes, and Related Topics*, Springer-Verlag, Berlin.

[1981] Smale, S., "Global Analysis and Economics," *Handbook of Mathematical Economics I*, K.J. Arrow and M.D. Intriligator, North-Holland, Amsterdam, pp. 331–370.

[1983a] Smale, S., "The Problem of the Average Speed of the Simplex Method," *Proceedings of the XIth International Symposium on Mathematical Programming*, A. Bachem, M. Grotschel, and B. Korte, eds., Springer-Verlag, New York, pp. 530–539.

[1983b] Smale, S., "On the Average Number of Steps of the Simplex Method of Linear Programming," *Mathematical Programming*, **27**, 241–262.

[1987] Smale, S., "Global Analysis in Economic Theory," *The New Palgrave, A Dictionary of Economics*, John Eatwell, Murray Milgate, and Peter Newman, eds., Macmillan, New York, Vol. 2, pp. 532–534.

[1972] Sonnenschein, H., "Market Excess Demand Functions," *Econometrica*, **40**, 549–563.

[1973] Sonnenschein, H., "Do Walras' Identity and Continuity Characterize the Class of Community Excess Demand Functions?," *Journal of Economic Theory*, **6**, 345–354.

[1964] Vind, K., "Edgeworth Allocations in an Exchange Economy with Many Traders," *International Economic Review*, **5**, 165–177.

[1874–77] Walras, L., *Eléments d'Economie Politique Pure*, L. Corbaz and Company, Lausanne.

[1975] Wan, H.Y., "On Local Pareto Optima," *Journal of Mathematical Economics*, **2**, 35–42.

15
Topology and Economics: The Contribution of Stephen Smale[1]

GRACIELA CHICHILNISKY

1. Introduction

Classical problems in economics are concerned with the solutions of several simultaneous nonlinear optimization problems, one for each consumer or producer, all facing constraints posed by the scarcity of resources. Often their interests conflict, and it is generally impossible to find a single real-valued function representing the interests of the whole of society. To deal with this problem, John Von Neumann introduced the theory of games. He also defined and established the existence of a general economic equilibrium, using topological tools [Von Neumann, 1938]. The work of Stephen Smale follows this tradition. He uses topological tools to deepen and refine the results on existence and other properties of another type of economic equilibrium, the Walrasian equilibrium (Walras [1874–77]), as formalized by Kenneth J. Arrow and Gerard Debreu [1954], and of non-cooperative equilibrium in game theory as formalized by Nash (1950). This article aims to show that topology is intrinsically necessary for the understanding of the fundamental problem of conflict resolution in economics in its various forms[2] and to situate Smale's contribution within this perspective. The study of conflicts of interests between individuals is what makes economics interesting and mathematically complex. Indeed, we now know that the space of all

---

[1] This is a revised version of the paper presented at Conference in Honor of Stephen Smale at the University of California Berkeley, August 1990. I thank those who commented on the presentation of this paper on August 5, 1990, and Ralph Abraham, Don Brown, Curtis Eaves, Geoffrey Heal, Morris Hirsch, Jerrold Marsden, Andreu Mas Colell and Dennis Sullivan for comments. I am particularly grateful to Kenneth J. Arrow for valuable suggestions. Hospitality and financial support from the Stanford Institute of Theoretical Economics at Stanford University is gratefully acknowledged.

[2] A similar statement is made by Von Neumann about his general equilibrium market solution: "The mathematical proof (of existence of an economic equilibrium) is possible only by means of a generalization of Brouwer's Fix-Point theorem, i.e. by the use of very fundamental *topological facts*" (Von Neumann, [1945–46], p. 1, parenthesis supplied).

individual preferences, which define the individual optimization problems, is topologically nontrivial, and that its topological complexity is responsible for the impossibility of treating several individual preferences as if they were one, i.e., aggregating them (Chichilnisky, 1980; Chichilnisky and Heal, 1983). Because it is not possible, in general, to define a single optimization problem, other solutions are sought. This article will develop three solutions, discussed below.

Because of the complexity arising from simultaneous optimization problems, economics differs from physics where many of the fundamental relations derive from a single optimization problem. The attempts to find solutions to conflicts among individual interests led to three different theories about how economies are organized and how they behave. These are *general equilibrium theory*, the *theory of games*, and *social choice theory*. Each of these theories leads naturally to mathematical problems of a topological nature. Steve Smale has contributed fruitfully to the first two theories: general equilibrium theory and the theory of games. I will argue that his work is connected also with the third approach, social choice theory, by presenting in Section 4 results which link closely, and in unexpected ways, two seemingly different problems: the existence of a general equilibrium and the resolution by social choice of the resource allocation conflict in economics (Chichilnisky, 1991).

2. A First Approach to Conflict Resolution in Economics: General Equilibrium Theory

One method of resolving social conflicts of interests is through market allocations or general equilibrium solutions. This is the area of economics to which Stephen Smale contributed most. General equilibrium theory explains how an economy behaves through a formal model of interacting markets. In particular, general equilibrium theory explains how society's resources will be allocated among the different individuals through the allocations which occur when markets clear. Such a solution is appealing because, under certain conditions, the general equilibrium allocation maximizes the Pareto order which ranks allocations among the different individuals. This is a statement of the "first theorem of welfare economics", which establishes conditions (Arrow and Hahn (1971)) under which Pareto optimality is achieved at the market clearing allocations. Pareto optimality of an allocation means that no other feasible allocation can be found at which all individuals are simultaneously better off. Pareto optimality provides a powerful rationale for studying market-induced solutions to the fundamental conflict of allocating scarce resources among several individuals.

General equilibrium theory starts with the assumption that there exist l commodities, so that a consumption bundle is described by a vector in R^l. A *pure exchange economy* is defined by three primitives, which are left unexplained:

1. Consumption sets X_i, nonempty subsets of R^l, for each agent $i = 1, \ldots, m$.
2. A vector $\mathbf{e} \in R^l$ describing a quantity of each commodity available to the economy as a whole. If the economy is *privately owned*, then there are vectors describing the initial allocations of the l commodities for each of the m-consumers or traders, $\mathbf{e}_i \in \mathbf{R}^l$, $i = 1, \ldots, m$, which describe each *individual's initial ownership of resources* or endowments, and satisfying $\sum_i e_i = e$.
3. For each trader i, a preference $\leq_i$ which is a total preorder (i.e., reflexive and transitive) on X_i.

A *pure exchange privately owned economy* is, therefore, a system, $E = \{(X_i), (\leq_i), (\mathbf{e}_i), i = 1 \ldots m\}$. The vector $\mathbf{e} = \sum_i \mathbf{e}_i \in R^l$ describes the scarcity of resources for the economy as a whole[3] and is called the initial endowment of the economy. The preference $\leq_i$ describes the criteria used by trader i to evaluate his/her allocation. These preferences are generally not linear, leading to a nonlinear optimization problem for each trader. The space of all possible preferences is an infinite-dimensional space related to, but quite different from, the space of real-valued functions on the commodity space R^l. Indeed, as it will be discussed in Section 4, the global topology of this space is responsible for the impossibility of constructing *one* preference $\leq$ which represents all of society, and which reduces the problem to a single optimization problem (Chichilnisky, 1980).

We now define in steps the general equilibrium solution. The index i describes consumers in the economy, $i = 1, \ldots, m$. Define an *attainable state s of the economy* as a list of consumption-vectors for each trader, $\{(x_i)_{i=1,\ldots,m}\}$, and a vector $\mathbf{p}$ in the positive orthant of euclidean space R^l, denoted R^{l+}, called a *price-vector*, satisfying the conditions:

(i) for every i, $x_i \in X_i$, and $\mathbf{p} \in R^{l+}$;
(ii) $\sum_i x_i \leq \sum_i \mathbf{e}_i$.

The price-vector $\mathbf{p}$ assigns to each commodity bundle z in R^l a "value" defined by the inner product $\mathbf{p} \cdot \mathbf{z}$. For each price $\mathbf{p}$, the *budget set* $B_i(\mathbf{p})$ *of the* ith *trader* is the set $B_i(\mathbf{p}) = \{x \in R^l / \mathbf{p} \cdot x \leq \mathbf{p} \cdot \mathbf{e}_i\}$.

An attainable state s is a *general equilibrium* if, for every i, the consumption vector x_i is best according to $\leq_i$, within the ith trader's budget set $B_i(\mathbf{p})$. An attainable state s is called *Pareto optimal* if no other attainable state s' exists at which all traders are better off in terms of their preferences.

A general equilibrium, therefore, proposes an interesting solution to the resource allocation problem. Total resources are scarce, constrained as in (ii)

[3] In more general terms, production is considered as well: Smale [1974a] treated production with considerable generality; see also Debreu [1991]. However, for our purposes it will suffice to study the pure exchange economy, in which individuals trade between themselves a fixed total stock $\mathbf{e} \in R^l$ of goods.

by the initial endowment of the economy $\mathbf{e}$. *In a general equlilibrium, each trader is allocated a share x_i of all l commodities in such a way that each trader i maximizes his/her preference $\leq_i$ within their budget sets $B_i(\mathbf{p})$, and without exceeding society's total resources $\sum_i x_i \leq \mathbf{e}$.*

Note that auxiliary (dual) variables have been introduced: prices $\mathbf{p}$ which are vectors in R^l (one for each commodity). These variables play an important role in the definition of an equilibrium. This was already noted by Von Neumann [1938] who wrote: "Another feature of our theory, so far without explanation, is the remarkable duality (symmetry) of the monetary variables (prices y_j, interest factor β) and the technical variables." The use of prices in defining equilibria and, thus, characterizing Pareto optimal allocations is one of the fundamental insights provided by welfare economics. This has also been the subject of Smale's work (e.g., Smale [1973, 1974b, 1974d, 1975, 1976d]). The general equilibrium solution to the resource allocation problem defined by Arrow and Debreu (1954) satisfies the following remarkable properties: It respects the value of private ownership because $\sum_i \mathbf{p}_i x_i \leq \sum_i \mathbf{p}_i \mathbf{e}_i$, and it optimizes the Pareto order on social allocations. This is the main claim for its use as an acceptable form of resource allocation.

Topology has long been connected with the general equilibrium of markets. The connection between topology and general equilibrium models in economics has been established in a number of pieces starting with Von Neumann's [1938]. As its title indicates, Von Neuman's article proved a generalization of Brouwer's Fixed-Point Theorem to establish the existence of an economic equilibrium. When the Walrasian model of a general equilibrium (Walras [1874–77]) was formalized by Arrow and Debreu [1954], a fixed-point theorem was also used to establish the existence of an equilibrium. Indeed, all proofs of existence of an economic equilibrium use topological tools.

Until Smale's contribution, all proofs of existence of a general equilibrium relied on fixed-point theorems, and convexity was generally required. As Debreu [1993] points out, Smale's work coincided with the return of the differentiable tradition in economics, a tradition that had been abandoned in the 1950s and 1960s in favor of the convex approach.

Smale's papers on general equilibrium use a different topological approach. The results of M. Hirsch [1963] on the nonretractibility of a cell onto its boundary, Lemke's [1965] algorithm for affine cases, and Eaves's extension to nonlinear cases [1971, 1972], provided topological methods for computing an equilibrium and a precedent for Smale's approach to proving the existence of an economic equilibrium. Existence is established by studying the behavior of certain one-dimensional manifolds starting at the boundary of a set (the price space) and tending to a general equilibrium. Smale's proof of the existence of a general equilibrium follows this general line, using Sard's theorem, the implicit function theorem, and the classification of one-dimensional manifolds.

As in Von Neumann [1938], Smale [1974a, 1975] reduces the problem to the existence of a zero of several simultaneous nonlinear equations, each describing the behavior of one market. The zeros represent the market-clearing positions in each market. This is an old approach to the problem to which he gives a new light. He considers the standard excess demand of the economy at each price $\mathbf{p}$, denoted $z(\mathbf{p})$. This is defined as the difference between two vectors: the *total demand* vector [the sum of the optimal choices of all the traders at their budget sets $B_i(\mathbf{p})$] and the *total supply* (the social endowment $\mathbf{e}$). Excess demand $z(\mathbf{p})$ is a correspondence defined on the *price space* which is the unit simplex in R^l, $\Delta_l = \{\mathbf{p} \in R^{l+}/\sum \mathbf{p}_j = 1\}$, a closed convex subset of R^l, and with values in the commodity space R^l. Smale makes assumptions which ensure that $z(\mathbf{p})$ is a well-defined C^2 function from the price space Δ_l into R^l. For each $\mathbf{p}$, the vector $z(\mathbf{p})$ is called the "excess demand" of the economy because it is the difference between what is demanded and what is supplied of each good.

The zeros of $z(\mathbf{p})$ are the general equilibria of the economy, called the "equilibrium prices." At an equilibrium price $\mathbf{p}$ [one satisfying $z(\mathbf{p}) = 0$], the market clears and traders are within their optimal behavior. Thus, Smale's interest in finding the zeros of a function from R^m into R^m (e.g., Hirsch and Smale [1979]).

Locating the zeros of the excess demand function is achieved by studying the behavior of a one-dimensional manifold. This manifold is the inverse image of a regular value of the function

$$\varphi^*(\mathbf{p}) = \varphi(\mathbf{p})/\|\varphi(\mathbf{p})\|$$

where

$$\varphi(\mathbf{p}) = z(\mathbf{p}) - \mathbf{p}\left[\sum_h z^h(\mathbf{p})\right].$$

The zeros of $z(\mathbf{p})$ are the same as the zeros of $\varphi(\mathbf{p})$, so that φ^* is not defined on the set of equilibria, denoted $\sum$. Thus, φ^* is defined only on $\Delta_1 - \sum$. Consider a regular value y of φ^*. A connected component of the one-dimensional manifold defined by the inverse image of y, starting at the boundary of Δ_l, must end outside the domain of definition of φ^*, and, thus, at a zero of $\varphi(x)$. Therefore, by following such a manifold, one finds an equilibrium (Smale, 1974a).

Smale's proof is interesting in that it can be extended in several ways to *compute* the equilibrium, by following related one-dimensional manifolds starting at the boundaries of the price space. Indeed, this became Smale's work on the Global Newton Method [1976b], related to his work with M. Hirsch "On Algorithms for solving $f(x) = 0$" (Hirsch and Smale, 1979).

Smale's work on existence of an equilibrium also provided very useful insights into the approximation of equilibrium through "nonequilibrium" paths (Smale, 1976a, 1976c, 1977, 1978). Here, Smale attacked also the problem of how the economy arrives at an equilibrium, a problem which is not yet fully solved but to which he contributed valuable technical insights. In addition, Smale explored fruitfully the question of trading outside of an equilibrium [1976c], the so called "non-tatonnement" model. This refers to a nontrivial variant of the Walrasian model which allows trading even though markets have not cleared (Arrow and Hahn, 1971).

Indeed, one of the difficulties with the Walrasian general equilibrium model is that trading takes place only at an equilibrium. An interpretation of this is that, in a Walrasian economy, there exists an "auctioneer" or "government" agent which controls the market in the sense of allowing "bets" at those prices that do not clear the market, but not allowing trading unless prices are achieved at which all markets clear. The interpretation of the auctioneer is problematic because it is at odds with the desire to view a market as a purely decentralized institution, one with no role for a "government" or other institution who knows everyone's position. Smale's work includes cases which go beyond this limitation, cases where trading takes place outside of an equilibrium. He studied conditions for the convergence of such trading to a general equilibrium [1976c].

Added to the above results, which use differentiable and algebraic topology in various forms, the Global Newton Method specified by Smale [1976b] modifies the standard Newton method (e.g., Arrow and Hahn [1971]) so that it becomes well-defined on the "singularities" where the Newton method is not, for generic sets of initial conditions. Again his approach uses topology to determine how to "continue" the Newton method through a singularity. His work on the computation of a zero of the excess demand functions led him, in turn, to develop in recent years a fruitful approach to the development of more general algorithms for the computation of zeros of a function of R^m into R^m.

2.1. Equilibrium in Nonconvex Economies

Smale also contributed to the understanding of the deterministic properties of equilibrium, i.e., to the number of equilibria of an economy. Using Sard's theorem, a technique used also by Debreu [1970], Smale [1974a] established the generic local uniqueness of equilibrium in rather general models, where production is allowed, even nonconvex production. This is a realistic assumption on the so-called "returns to scale" of the economy, which allows for "increasing" as well as "decreasing returns in production." The convexity of production sets derives from the assumption that added units of inputs produce proportionally fewer output. Instead, *increasing* returns means that production becomes *more efficient* when the quantity produced increases: An added unit of input produces proportionally a higher amount of output. With increasing returns, production sets are not convex. Economic production with informational inputs (such as, for example, those involving communications) typically exhibit increasing returns to scale, as do all industries with large initial fixed costs of production (such as R&D). Indeed, some of the most interesting and productive new technologies have this property.

Smale established properties of such nonconvex economies with simple methods, in an area which subsequently became important in economics in terms of its ability to yield new insights on the welfare properties of equilibria. With nonconvex technologies, a general equilibrium reinterpreted as Smale does, in differentiable terms, becomes essentially what is usually called a *marginal cost pricing equilibrium.*[4] Smale does not call attention to this fact, which has largely been overlooked in the literature.

Smale's framework [1974a] is as follows. He works with variable endowment-vectors $\mathbf{e}_i$ for traders, $i = 1, \ldots, m$. Each individual has a fixed consumption set denoted $P \subset R^l$ (previously X_i). The economy is thus defined by one vector $\mathbf{e} = (\mathbf{e}_1, \ldots, \mathbf{e}_m) \in P^m$, and by allocations of total resources, namely, of $\sum_i \mathbf{e}_i$, to the consumers denoted $x = (x_1, \ldots, x_m)$ such that $\sum_i x_i = \sum_i \mathbf{e}_i$. A trader's preference is defined by a C^2 real-valued function u_i defined on P, with no critical points, and thus having a well-defined gradient vector $g_i(x_i)$, $\forall x_i \in P$. Smale's definition of the set Σ of equilibria is then

$$\Sigma = \left\{(\mathbf{e}, x) \in P^m \times P^m / \forall i,\ g_i(x_i) = \mathbf{p} \text{ and } \mathbf{p} \cdot x_i = \mathbf{p} \cdot \mathbf{e}_i, \sum_i x_i = \sum_i \mathbf{e}_i\right\}.$$

Smale's definition corresponds to the previously given definition of a general equilibrium (which is, in turn, in Debreu's [1993]) when preferences are concave and the trader's endowments $\mathbf{e}_i$ are fixed.

Smale replaces the optimization of preferences by a "common gradient condition": $\forall i$, $g_i(x_i) = \mathbf{p}$. This condition is equivalent to the condition of optimizing traders' preferences on their budget sets *when preferences are concave*, but not otherwise.

Smale's definition of a general equilibrium is extended to a similar definition with production, in which case it implies optimization (of profits by producers) only when there is convexity (of production sets). With general production sets (not just convex) as considered by Smale, his definition becomes essentially equivalent to the notion of "marginal cost pricing equilibrium" (Brown and Heal, 1979). This simply means that at an equilibrium the "equal gradient condition" is satisfied, although no optimization of profits is assured. Marginal cost pricing is a well-established principle for economies with nonconvex production, and it is widely used in practice for regulating industries with "increasing returns to scale," i.e., with nonconvex production. In addition, marginal cost pricing can be shown to be a necessary (but not sufficient) condition for efficiency of an equilibrium.

Obviously, marginal cost pricing equilibria need not satisfy Pareto optimality conditions in nonconvex economies. Such cases are included in Smale's existence work, as well as in his work on the local uniqueness of such equilibria [1974a], although he does not discuss the welfare properties of the nonconvex cases.

---

[4] When an additional condition ensuring positive incomes is added. Such a condition is not necessary in convex economies.

Since Smale's existence work on economies with nonconvex production, we have learned that this failure of optimality can be quite serious: there exist nonconvex economies with several (locally unique) equilibria as defined by Smale, where all the equilibria fail to be Pareto optimal. This was proved by Guesnerie [1975] and by Brown and Heal [1979], following parallel lines which did not make contact with Smale's work. More recently, we have learned that this problem of failure of optimality is generic. Generically in nonconvex economies, all marginal cost pricing equilibria fail to be Pareto optimal (Chichilnisky, 1990).

Smale's work on nonconvex economies is not concerned with Pareto optimality. It appears to emerge from a natural mathematical position with respect to the limitations of convexity. While powerful, convexity is a rather special case, and it cannot be justified from the observation of economic reality. Smale's work, which focuses on a differentiable approach, naturally included convex as well as nonconvex cases. This simpler and more general approach to general equilibrium—which, as we noted, leads to unexpected failures of the optimality of equilibrium—is possibly one of the most interesting derivatives of Smale's work in general equilibrium theory. It seems fair to say that his work in this area has hardly been utilized in the economic literature, so that we may expect many more ramifications in the future.

Smale followed the path started by Von Neumann of utilizing topological methods to resolve simultaneous, conflicting, optimization problems arising from economics. Whereas Von Neumann introduced new models in economics, such as the Theory of Games and General Equilibrium Theory, Smale deepened and refined our understanding of existing models and open problems. He also provided a fruitful introduction of differential topology methods and showed the value of this machinery both for theoretical and applied purposes.

3. A Second Approach to Conflict Resolution in Economics: The Theory of Games

Smale also contributed to the theory of games, particularly in one paper studying the solutions of the "prisioner dilemma" game [1980]. Game theory provides a different approach to the resource-allocation problem. It typically assumes that the players have certain "strategies" available to them. The game is defined by specifying the outcomes which obtain when each player is playing a given strategy. This is called a "payoff function." A widely used solution concept is that of a Nash equilibrium (Nash, 1950). Here, each player chooses a strategy which maximizes his/her preferences for outcomes, *given the strategies played by the other players.* The fact that other player's strategies are taken as given is usually described by saying that the equilibrium is noncooperative. Instead, *cooperative solutions* involve communication between players, so that players influence each other strategies and, in this sense, coalitions are formed which decide the outcomes.

Game theoretical models differ from market models in that they have much less structure. This gives more flexibility to the theory, so that more situations can be represented, but obviously, fewer general results obtain. Because the problem is posed in a very wide framework, particular assumptions are made in each case to prove results, with the consequence that the results are far less general than those of the general equilibrium theory. In particular, Pareto efficiency is generally lost in noncooperative solutions. The trade-off is in the structure of the problem: market theory has far more structure and, therefore, far more results.

Yet, game theory is closely connected with the market-allocation problem. Indeed, one of the first existence theorems produced by G. Debreu for a general equilibrium proved the existence of a market equilibrium by using a proof due to Nash (1950) to establish the existence of an equilibrium of a game in which, in addition to the traders defined in Section 2, a "government" was one of the players (Debreu, 1952). In addition, a widely used concept of a cooperative game solution, the *core* of a market game, is closely associated with the market equilibrium. In an appropriately defined limiting sense, the two concepts coincide.

The problem tackled by Smale in game theory is the Pareto optimality of solutions to the prisioner's dilemma, the archetype example of a game where noncooperative solutions, the Nash equilibria, are Pareto inferior. A longstanding intuition about this game is that if the players played repeatedly, they would "learn" that the noncooperative solution is inferior and alter their behavior appropriately. The question is to formalize this learning process so that a new solution may be defined that leads to Pareto optimal outcomes.

Smale studies that questions by defining a repeated game, i.e., one which is played time and again, with finite memory. This is formalized by defining a time-dependent map from the set S of convex combination of all possible strategies to itself, $\beta_T: S \to S$, where $T \in N$ denotes time. The idea is to represent memory of the past by the average of past strategies. Given an initial strategy at time $T = 0$, at each T the map, β_T summarizes a time-dependent sequence of strategies, each generated by the average of the previous ones, thus defining dynamical systems associated to the noncooperative game. The interest of his approach is that he obtains Pareto efficient solutions asymptotically, the type of solutions which one may expect from a cooperative approach, by studying noncooperative solutions when the agents have memory. Again using topological tools, Smale uses his formulation to prove, under certain conditions, the existence of Pareto optimal solutions which are globally stable as usually defined in dynamical systems theory.

4. A Third Approach to Conflict Resolution: Social Choice Theory

Our last task is to explain the relation between Smale's work and social choice theory, an area in which Smale did not contribute directly.

Social choice theory starts by the assumption that there exists a set X of all possible choices or social states, which could be interpreted as the space of all possible allocations of the total social endowments among all m individuals. X is generally a closed, convex subset of R^l. Each individual i has a preference p_i defined on X, $i = 1, \ldots, m$, (see definition of $\leq$ in Section 2). The space of all individual preferences is denoted P; this is an important space with a non-trivial topological structure. As seen below, the solution of the social choice allocation problem hinges on its topology. The social choice problem is to find a "social preference" p, which ranks the elements in X, and, thus, can choose the social optimum among the feasible allocations in X.

It is desirable, and thus assumed, that the social preference depends on the individual preferences p_i, $i = 1, \ldots, m$, i.e., $p = \phi(p_1, \ldots, p_m)$. Equivalently, there must exist a function $\phi: P^m \to P$. The properties of the function ϕ determine the manner in which the social preference depends on the individual preferences. Ethical principles dictate that ϕ should be symmetric on the individuals, an "equal treatment" property, (i.e., *respect anonymity*), and that ϕ should be the identity map on the diagonal of P^m (meaning that ϕ *respects unanimity*). Practicality requires that ϕ be *continuous*; this assures the existence of sufficient statistics, namely, "polls" taken from finite samples which yield information approximating the true function. These three axioms: continuity, anonymity, and respect of unanimity were introduced in Chichilnisky [1980]; they relate to, but are different from, other axioms used in the Social Choice literature commenced by Kenneth Arrow's work.

In general, the three axioms, continuity, anonymity and respect of unanimity, are inconsistent with each other (Chichilnisky, 1980) and certain restrictions are necessary for social choice allocations satisfying these axioms to be possible (Chichilnisky and Heal, 1983). This means that a function $\phi: P^m \to P$ respecting all three axioms does not exist when P is the space of all preferences Chichilnisky (1980). The nonexistence result hinges on the topological structure of the space of preferences P. This became clear when, after a substantial literature developed on this problem, Chichilnisky and Heal (1983) proved that for any given (CW complex) space X a map $\phi: X^m \to X$ exists for all m if and only if the homotopy groups $\Pi_j(X) = 0$ for all $j \geq 1$ (Theorem 3 below). This is equivalent to X being contractible, i.e., topologically equivalent to a point: therefore, the allocation problem in economics, posed in terms of a social choice, is essentially a topological issue, one which can be resolved if and only if there is no topological complexity.

This section will show that the existence of the competitive equilibrium has the same characteristics as the social choice allocation problem: It will be shown that the topological obstruction for the social choice resolution and for the existence of an equilibrium are one and the same: They lie in the topology of a family of cones which are naturally associated to the economy's preferences..

We need some notation, in particular we must define the space of preferences and explain how it differs from a vector space of real-valued functions. The latter, being linear, is, of course, topologically trivial.

The following is a definition of preferences equivalent to that of Section 2.

Consider the space of real-valued functions on X, $f: X \to R$, with the equivalence relation $f \approx g$ if and only if $\forall x, y \in X, f(y) \geq f(x) \leftrightarrow g(y) \geq g(x)$. The space P of all preferences is then the quotient space of the space F of all real-valued functions on X under the equivalence relation $\approx$, $F/\approx$. With appropriate smoothness and regularity restrictions on F ($f \in F \to f$ is C^2 and has no critical points) such as those used is Smale's work, one considers P as the space of all C^1 unit vector fields $v: X \to R^l$ which are globally integrable, i.e., essentially the space of codimension—one C^1 oriented foliations of X without singularities (Chichilnisky, 1980).

Recent results have been used to establish the topological connection between the problem of existence of a general equilibrium to which Smale contributed extensively, and the social choice solution of the resource allocation conflict in economics in Chichilnisky [1991]. The rest of this section will explain this connection.

4.2. Social Choice and the Existence of a Competitive Market Equilibrium

Consider the exchange general equilibrium model defined in Section 2. Typically, it is assumed that consumption sets X_i are positive orthants. Here, instead, we assume that the individual consumption sets X_i are all of R^l; this is elaborated further below. In this case, it is possible to give necessary and sufficient condition a simple for the existence of a general equilibrium (Chichilnisky, 1992), defined on the properties of a family of cones which represent the relationships between the different individual preferences. These necessary and sufficient conditions are in turn also necessary and sufficient for the resolution of the social choice problem as defined by Chichilnisky [1980] and Chichilnisky and Heal [1983].

Consider the pure exchange general equilibrium model as defined in Section 2 with one further specification: for each consumer i, the consumption set X_i, which is the set of all consumption vectors which are feasible to the consumer, is the whole commodity space R^l. Such consumption sets are familiar in financial markets: they include "short" sales of commodities, which are sales involving a quantity larger than what is owned by the seller. When there are no limitations on the coordinates of the consumption vectors,[5] x_i, $\forall i, X_i = R^l$.

---

[5] Debreu [1993] mentions that for each consumer, the consumer inputs are positive numbers, and outputs are negative numbers. The hth coordinate of a vector x_i in the consumption set X_i is then the quantity of the hth commodity that he/she consumes if $x_i^h > 0$, or the negative of the quantity he/she produces if $x_i^h < 0$. For example, if a consumer's output is the amount of labor he/she contributes, because labor output is limited to 24 hours a day, then the negative coordinates of the consumption set X_i are bounded below. Such restrictions are however not applicable to financial markets, where in principle X_i could be unbounded below, and, in fact, the whole of the space R^l, see also Chichilnisky and Heal [1984, 1993].

When each consumer's consumption set X_i is the whole commodity space R^l rather than R^{l+}, certain conditions on individual preferences are needed for the existence of a competitive equilibrium without which it fails to exist. These conditions replace the boundary conditions used by Smale to prove existence. Arrow–Debreu consider for example the model defined in Section 2, where commodities are also indexed by date and state of nature so that, for example, an apple tomorrow if it rains is different from an apple today, and also different from an apple tomorrow if it does not rain, and they all have potentially different prices. In such models, consumer preferences include their subjective probabilities for the states of the world in the next period. It is clear that with two traders who have the opposite expectations about the price of apples tomorrow (for example, one expecting that with probability one the price of apples will increase, and the other the opposite, also with probability one), they will always increase their utility by trading increasing amounts in the opposite direction for delivery tomorrow. The trader who expects prices to drop will want to sell short, any amount, and the trader who expects prices to increase will want to buy more, any amount. Thus, in this two-person economy, an equilibrium will not exist unless an exogenous bound is introduced on "short" apple sales, such as, for example, a lower bound on the consumption sets X_i of the traders.

Such lower bounds are obviously artificial. In addition, the equilibria found by imposing exogenously such lower bounds, depend, of course, on the chosen bound, leading to indeterminacy of the model (i.e., which bound to choose?).

For this reason, it is preferable to postulate conditions on preferences, which are the primitives of the model, to assure the existence of an equilibrium. This is what has been done in Chichilnisky and Heal [1984, 1993]. We pursue a similar approach here, except that here we shall provide conditions which are "tighter" being both necessary and sufficient for the existence of an equilibrium.

We need some definitions. Let $E = \{(X_i), (\leq_i), (e_i), i = 1, \ldots, m\}$ be an economy as in Section 2, with $\forall i$, $X_i = R^l$. Define the ith trader *preferred cone* $A_i = \{v \in R^l\colon \forall x \in R^l \exists \lambda > 0\colon (e_i + \lambda v) >_i (x_i)\}$, and its *dual* $D_i = \{v \in R^l\colon \forall y \in A_i, \langle y, v \rangle > 0\}$. We shall say that a smooth preference p is consistent with the cone A_i if $\forall x_i \in R^l$ and $\forall y \in A_i$, $\langle p(x_i), y \rangle > 0$, where $p(x)$ is the vector field defining the preference p. This means that the preference p increases in the direction of the preferred cone A_i. For any subset of traders $\theta \subset \{1, \ldots, m\}$, a smooth preference p is similar with those of the *subeconomy* $E_\theta = \{(X_i), (\leq_i), (e_i), i \in \theta\}$ if it is consistent with the preferred cone A_i for some $i \in \theta$. The space of all preferences similar to those of the subeconomy E_θ is denoted P_θ. Define now the following conditions:

(A) The intersection of the duals $\bigcap_{i=1}^m D_i$ is not empty.
(B) The union $\bigcup_{i \in \theta} D_i$ is contractible for all $\theta \subset \{1, \ldots, m\}$.

Then we have the following results:

Theorem 1. *Conditions* (A) *and* (B) *are equivalent* (*Chichilnisky*, 1981, 1991).

Theorem 2. *Consider the pure exchange economy* $E = \{(X_i), (\leq_i), (e_i), i = 1, \ldots, m\}$ *of Section* 2. *Assume that all standard conditions on preferences are satisfied*: $\forall i$, $\leq_i$ *is smooth, concave and increasing. The consumption sets are* $X_i = R^l$ *for all i. Then condition* (A) *is necessary and sufficient for the existence of a competitive equilibrium* (*Chichilnisky*, 1992).

If X is a CW complex:

Theorem 3. *A continuous anonymous map* $\Phi: X^k \to X$ *respecting unanimity exists for all* $k \geq 1$ *if and only if* $\pi_j(X) = 0$ *for all j. In particular, X is contractible.*

From Theorems 1, 3 and Chichilnisky (1991):

Theorem 4. *For each subset of traders* $\theta \subset \{1, \ldots, m\}$, *consider the space* P_θ *of all smooth preferences on* R^l *which are similar to those in the subeconomy* E_θ. *Then there exists a continuous anonymous map* $\Phi: P_\theta^k \to P_\theta$ *which respects unanimity for all* $k \geq 1$ *and* θ, *if and only if* (B) *is satisfied.*

Theorems 1, 2 and 4 imply:

Theorem 5. *The economy* $E = \{(X_i), (\leq_i), (e_i), i = 1, \ldots, m\}$ *of Theorem* 2 *has a competitive equilibrium if and only if for each subset of traders* $\theta \subset \{1, \ldots, m\}$ *the space of preferences* P_θ *similar to those of the subeconomy* E_θ *admits a continuous anonymous map* $\Phi: P_\theta^k \to P_\theta$ *respecting unanimity*, $\forall k \geq 1$.

This last theorem proves an equivalence between two topological problems, the existence of a general equilibrium and the resolution of the social choice problem, and completes our arguments.

References

Arrow, K. and G. Debreu [1959], "Existence of an Equilibrium for a Competitive Economy," *Econometrica* **22**, 264–290.

Arrow, K. and F. Hahn [1971], *General Competitive Analysis*, Holden-Day, San Francisco.

Brown, D. and G.M. Heal [1979], "Equity Efficiency and Increasing Returns," *Review of Economic Studies* **46**, 571–585.

Chichilnisky, G. [1980], "Social Choice and the Topology of Spaces of Preferences," *Advances in Mathematics*, **37**(2), 165–176.

Chichilnisky, G. [1981], "Intersecting Families of Sets: A Topological Characterization," Working Paper No 166, Essex Economic Papers, University of Essex.

Chichilnisky, G. [1990], "General Equilibrium and Social Choice with Increasing Returns," *Annals of Operations Research* **23** 289–297.

Chichilnisky, G. [1991], "Markets, Arbitrage and Social Choice," Working paper, Columbia University.

Chichilnisky, G. [1992], "Limited Arbitrage is Necessary and Sufficient for the Existence of a Competitive Equilibrium," Working paper, Columbia University.

Chichilnisky, G. and G.M. Heal [1983], "Necessary and Sufficient Conditions for a Resolution of the Social Choice Paradox," *Journal of Economic Theory* **31**(1), 68–87.

Chichilnisky, G. and G.M. Heal [1984, revised 1993], "Existence of a Competitive Equilibrium in Sobolev Spaces without bounds on short sales," *Journal of Economic Theory.*

Debreu, G. [1952], "A Social Equilibrium Existence Theorem," *Proceedings of the National Academy of Sciences.*

Debreu, G. [1970], "Economies with a Finite Set of Equilibrium," *Econometrica* **38**, 387–392.

Debreu, G. [1993]: "Stephen Smale and the Economic Theory of General Equilibrium," this volume.

Eaves, B.C. [1971], "On the Basic Theorem of Complementarity," *Mathematical Programming,* **1**(1), 68–75.

Eaves, B.C. [1972], "Homotopies for Computation of Fixed Points," *Mathematical Programming,* **3**(1).

Guesnerie, R. [1975], "Pareto Optimality in Nonconvex Economies," *Econometrica* **43**, 1– 29.

Hirsch, M. [1963], "A Proof of the Non–Retractibility of a Cell onto its Boundary," *Proceedings of the American Mathematical Society* **14**, 364–365.

Hirsch, M. and S. Smale [1979], "On Algorithms for Solving $f(x) = 0$," *Communications on Pure and Applied Mathematics* **32**, 281–312.

Lemke, C.E. [1965], "Bimatrix Equilibrium Points and Mathematical Programming," *Management Science* **11**(7).

Lemke, C.E. "Recent Results on Complementarity Problems," Math. Dept., Rensselaer Polytechnic Institute.

Nash, J.F. [1950], "Equilibrium Points in n-Person Games," *Proceedings of the National Academy of Sciences (USA)* **36**, 48–49.

Smale, S. [1973], "Global Analysis and Economics I, Pareto Optimum and a Generalization of Morse Theory," in *Dynamical Systems,* M. Peixoto, ed., Academic Press, New York.

Smale, S. [1974a], "Global Analysis and Economics IIA, Extension of a theorem of Debreu," *Journal of Mathematical Economics* **1**, 1–14.

Smale, S. [1974b], "Global Analysis and Economics III, Pareto Optima and Price Equilibrium," *Journal of Mathematical Economics* **1**, 107–117.

Smale, S. [1974c], "Global Analysis and Economics IV, Finiteness and Stability of Equilibria with General Consumption Sets and Production," *Journal of Mathematical Economics* **1**, 19–127.

Smale, S. [1974d], "Global Analysis and Economics V, Pareto Theory with Constraints," *Journal of Mathematical Economics* **1**, 213–221.

Smale, S. [1975], "Optimizing Several Functions," *Manifolds Tokyo 1975*, University of Tokyo Press, Tokyo, pp. 69–75.

Smale, S. [1976a], "Dynamics in General Equilibrium Theory," *American Economic Review* **66**, 288–294.

Smale, S. [1976b], "A Convergent Process of Price Adjustment and Global Newton Method," *Journal of Mathematical Economics* **3**, 107, 120.

Smale, S. [1976c], "Exchange Processes with Price Adjustment," *Journal of Mathematical Economics* **3**, 211–226.

Smale, S. [1976d], "Global Analysis and Economics VI, Geometric Analysis of Pareto Optimum and Price Equilibrium under Classical Hypothesis," *Journal of Mathematical Economics 3*, 1–14.

Smale, S. [1977], "Some Dynamical Questions in Mathematical Economics," *Colloques Internationaux du Centre National de la Recherche Scientifique. N. 259: Systemes Dynamiques et Modeles Economiques.*

Smale, S. [1978], "An Approach to the Analysis of Dynamic Processes in Economic Systems," *Equilibrium and Disequilibrium in Economic Theory*, G. Schwodiauer, ed. D. Reidel Publishing Co., Boston, pp. 363–367.

Smale, S. [1980], "The Prisioner's Dilemma and Dynamical Systems Associated to Non-Cooperative Games," *Econometrica* **48**, 1617–1634.

Von Neumann, J. [1938], "Uber ein Okonomisches Gleichungsystem und ein Verallgemainerung des Brouwerschen Fixpunktsatzes," *Ergebnisse eines Mathematischen Seminars*, K. Mengen, ed., Vienna.

Von Neumann, J. [1945–46], "A Model of General Economic Equilibrium" (translated into English by G. Morgenstern), *Review of Economic Studies* **13**(1) 33, 1–9.

Walras, L. [1874–77], *Elements d'Economie Politique Pure*, L. Corbaz and Company, Lausanne.

16
Comments

Y.-H. WAN

Yieh-Hei Wan: I would like to make a brief comment on a mathematical problem motivated by economics. Probably I am the first student of Stephen Smale in mathematical economics. My thesis was on Morse theory for two functions. The Morse theory for a single function on a manifold M can be regarded as a globalization of the studies on optimizing this single function. One obtains the so-called Morse inequalities. Following this viewpoint, one can also get a Morse theory for two functions $f = (f_1, f_2)\colon M \to R^2$ as a globalization of the studies on optimizing two functions. In particular, one can obtain a suitable version of the Morse inequalities. Clearly, Smale's idea about Morse theory for two functions shows a connection between this morning's talks on differential topology and this afternoon's talks on mathematical economics.

Part 5
Dynamical Systems

DISCUSSION

CONTRIBUTED PAPERS

Dynamical Systems

17
On the Contribution of Smale to Dynamical Systems

J. PALIS

It is rare, and, in fact, one may not ever be so lucky in one's lifetime, to follow rather closely a dramatic change, an absolute upgrading of a branch of Science. In our case, Mathematics, and, more specifically, Dynamical Systems, whose foundations were set by Poincaré about 100 years ago and enriched in the next decades by Birkhoff, Lyapunov, Hadamard, Perron, Morse, Fatou, Julia, and others: One aims at the description of the asymptotic behavior of solutions of differential equations (or trajectories of iterated maps) without necessarily integrating them so as to find explicit expressions for the solutions. Also, in the appropriate setting, one looks for the behavior of solutions of "typical" equations and of "typical" solutions.

After a noble infancy, an upturn in the subject took place in the 1960s with two main developments: the construction of the hyperbolic theory for general systems, led by Smale, and the KAM theory (after Kolmogorov, Arnold, and Moser) for conservative systems, which incidentally is mentioned here only briefly.

The hyperbolic theory stands today as a very solid and rather well-understood region of the large world of dynamical systems. It has, in the tradition of the founders Poincaré and Birkhoff and almost by conception, a global character: After all, one searchs for the behavior at large of many trajectories of a system. But perhaps due to his topology background, Smale excelled at that: His vector fields were always defined on manifolds (their solutions often wrapping around in the phase space), his diffeomorphisms were global Poincaré maps of vector fields defined on one higher dimensional manifolds, his stable and unstable manifolds (of points, later of sets of points or "foliations") were always globally defined. Of course, the hyperbolic theory, and particularly Smale's contribution, goes much beyond its global character. Yet, one may ask nowadays, how could dynamical systems be otherwise studied? Well, if one looks at the literature of the 1950s, one may not see this global setting that often or that sharply as offered by Smale from the beginning, even when he was just a newcomer to the subject, and yet dared to propose what he thought should be some of the main goals of the theory. Indeed, sometimes one has the feeling that if this perspective had been present, some of the good work developed at that period could have been at least better focused or better integrated with other current research.

But let us go back to the hyperbolic theory and how its construction in the 1960s was orchestrated by Smale with the help of, and in fact with contributions of fundamental importance from, colleagues, the Russian School, his students, students of his students, and so on.

In 1937, Andronov and Pontryagin introduced the notion of *topological equivalence* (*same dynamical behavior*) for vector fields: One is equivalent to the other if there is a homeomorphism of the ambient space which sends trajectories of the first onto trajectories of the second. *A vector field is structurally stable if it is equivalent to any C^k-small perturbation of it* (*the vector fields are taken to be C^k with $k \geq 1$*). Similar definitions can be given for maps of a manifold into itself: *f is stable if it is topologically conjugate to C^k-small perturbations of it, i.e., for any g which is C^k-close to f there is a homeomorphism h of the ambient manifold such that $hf = gh$.* In that early work, Andronov and Pontryagin considered vector fields on the two-dimensional disk, say transversal to the boundary, from the point of view of (structural) stability. They stated that such a vector field is stable if and only if

—it has only finitely many singularities and periodic orbits, all of them hyperbolic,
—it exhibits no saddle-connections.

Recall that a singularity of a vector field is hyperbolic if no eigenvalue of its linear part (Lyapunov exponent) is pure imaginary. When some eigenvalue has positive real part and some negative real part, the singularity is called a saddle, otherwise a source or a sink. A saddle-connection means an orbit joining two saddles, i.e., having one of them as its ω-limit set and the other as its α-limit set. Notice that on the 2-disk or the 2-sphere, the two conditions of Andronov and Pontryagin imply a third one: *every orbit has a singularity* (*fixed point*) *or a periodic orbit as its ω-limit set and as its α-limit set.* This is a consequence of the Poincaré–Bendixson Theorem.

Adding this third condition, Peixoto generalized the statement of Andronov and Pontryagin to any two-dimensional compact orientable surface, proving in 1962 that any vector field on the surface can be approximated by one of these. The relevant case of vector fields on the 2-torus without singularities was independently done by Pliss and by Arnold.

Back in 1959, when he got acquainted with the work on two-dimensional flows that Peixoto was then developing, and inspired by it, Smale immediately defined a higher-dimensional version of the set first introduced by Andronov and Pontryagin on the 2-disk: *the limit set is formed by a finite number of hyperbolic singularities and periodic orbits and their stable and unstable manifolds intersect transversally.* This was quite a step: Instead of *no saddle connections*, in two dimensions, he was talking about *transversality between stable and unstable manifolds* in a much more general setting. Also, he defined a corresponding set of diffeomorphisms with similar dynamic conditions. Note that instead of the limit set we can define these systems using the (usually larger) *nonwandering set*: A point is nonwandering if any neighborhood of it intersects some of its positive or negative iterates. It is a bit puzzling why Smale changed from one concept (limit set) to the other (nonwandering set) to define these systems. My guess is that he became more acquainted with the work of Birkhoff who often deals with the nonwandering set. Also, Smale played on the safe side from the point of view of stability when he imposed more structure to the system (the nonwandering set is usually larger than the limit set). Anyhow, he never really proved that the two sets of systems using the two different concepts are the same (the reader may check that it is not that simple).

For such a vector field or diffeomorphism, the ambient manifold become decomposed in a finite number of stable manifolds, i.e., embedded submanifolds of several dimensions (indices) from possibly zero (point repellers) to the top one (attracting orbits). To Smale, this cried out for the topologically fundamental Morse inequalities, done classically for nondegenerate gradient flows and, indeed, this was his first result in dynamics! Thom, from whom Smale had learned transversality not long before, christened the set of flows or diffeomorphisms as above Morse–Smale systems. Perhaps Andronov and Pontryagin should also be remembered in connection with these systems, but the grandiose view about their place in dynamics was due to Smale.

In fact, by 1959 he was *conjecturing* that on any closed (compact, boundaryless) manifold the Morse–Smale systems formed an *open* and *dense* subset of all systems (flows, diffeomorphisms) of class C^k, $k \geq 1$, in the C^k topology. Moreover, each of its elements should be structurally stable (dynamically robust). Since these systems are dynamically rather simple, he was just proposing the world (of dynamics) to be mostly simple: *Any system could be approximated by a Morse–Smale one* (and thus with a limit set consisting only of finitely many singularities and periodic orbits) *and any Morse–Smale system would be dynamically robust.*

In his own words, he was at the time extremely naive about differential equations and also extremely presumptuous and "*if I had been at all familiar with the work of Poincaré, Birkhoff, Cartwright and Littlewood I would have seen how crazy this idea was,*" refering to the possibility of approximating any system by a simple one. Yet he was on the threshold of his fundamental contribution to Dynamical Systems.

First of all, although the robustness (or stability) of Morse–Smale systems was only proved about 7 years later, it was clear to him that his program made a lot of sense for gradient systems. In this context, he was even doing some bifurcation theory by cancelling singularities (handles) with indices differing by 1 through what we call saddle-node bifurcations. In particular, he showed that every gradient field is isotopic to one exhibiting only one source (point repeller) and one sink (point attractor). And this was a step toward his famous solution in 1960 of the generalized Poincaré Conjecture in dimensions greater than four: *A smooth closed manifold of the homotopy type of the sphere is homeomorphic to the sphere.*

At this point, he was visiting the Instituto de Matemática Pura e Aplicada (IMPA) at Rio de Janeiro in 1960, when he got a (dramatic for him) letter of Levinson. *There were certainly robust examples of flows and diffeomorphisms in three and two dimensions, respectively, for which the limit set contained infinitely many coexisting periodic orbits*! This was clear from the existence of *transversal homoclinic orbits* in the work of Levinson and that of Cartwright and Littlewood on the Van der Pol's equation with friction ("real" world) and, coming straight to the point, a result of Birkhoff in 1935 stating that a transversal homoclinic point of a surface diffeomorphism is an accumulation point of periodic orbits. Amazingly enough, 70 years before Poincaré was receiving a dramatic letter saying that there was something wrong concerning *homoclinic orbits* in the last part of his prize memoir (King Oscar of Sweden) on the stability of the solar system. Before the final publication of his essay, in fact that issue of the Acta Mathematica was already printed and about to be distributed, he got it straightened up: A second corrected printing was made! Of course, Poincaré still deserved that and all other possible prizes. This incident may have been in his mind when he wrote in *Les Méthodes Nouvelles de la Mécanique Céleste*: "On sera frappé de la complexité de cette figure, que je ne cherche même pas à tracer. Rien n'est plus propre à nous donner une idée de la complication du problème des trois corps et en général de tous les problèmes de Dynamique ... ."

Homoclinic orbits, that puzzled both Poincaré in 1890 and Smale in 1960, are orbits of intersection of the stable and unstable manifolds of saddles (excluding the saddles themselves). Again curiously (symptomatically?), Lit-

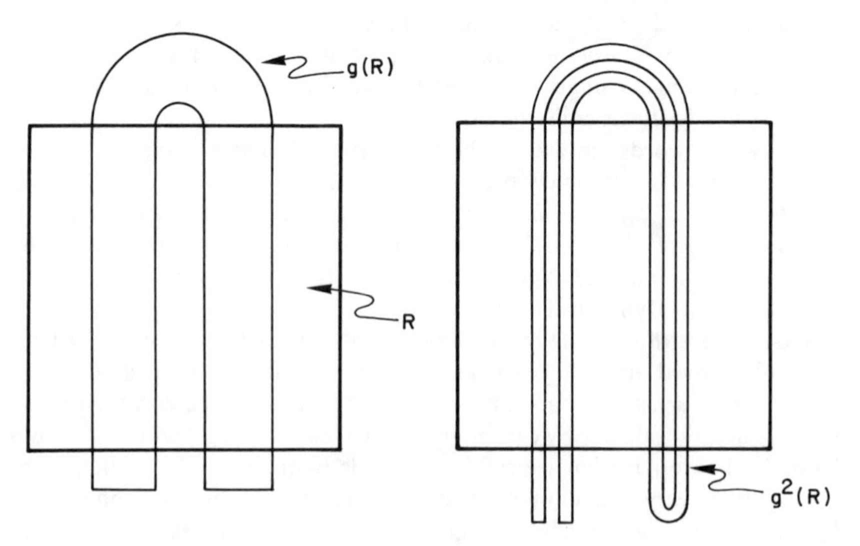

FIGURE 1

tlewood refers to his work on the subject, done partially with Cartwright, as the "monster." Studying Levinson's work, Smale considered the Poincaré transformation, say f on a section to the flow. He *abstracted* from a transversal homoclinic orbit of this transformation *the horseshoe map*: a map g sending a rectangle R into a horseshoe as in Fig. 1.

The horseshoe map can be taken as (or it is conjugate to) a high power of the original transformation f restricted to a rectangle R^* containing the local stable manifold of the associated saddle. See Fig. 2. So he was indeed showing that *if a diffeomorphism exhibits a transversal homoclinic orbit, then this orbit* is contained in an *invariant hyperbolic Cantor set, like $\Lambda = \bigcap_{n \in \mathbf{Z}} g^n(R)$, and a power of the original map restricted to the Cantor set is conjugate to the shift on two symbols.* The g-invariant set Λ is hyperbolic in the sense that there are two continuous bundles over Λ in which the derivative dg acts as a contraction and as an expansion, respectively.

The corollaries were remarkable: In any dimension, a transversal homoclinic orbit implies a rich dynamics, including nontrivial minimal sets, infinitely many coexisting periodic saddles, infinitely many coexisting transversal homoclinic orbits, all "living together" densely in an invariant Cantor set. And, as it is clear from the previous statement, this dynamic phenomenon was not *pathological*; much to the contrary, it was robust (open) and structurally stable. Moreover, it lived in any manifold of dimension two or larger in the case of diffeomorphisms or, via suspension, for flows in dimension three or larger. It was also realized later that this structure (the horseshoe) also implied positive topological entropy.

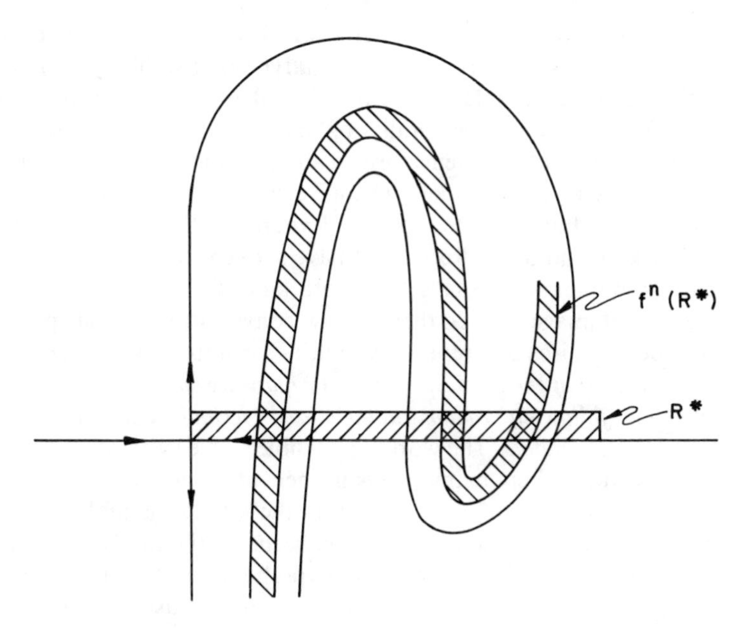

FIGURE 2

Let me make a little disgresion as to how often this phenomenon occurs. *I have been conjecturing in the past 5 years or so, that the systems exhibiting a horseshoe (or a transversal homoclinic orbit) together with the Morse–Smale ones form an open and dense subset of all C^k systems, $k \geq 1$. (Actually, this conjucture is implied by one stating that the hyperbolic systems, or Axiom A systems to be defined later, together with the ones exhibiting a homoclinic bifurcation form a C^k-dense subset).* Noteworthily, the (stronger) conjecture has been proved for C^1 surface diffeomorphisms by Aranjo and Mañé. It can be interpreted as a bit of vindication for Smale's original proposal (Morse–Smale systems are dense) or its later modifications (stable or limit set-stable systems are dense) as we shall discuss in the sequel. Still, nowadays we are much focused on *attractors*, i.e., sets attracting an open or at least a positive Lebesgue measure set of orbits. *And, worth noting, the set of orbits attracted to a horseshoe has measure zero, at least if the system is differentiable of class C^2.* (Ten years later, Bowen provided examples of C^1 horseshoes with positive Lebesgue measure.)

Concerning the horseshoe, another important fact is that *every point of the Cantor set has stable and unstable manifolds* (locally embedded Euclidean spaces), defined as the set of points whose orbits are positively and negatively asymptotic to the orbit of the initial point. They form a net: *The Cantor set is locally the product of two Cantor sets*, which are the closure of (transversal) homoclinic points with the same local stable and unstable manifolds, respectively. By *abus de langage*, we refer to them as the *stable and unstable foliations* of the Cantor set: The leaves are, respectively, the stable and unstable manifolds of the points of the Cantor set.

Now armed with Morse–Smale systems, which were still only conjecturally stable, and the horseshoe, a prototype of a large set of stable systems with a much richer limit set, Smale began an admirable battle to have a global description (even if only conjecturally) of "most" of the world of dynamics, still hoping that the stable systems formed an open and dense subset of it. He then went to Moscow and Kiev, where he met Kolmogorov and, in his words, an extraordinarily gifted group of young mathematicians: Anosov, Arnold, Novikov, and Sinai. Recently, I have spoken to Arnold and Anosov about this encounter with Smale, and both (and probably the others too) have a very vivid image of it. In their intense conversations, Smale presented the horseshoe and discussed other possibilities for stable systems. *In particular he conjectured that the geodesic flows on negatively curved closed manifolds were stable and, following Hadamard, the closed (periodic) geodesics would be dense in the whole manifold.* The world of possible stable systems was growing from the ones with finitely many orbits in their limit set (Morse–Smale systems) to some for which the limit set was the whole manifold, as in the geodesic flows above. And somewhat in between, lay the surely stable horseshoe. Thom was also asking about the hyperbolic linear toral diffeomorphisms: They were *globally hyperbolic*, their periodic orbits being dense in the whole manifold and these properties were persistent under perturbations.

Were these maps stable? The visit of Smale to Russia and Ukraine was most fruitful: Not long in the future, Anosov would announce his fundamental results answering positively the last questions, as we shall mention in the sequel. That was perhaps the most visible mathematical consequence of the trip, and a remarkable one.

There is a paper of Smale that I particularly like to discuss: his address at the International Congress of Mathematicians at Stockholm in 1962. When I first read it about 3 years later, I was initially amazed and then taken by it. It looked to me like an unfinished painting with several superposed sketches: Was that allowed in Mathmatics? It was certainly very inspiring then as well as three (and many more) years later. *Some new results were announced in the paper, but mostly it made transparent his struggle at the time to reach the right concept of hyperbolic systems* (*to be called later Axiom A systems by him*), *still dreaming the semi-impossible dream that they could be both* (*structurally*) *stable and form a dense subset of all systems.* It is curious to remark that at that very moment, a remarkable work that would remain unknown to us for the next 10 years was being developed: *Lorenz was proposing the impossibility of the second half of the dream through his still experimental robust* (*open*) *strange attractors* (*unstable*). *Unaware of that, Smale himself would show in* 1965 *that the stable systems are not dense.* And, in the seventies, we became definitively aware that the world of dynamics is much bigger (and fascinating and perhaps not so indomitable) than the set of hyperbolic systems and its boundary.

Back to Smale's address at the International Congress of 1962. *Smale offered a possible set of axioms and a variation of it to be satisfied by the "hyperbolic systems,"* but the definite concept would be achieved only a bit later.

Among the novelties, he announced that for the elements of a residual subset of C^k systems, any $k \geq 1$, the fixed and periodic orbits are hyperbolic and their stable and unstable manifolds all mutually transverse. This was also and independently proved by Kupka. *Smale then suggested the question whether such a diffeomorphism could exist on the 2-sphere without displaying a sink or a source.* The question has been considered by a string of mathematicians in the span of 25 years: Bowen and Franks, Franks and Young, Gambaudo and Tresser and van Strien, with a positive answer for C^1, C^2, and C^∞ diffeomorphisms, respectively. *Is there an analytic example*?

As before, let us call an invariant set for a dynamical system an *attractor* if it contains the ω-limit set of all points in an open (or a positive Lebesgue measure) set in the phase space. Another question that one may say is certainly inspired by the same paper is: *Can a diffeomorphism on S^2 exhibiting some attractor always be approximated by one exhibiting a sink?* An important result of Plykin in the mid-seventies showed that this was not so: There are robust or fully persistent (under C^k small perturbations) attractors for diffeomorphism on S^2. The Plykin attractors are, in fact, "hyperbolic," as we shall define later, and (structurally) stable.

Smale reminds the reader that the stability of the Morse–Smale systems still remained an unsolved problem, stressing that they exist on every manifold and that they satisfy the Morse inequalities.

He also hints at the *centralizer problem*: *Given a diffeomorphism, say of a closed manifold, how "large" is the group (under composition) of diffeomorphisms that commute with it?* In particular, is it common for a diffeomorphism to embed in a flow, i.e., to be the time-one map of a flow? One of the first results in this direction, a rigidity result, is due to Bochner and Montgomery and concerns the differentiability of a flow, i.e., a one-parameter group of C^k diffeomorphisms depending C^r on the parameter, $r \leq k$: The result states that $r = 0$ implies $r = k$. Again, a number of mathematicians have since worked on the problem; for example, Kopell, Lam, Mather (responsible for synthesizing an obstruction invariant to imbeddability in a flow), myself, Brin, Anderson, Walters, Morimoto, Sad, Yoccoz and myself, Rocha, showing in several contexts the *rigidity of centralizers*: *They are commonly (open and dense or residual subsets) very small, consisting just of integer powers of the initial map (or constant multiples of the initial vector field).* Rigidity of group actions has become a rather large topic of research (Katok, Hurder, Zimmer, Ghys, etc.).

Smale also asks if any diffeomorphism (flow) can be C^k approximated by one with a periodic orbit for $k \geq 1$, a form of the difficult and important "closing lemma" proved by Pugh 4 years later in the C^1 category: Residually the nonwandering set is the closure of the periodic orbits.

The paper mentions the horseshoe and that the stability of the hyperbolic linear automorphisms has been resolved by him (unpublished), by Arnold and Sinai ("my only paper with a wrong proof," Arnold tells me), and, it announces the fundamental results of Anosov in that line without any further detail.

The last paragraph of the 1962 paper shows his enthusiasm: "*Certainly the main problems stated here are very difficult. On the other hand, it seems quite possible to us that this field may develop rapidly and already as indicated here, there have been some initial steps in this direction.*

By 1965, Smale cleverly combined two diffeomorphisms with a very different dynamical behavior, that he knew were persistent under small perturbations and, in fact, stable, to provide *an open set of unstable diffeomorphisms*! As a consequence, his dream that stability would be a property displayed by the elements of a dense subset of systems was shattered (*was it*?) and shattered by himself. He wrote at the end of the paper: "The example in this paper surely reduces the importance of the notion of structural stability." Still he also manifested his usual tenacity in creating new and more complex scenarios as needed, promising a paper "based on an axiom, which we consider to be of central importance, axiom A ... ." In fact, he was still hoping that some kind of stability, say limit set-stability or nonwandering set-stability, was a property displayed by a dense (and open) subset of dynamical systems. More importantly, he had finally the set of axioms, now synthesized in the letter A, he was so much looking for: The concept of *hyperbolic systems. Thus, the hyperbolic theory was about to get its contour.*

The example introduced by Smale was based on the product of the north pole–south pole diffeomorphism on the circle S^1 (a source and a sink) and the linear hyperbolic authomorphism of the torus $\mathbf{T}^2$ mentioned above. Thus, the example was constructed just on $\mathbf{T}^3$. He then produced a "bump" to provide a *persistent lack of transversality* (*sometimes a tangency*) between stable and unstable manifolds. That is, he provided an open set of C^k diffeomorphisms on $\mathbf{T}^3$, $k \geq 1$, such that the elements of a dense subset of it exhibit an orbit of nontransversal intersection between stable and unstable manifolds of periodic orbits. (Recall that residually on the space of diffeomorphisms, stable and unstable manifolds of periodic orbits are in general position.) This fact was clearly an obstruction to stability: Topological conjugacies must send periodic orbits and their invariant manifolds (and thus their intersection) onto similar orbits and invariant manifolds (and their intersection). Later, Williams and Robinson used the same principle to provide an open set of unstable two-dimensional diffeomorphisms on the bitorus: It was like a "north pole–south pole" example, the attractor and repeller being the (nontrivial) ones constructed by Smale in what he called *diffeomorphisms derived from Anosov* (in particular, derived from hyperbolic torus automorphisms), to be again mentioned in the sequel. And that is not all: Again and again the same principle would be used to shatter the "modified dream" that the limit set-stability or nonwandering set-stability would be a dense property for diffeomorphisms in two or more dimensions (Abraham and Smale, Newhouse, Simon, Shub, Mañé, and so on). But let us not invert history and come back to this point later. A last word here on the "constructive" side: Several of these "shattering" examples have recently inspired some insights in the *dark realm of dynamics* (complement of the hyperbolic systems).

We are now arriving at a masterpiece of the mathematical literature: Smale's celebrated paper of 1967 "Differentiable Dynamical Systems." Nothing else synthesizes better his permanent legacy to dynamics. Like the 1962 paper, this was exuberant in new ideas, new problems, new or unfinished paths of research and, in fact, so much more. Yet it was also much more solid, coherent, overwhelming: a theory, the theory of hyperbolic systems. Some of us were much aware that he was preparing an extensive paper, but the result surpassed the expectations for the varied scenes and the global scenario of dynamics it presented. Still dreaming that nonwandering set-stability could be shared by a dense subset of systems (he called it Ω-stability because in his writings Ω always represented the nonwandering set). But this was not emphasized as much any more: The theory of hyperbolic systems was now standing by itself, solid and fundamental. (And just a year later, the dream was finally shattered by Abraham and himself in higher dimensions (≥ 4): again an example displaying a persistent orbit of nontransversal intersection of stable and unstable manifolds was provided, but this time the phenomenon occurring inside the nonwandering set!).

A diffeomorphism f is called hyperbolic or satisfies Axiom A if its non-

wandering set $\Omega(f)$ is hyperbolic and its set of periodic orbits is dense in $\Omega(f)$. We say that an invariant set is hyperbolic if the tangent bundle of M restricted to it splits into two subbundles, in which the derivative of the map operates as a uniform contraction and expansion, respectively. In particular, the hyperbolic torus automorphisms as well as the Morse–Smale diffeomorphisms satisfy Axiom A. One can also construct an Axiom A diffeomorphism of the 2-sphere with its nonwandering set consisting of one source, one sink and a horseshoe. In the first case, the whole manifold is a hyperbolic set for the diffeomorphism: We call such a diffeomorphism globally hyperbolic. In the other "extreme," the Morse–Smale diffeomorphisms have their nonwandering set consisting of finitely many hyperbolic periodic orbits. It is to be noted that in the first case the indices (dimension of the stable manifold) are the same for all periodic orbits, whereas in the other cases they vary: In the Morse–Smale case, they go from zero to the dimension of the ambient manifold.

The previously mentioned result of Anosov states that *globally hyperbolic diffeomorphisms are structurally stable. Also, the periodic orbits form a dense subset of the nonwandering set.* In proving these results, Anosov constructed stable and unstable manifolds for all points in the ambient: They consist of the set of points whose orbits are, respectively, positively or negatively asymptotic to the positive or negative orbit of the initial point.

By 1966, Moser provided a different and elegant proof of Anosov's Theorem, and this was the theme of Mather's Appendix to Smale's paper.

Smale was much inspired by the work of Anosov to finally achieve the concept of a hyperbolic diffeomorphism. Thus, he named the globally hyperbolic diffeomorphisms after Anosov. He now saw the nonwandering set for an Axiom A diffeomorphism decomposed into "Anosov pieces" assembled together somewhat as in the Morse–Smale case. Each of these pieces being transitive (dense orbit) is called a *basic set.* This is the content of Smale's Spectral Decomposition Theorem: *The nonwandering set of an Axiom A diffeomorphism can be decomposed into a finite number of basic sets.*

It was now clear to him that by imposing some global condition on the basic sets, to avoid "Ω-explosions," the diffeomorphism would be Ω-stable, i.e., its restriction to the nonwandering set would be structurally stable. He introduced another axiom, Axiom B: *If for a pair of basic sets Λ_1 and Λ_2, the stable manifold $W^s(x_1)$ intersects the unstable manifold $W^u(x_2)$, for some $x_1 \in \Lambda_1$ and $x_2 \in \Lambda_2$, then $W^s(x_1^*)$ has an orbit of transversal intersection with $W^u(x_2^*)$ for some $x_1^* \in \Lambda_1$ and $x_2^* \in \Lambda_2$.* He then announced: Diffeomorphisms satisfying Axioms A and B are Ω-stable.

Not so untypical of Smale at the time (somehow his lectures now seem much more "organized"), he was once speaking on this theorem to a rather large audience in Berkeley and in a kind of noisy atmosphere with Moe Hirsch going often to the blackboard to help him out. Terry Wall was quietly attending the lecture, when Smale declared: *In the presence of Axiom B the basic sets are partially ordered, i.e., if $W^s(x_1)$ intersects $W^u(x_2)$ and $W^s(x_2^*)$*

intersects $W^u(x_3)$, *for* $x_1 \in \Lambda_1$, x_2, $x_2^* \in \Lambda_2$ *and* $x_3 \in \Lambda_3$, *for basic sets* Λ_1, Λ_2, *and* Λ_3, *then* $W^u(x_3^*)$ *does not intersect* $W^s(x_1^*)$ *for any* $x_3^* \in \Lambda_3$ *and* $x_1^* \in \Lambda_1$. At this point, Wall asked (his only question): Is it reasonable to assume just the partial ordering of the basic sets? Smale shrugged his shoulders and went on lecturing. Before the end of the exposition, I was suggesting that the no-cycle condition, as defined above, should be adopted instead of Axiom B because this was *necessary to avoid* Ω-*explosions* under small perturbations of the map (Rosenberg had 10 years earlier suggested the no-cycle condition in the context of the Morse inequalities). And so the Ω-Stability Theorem of Smale when published in 1970 in the Proceedings of the 1968 Global Analysis meeting at Berkeley states: *an Axiom A diffeomorphism satisfying the no-cycle condition is* Ω-*stable*. We also knew that in the presence of Axiom A, the no-cycle was a necessary condition for the result.

In another development, as his Ph. D. student at the same time as Nancy Kopell and Michael Shub (the first three except for Moe Hirsch, an "informal" student back in their Chicago time) I had proved that the Morse–Smale systems are stable in low dimensions (less than four). Kopell proved very nice results concerning *centralizers of diffeomorphisms, especially for maps of the circle*. And Shub proved definitive results about the *stability of expanding maps* (derivative with norm bigger than one everywhere) and initiated their classification, which was finally and beautifully done by Gromov many years later, showing in particular that they *only exist* on infranilmanifolds. After many discussions with Smale (always good humored and loud—once an office neighbor asked us to be quieter!), we succeed in proving: *The Morse–Smale systems are stable*. In particular, an open and dense set of gradient systems are stable (a restricted true piece of the "dream"). We then dared to state the Stability Conjecture: *A system is* C^k *stable*, $k \geq 1$, *if and only if it is hyperbolic* (*Axiom A*) *and satisfies the transversality condition* (*all stable and unstable manifolds are in general position*). *For* Ω-*stability one imposes the no-cycle instead of the transversality condition.*

Notice that if the limit set of a diffeomorphism is hyperbolic, then the periodic orbits are automatically dense on it. So the Spectral Decomposition Theorem is valid for the limit set in this case (Newhouse). And if the no-cycle condition is also valid, the limit set is the same as the nonwandering set, and the diffeomorphism satisfies Axiom A. On the other hand, the nonwandering set may be hyperbolic without the periodic points being dense on it (Dankner); such a phenomenon cannot occur for surface diffeomorphisms as shown by Newhouse and myself.

Let me say that these concepts and results concerning diffeomorphisms were also established for flows (or differential equations as Smale liked to stress). The hyperbolicity of a flow φ_t, $t \in \mathbf{R}$, requires that the tangent bundle to the ambient manifold restricted to the nonwandering set splits into three subbundles, one along the flow (vector field) and the others contracting and expanding, respectively, by the derivative of φ_t, $t > 0$. However, about the Ω-Stability Theorem for flows, Smale declared that it was very likely to be true using the same methods of proof. Well, this was only done a few years later by Pugh and Shub, the methods being a bit more involved.

In the direction of the Stability Conjecture, Robbin for C^2 diffeomorphisms, de Melo for C^1 surface diffeomorphisms, and Robinson for C^1 diffeomorphisms in general proved in the early seventies the direct implication: *Hyperbolicity and transversality implies stability.* Robinson also proved the corresponding result for flows. Around 1980, Liao, Mañé, and Sannami proved the converse for C^1 surface diffeomorphisms. Finally, in a remarkable paper, Mañé, 20 years after it was stated, proved *the C^1 Stability Conjecture for diffeomorphisms in all dimensions.* I was able to extend slightly his formidable work to prove the C^1 Ω-Stability Conjecture. The question for flows remains essentially unsolved.

Smale also asked for the *classification of Anosov diffeomorphisms and flows.* For instance, we still do not know if the periodic orbits of an Anosov diffeomorphism are dense in the ambient manifold except in the codimension-one case (Newhouse), i.e., when the dimension of the unstable (or the stable) manifold is 1. Ten years later, Franks and Williams showed that this is *not* the case for Anosov flows even in three dimensions. On the positive side, there are the results of Franks, especially in the codimension-one case, and Manning for Anosov diffeomorphisms on tori, nil and infranilmanifolds. About Anosov flows, there are the works of Tomter, Plante, and Verjovsky (codimension-one case in dimension larger than three).

Notably, Smale constructed hyperbolic attractors which were neither sinks nor the whole manifold (Anosov diffeomorphisms). The main ones were the *solenoid* and, perhaps more unexpected, *the DA* (derived from Anosov) *attractors*, the last ones being obtained through a saddle-node bifurcation of, say, a hyperbolic torus automorphism. These and the Anosov ones can also be "*multiplied*" *by strong normal contractions* to generate nontrivial attractors in higher dimensions. A very relevant contribution to the study of hyperbolic attractors is the work of Williams. Hyperbolic attractors were used a bit later to construct examples of *persistent lack of transversality* (in particular, *persistent tangency*) inside the nonwandering set or even the limit set, as first done by Abraham and Smale. As mentioned before, these examples proved that the Ω-stable or limit set-stable systems are not C^k dense in dimensions greater than two and any $k \geq 1$. It remained the case of diffeomorphisms on surfaces. This was settled with the same answer and around the same time for C^2 surface diffeomorphisms by Newhouse (also a student of Smale, like Franks) in a much more subtle way: he used the arithmetic difference of subsets of the line and a fractal dimension (thickness) to show that often two Cantor sets in the line intersect each other and do so in a robust way when we slightly change the Cantor sets metrically (this corresponds to *persistent tangencies* in the creation and unfolding of a two-dimensional homoclinic tangency). Still we do not know the answer for C^1 surface diffeomorphisms.

Smale talks again about the *centralizer problem*, now in more precise terms: For an open and dense set of hyperbolic diffeomorphisms or even for general diffeomorphisms they should be trivial (conjecture). As mentioned before, this was solved by Kopell for general diffeomorphisms of S^1 and for smooth hyperbolic diffeomorphisms in higher dimensions by Yoccoz and myself (Rocha in the analytic case). Smale also discusses hyperbolicity and stability of group actions and this was the theme of Camacho's thesis, also his student.

Similarly, he posed the question of the *rationality of the zeta function* for a residual set of diffeomorphisms (which turned out to be false by Simon's work) and for hyperbolic diffeomorphisms (which is true by Williams and Guckenheimer). He also pointed in the direction of ergodic theory to be brillantly followed by Bowen, who worked very closely with Ruelle and, although far apart, with Sinai (he also did some initial work close to Lanford). A high point was the construction of the Bowen–Ruelle–Sinai measure for hyperbolic attractors about 10 years later.

I could go on mentioning other topics and the work of other of his students at the time or a bit afterward: Fried, Guckenheimer, Nitecki, Schneider, Shahshahani, and so on.

But when the atmosphere in Berkeley seemed most extraordinarily exciting, Smale was preparing to leave the subject! He became more interested in Mechanics, then Economics, then Theory of Computation (Complexity of Algorithms), and then Numerical Analysis, and so on. I particularly like to detect some of the same old spirit of dynamics in his work in these different areas, but maybe it is just my passion for dynamics! I also like to remember the many discussions we had about dynamics and mathematics in general, almost never about details and almost always about ideas, directions, theories and this in a concrete way, nothing like abstract supermachines. He certainly influenced many of us in these ways.

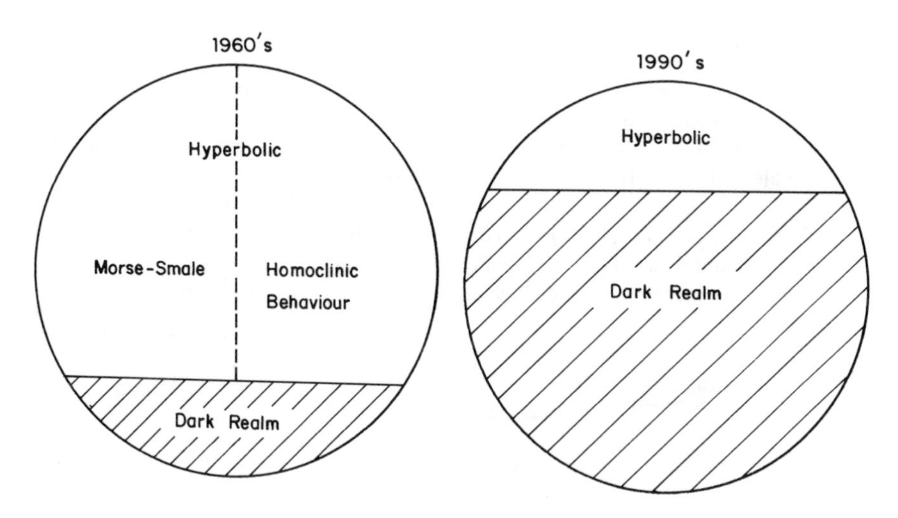

FIGURE 3

But the pace of research in dynamics continued unabated for us and still to a large extent for him up to the Global Analysis meeting that he has organized with Chern in the Summer of 1968. *Yet somehow there seemed to be a sense of hurry to "complete" the hyperbolic theory.* The decade was closing and Smale was leaving. We still had to fix a piece of his Ω-Stability Theorem, concerning the stable set and the union of the stable manifolds of the points in a basic set. Hirsch, Pugh, Shub, and myself were quite busy doing that in the Warwick meeting of 1969. And Smale calmly but frequently would come around to see how that was going.

The seventies would reserve a big surprise for us, happy builders of the hyperbolic theory of the sixties. The Dark Realm of dynamics was much "bigger" than we thought, as indicated to us mostly by physicists, biologists, astronomers, and so on doing experimental work using dynamical systems models. Bigger and also very challenging and exciting. In particular, Feigenbaum made his celebrated discovery on cascades of period doubling bifurcations (Coullet and Tresser are to be mentioned in this context too) after listening to a lecture of Smale on quadratic maps of the interval. We are indeed lucky, as I mentioned in my first sentence, to be part of one revolution in the subject: Imagine two! It is my belief that it is just happening .... Of course, the hyperbolic theory stands as solid as ever and is the starting point to look across its boundaries. Neverthless, the methods and views are different, more analytic perhaps, the sense of *prevalence* of a dynamical phenomenon being more of a measure theoretical sense in a parametrized form (positive measure, positive density, total measure) than residuality. But this is a topic of another discussion and I shall finish by just presenting my pictorial view (Fig. 3) of how hyperbolicity stands vis-à-vis its complement (the Dark Realm) in the World of Dynamical Systems in the sixties and nineties.

References

1. S. Smale, Dynamical systems and the topological conjugacy problem for diffeomorphisms. In: Proceedings of the International Congress of Mathematican 1962. Institut Mittag-Leffler, Djursholm Sweden, 1963. Editor: V. Stenström, pp. 490–496.
2. S. Smale, Differentiable dynamical systems, *Bull. Amer. Math. Soc.* **73** (1967), 747–817.
3. S. Chern and S. Smale, eds., Proceedings of the Symposium on Pure Mathematics, Vol. XIV, American Mathematical Society, Providence, RI, 1970.
4. S. Smale, *The Mathematics of Time*, Springer-Verlag, New York, 1980.

18
Discussion

S. NEWHOUSE, R.F. WILLIAMS, AND AUDIENCE MEMBERS

R.F. (Bob) Williams: Sheldon and I have been thinking about this discussion for about 45 minutes. It's not well prepared, but we agreed a moment ago that I would talk a little bit about the work of Rufus Bowen, John Franks, John Guckenheimer, and others on zeta functions, since that was something that Jacob did not have a chance to talk about. It was, of course, one important thing about the 1967 paper which many of us were working hard on in 1966, because it came out in preprint. It was very much an outline. In it, Smale did things that he's done so many times that have been so helpful to us. He's laid out ideas with strong statements, conjectures if you wish, which he didn't feel all that secure about, but he's brave enough to make the conjectures, so we got to play with them. Two kinds of zeta functions were defined. Some proofs were given, some computations were made. I remember very well correcting a slight error of his in this paper, and pointing it out, and of course when it came out, not only was the error corrected but he used my own notations .... Anyway, he was brave enough to conjecture something that took years to prove by John Guckenheimer—first Mañé—the rationality of the zeta functions.

Jacob Palis: It's Manning.

Bob Williams: Did I say Manning?

Jacob Palis: You said Mañé.

Bob Williams: Yes, the first proof of the rationality was given by John Guckenheimer. and then Manning in England gave the second proof. Manning proved this slightly stronger theorem.

Jacob Palis: You're being shy about your own work.

Bob Williams: Well, that's all right. I haven't always been.

Jacob Palis: Their work was just an improvement of yours.

Bob Williams: It's true that I was very struck by this strange idea that showed the zeta function was rational. I worked very hard on this myself. An offshoot of my work was a seminar I had where we went through the '67 paper. And one of the things that wasn't hinted at, I think, was Paul Schweitzer's work. He showed the Seifert conjecture was false, for C^1 vector fields. Paul was part of the seminar that I ran and therefore his work was very much an offshoot of Smale's ideas. Schweitzer produced a counterexample to the Seifert conjecture. In addition, it also related to my own work. It's the work on symbolic dynamics. As I understand it—and I should have, during this 45 minutes, looked up the literature—there are two definitions of what we now call subshifts of finite type. One was given earlier by Parry, but the definition given by Smale is in some sense the proper one. In particular, the name that is now used, the name that Smale chose, is "subshifts of finite type." Bill Parry called them "Topological Markov Chains." This was an important idea. It came from the knowledge of horseshoes and related systems, and it immediately was taken up by students of his, Bowen and Lanford. They showed that the zeta functions of subshifts of finite type were rational. They introduced the whole terminology to make that easy to see. Rufus proceeded to use this concept, symbolic dynamics, to set up the matrix theory. In addition, there's a slightly tighter collection of work that stems from this; that is to say, a more geometric aspect of subshifts of finite type pursued very effectively by me and many young people, and I can still call John Franks young since I'm so old. John Franks, Mike Boyle, Bruce Kitchens, a large collection of people, have made a very pretty theory of this. In large part, one can say this stems from Smale's ideas about horseshoes. There's something kindly called a Williams conjecture. I thought I proved it .... It still occupies a lot of thought.

John Franks: Do you call it the Williams conjecture or the Williams problem?

Bob Williams: Well, I certainly have a problem with it.

John Franks: Would you care to venture a prediction on whether or not it's true?

Bob Williams: I still think it's true.

John Franks: Would you say it's true?

Bob Williams: Uh, yes. Possibly. Let's see, it concerns matrices A and B whose entries are non-negative integers. One defines two equivalence relations, one of which is very natural, and one of which is correct .... The natural one, one says A is shift equivalent to B, provided there are three diagrams, A takes place across the top and B on the bottom, and there are matrices R

which form a commutative diagram. If you think of these as matrices, you hardly need to say what their corners are:

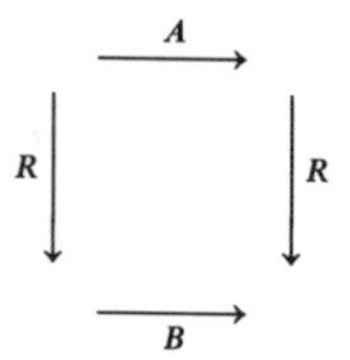

That's one diagram. There's a similar diagram with a matrix S going up:

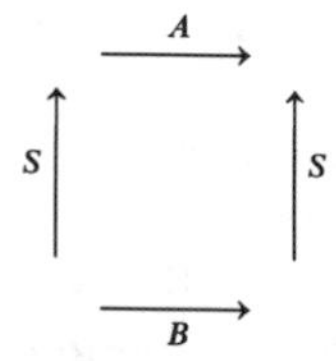

Then, finally, if these were to be conjugated, one would want to know that R was the inverse of S. R and S put together is the identity. But one has the weaker assumption that R followed by S is A to somepower, and S followed by R is B to some power:

$$SR = A^m, \qquad RS = B^m.$$

The point is, that's very natural. It would take a while to show how natural it is, but it is natural. The other notation is sort of crazy, but it's correct. And it says that A is equivalent to B provided you can proceed by a finite string. A is equal to $R_1 S_1$. A_1 is equal to $S_1 R_1$. But A_1 is also $R_2 S_2$. Then A_2 is $S_2 R_2$, etc. In other words, you proceed by finitely many steps.

$$\begin{aligned} A &= R_1 S_1, \qquad S_1 R_1 = A_1; \\ A_1 &= R_2 S_2, \qquad S_2 R_2 = A_2; \\ &\vdots \\ A_{n-1} &= R_n S_n, \qquad S_n R_n = B. \end{aligned}$$

Eventually you get to B. Now the difficulty with this problem, assuming these are equal, is that the non-negativeness of the integers involved. If one does this over any other category, it's easy to see these are essentially equivalent; very easy arguments have been given that work in any other sort of category.

Well, that's certainly enough about me. Did I miss something about Rufus?

John Franks: There are many names that deserve more praise than you and I can give them today. But the thing that hasn't been mentioned is the relationship between this and the other things we've been talking about today, in particular hyperbolic dynamics. The theorem on Markov partitions basically says all of the hyperbolic systems that Jacob told us about today can be understood in terms of symbolic dynamics. And there's a great deal of information to be gleaned about these hyperbolic systems from this symbolic dynamics. The existence of that connection is the existence of Markov partitions, in the development of which Rufus was perhaps the most important figure.

Bob Williams: Well, I've left to Sheldon the enormous task of describing developments in ergodic theory.

Moe Hirsch: Before we get to that, I'd like to give a comment about the hyperbolic theory, and that is that it's had an enormous influence in applied dynamics. Now, we don't talk much about applied dynamics in this conference, but anyone who deals with applied dynamics would know the extreme importance Smale's work on hyperbolicity has had in giving rigorous examples of chaotic systems. There, I said the "c" word; nobody has said "chaos" yet. But as you know from reading the popular magazines and the press, chaos is a very popular word, and we all appreciate—but a lot of people who are novices don't appreciate—how important the hyperbolic systems are in rigorously demonstrating the existence of chaotic systems and the importance of horseshoes. I don't know if Jacob said that one of the things Smale proved is that when you have a horseshoe, you can model it with symbolic dynamics.

Jacob Palis: I did.

Moe Hirsch: Did you say that?

Jacob Palis: Why, of course.

Moe Hirsch: Okay. It's extremely important. And in a sense, this responds to Poincaré's despair in understanding horseshoes, because you can understand them now in the sense that we have labeled them by symbols from a finite alphabet, and we can understand how that goes. So, this is a far-reaching offshoot with all sorts of people who could not tell you what a nonwandering point is, but have to deal everyday with applied systems. They can understand, and you can do the same. A very important influence in applied mathematics.

Jacob Palis: Yes, I did not use the word chaos, because I was afraid of René Thom.

Mike Shub: It is important to say that the first complicated structurally stable attractors were introduced in the 1967 paper. Up to that point the only structurally stable attractors were points and periodic solutions. The 1967 paper introduced a whole new class of structurally stable attractors. Takens and Ruelle adopted those attractors as an alternative model of turbulence and chaotic behavior.

Bob Williams: Yeah, I claim even the word "attractor" was invented by these guys, Thom and Smale. Each says the other invented it.

Mike Shub: It is also important to stress that the whole thing is expressed in terms of topological conditions—there are no equations. It had the effect of restructuring the entire subject.

Sheldon Newhouse: Yeah, let me just add—I think Mike hit on the fundamental thing, at least from my point of view—one of the great influences that Steve had, at least on me and I think many others, aside from excitement, was the whole idea that one might get a structure theory for general dynamical systems. That such a general structure theory for most systems could exist I think was a goal of Poincare and Birkhoff, and somehow in the intervening years, it was lost sight of by people working on specialized problems. The idea of getting a global structure of all systems has spawned a lot of development in the past, and it is still going on.

I think Jacob did an excellent job of describing the atmosphere that existed in the sixties, with all the excitement and developments happening virtually every other week, in the seminar, out of the seminar. I want to give a little personal perspective of what it was like to be a rather ignorant graduate student at that time, with all of this stuff going on, not understanding much of anything. In particular, I remember one aftemoon in a seminar, John Franks was doing what you would call nowadays a thesis defense. So you know what a graduate student feels like at his first big lecture. He was talking about some theorems in his thesis and then Moe asked him a question, "Well, how do you prove that?" And John said, "Well, that's in Steve's paper." And then Steve piped up, "Yeah, but I don't think I know how to prove that anymore!" John just sort of didn't know what to say. I think he worked out a proof. How long did it take?

John Franks: About two days.

Sheldon Newhouse: About two days. So, there was a very kind of cavalier idea about trying to develop a subject, and the idea always was "What are the general sort of structures? And let's not worry about the details." And that got even to the level of the graduate student. So, another historical thing that René Thom was just telling me today which I think is quite interesting is that the

$$\begin{pmatrix} 2 & 1 \\ 1 & 1 \end{pmatrix}$$

map actually occurred in response to a question that Steve asked about whether every diffeomorphism in the 2-sphere had an attracting periodic orbit. Well, obviously, it couldn't be that statement, right? Because an area-preserving diffeomorphism doesn't have it. He was obviously thinking about gradient systems and dissipative dynamical systems. That question in itself led to the

$$\begin{pmatrix} 2 & 1 \\ 1 & 1 \end{pmatrix}$$

map on the torus, which was not a counterexample for the 2-sphere; it was a counterexample to the general kind of question about whether one always had attracting periodic orbits.

John Franks: I don't know the details, but I would like to make a comment about your historical remarks about being a graduate student. It was a strange period, even in retrospect. The people who were there as graduate students include many of the people who are speaking at this conference and in this room today: Jacob certainly, Mike Shub, Nancy Kopell, Sheldon was here, John Guckenheimer, Rufus Bowen, Oscar Lanford was an instructor, I believe. Anyway, with this large group of people, mostly graduate students, it was a very heady time, but we didn't really realize that, I think. Being naive graduate students, as he said, we assumed this is the way it always was for students in graduate school. It was a period when we learned a lot; we started with a clean slate, in a way. I think many of us—certainly I did—learned dynamics from Steve's '67 paper. If you knew nothing and you took that paper, you could become an expert by reading through it. At least, that's the way we viewed it in our naive way. I think that Steve's success as a thesis advisor is illustrated in that group and the 30 other people who will go at the top of Steve's family tree.

But I'd like to make one further comment. While I think Steve's done an amazing job as a thesis advisor, he shouldn't get all the credit. A great deal of praise goes to two people, Moe Hirsch and Charles Pugh, who haven't gotten mentioned very prominently yet in this meeting, but were a real essential part of my education and I think the education of many of the rest of us here today. They were the people creating a lot of the ferment and a lot of the talk, a lot of the chaos, if you will, that was flowing around. It was being churned up by them.

Sheldon Newhouse: I would second that thought. Thank you.

Nancy Kopell: I'd like to comment also on Moe's point about Steve's influence on applied mathematics, which I think has been enormous. It's not just in chaos or horseshoe dynamics, or particular things that he's done, but rather it's the whole qualitative point of view. And this is extremely important, especially in subjects unlike physics where you don't know these exact equations, but instead you have to reason with qualitative properties. In particular, mathematical biology has been strongly influenced by this point of view precisely because one has only some qualitative properties to begin with, and one has to reason about the consequence of those qualitative properties. And it turns out that dynamical systems is a wonderful framework in which to do that.

John Guckenheimer: I'd like to add to that Steve's influence even in the realm of physics, and the attention drawn to chaos in recent years. One was able to draw very specific conclusions from qualitative models. Feigenbaum says that he was motivated to begin looking at the dynamics of maps of the interval by a lecture that Steve gave in Aspen at the Center for Physics, I think in 1975, and that this was oriented to properties of physics and population models. I think that the one event which, in some sense, was the key to catching the attention of physicists that this was real stuff was in the experiment of Libchaber, in which he took this phenomenon of period doubling and saw it in a real fluid experiment. It is totally inconceivable to me that anyone could have predicted, by staring at the Navier–Stokes equations, that something like this would be found in the real world.

Christopher Zeeman: It's extremely plausible that you could stare at them forever.

Sheldon Newhouse: Well, perhaps we ought to have some refreshments.

19
Recurrent Sets for Planar Homeomorphisms

Marcy Barge and John Franks

In this article, we investigate fixed-point-free orientation-preserving homeomorphisms of $\mathbb{R}^2$ and the recurrence of connected subsets of the plane under them. In Section 1, we give a short elementary proof that an orientation-preserving homeomorphism of the plane with a periodic point must have a fixed point.

In Section 2, we consider X, a connected set which is open or closed, and let $E(X,f) = \{n > 0 | f^n(X) \cap X = \varnothing\}$. We prove that $E(X,f)$ is closed under addition. Moreover, we show that for any set of positive integers S which is closed under addition, there is an open disk D such that $S = E(D,T)$, where $T(x,y) = (x+1,y)$.

1. Recurrence Implies Fixed Points

Perhaps the most striking dynamical property of orientation-preserving homeomorphisms of $\mathbb{R}^2$ is that almost any kind of recurrence implies the existence of a fixed point. A fairly general formulation of this property [Proposition (1.5) below] is a very important tool in the study of the dynamics of surface homeomorphisms. We begin with a proof that the existence of a periodic point for an orientation-preserving homeomorphism $f\colon \mathbb{R}^2 \to \mathbb{R}^2$ implies the existence of a fixed point. This beautiful result has its origins in the works of Brouwer on planar homeomorphisms. A simple modern proof due to M. Brown can be found in [Br] and A. Fathi [Fa] has given another very nice proof. The proof we give below is similar to the proof of Fathi.

(1.1) Theorem. *Let $f\colon \mathbb{R}^2 \to \mathbb{R}^2$ be an orientation-preserving homeomorphism which possesses a periodic point p which is not a fixed point. Then there is a fixed point for f. If the fixed points of f are isolated, there is one z which has positive index $I(z,f)$.*

We first prove the special case when f has a point of period 2.

(1.2) Lemma. *Let $f: \mathbb{R}^2 \to \mathbb{R}^2$ be an orientation-preserving homeomorphism which possesses a periodic point p of least period 2. Then there is a fixed point for f. If the fixed points of f are isolated, there is one z which has positive index $I(z, f)$.*

PROOF. Let S^2 be the one-point compactification of $\mathbb{R}^2$ and let f_0 denote the extension of f to S^2 obtained by setting $f(\infty) = \infty$. Because p is a point of period 2, it is possible to alter f_0 in an arbitrarily small neighborhood of the points p and $f(p)$ to obtain a homeomorphism f_1 which has the same fixed points as f_0, but has the property that f_1^2 is the identity on a neighborhood of p. A proof that this is possible is straightforward using the fact that if $D_1 \subset D_2$ are closed disks in the plane, then any orientation-preserving embedding of D_1 in the interior of D_2 can be extended to a homeomorphism of D_2 which is the identity on the boundary.

We can now choose a small open disk U containing p on which f_1^2 is the identity. If we remove U and $f(U)$, we are left with an annulus $A = S^2 \backslash (U \cup f(U))$ which is invariant under the homeomorphism f_1. Note that there is a well-defined correspondence between points of $\mathbb{R}^2$, other than those in U and $f(U)$, and points of A other than the point ∞ (henceforth, we will denote the point $\infty \in A$ by q). In particular, any fixed points of f_1 other than q will correspond to fixed points of f and will have the same index.

Thus, it suffices to prove that f_1 has a fixed point z other than q and this point has index $I(z, f_1) > 0$. The sum of the indices of all fixed points of $f_1: A \to A$, which we will denote $\mathscr{I}(f_1)$, is 2. In fact, this is true for any map homotopic to f_1. This follows from the Lefschetz fixed-point theorem. [Note that in this case, one only needs to know that $\mathscr{I}(f_1)$ depends only on the homotopy type of f_1 because f_1 is homotopic to the homeomorphism of the annulus which sends (θ, t) to $(-\theta, -t)$ in A, and this homeomorphism has two fixed points each of index 1.]

Consider the twofold covering $\pi: A \to A$ given by $\pi(\theta, t) = (2\theta, t)$ and let $\pi^{-1}(q) = \{q_1, q_2\}$. Let $F: A \to A$ be the lift of f_1 for this covering which satisfies $F(q_1) = q_2$. Then because F is homotopic to f_1, we have $\mathscr{I}(F) = 2$. It follows that F must have a fixed point z_0 with $I(z_0, F) > 0$. Because $\pi \circ F = f_1 \circ \pi$, we conclude that $z = \pi(z_0)$ is a fixed point of f_1. Also $I(z, f_2) = I(z_0, F)$. ■

(1.3) Definition [F]. A periodic disk chain for f is a collection $\{D_i\}_{i=1}^n$ of open topological disks in $\mathbb{R}^2$ such that:

(i) $D_i \cap D_j = \varnothing$ for $i \neq j$;
(ii) $f(D_i) \cap D_i = \varnothing$ for $i = 1, \ldots, n$; and
(iii) there are positive integers $k_1, k_2, \ldots, k_n$ such that $f^{k_i}(D_i) \cap D_{i+1} \neq \varnothing$ for $i = 1, \ldots, n-1$ and $f^{k_n}(D_n) \cap D_1 \neq \varnothing$.

We say that $m = \sum k_i$ is the period of the disk chain.

(1.4) Lemma. *If f is an orientation-preserving homeomorphism of the plane for which there exists a periodic disk chain of period m, then there is an orientation-preserving homeomorphism, g, of the plane with a point of period m such that the fixed point sets* $\mathrm{Fix}(f)$ *and* $\mathrm{Fix}(g)$ *are equal and $f = g$ on a neighborhood of this set.*

PROOF. The idea of the proof is to alter the homeomorphism f to g in such a way that f has the same fixed-point set as f and so that g has a periodic point.

Let $D_1, \ldots, D_{n-1}$ be a periodic disk chain for f. Let x_{i+1} be a point in $f^{k_i}(D_i) \cap D_{i+1}$ [and $x_1 \in f^{k_n}(D_n) \cap D_1$]. Let $z_i \in D_i$ be defined by $z_i = f^{-k_i}(x_{i+1})$, $1 \leq i \leq n-1$, and $z_n = f^{-k_n}(x_1)$. Let $P = \bigcup_{i=1}^{n} \bigcup_{j=1}^{k_i-1} f^j(z_i)$, so P is the union of the partial orbits starting at $f(z_i)$ and going to $f^{-1}(x_{i+1})$. Now for each i, choose a polygonal arc Γ_i from x_i to z_i such that $\Gamma_i \subset D_i$ and $\Gamma_i \cap P = \emptyset$.

Let V_i be a neighborhood of Γ_i which is contained in D_i and disjoint from P. By means of an isotopy with support in $\bigcup V_i$ which pushes x_i along Γ_i to z_i, we can construct a homeomorphism $h: \mathbb{R}^2 \to \mathbb{R}^2$ supported in $\bigcup V_i$ and satisfying $h(x_i) = z_i$ for $1 \leq i \leq n$.

Let $g = f \circ h$ and note that $g(x_i) = f(h(x_i)) = f(z_i)$ and $g^{k_i-1}(f(z_i)) = f^{k_i-1}(f(z_i)) = f^{k_i}(z_i) = x_{i+1}$. Hence, $g^{k_i}(x_i) = x_{i+1}$ [and likewise $g^{k_n}(x_n) = x_1$] so g is an orientation-preserving homeomorphism with a periodic point whose period is the same as that of the disk chain for f.

We must show that g and f have the same fixed points. But, clearly, because $V_i \subset D_i$ and $h(V_i) = V_i$, it follows that $f(D_i) = g(D_i)$. Because $f(D_i)$ is disjoint from D_i, neither f nor g have any fixed points in D_i. Outside of $\bigcup D_i$, however, $f = g$. Thus, f and g have the same fixed points and agree on an open set containing their common fixed-point set. ■

PROOF OF THEOREM 1.1. We can assume that any fixed points f may have are isolated. Given a point p of period n for f, we will show the existence of a homeomorphism g with a point of period $n-1$ and the same fixed points as f. In fact, we will have $f = g$ on an open set containing the fixed points of both. Applying the same process to g, we obtain a homeomorphism with a point of period $n-2$. Repeating this ultimately results in a homeomorphism with a point whose period is 2 and which has the same fixed point set as f. Because we showed above that a homeomorphism with a point whose period is 2 must have a fixed point, it follows that f must have a fixed point. Because the constructed homeomorphism agrees with f on a neighborhood of the fixed-point set, the existence of a fixed point with positive index also follows. Thus, to complete the proof we need only show it is always possible to find a homeomorphism g as described above.

To construct g, first choose an arc joining p to $f(p)$ which is disjoint from the fixed point set of f. Let U be an open neighborhood of this arc which is also disjoint from the fixed-point set of f. Let V_t, $t \in [0, 1]$ be a continuously

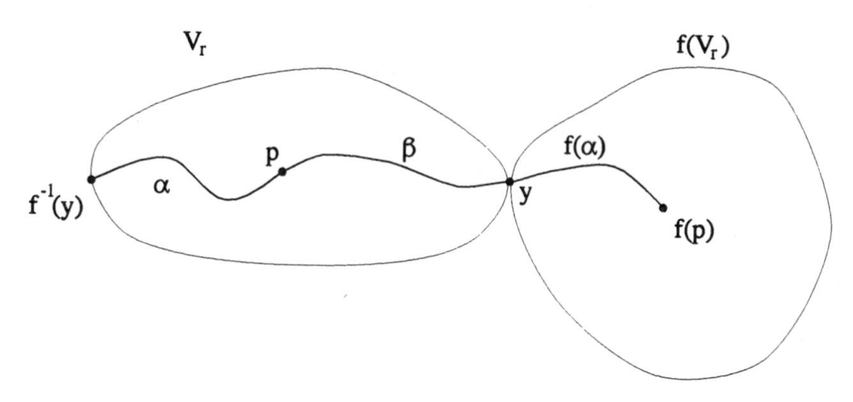

FIGURE 1

varying family of closed topological disks satisfying $p \in V_0$, $V_s \subset V_t$ whenever $s < t$, and $f(p) \in V_1 \subset U$. Let $r = \sup\{t \mid V_t \cap f(V_t) = \emptyset\}$ and let $y \in V_r \cap f(V_r)$. (See Fig. 1.) The points y, $f^{-1}(y)$, and $f(y)$ can be assumed to be distinct because otherwise y would be a point of period 2 (it cannot be fixed as it lies in U).

Note that the interiors of $V_r, f(V_r), f^2(V_r), \ldots, f^{n-1}(V_r)$ can be assumed to be pairwise disjoint because otherwise a subset of them would form a periodic disk chain of period less than n, and by Lemma 1.4, the homeomorphism g exists as desired.

Choose an arc α from $f^{-1}(y)$ to p which lies in V_r and another β from p to y (see Fig. 1). Let γ be the arc from p to $f(p)$ formed by joining β and $f(\alpha)$. Let $\gamma_i = f^i(\gamma)$, $0 \le i \le n-1$. The collection $\{\gamma_i\}$ forms a closed curve in the plane. If this closed curve is not simple (i.e., if it intersects itself), then there is $j \le n-1$ such that γ and $f^j(\gamma)$ have an interior point of intersection. In this case, there is a compact subarc ζ of γ with $\zeta \cap f^j(\zeta) \neq \emptyset$. Then a small disk neighborhood of ζ will be a periodic disk chain for f of period $j < n$ so once again we can apply Lemma 1.4 to obtain the result we want.

Thus, we can assume that the interiors of the arcs γ_i are all pairwise disjoint.

Next choose a short arc W_1 in $\gamma_{n-1} \cup \gamma_0$ which forms a neighborhood of p in the closed curve formed by the arcs $\{\gamma_i\}$ and with the property that $f(W_1) \cap W_1 = \emptyset$. (See Fig. 2.) Choose a closed proper subarc W_2 of γ_1 with the properties that $f(W_1) \cap W_2 \neq \emptyset$ and $f^{n-2}(W_2) \cap W_1 \neq \emptyset$. If D_1 and D_2 are sufficiently small open neighborhoods of W_1 and W_2, then they form a periodic disk chain. Because $f(D_1) \cap D_2 \neq \emptyset$ and $f^{n-2}(D_2) \cap D_1 \neq \emptyset$, this chain has period $n-1$. Thus, once again the existence of the desired g follows from Lemma 1.4. ■

The following proposition is now immediate from Theorem 1.1 and Lemma 1.4.

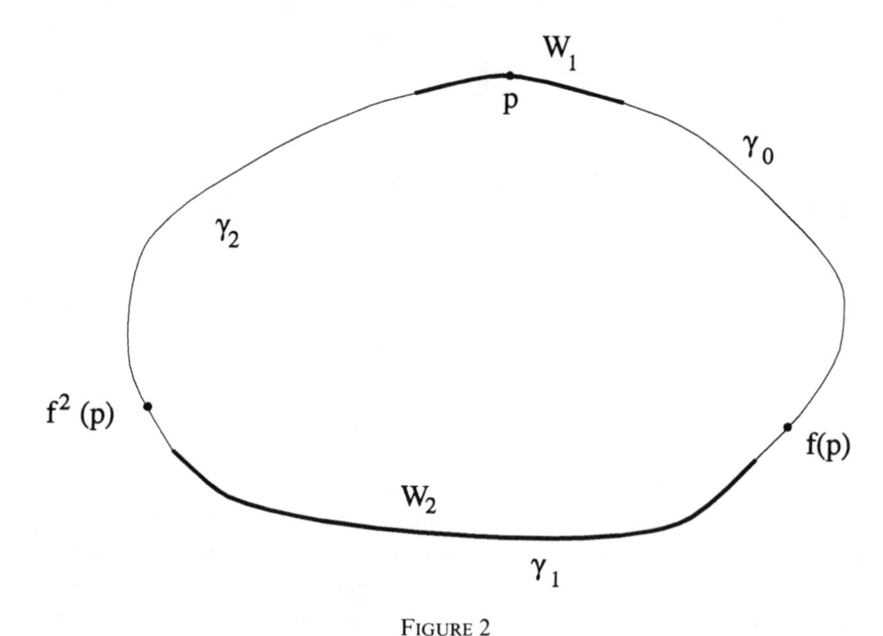

FIGURE 2

(1.5) Proposition. *If f is an orientation-preserving homeomorphism of the plane for which there exists a periodic disk chain, then f has a fixed point. If the fixed points of f are isolated, there is a fixed point of positive index.*

Remark. The fact that the D_i in the definition of disk chain are topological disks is never used in the proof. The only topological properties we used were that the D_i be open and connected. Of course, if the D_i are closed connected sets, then they have open neighborhoods which satisfy the requisite properties. It is sometimes useful to consider "disk chains" where the D_i are, say, compact arcs. The theorem remains valid in this case.

2. Non-return Times of Connected Sets

Given a homeomorphism $f: \mathbb{R}^2 \to \mathbb{R}^2$ and a set $X \subset \mathbb{R}^2$, let

$$E(X,f) = \{n > 0 | f^n(X) \cap X = \varnothing\}.$$

The main results of this section characterize the sets of integers $E(D,f)$ which are possible when D is a disk and f is orientation preserving and fixed-point-free.

(2.1) Theorem. *If f is a fixed-point-free orientation-preserving homeomorphism of the plane and X is a connected subset of the plane which is open or closed, then $E(X,f)$ is closed under addition.*

There is also a converse to this result.

(2.2) Theorem. *If E is any set of positive integers that is closed under addition, then there is an open topological disk D in the plane such that $E(D, T) = E$ where $T(x, y) = (x + 1, y)$ is translation. If $\mathbb{Z}^+ \setminus E$ is finite, D may be taken to be bounded.*

Proof of Theorem 2.1. Consider first the case that $X = D$ is an open connected subset of the plane. Suppose that $E = E(D, f)$ is not closed under addition and let m be the smallest positive integer such that $m \notin E$ and $m = k + l$ for some k and l in E. We will show that f must have a fixed point. Let

$$k_1 \leq k_2 \leq \cdots \leq k_n$$

be elements of E with the properties (i) $m = k_1 + k_2 + \cdots + k_n$ and (ii) if $m = l_1 + l_2 + \cdots + l_r$ with $l_i \in E$ for $i = 1, \ldots, r$, then $r \leq n$. Let $g = f^{k_1}$, let $D_1 = D$ and, for $i = 2, \ldots, n$, let $D_i = f^{k_2 + \cdots + k_i}(D)$. We will prove that $\{D_i\}_{i=1}^n$ is a periodic disk chain for g. (Recall that by the remark at the end of Section 1, it is not necessary that D_i be a disk, only that it be open and connected.)

For $i > 1$, $D_1 \cap D_i = D \cap f^{k_2 + \cdots + k_i}(D) = \varnothing$ because $k_2 + \cdots + k_i < m$ and m is the smallest sum of elements of E that is not in E. For $1 < i < j \leq n$, $D_i \cap D_j = f^{k_2 + \cdots + k_i}(D) \cap f^{k_2 + \cdots + k_j}(D) = f^{k_2 + \cdots + k_i}(D \cap f^{k_{i+1} + \cdots + k_j}(D)) = \varnothing$ because $k_{i+1} + \cdots + k_j \in E$, again by minimality of m. Thus, $\{D_i\}_{i=1}^n$ is a disjoint collection.

Because $k_1 \in E$, $g(D_1) \cap D_1 = f^{k_1}(D) \cap D = \varnothing$ and, for $2 \leq i \leq n$, $g(D_i) \cap D_i = f^{k_1 + \cdots + k_i}(D) \cap f^{k_2 + \cdots + k_i}(D) = f^{k_2 + \cdots + k_i}(f^{k_1}(D) \cap D) = \varnothing$. Thus, condition (ii) in the definition of periodic disk chain is satisfied.

Now note that if $k_2 > k_1$, then $D \cap f^{k_2 - k_1}(D) \neq \varnothing$ for otherwise $k_2 - k_1 \in E$ and

$$k_1, k_1, k_2 - k_1, k_3, k_4, \ldots, k_n$$

would be a collection of $n + 1$ elements of E whose sum is m, contradicting the maximality property (ii) of n. Thus, $g(D_1) \cap D_2 = f^{k_1}(D) \cap f^{k_2}(D) = f^{k_1}(D \cap f^{k_2 - k_1}(D)) \neq \varnothing$. A similar argument shows that $g(D_i) \cap D_{i+1} \neq \varnothing$ for $2 \leq i < n$. Finally, $g(D_n) \cap D_1 = f^{k_1 + k_2 + \cdots + k_n}(D) \cap D \neq \varnothing$ because $k_1 + k_2 + \cdots + k_n = m \notin E$.

Thus, $\{D_i\}_{i=1}^n$ is a periodic disk chain for $g = f^{k_1}$ and g must have a fixed point. Then f has a periodic point and f must also have a fixed point.

In case the set X is a closed connected set, we observe that the argument given above involved only the first m iterates of D, where m is the smallest positive integer such that $m \notin E$ and $m = k + l$ for some k and l in E. If we define m similarly for $E = E(X, f)$, then we can choose an open connected neighborhood D of X such that the sets $E(X, f)$ and $E(D, f)$ have precisely the same elements $\leq m$. Thus, the case when X is closed reduces to the previous case. ■

PROOF OF THEOREM 2.2. Let E be a subset of the positive integers that is closed under addition and let $\mathcal{S} = \{n_1 < n_2 < \cdots\} = \{n \geq 0 | n \in E \text{ and } n+1 \notin E \text{ or } n \notin E \text{ and } n+1 \in E\}$. We first treat the case of $\mathcal{S}$ being infinite. Note that in this case, $n_1 > 0$ (otherwise $E = \mathbb{Z}^+$ and $\mathcal{S} = \{0\}$), $n_{2i-1} \notin E$, and $n_{2i} \in E$ for $i = 1, 2, \ldots$. Define horizontal line segments u_i and l_i and vertical line segments v_i as follows:

$$u_0 = \left[0, n_1 + \frac{1}{2}\right] \times \{1\} \text{ and, for } i > 0,$$

$$u_i = \left[n_{2i} + \frac{2i}{2i+1}, n_{2i+1} + \frac{1}{2i+2}\right] \times \left\{\frac{1}{i+1}\right\};$$

$$l_i = \left[n_{2i-1} + \frac{1}{2i}, n_{2i} + \frac{2i}{2i+1}\right] \times \left\{\frac{1}{i} - 2\right\};$$

$$v_0 = \{0\} \times [0, 1],$$

$$v_{2i-1} = \left\{n_{2i-1} + \frac{1}{2i}\right\} \times \left[\frac{1}{i} - 2, \frac{1}{i}\right], \qquad i = 1, 2, \ldots,$$

and

$$v_{2i} = \left\{n_{2i} + \frac{2i}{2i+1}\right\} \times \left[\frac{1}{i} - 2, \frac{1}{i+1}\right], \qquad i = 1, 2, \ldots.$$

Let

$$A = \left(\bigcup_{i=0}^{\infty} u_i\right) \cup \left(\bigcup_{i=1}^{\infty} l_i\right) \cup \left(\bigcup_{i=0}^{\infty} v_i\right).$$

Then A is an embedded half-line and we will prove that, for $n > 0$, $T^n(A) \cap A = \emptyset$ if and only if $n \in E$. That is, $E(A, T) = E$.

For convenience, let $E^c = \mathbb{Z}^+ \backslash E$. Note that $[0, n_1] \cap E = \emptyset = [n_{2i} + 1, n_{2i+1}] \cap E$ for $i = 1, 2, \ldots$ and that $[n_{2i-1} + 1, n_{2i}] \cap E^c = \emptyset$ for $i = 1, 2, \ldots$. For $0 < m \leq n_1$, $T^m(u_0) \cap u_0 \neq \emptyset$ and, if $m \in [n_{2i} + 1, n_{2i+1}]$, $T^m(v_0) \cap u_i \neq \emptyset$. Thus, $T^m(A) \cap A \neq \emptyset$ for $m > 0$ and not in E.

To see that $T^m(A) \cap A = \emptyset$ for $m \in E$, we will check the nine possible types of intersection.

(1) $T^m(u_i) \cap l_j = \emptyset$ for $i \geq 0$, $j \geq 1$. This is clear.
(2) $T^m(l_i) \cap u_j = \emptyset$ for $i \geq 1$, $j \geq 0$. This is also clear.
(3) $T^m(u_i) \cap u_j = \emptyset$ for $i, j \geq 0$. First note that since $m \in E$, $m > n_1$. Thus, $T^m(u_0) \cap u_0 = \emptyset$. Furthermore, $n_{2i+1} - n_{2i} \leq n_1$ for all $i \geq 1$ for otherwise there would be an integer k such that $k(n_1 + 1) \in [n_{2i} + 1, n_{2i+1}]$. Since $n_1 + 1 \in E$ and $[n_{2i} + 1, n_{2i+1}]$ is disjoint from E, this would contradict the additive closure of E. Thus, $\text{length}(u_i) < n_{2i+1} - n_{2i} \leq m$ and $T^m(u_i) \cap u_i = \emptyset$ for $i = 1, 2, \ldots$. That $T^m(u_i) \cap u_j = \emptyset$ for $i \neq j$ is clear.
(4) $T^m(l_i) \cap l_j = \emptyset$ for $i, j \geq 1$. If $\text{length}(l_i) = (n_{2i} - n_{2i-1}) + (2i/(2i+1) - 1/2i) \geq m$, it would follow that $(n_{2i} + 1) - m > n_{2i-1}$ so that $(n_{2i} + 1) -$

$m \in E$. Then $m + ((n_{2i} + 1) - m) = n_{2i} + 1 \notin E$ would contradict the closure of E under addition. Thus, length$(l_i) < m$ so that $T^m(l_i) \cap l_i = \emptyset$ for $i \geq 1$. That $T^m(l_i) \cap l_j = \emptyset$ for $i \neq j$ is clear.

(5) $T^m(v_i) \cap v_j = \emptyset$ for $i, j \geq 0$. This is clear.

(6) $T^m(v_i) \cap u_j = \emptyset$ for $i, j \geq 0$. $T^m(v_0) \cap u_j = \emptyset$ since $m \in E$, $m > n_1$, and $[n_{2i} + 1, n_{2i+1}]$ is disjoint from E. For $i \neq 0$, we consider two cases. If $i = 2S$ is even and $T^m(v_{2S}) \cap u_j \neq \emptyset$, then

$$n_{2S} + m + \frac{2S}{2S+1} \in \left[n_{2j} + \frac{2j}{2j+1}, n_{2j+1} + \frac{1}{2j+2}\right].$$

Since $m > 1$ (otherwise $\mathscr{S}$ would be finite), we see that $2S \leq 2j$ and then

$$n_{2S} + m \in [n_{2j} + 1, n_{2j+1}].$$

But $n_{2S} \in E$, $m \in E$, and $[n_{2j} + 1, n_{2j+1}]$ is disjoint from E so that closure of E under addition has been violated.

If $i = 2S - 1$ is odd and $T^m(v_i) \cap u_j \neq \emptyset$, then

$$n_{2S-1} + m + \frac{1}{2S} \in \left[n_{2j} + \frac{2j}{2j+1}, n_{2j+1} + \frac{1}{2j+2}\right].$$

Thus, $2S - 1 < 2j + 2$ and $n_{2S-1} + m + 1/2S \leq n_{2i+1} + 1/(2j+2)$. It follows that $(n_{2S-1} + 1) + m \in [n_{2j} + 1, n_{2j+1}]$, again in contradiction to the additive closure of E.

(7) $T^m(v_i) \cap l_j = \emptyset$ for $i \geq 0$, $j \geq 1$. This is clear unless $i = 2S - 1$ and $j = S$ for some $S \geq 1$. But $T^m(v_{2S-1}) \cap l_S \subset T^m(l_0) \cap l_S = \emptyset$ by (4).

(8) $T^m(u_i) \cap v_j = \emptyset$ for $i, j \geq 0$. This is clear unless $j = 2S + 1$ and $i = S$ for some $S \geq 0$. But $T^m(u_S) \cap v_{2S+1} \subset T^m(u_S) \cap u_S = \emptyset$ by (3).

(9) $T^m(l_i) \cap v_j = \emptyset$ for $i \geq 1$, $j \geq 0$. We have shown that $T^m(l_i) \cap l_i = \emptyset$ and, consequently, $T^m(l_i) \cap v_{2i} = \emptyset = T^m(l_i) \cap v_{2i-1}$. If $T^m(l_i) \cap v_{2S} \neq \emptyset$ for some $S \geq 0$, then $S \neq i$ and $n_{2i-1} + m + 1/2i \leq n_{2S} + 2S/(2S+1) \leq n_{2i} + m + 2i/(2i+1)$. It follows that $n_{2i-1} < n_{2S}$ so that $2S > 2i$. Now

$$\begin{aligned} n_{2i-1} + 1 + \left(\frac{1}{2i} - \frac{2S}{2S+1}\right) &\leq (n_{2S} + 1) - m \\ &\leq n_{2i} + 1 + \left(\frac{2i}{2i+1} - \frac{2S}{2S+1}\right) \end{aligned}$$

and, since $2i/(2i+1) - 2S/(2S+1) < 0$,

$$n_{2i-1} \leq (n_{2S} + 1) - m \leq n_{2i} + 1.$$

Then $(n_{2S} + 1) - m \in E$, $m \in E$, $n_{2S} + 1 \notin E$ contradicts the additive closure of E.

If $T^m(l_i) \cap v_{2S-1} \neq \emptyset$ for some $S \geq 1$, then $S \neq i$ and $n_{2i-1} + m + 1/2i \leq n_{2S-1} + 1/2S \leq n_{2i} + m + 2i/(2i+1)$. It follows that $S > i$ and that $n_{2i-1} + m + 1 \leq n_{2S-1} \leq n_{2i} + m$. But then $n_{2S-1} - m \in E$, $m \in E$, and $n_{2S-1} \notin E$ contradicts the additive closure of E.

Thus, $T^m(A) \cap A = \varnothing$ for $m \in E$ and we have $E(A, T) = E$ as desired. In case $\mathscr{S}$ is empty, let $A = \{(0,0)\}$. If $\mathscr{S}$ is finite and nonempty, then $\mathscr{S} = \{n_1 < n_2 < \cdots < n_{2k+1}\}$ for some $k \geq 0$. Let $u_0, u_1, \ldots, u_{k-1}, l_1, l_2, \ldots, l_k, v_0, v_1, \ldots, v_{ik}$ be as before but define

$$u_k = \left[n_{2k} + \frac{2k}{2k+1}, n_{2k+1}\right] \times \left\{\frac{1}{k+1}\right\}$$

and

$$v_{2k+1} = \{n_{2k+1}\} \times \left[0, \frac{1}{k+1}\right].$$

The reader can check (this is virtually the same as in the infinite $\mathscr{S}$ case) that if A is the arc

$$A = \left(\bigcup_{i=0}^{k} u_i\right) \cup \left(\bigcup_{i+1}^{k} l_i\right) \cup \left(\bigcup_{i=0}^{2k+1} v_i\right),$$

then $E(A, T) = E$.

Now given an additively closed E, we may fatten the A constructed above (less as we move out along the x-axis) to obtan an open topological disk D with $E(D, T) = E$. ■

Remark. If f is an arbitrary orientation-preserving fixed-point-free homeomorphism of the plane, we do not know whether E and $\mathbb{Z}^+ \setminus E$ both being infinite forces any disk D with $E(D, f) = E$ to be unbounded. It seems plausible that it does. But we do not know, for example, whether there can be a bounded disk D and homeomorphism f as above with $E(D, f)$ equal to the positive even integers.

Suppose X is any compact connected set in the plane and v is any fixed nonzero vector in $\mathbb{R}^2$. A result due to H. Hopf [H] asserts that if $T_s(x) = x + sv$, $s \in (0, \infty)$, then the set $\mathscr{S}$ of s such that $T_s(X) \cap X = \varnothing$ is a semigroup, i.e., closed under addition.

Theorem 2.1 extends this to the case of an arbitrary fixed-point-free flow on $\mathbb{R}^2$ instead of just translations.

(2.3) Corollary. *Suppose that X is a compact connected set in the plane and f_t is a fixed-point-free flow on the plane defined for all $t \in \mathbb{R}$. Then the set $\mathscr{S}$ of $s \in (0, \infty)$ such that $f_s(X) \cap X = \varnothing$ is a semigroup.*

PROOF. We must show that if $s_1 \in \mathscr{S}$ and $s_2 \in \mathscr{S}$, then $s_0 = s_1 + s_2 \in \mathscr{S}$. The set $\mathscr{S}$ is open by the continuity of f_t and compactness of X. Hence, there are neighborhoods U_i, $i = 1, 2$, of s_i in $\mathbb{R}$ such that $U_i \subset \mathscr{S}$. If N is sufficiently large, there will exist integers n and m such that $n + m = N$,

$$\frac{n}{N}s_0 \in U_1,$$

and

$$\frac{m}{N}s_0 \in U_2.$$

Let $h = f_t$ where $t = s_0/N$. Then $h^n(X) \cap X = \varnothing$ because

$$\frac{n}{N}s_0 \in \mathscr{S}.$$

Likewise $h^m(X) \cap X = \varnothing$. It then follows from Theorem 2.1 that $h^{n+m}(X) \cap X = \varnothing$, but $h^{n+m}(X) = f_{s_0}(X)$, so $s_0 \in \mathscr{S}$. ■

References

[Br] Morton Brown, A new proof of Brouwer's lemma on translation arcs, *Houston J. Math.* **10** (1984), 35–41.

[Fa] Albert Fathi, An orbit closing proof of Brouwer's lemma on translation arcs, *L'enseignement Mathématique* **33** (1987), 315–322.

[F] John Franks, Generalizations of the Poincaré–Birkhoff Theorem, *Ann. Math.* **128** (1988), 139–151.

[H] Heinz Hopf, Uber die Sehnen ebener Kontinuen und die Schleifen geschlossener Wege, *Comment. Math. Helv.* **9** (1937), 303–319.

20
Convergence of Finite-Element Solutions for Nonlinear PDEs

XIAOHUA XUAN

1. Introduction

The aim of this note is to prove the convergence of FEM solutions (finite-element-method solutions) in solving

$$\begin{aligned} \Delta u &= b(x, u, Du) \quad \text{in } \Omega, \\ u &= 0 \quad \text{on } \partial\Omega. \end{aligned} \tag{1}$$

Here the domain $\Omega \subset \mathbb{R}^n$ ($n = 2$) is a bounded convex polygon and the right-hand side $b(x, z, p)$, smooth, has at most quadratic gradient growth.[1] Indeed, the proof below will also apply to second-order quasilinear boundary value problems in the divergence form

$$\begin{aligned} -\text{div } A(x, u, Du) &= b(x, u, Du) \quad \text{in } \Omega, \\ u &= 0 \quad \text{on } \partial\Omega, \end{aligned}$$

with the uniform ellipticity of the operator A. Such convergence is important for the study of feasibility and complexity of finite-element methods for nonlinear boundary value problems.

Let the Sobolev space $W^{k,p}(\Omega)$ with norm $\|\cdot\|_{k,p}$, the subspace $W_0^{k,p}(\Omega)$ of $W^{k,p}(\Omega)$, and FEM solutions $\{u_h\}$ be defined in the usual way (see precise definitions in Section 2). We make the following two assumptions.

Assumption 1. (Boundedness). The FEM solutions $\{u_h\}$ is uniformly bounded in $W^{1,\infty}(\Omega)$, i.e.,

$$\sup_h \|u_h\|_{1,\infty} < \infty. \tag{2}$$

---

[1] The $b(x, z, p)$ is said to have at most quadratic gradient growth if $\exists K \in \mathbb{R}$ such that

$$|b(x, z, p)| \le K(1 + |p|^2) \quad \text{for all } x \in \Omega,\ z \in \mathbb{R},\ p \in \mathbb{R}^n.$$

In fact, the proof will only need the following inequality.

$$|b(x, u, Du| \le K(M)(1 + |Du|^2) \quad \text{for all } u \in W^{1,2}(\Omega) \text{ satisfying } |u| \le M \text{ a.e.}$$

From the boundedness (2), one can extract a subsequence of $\{u_h\}$ converging weakly in $W^{1,2}(\Omega)$ and strongly in $W^{0,2}(\Omega) = L^2(\Omega)$.

Assumption 2 (Regularity). The FEM solutions $\{u_h\}$ converge to a function u weakly in $W^{1,2}(\Omega)$ and strongly in $L^2(\Omega)$, and

$$u \in W^{k,2}(\Omega) \cap W^{1,\infty}(\Omega) \quad \text{with } k = 2. \tag{3}$$

We can now state our result.

Main Theorem. *Under the above assumptions, the FEM solutions $\{u_h\}$ converge to u (strongly) in $W^{1,2}(\Omega)$, i.e.,*

$$\|u_h - u\|_{1,2} \to 0, \quad \text{as } h \to 0, \tag{4}$$

and the function u satisfies the following equation (weak solution):

$$\int_\Omega Du\, Dv\, dx = \int_\Omega b(x, u, Du) v\, dx \quad \text{for all } v \in W_0^{1,2}(\Omega). \tag{5}$$

Remark on the Assumptions. The main theorem holds for nonconvex polygons, but the statements in the lemmas of the next section needs changes ($h \to h^\alpha$ with $\alpha < 1$). The restriction on dimension $n = 2$ could also be easily removed with a possible increase of k in Assumption 2 and the main theorem. Whereas Assumption 1 seems able to be verified in applications with the help of explicit nonlinear structures, Assumption 2 is difficult to verify. Often one proves such regularity (3) after proving that u is a weak solution of (1). An modification of the main theorem without this regularity assumption is under investigation.

The material of this note is mainly from my Ph.D. thesis supervised by Steve Smale, to whom I am grateful for his support and encouragement. The goal there, as proposed in Smale [7, 8], is to do global analysis of computational cost for solving nonlinear partial differential equations. The result here is also motivated by the work of Frehse [4] on the existence of a solution for problem (1). The approach, different from typical approaches in Bank [2], Mansfield [5], and Xuan [9], is that only uniform ellipticity is found in the equation, instead of the stronger ellipticity (called V-monotonicity, or V-ellipticity). This lack of the stronger ellipticity causes difficulty in proving the convergence of the FEM solution. An important work for linear equations using such uniform ellipticity in the principal part is Schatz [6].

2. Proof

2.1. *Definition*

Let the function spaces $C(\Omega)$, $C(\overline{\Omega})$, $C^k(\Omega)$ ($k \in \mathbb{N} = \{1, 2, \ldots\}$), and $L^p(\Omega)$ ($1 \leq p \leq \infty$) be defined in usual way. We define

$$C_0^\infty(\Omega) = \{v: v \in C^k(\Omega), \forall k \in \mathbb{N} \text{ and } v \text{ has compact support}\}.$$

By making $W^{0,p} = L^p(\Omega)$ and $\|\cdot\|_{0,p} = \|\cdot\|_{L^p(\Omega)}$, we can give the definition of the Sobolev spaces $W^{k,p}(\Omega)$ and its norm $\|\cdot\|_{k,p}$.

Definition 2.1. For $k \in \mathbb{N}$ and $1 \le p \le \infty$, we define

$$W^{k,p}(\Omega) = \{v \in L^p(\Omega): D^\alpha v \in L^p(\Omega) \text{ for all } |\alpha| \le k\},$$

$$\|v\|_{k,p} = \left(\sum_{|\alpha| \le k} \|D^\alpha v\|_{0,p} \right)^{1/p},$$

where the derivative $D^\alpha v$ is understood in the distribution sense. Also, we define

$$W_0^{k,p}(\Omega) = \text{closure of } C_0^\infty(\Omega) \text{ in } W^{k,p}(\Omega).$$

In the case $p = 2$, we set

$$H^k(\Omega) = W^{k,2}(\Omega), \qquad \|\cdot\|_k = \|\cdot\|_{k,2}, \qquad H_0^k(\Omega) = W_0^{k,2}(\Omega).$$

Our next task is to define the finite-element space and the FEM solutions. To do that, we assume that $\{\pi_h\}$ is a sequence of triangulations of the domain Ω; those triangles have the largest side h and the smallest angle $\theta \ge \theta_0 > 0$.

Definition 2.2. For each such triangulation $\{\pi_h\}$, we define $S^h(\Omega)$ to be the space of all continuous functions on the domain Ω which are linear on each triangle, and $S_0^h(\Omega)$ to be the subspace of $S^h(\Omega)$ consisting of those functions vanishing on the boundary of the domain Ω, i.e.,

$$S_0^h(\Omega) = \{v \in S^h(\Omega): v = 0 \text{ on } \partial\Omega\}$$

The finite-element method is to solve the boundary value problem (1) in a finite-dimensional space $S_0^h(\Omega)$ of $H_0^1(\Omega)$ in a weak sense.

Definition 2.3. The FEM solution u_h is characterized as the following equation:

$$\int_\Omega Du_h\, Dv\, dx = \int_\Omega b(x, u_h, Du_h) v\, dx \quad \text{for all } v \in S_0^h(\Omega). \tag{6}$$

We will also be interested in a particular function of $S_0^h(\Omega)$ associated with u (called interpolant).

Definition 2.4. Let $u \in C(\overline{\Omega})$; the interpolant $u_I \in S^h(\Omega)$ of u for a triangulation π_h is defined to be the unique piecewise linear function which is equal to u at the nodal points of the triangulation.

2.2. *Approximation Inequalities*

Since the functions in $S_0^h(\Omega)$ are to be used to approximate the solution of the original problem. The approximation inequalities about the space $S_0^h(\Omega)$ play importance role in proving convergence. Within the following lemmas, the notation $C(\cdot, \cdots)$ conventionally represents a constant dependent of its arguments.

The following lemma is well know, e.g., see Babuška [1], Ciarlet [3].

Lemma 2.1. *If $u \in H^2(\Omega)$, then the interpolant u_I of u satisfies*

$$\|u - u_I\|_1 \le C(u, \Omega)h, \tag{7}$$

We now turn to state some other approximation inequalities. The proofs of these inequalities are omitted here. Recall the *Assumption 1 and 2* in the previous section.

Lemma 2.2. *Under the* Assumption 1 and 2, *we have*

i) $\|e^{\lambda(u_h - u)} - e^{\lambda(u_h - u_I)}\|_1 \le C(\lambda, u, \Omega)h.$
ii) $\|e^{\lambda(u_h - u_I)} - (e^{\lambda(u_h - u_I)})_I\|_1 \le C(\lambda, u, \Omega)h.$

for any given real number λ.

Lemma 2.3. *Under the* Assumption 1 and 2, *there exist $s_h(x) \in S_0^h(\Omega)$ such that*

$$\|\sinh(\lambda(u_h - u)) - s_h(x)\|_1 \le C(\lambda, u, \Omega)h. \tag{8}$$

for any given real number λ.

2.3. *Proof of Main Theorem*

The main theorem can now be proved.

Proof. Since the finite element solution $u_h(x)$ solves (6), we first have

$$\int_\Omega Du_h Dv\,dx = \int_\Omega b(x, u_h, Du_h)v\,dx. \tag{9}$$

Taking $v(x) = s_h(x)$ defined in lemma 2.3, and substituting in the above equation, we obtain

$$\int_\Omega Du_h Ds_h\,dx = \int_\Omega b(x, u_h, Du_h)s_h\,dx. \tag{10}$$

Let $\phi(x) = \sinh(\lambda(u_h - u))$ where the real number λ will be chosen later. It then follows from lemma 2.3 that

$$\int_\Omega Du_h D\phi\,dx = \int_\Omega b(x, u_h, Du_h)\phi\,dx + o(1).$$

Using the fact that $u_h \to u$ strongly in $L^2(\Omega)$, we have

$$\int_\Omega D(u_h - u)D\phi\, dx = \int_\Omega b(x, u_h, Du_h)\phi\, dx + o(1).$$

Now following the same estimate in Frehse [4], we have

$$\lambda \int_\Omega (Du_h - Du)^2\, dx \le K \int_\Omega (Du_h - Du)^2\, dx + o(1), \tag{11}$$

i.e.,

$$(\lambda - K) \int_\Omega (Du_h - Du)^2\, dx \le o(1),$$

If the real number λ is chosen such that $\lambda > K$, we have

$$\|u_h - u\|_{1,2} \to 0, \qquad \text{as } h \to 0, \tag{12}$$

proving the convergence of FEM solutions $\{u_h\}$ to u. Now passing the limit in equality (9), and then using the fact that $\bigcup_h S_0^h(\Omega)$ is dense in $W_0^{1,2}(\Omega)$, we obtain the conclusion (5).

References

[1] Babuška I., and Aziz K., "Survey lectures on the mathematical foundations of the finite element methods, "*The Mathematical Foundations of the Finite Element Method, With applications to Partial Differential Equations* (edit by A.D. Aziz), Academic Press, New York and London 1972.

[2] Bank R., "Analysis of a multilevel iterative methods for nonlinear finite element equations," *Math. Comp.*, 39 (1982), 453–465.

[3] Ciarlet P.G., *The Finite Element Method for Elliptic Problems*, North-Holland Publ., Amsterdam. Bifurcation Theory, Vol. 2, Springer Verlag: New York, 1982.

[4] Frehse J., "Existence and perturbation theorems for nonlinear elliptic systems," *Nonlinear PDE and Their Applications: Collège de France Seminar Vol. IV*, Pitman, New York, 1983.

[5] Mansfield L., "On the solution of nonlinear finite element system," *SIAM J. Numer. Anal.*, **17** (1980), 752–765.

[6] Schatz A.H., "An observation concerning Ritz-Galerkin methods with indefinite bilinear forms," *Math. Comp.*, **28** (1974), 959–962.

[7] Smale S., "On the efficiency of algorithms of analysis, "*Bull. Amer. Math. Soc.*, **13** (1985), 87–121.

[8] Smale S., "Some remarks on the foundations of numerical analysis," *SIAM Review*, **32** (1990), 211–220.

[9] Xuan Xiaohua, *Nonlinear boundary value problems: computation, convergence and complexity*, Ph. D. Dissertation, University of California, 1990.

21
Ergodic Theory of Chaotic Dynamical Systems*

L.-S. Young

A dynamical system in this article consists of a map of a manifold to itself or a flow generated by an autonomous system of ordinary differential equations. For definiteness, we shall discuss the discrete-time case, although most things here have their continuous-time versions. We shall not attempt to define "chaos," except to mention two of its essential ingredients:

1. exponential divergence of nearby orbits,
2. lack of predictability.

Ingredient 1 is a geometric condition. Ingredient 2 says that there is some "randomness" in the system in the sense that it is difficult to predict future behavior based on information from the past. These ideas will be made precise later on.

A class of dynamical systems having these properties was axiomatized and systematically studied for the first time in the sixties, following the groundbreaking papers of Smale [Sm] and Anosov [A]. They are the uniformly hyperbolic or *Axiom A* systems. Roughly speaking, a diffeomorphism is uniformly hyperbolic on a compact invariant set if, at every point in the set, the map has saddlelike behavior with coefficients of expansion and contraction uniformly bounded away from 1. The topological and ergodic theories of Axiom A systems flourished in the late sixties and early seventies.

There are, however, many examples that are clearly "chaotic" but are not Axiom A: the periodically forced Duffing equation, Lorenz attractor [Lo], Henon mappings, etc. (See [GH] for more information and references.) See also [N]. As time went on, it became increasingly clear that it would be desirable to have a theory that applies to a broader class of chaotic systems.

In [O], Oseledec introduced the notion of Lyapunov exponents. This, together with the local analysis developed by Pesin [Pe1], laid the foundation for a *nonuniform hyperbolic theory*. The setting here consists of a dy-

---

* This is an expanded version of a lecture presented in Berkeley in August 1990, in a conference honoring Steve Smale on his 60th birthday. The author is partially supported by NSF.

namical system with an invariant probability measure. We assume that at almost every point x with respect to this measure, the derivative of the nth iterate of the map expands and contracts exponentially as n tends to infinity. The onset of hyperbolicity is not necessarily uniform in x, nor are the rates of expansion and contraction. As we shall see, these assumptions give rise to fairly rich statistical properties in a dynamical system.

In this chapter I would like to share with the reader what I learned in this subject. I have aimed the exposition at nonspecialists, assuming only some acquaintance with "horseshoes," etc. In Section 1, two examples are used to introduce the notion of nonuniform hyperbolicity. The formal setting of this theory is presented in Section 2. In Section 3, we discuss the relation among Lyapunov exponents, entropy, and Hausdorff dimension. This leads naturally to a notion of "attractors," which is the topic of Section 4. Section 5 concerns questions of stability when the system is perturbed by random noise.

Obviously, these notes do not give a complete picture of nonuniformly hyperbolic systems. We have tried to explain a few ideas and have neglected many others. We refer the reader to [Si4] for a survey of ergodic theory and its applications. We recommend also [ER], which overlaps with this chapter and contains, among other things, physical motivations and numerical implementations of many of the ideas discussed here.

1. Examples

In this section, we discuss *very informally* two examples that appear to have exponential divergence of nearby orbits. Precise results are stated at the end of Section 2.

EXAMPLE 1. Let $f\colon [-1,1] \circlearrowleft$ be defined by $f(x) = 1 - 2x^2$. This example was first studied by von Neumann and Ulam in the 1940s. If we make the change of coordinates $h(x) = \sin[(\pi/2)\theta]$, then an easy calculation shows that $g = h^{-1} \circ f \circ h = (2/\pi)\arcsin\{1 - 2\sin^2[(\pi/2)\theta]\}$ has derivative $g'(\theta) = \pm 2$. So if $f^n x \neq 0$, $\forall n$, then

$$|(f^n)'x| \sim 2^n.$$

Note that the growth of $(f^n)'$ cannot be uniform in x because there are always points which return arbitrarily close to 0.

To prepare ourselves for more general situations, let us not make this change of coordinates and try to understand, at least qualitatively, why $(f^n)'$ is expected to grow. Naturally we are concerned with x near 0, for there the derivative is small. Note, however, that the critical point 0 is preperiodic, with $f(0) = 1$ and $f^n(0) = -1$ $\forall n > 1$. So for x near 0, the orbit of x will also stay near ± 1 for several iterates; and near ± 1, $|f'| \sim 4$. Indeed, the closer x is to 0, the longer the orbit of x will stay near -1. With a small amount of fiddling, one can show that there exist $\delta > 0$ and $\lambda > 1$ s.t. $\forall x \notin (-\delta, \delta)$, $|f'x| \geq \lambda$, and $\forall x \in (-\delta, \delta)$, $\exists n(x)$ s.t. $f^k x$ is near ± 1 for $1 \leq k < n(x)$ and $|(f^{n(x)})'x| \geq \lambda^{n(x)}$. After time $n(x)$, the orbit of x may stay outside $(-\delta, \delta)$ for a number of iterates, with $(f^n)'$ continuing to grow. The argument above is repeated when it returns to $(-\delta, \delta)$.

EXAMPLE 2. Consider a billiard table consisting of a square with a (round) disk in the middle removed, and a billiard ball moving without friction on this table. We assume that the ball bounces elastically off the edges—so that the angle of reflection equals the angle of incidence.

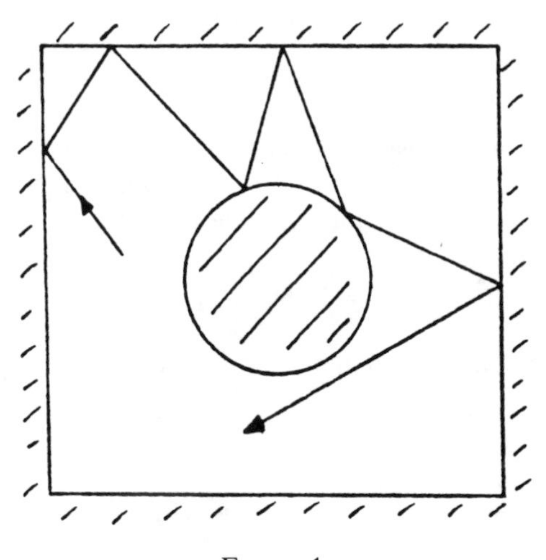

FIGURE 1

It is easy to see that if two nearby trajectories are diverging, then upon reflecting off a concave edge, they will continue to diverge—in fact with a greater angle than before.

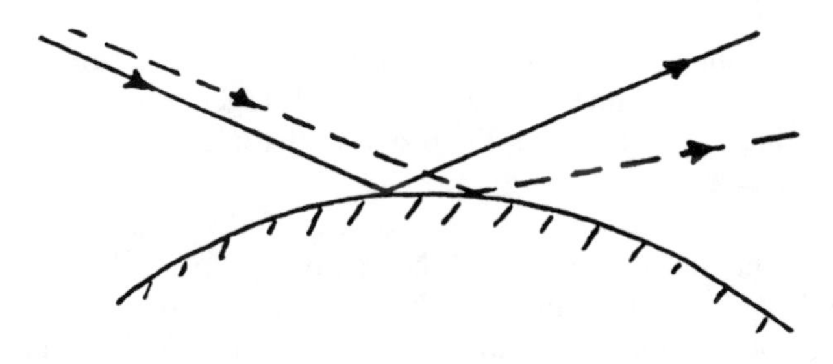

FIGURE 2

This is an example of a "semi-dispersing" billiard system, first studied by Sinai in [Si2]. We have not proved this, but two nearby, initially parallel billiard trajectories will separate exponentially fast. It is easy to see that the rates with which they separate cannot be uniform, for there is a band of periodic orbits running horizontally between the sides of square without hitting the round disk. These periodic orbits are clearly nonhyperbolic. True, they form a set of measure zero, but orbits near them and nearly horizontal also spend a long time bouncing between the sides of the square before they experience the scattering action described above.

2. General Setting and Basic Facts in the Theory of Nonuniformly Hyperbolic Systems

A. Uniformly Hyperbolic Sets

Let f be a C^1 diffeomorphism of a compact manifold onto itself, and let Λ be a compact f-invariant set. We say that f is *uniformally hyperbolic* on Λ if there is a continuous splitting of the tangent bundle over Λ into a direct sum of two Df-invariant subbundles, written

$$T\Lambda = E^u \oplus E^s,$$

so that for all $x \in \Lambda$ and $n > 0$, the following hold:

$$v \in E^u(x) \quad \Rightarrow \quad |Df_x^n v| \geq C^{-1}\lambda^n|v|$$

and

$$v \in E^s(x) \quad \Rightarrow \quad |Df_x^n v| \leq C\lambda^{-n}|v|,$$

where $\lambda > 1$ and $C > 0$ are constants independent of x.

The following terminologies were introduced in [Sm]. We say that f is an *Anosov diffeomorphism* if it is uniformly hyperbolic on all of M. A compact invariant set Λ is called an *Axiom A basic set* if

1. $f|\Lambda$ is uniformly hyperbolic;
2. there is an open neighborhood U of Λ s.t. $\Lambda = \{x: f^n x \in U \ \forall n \in \mathbf{Z}\}$;
3. $f|\Lambda$ has a dense orbit.

A diffeomorphism $f: M \circlearrowleft$ is said to satisfy *Axiom A* if M can be written as the disjoint union of a finite number of basic sets together with a "wandering" part, consisting of points x with the property that for some neighborhood V of x, $f^n V \cap V = \phi$, $\forall n \neq 0$.

If Λ is a nontrivial basic set, i.e., if it consists of more than a periodic orbit, then $f|\Lambda$ is "chaotic" in the sense of the first paragraph of this paper. Horseshoes and hyperbolic toral automorphisms (e.g., the 2-1-1-1 map) are prototypical examples of nontrivial Axiom A basic sets and diffeomorphisms. (See [Sm].) Geodesic flows on manifolds of negative curvature are prime examples of Anosov flows (see [A]).

The ergodic theory of Anosov diffeomorphisms was developed by Sinai (see, e.g., [Si3]). This theory was subsequently extended to Axiom A attractors by Bowen and Ruelle (see, e.g., [Bo]). Their main tools were Markov partitions, which allowed them to transform their problems into problems on shift spaces. They then handled these problems using methods from statistical mechanics. We shall discuss this work in more detail in Sections 3 and 4.

B. Linear Theory: Lyapunov Exponents

Let $A \in GL(m, \mathbf{R})$, and let v be a vector in $\mathbf{R}^m$. We are interested in the rate of growth of $|A^n v|$ as n tends to infinity. It is easy to see, by writing A in its real Jordan form, for instance, that $\mathbf{R}^m$ can be decomposed into

$$\mathbf{R}^m = E_1 \oplus \cdots \oplus E_r$$

and that for each i, there is a number λ_i such that for all $v \in E_i$,

$$\lim_{n\to\infty} \frac{1}{n} \log|A^{\pm n} v| = \pm\lambda_i.$$

The λ_i, of course, are the log of the moduli of the eigenvalues of A.

Next we consider a bi-infinite sequence of matrices $\ldots, A_{-1}, A_0, A_1, \ldots$ and ask about the rates of growth of $|A_n \cdots A_0 v|$. We say that the sequence $\{A_n\}$ is *regular* if the following holds: Write $A^{(n)} = A_{n-1} \cdots A_0$ and $A^{(-n)} = (A_{-1} \cdots A_{-n})^{-1}$. Then there is a decomposition of $\mathbf{R}^m$ into $\mathbf{R}^m = E_1 \oplus \cdots \oplus E_r$ such that

1. for each i, there is a number λ_i s.t. for all $v \in E_i$,

$$\lim_{n\to\infty} \frac{1}{n} \log|A^{(\pm n)} v| = \pm\lambda_i;$$

2. for every $j \neq k$,

$$\lim_{n\to\infty} \frac{1}{n} \log \angle(A^{(\pm n)} E_j, A^{(\pm n)} E_k) = 0,$$

where $\angle(E_j, E_k)$ denotes the minimum angle between the subspaces E_j and E_k. (In the literature, there are various notions of "regularity," not all of which are equivalent.) The following theorem is the matrix version of the Birkhoff ergodic theorem and is proved by Oseledec in 1968:

Theorem [O]. *Let (X, μ) be a probability space, and let $f: X \circlearrowleft$ be an invertible measure-preserving transformation. Let $A: X \to GL(m, \mathbf{R})$ be a measurable mapping satisfying*

$$\int \log^+ \|A\| \, d\mu, \quad \int \log^+ \|A^{-1}\| \, d\mu \quad < \quad \infty,$$

where $\log^+ a = \max(0, \log a)$. *Then, for* μ*-a.e.* x, *the sequence* $\{A(f^n x)\}$ *is regular.*

There now exist more than one proof of Oseledec's theorem in the literature. See, e.g., [Ru3].

This theorem is easily adapted to apply to diffeomorphisms preserving a Borel probability measure with $A(x) = Df_x$. It is easy to see that if

$$T_x M = E_1(x) \oplus \cdots \oplus E_{r(x)}(x)$$

is the decomposition of the tangent space at x and $\lambda_i(x)$ is the rate of growth for vectors in $E_i(x)$, then $Df_x E_i(x) = E_i(fx)$ and $\lambda_i(fx) = \lambda_i(x)$. It follows from this that if (f, μ) is ergodic, then the functions $x \mapsto \lambda_i(x)$ and $\dim E_i(x)$ are constant μ-a.e.

The numbers $\lambda_1(x), \ldots, \lambda_{r(x)}(x)$ together with their multiplicities are called the *Lyapunov exponents* at x. If (f, μ) is ergodic, then $\lambda_1, \ldots, \lambda_r$ and their multiplicities are called the Lyapunov exponents of the dynamical system (f, μ).

We remark that, in general, Lyapunov exponents are only defined almost everywhere and that the functions $x \mapsto \lambda_i(x)$ and $E_i(x)$ are only Borel measurable. Moreover, there is no lower bound on the angles between the various Oseledec subspaces, and, depending on x, it could take arbitrarily long for a tangent vector to attain its supposed rate of growth.

A diffeomorphism preserving a Borel probability measure μ is sometimes called *nonuniformly hyperbolic*[1] if μ-a.e. we have $\lambda_i \neq 0$ for every i. The pair (f, μ) is sometimes called *partially nonuniformly hyperbolic* if on a set of positive μ-measure, we have $\lambda_i \neq 0$ for some i.

C. Nonlinear Theory and Approximation by Uniformly Hyperbolic Sets

The material in this section is due to Pesin, who first developed a machinery for translating the linear theory of Lyapunov exponents into nonlinear results in neighborhoods of typical trajectories. He proved, in particular, the existence of stable and unstable manifolds for nonuniformly hyperbolic systems. These results are contained in [Pe1]. Similar results are obtained by Ruelle using different techniques. See [Ru3].

Throughout this section, let f be a C^2 diffeomorphism of a compact Riemannian manifold preserving a Borel probability measure μ. For simplicity of exposition, let us assume that (f, μ) is ergodic and nonuniformly hyperbolic, i.e., $\lambda_i \neq 0$ for all i, although more general situations can be handled

[1] Throughout this chapter, and in most of the literature for that matter, the term "nonuniform hyperbolicity" really refers to hyperbolicity that is *not necessarily* uniform.

easily with a small amount of fuss. We begin by defining a parameter that measures the "strength" of hyperbolicity of the map at each point:

Let $\lambda^+ = \min_{\lambda_i>0} \lambda_i$ and $\lambda^- = \max_{\lambda_i<0} \lambda_i$. Wherever it makes sense we let $E^u(x) = \bigoplus_{\lambda_i>0} E_i(x)$ and $E^s(x) = \bigoplus_{\lambda_i<0} E_i(x)$. Fix ε with $0<\varepsilon \ll \min(\lambda^+, -\lambda^-)$ and define

$$\ell(x) = \operatorname{Inf}\left\{ \ell\colon \begin{array}{ll} |Df_x^{-n}v| \le \ell e^{-(\lambda^+-\varepsilon)n}|v|, & \forall v \in E^u(x) \text{ and } n \ge 0 \\ |Df_x^{n}v| \le \ell e^{(\lambda^-+\varepsilon)n}|v|, & \forall v \in E^s(x) \text{ and } n \ge 0 \\ \angle(E^u(f^n x), E^s(f^n x)) \ge \ell^{-1}e^{-\varepsilon|n|}, & \forall n \in \mathbf{Z}. \end{array} \right\}$$

Oseledec's theorem guarantees that $\ell(x) < \infty$ for μ-a.e. x. For $k = 1, 2, \ldots$, let

$$\Lambda_k = \{x \in M : \ell(x) \le k\}.$$

It is easy to see that each Λ_k is a closed set, usually noninvariant, and that restricted to each Λ_k, f^n is uniformly hyperbolic. Moreover, $\Lambda_1 \subset \Lambda_2 \subset \cdots$, and the "strength" of hyperbolicity on Λ_k weakens as k increases. Clearly, $\mu(\bigcup \Lambda_k) = 1$.

One can, in fact, make nonautonomous changes of coordinates a.e. so that in these "Lyapunov charts," the map appears to be uniformly hyperbolic. More precisely, at μ-a.e., x one can construct an embedding Φ_x taking a neighborhood N_x of 0 in $\mathbf{R}^{u+s}$ into M. (Here $u = \dim E^u$, $s = \dim E^s$.) These embeddings have the property that $\Phi_x(0) = x$, $D\Phi_x(0)(\mathbf{R}^u \times \{0\}) = E^u(x)$, $D\Phi_x(0)(\{0\} \times \mathbf{R}^s) = E^s(x)$, and if $\tilde{f}_x = \Phi_{fx}^{-1} \circ f \circ \Phi_x$, then

(i) $|D\tilde{f}_x(0)v| \ge e^{\lambda^+-\varepsilon}|v|$, $\forall v \in \mathbf{R}^u \times \{0\}$
(ii) $|D\tilde{f}_x(0)v| \le e^{\lambda^-+\varepsilon}|v|$, $\forall v \in \{0\} \times \mathbf{R}^s$
(iii) $\operatorname{Lip}(\tilde{f}_x - D\tilde{f}_x(0)) \le \varepsilon$.

The diameter of N_x is $\sim\ell(x)^{-2}$, and each Φ_x carries the Euclidean metric on N_x to a metric equivalent to the Riemannian metric on M by a constant $\sim\ell(x)^2$.

Recall that to construct a local unstable manifold at x, we put a "trial disk" V_n at $f^{-n}x$, tangent to $E^u(f^{-n}x)$, and show that $f^n V_n$ appropriately restricted converges to $W^u_{\text{loc}}(x)$. Because we only have uniform estimates for $\{\tilde{f}_x\}$, it is convenient to carry out our construction in charts. This requires that we have some control on how $\operatorname{diam}(N_x)$ varies along typical orbits. One possibility is to replace ℓ in the analysis above by a function $\tilde{\ell}$ satisfying $\ell \le \tilde{\ell} < \infty$ a.e. and $\tilde{\ell}(x)e^{-\varepsilon} \le \tilde{\ell}(fx) \le \tilde{\ell}(x)e^{\varepsilon}$. It is shown in [Pe1] that such a function exists.

Having constructed $W^u_{\text{loc}}(x)$, we define the *global unstable manifold*

$$W^u(x) := \bigcup_{n\ge 0} f^n W^u_{\text{loc}}(f^{-n}x).$$

As in the uniform case, $W^u(x)$ is an injectively immersed copy of $\mathbf{R}^u$ tangent at x to $E^u(x)$. It has the characterization

$$W^u(x) = \{y \in M : d(f^n x, f^n y) \le e^{-(\lambda^+-\varepsilon)n} \text{ for all sufficiently large } n\}.$$

Stable manifolds are defined analogously. For more details on stable and unstable manifolds in this setting, see [FHY, Pe1, or Ru3].

We close this section by mentioning a result of Katok's, which states that at least in dimension 2, all the randomness in a dynamical system is "caused" by the presence of horseshoes:

Theorem [Ka2]. *Let f be a C^2 diffeomorphism of a compact surface M with positive topological entropy. Then given any $\varepsilon > 0$, there exist $n \in \mathbf{Z}^+$ and a compact f^n-invariant set $\Lambda \subset M$ s.t.*

1. *$f^n|\Lambda$ is uniformly hyperbolic;*
2. *$f^n|\Lambda$ is topologically conjugate to a full shift on m symbols;*
3. $$\frac{1}{n} h_{\text{top}}(f^n|\Lambda) > h_{\text{top}}(f) - \varepsilon,$$

 where $h_{\text{top}}(f)$ denotes the topological entropy of f.

In fact, Katok showed that on a manifold of any dimension, if (f, μ) is nonuniformly hyperbolic, i.e., if all the exponents are different from zero, then most of the metric entropy is captured on uniformly hyperbolic sets.

D. Examples

EXAMPLE 1. Consider the 1-parameter family f_a: $[-1, 1]$ defined by

$$f_a(x) = 1 - ax^2.$$

We quote a well-known theorem of Yakobson's:

Theorem [J]. *There is a positive Lebesgue measure set Δ in parameter space with the property that for all $a \in \Delta$*

1. *f_a admits an invariant probability measure μ_a that is absolutely continuous with respect to Lebesgue;*
2. *the Lyapunov exponent of (f_a, μ_a) is positive.*

The existence of an absolutely continuous invariant probability implies, in fact, a positive exponent for certain classes of maps. (See [BL].) Since Jakobson published his result, other proofs (e.g., [BC1, Ry]) have appeared. I will give the version I am most familiar with. Here are the main steps:

Step I. Understand the construction of smooth invariant measures for uniformly expanding maps. This can be done using the Perron–Frobenius operator. Let m denote Lebesgue measure on the interval and let $f_* m$ be the measure defined by $f_* m(E) = m(f^{-1}E)$. Prove that the densities of $(f^n)_* m$ are uniformly bounded and obtain as an invariant measure an accumulation point of $(1/n) \sum (f^i)_* m$.

Step II. Fix a neighborhood V of the critical point 0. For a sufficiently near 2, f_a is essentially expanding outside of V. Fix a so that $f = f_a$ has the property that the orbit of 0 stays away from 0 for a very long time and is not allowed to approach 0 faster than a certain rate. Call this condition (*).

- Show with an argument similar to that in Example 1 of Section 1 that even though f is not expanding uniformly, its derivatives grow in a very controlled way.
- Argue further that this controlled expansion allows us to mimick the construction of invariant measures in the uniformly expanding case.

Step III. Prove that for a near 2,

$$\frac{d}{da} f_a^n(0) \qquad \text{and} \qquad \frac{d}{dx} f_a^{n-1}(f0)$$

are comparable as long as $(f_a^{n-1})'(f0)$ grows fast. This allows us to estimate the measure of the set of parameters a that fail to satisfy condition (*) above.

(This is a sketch of [BC1], a revised version of which appears in [BY1].)

EXAMPLE 2. Billiards

Let $\Omega \subset \mathbf{R}^2$ be a compact region with piecewise smooth boundaries. Consider the flow generated by the uniform motion of a point mass in Ω with elastic reflections at $\partial\Omega$. Let $S = \partial\Omega \times [0, \pi]$, and for $(x, \theta) \in S$, define $(x', \theta') = f(x, \theta)$ as follows: Let ℓ be the directed line segment starting at $x \in \partial\Omega$, pointing into the interior of Ω and making an angle of θ with $\partial\Omega$. Then x' is the first point of contact of ℓ with $\partial\Omega$, and if ℓ' is the reflected line segment at x', then θ' is the angle ℓ' makes with $\partial\Omega$.

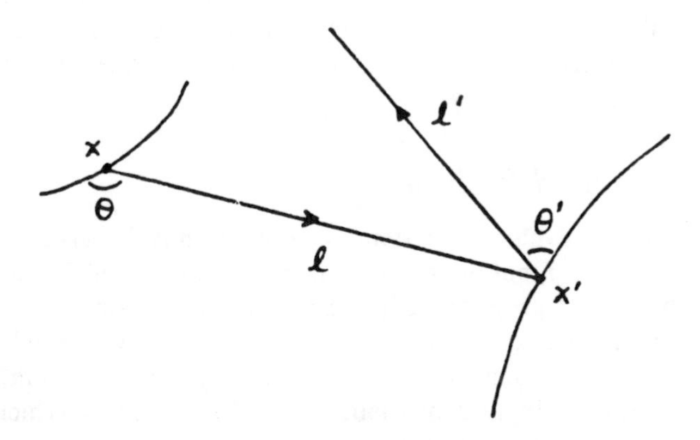

FIGURE 3

Let $\mu = \sin\theta \, dx \, d\theta$ normalized. It is easy to see that f preserves μ. One way of detecting positive Lyapunov exponents is to look for invariant cones. Here is how it is explained in [W1]:

Lemma. *Suppose f has a measurable family of strictly invariant cones, i.e., for a.e. (x, θ), there is a closed sector $C(x, \theta)$ in the tangent space at (x, θ), such that $Df_{(x,\theta)}C(x, \theta) \subset C(f(x, \theta))$ and $Df^{n(x,\theta)}_{(x,\theta)}$ takes $C(x, \theta)$ into the interior of $C(f^{n(x,\theta)}(x, \theta))$ for some $n(x, \theta) > 0$. Then f has a positive Lyapunov exponent a.e.*

The idea, then, is to look for geometric conditions on $\partial\Omega$ so that Df would preserve natural families of cones. Observe that a tangent direction at (x, θ) can be represented by a curve in S through (x, θ), which, in turn, can be thought of a smooth family of lines $\{\ell(s), |s| < \varepsilon\}$, with $\ell(0) = \ell_{(x,\theta)}$, the oriented line through x with direction θ. Because $\{\ell(s)\}$ focuses (in linear approximation) at some point in $\ell_{(x,\theta)} \cup \{\infty\}$, we can identify the projective coordinates at (x, θ) with $\ell_{(x,\theta)} \cup \{\infty\}$.

Suppose, for instance, that $\partial\Omega$ consists entirely of concave pieces and that $C(x, \theta)$ consists of families of diverging rays, i.e., rays that focus "behind" x in $\ell_{(x,\theta)}$. Then these cones are strictly invariant and we conclude immediately that there is a positive Lyapunov exponent a.e. The same cones work for Example 2 in Section 1. These are the "dispersing" and "semidispersing" billiards first studied by Sinai [Si2].

Billiards on domains with convex or focusing components were first studied in [Bu]. Conditions that guarantee invariant cones are presented in [W1].

We mention a few other examples of (partially) nonuniformly hyperbolic systems that have been successfully analyzed: (1) Systems with more degrees of freedom, such as [CS, KSS], Wojtkowski's bouncing balls [W2], etc. (2) Geodesic flows on manifolds of nonpositive curvature. If, for instance, all the sectional curvatures at one point are <0, then there is a set in the unit tangent bundle with positive Liouville measure on which the geodesic flow is nonuniformly hyperbolic. See [BB]. (3) Katok's surface diffeomorphisms (see [Ka1]).

E. An Unresolved Problem

Whether or not a dynamical system is nonuniformly hyperbolic depends not only on the map f, but on its invariant measure μ as well. When we start to discuss the ergodic properties of (f, μ), the dependence on μ becomes even more obvious. Dynamical systems, however, are usually "given" to us as maps or in terms of systems of equations, and they typically admit uncountably many mutually singular invariant probability measures. Which ones of these measures are more relevant than others?

In each on the examples mentioned in the last subsection, there is an invariant measure that is related to Lebesgue. For many purposes it is probably reasonable to think of this as the *natural* invariant measure, although, depending on one's motives, other invanant measures can be quite natural as well. Measures that maximize entropy, or equilibrium states for certain potentials, etc., all have their role in dynamical systems.

Going back to the view that Lebesgue measure has an elevated status—which seems to be reasonable from the point of view of physical observations—we are left with the following question: Consider, for instance, a volume-decreasing diffeomorphism with an attractor Λ. All invariant measures on Λ are necessarily singular with respect to Lebesgue. What role, if any, does Lebesgue measure play in the ergodic theory of these systems? An attempt to answer this question is made in Section 4.

3. Entropy, Lyapunov Exponents, and Hausdorff Dimension

A. Metric Entropy

This well-known invariant in ergodic theory is due to Kolmogorov and Sinai. It is difficult to give its complete definition and properties in this limited space, so we shall try only to explain its intuitive meaning here. For a more formal introduction to metric entropy, we refer the reader to an ergodic theory text.

Let T be a measure-preserving transformation of a probability space $(X, \mathscr{B}, m)$. In this subsection, α will always denote a finite partition of X. The metric entropy of T, $h_m(T)$, is defined to be the supremum over all α of the number

$$h_m(T, \alpha) := \lim_{n \to \infty} H\left(T^{-n}\alpha \,\middle|\, \bigvee_{i=0}^{n-1} T^{-i}\alpha \right).$$

This number has the following interpretation. Suppose that we are given the α-addresses of x, $Tx, \ldots,$ up to $T^{n-1}x$ for some very large n, and try to guess the α-address of $T^n x$ based on this information alone. The number $h_m(T, \alpha)$ measures the amount of uncertainty, averaged over all x, in making this prediction.

The theorem of Shannon–Breiman–McMillan gives another view of the same invariant. Assume that (T, m) is ergodic. Let α be a finite partition and write $h = h_m(T, \alpha)$. Then for all sufficiently large n, there is a subset X_n of X with $mX_n > 1 - \varepsilon$ such that X_n is made up of $\sim e^{nh}$ elements of $\bigvee_0^{n-1} T^{-i}\alpha$ each having measure $\sim e^{-nh}$. (Two points x and y are in the same element of $\bigvee_0^{n-1} T^{-i}\alpha$ if $T^i x$ and $T^i y$ have the same α-address for $i = 0, \ldots, n-1$.)

B. Statements of Results

We have seen from Section 2 that Lyapunov exponents, in particular positive Lyapunov exponents, give the rates at which nearby orbits diverge. They provide us with a *geometric* measurement of chaos. Metric entropy, on the other hand, is a purely *probabilistic* notion. It measures randomness in the sense of information and predictability. We compare these two invariants:

Theorem. *Let f be a C^2 diffeomorphism of a compact Riemannian manifold M, and let μ be an f-invariant Borel probability measure. We let $\{\lambda_i\}$ denote the*

Lyapunov exponents of (f, μ) *and* $\{E_i\}$ *the corresponding subspaces. Then*

$$h_\mu(f) \le \int \sum \lambda_i^+ \dim E_i \, d\mu,$$

where $a^+ = \max(a, 0)$. *Equality holds if and only if* μ *has absolutely continuous conditional measures on unstable manifolds.*

The inequality part of this theorem holds, in fact, for C^1 mappings that are not necessarily invertible. We define what it means for μ to have absolutely continuous conditional measures on unstable manifolds. Let A be a set with $\mu A > 0$, and suppose that there is a measurable partition η of A into open pieces of local unstable manifolds. Then, for every x, $\eta(x)$ is a Riemannian manifold and we let $m_\eta(x)$ denote its Riemannian measure. Disintegrate μ into a family of conditional probability measures $\{\mu_\eta(x), x \in A\}$ on elements of η. Our condition demands that for a.e.-x, $\mu_\eta(x)$ is absolutely continuous with respect to $m_\eta(x)$.

Many people contributed to this theorem in its present form. The *Axiom A* case is contained in the work of Sinai, Bowen, and Ruelle. See [A, Bo, Ru1, Si1, and Si2]. The next breakthrough came when these ideas were brought to the level of Lyapunov exponents. Pesin proved the entropy formula for arbitrary diffeomorphisms preserving smooth measures [Pe2], and Ruelle proved the inequality with no assumptions on the invariant measure [Ru2]. Margulis independently obtained a similar result. See also [M1]. In the context of nonuniformly hyperbolic systems, the characterization of measures for which equality holds is due largely to Ledrappier, following earlier ideas of Sinai. This characterization is later shown to be valid even in the presence of zero exponents. See [LS, Le1, and Part I of LY1].

In light of the above theorem, it seems reasonable to try to explain the "gap" in Ruelle's inequality in terms of geometric properties of the invariant measure. We make the following definitions:

Let X be a metric space, and let m be a Borel probability measure on X. For $x \in X$ and $\rho > 0$, we let $B(x, \rho)$ denote the ball of radius ρ centered at x. We say that the dimension of m, written $\dim(m)$, exists and is equal to α if for m-a.e. x,

$$mB(x, \rho) \sim \rho^\alpha$$

as $\rho \to 0$. The Hausdorff dimension of the measure m is defined to be

$$HD(m) = \inf_{\substack{Y \subset X \\ mY = 1}} HD(Y).$$

It is easy to see that if $\dim(m)$ exists, then it is equal to $HD(m)$.

Theorem (Part II of [LY1]). *All hypotheses are as in the last theorem. We assume in addition that* (f, μ) *is ergodic. Then*

$$h_\mu(f) = \sum \lambda_i^+ \sigma_i,$$

where σ_i satisfies $0 \le \sigma_i \le \dim E_i$ and is, in some sense, the dimension of μ in the directions of the subspaces corresponding to E_i.

We shall postpone the precise definition of σ_i to the next subsection, where the ideas behind a proof of this theorem are discussed. In the meantime, suffice it to say that because $0 \le \sigma_i \le \dim E_i$, this formula agrees with Ruelle's inequality in general, and that when μ is smooth on unstable manifolds, then $\sigma_i = \dim E_i$, and we recover the entropy formula in the previous theorem.

Given that entropy can be expressed in terms of Lyapunov exponents and dimension, one might ask if $HD(\mu)$ can be expressed in terms of the other two invariants. This, in fact, has some practical consequence, because the Hausdorff dimension of a set tells us effectively how many variables we need to describe it completely. (See [M2] for a precise statement.) For each $\lambda_i > 0$, let σ_i be defined as in the last theorem (or rather, in the next subsection). For $\lambda_i < 0$, we can define σ_i similarly by considering f^{-1}. Suppose now that there are no zero exponents. It would be reasonable to expect that $HD(\mu) = \dim(\mu) = \sum \sigma_i$, where the sum is taken over all i. So far, this has only been proved when M has dimension 2 [Y1] or when μ has absolutely continuous conditional measures on unstable manifolds [Le2]. In the latter case Yorke et al. have conjectured that, "typically," $HD(\mu)$ is equal to a quantity they call "Lyapunov dimension" [FKYY].

C. Sketch of Proof of Dimension Formula

Let us first consider the case when all the exponents of (f, μ) are identically equal to some $\lambda > 0$. (Think of f as not being injective, otherwise this forces μ to live on a repelling periodic orbit.) Let $B(x, \varepsilon; n) = \{y \in M : d(f^k x, f^k y) < \varepsilon, \forall k \le n\}$. Pretending that locally f is a dilation by e^λ, we have

$$B(x, \varepsilon; n) \sim B(x, \varepsilon e^{-\lambda n}).$$

Estimates of this type are easily carried out using Lyapunov charts (see Section 2C) and are not far from being accurate. Using a result in [BK], which is a modification of the Shannon–Breiman–McMillan theorem, we also have

$$\mu B(x, \varepsilon; n) \sim e^{-nh},$$

where $h = h_\mu(f)$. Comparing these two expressions and letting $\rho = e^{-\lambda n}$, we have

$$\mu B(x, \rho) \sim \rho^{h/\lambda},$$

and so $\dim(\mu)$ exists and is equal to h/λ.

Note that if more than one positive exponent is present, then the "eccentricity" of $B(x, \varepsilon; n)$ approaches ∞ as n gets large. $\dim(\mu)$, as with Hausdorff dimension, cannot be estimated using only sets that are unboundedly eccentric, and so the argument in the last paragraph fails.

To handle this more general situation, let $\lambda_1 > \cdots > \lambda_u$ be the positive exponents of (f, μ). For each i, let $W^i(x)$ denote the unstable manifold of f at x corresponding to $\bigoplus_{j \le i} E_j(x)$. We then have a nested family of "foliations"

$$W_1 \subset W_2 \subset \cdots \subset W^u.$$

We introduce a notion of entropy along W^i, written h_i, measuring the randomness of f along the leaves of W^i and ignoring what happens in transverse directions. We say that the conditional measures of μ on W^i have a well-defined dimension, called δ_i, if for every measurable set made up of local W^i-leaves, the dimensions of the conditional measures exist and are equal a.e. to δ_i. The following is proved in [LY1]:

(i) $h_1 = \delta_1 \lambda_1$;
(ii) for $i = 2, \ldots, u$, δ_i is well-defined and satisfies

$$h_i - h_{i-1} = (\delta_i - \delta_{i-1})\lambda_i;$$

(iii) $h_u = h_\mu(f)$.

The proof of (i) is the same as before because it only involves one exponent. Let us try to explain (ii). Consider the action of f on the leaves of W^i, quotient out by the leaves of W^{i-1}. This "quotient" dynamical system has exactly one Lyapunov exponent, namely, λ_i. It behaves as though it leaves invariant a measure with dimension $\delta_i - \delta_{i-1}$ and has entropy $h_i - h_{i-1}$. Once we make sense out of these statements, we can appeal again to the single-exponent argument to obtain (ii).

This is what is done in the real proof, modulo a technical mess. Letting $\sigma_1 = \delta_1$ and $\sigma_i = \delta_i - \delta_{i-1}$ for $i = 2, \ldots, u$, and summing the equations in (ii) over i, we obtain $h_u = \sum_{i=1}^{u} \sigma_i \lambda_i$. The last step is a statement on the non-involvement of zero exponents. With that, we are done.

4. Attractors

A. Definitions and Properties

One way of looking at the entropy inequality

$$h_\mu(f) \le \int \sum \lambda_i^+ \dim E_i \, d\mu \qquad (*)$$

is that randomness or uncertainty in a dynamical system is created by the exponential divergence of nearby orbits. In particular, entropy is zero when all the exponents are ≤ 0. One could also think that when $(*)$ is a strict inequality, then some of the expansion is "wasted" in the sense that there is some "leakage" from the system. This line of reasoning would lead us to associate measures having absolutely continuous conditional measures on unstable manifolds with either "attractors" or "conservative systems."

Definition. *Let (f, μ) be a dynamical system. We call μ a* Sinai–Bowen–Ruelle (*SBR*) measure *if there is a positive exponent μ-a.e. and μ has absolutely continuous conditional measures on unstable manifolds.*

Whereas the remarks in the first paragraph are largely heuristic, there are rigorous results for Axiom A systems. An Axiom A basic set Λ is called an *attractor* if there is an open neighborhood U of Λ s.t.

$$\Lambda = \bigcap_{n \geq 0} f^n U.$$

(Note that this includes the case where Λ is the entire manifold.) Define

$$P := \sup \left\{ h_\nu(f) - \int \sum \lambda_i^+ \dim E_i \, d\nu \right\},$$

where the supremum is taken over all f-invariant measures supported on Λ. For the rest of this Chapter, let m denote the Riemannian measure on our manifold M.

Theorem [Bo, Ru1, Si3]. *Let Λ be an Axiom A basic set for a C^2 diffeomorphism f.*

1. *If Λ is an attractor, then $f | \Lambda$ admits an SBR measure.*
2. *If Λ is not an attractor, then $P < 0$, and if U is a small neighborhood of Λ, then*

$$m\{x\colon x, fx, \ldots, f^n x \in U\} \sim e^{Pn}.$$

Next we mention some properties of SBR measures. The Dirac measure at $x \in M$ is written δ_x. If μ is a probability measure, we say that x is *generic* with respect to μ if

$$\frac{1}{n} \sum_{i=0}^{n-1} \delta_{f^i x} \to \mu \quad \text{(weak convergence)}.$$

Theorem [Pe2, Le1, etc]. *Let (f, μ) be a nonuniformly hyperbolic system on a manifold M, and assume that μ is an SBR measure.*

1. *Ergodic properties of (f, μ). There is at most a countable number of sets Λ_1, $\Lambda_2, \ldots$ with $\mu\Lambda_i > 0$ and $\mu(\bigcup \Lambda_i) = 1$ s.t. for each i, $(f | \Lambda_i, \mu | \Lambda_i)$ is ergodic, and $\mu | \Lambda_i$ is SBR. Furthermore, each Λ_i can be decomposed into $\Lambda_i = \Lambda_i^1 \cup \cdots \cup \Lambda_i^{n_i}$ s.t. $f \Lambda_i^j = \Lambda_i^{j+1}$, $f \Lambda_i^{n_i} = \Lambda_i^1$, and $(f^{n_i} | \Lambda_i^1, \mu_i | \Lambda_i^1)$ is measure theoretically isomorphic to a Bernoulli shift.*
2. *Assume that (f, μ) is ergodic. Then there is a Borel set $X \subset M$ with $m(X) > 0$ such that every $x \in X$ is generic wrt μ.*

Property 2 is what gives SBR measures their unique place of importance among all invariant measures. It says that they can be observed physically. (See Section 2E.) This property is, in fact, an easy consequence of another important property called the *absolute continuity of the stable foliation*, which we now describe. Let Σ_1 and Σ_2 be transversals to a family of local stable disks $\{D^u_\alpha\}$, with each D^u_α intersecting both Σ_1 and Σ_2. For $i = 1, 2$, let m_i denote the Riemannian measure on Σ_i. Then

$$m_1\left(\Sigma_1 \cap \bigcup_\alpha D^u_\alpha\right) > 0 \quad \text{iff} \quad m_2\left(\Sigma_2 \cap \bigcup_\alpha D^u_\alpha\right) > 0.$$

(See, e.g., [KS, Pe2, or PuSh] for more details.)

PROOF of Property 2 in the theorem assuming the absolute continuity of W^s. It is easy to see that if a point x is generic wrt μ, then so is y for every y in $W^s(x)$. Now let L be a piece of unstable manifold and let L' be the set of points in L that are generic wrt μ. Because μ is SBR, we can choose L so that $m_L(L') > 0$, where m_L denotes the induced Riemannian measure on L. It follows from the absolute continuity of the stable foliation that

$$m\left(\bigcup_{x \in L'} W^s(x)\right) > 0,$$

and every point in this set has the desired property. ∎

The discussion in this section seems to suggest the following:

Definition. *We say that* (f, μ) *is a* nonuniformly hyperbolic attractor *if*

(i) (f, μ) *is nonuniformly hyperbolic and ergodic,*
(ii) μ *is an SBR measure.*

B. Examples of Nonuniformly Hyperbolic Attractors

EXAMPLE 1. Axiom A attractors.

Let f be a C^2 diffeomorphism and let Λ be an Axiom A attractor. Here is one way of constructing an SBR measure on Λ.

Let L be a piece of local unstable manifold, and let m_L denote the normalized Riemannian measure on L. We transport m_L forward by f^n and let μ be an accumulation point of $(1/n) \sum (f^i)_* m_L$. To see that μ has the SBR property, let $\{D^u_\alpha\}$ be a continuous family of local unstable disks of size δ. If $f^n L$ contains one of these disks, say D^u_α, and ρ_n is the density of $f^n_* m_L$ with respect to the Riemannian measure on D^u_α, then for x and $y \in D^u_\alpha$,

$$\frac{\rho_n(x)}{\rho_n(y)} = \prod_{i=1}^{n} \left| \frac{\det Df \mid E^u(f^{-i}y)}{\det Df \mid E^u(f^{-i}x)} \right|.$$

A standard distortion estimate for C^2 expanding maps tells us that this quotient is bounded above and below by two constants that depend only on f and δ, and not on D^u_α or on n. These uniform bounds on the densities of $f^n_* m_L$ are passed on to μ.

To see that μ is ergodic, we first use the absolute continuity property of the stable foliation to show that for a.e. ergodic component μ_e of μ, the set of μ_e-generic points is, modulo a set of m-measure zero, open in M. The fact that $f|\Lambda$ has a dense orbit is then used to show that there can only be one such ergodic component.

Because the basin of the attractor U is equal to

$$\bigcup_{x \in \Lambda} W^s(x),$$

we have, in fact, shown that m-a.e. x in U is generic wrt μ.

EXAMPLE 2. Partially uniformly hyperbolic and piecewise hyperbolic attractors.
It is clear from our construction in the last paragraph that if all we want is an SBR measure—with no requirements on uniqueness—then all that is needed is that $f|\Lambda$ has a continuous expanding subbundle. See, e.g., [PeSi]. In fact, the construction above, with some modification, works for maps that have mild singularities but are otherwise uniformly hyperbolic. See, e.g., [CL] and [Y3] for the ergodic properties of Lozi attractors.

EXAMPLE 3. Systems with smooth invariant measures.
If (f, μ) has positive exponents and μ is absolutely continuous wrt m, then μ is SBR. This applies, for instance, to the examples in Section 2D.

EXAMPLE 4. The Henon mappings.
This family of mappings of the plane is given by

$$T_{a,b}\colon \begin{pmatrix} x \\ y \end{pmatrix} \mapsto \begin{pmatrix} 1 - ax^2 + y \\ bx \end{pmatrix}.$$

Numerical studies show that for parameter values near $a = 1.4$ and $b = 0.3$, the mapping $T_{a,b}$ has what appears to be an attractor with chaotic dynamics. The fact that randomly picked trajectories seem to have the same asymptotic distribution lends credence to the existence of an SBR measure. I know of no rigorous results supporting or disproving these conjectures.

In a different parameter range, Benedicks and Carleson recently developed some machinery for analyzing the dynamics of certain Henon maps that are small perturbations of one-dimensional maps. The following theorem relies heavily on their machinery.

Theorem [BC2, BY2]. *There is a set $\Delta \subset \mathbf{R}^2$ with* Leb$(\Delta) > 0$ *s.t. for all $(a, b) \in \Delta$, the following is true for $T = T_{a,b}$:*

1. *There is a compact invariant set $\Lambda = \Lambda_{a,b}$ and an open neighborhood U of Λ s.t. for every $z \in U$, $T^n z \to \Lambda$ as $n \to \infty$;*
2. *$T|\Lambda$ admits a unique SBR measure; moreover,* supp $\mu = \Lambda$ *and (T, μ) is Bernoulli.*

Let us first state some easy facts. For all $a < 2$ and ≈ 2, $b > 0$ and small, $T = T_{a,b}$ has a unique fixed point $\hat{z} = \hat{z}_{a,b}$ in the first quadrant. This fixed point is hyperbolic; its expanding direction is roughly horizontal and the eigenvalues of $DT_{\hat{z}}$ are ≈ -2 and $b/2$. Let W denote the global unstable manifold at $\hat{z}$, and let Λ be the closure of W. It is not hard to see that given $a_0 < a_1 < 2$, there exists $b_0 > 0$ s.t. for all $(a,b) \in [a_0, a_1] \times (0, b_0]$, Λ is a compact set that attracts all points in some neighborhood. One checks easily also that for $\delta \gg b$, all invariant subsets of Λ that are outside of $(-\delta, \delta) \times \mathbf{R}$ are uniformly hyperbolic, but that the angles between the stable and unstable directions deteriorate as δ gets small. The attracting set Λ is definitely *not* an Axiom A attractor.

For each sufficiently small b, Benedicks and Carleson inductively constructed a positive measure set of a's for which $T_{a,b}$ has certain nice properties. Their induction is far too involved to be discussed here, but we will try to indicate what the picture looks like for their chosen parameters. For the rest of this section, let $T = T_{a,b}$ where (a,b) is a pair of "good" parameters.

The idea in [BC2] is to model the analysis of $T|W$ on that of a certain class of one-dimensional maps. First, we describe the geometry of W. Let W_1 be the maximal piece of $W^u_{\text{loc}}(\hat{z})$ that is roughly horizontal. Remembering that $a \approx 2$ and b is very small, it is easy to believe that W_1 stretches from approximately $(-1, 0)$ to $(1, 0)$, and that for small $n > 0$, $T^n W_1$ contains 2^n roughly horizontal pieces much like W_1, joined together by sharp turns near $(\pm 1, 0)$. Because Λ is not a full horseshoe, this picture is not maintained as n gets large: Some of the horizontal pieces are less than full length; they turn around at random places.

The role of the critical point in one dimension is played by a subset $\mathscr{C}$ of W called the *critical set*. As a first approximation, we can think of $\mathscr{C}$ as $W \cap \{x = 0\}$, although $\mathscr{C}$ does not really lie on any smooth curve. Intuitively, we want to say that a point $z \in W$ is in $\mathscr{C}$ if it lies in a flat piece and is mapped into a "turn"—but it is not so easy to define a "turn" mathematically. The characterization of $\mathscr{C}$ used in [BC2] is a dynamical one. Let u denote a unit tangent vector to W. Then $z \in \mathscr{C}$ iff $|DT^n_z u| \leq (\text{const} \cdot b)^n$, $\forall n \geq 0$. This is about as fast as any vector can shrink under DT.

Through careful parameter selection, our "good" map T has the property that all $z \in \mathscr{C}$ are uniformly hyperbolic in their future iterates, and the orbits of $\mathscr{C}$ do not approach $\mathscr{C}$ faster than a certain rate. These conditions are reminiscent of condition (*) in Example 1 of Section 2. Indeed, one proves that for arbitrary $z \in W$, $|DT^n_z u|$ behaves qualitatively very much like $|(f^n)'x|$ for $x \in [-1, 1]$ and f satisfying (*). That is to say, $DT^n_z u$ grows exponentially fast, but this growth is not uniform because $DT|W$ contracts sharply for a number of iterates every time the orbit of z comes near a "turn" of W. The estimates here are considerably more delicate because they are two dimensional in nature: we have to keep track of *angles* as well as lengths.

With this machinery in place, we now carry out the construction of SBR measures as is done in Example 1 of this subsection. Using the notation there, we take $L \subset W$ and push m_L forward. Because the expansion along W is not uniform, in order to get uniform estimates at the nth stage, we must abandon part of $T_*^n m_L$. In fact, it is sufficient to prove that the forward images of a positive percentage of m_L have uniformly bounded densities. The uniqueness and, hence, ergodicity of our SBR measure follows from the fact that $\overline{W} = \Lambda$.

5. Random Perturbations and Stability

A. Some Notions of Stability

One of the successes of the program begun in [Sm] is the introduction of the notion of *structural stability* and the complete characterization of dynamical systems that are structurally stable. Recall that a diffeomorphism $f\colon M \circlearrowleft$ is said to be C^r-structurally stable if every g that is C^r-near f is equal to f up to a continuous change of coordinates, i.e., $g = hfh^{-1}$, where h is a conjugating homeomorphism.

Theorem. *A diffeomorphism f of a compact Riemannian manifold is C^1-structurally stable iff f is Axiom A and $W^u(x)$ meets $W^s(y)$ transversally for every x, y in the nonwandering set.*

The "if" part of this theorem is proved by Robbin [Robb] and Robinson [Robi], the converse by Mañé [M3]. There is also the notion of Ω-stability, where we require only that the structure of the nonwandering set remain unchanged when f is perturbed. For the corresponding theorem, see [Pa].

Consider now a nonuniformly hyperbolic attractor (f, μ) with supp $\mu = \Lambda$. According to the theorems of Mañé and Palis, if $f|\Lambda$ is not Axiom A, then there is no chance that the orbit structure of f on Λ can be stable in the sense described above. This leads one to think that perhaps structural stability is too much to ask for in this nonuniform context. A weaker notion of stability might be that small perturbations do not drastically change the asymptotic distribution of typical orbits. We formulate something along these lines.

Let U be an open set in M, and let f be a map with $fU \subset U$. Our random perturbations of f will be given by a Markov chain with state space U. For each x, let $P(\cdot|x)$ be a probability measure on U. Think of $P(\cdot|x)$ as giving the distribution of the next iterate if one starts at x. Consider the Markov chain $\mathscr{X}$ with $\{P(\cdot|x)\}$ as transition probabilities. A probability measure μ is called an invariant measure for $\mathscr{X}$ if for every Borel subset E of $\overline{U}$,

$$\mu(E) = \int P(E|x)\, d\mu.$$

Invariant measures always exist if $x \mapsto P(\cdot|x)$ is continuous and U is bounded. We are particularly interested in the case where, for every x, $P(\cdot|x)$ is absolutely continuous wrt m (written $\ll m$). In this case, all invariant measures of $\mathscr{X}$ are also $\ll m$. We say that the 1-parameter family of Markov chains $\{\mathscr{X}^\varepsilon, \varepsilon > 0\}$ converges to f if, for every $x \in U$, $P^\varepsilon(\cdot|x)$ converges weakly to δ_{fx} as $\varepsilon \to 0$ and this convergence is uniform in x. If μ^ε is an invariant measure for $\mathscr{X}^\varepsilon$, then it is easy to see that any accumulation point of μ^ε as $\varepsilon \to 0$ is an invariant measure for the map f.

Definition. We say that the dynamical system (f, μ) is stochastically stable if there is an open set U containing the support of μ on which the following hold.

1. $f\overline{U} \subset U$;
2. if $\{\mathscr{X}^\varepsilon, \varepsilon > 0\}$ is a family of Markov chains on U satisfying certain regularity conditions and converging to f, and if μ^ε is an invariant measure for $\mathscr{X}^\varepsilon$, then $\mu^\varepsilon \to \mu$ as $\varepsilon \to 0$.

First, we explain what we mean by "regularity conditions." For continuous-time dynamical systems, the only reasonable models are probably those given by differential equations plus some Brownian motion terms. In the discrete-time case, it is less clear what reasonable assumptions are. At a minimum, we probably want to assume that every $P^\varepsilon(\cdot|x)$ is $\ll m$, and that the densities of $P^\varepsilon(\cdot|x)$ should not become increasingly wild as $\varepsilon \to 0$. If, for instance, $U \subset \mathbf{R}^n$ and $P^\varepsilon(\cdot|x)$ is the uniform distribution on the ε-disk centered at fx, then $\{\mathscr{X}^\varepsilon\}$ is definitely "regular." In general, we might want to assume that the densities of $P^\varepsilon(\cdot|x)$ are not too far from some fixed function scaled appropriately.

Suppose now that (f, μ) is a nonuniformly hyperbolic attractor, and that m-a.e. x in a neighborhood of the support of μ is generic wrt μ. Suppose also that (f, μ) is stochastically stable. Under some mild assumptions of transitivity, μ^ε is unique and ergodic. By the ergodic theorem for Markov chains, if $x_0, x_1, \ldots$ is a typical trajectory of $\mathscr{X}^\varepsilon$, then $(1/n)\sum \delta_{x_i} \to \mu^\varepsilon$. What this all means is that for small $\varepsilon > 0$,

$$\frac{1}{n}\sum_{i=0}^{n-1} \delta_{x_i} \sim \frac{1}{n}\sum_{i=0}^{n-1} \delta_{f^i x},$$

where $x_0, x_1, \ldots$ is a typical orbit of $\mathscr{X}^\varepsilon$ and x is an m-typical point. Thus, small random perturbations of f do not disturb very much the asymptotic distribution of typical orbits.

As far as I know, this notion of stochastic stability goes back to Kolmogorov—although we may have formulated things a little differently. The theory for flows with simple nonwandering sets was developed in [VF]. The first results for chaotic dynamical systems that I know of are in [Ki1] and [Si3]. [Ki3] is a good reference for what is discussed here.

B. Examples of Stable Systems

If μ lives on an attracting set Λ that admits no other invariant measure, then clearly (f, μ) is stochastically stable. This is the case when Λ is an attractive periodic orbit, or the Feigenbaum attractor, etc. Aside from these, most of the systems proved to be stable have μ as their SBR measures.

Kifer first gave a detailed proof of the stochastic stability of Axiom A attractors [Ki1, Ki2]. For a different approach using random maps, see [Y1]. Let us indicate why this is true assuming, say, that $P^\varepsilon(\cdot|x)$ is the uniform distribution on the ε-disk centered at fx. For $\varepsilon > 0$, consider the operator $\mathscr{P}^\varepsilon$ on the space of probability measures on U defined by

$$\mathscr{P}^\varepsilon(\nu)(E) = \int P^\varepsilon(E|x)\, d\nu.$$

Fixed points of $\mathscr{P}^\varepsilon$ are precisely the invariant measures of $\mathscr{X}^\varepsilon$. In this case, there is only one. As before, let m_L be the normalized Riemannian measure on an unstable leaf. Since $\mathscr{P}^\varepsilon m_L$ is essentially $f_* m_L$ convolved with the uniform distribution on an ε-disk, it is easy to see that $\mathscr{P}^\varepsilon m_L$ has bounded densities in unstable directions—meaning that if we take a foliation with leaves roughly parallel to W^u and disintegrate $\mathscr{P}^\varepsilon m_L$ on its leaves, then we get nice bounded densities. The same is true of $(\mathscr{P}^\varepsilon)^k m_L$ for all k because at each step, $\mathscr{P}^\varepsilon$ stretches in the unstable directions and then averages. These uniform bounds are passed on to $\mu^\varepsilon = \lim[(1/n) \sum (\mathscr{P}^\varepsilon)^k m_L]$. We conclude that any accumulation point of μ^ε as $\varepsilon \to 0$ must also have absolutely continuous conditional measures on unstable manifolds.

The argument in the last paragraph tells us that as long as $f|\Lambda$ has a continuous or somehow recognizable expanding subbundle, then any accumulation point of μ^ε as $\varepsilon \to 0$ is necessarily an SBR measure. This applies to partially uniformly hyperbolic attractors, or piecewise hyperbolic ones. It also proves the stochastic stability of ergodic (but not necessarily hyperbolic) toral automorphisms.

As a final example, let us go back to the quadratic family $f_a\colon x \mapsto 1 - ax^2$ on $[-1, 1]$. As the parameter a varies from 0 to 2, the family $\{f_a\}$ goes through many different dynamic types. Let $\mathscr{A}$ consist of those parameters a for which f_a is Axiom A, i.e., the nonwandering set of f_a is made up of a uniformly expanding periodic orbit or cantor set together with a hyperbolic periodic sink. These maps are structurally as well as stochastically stable, with almost every trajectory going to a sink. The parameter set $\mathscr{A}$ is open and believed (though not yet proven) to be dense.[(1)] Now let $\mathscr{B}$ be the set of parameters satisfying condition $(*)$ in our proof of Jakobson's theorem (see Section 2D). Then $\mathscr{B}$ has positive Lebesgue measure, and for $a \in \mathscr{B}$, f_a has an invariant probability measure μ_a that is $\ll$Lebesgue. Moreover, (f_a, μ_a) has a

---

[(1)] Added in proof. The density of $\mathscr{A}$ has recently been proved by G. Swiatek.

positive exponent. This system has been shown to be stochastically stable [BY1, C]. See also [KK].

Perhaps it is worth noting that here we have two very different kinds of dynamical systems living in close proximity to each other. Yet both kinds of dynamics are robust, and both are observed numerically.

C. Stability of Lyapunov Exponents

So far, we have focused on the stability of asymptotic distributions. One could ask the more general question of how noise affects other dynamical invariants as well. Consider, for instance, the case of Lyapunov exponents.

Let $f: X \circlearrowleft$ be a self-map of some space preserving a probability measure μ, and let $A: X \to GL(d, \mathbf{R})$. It is easy to cause the exponents of $(f, \mu; A)$ to jump by tempering slightly with A. Suppose, for instance, that A has one positive and one negative exponent, and the angles between the expanding and contracting directions are not bounded away from zero. The following precedure will ruin the exponents at x. Allow $v \in E^u(x)$ to grow for a while. Suppose that at time n, $\angle(E^u(f^n x), E^s(f^n x))$ is very small. Rotate $Df_x^n v$ slightly so that the resulting vector is in $E^s(f^n x)$ and will start to shrink. Allow the vectors in $E^u(f^n x)$ to grow for some time. Then repeat the same operation, and so on. This argument can be made precise. In fact, Mañé used fancier methods based on the same idea to show that if f is an area-preserving surface diffeomorphism that is not Anosov, then it can be C^1-perturbed into another diffeomorphism all of whose exponents are zero a.e. (This work is unpublished.)

The perturbations discussed in the last paragraph are *deterministic* or *deliberate* perturbations. Let us consider now the effects of *random* noise. Here is one of the situations we can handle.

Let f be a volume-preserving diffeomorphism of a torus $\mathbf{T}$, and let $\lambda_1 \geq \cdots \geq \lambda_d$ denote its integrated Lyapunov exponents with respect to the Riemannian measure m. For $\varepsilon > 0$, we define a Markov chain $\mathscr{X}^\varepsilon$ as follows. Let p_ε be a probability on $\mathbf{T} = \mathbf{R}^d/\mathbf{Z}^d$ with p_ε (the ε-disk at 0) $> 1 - \varepsilon$, and let $P^\varepsilon(E|x) = p_\varepsilon(E - fx)$. Then m is also invariant under $\mathscr{X}^\varepsilon$. (We need this.) Let $x_0, x_1, \ldots$ be a typical trajectory of $\mathscr{X}^\varepsilon$. For each x_i, we let ν_{x_i} be a probability measure on $GL(d, \mathbf{R})$ that is not far from the uniform distribution on a ε-disk centered at Df_{x_i}, and pick $A_0, A_1, \ldots$ independently with laws $\nu_{x_0}, \nu_{x_1}, \ldots$, respectively. Lyapunov exponents are then computed using $\{A_n\}_{n=0,1,\ldots}$. Let $\lambda_1^\varepsilon \geq \cdots \geq \lambda_d^\varepsilon$ be the expected values of the exponents of this ε-perturbed process.

Theorem [LY2]. *$\lambda_i^\varepsilon \to \lambda_i$ as $\varepsilon \to 0$, $i = 1, \ldots, d$.*

To summarize, if f is not uniformly hyperbolic and we wish to cause its exponents to jump, it can always be done by arbitrarily small perturbations. In this sense, Lyapunov exponents are highly unstable. However, small errors that are genuinely random and of the type described above are not likely to drastically affect the accuracy of our computations.

A Final Note

Parts of Sections 4 and 5 contain material that is still being developed at the writing of this Chapter. Some of the ideas have not yet been carefully thought out, and many questions remain. We mention some obvious ones.

We have tried to stress the importance of SBR measures and to define a notion of nonuniformly hyperbolic attractors. But *how prevalent are these objects*? Physicists tend to take them for granted; numerical studies rely on and to some extent verify their existence. Yet there are very few non-Axiom A dissipative systems that have been *proved* to have SBR measures. To me, this is one of more pressing questions in this subject.

Concerning the notion of stochastic stability, all that we really have at the present time is the vague notion that random noise with its averaging properties should be relatively harmless, together with a handful of examples which seem to suggest the stability of SBR measures and Lyapunov exponents. We do not know in what generality these statements are true, or what types of conditions will guarantee stability.

Incomplete and imperfect as it is in its present state, the theory of nonuniformly hyperbolic systems probably has potential applications. In the hope of bringing it to researchers in other areas, I have tried to communicate these ideas in as nontechnical a fashion as I know how. This, of course, has already been done in many books and review articles. See, e.g., [ER] for a list of references.

Acknowledgments. Part of this Chapter was written while I was visiting MSRI. I wish also to thank F. Ledrappier for reading over this manuscript.

References

[A] Anosov, D., Geodesic flows on closed Riemann manifolds with negative curvature, *Proc. Steklov Inst. Math.* No. 90 1967. Amer. Math. Soc. 1969 (translation). pp. 1–235.

[BB] Ballman, W. and Brin, M., On the ergodicity of geodesin flows, *Ergod. Theory Dynam. Syst.* **2** (1982), 311–315.

[BC1] Benedicks, M. and Carleson, L., On iterations of 1-ax^2 on $(-1, 1)$, *Ann. Math.* **122** (1985), 1–25.

[BC2] ——— and ——— The dynamics of the Henon map, Ann. Math. **133** (1991), 73–169.

[BY1] ——— and Young, L.-S., Absolutely continuous invariant measures and random perturbations for certain one-dimensional maps, Erg. Th. & Dynam. Sys. **12** (1992), 13–38.

[BY2] ——— and ———, SBR measures for certain Henon maps, to appear in Inventiones (1993).

[BK] Brin, M. and Katok, A., On local entropy, *Geometric Dynamics*, Springer Lecture Notes in Mathematics, No. 1007, Springer-Verlag, Berlin, 1983, pp. 30–38.

[BL] Blokh, A.M. and Lyubich, M. Yu., Measurable dynamics of S-unimodal maps of the interval, preprint, 1990.

[Bo] Bowen, R., *Equilibrium States and the Ergodic Theory of Anosov Diffeomorphisms*, Springer Lectures Notes in Mathematics No. 470, Springer-Verlag, Berlin, 1975.

[Bu] Bunimovich, L.A., On the ergodic properties of nowhere dispersing billiards, *Commun. Math. Phys.* **65** (1979), 295–312.

[C] Collet, P., Ergodic properties of some unimodal mappings of the interval, preprint, 1984.

[CL] ——— and Levy, Y., Ergodic properties of the Lozi mappings, *Commun. Math. Phys.* **93** (1984), 461–482.

[CS] Chernov, N.I. and Sinai Ya.G., Ergodic properties of some systems of 2-dimensional discs and 3-dimensional spheres, *Russ. Math. Surveys* **42** (1987), 181–207.

[ER] Eckmann, J.-P. and Ruelle, D., Ergodic theory of chaos and strange attractors, *Rev. Mod. Phys.* **57** (1985), 617–656.

[FHY] Fathi, A., Herman, M., and Yoccoz, J.-C., A proof of Pesin's stable manifold theorem, Springer Lecture Notes in Mathematics. No. 1007, Springer-Verlag, Berlin, 1983, pp. 117–215.

[FK] Furstenberg, H. and Kifer, Yu., Random matrix products and measures on projective spaces, *Israel J. Math.* **46** (1-2) (1983), 12–32.

[FKYY] Frederickson, P., Kaplan, K.L., Yorke, E.D. and Yorke, J.A., The Lyapunov dimension of strange attractors, J. Diff. Eq. **49** (1983) 185–207.

[GH] Guckenheimer, J. and Holmes, P., *Nonlinear Oscillations, Dynamical Systems, and Bifurcations of Vector Fields*, Springer-Verlag, New York, 1983.

[J] Jakobson, M., Absolutely continuous invariant measures for one-parameter families of one-dimensional maps, *Commun. Math. Phys.* **81** (1981), 39–88.

[Ka1] Katok, A., Bernoulli diffeomorphisms on surfaces, *Ann. Math.* **110** (1979), 529–547.

[Ka2] ———, Lyapunov exponents, entropy and periodic orbits for diffeomorphisms, *Publ. Math. IHES* **51** (1980), 137–174.

[KK] ——— and Kifer, Yu., Random perturbations of transformations of an interval, *J. Analy. Math.* **47** (1986), 194–237.

[KS] ——— and Strelcyn, J.M., *Invariant Manifolds, Entropy and Billiards; Smooth Maps with Singularities*, Springer Lecture Notes in Mathematics No. 1222, Springer-Verlag, Berlin, 1986.

[Ki1] Kifer, Yu., On small random perturbations of some smooth dynamical systems, *Math. USSR Ivestjia* **8** (1974), 1083–1107.

[Ki2] ———, General random perturbations of hyperbolic and expanding transformations, *J. Anal. Math.* **47** (1986), 111–150.

[Ki3] ———, *Random Perturbations of Dynamical Systems*, Birkhauser, Basel, 1986.

[KSS] Kramli, A., Simanyi, N., and Szasz, D., The K-property of three billiard balls, *Ann. Math.* **133** (1991), 37–72.

[Le1] Ledrappier, F., Preprietes ergodiques des mesures de Sinai, *Publ. Math. IHES* **59** (1984), 163–188.

[Le2] ———, Dimension of invariant measures, *Teubner-Texte zur Math.* **94** (1987), 116–124.

[LS] ——— and Strelcyn, J.-M., A proof of the estimation from below in Pesin entropy formula, *Ergod. Theory Dynam. Syst.* **2** (1982), 203–219.

[LY1] ——— and Young, L.-S., The metric entropy of diffeomorphisms, *Ann. Math.* **122** (1985), 509–574.

[LY2] ——— and ———, Stability of Lyapunov exponents, Ergod. Th. & Dynam. Sys. **11** (1991), 469–484.

[Lo] Lorenz, E., Deterministic nonperiodic flow, *J. Atmos. Sci.* **20** (1963), 130–141.

[M1] Mãné, R., A proof of Pesin's formula, *Ergod. Theory Dynam. Syst.* **1** (1981), 95–102.

[M2] ———, On the dimension of compact invariant sets of certain non-linear maps, Springer Lecture Notes in Mathematics No. 898, Springer-Verlag, Berlin, 1981, pp. 230–241 (see erratum).

[M3] ———, A proof of the C^1-stability conjecture, *Publ. Math. IHES* **66** (1988), 160–210.

[N] Newhouse, S., *Lectures on Dynamical Systems*, Birkhauser, Basel, (1980), 1–114.

[O] Oseledec, V.I., A multiplicative ergodic theorem: Liapunov characteristic numbers for dynamical systems, *Trans. Moscow Math. Soc.* **19** (1968), 197–231.

[Pa] Palis, J. On the C^1-stability conjecture, *Publ. Math. IHES* **66** (1988), 211–215.

[Pe1] Pesin, Ya.B., Families if invariant manifolds corresponding to non-zero characteristic exponents, *Math. USSR Izvestjia* **10** (1978), 1261–1305.

[Pe2] ———, Characteristic Lyapunov exponents and smooth ergodic theory, *Russ. Math. Surveys* **32** (1977), 55–114.

[PeSi] ——— and Sinai, Ya.G., Gibbs measures for partially hyperbolic attractors, *Ergod. Theory Dynam. Syst.* **2** (1982), 417–438.

[PuSh] Pugh, C. and Shub, M., Ergodic attractors, *Trans. AMS* **312** (1) (1989), 1–54.

[Robb] Robbin, J., A structural stability theorem, *Ann. Math.* **94** (1971), 447–493.

[Robi] Robinson, C., Structural stability of C^1 diffeomorphisms, *J. Diff. Eqs.* **22** (1976), 28–73.

[Ru1] Ruelle, D., A measure associated with *Axiom A* attractors, *Amer. J. Math.* **98** (1976), 619–654.

[Ru2] ———, An inequality of the entropy of differentiable maps, *Bol. Sc. Bra. Mat.* **9** (1978), 83–87.

[Ru3] ———, Ergodic theory of differentiable systems, *Publ. Math. IHES* **50** (1979), 27–58.

[Ry] Rychlik, M., A proof of Jakobson's theorem, *Ergod. Theory Dynam. Syst.* **8** (1988), 93–109.

[Si1] Sinai, Ya.G., Markov partitions and C-diffeomorphisms, *Funct. Anal. Appl.* **2** (1968), 64–89.

[Si2] ———, Dynamical systems with elastic reflections: ergodic properties of dispersing billiards, *Russ. Math. Surveys* **25** (2) (1970), 137–189.

[Si3] ———, Gibbs measures in ergodic theory, *Russ. Math. Surveys* **27** (4) (1972), 21–69.

[Si4] ——— (ed.), *Dynamicals Systems II*, Springer-Verlag, Berlin, 1989.

[Sm] Smale, S., Differentiable dynamical systems, *Bull. Amer. Math. Soc.* **73** (1967), 747–817.

[VF] Ventsel, A.D. and Freidlin, M.I., Small random perturbations of dynamical systems, *Russ. Math. Surveys* **25** (1970), 1–55.

[W1] Wojtkowski, M., Principles for the design of billiards with nonvanishing Lyapunov exponents, *Commun. Math. Phys.* **105** (1986), 391–414.

[W2] ———, A system of one-domensional balls with gravity, *Commun. Math. Phys.* **126** (1990), 507–533.

[Y1] Young, L.-S., Dimension, entropy and Lyapunov exponents, *Ergod. Theory Dynam. Syst.* **2** (1982), 109–129.

[Y2] ———, Stochastic stability of hyperbolic attractors, *Ergod. Theory Dynam. Syst.* **6** (1986), 311–319.

[Y3] ———, Bowen-Ruelle measures for certain piecewise hyperbolic maps, Trans. AMS, **287**, No. 1 (1985), 41–48.

22
Beyond Hyperbolicity: Expansion Properties of One-Dimensional Mappings

JOHN GUCKENHEIMER and STEWART JOHNSON

Smale's definition of Axiom A [26] was a brilliant insight into the structure of dynamical systems that led to the geometric characterization of structurally stable diffeomorphisms. "Real-world" dynamical systems are often found with attractors that are structurally unstable, and these have been described as belonging to the "dark realm" [22]. The iterations of one-dimensional mappings provide one route to illuminating a small corner of the dark realm, and much has been discovered about these iterations during the past 15 years. The keys to understanding the dynamics of these transformations lies in

1. rigid topological structure described by kneading theory,
2. distortion properties of iterates on intervals of monotonicity,
3. properties of expansion that are determined by growth rates of the derivative along trajectories of critical points.

Large sets of these transformations have positive Lyapunov exponents and behave in many ways as if they satisfied Axiom A despite the presence of critical points, but the presence of positive Lyapunov exponents is not a structurally stable phenomenon. This Chapter describes the growth rate properties of derivatives along trajectories for unimodal transformations with negative Schwarzian derivative.

The Schwarzian derivative Sf of a function $f: I \to I$ is defined by the formula

$$Sf(x) = \frac{f'''(x)}{f'(x)} - \frac{3}{2}\left(\frac{f''(x)}{f'(x)}\right)^2.$$

The Schwarzian derivative can be interpreted geometrically as the infinitesimal distortion of cross-ratios or as a measure of the infinitesimal deviation of a function from being a fractional linear transformation. A direct calculation shows that the class of functions whose Schwarzian derivative is negative everywhere is closed under composition.

Definition. Let $I \subset R$ be an interval. The function $f: I \to I$ is an S-unimodal map if it has the following properties:

1. f has a single critical point c that is nondegenerate.
2. f is C^3 and its Schwarzian derivative is negative at all points except c.
3. $f(\partial I) \subset \partial I$ and the fixed point in ∂I is expanding.

The class of S-unimodal maps is clearly special, but many of the properties that we describe for these maps have been extended to far more general classes of maps with little change in the underlying geometric insight. Working in the class of S-unimodal maps eliminates some technical considerations that complicate the statements of results and the proofs of theorems. General procedures for extending the results to C^2 maps have been described by de Melo and van Strien [17], Sullivan [27], and by Swiatek [28].

One-parameter families of unimodal maps display remarkable features of geometric rigidity that have been described as universal. The topological types of S-unimodal maps are all displayed in one-parameter families. Indeed, every S-unimodal map is topologically equivalent to a quadratic map of the form $f(x) = a - x^2$. The topological equivalence classes can be divided into three types: those with stable periodic orbits, those that are infinitely renormalizable, and those that have sensitive dependence to initial conditions.

Definition. The unimodal transformation f is renormalizable if there is an integer n and a neighborhood J of its critical point such that the intervals $f^i(J)$ are disjoint for $i < n$, $f^n(J) \subset J$ and $f^n(\partial J) \subset \partial J$. Then $f^n|_J$ is a unimodal map and is called a renormalization of f.

The maps with stable periodic orbits satisfy Axiom A and are well understood. The maps that are infinitely renormalizable have nonwandering sets that consist of an infinite number of hyperbolic basic sets and an attracting Cantor set of Lebesgue measure zero on which the transformation acts as a homeomorphism. Much of the metric structure of these Cantor sets is determined by the topological equivalence class of the transformation [27].

The geometric structure of transformations that have sensitive dependence to initial conditions is less fully understood than that of the first two types of transformations. For some of these transformations, the trajectory of the critical point remains a finite distance from the critical point. These transformations of Misiurewicz type also behave very much like ones that satisfy Axiom A. They have induced transformations which are expanding, positive Lyapunov exponent for the trajectory of the critical value and absolutely continuous invariant measures [18]. The set of parameter values giving transformations of Misiurewicz type apparently has Lebesgue measure zero and is much smaller than the full set of parameter values giving transformations that have sensitive dependence to initial conditions in typical one-parameter families. Yakobson [10] proved that the set of parameter values giving rise to transformations with absolutely continuous invariant measures has positive measure, but Johnson [12] and Keller [15] have given examples of transformations which have sensitive dependence to initial conditions but no finite absolutely continuous invariant measure.

There are several approaches to the identification of conditions that guarantee the existence of absolutely continuous invariant measures. One approach to the construction of invariant measures is to look at the forward images $f_*^n\lambda$ of Lebesgue measure λ and their averages:

$$\frac{1}{n}\sum_{i=0}^{n-1} f_*^i\lambda$$

or at the limits of Birkhoff sums

$$\frac{1}{n}\sum_{i=0}^{n-1} \delta_{f^i(x)}$$

for typical $x \in I$. In "good" cases, the sequence of forward images of Lebesgue measure converges and the limit is the unique absolutely continuous invariant measure of a transformation. These take (at least) the following forms:

1. Invariant measures for expanding, induced transformations are constructed and pulled back to invariant measures for the transformation f [10, 11, 23].
2. If $|f^n(c) - c| > \exp(-\sqrt{n})$, then there is an absolutely continuous invariant measure [2].
3. If the rate of growth of $(f^n)'(f(c))$ is exponential, then there is an absolutely continuous invariant measure [5, 18].
4. Keller gives a criterion for the existence of absolutely continuous invariant measures that is expressed in terms of the metric structure of the forward trajectory of the critical point, but does not explicitly involve the number of iterates associated to each point [16].

These conditions may well define distinct classes of transformations. Condition 2 gives necessary conditions for the existence of absolutely continuous invariant measures. Blokh and Lyubich [4] state that transformations with absolutely continuous invariant measures have positive entropy and Ledrappier showed that transformations with absolutely continuous invariant measures of positive entropy satisfy Pesin's formula that

$$h(\mu) = \int |f'|\, d\mu.$$

Combined with the ergodic theorem, this yields a positive Lyapunov exponent for f.

To obtain a comprehensive understanding of unimodal transformations, we need more information about unimodal transformations that are finitely renormalizable, but have no absolutely continuous invariant measure. These are the transformations that are understood least well from a measure theoretic point of view. They inhabit remote corners of the dark realm. The topological attractors in these cases are finite union of intervals, but the metric attractors may not be. If the metric and topological attractors do not coincide, then the omega limit set of typical points is a nonhomeomorphic Cantor set with interesting topological properties. These Cantor sets are called *absorbing* Cantor sets and are further explored in [8, 13]. The existence of absorbing Cantor sets is doubtful for S-unimodal maps [28], but non-smooth examples are readily constructed. No smooth examples are known.*

If the metric attractor coincides with the topological attractor, then this finite union of intervals may support a finite absolutely continuous invariant measure, an infinite but σ-finite continuous invariant measure, or will support no continuous invariant measure. The latter case is called type III, and no S-unimodal examples are known. Katznelson [14] gave an example of a C^∞ circle homeomorphism of type III. Using similar techniques and a recent construction of a type III example for an exact endomorphism [24] it may be possible to construct a type III example for an S-unimodal map. This remains an open question.

A key to understanding the metric properties of these maps is the growth along the trajectory of the critical point. We consider this growth in terms of induced maps.

Definition. An *induced map* of f is a map g with $g(x) = f^{k(x)}(x)$ for some integer valued function $k: \mathbf{I} \to \mathbf{Z}$. The function $k(x)$ is called the *stopping rule of* g.

Induced maps are useful as a tool for studying properties of f because the itineraries of points for induced maps of f are subsets of the itineraries for f. Thus, one can think of an induced map g as a map which is related to f by a change of time scale which depends upon the point being studied.

Our first observation is the following proposition:

Proposition. *If* $\gamma: Z^+ \to R^+$ *satisfies* $\gamma(n) \to \infty$ *as* $n \to \infty$, *then there is an* S-*unimodal map* f *that has no stable periodic orbit, is not renormalizable, and has the property that* $|(f^n)'(f(c))| < \gamma(n)$ *for* n *sufficiently large.*

Given a neighborhood U of the critical point c, the set of points whose trajectory avoids U forms a subshift of finite type. Thus, examples of transformations with a slow growth rate for the derivative of the trajectory of the critical value must have returns of the critical point to a neighborhood that occur frequently and close relative to the number of iterations. One circumstance in which such close returns occur repeatedly is given by a transforma-

* Since this paper was written, Lyubich has constructed a proof that absorbing Cantor sets do not exist (M. Lyubich, (1992), Combinatories, Geometry and Attractors of Quasi-quadratic Maps, SUNY Stony Brook Institute for Mathematical Sciences Preprint 1992/18).

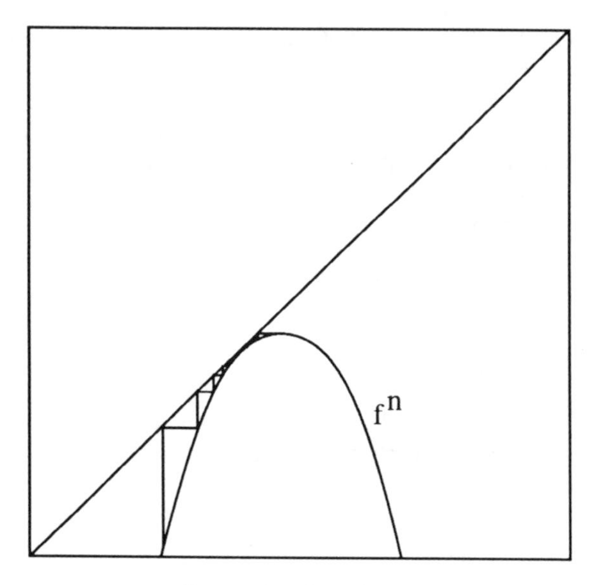

FIGURE 1. A funnel for f.

tion that is close to one having an indifferent period orbit of period n. In this situation, f^n near the critical point is a parabolic-like transformation whose graph has a section that is almost tangent to the identity. Trajectories of f^n starting near the critical point then spend many iterates passing through the narrow funnel between the graph of f^n and the graph of the identity function. See Fig. 1. During the passage through the funnel, there is no net growth of the derivative along the trajectory. This is the basic idea underlying the proof of the proposition.

To be more specific, consider an S-unimodal transformation f close to one that has an indifferent periodic orbit of period 3, but that has no periodic orbit of period 3. Let e be the minimum of $f^3(x) - x$ in the interval about the critical point where there is nearly a fixed point. If e is small, then f^3 can be approximated near this point by transformation of the form $-e+(x-x_0)+a(x-x_0)^2+b(x-x_0)^3$. As $e \to 0$, the number of iterates that it takes for points to transit a fixed neighborhood of x_0 grows without bound. On the other hand, if $f^{3n}(x_0+\varepsilon) \approx x_0 - \varepsilon$, the derivative of f^{3n} is bounded independent of n. Thus, one can find a transformation in which the derivatives of the first $3n$ iterates are bounded independent of n as they proceed through the "funnel" near x_0. Thus, it is clear that being trapped in small funnels for long periods of time will destroy any previous growth. By forcing the trajectory of the critical point into a sequence of smaller and more restrictive funnels, we can create as slow a growth rate as we wish. For this, it must be shown that we can trap the critical point in a funnel for sufficiency long to reduce any growth that may occur after escaping from that funnel and before being trapped in the next smaller funnel. The key observation to this and similar arguments is that for close returns of the critical point, the dynamics near the critical point change dramatically for small perturbations in the parameter value, whereas the dynamics away from the critical are stable.

To see how this works more clearly, suppose that $T_{\lambda_0} = f_{\lambda_0}^n$ has an indifferent fixed point at x_0. WIthout loss of generality, we assume $x_0 < c$ and $T_{\lambda_0}''(c) < 0$, and we assume the fixed point at x_0 disappears for $\lambda < \lambda_0$. For any $\varepsilon > 0$, we may find points

$$x_0 - \varepsilon < a < b < x_0$$

such that

$$T_{\lambda_0}^k(a) < c < T_{\lambda_0}^k(b) \qquad \text{and} \qquad T_{\lambda_0}^{k'}|_{(a,b)} > 0$$

for some k. The existence of these points and their behavior for the first k iterates under T_{λ_0} is stable for small perturbations in λ. That is, for λ sufficiently close to λ_0,

$$x_0 - \varepsilon < a < b < x_0,$$

$T_\lambda^k(a) < c < T_\lambda^k(b)$, and $T_\lambda^{k'}|_{(a,b)} > 0$ and for $x \in (a, b)$

$$\max\{|T_\lambda'(x) - T_{\lambda_0}'(x)|, \ldots, |T_\lambda^{k'}(x) - T_{\lambda_0}^{k'}(x)|\} < \varepsilon.$$

Now, we can take $\lambda < \lambda_0$ arbitrarily close to λ_0 such that an iterate of c proceeds through the funnel and lands anywhere we want in (a, b). Thus, for any $\delta > 0$, there is a $\lambda < \lambda_0$ arbitrarily close to λ_0 and a k_λ such that

$$T_\lambda^{k_\lambda}(c) \in (a, b)$$

and

$$0 < c - T_\lambda^{k_\lambda + k}(c) < \delta.$$

It follows that the map $T_\lambda^{k_\lambda + k}$ has a new funnel, and because δ was chosen independently of k, the new funnel is as restrictive as we would like. We choose λ sufficiently close to λ_0 so as to reduce any growth that may be contributed along the trajectory $T^{k_\lambda}(c), \ldots, T^{k_\lambda + k - 1}(c)$.

The passage through funnels described above is the most severe obstruction to obtaining exponential growth of the derivatives along the trajectory of the critical value. There is a second, weaker obstruction that occurs when the critical point returns close to itself after n iterates with a distance that is comparable to $(f^{n-1})'(f(c))$. One then has an "almost restrictive" return in which there is a periodic point p of period n such that f^n maps the interval $(p, f^{-1}(f(p)))$ across itself (folding just once) with an image that is comparable to the length $(p, f^{-1}(f(p)))$. See Fig. 2. Assuming distortion bounds of the type proved in [8], the derivative $(f^n)'(p)$ will be bounded independent of n. If n is large, the derivatives of the trajectory of the critical value will grow slowly if they spend a long time in the vicinity of the orbit of p.

The metric structure of a unimodal map is determined by the locations of points with specific itineraries. Replacing a unimodal map with an induced map throws away some of the information in each itinerary, but does not

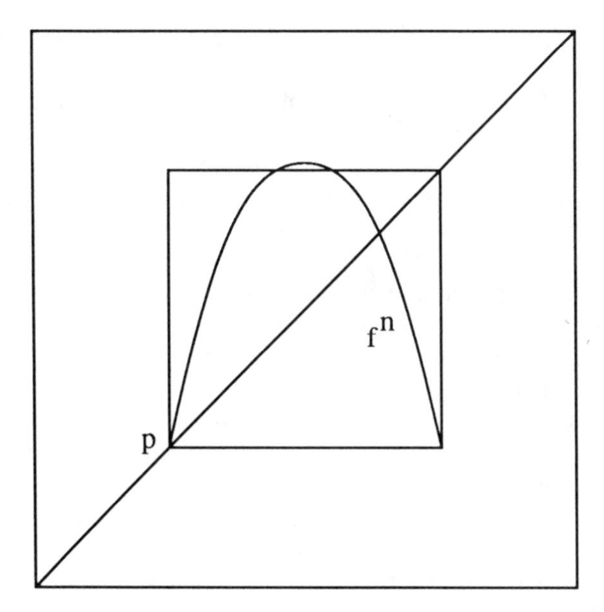

FIGURE 2. An almost restrictive return.

change the locations of a point associated to an itinerary. To recover some concept of expansion for a unimodal map f with sensitive dependence on initial conditions, one can seek to replace f with an induced map g that has better expansion properties but retains sufficient recurrence that absolutely continuous invariant measures can be constructed. If induced maps are constructed according to the stopping rules formulated in [8], then the weaker obstruction to expansion described above disappears. Specifically, in the class of S-unimodal maps studied in [8] that are not renormalizable, there is a universal lower bound larger than one to the derivative along any periodic orbit. Funnels still occur in the induced maps studied there, however. They can be eliminated by changing the stopping rules associated with the construction of induced maps. The following theorem can be proved by using this analysis.

Theorem. *The attracting Cantor set of an S-unimodal transformation that is infinitely renormalizable has Hausdorff dimension smaller than one.*

We assume that an attracting Cantor set is constructed as $\bigcap_{p=1}^{\infty} \bigcup_{i=n}^{n_p} f^i(J_p)$, where $f^{n_p}|_{J_p}$ for $p = 1, 2, \ldots$ is a sequence of renormalizations of f. We remark that if the return time from one renormalization to the next is bounded, i.e., n_{p+1}/n_p is bounded, then the result is known [24]. The proof in the general case requires that one determine metric information about the intervals $f^i(J_p)$ for $i = 1, 2, \ldots, n_p$. As return patterns are examined with increasing p, there are two possibilities for the behavior of the intervals $f^i(J_p)$: either they pass through funnels or their lengths are exponentially decreasing in n_p with a universal rate of decrease. This latter possibility is sufficient to bound the Hausdorf dimension less that 1. If the interval J_p does pass through a funnel, then one can calculate that its length decreases at a rate proportional to $1/k^2$ going into the funnel and increases at a similar rate coming out of the funnel, where k is the number of iterates required to pass through the funnel. As long as there is a definite proportion of each interval $f^i(J_p)$ that is not contained in $\bigcup_{i=1}^{p_{n+1}} f^i(J_{p+1})$, then one can cover $\bigcup_{i=1}^{p_n} f^i(J_p)$ with intervals of length l_k where $\sum l_k < r < 1$ (independent of p, n_p) and l_k decreasing either exponentially or as fast as $1/k^2$. This is sufficient to bound the Hausdorff dimension of the resulting Cantor set less than 1.

One of the fundamental outstanding problems for unimodal maps of the interval involves quantitatively estimating the way in which the dynamics change with changes in the parameters. The kneading theory [17] describes in topological terms an ordering of topological equivalence classes, but it gives no indication as to how abundant (in measure) different types of topological equivalences are or whether topological equivalence classes are traversed in a monotone manner. Myberg [21] suggested a direct approach to the monotonicity question through the use of chain rule formulas that express the rate of change of the iterates of the critical point with respect to the parameter. In all proofs of Yakobson's Theorem, these formulas are used to relate the changes in the iterates of the critical point to the amount of expansion occurring along the trajectory of the critical value. The leading term of the chain rule formula dominates in this situation. In other circumstances, useful estimates based upon the chain rule formula have not been found.

More subtle approaches based upon extending real mappings to quasiconformal maps of the complex plane and using quasiconformal deformation theory have yielded a number of results about the dependence of unimodal maps upon parameters [17, 24]. The quasiconformal theory motivates questions about the regularity of the conjugacies among unimodal maps. In particular, a plausible conjecture is that the conjugacies between S-unimodal maps are quasisymmetric [24].* Conjugacies between unimodal maps depend only on the labeling of points by their symbolic sequences, so induced maps can be used in studying the properties of conjugacies without further attention to the number of iterates used to produce each branch of the induced map. We make the following conjecture:

Conjecture. *The topological conjugacy h between two topologically equivalent S-unimodal maps f and g is quasisymmetric if and only if there is a constant C such that for every pair of periodic points x and y of period n with the same itineraries for f and g, respectively, the following formula holds*

---

* Combined with recent results of Yoccoz, Swiątek has announced a proof of this conjecture for quadratic maps (G. Swiątek, Hyperbolicity is Dense in the Real Quadratic Family, SUNY Stony Brook Institute for Mathematical Sciences, Preprint 1992/10).

$$\frac{1}{C} < \frac{\log|Df^n(x)|}{\log|Dg^n(y)|} < C.$$

It is readily proved that this formula is a necessary condition for quasi-symmetric conjugacy. Note also that the ratio of derivatives in the formula is unchanged if one replaces n by a multiple of n. Our hope is that further analysis of the derivatives along the trajectory of the critical value for unimodal maps can be used with our earlier estimates of distortion to prove that topologically equivalent S-unimodal maps satisfy the inequality in this conjecture.

The theory of Axiom A diffeomorphisms and flows is a striking story from the pages of mathematical history in the past 30 years. Smale's insight into the relationship between the theory of structural stability and geometric conditions of hyperbolicity of invariant sets is truly remarkable. As dynamical systems, theory has sought to understand the behavior of systems beyond those that are hyperbolic; the theory of iterations of one-dimensional mappings has provided a context within which a rather comprehensive theory has developed. This story is not complete, but it has already illuminated some parts of the dark realm that is populated by structurally unstable dynamical systems.

Acknowledgments. This research was partially supported by the National Science Foundation and the Department of Energy. We thank the Institute for Advanced Study for their hospitality during the academic year 1988–89 when the ideas underlying this Chapter were formulated.

References

1. M. Benedicks and L. Carleson (1983), On iterations of $1 - ax^2$ on $(-1, 1)$, *Ann. Math.* **122**, 1–25.
2. M. Benedicks and Lai-Sang Young (1990), Absolutely continuous invariant measures and random perturbations for certain one-dimensional maps, preprint.
3. A.M. Blokh and M.Yu. Lyubich (1987), Attractors of transformations of an interval, *Funct. Anal. Appl.* **21**, 148–150.
4. A.M. Blokh and M.Yu. Lyubich (1989), Measure and dimension of solenoidal attractors of one dimensional dynamical systems, preprint.
5. P. Collet and J.P. Eckmann (1981), Positive Lyapunov exponents and absolute continuity for maps of the interval, *Ergod. Theory Dynam. Syst.* **3**, 13–46.
6. J. Guckenheimer (1979), Sensitive dependence to initial conditions for one dimensional maps, *Commun. Math. Phys.* **70**, 133–16.
7. J. Guckenheimer (1987), Renormalization of one dimensional mappings and strange attractors, Lefschetz Centennial Conference, part III, 1984, A. Verjovsky, ed., *Contemp. Math.* **58**, 143–160.
8. J. Guckenheimer and S. Johnson (1990), Distortion of S-unimodal maps, *Ann. Math.* **132**, 71–130.

9. M. Jakobson (1980), *Construction of Invariant Measures Absolutely Continuous* with respect to dx for Some Maps of the Interval, Lecture Notes in Mathematics No. 819, Springer-Verlag, Berlin, pp. 246–257.
10. M. Jakobson (1981), Absolutely continuous invariant measures for interval maps, *Commun. Math. Phys.* **81**, 39–88.
11. S. Johnson (1989), Continuous measures in one dimension, *Commun. Math. Phys.* **122**, 293–320.
12. S. Johnson (1987), Singular measures without restrictive intervals, *Commun. Math. Phys.* **110**, 185–190.
13. S. Johnson (1990), Absorbing Cantor sets and trapping structures, preprint.
14. Y. Katznelson (1977), Smooth mappings of the circle without sigma-finite invariant measures, *Sympos. Math.* **22**, 363–369.
15. G. Keller (1987), Invariant measures and expansion along the critical orbit for S-unimodal maps, preprint.
16. G. Keller (1988), Quadratic maps without asymptotic measures, preprint.
17. W. de Melo and S. van Strien (1986), A structure theorem in one dimensional dynamics, *Ann. Math.* **129**, 519–546.
18. T. Nowicki (1985), Symmetric S-unimodal maps and positive Lyapunov exponents, *Ergod. Theory Dynam. Syst.* **5**, 611–616.
19. J. Milnor and W. Thurston (1988), *On Iterated Maps of the Interval*, Lecture Notes in Mathematics No. 1342, Springer-Verlag, Berlin.
20. M. Misiurewicz (1981), Absolutely continuous invariant measures for certain maps of an interval, *Publ. Math. IHES* **53**, 17–51.
21. P.J. Myrberg (1963), Iteration der reelen polynome zweiten grades III, *Annales Acad. Sci. Fennicae*, **336** (33), 1–18.
22. J. Palis (1990), Lecture at Smale Symposium, Berkeley, California. Chapter 17, this volume.
23. M. Rychlik (1988), Another proof of Jakobson's theorem and related results, *Ergod. Theory and Dynam. Syst.* **8**, 93–110.
24. C. Silva (1990), private communication.
25. D. Singer (1978), Stable orbits and bifurcations of maps of the interval, *SIAM J. Appl. Math.* **35**, 260–267.
26. S. Smale (1967), Differentiable dynamical systems, *Bull. Am. Math. Soc.* **73**, 747–817.
27. D. Sullivan (1990), On the structure of infinitely many dynamical systems nested inside or outside a given one, Centennial volume, American Mathematical Society, Providence, RI.
28. G. Swiatek (1990), Bounded distortion properties of one-dimensional maps, preprint.
29. M. Yakobson (1990), private communication.

23
Induced Hyperbolicity, Invariant Measures, and Rigidity

M. Jakobson*

1. The study of nonhyperbolic attractors is one of the interesting topics which originated from the theory of hyperbolic systems. Several results of the seventies lie in its background.

When studying one-parameter families of dissipative C^r-diffeomorphisms, $r \geq 2, f_t \colon M^2 \to M^2$, Newhouse [N] proved the persistence of homoclinic tangencies between stable and unstable manifolds of hyperbolic sets and the existence of a residual set of diffeomorphisms with infinitely many sinks. At the same time, Henon [H] discovered his famous farmly of attractors exhibiting chaotic behavior. The work of Henon and subsequent computer-assisted studies showed no trace of attracting periodic orbits for typical parameter values. That controversy has been recently clarified by Benedicks and Carleson [BC2]. They proved that for a positive measure set of parameter values in the Henon family, the corresponding diffeomorphisms have a unique chaotic attractor. The ergodic properties of Henon attractors seem to be similar to those of hyperbolic attractors carrying Bowen–Ruelle–Sinai measures.

The diffeomorphisms studied in [BC2] (see also [MV]) have strong contraction in one direction and could be considered as small perturbations of noninvertible one-dimensional maps.

So far, the global structure of Henon diffeomorphisms with the values of Jacobians not too close to zero (for example, close to the Henon values $b = 0.3$) has not been well understood. We can look at it as at a particular case of a more general problem: to find some reasonable sufficient conditions for the existence of global strange attractors in families of C^2-diffeomorphisms exhibiting homoclinic tangencies. A method of induced hyperbolicity similar to the method of induced expansion which we describe below could be useful for that purpose.

---

* Partially supported by NSF Grant DMS-9001631.

2. One-dimensional unimodal maps which initiated the recent study of Henon-like attractors still provide a lot of interesting questions. In the first work concerning abundance of stochastic behavior [J, BC1, R], a C^2-smoothness was essentially used and complicated estimates were involved. To satisfy those estimates, iterates of the critical point are required to stay far enough from that point. As a result, many maps remain out of consideration. So we still encounter the problem of a deeper understanding of the dynamics and its dependence on the parameter at least for some class of unimodal maps containing quadratic polynomials.

A natural class discovered by Singer [Si] consists of C^3-maps with negative Schwarzian derivative, or S-unimodal maps; see [M] for the main properties of such maps.

Here we formulate two recent results concerning S-unimodal maps. The first result [JSI] gives some sufficient conditions for such a map to be expansion-inducing. The second [JS2] is a kind of rigidity result. For two expansion-inducing S-unimodal maps, which are topologically conjugate, some sufficient conditions are given which imply that the conjugacy is quasisymmetric.

3. Let $f\colon[-1,1]\to[-1,1]$ be a nonrenormalizable S-unimodal map satisfying $f(-1)=f(1)=-1$, $Df(-1)>1$, $Df(0)=0$, $D^2f(0)\neq 0$. We call f *expansion-inducing* if there exists an interval $I=[-q,q]$ and a partition ξ of I into a disjoint union of intervals Δ_i such that

$$\operatorname{meas}(I\setminus\bigcup\Delta_i)=0$$

and the following properties I–III hold.

I. Each Δ_i is mapped onto I by some iterate f^{n_i} of the initial map f so that $f^{n_i}|\Delta_i$ is a diffeomorphism.

II. The "induced" map F defined by

$$F|\Delta_i=f^{n_i}|\Delta_i$$

is expanding, which means that there exist $n_0\in\mathbb{N}$ and $C_0>1$ such that

$$\left|\frac{dF^{n_0}}{dx}\right|>C_0.$$

III. The iterates $f^{n_i}|\Delta_i$ have uniformly bounded distortions

$$\sup_i\sup_{x_1,x_2\in\Delta_i}\left|\frac{df^{n_i}/dx(x_1)}{df^{n_i}/dx(x_2)}\right|<C_1.$$

The so-called "Folklore theorem" states that a C^2-map F satisfying I–III has an absolutely continuous invariant measure μ_F with strong mixing properties. If *summability condition*

$$\sum_i n_i|\Delta_i|<\infty$$

holds, then μ_F generates an f-invariant measure μ_f which also has strong mixing properties [L].

The *Koebe distortion property* of S-unimodal maps implies that the following is sufficient for expansion-inducing.

Extensibility Property. *There exists an interval J, such that* $\operatorname{int} J \supset I$ *and every* $f^{n_i}: \Delta_i \to I$ *can be extended as a diffeomorphism onto an interval* $\hat{\Delta}_i \supset \Delta_i$ *so that* f^{n_i} *maps* $\hat{\Delta}_i$ *onto* J.

To satisfy the extensibility property, it is enough to organize a kind of topological version of the *inducing construction* from [J].

Namely, we start by constructing an initial induced map $F_0|I$ which has several monotone extendable branches and a central parabolic branch defined in $\delta_0 \ni 0$. Then assuming that the critical value belongs to a domain of some monotone branch, we pull back with respect to the parabolic branch. As a result the former central domain, δ_0 is partitioned into domains of new monotone extendable branches, a new parabolic branch defined on a smaller central domain δ_1 and diffeomorphic preimages of δ_0. That can be written as

$$I = \left(\bigcup \Delta\right) \cup \delta_1 \cup \left(\bigcup \delta_0^{-1}\right),$$

where Δ is a common notation for domains of monotone extendable branches. As δ_0^{-1} are preimages of δ_0, we can pull back the partition of δ_0 onto δ_0^{-1} and obtain

$$\delta_0^{-1} = \delta_1^{-1} \cup \left(\bigcup \Delta\right) \cup \left(\bigcup \delta_0^{-2}\right).$$

Thus, we can represent I as

$$I = \left(\bigcup \Delta\right) \cup \delta_1 \cup \left(\bigcup \delta_1^{-1}\right) \cup \left(\bigcup \delta_0^{-2}\right).$$

By continuing the procedure of filling preimages δ_0^{-k}, we obtain

$$I = \left(\bigcup \Delta\right) \cup \delta_1 \cup \left(\bigcup \delta_1^{-1}\right) \cup \cdots \cup \left(\bigcup \delta_1^{-k}\right) \cup \left(\bigcup_{m>k} \delta_0^{-m}\right).$$

When k tends to infinity, the last term tends to

$$\bigcap_{n=1}^{\infty} \bigcup_{k=n}^{\infty} \delta_0^{-k}$$

which has measure zero. So we get a partition

$$\xi_1 : I = \left(\bigcup \Delta\right) \cup \delta_1 \cup \left(\bigcup_{k=1}^{\infty} \delta_1^{-k}\right) \cup T_1,$$

where δ_1 is the domain of parabolic branch h_1, δ_1^{-k} are diffeomorphic preimages of and meas $T_1 = 0$.

Then we proceed by induction in a similar way.

At every step of the construction, we check the position of the critical value $h_n(0)$. It can belong either to a domain Δ of some monotone branch, or to a preimage δ_n^{-k} of the central domain, or to a complement Cantor set of zero measure. That last possibility corresponds to "Misiurewicz case" when f has absolutely continuous measure [M]. We call the first possibility

$$h_n(0) \in \Delta$$

the *basic situation.* For the maps from [J], only the basic situation occurs and additionally $|h_n(0)|$ is required to be exponentially larger than the size of the domain of the parabolic branch.

For S-unimodal maps, we allow the critical value to approach zero as close as is consistent with the nonrenormalizable property of the initial map f.

Let us call f *Chebyshev-like* if f is topologically conjugate to $x \to 1 - 2x^2$. Using a more elaborate construction with no topological restrictions for the position of the critical value, we prove

Theorem I [JS1]. *For a given Chebyshev-like S-unimodal map f_0 there exists a C^3-neighborhood U, such that any nonrenormalizable $f \in U$ is expansion-inducing.*

By a theorem of Johnson [Jo], there exist nonrenormalizable quadratic maps f in any neighborhood of the Chebyshev polynomial which do not admit absolutely continuous invariant measure (see also [HK]). So we obtain

Corollary 1. *The maps f from Johnson's examples which are close enough to Chebyshev fail to satisfy the summability condition.*

Conjecture 1. *Every nonrenormalizable S-unimodal map is expansion-inducing.*

See also [GJ] for relevant discussion.

4. The recent results of Sullivan [S] and Yoccoz [Y] are the major steps toward a solution of the famous Fatou conjecture: the density of hyperbolic polynomials in the space of all polynomials of degree n for $n \geq 2$. These results prove the rigidity of certain classes of quadratic-like mappings. The following theorem can be also considered as a kind of rigidity result for S-unimodal maps.

Theorem 2 [JS2]. *Let f, f' be two topologically conjugate nonrenormalizable S-unimodal maps, and let φ: $[-1, 1] \to$ be the corresponding conjugacy. If at every step of inducing construction $h_n(0)$ belongs to a domain of a monotone branch ("basic situation"), then φ is a quasisymmetric homeomorphism.*

To prove Theorem 2, a sequence of partition ξ_n similar to ξ_1 is constructed such that ξ_n refine ξ_{n-1} and converge to the trivial partition into the points.

Every ξ_n is a partition into domains of some iterates of f. Let ξ'_n be the corresponding partition for f'. A sequence of piecewise smooth homeomorphisms φ_n which map elements of ξ_n into the corresponding elements of ξ'_n is constructed which converges to φ. The key problem is to prove uniform estimates for quasisymmetric distortions of φ_n. To do that, the whole construction is complexified and φ_n are extended up to quasiconformal homeomorphisms Φ_n of the plane. Then it is proved that quasiconformal distortions of Φ_n are uniformly bounded by a constant which depends only on the initial geometry of f and f'.

The maps from [J] satisfy the conditions of Theorem 2, but there are many more because "basic situation" is defined in purely topological terms and it does not include restrictions on the distances between the critical point and its iterates.

In fact, inducing construction from Section 3 works for arbitrary S-unimodal maps and could be useful for the study of rigidity in general case.

References

[BC1] M. Benedicks and L. Carleson, "On iterations of $1 - ax^2$," *Ann. Math.* **122** (1985), 1–25.

[BC2] M. Benedicks and L. Carleson, "The dynamics of the Henon map," preprint (1989).

[GJ] J. Guckenheimer and S. Johnson, "Distortion of S-unimodal maps," *Ann. Math.* **132** (1990), 71–130.

[H] M. Henon, "A two dimensional mapping with a strange attractor," *Commun. Math. Phys.* **50** (1976), 69–77.

[HK] F. Hofbauer and G. Keller, "Quadratic maps without asymptotic measure," preprint (1990).

[J] M. Jakobson, "Absolutely continuous invariant measures for one parameter families of one dimensional maps," *Commun. Math. Phys.* **81** (1981), 39–88.

[JS1] M. Jakobson and G. Swiatek, "Metric properties of non-renormalizable S-unimodal maps, I. Induced expansion and invariant measures," preprint (1990).

[JS2] M. Jakobson and G. Swiatek, "Metric properties of non-renormalizable S-unimodal maps, II. Quasisymmetric conjugacies," to appear.

[Jo] S. Johnson, "Singular measures without restrictive intervals," *Commun. Math. Phys.* **110** (1987), 185–190.

[L] F. Ledrappier, "Some properties of absolutely continuous invariant measures on an interval," *Ergod. Theory Dynam. Syst.* **1** (1981), 77–94.

[M] W. de Melo, *Lectures on one-dimensional dynamics*, IMPA, Rio de Janeiro (1989).

[Mi] M. Misiurewicz, "Absolutely continuous invariant measures for certain maps of an interval," *Publ. Math. IHES* **53** (1981), 17–51.

[MV] L. Mora and M. Viana, "Abundance of strange attractors," preprint (1990).

[N] S. Newhouse, "The abundance of wild hyperbolic sets and non-smooth stable sets for diffeomorphisms," *Publ. Math. IHES* **50** (1979), 101–152.

[R] M. Rychlik, "Another proof of Jakobson's theorem and related results," *Ergod. Theory Dynam. Syst.* **8** (1988), 93–109.

[S] D. Sullivan, "On the structure of infinitely many dynamical systems nested inside or outside a given one," preprint (1990).

[Si] D. Singer, "Stable orbits and bifurcations of maps of the interval," *SIAM J. Appl. Math.* **35** (1978), 260–267.

[Y] J.C. Yoccoz, "On local connectivity of the Mandelbrot set," to appear.

24
On the Enumerative Geometry of Geodesics

IVAN A.K. KUPKA and M.M. PEIXOTO

1. Introduction

The present work has its origin in a recent paper of Peixoto and Thom [7].

We begin by reformulating the Theorem of this paper (p. 639) in a way that renders its statement and proof more simple and conceptual. This is done by the use of the concept of *graph of a foliation*, introduced by Thom [9, p. 173] as follows: The graph of a foliated manifold F is the set of points $(p, q) \in F \times F$ such that p and q are on the same leaf.

Consider a second-order differential equation E, not necessarily autonomous, defined on a smooth manifold M of dimension m [3, p. 17]. Then E defines a foliation of codimension $2m$ on $\mathbf{R} \times TM$. Its graph $\Omega = \Omega(E)$ is a smooth closed submanifold of $(\mathbf{R} \times TM) \times (\mathbf{R} \times TM)$ with the same codimension $2m$ [10, p. 51]. The canonical projection $TM \to M$ induces a projection

$$\pi\colon (\mathbf{R} \times TM) \times (\mathbf{R} \times TM) \to (\mathbf{R} \times M) \times (\mathbf{R} \times M).$$

Let $\Sigma_i \subset (\mathbf{R} \times M) \times (\mathbf{R} \times M)$ be the set of all couples (t_1, x_1), (t_2, x_2), t_1, $t_2 \in \mathbf{R}$, $x_1, x_2 \in M$, such that the 2-point problem

$$x(t_1) = x_1, \qquad x(t_2) = x_2$$

for the equation E has exactly i solutions, $i = 0, 1, \ldots, \infty$. Let δ be the diagonal of $(\mathbf{R} \times M) \times (\mathbf{R} \times M)$ defined by $t_1 = t_2$.

Clearly, the sets $\Sigma_i - \delta$ are disjoint, i.e., they define a partition of $(\mathbf{R} \times M) \times (\mathbf{R} \times M) - \delta$. We propose the following version for the above-mentioned theorem.

Theorem 0 (Peixoto–Thom). *The graph $\Omega = \Omega(E)$ has the two properties*:

(a) $(t_1, x_1, t_2, x_2) \in \Sigma_i$ *if and only if* $(\pi/\Omega)^{-1}(t_1, x_1, t_2, x_2)$ *consists of i points,* $i = 0, 1, \ldots, \infty$.

(b) *If M and E are analytical and π/Ω is proper on* $(\mathbf{R} \times M) \times (\mathbf{R} \times M) - \delta$, *then there exists a Whitney stratification of* $(\mathbf{R} \times M) \times (\mathbf{R} \times M) - \delta$ *which is finer than the partition of* $(\mathbf{R} \times M) \times (\mathbf{R} \times M) - \delta$ *into the sets* $\Sigma_i - \delta$.

One sees then that the σ-decomposition of $B = (\mathbf{R} \times M) \times (\mathbf{R} \times M) - \delta$ introduces itself naturally in the description of the natural projection into B of the graph $\Omega = \Omega(E)$, itself a canonical object associated to the second-order equation E.

Although Theorem 0 covers a fairly general situation concerning the 2-point problem for a second-order differential equation E, it is natural also to put into this picture the variation of the equation E itself, within a convenient space. We will not pursue this direction here but only remark that for second-order differential equations on the line,

$$x'' = f(t, x, x'), \quad t, x, x' \in \mathbf{R},$$

the 2-point problem was considered, allowing a fairly general variation of f. Some kind of the Kupka–Smale theorem was then obtained by one of us [6]. This subject is related to the Dirichlet problem, as treated by Smale in [8].

Ending this digression, we consider from now on the special case where M is a complete Riemannian manifold, and as a second-order differential equation on it, we consider the geodesic flow.

The above considerations can be put on a more conceptual basis through the definition below.

Let $v \in TM$, $v \neq 0$, $p = \pi(v)$, where $\pi \colon TM \to M$ is the canonical projection. Call $I(v)$ the number of geodesics of length $\|v\|$ which connect p and $\exp_p(v)$ and $\Sigma_i = \{v \in TM | I(v) = i\}$, $i = 0, 1, 2, \ldots, \infty$. The zero-section $M \subset TM$ is said to belong to Σ_∞. Since M is complete, $\Sigma_0 = \phi$.

Definition. The σ-decomposition of TM is the partition of TM into the sets Σ_i, $i = 1, 2, \ldots, \infty$.

Clearly, this σ-decomposition of TM depends only on the metric of M. The enumerative point of view applied to geodesics leads to the formulation of the following:

Fundamental Problem. *For each Riemannian metric of M: to study the resulting σ-decomposition of TM into the sets Σ_i; in a more general way, to study the corresponding σ-decomposition of $\mathcal{R} \times TM$, $\mathcal{R}$ being the space of all Riemannian metrics on M, with a convenient topology.*

For a given metric, if $p \in M$ and $\sigma_i = \Sigma_i \cap T_pM$, the study of the decomposition of T_pM into the sets σ_i constitutes the *restricted problem with base point p*.

In this chapter, we show in Theorem 1 the existence of a Whitney stratification of TM which is finer than the partition of TM by the sets Σ_i. Similarly, on each fiber T_pM, $p \in M$, a Whitney stratification is obtained by refining its partition by the sets $\sigma_i = \Sigma_i \cap T_pM$.

The Angle Lemma gives an equiangular property satisfied by a family of geodesics of the same length connecting two points of M. This gives some insight on the relation between the integer i and the dimension of a stratum which is part of σ_i.

Finally, we study the σ-decomposition for flat tori. This turns out to be a surprisingly rich problem. In particular, it contains a natural generalization and geometrization of the problem of representation of integers by positive definite quadratic forms. For the form $x^2 + y^2$, the σ-decomposition exhibits also a close relationship with the Brillouin zones of Solid State Physics.

Figures 1 and 2 were made on a Sun 3/50 using a program written in C, by Stuart Cowan a graduate student at Berkeley. We thank him for that. The second author is grateful to the I.H.E.S. at Bures sur Yvette, France, for kind hospitality during the preparation of this work.

2. A Stratification Theorem

Let M be a complete analytical Riemannian manifold. We want to stratify TM by refining its partition by the sets Σ_i.

(2.1) **Theorem 1.** *There is a Whitney stratification of TM which is finer than the partition of TM by the sets Σ_i. Similarly, there is a Whitney stratification of T_pM, $p \in M$, finer than the partition of T_pM defined by the sets $\sigma_i = \Sigma_i \cap T_pM$.*

Proof. Let $\mathbf{R}_+ = \{r | r \geq 0\}$, $\pi\colon TM \to M$ be the canonical projection, $v \in TM$, $\pi(v) = p$, $q \in M$, $r \in \mathbf{R}_+$.

Consider the map

$$f\colon TM \to M \times M \times \mathbf{R}_+ \tag{2.2}$$

defined by

$$f(v) = (p, \exp_p v, \|v\|^2).$$

If we call $\Sigma_i' \subset M \times M \times \mathbf{R}_+$ the set of points (p, q, r) for which the cardinality of the fiber $f^{-1}(p, q, r)$ is $i > 0$, then, by the very definition of Σ_i,

$$\Sigma_i = f^{-1}(\Sigma_i'). \tag{2.3}$$

Observe that $\Sigma_0 = \phi$, whereas $\Sigma_0' \neq \phi$ and because of this we assume $i > 0$.

Now it is quite clear that f is both analytical and proper. From a general theorem [4, pp. 42, 43], it follows that this map can be stratified, i.e., one can find Whitney stratifications $\mathscr{A}$ of TM and $\mathscr{B}$ of $M \times M \times \mathbf{R}_+$ such that f maps submersively strata of $\mathscr{A}$ onto strata of $\mathscr{B}$ in such a way that over any stratum of $\mathscr{B}$ the fibers at any two points are homeomorphic. So the corresponding stratum of $\mathscr{A}$ belongs to the same Σ_i. Put another way, every Σ_i is the union of strata of $\mathscr{A}$ or equivalently the stratification $\mathscr{A}$ is obtained by refinement of the covering of TM by the sets Σ_i.

Consider now the case of a fixed base point. Similarly, for any point p in M there exists Whitney stratifications $\mathscr{A}_p$, $\mathscr{B}_p$ of T_pM and $M \times \mathbf{R}_+$, respectively such that:

$\mathscr{A}_p$ is a refinement of the partition of T_pM by the sets $\sigma_i = \Sigma_i \cap T_pM$, the mapping f_p: $T_pM \to M \times \mathbf{R}_+$,

$$f_p(v) = (\exp_p v, \|v\|^2), \tag{2.4}$$

is stratified with respect to $\mathscr{A}_p$ and $\mathscr{B}_p$. So we get a Whitney stratification of T_pM by refining its partition into the sets σ_i.

3. The Angle Lemma

As before, M is an analytical complete Riemannian manifold. In this section, we consider a fixed point $p \in M$ and a curve $q = q(\lambda)$: $\lambda \in \Lambda \to q(\lambda) \in M$, where Λ is some open interval of the line. We assume that the number of geodesics, going from p to $q(\lambda)$ and having length $\tau(\lambda)$, is constant for all $\lambda \in \Lambda$. We want to prove that under these circumstances all these geodesics make the same angle $\alpha(\lambda)$ with $q(\lambda)$.

This will not follow exactly from Theorem 1 but from its proof, in the case of a fixed base point p.

To make a precise statement, let $p \in M$ and Λ be an open interval. Consider a C^1-curve

$$(q, \tau): \Lambda \to M \times \mathbf{R}_+ \tag{3.1}$$

such that

$$\frac{dq(\lambda)}{d\lambda} \neq 0, \quad \lambda \in \Lambda, \tag{3.2}$$

(3.3) the number of geodesics of length $\tau(\lambda)$ joining p to $q(\lambda)$ *is finite* and the same for all $\lambda \in \Lambda$.

(3.4) any geodesic of length $\tau(\lambda)$ joining p to $q(\lambda)$ is the limit of geodesics of length $\tau(\mu)$, $\mu \in \Lambda$, joining p to $q(\mu)$ as $\mu \to \lambda$.

Assuming (3.1)–(3.4), we then have

(3.5) **Angle Lemma.** *The geodesics joining p to $q(\lambda)$ make the same angle $\alpha(\lambda)$ with the curve q at $q(\lambda)$ and*

$$\cos \alpha(\lambda) = \frac{1}{\|dq/d\lambda\|} \frac{d\tau(\lambda)}{d\lambda}. \tag{3.6}$$

PROOF. In the proof of Theorem 1 we saw that associated to the map

$$f_p: T_pM \to M \times \mathbf{R}_+$$

defined by (2.4) we have stratifications $\mathscr{A}_p$ of T_pM and $\mathscr{B}_p$ of $M \times \mathbf{R}_+$ with respect to which f_p is a stratified map and in such a way that every *stratum A of* $\mathscr{A}_p$ belongs to some σ_i and is mapped submersively by f_p onto some stratum B of $\mathscr{B}_p$.

Consider now the curve (q, τ) at a point $\lambda_0 \in \Lambda$ so that $(q(\lambda_0), \tau(\lambda_0)^2)$ belongs to some stratum B of the stratification $\mathscr{B}_p$ of $M \times \mathbf{R}_+$. If there is some open interval Δ, $\lambda_0 \in \Delta$, such that whenever $\lambda \in \Delta$, $(q(\lambda), \tau(\lambda)^2) \in B$, then we say that λ_0 is a *locally constant point* of Λ. Call $L \subset \Lambda$ the totality of all locally constant points.

We now show that L is open dense in Λ. To see this, first remark that the strata of any stratification, in particular of $\mathscr{B}_p$, are endowed with a natural partial order: If X and Y are strata, then $X > Y$ if and only if $\overline{X} \supset Y$.

If we put $\varphi(\lambda) = (q(\lambda), \tau(\lambda)^2)$ and identify $\varphi(\lambda)$ with the stratum to which it belongs, then there is a neighborhood Δ of λ such that $\varphi(\mu) > \varphi(\lambda)$. In other words, by a small variation of λ, $(q(\lambda), \tau(\lambda)^2)$ does not leave the star of the stratum where it was originally.

It is now easy to see that L is open dense in Λ. Take any open interval $I \subset \Lambda$ and consider the totality of strata

$$\{\varphi(\lambda) | \lambda \in I\}. \tag{3.7}$$

Because strata are partially ordered, by Zorn's lemma we choose, among those present in (3.7), a maximal chain. Because of the local finiteness character of strata, this chain has a maximum element $m(I)$. Therefore, the set

$$\{\mu \in I | \varphi(\mu) = m(I)\}$$

is open and nonempty in I. This shows that L is open dense in Λ.

We now prove (3.6) for any $\lambda_0 \in L$. From the definition of L, there is an open interval $\Delta \subset \Lambda$ such that $(q(\lambda), \tau(\lambda)^2)$ belongs to the same stratum $\varphi(\lambda_0)$ for all $\lambda \in \Delta$.

Consider now any geodesic γ of length $\tau(\lambda_0)$ joining p to $q(\lambda_0)$. Then there is a vector $v \in T_pM$ such that γ is the curve

$$\gamma: t \in [0, 1] \to \exp_p(tv), \qquad \|v\| = \tau(\lambda_0).$$

Now this vector v belongs to some stratum A of the stratification $\mathscr{A}_p$ of TM mentioned at the beginning, which is mapped submersively onto the stratum $\varphi(\lambda_0)$. From this and the very definition of exponential mapping, it follows that for a small $\hat{\Delta}$, $\lambda_0 \in \hat{\Delta} \subset \Delta$, there is a curve $\hat{v}: \hat{\Delta} \to T_pM$ such that $\hat{v}(\lambda_0) = v$, $f(\hat{v}(\lambda)) = (q(\lambda), \tau(\lambda)^2)$. In other terms

$$\exp_p(\hat{v}(\lambda)) = q(\lambda), \qquad \|\hat{v}(\lambda)\| = \tau(\lambda),$$

whenever $\lambda \in \hat{\Delta}$.

We now prove (3.6) for $\lambda \in \hat{\Delta}$. Let us denote by $w(\lambda)$ the tangent vector to the geodesic $\gamma(\lambda)$: $[0, 1] \to \exp(t\hat{v}(\lambda))$ at $t = 1$. Clearly,

$$\|w(\lambda)\|^2 = \|\hat{v}(\lambda)\|^2 = \langle \hat{v}(\lambda), \hat{v}(\lambda) \rangle = \tau(\lambda)^2, \tag{3.8}$$

$$\cos(\alpha(\lambda)) = \left\langle \frac{w(\lambda)}{\|w(\lambda)\|}, \frac{dq(\lambda)/d\lambda}{\|dq/d\lambda\|} \right\rangle. \tag{3.9}$$

But from Gauss' Lemma and (3.8), we have

$$\left\langle w(\lambda), \frac{dq(\lambda)}{d\lambda} \right\rangle = \left\langle \hat{v}(\lambda), \frac{d\hat{v}(\lambda)}{d\lambda} \right\rangle = \frac{1}{2}\frac{d\tau(\lambda)^2}{d\lambda}. \tag{3.10}$$

From (3.8)–(3.10), we finally get expression (3.6) for $\cos(\alpha(\lambda))$, proving (3.6) whenever $\lambda \in L$. Because L is open dense in Λ, from (3.4) and a passage to the limit we get (3.6) valid for $\lambda \in \Lambda$, as a consequence of assumption (3.4). This ends the proof of the Angle Lemma.

(3.11) *The Dimension of a Stratum in* σ_i

We now make an application of the Angle Lemma. We have the same situation as before; we are dealing with the stratification on T_pM, $p \in M$. Let s_i be one stratum contained in σ_i. To this stratum, we associate an integer $s = s(s_i)$ defined as follows. For any $q \in \exp_p(s_i)$, let $w_1(q), \ldots, w_i(q)$ be the i tangent vectors at q of the i geodesics joining p to q and whose length is the distance $d(p, \exp_p^{-1}(q))$ measured on T_pM. Let $W(q)$ be the subspace of T_qM generated by the differences $w_j(q) - w_k(q)$, $1 \leq j, k \leq i$. Call

$$s = s(s_i) = \sup\{\dim W(q) | q \in \exp_p(s_i)\}. \tag{3.12}$$

Clearly, $s \leq i - 1$ and we call it the *nondegenerate* case when $s = i - 1$.

From the Angle Lemma we derive now the following:

(3.13) **Corollary.** *If* s_i *is any stratum of* σ_i, *then*

$$\dim \exp_p(s_i) \leq \dim M - s,$$

where s *is given by* (3.12).

Proof. Let

$$q\colon (-\varepsilon, \varepsilon) \to \exp_p(s_i), \quad -\varepsilon < \lambda < \varepsilon, \tag{3.14}$$

be a C^1-curve such that

$$q(0) = q_0, \qquad \left(\frac{dq}{d\lambda}\right)_0 = v_0,$$

where $v_0 \neq 0$ is arbitrarily chosen in $T_{q_0}(\exp_p(s_i))$. We can choose (3.14) in such a way that all points $q(\lambda)$ are regular values of $\exp_p|s_i$. Choosing $\varepsilon > 0$ small enough, all assumptions of the Angle Lemma are met. Hence, if $\gamma_1, \ldots, \gamma_i$ are the i geodesics joining p to q_0 corresponding to the i vectors in $\exp_p^{-1}(q_0) \cap s_i$, then $v_0 = (dq/d\lambda)_0$ make the same angle with the vectors $w_1, \ldots, w_i$ tangent to $\gamma_1, \ldots, \gamma_i$, respectively, at q_0. Hence $T_{q_0}(\exp_p(s_i))$ is contained in the set V of all vectors v in $T_{q_0}M$ making the same angle with $w_1, \ldots, w_i$. Because for a generic q_0 the linear span of the differences $w_k - w_j$, $1 \leq k, j \leq i$, has dimension s, V is a linear space of dimension $\dim M - s$.

4. Some Trivial Examples

The determination of the σ-decomposition relative to a complete Riemannian manifold M is, in general, a difficult problem because it requires a very precise knowledge of the behavior of all the geodesics passing through a point $p \in M$. When $M = \mathbf{R}^n$, for the restricted problem with base point $p \in M$, $T_pM = \mathbf{R}^n$, $p \in \sigma_\infty$, all other points of T_pM belong to σ_1. If $M = S^n$ with the usual metric and radius $1/\pi$, the spheres centered at the base point p and integral radii belong to σ_∞ and all other points of the tangent space different from p belong to σ_1.

The same σ-decomposition appears when M is the projective space P^n with the usual metric.

When M is the cylinder $S^1 \times \mathbf{R}$, the σ-decomposition is made up of equidistant parallel lines, belonging to σ_2, all other points different from the base point p belong to σ_1 and $p \in \sigma_\infty$ by definition. The Möbius band exhibits the same σ-decomposition.

Another trivial example corresponds to the case when M is a simple-connected complete Riemannian manifold with negative sectional curvature: Any two geodesics starting from the base point p will never meet again so that σ_1 is the whole tangent space, with the exception of $p \in \sigma_\infty$.

5. The σ-Decomposition of Flat Tori

We consider now the simpler among the nontrivial examples of σ-decomposition. It turns out to be surprisingly rich, establishing a connection of our subject with Arithmetic and with Brillouin zones of Solid State Physics.

Consider the flat torus $T^2 = \mathbf{R}^2/\mathbf{Z}^2$ with the usual metric, induced by the canonical metric of $\mathbf{R}^2$. This metric is then canonically associated to the quadratic form $x^2 + y^2$.

Our problem is to determine the corresponding σ-decomposition of T_pT^2, relative to a point $p \in T^2$.

The tangent plane at p is canonically identified with $\mathbf{R}^2$ so that we can take p to be the origin. Besides, the exponential mapping is identified with the canonical projection $\pi\colon \mathbf{R}^2 \to T^2$.

To determine the σ-decomposition relative to the origin, one has to give a rule that associates to every $v \in \mathbf{R}^2$ its index $i(v)$. Clearly, $i(v)$ is the cardinality of the set of vectors u such that

$$\|u\| = \|v\|, \quad u - v \in \mathbf{Z}^2.$$

In coordinates, if $v = (a, b)$, $i(v) = i(a, b)$ will be the cardinality of the lattice points $(m, n) \in \mathbf{Z}^2$ such that

$$(a + m)^2 + (b + n)^2 = a^2 + b^2, \tag{5.1}$$

i.e.,

$$m^2 + n^2 + 2am + 2bm = 0. \tag{5.2}$$

Of course, $m = n = 0$ is always a solution of (5.2). The number $i(a, b)$ of integral solutions of (5.2) is the number of lattice points (m, n) situated on the circle $C(-a, -b)$ centered at $(-a, -b)$ and passing through the origin. Now the equation of such circle is exactly (5.2), a, b being constants and m, n being considered as variable.

So the determination of $i(a, b)$ is reduced to a precise arithmetic question, namely, the determination of the number of integral solutions (m, n) of (5.2). If (a, b) is a rational point, in general, there does not seem to exist in the literature a precise formula for the determination of the number $i(a, b)$ of integral solutions of (5.2).

But if $(a, b) \in \mathbf{Z}^2$, then $i(a, b)$ is the number of lattice points on the circle centered at the origin and passing through (a, b), i.e., the number of integral solutions of the equation

$$x^2 + y^2 = a^2 + b^2 = N. \tag{5.3}$$

Now a classical theorem of Jacobi [5, p. 242] tells us that the number of such solutions is given by the arithmetical function $r(N)$,

$$i(a, b) = r(a^2 + b^2) = r(N) = 4[d_1(N) - d_3(N)], \tag{5.4}$$

where $d_j(N)$, $j = 1, 3$, is the number of divisors of N congruous to j mod 4.

Thus, one gets a very precise formula giving $i(a, b)$ for $(a, b) \in \mathbf{Z}^2$.

Now the arithmetical function $r(N)$ considered above is defined for every positive integer N and vanishes when N is not the sum of two squares, i.e., when $d_1(N) = d_3(N)$.

For our purpose here, it is convenient to consider the function

$$\bar{r}\colon \mathbf{Z}^2 \to \mathbf{Z},$$

$$\bar{r}(a, b) = r(a^2 + b^2), \quad (a, b) \in \mathbf{Z}^2,$$

which never vanishes.

Our index i which is defined on the whole plane $\mathbf{R}^2$ and agrees with $\bar{r}$ on $\mathbf{Z}^2$ can then be considered as a natural extension of the original function r to the whole plane.

These arithmetical considerations do not allow us to determine the geometric structure of the σ-decomposition that we are looking for.

(5.5) *The Lines $L(m, n)$*

One has a good picture of the σ-decomposition if one looks at (5.2) from a different point of view. To this effect, consider, in (5.2), (m, n) as fixed and (a, b) as variable. Then for each pair (m, n) of nonvanishing integers, (5.2) represents the line $L(m, n)$ passing through $(-m/2, -n/2)$ and perpendicular to the line joining this point to the origin. Call

$$\mathscr{L} = \{L(m, n)\}$$

the totality of all such lines.

Now suppose that a point (a, b) belongs to no line of $\mathscr{L}$. Then (5.2) is satisfied only when $m = n = 0$ so that $i(a, b) = 1$, i.e., $(a, b) \in \sigma_1$. If (a, b) belongs to exactly one line of the family $\mathscr{L}$, then $(a, b) \in \sigma_2$ and, in general, if (a, b) belongs to exactly p lines of $\mathscr{L}$, $(a, b) \in \sigma_{p+1}$. In particular, if $(a, b) \in \mathbf{Z}^2$, the number of lines of the family $\mathscr{L}$ passing through it is obtained by subtracting one from $r(a^2 + b^2)$ given by (5.4).

So the lines $\mathscr{L}$ provide a way to look at the Diophantine equation

$$x^2 + y^2 = N$$

as being extended to the whole plane. From this point of view, the "multiple points," i.e., the points where two or more lines $L(m, n)$ meet, should be regarded on equal footing with the lattice points themselves.

Concerning multiple points, two preliminary problems present themselves: (a) to give some arithmetical characterization of these rational points; (b) to decide whether or not there are multiple points, not of the form $(m/2, n/2)$, m, $n \in \mathbf{Z}$, and exhibiting an arbitrarily high multiplicity.

In Fig. 1, one finds all lines $L(m, n)$

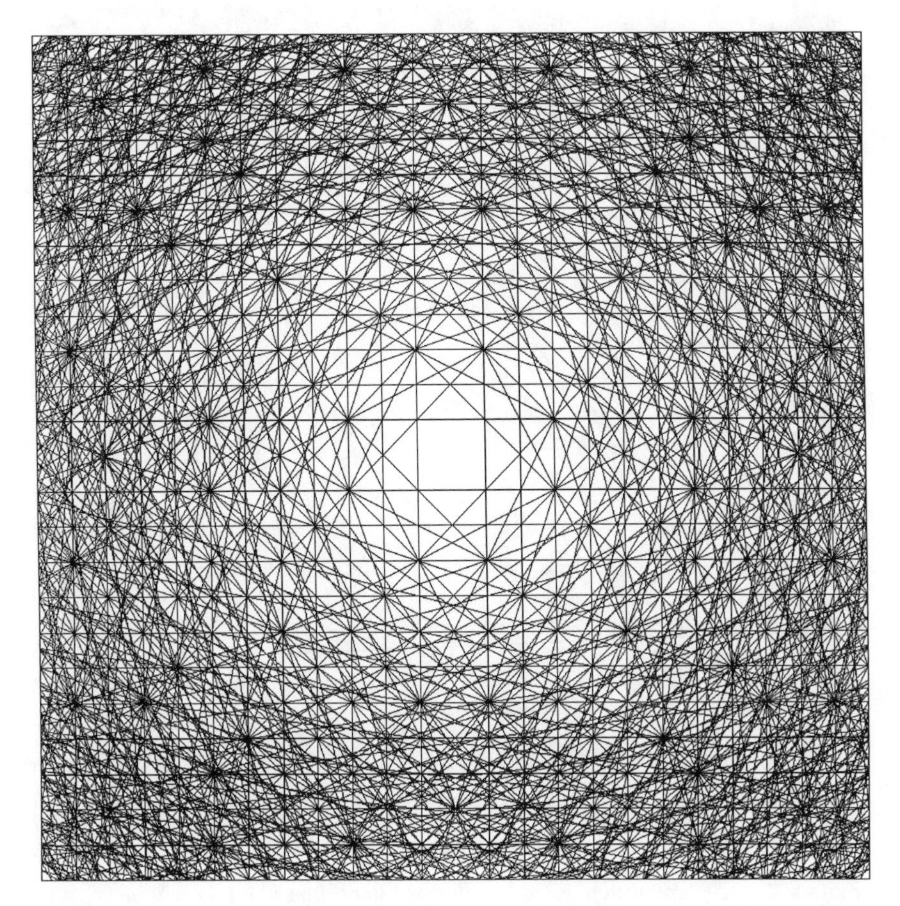

FIGURE 1. σ-decomposition for $x^2 + y^2$, lines $L(m, n)$ within $|x| \leq 6$, $|y| \leq 6$ (after S. Cowan).

(5.6)
$$m^2 + n^2 + 2mx + 2ny = 0$$

meeting the square $|x| \leq 6, |y| \leq 6$.

One bewildering feature is the presence of so many "envelopes." Do they have an arithmetical meaning?

(5.7) *The Brillouin Zones*

The lines $L(m,n)$ appears also in quite a different context. In fact, the so-called Brillouin zones of a cubic plane crystal are areas bounded by lines $L(m,n)$ of our family $\mathscr{L}$ [1, p. 106; 2, p. 39]. At first sight, there seems to be little in common between our geodesic interpretation of the lines $L(m,n)$ and the physical one that leads to the Brillouin zones.

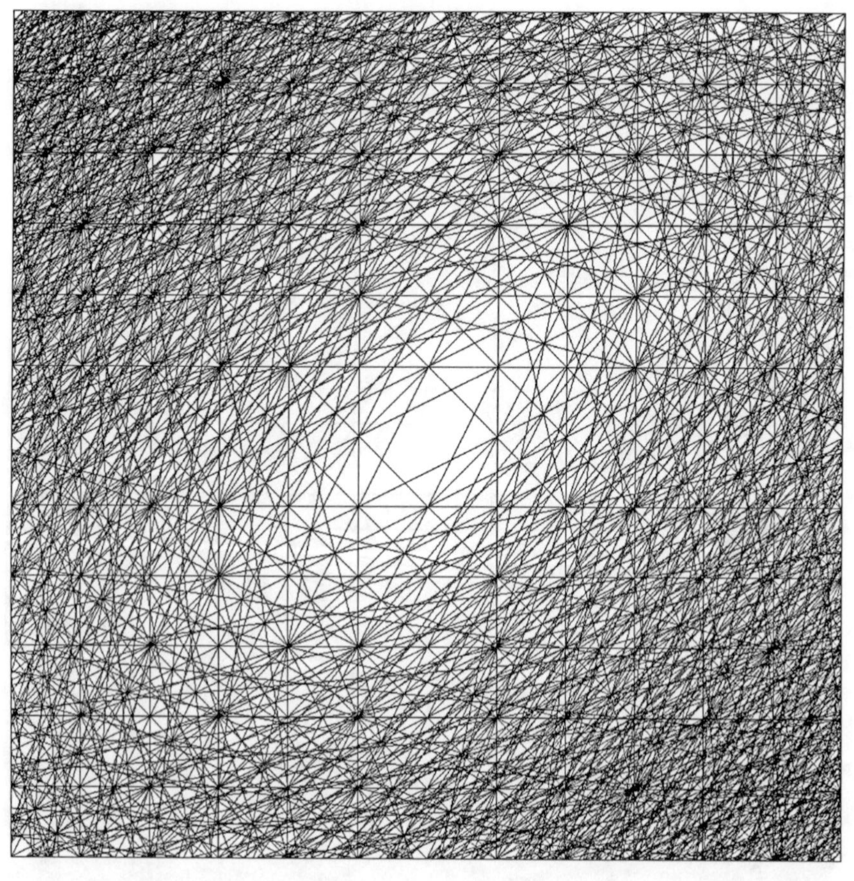

FIGURE 2. σ-decomposition for $x^2 + xy + y^2$, lines $L(m,n;1,1,1)$ within $|x| \leq 6$, $|y| \leq 6$ (after S. Cowan).

(5.8) *Flat Metric Associated to a Positive Defined Quadratic Form*

Given a positive definite quadratic form with integral coefficients

$$Ax^2 + Bxy + Cy^2, \qquad B^2 - 4AC < 0,$$

it determines a flat metric on T^2.

Then the problem arises of determining its σ-decomposition.

Proceeding as before, we get, instead of $L(m, n)$ defined by (5.6), a family of lines $L(m, n; A, B, C)$ defined by

$$Am^2 + Bmn + Cn^2 + (2Am + Bn)x + (Bm + 2Cn)y = 0.$$

The index of any point in the plane is the number of lines $L(m, n; A, B, C)$ passing through it plus one. (Fig. 2)

(5.9) *The σ-Decomposition for High-Dimensional Flat Tori*

All the previous considerations apply, with the obvious modifications, to flat tori in any dimension. In particular, for a flat metric of T^3 associated to the positive definite quadratic form $x^2 + y^2 + z^2$, we get the same kind of relationship between Brillouin zones and σ-decomposition.

References

[1] Brillouin, L., *Wave Propagation in Periodic Structures.* 2nd ed. Dover, New York (1953).

[2] Busch, G. and Schade, H., *Lectures on Solid State Physics.* Pergamon Press, Oxford (1976).

[3] Dieudonné, J. *Treatise on Analysis.* Academic Press, New York (1974), Volume 4.

[4] Goresky, M. and MacPherson, R., *Stratified Morse Theory.* Springer-Verlag, Berlin (1988).

[5] Hardy, G.H. and Wright, E.M., *An Introduction to the Theory of Numbers.* 5th ed. Clarendon Press, Oxford (1989).

[6] Peixoto, M.M., *On a Generic Theory of End Point Boundary Value Problems.* An. Acad. Bras. Cienc. 41 (1969), 1–6.

[7] Peixoto, M.M. and Thom, R., *Le point de vue énumératif dans les problèmes aux limites pour les équations différentielles ordinaires.* C.R. Acad. Sc. Paris, t. 303, Série I, n° 13 (1986), 629–632 and t. 307, Série I (1988), 197–198 (Erratum) and t. 303, Série I, n° 14 (1986), 693–698.

[8] Smale, S., *An infinite dimensional version of Sard's theorem.* Amer. J. Math. 87 (1965), 861–866.

[9] Thom, R., *Généralisation de la Théorie de Morse aux Variétés Feuilletées.* Ann. Inst. Fourier, Grenoble 14 (1964), 173–190.

[10] Winkelnkemper, H.E., *The Graph of a Foliation.* Ann. Glob. Analysis and Geometry, v. 1, n. 3 (1983), 51–75.

25
A Relation Between Newton's Method and Successive Approximations for Quadratic Irrationals

JULIAN PALMORE

A *quadratic irrational*,[1] ζ, has the form $(a + b^{1/2})/c$, where a, b, c are integers, $b > 0$, b is not the square of an integer, and $c \neq 0$. Let ζ^+ and ζ^- be quadratic irrationals corresponding to $\zeta^+ = (a + b^{1/2})/c$ and $\zeta^- = (a - b^{1/2})/c$. A quadratic polynomial that has these roots is $p(z) = (z - \zeta^+)(z - \zeta^-)$, or

$$p(z) = z^2 - \frac{2a}{c}z + \frac{a^2 - b}{c^2}.$$

We introduce two iterative processes for computing these algebraic numbers. We prove a relation between the convergent sequences produced by each process. Our point of view is dynamical systems and, in particular, complex analytic dynamics. For this reason, we use the extended complex plane, $\mathbb{C} \cup \{\infty\}$, for arithmetic when iterating rational functions.

Newton's map of $p(z)$, $g(z)$, is defined by

$$g(z) = z - \frac{p(z)}{p'(z)},$$

where $z \in \mathbb{C} \cup \{\infty\}$. Thus, $g(z)$ takes the simplified form

$$g(z) = \tfrac{1}{2}\left(z + \frac{a}{c}\right) + \frac{\tfrac{1}{2}(b/c^2)}{z - a/c}.$$

Newton's method has two superstable fixed points, ζ^+ and ζ^-. By starting in the basin of attraction[2] of either fixed point, an open half-plane, the sequence of iterates converges quadratically to the fixed point.

A second procedure is called successive approximations.[3] If we write $p(z) = 0$ in the form $z(z - (\zeta^+ + \zeta^-)) = -\zeta^+\zeta^-$, then we can define a map h on the extended complex plane as

$$h(z) = -\frac{\zeta^+\zeta^-}{z - (\zeta^+ + \zeta^-)}.$$

Consequently, successive approximations for $p(z)$ takes the form

$$h(z) = -\frac{(a^2 - b)/c^2}{z - 2a/c}.$$

There are two points fixed by h. These are ζ^+ and ζ^-. Provided $a \neq 0$, one fixed point is stable and the other is unstable.

We can always choose $a \neq 0$ by adding an arbitrary integer to our choice of quadratic irrationals. For the remainder of the chapter, without loss of generality, we adopt the following convention: $a < 0$. For $a < 0$, ζ^+ is a stable fixed point of $h(z)$. The fixed point ζ^- is unstable. The stable fixed point is a global attractor of h. This means that with the exception of the unstable fixed point, ζ^-, every point in the extended complex plane, as an initial point of iteration by h, is attracted to ζ^+.

The basin of attraction of ζ^+ for Newton's method and the basin of attraction of ζ^+ for successive approximations each contain $z = 0$.

For Newton's method, this is because $z = 0$ lies nearer ζ^+ than ζ^-.

Let $\{g^{(k)}(\xi)\}_{k \geq 0}$ and $\{h^{(k)}(\xi)\}_{k \geq 0}$ denote the sequences of iterates by Newton's method and by successive approximations, respectively, when initialized at $z = \xi$. In particular, $g^{(0)}(z) = z$, $g^{(1)}(z) = g(z)$, and $g^{(k)}(z) = g(g^{(k-1)}(z))$, $k \geq 1$. We find that for $p(z)$ above,

$$g(0) = \tfrac{1}{2}\frac{a}{c} - \tfrac{1}{2}\frac{b}{ac} = h(0).$$

We know that Newton's method converges quadratically to the fixed point ζ^+ and that successive approximations converges only linearly as a contraction mapping to ζ^+. Thus, we expect that Newton's method, when initialized with successive approximations as above, generates a subsequence of the convergent sequence produced by successive approximations.

Theorem. *Let the quadratic irrationals ζ^+ and ζ^-, the polynomial p and the rational functions g (Newton's map) and h (successive approximations) be defined as above. Then for all $n \geq 0$, we have*

$$g^{(n)}(0) = h^{(\alpha)}(0), \quad \textit{where } \alpha = 2^n - 1.$$

Outline of the Proof

We give a proof by induction. We initialize the sequences at $z = 0$, $g^{(0)}(0) = 0 = h^{(0)}(0)$. Assume $g^{(n)}(0) = h^{(\alpha)}(0)$, where $\alpha = 2^n - 1$ holds for $0 \leq n \leq N - 1$. We shall conclude that the statement holds for $0 \leq n \leq N$ and, consequently, for all $n \geq 0$.

For quadratic irrationals, the mapping $h(z)$ of successive approximations is a linear rational function. The iterates of $h(z)$, $h^{(\alpha)}(z)$, $\alpha \geq 0$, are linear rational functions. One fixed point of $h(z)$ is stable; the other fixed point is unstable.

Consequently, it follows that h is conjugate by a Möbius transformation to $f(w) = Kw$, constant K. We will use this conjugacy to establish the inductive step of the proof.

Illustration of the Method of Proof

Let $p(z) = z^2 + 2z - 1$ so that the zeros ζ^+ and ζ^- are $-1 \pm 2^{1/2}$. We have for Newton's method, the map

$$g(z) = \tfrac{1}{2}\frac{z^2 + 1}{z + 1}.$$

For successive approximations, we have the map $h(z) = 1/(2 + z)$. The fixed point $\zeta^+ = -1 + 2^{1/2}$ is the only stable fixed point for the map $h(z)$. For Newton's method, both fixed points are superstable.

We build a Möbius transformation so that the stable fixed point ζ^+ corresponds to $w = 0$ and the unstable fixed point ζ^- corresponds to $w = \infty$. In this way, we show that $h(z)$ is conjugate to $f(w) = Kw$ by a linear fractional transformation. Thus, we have the following commutative diagram:

$$\begin{array}{ccc} f\colon \mathbb{C} \cup \{\infty\} & \longrightarrow & \mathbb{C} \cup \{\infty\} \\ M\big\downarrow & & \big\downarrow M \\ h\colon \mathbb{C} \cup \{\infty\} & \longrightarrow & \mathbb{C} \cup \{\infty\}. \end{array}$$

The Möbius transformation $z = M(w)$ is $M(w) = (Aw + B)/(Cw + D)$, where the values of parameters are $A = 1$, $B = 1$, $C = 1 - 2^{1/2}$, and $D = 1 + 2^{1/2}$. The inverse, $w = M^{-1}(z)$, is $M^{-1}(z) = (Dz - B)/(-Cz + A)$. The constant K is determined by conjugacy to be $K = 2\cdot 2^{1/2} - 3$. We have the conjugacy, $h(z) = M \circ f(w) = (2\cdot 2^{1/2} - 3)w$. Consequently, for all $\alpha \geq 0$, $h^{(\alpha)}(z) = M \circ f^{(\alpha)} \circ M^{-1}(z) = M(K^\alpha w)$.

To prove by induction that $g^{(n)}(0) = h^{(\alpha)}(0)$, $\alpha = 2^n - 1$, $n \geq 0$, we start with $g^{(0)}(0) = 0 = h^{(0)}(0)$. We assume that the result holds for $0 \leq n \leq N - 1$. We prove that, $g^{(N)}(0) = g(g^{(N-1)}(0)) = g(h^{(\alpha)}(0)) = g(M \circ f^{(\alpha)} \circ M^{-1}(0)) = h^{(2\alpha+1)}(0)$, for $\alpha = 2^{N-1} - 1$. This step completes the induction.

In all that follows, we let $\alpha = 2^{N-1} - 1$. We denote K^α by K_α. The inductive step starts by evaluating $g(h^{(\alpha)}(0))$. We have

$$g(h^{(\alpha)}(0)) = \tfrac{1}{2}\frac{[h^{(\alpha)}(0)]^2 + 1}{h^{(\alpha)}(0) + 1}.$$

Writing

$$h^{(\alpha)}(0) = \frac{1 - K_\alpha}{[1 + 2^{1/2}] - [1 - 2^{1/2}]K_\alpha},$$

making a substitution into $g(h^{(\alpha)}(0))$, and simplifying, we find

$$g(h^{(\alpha)}(0)) = \frac{(2 + 2^{1/2})(1 - KK_\alpha^2)}{[4 + 3 \cdot 2^{1/2}] + 2^{1/2}KK_\alpha^2},$$

$$g(h^{(\alpha)}(0)) = \frac{1 - KK_\alpha^2}{[1 + 2^{1/2}] - [1 - 2^{1/2}]KK_\alpha^2}$$

$$= h^{(2\alpha+1)}(0).$$

Thus, the result is proved by induction.

Proof of the Theorem

The general case, $p(z) = z^2 - (2a/c)z + (a^2 - b)/c^2$, with $\zeta^+ = (a + b^{1/2})/c$ and $\zeta^- = (a - b^{1/2})/c$, can be proved easily by following the outline elaborated above.

We conjugate $h(z)$ to a linear mapping, $f(w) = Kw$, by a linear fractional transformation, that maps the stable fixed point of h, ζ^+, to $w = 0$ and maps the unstable fixed point of h, ζ^-, to $w = \infty$. It is convenient to choose $A = 1$ and $B = 1$ in the linear fractional transformation and to evaluate C and D directly as functions of a, b, and c, or equivalently, as functions of ζ^+ and ζ^-. Thus, from the relation, $M^{-1}(\zeta^+) = 0$, we find $D = (\zeta^+)^{-1}$. From the relation, $M^{-1}(\zeta^-) = \infty$, we find $C = (\zeta^-)^{-1}$. In particular, we find that K, evaluated as a function of a, b, and c, is $K = (a + b^{1/2})^2/(a^2 - b)$.

Following precisely the induction used in the example, we assume the truth of the result for $0 \le n \le N - 1$ and show that this implies the result for $0 \le n \le N$.

The form of the proof is identical to that above except that numerical values are replaced by

$$g(h^{(\alpha)}(0)) = \frac{1 - KK_\alpha^2}{D - CKK_\alpha^2} = h^{(2\alpha+1)}(0).$$

This completes the induction and the proof of the theorem.

Remark. Successive approximations, beginning with $z = 0$, is, in reality, a generator of a continued fraction expansion for a quadratic irrational. In the example above, $h(z) = 1/(2 + z)$, the sequence of iterates is

$$\{0, 1/2, 2/5, 5/12, 12/29, 29/70, 70/169, 169/408, \ldots\}.$$

Newton's method, as we have proved, generates a quadratically convergent subsequence

$$\{0, 1/2, 5/12, 169/408, 195025/470832, \ldots\}.$$

This illustrates the correspondence established by the theorem,

$$g^{(0)}(0) = 0 = h^{(0)}(0), \qquad g^{(1)}(0) = 1/2 = h^{(1)}(0), \qquad g^{(2)}(0) = 5/12 = h^{(3)}(0),$$

$$g^{(3)}(0) = 169/408 = h^{(7)}(0), \qquad g^{(4)}(0) = 195025/470832 = h^{(15)}(0).$$

In fact, at the 5th iteration, we have obtained by Newton's method a highly accurate approximation to $-1 + 2^{1/2}$, given as a rational,

$$g^{(5)}(0) = 259717522849/627013566048 = h^{(31)}(0).$$

It is interesting that dynamical systems theory in the context of complex analytic dynamics brings together various aspects of number theory and analysis.

References

[1] Ivan Niven, *Irrational Numbers*, Carus Mathematical Monographs, Mathematical Association of America, 1956.

[2] Harold Benzinger, Scott Burns, and Julian Palmore, "Complex analytic dynamics and Newton's method, *Phys. Lett. A* **119** (1987), 441–446.

[3] Julian Palmore, "Exploring chaos: science literacy in mathematics," *Proceedings of the Symposium on Science Learning in the Informal Setting*, The Chicago Academy of Sciences, Chicago, 1988, pp. 183–200.

26
On Dynamical Systems and the Minimal Surface Equation

PER TOMTER

0. Introduction

The relationship between the theory of dynamical systems and differential geometry is a long-standing and profound one. It has largely been focused around the geodesic flow on the unit tangent bundle of a Riemannian manifold. One may recall early work by Poincaré and Birkhoff for the case of convex surfaces, Morse theory, contributions by Hadamard, and, more recently, by Anosov, in the case of negatively curved manifolds—to register the decisive influence that has been asserted by this particular example in the development of the general theory of dynamical systems. Conversely, analysis of the geodesic flow has been useful to study problems in Riemannian geometry; one may point to the recent successful structure theory for Hadamard manifolds by Ballmann, Eberlein, Gromov, Schroeder, and others. Here we will be concerned with a different, more unexpected role that dynamical systems have recently played in the very active field of partial differential equations that arise in Riemannian geometry, notably the minimal surface equation. Briefly, in the presence of a suitably chosen Lie group G of isometries, one studies the G-invariant solutions. A simple case of this idea goes back to Delaunay's classification of rotationally invariant constant mean curvature surfaces [D]; more recently, the celebrated work of Bombieri, de Giorgi, and Giusti on the Bernstein problem may also be viewed in this context [BGG]. The ideas were developed in a more systematic way as a general program of "equivariant geometry" in a seminal paper by W.Y. Hsiang and B. Lawson [HL], and has since been developed, especially by Hsiang, to solve some long-standing open problems in Riemannian geometry [H1, H2, H3].

In the case when the principal orbits of G have codimension two in the manifold M, the reduced minimal equation in the orbit space M/G becomes a dynamical system; in fact, for a suitably modified metric on M/G, we are again faced with the geodesic flow. Because M/G has singular strata, where the curvature blows up, there are new features here, and some hard analysis is usually required to find closed solutions in each particular case. Most interesting solutions which have been found (minimal hyperspheres in various symmetric spaces, including counterexamples to the "spherical Bernstein problem" posed by S.S. Chern [H1, H2, T1]), have been obtained from geodesics in M/G which are "closed" only in the more general sense that their endpoints are on singular strata.

After giving a brief survey of generalities, we outline, in Section 1, the computation of the relevant orbital invariants for actions of suitable isometry groups on the projective planes over the complex numbers **C**, the quaternions **H**, and the Cayley numbers *Ca*, respectively. In Section 2, we investigate the resulting dynamical system, and in Section 3, we prove the following:

Theorem. *Let* $\mathbf{C}P(2)$, $\mathbf{H}P(2)$, $Ca(2)$ *be the projective planes over* **C**, **H**, *and Ca, respectively. In each case, there exists an embedded closed minimal hypersurface of generalized torus type, namely*:

(i) *for* $\mathbf{C}P(2)$: $S^1 \times S^1 \times S^1$*-type;*
(ii) *for* $\mathbf{H}P(2)$: $S^1 \times S^3 \times S^3$*-type;*
(iii) *for* $Ca(2)$: $S^1 \times S^7 \times S^7$*-type.*

This is an example of the type of question that has come up in Hsiang's program, and, in this case, the solutions are obtained by finding an honestly closed geodesic in the regular part of the orbit space. For dimension two (i.e., projective planes), the analysis is considerably simplified by the existence of a finite reflection group of symmetries in orbit space; in the case of the Cayley projective planes, these arise from triality.

In Section 4, we illustrate this by showing some solution curves obtained by numerical computations. These curves suggest the existence of many further closed solutions; moreover, they appear compatible with orbital stability. The curvature of M/G is positive, so some closed geodesics may be expected to be elliptic. This raises the question of developing the theory of geodesic flow for this type of stratified manifolds (in particular, to which extent the Kolmogorof–Arnold–Moser theory is applicable to prove the existence of many closed solutions near a closed solution, possibly of singular type, as described above). In some cases, many "closed" solutions of singular type have already been found by more ad hoc methods [HT, HHT], but the natural way in which these types of dynamical systems with singularities make their appearance, as well as the far-reaching results already obtained for specific cases, suggests further study of the general properties of such systems.

The computer computations were done in collaboration with Gil Bor at the University of California, Berkeley.

1

Let G be a compact Lie group acting isometrically on a closed Riemannian manifold M. The orbit projection $\pi: M \to M/G$ is then a Riemannian submersion when restricted to the open dense set M^0 of points on principal orbits; here, we use the orbital distance metric ds^2 on M^0/G. To reduce the study of the minimal equation in M to a problem in M^0/G, the only invariant needed from the geometry of the fibers of π is the function W on M^0/G which records the volume of the orbits. For a curve γ in M^0/G, the volume of $\pi^{-1}(\gamma)$ in M is then the integral of W along γ, or, equivalently, the arc length of γ with respect to the modified metric $ds^{-2} = W^2\, ds^2$ on M^0/G. It follows that $\pi^{-1}(\gamma)$ is minimal in M iff γ is a geodesic in $(M^0/G, ds^{-2})$; and the geodesic equation is as follows [HL]:

$$k(s) - \frac{d}{d\bar{n}} \ln W(\gamma(s)) = 0, \tag{1.1}$$

where $\gamma(s)$ is the curve parameterized by arc length relative to ds^2, k is the curvature of γ, and $d/d\bar{n}$ is the directional derivative of W in the normal direction of γ compatible with a chosen orientation. If we now assume that G has cohomogeneity two, i.e., $\dim M^0/G = 2$, then $\pi^{-1}(\gamma)$ is a minimal hypersurface in M iff γ is a solution of (1.1). Hence, the invariants needed to perform the reduction in a concrete case is the orbit space with its orbital distance metric and the volume function W.

It is, of course, only feasible to compute these invariants in simple cases. It is straightforward for cohomogeneity-two orthogonal actions on Euclidean spaces and spheres, and is easy for complex and quaternionic projective spaces by lifting in the Hopf fibration (see below). For the isotropy action on compact rank 2 symmetric spaces, the Cartan–Weyl theory will give these invariants: The orbit space is a flat Cartan polyhedron and the volume function W is given in terms of the restricted roots by Weyl's volume formula for the relative case [HHT].

1.1. Complex Projective Space: $\mathbf{C}P(n) = U(n+1)/U(1) \times U(n)$ (see [HHT].)

Let $G = U(1) \times U(1) \times U(n-1)$ act on $S^{2n+1} \subseteq \mathbf{C}^{n+1}$ in the standard way; the action of $U(1)$ sitting diagonally in G defines the Hopf fibration $p\colon S^{2n+1} \to \mathbf{C}P(n)$. Because the fibers of p are totally geodesic, p defines a one-to-one correspondence between $U(1)$-invariant minimal hypersurfaces of S^{2n+1} and minimal hypersurfaces of $\mathbf{C}P(n)$. Hence, it is sufficient to study the orbit data for S^{2n+1}. The G-orbits of $\mathbf{C}^{n+1} = \mathbf{C} + \mathbf{C} + \mathbf{C}^{n-1}$ are of the type $S^1(x) \times S^1(y) \times S^{n-3}(z)$, and are characterized by the three non-negative radii x, y, z. Hence, the orbit space for the restricted action on S^{2n+1} is the spherical triangle $\{(x, y, z) | x^2 + y^2 + z^2 = 1,\ x, y, z \geq 0\}$. Introducing spherical polar coordinates: $x = \sin r \cos\theta$, $y = \sin r \sin\theta$, $z = \cos r$, we have the following:

(i) the orbit space is parameterized by $(r, \theta) \in [0, \pi/2] \times [0, \pi/2]$;
(ii) the orbital distance metric is given by $ds^2 = dr^2 + \sin^2 r\, d\theta^2$.
(iii) $W(r, \theta) = xyz^{2n-3} = \frac{1}{2}\sin^2 r \cos^{2n-3} r \sin 2\theta$.

(Note that we may normalize the volume function by any chosen constant factor.)

1.2. *Quaternionic Projective Space:* $\mathbf{H}P(n) = Sp(n+1)/Sp(1) \times Sp(n)$

By a similar argument with $G = Sp(1) \times Sp(1) \times Sp(n-1)$ and lifting the G-action in the Hopf fibration $S^3 \to S^{4n+3} \to \mathbf{H}P(n)$, one obtains:

the orbit space is given as in (i), (ii) above;

the volume function $W(r,\theta) = \frac{1}{8}\sin^6 r \cos^{4n-5} r \sin^3 2\theta$.

1.3. *The Cayley Projective Plane:* $Ca(2) = F_4/\text{Spin } 9$

In this case, there does not exist any "Hopf-fibration" $S^7 \to S^{23} \to Ca(2)$, and it is a more challenging task to compute the orbital invariants. We refer to a forthcoming paper [T3] for the details of this; the methods developed there may also be useful in more general symmetric spaces. All three cases are treated uniformly in [T3] as follows: M is a two-point homogeneous space, hence the orbit space under the isotropy action may be represented by a geodesic segment from a point P to the cut locus, represented by A; we may parameterize this segment by $r \in [0, \pi/2]$. G is the subgroup of the isotropy group G_P that fixes a point on the cut locus; then G acts on M with cohomogeneity two. The G_P-orbits corresponding to interior points on PA are Berger-type spheres. Their isotropy representations have two irreducible components; hence, their homogeneous metrics are determined by two scaling factors along PA, which are well-known to be $\frac{1}{2}\sin 2r$ and $\sin r$. They are obtained by studying the lengths of suitable Killing–Jacobi fields along the geodesic. In case 3, we have $G_P = \text{Spin } 9$, $G = \text{Spin } 8$, the principal G_P-orbits are of type $\text{Spin } 9/\text{Spin } 7 = S^{15}$; the volume of these geodesic spheres varies with r as $2^{-7}\sin^7 2r \sin^8 r$. Under the G-action, each G_P-orbit has one-dimensional orbit space, which may again be taken as a geodesic segment parameterized by $\theta \in [0, \pi/2]$. The fact that the G_P-orbits are not canonically reductive as homogeneous spaces makes computations rather complicated. In case 3, the orbits are of type $S^7 \times S^7$, with a complicated metric (spheres not orthogonal). However, because G still defines an ample space of Killing fields, it turns out to be possible to compute the scaling of the volume of G-orbits along the geodesic segment $\theta \in [0, \pi/2]$; we refer to [T3] for the details of this. The orbit space is given by the same spherical triangle as in (i), (ii) above, the volume function is given by $W(r,\theta) = 2^{-7}\sin^{14} r \cos^7 r \sin^7 2\theta$.

Remark. This is the same result as one would obtain formally by assuming the existence of a "Hopf-fibration" $S^7 \to S^{23} \to Ca(2)$ where the G-action could be lifted.

2

Let $P(a+1)$ be the projective plane over $\mathbf{C}$ for $a = 1$, over $\mathbf{H}$ for $a = 3$, and over Ca for $a = 7$. Let Δ be the intersection of the unit sphere with the first octant in $\mathbf{R}^3$, parameterized by spherical polar coordinates (r, θ). We summarize some results from the last section.

Proposition 1. *Let $M = P(a+1)$, let $P \in M$, and let A be the first point on the cut locus along a geodesic from P. Let G_P be the isotropy group of P, and let G be the stabilizer subgroup of G_P at A. Then the orbit space M/G, considered in the orbital distance metric, is isometric to Δ. The volume function is given by $W(r,\theta) = \sin^{2a} r \cos^a r \sin^a 2\theta$ (when suitably normalized).*

We will now use the orbital distance metric $ds^2 = dr^2 + \sin^2 r\, d\theta^2$ unless specified otherwise. Let $\gamma(s)$ be a curve in int Δ parameterized by arc length, and let α be the angle between the tangent vector $\dot{\gamma} = \dot{r}\, \partial/\partial r + \dot{\theta}\, \partial/\partial\theta$ and $\partial/\partial r$, i.e., $\dot{r} \doteq \cos\alpha$, $\dot{\theta} = \sin\alpha \sin^{-1} r$.

Proposition 2. *Let $\gamma(s)$ be a curve in int Δ as above. Then $\pi^{-1}(\gamma)$ is a minimal hypersurface in M iff γ satisfies the following:*

$$(*) \qquad \begin{aligned} \dot{r} = \cos\alpha, \qquad \dot{\theta} &= \sin\alpha \sin^{-1} r, \\ \dot{\alpha} = K_r \sin\alpha &+ 2a\cos\alpha \sin^{-1} r \cot 2\theta, \end{aligned}$$

where $K_r = a\tan r - (2a+1)\cot r$.

PROOF. The normal of γ is $\bar{n} = -\sin\alpha(\partial/\partial r) + \cos\alpha \sin^{-1} r(\partial/\partial\theta)$, hence $d/d\bar{n} \ln W = -2a\sin\alpha \cot r + a\sin\alpha\tan r + 2a\cos\alpha\sin^{-1} r \cot 2\theta$. Furthermore, the curvature of γ in Δ is easily computed by Liouville's formula to be $\dot{\alpha} + (\cos r)\dot{\theta}$, and $(*)$ follows by substitution into (1.1). Q.E.D.

Proposition 3. *Let S_3 be the symmetric group on three letter, which acts on the orbit space by reflections across the bisectors of Δ. Then S_3 is a symmetry group for the dynamical system $(*)$.*

PROOF. These arise from the isometry groups on M, for example, in the case of $\mathbf{C}P(2) = U(3)/U(1) \times U(1) \times U(1)$ from the Weyl group permuting the three $U(1)$'s. In case $M = Ca(2)$, this comes from the triality principle: In this case, the three edges of Δ are the fixed point sets of the three nonconjugate subgroups of Spin 7 type in Spin 8, which are permuted by the triality outer automorphisms. These outer automorphisms of Spin 8 may, in fact, be represented as restrictions of conjugations by elements of F_4, and define isometries (see [T3]).

Corollary 1. *Let $\gamma(s)$ be a solution curve for $(*)$ which intersects two bisectors of Δ orthogonally. Then $\gamma(s)$ may be extended to a periodic solution curve.*

Proof. This follows from Proposition 3 by repeated reflections across bisectors.

Proposition 4. *The following curves are solutions for* $(*)$:

(a) $r = r_0 = \arctan\sqrt{\dfrac{2a+1}{a}}, \quad \alpha = \pm\dfrac{\pi}{2}$ (equator).

(b) $\theta = \theta_0 = \dfrac{\pi}{4}, \quad \alpha = 0$ or $\alpha = \pi$ (meridian).

(c) $\sin r \sin\theta = \sqrt{\dfrac{a}{3a+1}}$ or $\sin r \cos\theta = \sqrt{\dfrac{a}{3a+1}}$.

(d) $\sin r \sin\theta = \cos r$ and $\sin r \cos\theta = \cos r$ (bisectors).

Proof. (a) and (b) are verified by substitution; (c) and (d) are images of (a) and (b), respectively, by reflections from S_3.

Remark 1. The equator solution (a) is the geodesic sphere in M around the point $r = 0$ with maximal volume. Note that $r = r_0$ is a "closed geodesic of singular type."

Remark 2. It follows that any other solution curve which goes from an interior point of one boundary arc to one on another boundary arc of Δ must also be the projection of a minimal hypersphere in M. Hence, the existence of such solution curves is central to the uniqueness aspect of Bernstein-type problems; they have been investigated in great detail [H1, H3, T1, HT].

3

In this section, we do the analytical work needed to prove the existence of the desired closed solution curves in int Δ.

Proposition 5. *Let* $\gamma(s) = (r(s), \theta(s), \alpha(s))$ *be a solution curve for* $(*)$. *Then we have*:

(a) *a critical point for* $\theta(s)$ *is a local maximum (resp. local minimum) in the region* $\theta > \pi/4$ *(resp.* $\theta < \pi/4$*)*;

(b) a critical point for $r(s)$ *is a local maximum (resp. local minimum) in the region* $r > r_0$ *(resp.* $r < r_0$*)*;

(c) *a critical point for* $\theta(s)$ *is nondegenerate except for the meridian solution* $\theta = \pi/4$, *a critical point for r is nondegenerate except for the equator solution* $r = r_0$.

Proof. By computation we get $\ddot{\theta} = 2a\sin^{-2} r\cos^2\alpha\cot 2\theta$ when $\dot{\theta} = 0$, *and* $\ddot{r} = -K_r \sin^2\alpha$ when $\dot{r} = \cos\alpha = 0$.

Proposition 6. *For each interior point p of an edge of Δ there exists a unique solution of $(*)$ with initial point at p. The initial direction of this solution is orthogonal to the edge.*

PROOF. This can be seen by some local analysis of $(*)$ near the boundary and an argument using formal power series expansion; see [HH, T2].

Proposition 7. *Let O be the intersection point of the three bisectors of Δ, then the coordinates of O are $(r, \theta) = (\arctan\sqrt{2}, \pi/4)$. Let B be the corner of Δ given by $(r, \theta) = (\pi/2, \pi/2)$. Then the angle α of the bisector going into B is in the interval $[\pi/4, \pi/2]$.*

PROOF. Straightforward.

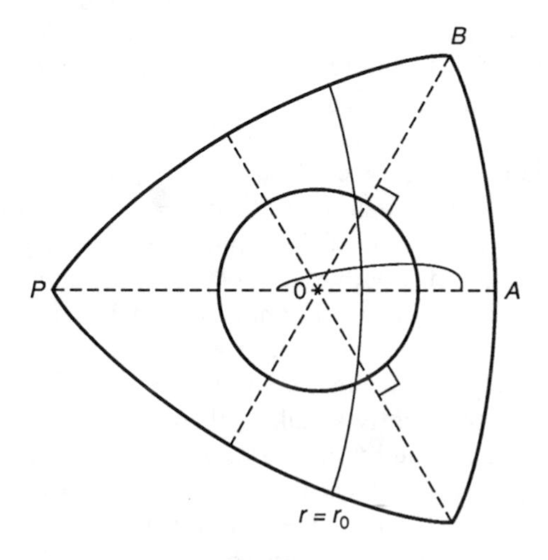

FIGURE 1

Definition. Let $\gamma_b(s) = (r_b(s), \theta_b(s), \alpha_b(s))$ be the solution curve of $(*)$ with initial conditions $r_b(0) = b$, $\theta_b(0) = \pi/4$, $\alpha_b(0) = \pi/2$. Let $s(b)$ be the smallest possible s such that $\dot{\theta}(s(b)) = 0$, i.e., the first local maximum for $\theta_b(s)$, $s > 0$.

Theorem. *There exists a $b \in (r_0, \pi/2)$ such that $\gamma_b(s)$ defines a simple, closed solution curve of $(*)$ in* int Δ.

PROOF. By Corollary 1, it is sufficient to show that there exists a $b \in (r_0, \pi/2)$ and a positive d such that $\gamma_b(s)$, $s \in [0, d]$, is a simple arc in the triangle OAB intersecting the bisector segment OB orthogonally. The equator solution γ_{r_0} intersects the bisector OB at an angle in $(0, \pi/2)$; by continuous dependence on initial conditions, this also holds for γ_b, $b \in (r_0, r_0 + \varepsilon)$, when ε is sufficiently small. It also follows that this intersection then occurs before $s = s(b)$.

Because the bisector and γ_b are both solutions of (*), they never become tangent, and it follows that the first intersection point must persist as b varies in $(r_0, \pi/2)$. (Note that they cannot intersect at the boundary edge, by Proposition 6.) Furthermore, as long as the intersection occurs before $s = s(b)$, it must be on the segment OB. Assume that it should occur after $s = s(b_1)$ for some b_1; then, by continuity, it must occur exactly at $s = s(b_2)$ for some $b_2 \in (r_0, b_1)$. But then, because $\alpha_{b_2}(s(b_2)) = \pi$ and by Proposition 7, the intersection angle between γ_{b_2} and OB is in $(\pi/2, 3\pi/4)$. From continuity, it would then follow that γ_b intersects OB orthogonally for some $b \in (r_0, b_2)$, and we would be done.

Hence, for the remainder of the proof we may assume that γ_b has its first intersection with OB for some $s < s(b)$, whenever $b \in (r_0, \pi/2)$. We wish to consider b close to $\pi/2$. Let $\bar{r} = -r$; from (*) we get: $d\alpha/d\bar{r} = K_{\bar{r}} \tan\alpha + 2a \sin^{-1} r \cot 2\theta$; hence $d\alpha/d\bar{r} > K_{\bar{r}} \tan\alpha$ in the region under consideration. Furthermore, $K_{\bar{r}} \tan\alpha > 0.9a \tan\bar{r} \tan\alpha$ for $r \in (r_1, \pi/2)$, with $r_1 = \arctan\sqrt{20 + 10/a}$. So in this region we now have $d\alpha/d\bar{r} > 0.9a \tan\bar{r} \tan\alpha$. The solution of the differential equation $d\alpha/d\bar{r} = 0.9a \tan\bar{r} \tan\alpha$ with the initial condition $\alpha(-b) = \pi/2$ is $\sin\alpha = \cos^{0.9a} b \cos^{-0.9a} \bar{r}$; comparing with this, we may conclude that $\sin\alpha_b < \cos^{0.9a} b \cos^{-0.9a} \bar{r}_b$ as long as $r_b > r_1$. Hence, for a given positive ε, we get $\sin\alpha_b < \varepsilon/2$ at $r = r_1$ by choosing b sufficiently close to $\pi/2$. This shows that such γ_b must turn sharply to $\alpha_b \in (\pi - \varepsilon, \pi)$ before $r = r_1$, and continue with $\alpha_b \in (\pi - \varepsilon, \pi)$ at least until $r = r_0$, because, by (*), $\dot{\alpha}_b$ is positive in this region, and, by assumption, α_b does not reach π [corresponding to $s = s(b)$] before crossing OB, say at $s =_b s$. Furthermore, unless α_b reaches a local maximum before $s =_b s$, the same holds until that crossing; for ε sufficiently small, γ_b then crosses OB with an angle in $(\pi/2, 3\pi/4)$, and we are done by the previous continuity argument. The final possibility that α_b reaches a local maximum before $s =_b s$ would give $\dot{\alpha}_b = K_{r_b} \sin\alpha_b + 2a \cos\alpha_b \sin^{-1} r_b \cot 2\theta_b = 0$; in this range, this would imply that θ_b at such a point would approach zero as $\varepsilon \to 0$. In that case, the initial conditions of γ_b at $r_b = r_0$ approach those of the meridian solution $\theta = \pi/4$, $\alpha = \pi$ as $\varepsilon \to 0$, so continuous dependence on initial conditions would again imply that α_b stays close to π until intersection with OB. Q.E.D.

4

In this section, we reproduce some solution curves γ_b which were numerically obtained on a computer for the case of the Cayley projective plane, $a = 7$. We remark that the curves suggest the existence of many more nonsimple, closed solution curves. Moreover, small perturbations of the initial conditions of some of these appear to produce solution curves that project into a ring-shaped domain around the closed solution, which contains other closed solutions of higher period.

Recalling that this is the geodesic flow of (Δ, ds^{-2}), it is Hamiltonian, and the

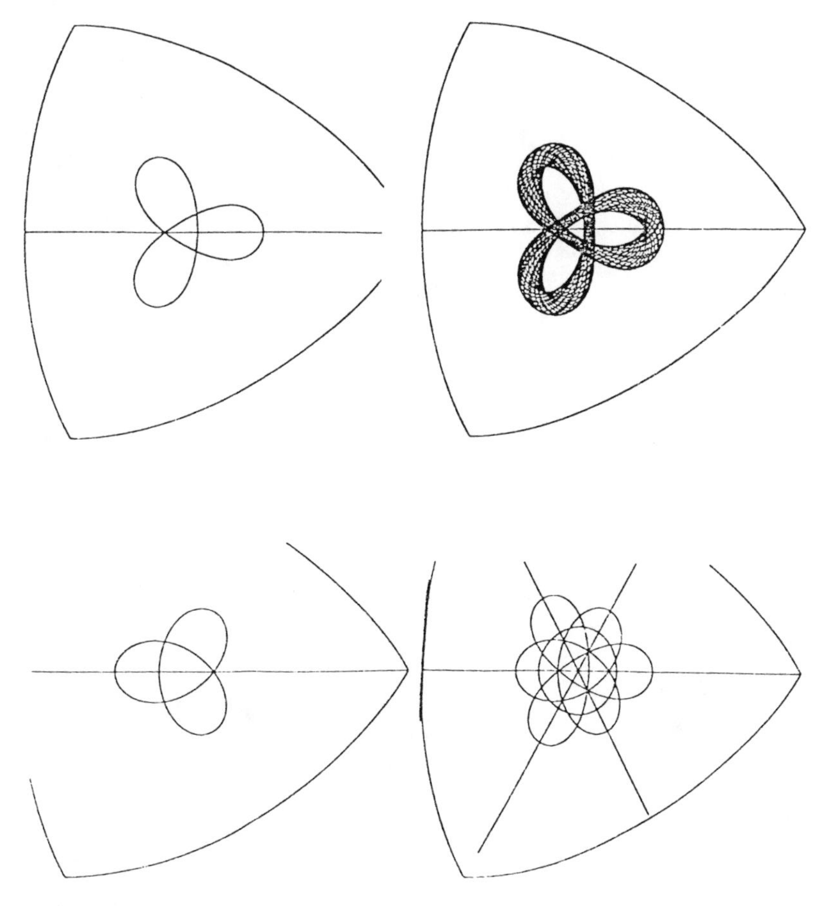

FIGURE 2

characteristic multipliers of a closed orbit are either λ, λ^{-1}, with real λ, $|\lambda| > 1$ (hyperbolic case) or $e^{i\alpha}$, $e^{-i\alpha}$ with real α (elliptic case [AM]). The curvature of (Δ, ds^{-2}) is computed to be $K = \sin^{-(2a+1)} r \cos^{-a} r \sin^{-a} 2\theta (2a \sin^{-1} r \sin^{-2} 2\theta + (a+1) \sin r + a \tan r + a \tan^2 r \sin r) > 0$; hence, elliptic periodic orbits are not unexpected. Analysis of the Poincaré map is of interest because periodic points correspond to new closed minimal hypersurfaces in M. In the generic elliptic case, the Moser twist stability theorem gives the existence of invariant tori containing periodic solutions of arbitrarily large period. The curves would fit in with ring-shaped domains obtained as projections on the configuration space of such invariant tori; there has not been any study of this aspect so far. More interesting might be the case of singular closed orbits (as the equator γ_{r_0}). Toward this, we only make the following observation:

Proposition 8. *There is a well-defined Poincaré map for a closed orbit γ of singular type.*

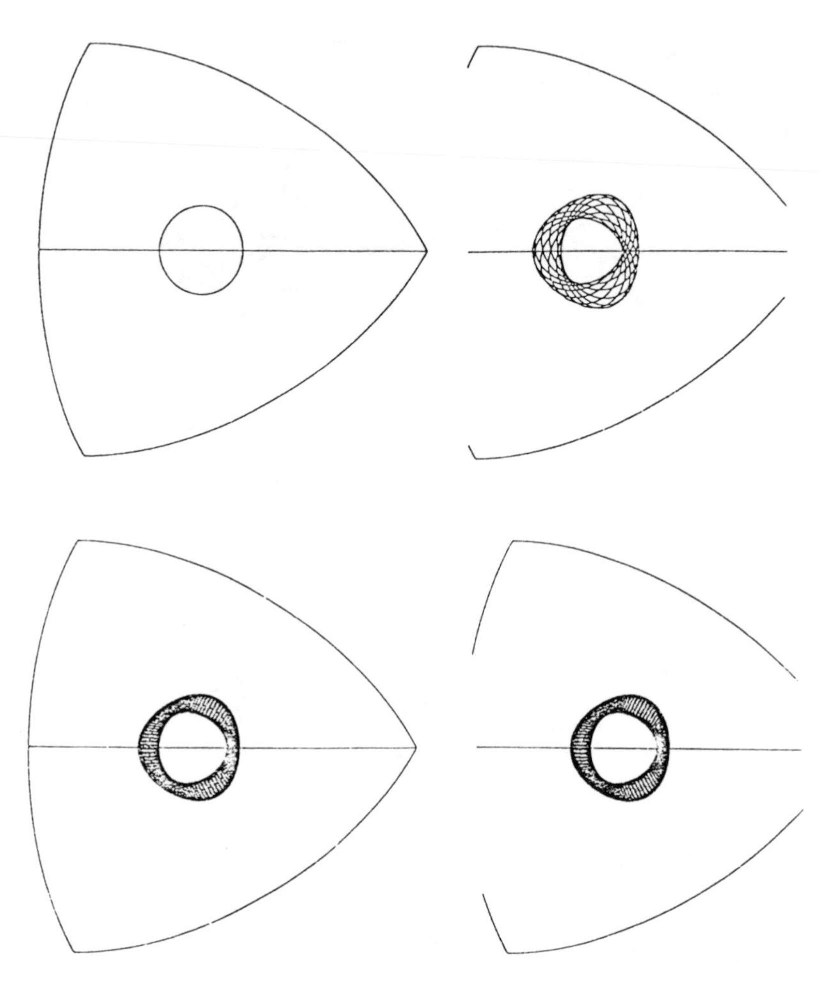

FIGURE 3

PROOF. We consider γ as defined globally by reversing direction from a boundary point; i.e., we oscillate back and forth along γ, with a discontinuity jump of π for α at each boundary intersection. The only problem is to establish continuous dependence on initial conditions for solutions starting out near γ, also beyond γ's intersection with the singular boundary arc. This is nontrivial, although one can prove (with estimates similar to the proof of the theorem in Section 3) that such a curve must follow γ toward the boundary, turn sharply there, and start out again with initial data close to γ on the way back. However, as we get closer to the boundary, there is no control over the Lipschitz constants of $(*)$. Nevertheless, it is known that continuous dependence on initial conditions does persist past intersections with the boundary; for a proof, see [HT].

For example, in the case of the standard sphere, one could consider the possibility of using the Poincaré map of singular closed solutions to obtain nearby minimal closed hypersurfaces. If they correspond to singular closed solution curves, these would be hyperspheres. (In the case of the equator hypersphere, however, by an extrinsic rigidity theorem [S], it is known that there does not exist C^1-close minimal hyperspheres.) Although the particular question of existence of infinitely many minimal hyperspheres has already been treated successfully both for spheres [HT] and projective spaces [T3], it appears that the Hamiltonian viewpoint could be relevant for further study. This might especially apply to generic density properties of closed minimal hypersurfaces with respect to the cohomogeneity-two metrics on M respecting G.

References

[AM] Abraham, R. and Marsden, J.E., *Foundations of Mechanics*, 2nd ed., The Benjamin/Cummings Publ. Co., Reading, MA, 1978.

[BGG] Bombieri, E., de Giorgi, E., and Giusti, E., Minimal cones and the Bernstein problem. *Ivent. Math.* **7** (1969), 243–268.

[D] Delaunay, C., Sur la surface de révolution dont la courbure moyenne est constante, *J. Math. Pures. Appl, Sér. 1.* **6** (1841), 309–320.

[H1] Hsiang, W.Y., Minimal cones and the spherical Bernstein problem I. *Ann. Math.* **118** (1983), 61–73.

[H2] Hsiang, W.Y., Minimal cones and the spherical Bernstein problem II. *Invent. Math.* **74** (1983), 351–369.

[H3] Hsiang, W.Y., New examples of constant mean curvature immersions of $(2k-1)$-spheres into Euclidean $2k$-spaces. *Ann. Math.* **117** (1983), 609–625.

[HH] Hsiang, W.T. and Hsiang, W.Y., On the existence of codimension one minimal spheres in compact symmetric spaces of rank 2, II. *J. Diff. Geom.* **17** (1982), 582–594.

[HHT] Hsiang, W.T., Hsiang, W.Y., and Tomter, P., On the existence of minimal hyperspheres in compact symmetric spaces. *Ann. Scient. E.N.S.* **21**, (1988), 287–305.

[HL] Hsiang, W.Y. and Lawson, H.B., Minimal submanifolds of low cohomogeneity. *J. Diff. Geom.* **5** (1970), 1–37.

[HT] Hsiang, W.Y. and Tomter, P., On minimal immersions of S^{n-1} into $S^n(1)$, $n \geq 4$. *Ann. Scient. E.N.S.* **20** (1987), 201–214.

[S] Simons, J., Minimal varieties in Riemannian manifolds. *Ann. Math.* (2) **88** (1986), 62–105.

[T1] Tomter, P., The spherical Bernstein problem in even dimensions and related problems. *Acta Math.* **158** (1987), 189–212.

[T2] Tomter, P., Existence and uniqueness for a class of Cauchy problems with characteristic initial manifolds. *J. Diff. Equations* **71** (1988), 1–9.

[T3] Tomter, P., Minimal hyperspheres in two-point homogeneous spaces. Preprint series No. 12 June 1992. Institute of Mathematics, University of Oslo.

27
A New Zeta Function, Natural for Links

R.F. WILLIAMS

We overcome a former frustration: A new "determinant" is defined, not requiring completely abelian matrix entries. In fact, non-abelian invariants do occur in dynamical systems, so that the usual determinant is not available. This "link-determinant" is invariant only under the group of permutations, not the general linear group. But permutations are the only pertinent transformations in the setting of symbolic dynamics. That is to say, permuting the elements of a Markov matrix makes sense; linear combinations of them do not. For our immediate purpose, keeping track of knots and links via their fundamental group words, the zeta function defined in terms of this new determinant seems ideal.

In fact, it seems felicitous to us that what one *can* save of the noncommuting nature of the entries is, in our setting, precisely the natural invariant in this setting. This good luck continues for a bit, and the Cayley–Hamilton theorem seems to be true [KSW] in this setting. It follows that the equation

$$\det(I - A) = \exp[-(\mathrm{tr}(A + A^2/2 + A^3/3 + \cdots)],$$

so important in symbolic dynamics, also holds. This, of course, makes it obvious that trace A^i is determined for all i once it is known for $i < \dim A$. Here, we restrict our attention to the Lorenz attractor, and to the special case in which there is a double saddle connection: We verify directly that the trace of all powers of A is determined by link-det$(I - A)$. As in an earlier work [W1], a matrix having as elements the generators x and y of a free group is introduced. The powers of such a matrix have entries in the group ring—no trouble so far. The trace of a power is, in turn, a sum over Z of words in x and y. Now each of these words is a representative of the value of a periodic orbit in the fundamental group of a neighborhood of the attractor. For example, $xxyxy$. But since there is no natural way to choose a starting point for a periodic orbit, $xyxyx$ is just as good; in fact, any cyclic permutation is okay, so that the natural invariant is the *cyclic permutation class* $(xxyxy)$ of such words, called the "free-knot symbol" below. But this information is easily retrieved from the combinatorics of the definition

$$(*) \qquad \det A = \sum_p \mathrm{sg}(p) A(1, p_1) A(2, p_2) \cdots A(n, p_n) \qquad (p \text{ a permutation}).$$

This is just the fact that permutations are uniquely factored into their cyclic parts. The order in which the the cyclic factors occur in (*) is *not* preserved by permutations of the basis. But one finds that when a product of two or more free-knot symbols arises in our new determinant, it refers to the disjoint union of the respective periodic orbits. These are "links" to a topologist; hence, the title of the chapter. Thus, it would not be natural to assign any order to such products anyway. So that the natural thing is to allow such products to commute, e.g.,

$$(**) \qquad (xxyxy)(xxyy) = (xxyy)(xxyxy).$$

So that again, the natural invariant is exactly what we can have in our new "determinant."

Topologists use the term link for the disjoint union of knots, so that we call the products as in (**) "free-link symbols."

This chapter is mostly self-contained. However, we are concerned here only with matrices of a certain type, to wit, those that were introduced in [W1, W2] (see the next section) associated with a Lorenz attractor with a double saddle connection. Note in particular, that the diagonal entries of these matrices are all 0. These restrictions were needed for the proofs in an earlier version of this chapter, but are now known to be unnecessary, by virtue of a Cayley–Hamilton theorem [KSW] proved in this setting. We retain these assumptions to make the chapter self-contained.

It was a pleasure to get, as it is to acknowledge, help from my colleagues David Saltman, Bill Schelter, Mat Stafford, and John Tate.

The Lorenz Template

In [W1, W2], a certain amount of machinery was set up to study the periodic orbits of the (geometrically defined) Lorenz attractor. This is an ordinary differential equation in R^3 which has infinitely many periodic orbits, mostly knotted and generally linked to one another. By a reduction process, they were shown to correspond up to isotopy to the periodic orbits on a branched surface, L (these are equipped with a semiflow), such as that pictured later in Example D. The hole on the left is called x, that on the right, y. Then the orbits are in 1-to-1 correspondence with the (aperiodic) words in x and y, such as $xxyxy$. In fact, with a bit of worry about a base point, the fundamental group of L is free on generators x and y and the periodic orbit corresponding to a word w is an embedded 1-sphere, and thus a "knot" K_w to a topologist. Then w represents the homotopy class of K_w in the fundamental group of L. See also [BW1, BW2]. Note that on the branch line, which is horizontal in our figure, the semiflow is not well-defined for any negative values because there are two flow lines coming into these points. The point 0 in the bottom center is a rest point of the flow and is hyperbolic. Leaving 0 in positive time are the two "unstable orbits," one to the right and the other to the left. Note that these two orbits return to 0 in positive time. Thus, they form what is called, variously, a double saddle connection or a double homoclinic orbit.

In [W1], we introduced matrices with entries x, y, and 0. These derive from a "countable Markov partition" of the branch line, though this name was not used in these earlier works. However, in the double saddle connection case, these reduce to the usual finite Markov partitions. In any case, tr A^i is a formal sum of a collection of words w of length i, corresponding to the periodic orbits of length i. Though determined only up to cyclic permutation, this suffices to distinguish x^3y^2 and x^2yxy, for example. (The second word corresponds to a trefoil knot, the first to an unknot.) This distinction is precisely what was needed for our characterization in [W2] and is preserved in our new zeta function.

Definitions

1. Link-ring. Given a group F and a matrix A with entries in F, we define the *link-ring* $R(A)$ as follows:

a. For each sequence (called a *cycle* below)

$$i_1, i_2, \ldots, i_k, \qquad i_r \neq i_s \quad \text{for } r \neq s,$$

such that the product

$$A(i_1, i_2)A(i_2, i_3)\cdots A(i_k, i_1) \neq 0,$$

let $(i_1, i_2, \ldots, i_k)$ be the equivalence class under cyclic permutations of this product. Terminology: These equivalence classes are called *free-knot symbols* and the indices $i_1, i_2, \ldots, i_k$ are called *nodes* of the free-knot symbol.

b. Define $R(A)$ to be the free abelian group generated by the free-knot symbols defined in (a).

Example. For the matrix A

$$\begin{matrix} 0 & x & 0 & 0 \\ 0 & 0 & x & x \\ y & y & 0 & 0 \\ 0 & 0 & y & 0 \end{matrix}$$

the link-ring $R(A)$ is free abelian on the generators

$$(xxy), \quad (xxyy), \quad (xy), \quad (xyy)$$

corresponding to the cycles

$$(123), \quad (1234), \quad (23), \quad (234).$$

In the special case treated here, the diagonal entries of A are all 0. However, for matrices with, say, $A(i, i) = x$, (i) would be a cycle and (x) would be a "free-knot symbol."

In the definitions below, except for 4 and 5, we have a fixed matrix A, with entries in our group F. It is only in this setting that the term "free-knot symbol" makes sense. In turn, the terminology is metaphorical, coming from the main example we have in mind, of a given flow and given Markov partition. The "places" or "nodes" 1 to n in the matrix correspond to the cells in a Markov partition. The "cycles" of the matrix correspond to periodic orbits of the Markov partition and, thus, to periodic orbits of our flow. These orbits are simple closed curves in 3-space and, thus, "knots"; hence, the term "free-knot symbol" for $(xxyxy)$. Topologists call the union of two (disjoint) knots a link; proceeding with our analogy, when such symbols as $(xxy)(xxyxy)$ can occur together, they are "free-link symbols." By occurring together, we mean that they have no node in common, just as two cycles for the matrix A can occur in a product of the determinant, only if they have no node in common, such as (126) (35847).

2. By a *free-link symbol* in the ring $R(A)$ is meant a product $x_1 x_2 \ldots x_l$ of free-knot symbols, no two of which have a node in common.

3. Define the determinantlike function

$$\text{link-det}(I - A) = \sum_{\text{free-link symbols}} (-1)^l x_1 x_2 \cdots x_l.$$

4. More generally, for a generic matrix A define

$$\text{(D4)} \qquad \text{cycle-det}(A) = \sum_p \operatorname{sign}(p) c_1 c_2 \cdots c_k$$

where the sum is over all permutations p of $\{1, 2, \ldots, n\}$, $n = \dim A$, and $c_1 c_2 \cdots c_k$ is a factorization of p into cycles. Here the c_i are unique, up to cyclic permutation; as there is no canonical order, we allow $c_i c_j = c_j c_i$.

Remark. These two are related as follows: for each cycle $c = (i_1, i_2, \ldots, i_k)$, let $x(c) = A(i_1, i_2) A(i_2, i_3) \cdots A(i_k, i_1)$, and extend x to be a ring homomorphism. Then

$$\text{(D4}') \quad \text{link-det}(I - A) = x(\text{cycle-det}(I - A)) = \sum_p \operatorname{sign}(p) x(c_1) x(c_2) \cdots x(c_k).$$

Proof. This is fairly clear except for two problems with the diagonal elements of $I - A$. First, the free-link symbols in (D4) were not required to have total length n. However, they can always be filled in to be of length n because each diagonal element of $I - A$ has a 1 as summand. Thus, (i) is a cycle for each i and $x((i)) = 1$. Second, if $A(i, i) = a \neq 0$, then there will be other kinds of terms. But they all belong, as one sees after a bit of reflection. This will be dealt with fully in [KSW].

5. Define the characteristic polynomial $\chi(A)$ by

$$\chi(A)(t) = \text{cycle-det}(tI - A).$$

In [KSW], it is shown that the "Cayley–Hamilton Theorem" holds:

$$\chi(A)(A) = 0.$$

Here we limit our discussion to the case of a double saddle connection: This means that we have a finite Markov partition of the branch set I of the Lorenz template L and an incidence map and incidence matrix of a certain special type. To make a long story short, our matrices are like the matrix A above or A', defined to be the following:

$$\begin{matrix}
0 & 0 & x & x & 0 & 0 & 0 & 0 & 0 & 0 & 0 & 0 \\
0 & 0 & 0 & 0 & x & x & 0 & 0 & 0 & 0 & 0 & 0 \\
0 & 0 & 0 & 0 & 0 & 0 & x & 0 & 0 & 0 & 0 & 0 \\
0 & 0 & 0 & 0 & 0 & 0 & 0 & x & x & 0 & 0 & 0 \\
0 & 0 & 0 & 0 & 0 & 0 & 0 & 0 & 0 & x & x & 0 \\
0 & 0 & 0 & 0 & 0 & 0 & 0 & 0 & 0 & 0 & 0 & x \\
y & 0 & 0 & 0 & 0 & 0 & 0 & 0 & 0 & 0 & 0 & 0 \\
0 & y & y & 0 & 0 & 0 & 0 & 0 & 0 & 0 & 0 & 0 \\
0 & 0 & 0 & y & y & 0 & 0 & 0 & 0 & 0 & 0 & 0 \\
0 & 0 & 0 & 0 & 0 & y & 0 & 0 & 0 & 0 & 0 & 0 \\
0 & 0 & 0 & 0 & 0 & 0 & y & y & 0 & 0 & 0 & 0 \\
0 & 0 & 0 & 0 & 0 & 0 & 0 & 0 & y & y & 0 & 0
\end{matrix}$$

In fact, the matrix A corresponds to the canonical Markov partition of a Lorenz attractor with a double saddle connection and A' to a subdivision of this partition formed by a (3–4 torus knot) periodic orbit. We will return to these two matrices in an example below.

Basic Results

Remark 1. The link-determinant is a generalization of the ordinary determinant, in that link-det$(I - A)$ becomes $\det(I - A)$ under abelianization.

PROOF. One of the standard definitions of the determinant is

$$\det(B) = \sum_p \operatorname{sign}(p) B(1, i_1) B(2, i_2) \cdots B(n, i_n)$$

where the sum is over all permutations $p = i_1, \ldots, i_n$, where n is the dimensionof B. Thus, each of our free-link symbols occurs in *some order*, as a *part* of one of the basic products. Then note that we can choose our favorite order for this product, and that we can always complete each partial product by choosing 1's from the main diagonal because we have allowed only 0's on the diagonal of A. Thus, it remains only to check the sign we have chosen. This sign has three ingredients: first, the sign of a cyclic permutation $(i_1, \ldots, i_k)$ is

$(-1)^{k+1}$; second, there is a minus in front of each nondiagonal term of $I - A$, giving a total of -1, for each free-knot symbol; finally, the sign of a product of permutations is the product of the individual signs and we are done.

Remark 2. Link-det$(I - A)$ is invariant under permutation of the basis elements of A, that is, more precisely

$$\text{link-det}(I - PAP^{-1}) = \text{link-det}(I - A)$$

for a permutation matrix P.

PROOF. Let p be the permutation of the bases of A induced by the matrix P. Then the permutation p permutes the elements of the free-knot symbols, which can change these symbols, but only by a cyclic permutation. But this does not change the cyclic equivalence class and, thus, *link-det* is unchanged.

Statement of Results

Theorem A. *For Lorenz attractors with a double saddle connection,*

$$\text{link-det}(I - A) = \sum_{\mathscr{L}} (-1)^{|L|} \operatorname{fls}(L),$$

where $\mathscr{L}$ is the collection of all links L in the attractor which have at most one point in each partition set, $|L|$ is the number of components in L, and fls(L) *means the free-link symbol of L.*

Corollary. *For the example of* [W1, pp. 108/10], link-det *distinguishes two systems that the ordinary determinant does not.*

The lack of commutativity for our matrices poses no problem when just taking powers A^i. The x's and y's come out just where they should be, so that tr A^i is just the "sum" of the words corresponding to periodic orbits of length i. Here x^3y^2 is clearly distinct from x^2yxy which is as it should be. Our main purpose here is to see that the infinite expression involving the traces of all the powers of A can be reduced to a finite expression, without losing this distinction.

Corollary C (to the Cayley–Hamilton Theorem [KSW]). *The following equation, important in symbolic dynamics, follows from the Cayley–Hamilton Theorem.*

(EF) $$\exp \sum_{i=0}^{\infty} -\operatorname{tr}(A^i)/i = \text{link-det}(I - A).$$

PROOF. The entries in A do not commute. But the diagonal entries of any power of A are just products of free-knot symbols, and thus commute. Thus, let $\hat{A}$ be the "companion matrix" of A. That is, $\hat{A}$ is a matrix with entries which commute, and by the Cayley–Hamilton Theorem, has the same trace as A for each power $i = 1, 2, \ldots$. Thus, (EF) holds for $\hat{A}$ in place of A, as one can see by checking it at a general point. At such a point, we can use eigenvalues and the usual proof works. But (EF) cannot distinguish between A and $\hat{A}$, so it holds for A as well.

For completeness, we include a direct proof of a corollary of this equation in our special case.

Theorem C. *For matrices A associated to the Lorenz attractor, with double saddle connection,* link-det$(I - A)$ *determines* trace A^n, *for all n.*

Proof. In fact, two of the free-knot symbols in link-det$(I - A)$ are the kneading sequences for this attractor. That is, the saddle connections themselves show up as "loops" or free-knot symbols. As such, they will be recognized as the first and last words in the lexicographical ordering. For example, for the matrix A given earlier, the kneading sequences are

$$k_\ell = xx, \qquad k_r = yy$$

because the first and last words are xxy, yyx. (This is according to the convention used in [W2, p. 327]; a simpler convention would give yxx, xyy, leaving the initial y and x in place.) But the kneading sequences determine the Markov partition and thus the matrix A, up to a permutation.

Though our invariant works well for the partitions as in Theorem A, it is less satisfactory under subdivisions of the canonical partition. In the usual treatment tr A^n gets replaced by tr A'^n − tr B^n. A computation based on the exponential formula leads to a quotient of two polynomials. But this reduces to the previous answer:

$$\frac{\det(I - B)}{\det(I - A')} = \frac{1}{\det(I - A)}.$$

In the current setting, this equality is a priori complicated by the fact that the link-ring of A' is larger than that of A. However, one can choose a projection of the larger to the smaller, guided by the fact that the larger Markov partition is a refinement of the smaller. (Note, in particular, that this equation does hold after abelianizing x and y.) Unfortunately, this does not work well here:

Example D. For the canonical partition of the Lorenz attractor described below, (see Fig. 1) and the indicated subdivision, one computes

$$\text{link-det}(I - B)\,\text{link-det}(I - A) \neq \text{link-det}(I - A').$$

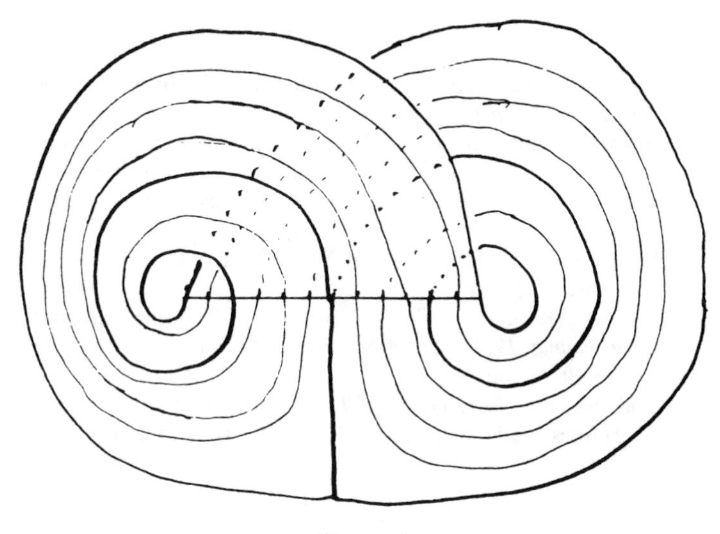

FIGURE 1

In fact, their difference is

$$(x^2y)(xy^2) - (xy)(x^2y^2).$$

Here, the terms involving A' and B are evaluated in the smaller ring under a chosen projection.

Remark. This example is unsatisfactory in several ways. These all seem to stem from the fact that there is no reasonable homomorphism from the group ring $Z[F(x, y)]$ on the free group with generators x and y, to the ring $R(A)$ we introduced. In particular, $R(A)$ can have more than two generators, so that there would be no onto homomorphism. However, the equation in Corollary C indicates that there must be some way to bypass the problem presented in Example D.

The matrices A dealt with in this chapter are quite sparse: They have at most one x and one y in each column. Thus, 3^n is an obvious upper bound to the number of nonzero terms in link-det$(I - A)$. However, this number seems to grow a lot slower, and for small values, to be less than exponential. In the following table we list, as the output of a computer program, the *smaller* number of *cycles* for a representative type of an n-dimensional matrix. Computing the cycles is the first step in computing link-det; we seem to have exponential growth:

n	5	11	21	25	31	35	41	45
	5	21	210	970	4422	8369	89653	168876

References

[BW1] Birman, J. and Williams, R., Knotted periodic orbits I: Lorenz knots, *Topology*, **22** (1983), 47–82.

[BW2] Birman, J. and Williams R., Knotted periodic orbits II: Fibered knots, *Low Dimensional Topology* (S. Lomonaco, ed.), Contemporary Mathematics, Vol. 20, American Mathematical Society, Providence, RI, 1983, pp. 1–60.

[KSW] Kennedy, S., Stafford, M., and Williams, R., *A new Cayley–Hamilton Theorem*, to appear.

[W1] Williams, R., Lecture VII: The structure of Lorenz attractors, *Turbulence Seminar, 1976/77*, Lecture Notes in Mathematics No. 615, Springer-Verlag, Berlin, pp. 96–112.

[W2] Williams, R., *The structure of Lorenz attractors*, Publications Institute des Hautes Etudes Scientifique, **50** (1979) 307–47, Paris.

Part 6
Theory of Computation

Theory of Computation

28
On the Work of Steve Smale on the Theory of Computation

MICHAEL SHUB*

The theory of computation is the newest and longest segment of Steve Smale's mathematical career. It is still evolving and, thus, it is difficult to evaluate and isolate the more important of Smale's contributions. I think they will be as important as his contributions to differential topology and dynamical systems. He has firmly grounded himself in the mathematics of practical algorithms, Newton's method, and the simplex method of linear programming, inventing the tools and methodology for their analysis. With the experience gained, he is laying foundations for the theory of computation which have a unifying effect on the diverse subjects of numerical analysis, theoretical computer science, abstract mathematics, and mathematical logic. I will try to capture some of the points in this long-term project. Of course, the best thing to do is to read Smale's original papers; I have not done justice to any of them.

Smale's work on the theory of computation begins with economics [Smale, 1976]. Prices $p = (p_1, \ldots, p_\ell) \in R^\ell_+$ for ℓ commodities give rise to demand and supply functions $D(p)$ and $S(p)$. The excess demand function $f(p) = D(p) - S(p) \in R^\ell$ has as the ith coordinate the excess demand for the ith good at prices p. A price equilibrium is a system of prices p for which $f(p) = 0$, that is, supply equals demand.

$$f\colon R^\ell_+ \to R^\ell$$

and the problem is to find a zero of f[1]. Given a C^2 function $f\colon M \to R^n$ defined on a domain $M \subset R^n$, Smale suggested using the "global Newton" differential equation

$$Df(x)\frac{dx}{dt} = -\lambda f(x)$$

* Partially supported by an NSF grant.

[1] Actually, f is assumed to be scale invariant $f(\lambda P) = f(P)$ for $\lambda > 0$ so f may be restricted to the unit sphere intersect R^ℓ_+, $S^{\ell-1}_+$. Walras' law is $p \cdot f(P) = 0$, so f is tangent to $S^{\ell-1}_+$ and the problem is to find a zero of this vector field.

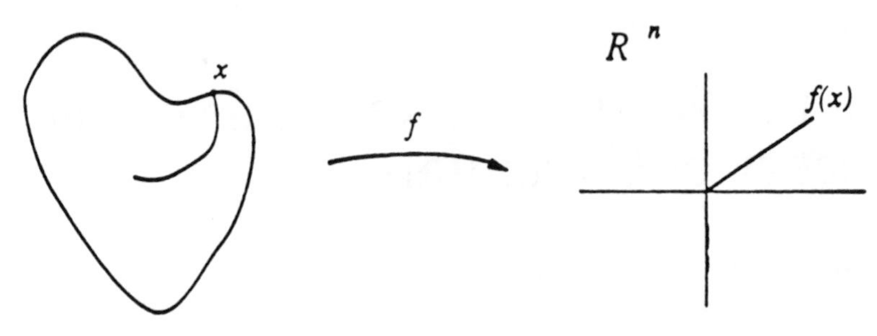

FIGURE 1

where λ is a real number depending on the sign of $\text{Det}(Df(x))$. The solution curves of this equation through points x with $f(x) \neq 0$ are inverse images of rays pointing to zero.

or what is the same, inverse images of points $x \in S^{n-1}$ for the function

$$g: M - E \to S^{n-1}$$

defined by $g(x) = f(x)/\|f(x)\|$ and where $E = \{x \in M \mid f(x) = 0\}$. With the right boundary conditions, the "global Newton" vector field is transverse to the boundary and g is nonsingular on the boundary.

By Sard's theorem, almost every value in S^{n-1} is a regular value, and for almost every $m \in \partial M$, $g^{-1}(g(m))$ is a smooth curve. If M is compact, this curve must lead to the set of zeros E.

Smale proved the existence of price equilibria this way. Differential equation solvers can then be used to locate these points. Smale pursued these ideas with Hirsch in [Hirsch-Smale, 1979] where they suggested various explicit algorithms for solving $f(x) = 0$. In particular, their work included polynomial mappings

$$f: \mathbb{C}^m \to \mathbb{C}^m \quad (\text{or } R^m \to R^m)$$

which are proper and have nonvanishing Jacobian outside of a compact set.

An interesting feature of these algorithms is that their natural starting points are quite far from the zero set; for example, in the discussion of global Newton above, they are in ∂M.

The global Newton differential equation had been considered previously and independently by Branin without convergence results. The polynomial system $g: \mathbb{R}^2 \to \mathbb{R}^2$ proposed by Brent,

$$g_1(x_1, x_2) = 4(x_1 + x_2),$$

$$g_2(x_1, x_2) = 4(x_1 + x_2) + (x_1 + x_2)((x_1 - 2)^2 + x_2^2 - 1),$$

is illustrated in [Branin, 1972] and reproduced below. The curve $|J| = 0$ is $\text{Det}(g) = 0$. Note the region of closed orbits and nonconvergence which are close to zero.

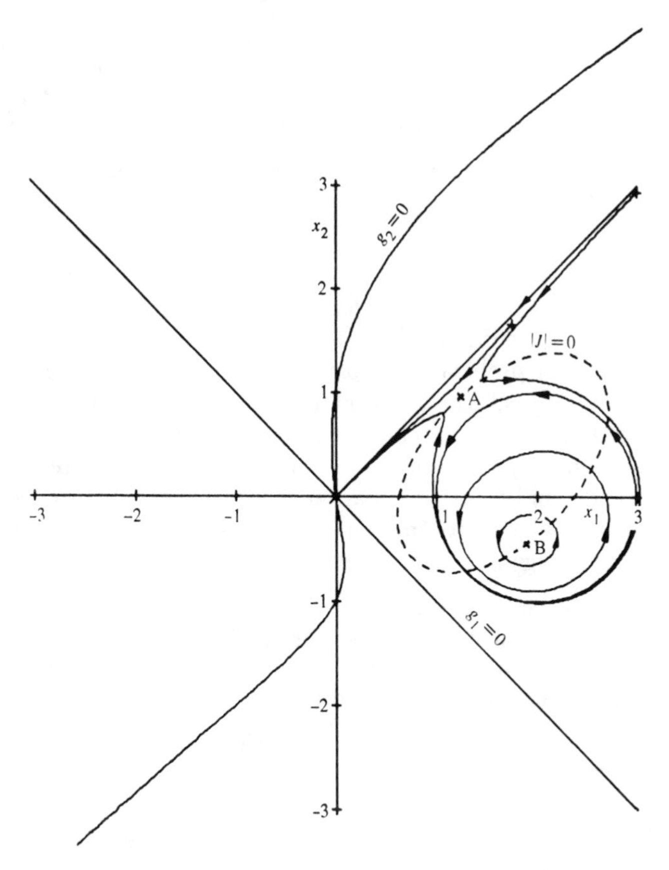

FIGURE 2

Other algorithms for the location of pure equilibria were known. Scarf, in particular, had developed simplicial algorithms; see [Smale, 1976] for a discussion of this. Having competing methodologies to solve the problem led Smale to develop a framework in which to compare their efficiency. Smale's article "The Fundamental Theorem of Algebra and Complexity Theory" [Smale, 1981] is a startling step forward. Consider the first three paragraphs of the paper:

> The main goal of this account is to show that a classical algorithm, Newton's method, with a standard modification, is a tractable method for finding a zero of a complex polynomial. Here, by "tractable" I mean that the cost of finding a zero doesn't grow exponentially with the degree, in a certain statistical sense. This result, our main theorem, gives some theoretical explanation of why certain "fast" methods of equation solving are indeed fast. Also this work has the effect of helping bring the discrete mathematics of complexity theory of computer science closer to classical calculus and geometry.

A second goal is to give the background of the various areas of mathematics, pure and applied, which motivate and give the environment for our problem. These areas are parts of (a) Algebra, the "Fundamental theorem of algebra", (b) Numerical analysis, (c) Economic equilibrium theory and (d) Complexity theory of computer science.

An interesting feature of this tractability theorem is the apparent need for use of the mathematics connected to the Bieberbach conjecture, elimination theory of algebraic geometry, and the use of integral geometry.

The scope of the undertaking is very large and the terrain unclear. It was the discrete theoretical computer scientists who had the most developed notion of algorithm, of cost, and of complexity as a function of input size. For them, the space of problems, complex polynomials of a given degree, so natural for a mathematician is not natural because it is not discrete. From another direction, numerical analysts have practical experience with root-finding, algorithms which are fast and algorithms which are sure, algorithms which are stable and those which are not. In [Smale, 1985], we see Smale grappling again with these questions. First, there is the quote from von Neumann quoted again in [Smale, 1990].

The theory of automata, of the digital, all or none type, as discussed up to now, is certainly a chapter in formal logic. It would, therefore, seem that it will have to share this unattractive property of formal logic. It will have to be from the mathematical point of view, combinatorial rather analytical.

... a detailed, highly mathematical and more specifically analytical, theory of automata and of information is needed.

and Chapter 2, Section 6

6. What is an algorithm?
PROBLEM 11. What is the fastest way of finding a zero of a polynomial? This is a kind of super-problem. I would expect contributions by several mathematicians rather than a single solution. It will take a lot of thought even to find a good mathematical formulation.

In some ways, one could compare this problem with showing the existence of a zero of a polynomial. The concept of complex numbers had to be developed first. For Problem 11, one must develop the concept of algorithm to deal with the kind of mathematics involved. Consistent with the von Neumann statement quoted in the introduction, my belief is that the Turing approach to algorithms is inadequate for these purposes.

Although the definitions of such algorithms are not available at this time, my guess is that some kind of continuous or differentiable machine would be involved. In so much of the use of the digital computer, inputs are treated as real numbers and the output is a continuous function of the input. Of course a continuous machine would be an idealization of an actual machine, as is a Turing machine.

The definition of an algorithm should relate well to an actual program or flow-chart of a numerical analyst. Perhaps one could use a Random Access Machine (RAM, see Aho–Hopcraft–Ullman) and suppose that the registers could hold real numbers.

Then one might with some care expand the list of permissible operations. There are pitfalls along the way and much thought is needed to do this right.

To be able to discuss the fastest algorithm, one has to have a definition of algorithm. I have used the word algorithm throughout this paper, yet I have not said what an algorithm is. Certainly the algorithms discussed here are not Turing machines; and to force them into the Turing machine framework would be detrimental to their analysis. It must be added that the idealizations I have suggested do not eliminate the study of round-off error. Dealing with such loss of precision is a necessary part of the program.

Problem 11 is not a clear-cut problem for various reasons. Factors which could affect the answer include dependence on the machine, whether one wants to solve one or many problems, time taken to write the program, whether polynomials have large or small degree, how the problem is presented, etc.

So in [Smale, 1981] he has launched into a discussion of the total cost of an algorithm without a precise notion of cost or algorithm available; these will come later! To have enough conviction that such a long-range project will work out is not uncharacteristic of Smale.

When Smale was awarded the Fields Medal in 1966 for his work in differential topology, René Thom wrote (translation my own):

... Smale is a pioneer, who takes his risks with calm courage, in a completely unexplored domain, in a geometric jungle of inextricable richness he is the first to have cleared a path and planted beacons. [Thom, 1966]

Returning to the 1981 paper, Smale restricted the class of functions for which roots are to be found to complex polynomials of one variable and degree d, normalized as $f(z) = \sum_{i=0}^{d} a_i z^i$ with $a_i \in \mathbb{C}$ $a_d = 1$ and $|a_i| \leq 1$ for $1 \leq i \leq d$; call this space $P_d(1)$. For complex polynomials f, the Jacobian is always ≥ 0, so global Newton can be taken as

$$\frac{dx}{dt} = -Df(x)^{-1}f(x).$$

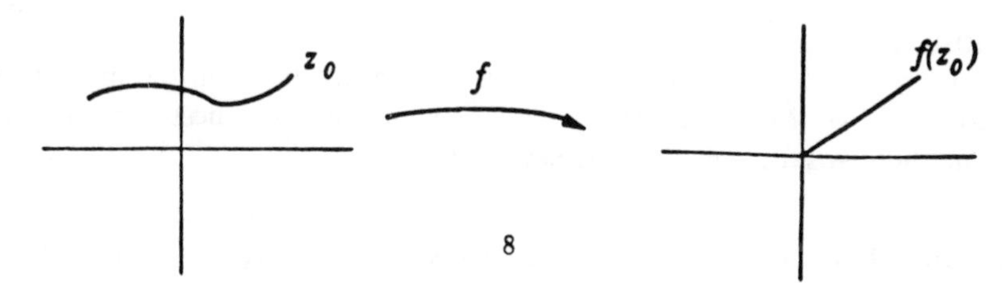

FIGURE 3

Now the solution curve through z_0 is the branch through z_0 of the inverse of image of the ray from $f(z_0)$ to 0, and because f is proper, it does lead to a zero of f with the exception of a finite number of rays which contain the critical values of f. Incidentally, this argument proves the fundamental theorem of algebra. To turn the proof into an algorithm, Smale considers the Euler approximation to the solution of the differential equation, for h a positive real,

$$N_h(f)(z) = z - hDf^{-1}(z)f(z).$$

When $h = 1$, this is Newton's method; we denote this also by $N(f)$. Newton's method gives quadratic convergence near simple zeros, so Smale defines an approximate zero of f as a point where Newton's method is converging quadratically.

Definition. z_0 is an approximate zero for f iff $|z_k - z_{k-1}| \leq (1/2)^{2^{k-1}-1}|z_1 - z_0|$, where $z_i = N(f)z_{i-1}$. The goal becomes to find an approximate zero of f.

Smale uses $x_0 = 0$ as a starting point for his algorithm and then iterates $N_h(f)$. The first main theorem using Lebesgue measure on $P_d(1)$ in [Smale, 1981] is:

Main Theorem. *There is a universal polynomial $S(d, 1/\mu)$, and a function $h = h(d, \mu)$ such that for degree d and μ, $0 < \mu < 1$, the following is true with probability $1 - \mu$. Let $x_0 = 0$. Then $x_n = N_h(f)(x_{n-1})$ is well-defined for all $n > 0$ and x_s is an approximate zero for f where $s = S(d, 1/\mu)$.*

More specifically, we can say, if $s \geq [100(d + 2)]^9/\mu^7$, then with probability $1 - \mu$, x_s is well-defined by the algorithm for suitable h and x_s is an approximate zero of f.

Borrowing the notion of polynomial cost computation from computer sciences, Smale has proven that approximate root-finding is "polynomial" in the degree d and one over the probability of failure.

The theorem reflects Smale's notion that fast algorithms fail sometimes (as Newton's method does) and are slow near where they fail, thus making a statistical analysis appropriate.

The proof of the main theorem is very long involving all the ingredients quoted above. Moreover, there are some outstanding questions on mean value theorems for polynomials. Smale proves:

Theorem 1. *Suppose $f(z)$ is a complex polynomial with $f(0) = 0$ and $f'(0) \neq 0$, then*

(a) *there is a critical point Θ (i.e., $f'(\Theta) = 0$) such that*

$$\frac{|f^{(k)}(0)|}{k!} \frac{|f(\Theta)|^{k-1}}{|f'(0)|^k} \leq 4^{k-1};$$

(b) *there is a critical point* Θ *with*

$$\frac{|f(\Theta)|}{|\Theta|}\frac{1}{|f'(0)|} \leq 4.$$

Smale raises the problem [Smale, 1981]:

Problem

(a) Can the 4 in (a) of Theorem 1 be reduced, perhaps to 1?
(b) Can the 4 in (b) of Theorem 1 be reduced to 1, in fact, to $1 - 1/d$?

I do not think there has been much progress on (a). There is some progress on (b) by Tischler [1989].

It was these problems that got me involved in analyzing Newton's method. On one of my many trips to Berkeley, Steve was working on Theorem 1 for his complexity analysis which was to become the 1981 paper. He asked me if I could prove something like Theorem 1 above for some constant; in fact, he thought the constant should be 1. I was able to see the first case of the inequality (here my memory is a little different than Steve's) for the second derivative by applying Bieberbach's estimate for a_2 to the inverse of f, which must be defined and injective on a disc of radius at least the smallest modules of a critical value. I told this to Steve and returned to New York. I had no idea what he wanted it for. The next year I was in Berkeley on a sabbatical, [Smale, 1981] was already written and Steve was teaching a course on it, which I took. Steve and I began collaborating during the semester; our work was finally published in [Shub-Smale, 1985] and [Shub-Smale, 1986a]. We extended the analysis that Steve did to higher-order methods which we called generalized Euler iterations. It was our impression at the time that using methods of order $\ln d$ for d degree polynomials might be the most efficient among "incremental" algorithms. We gave a lower bound estimate for the area of the approximate zeros in $P_d(1)$ across the unit disc which was later improved; see [Friedman, 1990; Smale, 1986]. By choosing starting points far from the roots, we were able to improve the cost estimates of [Smale, 1981] for finding approximate zeros. We also studied the problem of finding small values of polynomials. Examining the foliation of the complex plane $\mathbb{C}$ by inverse images of rays for a polynomial f, we found that the inverse images of rays with small angle with respect to critical rays occupy about the same proportion (i.e., angle) of a large circle. For algorithms of Newton–Euler type we proved:

Theorem A. *For each* f, ε, *there is a Newton–Euler Algorithm which terminates with probability* 1 *and produces a* z *with* $|f(z)| < \varepsilon$. *The average number of iteration is less than* $O(d + |\log \varepsilon|)$.

Theorem B. *There is a Newton–Euler algorithm which produces an approximate zero for $f \in P_d(1)$ with probability* 1 *and average number of iterations* $O(d \log d)$.

Theorem B improved the main theorem in [Smale, 1981] in the sense that the estimate given there does not produce a finite average cost algorithm. Theorem A was improved in [Smale, 1985a; Kim, 1988a; and [Renegar, 1987b], where $|\log \varepsilon|$ is replaced by $\log|\log \varepsilon|$. Also Schonhage and others have zero finding algorithms of different flavor. Most recently, Neff [1990] has made a good contribution. Renegar [1987a] proved and an n-variable analogue of Smale's 1981 main theorem, and Canny [1988, 1990], Renegar [1989], and others have studied more algebraic approaches to the n-variable root-finding problem. Recently, Sutherland [1989] has an interesting result about the convergence of Newton's method itself on large circles.

Already in [Smale, 1981], the average case analysis of the Dantzig's simplex method for linear programming is cited as Problem 6. Consonant with Smale's perspective that fast algorithms are not always fast the simplex method was known to be a worst-case exponential but practically highly efficient. The linear programming problem (LPP) is: Given $m \times n$ matrices A and vectors $b \in R^m$, $c \in R^n$, determine if the function cx has a minimum on $Ax \geq b$ and $x \geq 0$. If it does, find a point x which minimizes it.

Taking a Gaussian distribution on $R^{mn} \times R^m \times R^n$ and letting $\rho(m, n)$ be the number of steps of Dantzig's self-dual method to solve LPP, Smale [1983a] proves that ρ is sublinear in the number of variables; see also [Smale, 1983b].

Theorem 2. *Let p be a positive integer. Then depending on p and m there is a positive constant* c_m *such that for all n*

$$\rho(m, n) \leq c_m n^{1/p}.$$

In [Smale, 1985a] which won the Chauvenet Prize of the Mathematical Association of America in 1988, Smale confronts some new issues. First is the problem of ill-posed problems. In the fall of 1983, Lenore Blum was visiting New York and we studied the problem of the average loss of precision (or significance) in evaluating rational functions of real variables. Let δ be the input accuracy necessary for desired output accuracy ε. Then $|\ln \delta| - |\ln \varepsilon|$ is the loss of precision (or significance for relative accuracy). We showed this loss was tractable on the average [Blum–Shub, 1986]; Smale focused on linear algebra. There the condition number, $K_A = \|A\| \, \|A^{-1}\|$, of a matrix A measures the worst-case relative error of the solution x of the equation $Ax = b$ divided by the relative error of the input b. Thus, $\log K_A$ measures the worst-case loss of significance. Smale wrote an explicit integral for the average of $\log K_A$ and Ocneanu, Kostlan, Renegar, and others made progress toward its estimation. Finally, Edelman [1988] has shown that up to an additive constant the average is $\ln n$. This result helps explain the success of fixed precision computers in solving fairly large linear systems. Demmel [1987a; 1987b] interprets the condition number as the inverse of the distance to the determinant zero variety, i.e., the singular matrices. Thus, there is an analogy between the success of fast algorithms and the intrinsic difficulty of robust computation. They are both measured in terms of distance to a subvariety of bad problems. This theme surfaces again in Smale's work, although the precise relationship remains somewhat mysterious.

The same 1985 paper dealt with two other problems. The efficiency of approximation of integrals: I will not say much about this except that Smale showed that the trapezoid rule is more efficient than Riemann integration on the average for fixed error on $\mathscr{H}^1$ functions, similarly Simpson's rule is more efficient for $\mathscr{H}^2$ functions. Newton's method is an example of a purely iterative algorithm for solving polynomial equations. A purely iterative algorithm is given as a rational endomorphism of the Riemann sphere which depends rationally on the coefficients of the polynomials (of fixed degree d) which are to be solved. A purely iterative algorithm is generally convergent if for almost all (f, x) iterating the algorithm on x, the iterates converge to a root of f. Smale [1986] conjectures that there are no purely iterative generally convergent algorithms for general d. McMullen [1988] proved this for $d \geq 4$ and produced a generally convergent iterative algorithm for $d = 3$. For $d = 2$, Newton's method is generally convergent. Doyle and McMullen [1979] have gone on to add to this examining $d = 5$ in terms of a Galois theory of purely iterative algorithms. In contrast, Steve and I showed in [Shub–Smale, 1986b] that if complex conjugation is allowed, then there are generally convergent purely iterative algorithms even for systems of n complex polynomials of fixed degree in n variables.

Smale has devoted a lot of effort to understanding Newton's method. These are recounted in [Smale, 1986], but let me mention a few of the results of this paper and [Smale, 1985b].

To generalize the one-variable theory, Smale considers the zero-finding problem for $f: E \to F$, where f is an analytic map of Banach spaces. Newton's method is the same $z' = N(f)z = z - Df(z)^{-1}f(z)$, and the definition of approximate zero is the same.

Definition. z_0 is an approximate zero for f iff $\|z_k - z_{k-1}\| \leq (1/2)^{2^{k-1}-1} \|z_1 - z_0\|$, where $z_i = N(f)z_{i-1}$.

Let $\beta(z,f) = \beta(z) = \|Df(z)^{-1}f(z)\|$, i.e., β is the norm of the Newton step $z' - z$. Let $\gamma(f,z) = \gamma(z) = \sup_{k\geq 2} \|(1/k!)Df(z)^{-1}D^k f(z)\|^{1/k-1}$ and let $\alpha(z,f) = \beta(z,f)\gamma(z,f)$; from [Smale, 1985a].

Theorem A. *There is a naturally defined number α_0 approximately equal to* 0.130707 *such that if $\alpha(z,f) < \alpha_0$ then z is an approximate zero of f.*

Thus, we have a test for approximate zeros in terms of data computed at the point z alone, and which is, hence, quite different from Kantorovich-type estimates for the domain of convergence of Newton's method that require estimates of derivatives on a neighborhood, and it is very useful. Smale uses it to get an estimate on global Newton. Given $f: E \to F$ and $z_0 \in E$, suppose $Df^{-1}(z)$ is defined on the whole ray $tf(z_0)$ for $0 < t \leq 1$, so that we may invert the ray. Call the inverse image σ. Let $M(z_0, f) = \max_{z \in \sigma} \alpha(z,f)/\|f(z)\|$ and ∞ if σ is not defined.

Theorem 3 [Smale, 1986] (The Speed of a Global Newton Method). *There exist (small) positive constants c real; ℓ an integer with this property. Let f: $E \to F$ be analytic, $z_0 \in E$ with $M(z_0, f) < \infty$. Suppose n is an integer*

$$n > c\|f(z_0)\| M(z_0, f), \quad \Delta = 1/n.$$

Let $w_i = (1 - i\Delta)f(z_0)$, $i = 0, \ldots, n$. Then, inductively, $z_i = N^{\ell}_{f-w_i}(z_{i-1})$ is well-defined and z_n is an approximate zero of f.

Renegar and I [Renegar–Shub, 1992] use a version of these theorems to give a simplified and unified proof of the convergence properties of several of the recent polynomial time linear programming algorithms. For one variable, Kim [1988a] also considered the algorithms of Theorem 3.

Smale [1986] also estimates $\gamma(f, z)$ for a polynomial f in terms of the norms of the coefficients, the norm of z, and the norm of the derivative of f at z. With this estimate, he was able to prove:

Theorem 4 [Smale, 1986]. *The average area of approximate zeros for $f \in P_d(1)$ is greater than a constant $c > 0$, where c is independent of d.*

Smale [1986] also dealt with a regularized version of the linear programming problem which I will not consider here. Smale [1985a] asserts:

> A study of total cost for algorithms of numerical analysis yields side benefits. It forces one to consider global questions of speed of convergence, and in so doing one introduces topology and geometry in a natural way into that subject. I believe that this will have a tendency to systematize numerical analysis. This development could turn out to be comparable to the systematizing effect of dynamical systems on the subject of ordinary differential equations over the last twenty-five years.

We have already seen geometry. In [Smale, 1987], he turns his attention to topology. To begin with, Smale defines a (uniform) algorithm for a problem. An algorithm is a rooted tree, the root at the top for input. Leaves are at the bottom for output. There is an input space $\mathscr{I}$ a state space $\mathscr{S}$, and an output space $\mathscr{O}$ which are finite-dimensional real vector spaces.

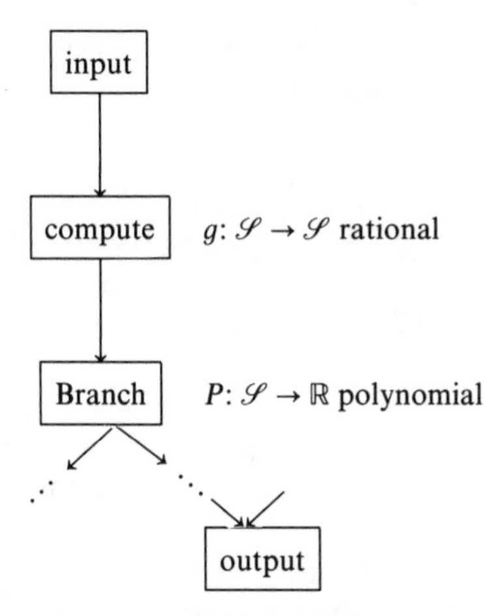

There are two additional types of nodes computation and branching nodes. Computations are rational functions from $\mathscr{S}$ to $\mathscr{S}$. The input and computation nodes have one existing edge. The branch nodes have two and which is chosen is determined by a polynomial inequality $p(x) \geq 0$ or $p(x) < 0$. The input and output functions are rational.

These algorithms are also called tame machines; they are limited, for example, by the fact that there are no loops.

A problem is a subset $X \subset \mathscr{I} \times \mathscr{O}$. Let $f: X \to Y$ be the restriction of the projection $\pi: \mathscr{I} \times \mathscr{O} \to \mathscr{O}$ to X and assume f is surjective. A solution of a problem is a section $\sigma: Y \to X$ such that $f \circ \sigma$ is the identity. A tame machine whose input–output map on inputs $y \in Y$ is a solution to a problem is said to solve the problem.

The minimum number of branch nodes of a tame machine which solves a problem is the branching complexity or topological complexity of the problem.

An example of a problem is the ε-all root problem for $f \in P_d$, the space of complex univariate monic polynomials.

$$X_\varepsilon \subset P_d \times \mathbb{C}^d = \{(f,(a_1,\ldots,a_d)) | f = \prod_{i=1}^{d} (z - r_i) \text{ and } |a_i - r_i| < \varepsilon\} \qquad \text{and}$$

$$Y = P_d.$$

These definitions of problem and algorithm are a large step in the program called for in [Smale, 1985a] for the definition of an algorithm. Given a problem $f: X \to Y$, let $K(f)$ be the kernel of $f^*: H^*(Y) \to H^*(X)$ and $K(f) = \{\gamma \in H^*(Y) | f^*(\gamma) = 0\}$, where $H^*(Y)$ is the singular cohomology ring of Y. The cup length of $K(f)$ is the maximum number of element $\gamma_1, \ldots, \gamma_k$ of $K(f)$ s.t. the cup product $\gamma_1 \cup \cdots \cup \gamma_k \neq 0$. Smale [1987] proves:

Theorem 5. *Let $f: X \to Y$ be a problem. The topological complexity of f is bigger than or equal to the cup length of $K(f)$.*

He applied this result to prove, via a complex algebraic topology computation, the main theorem of Smale [1987].

Theorem 6. *There is a* $\varepsilon(d) > 0$ *such that for all* $0 < \varepsilon < \varepsilon(d)$ *the topological complexity of the* ε*-all root problem for* P_d *is greater than* $(\log_2 d)^{2/3}$.

This result has been vastly improved by Vasiliev, [1988]. Levine [1989] has also worked on n-dimensional analogues.

In the fall of 1987, Lenore Blum and Steve Smale were visiting at Watson. Steve began extending his model of computation from tame machines to allow loops and to work over ordered rings. Soon all three of us were involved.

First, we specify a ring, the functions we compute on the ring, and the branching structure. Our functions are polynomial or rational, involving only a fixed finite number of variables, and having a finite number of nontrivial coordinates. We branch on $\neq 0$ or $= 0$ and ≥ 0 or < 0. Examples are:

1. the integers $\mathbb{Z}$ with polynomial functions (i.e., with $+$, $-$, $\times$) and branching on ≥ 0 or < 0;
2. the integers $\mathbb{Z}$ with polynomial functions and branching on $\neq 0$ or $= 0$;
3. the reals $\mathbb{R}$ with rational functions and branching on ≥ 0 or < 0;
4. the complexes $\mathbb{C}$ with rational functions and branching on $\neq 0$ or $= 0$.

Frequently, we suppress the functions or the branching structure. If we discuss machines over $\mathbb{Z}$, $\mathbb{R}$, or $\mathbb{C}$ without further qualification, we mean 1, 3, and 4, respectively.

Our machines now are finite-directed graphs with one input node, computation nodes, branch nodes, output nodes, and a certain fifth node. The input and output spaces are the infinite direct product of the ring with itself, R^∞, and the state space is $\mathbb{Z}_+ \times \mathbb{Z}_+ \times R^\infty$. All computations only involve and affect a finite number of the coordinates of R^∞. The two $\mathbb{Z}_+$'s are like counters and a fifth node will copy the contents of the jth coordinate to the ith of R^∞ if the first two coordinates are (i, j).

This is all worked out formally in [Blum–Shub–Smale, 1989] except that there we always assume the ring ordered and branching done on ≥ 0 or < 0.

By doing things in an integrated way, we hope that the various settings will illuminate one another, that the theory of recursive functions and complexity over $\mathbb{R}$ or $\mathbb{C}$, for example, will benefit from the more developed discrete theory. On the other hand, we do not want to be so general that we lose the basic contacts with algebra and the geometry and topology of the reals and complexes.

A simple example of a machine over $\mathbb{R}$ is given by the iterates of the complex polynomial $g(z) = z^2 + 1$.

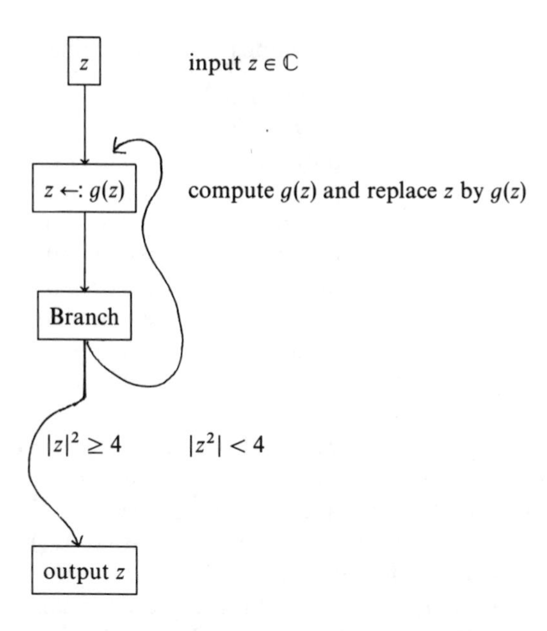

This machine halts on the basin of infinity and fails to halt on its complement, the Julia set.

The computable functions over the integers are the usual computable functions, while over the reals, most Julia sets are not halting sets. Universal machines are constructed over any ring.

To study the complexity of a problem, two additional pieces of data are required: the input size of the problem and the cost of the computation.

The *size* of an element is the *length* plus the *height*. The *length* of an element is the least k such that $x = (x_1, \ldots, x_{k-1}, 0, \ldots, 0, \ldots)$ and $x_i = 0$ for $i \geq k$. The *height* of $x = (x_1, \ldots, x_i, \ldots)$ is max *height* x_i over all i. Now we need to define the height of an element of the ring R. For the integers $\mathbb{Z}$, height $x = \log(|x| + 1)$. With this notion of height, we will call the size the *bit size*. For the reals $\mathbb{R}$ or complexes $\mathbb{C}$, the height is perhaps most naturally taken as one for any element. Then the size is the dimension. But one might also want the height to reflect the size of the number and then one might take height $x = \log(|x| + 1)$ in analogy with the integers. I will call this logarithmic height and the corresponding size logarithmic size. Given a notion of height, there is a natural notion of cost of a computation. Suppose on input $y \in \mathscr{I}$, machine M visits $(m - 1)$ nodes in succession and finally arrives at an output node. Then we say the time of the computation $Y_M(y) \equiv m$ and the cost $C_M(y)$ is the time $_M(y) \times$ max height generated in the computation. For the integers and logarithmic height, we call the cost the bit cost. For height identically one, we call the cost the algebraic cost.

Thus, we have various settings in which to discuss complexity. I list a few.

Input size	Ring	Functions	Branching	Cost
1) bit	$\mathbb{Z}$	polynomial	≥ 0 or <0	bit
2) bit	$\mathbb{Z}$	polynomial	≥ 0 or <0	algebraic
3) bit	$\mathbb{Z}$	polynomial	$=0$ or $\neq 0$	bit
4) bit	$\mathbb{Z}$	polynomial	$=0$ or $\neq 0$	algebraic
5) dimension	$\mathbb{R}$	rational	≥ 0 or <0	algebraic
6) logarithmic	$\mathbb{R}$	rational	≥ 0 or <0	logarithmic
7) dimension	$\mathbb{C}$	rational	$\neq 0$ or $=0$	algebraic
8) logarithmic	$\mathbb{C}$	rational	$\neq 0$ or $=0$	logarithmic
9) bit	$\mathbb{Z}$	linear	$\neq 0$ or $=0$	bit
10) dimension	$\mathbb{R}$ or $\mathbb{C}$	linear	$=0$ or $\neq 0$	algebraic

Given a problem $X \subset \mathscr{I} \times \mathcal{O}$, $f: X \to Y$, we say that f is in class P if there is a machine M which solves the problem, a real constant $c > 0$, and a positive integer q such that for input $y \in Y$, $C_M(y) \leq c(\text{size } y)^q$.

A special class of problems are decision problems (Y, Y_{yes}): Given a set of inputs Y and a subset Y_{yes}, determine if $y \in Y$ is in Y_{yes}.

A decision problem (Y, Y_{yes}) is in NP if there are constants $c > 0$, $q \in \mathbb{Z}_+$ and a machine M with input space $\mathscr{I} \times \mathscr{I}'$ such that on inputs $(y, y') \in Y \times \mathscr{I}'$

(a) M outputs 1 (yes) or 0 (no);
(b) M outputs 1 only if $y \in Y_{\text{yes}}$;
(c) if $y \in Y_{\text{yes}}$, then there is a $y' \in \mathscr{I}'$ such that M outputs 1 on (y, y') and $C_M(y, y') \leq c(\text{size } y)^q$.

This is the formalism of NP completeness theory (see [Garey–Johnson, 1979]). In analogy with the standard problem in setting (1), one may ask for decision problems:

Problem. Does $P \neq NP$?

A decision problem (Y, Y_{yes}) is called NP complete if given any other decision problem (Y', Y'_{yes}) there is a machine M which maps (Y', Y'_{yes}) to (Y, Y_{yes}) faithfully (i.e., M takes an input $y' \in Y'$ into Y_{yes} iff $y' \in Y'_{\text{yes}}$) and $\exists c > 0$, $q \in \mathbb{Z}_+$ such that $C_M(y') \leq c(\text{size } y')^q$ for all $y' \in Y'$.

Thus, $P \neq NP$ iff any NP-complete problem is not in P.

For each complexity setting (1)–(10) above, we may consider the existence of NP complete problems and the question: Is $P \neq NP$? In [Blum–Shub–Smale, 1989], we mainly considered case (5) which we simply refer to as over $\mathbb{R}$. Similarly, the standard setting (1) is called over $\mathbb{Z}$ and (7) is called over $\mathbb{C}$. The next theorem is proved in [Blum–Shub–Smale, 1989] over $\mathbb{R}$, but the arguments apply more generally; so I state it over $\mathbb{C}$ as well.

Theorem 7. (1) *Let* 4-*satisfiability be the decision problem* $(\mathscr{F}, \mathscr{F}_{\text{yes}})$ *where* $\mathscr{F}$ *is the set of all* 4th *degree polynomials* $f: \mathbb{R}^n \to \mathbb{R}$ *for all* $n \in \mathbb{Z}_+$ *and* $\mathscr{F}_{\text{yes}}$ *are* *those f such that $\exists x \in \mathbb{R}^n$ with $f(x) = 0$. Then 4-satisfiability is NP complete over R.*

(2) *Let Hilbert Nullstellensatz (HN) be the decision problem $(\mathcal{F}, \mathcal{F}_{\text{yes}})$ where $\mathcal{F}$ is all sets of k polynomials of arbitrary degree*

$$f_i: \mathbb{C}^n \to \mathbb{C}, \quad i = 1, \ldots, k,$$

for all n and $k \in \mathbb{Z}_+$ and $\mathcal{F}_{\text{yes}}$ are those for which the algebraic set they define is nonempty, i.e., $\exists x \in \mathbb{C}^n$ s.t. $f_i(x) = 0$, $\forall i = 1, \ldots, k$, or what is the same set of those for which 1 is not in the ideal generated. Then HN is NP-complete over $\mathbb{C}$.

The $\mathcal{F}$ in HN could also be replaced by systems of degree $\leq d$ where $d \geq 2$.

In Rio de Janeiro, last January, Smale discussed the problem of the existence of NP-complete problems and the question $P \neq NP$ in various settings. Megiddo [1993] proves a general NP-completeness theorem. The question $P \neq NP$ does not make much sense in all of our settings because in settings (2) and (4), there are problems in NP which are not even decidable. In Rio, Smale proved that $P \neq NP$ in settings (9) and (10) and presented some ideas on (3), (6), and (8). In [Shub, 1993], I give a simple argument for setting (3).

To extend the notion of the polynomial time algorithm to the numerical analysis context requires incorporating approximate solutions and round-off error into the problem. It is not exactly clear in all cases how to do this. A beginning was made in [Blum–Shub–Smale, 1989; Smale, 1990] went further.

In [Blum–Shub–Smale, 1989, we consider ε as a variable and the effect of scaling on approximate solutions. We take as a new size the logarithmic size $+|\ln \varepsilon|$, and as cost function the algebraic cost. In [Smale, 1990], the round-off error is incorporated and probabilistic algorithms. For round-off error, the size is taken as dimension $+\ln|\varepsilon| + \ln W$, where W is a weight which might reflect the logarithmic size and the inverse of the distance to the ill-posed problems. The cost reflects the admissible round-off error and the polynomial class is called numerically stable. Whereas the notion of numerical stability is good for computations, it does not seem good for algorithms because according to the definition there is no numerically stable algorithm for the square root. So here I will make another attempt which is slightly different in the definition of admissible error and uses a different cost function. The computation of a machine M on input y is described by the computing endomorphism $(n_i, x_i) = H_M(n_{i-1}, x_{i-1})$, where essentially (n_{i-1}, x_{i-1}) is the (node, state) at time $i - 1$ and (n_i, x_i) is the (next node, next state) with the further conditions that n_0 is the input node, $x_0 = y$, $n_{T_M(y)}$ is an output node, and M output $x_{T_M(y)}$.

We will say that the sequence (n_i, x_i) is a *δ-pseudo-computation* if n_0 is the input node, and for $i \geq 1$ $|(n_i, x_i) - H_M(n_{i-1}, x_{i-1})| \leq \delta$.

Now given a problem $f: X \to Y$, a machine M, an input $y \in Y$, and an $\varepsilon > 0$, then δ is an *admissible error* (*round-off and input error*) for (ε, y, M) if for any δ-pseudo-computation x_i with $|y - x_0| \le \delta$ there is a first time I such that n_I is an output node and x_I is within ε of a solution to problem instance y, i.e., $\exists x \in f^{-1}(y)$ such that the distance from x_I to x is less than or equal to ε.

Let $\delta(\varepsilon, y) = \delta(\varepsilon, y, M)$ be the maximum admissible round-off error. The round-off cost

$$C_R(\varepsilon, y) = \min_{0<\delta\le\delta(\varepsilon,y)} \text{max over } \delta\text{-pseudo-computations of}$$

$$I \times \left(\max_{0\le i\le I} \text{logarithmic height } (x_i) + |\log^- \delta(\varepsilon, y)| \right),$$

where $\log^-$ is $\min(\log, 0)$.

The input size of input (ε, y) as in [Smale, 1990] is taken as

$$S(\varepsilon, y) = \text{dimension} + \log|\varepsilon| + \log W,$$

where W is a weight representing the size of y, the inverse of the distance of y to the ill-posed problems, or something like the condition of the problem instance y; see [Smale, 1990] for a discussion of this. The polynomial class are those problems for which there are algorithms (machines) for which $\exists c > 0$ and $q \in \mathbb{Z}_+$ such that

$$C_R(\varepsilon, y) < cS(\varepsilon, y)^q.$$

For univariate polynomial root-finding, Myong-Hi Kim has found such algorithms [Kim, 1988b]. When round-off error is not taken into account, $|\log|\log(\varepsilon)||$ is perhaps more appropriate than $|\log \varepsilon|$; see [Renegar, 1987b]. The role of the distance to the ill-posed problems is less clear, although size of coefficients must play a role because of scaling.

The work of Smale in conjunction with Blum [Blum–Smale, 1993] on Gödels theorem is in this volume so I will not comment on it. Also to be mentioned are the expository articles [Smale, 1988; 1989] where some of Smale's philosophy on real number machines are exposited for a general audience.

Appendix: Personal Reminiscences by Mike Shub

I first met Steve Smale during the 1961–62 academic year. I was a sophomore at Columbia College. Some of my older roommates had a copy of one of Steve's papers on structural stability and they couldn't make out some of the subtleties of the definition. I think it had to do with epsilons and deltas. Somehow they put me up to going to see Steve and asking about it. Steve answered my question rapidly and succinctly but at a depth I didn't even know existed. Apparently I had phrased my elementary question in terms close to problems he was thinking about. He must have thought me quite presumptuous. I am still frequently shocked when Steve answers some question of mine or a colloquium question at a depth I didn't imagine.

The next year my friends advised me to try to take Steve's graduate course in differential topology since he was a famous topologist who had proven a great theorem. In those days Columbia College didn't have much of an undergraduate mathematics curriculum. Many math majors vied in taking graduate courses which we were frequently hopelessly unprepared for. I enrolled for Steve's course. The very first day, he arrived and announced that the course would be about infinite-dimensional differential topology because that was where the most interesting work was to be done. In the class the first day was Sammy Eilenberg and other luminaries of the Columbia math department. Steve began by defining the derivative in Banach space. He didn't quite get it right, and the class degenerated as various of the luminaries shouted out suggested corrections. I ran out and bought Lang's infinite-dimensional *Introduction to Differential Topology* downtown at the publisher's office (it wasn't in the bookstores yet), and began struggling with Lang and Ralph Abraham's course notes which were trailing the lectures. I was always behind, but some of the seniors could follow. Sometimes as I was sitting through a lecture which I couldn't understand I broke into giggles as Steve would get confused at the blackboard only to be saved by an undergraduate. One day, Serge Lang took David Frank who was also enrolled in Steve's course and me to lunch. He explained that while the undergraduates were locally correct, Steve was almost always locally wrong but globally correct. Actually, in the many courses I have taken from Steve since I haven't noticed so many local errors, but Serge's hyperbole was comforting at the time. Steve left Columbia after the spring semester 1964 for Berkeley. David Frank and I decided to go to Berkeley for graduate school. Steve and Clara offered us their car to drive across the country. David and Kathy Simon were getting married. So the three of us first drove to Pittsburgh for their wedding and then on to California. Beth Pessen and I got married that September.

The fall semester at Berkeley that year was dominated by the Free Speech Movement. David, Kathy, Beth, and I were loyal foot soldiers in the movement. Steve Smale and Moe Hirsch were prominent faculty supporters. The Free Speech Movement had a very good effect on faculty–student relations in general on the Berkeley campus. The faculty and students were thrown together and made common cause on a political matter where they were more equals than in academic disciplines. The faculty became more aware of student concerns and reached out to accommodate a spirit of reform and even revolution. David, Kathy, Beth, and I were arrested in Sproul Hall. David and I were, I believe, the only math graduate students among the 800 or so students arrested there on December 8. David was taken to the Berkeley jail. I to Santa Rita. Steve actually went down to the Berkeley jail and bailed David out. I was released early the next morning when the faculty raised the funds to bail us out en masse.

That spring Steve was co-chairman of the Viet Nam Day Committee with Jerry Rubin. I was on the steering committee. Sometime that year I remember Steve telling Charles Pugh that he had proven that structurally stable systems were not dense. I was amazed by his ability to do research in the midst of all the turmoil. Beth and I were frequently at Steve and Clara's for dinners and parties that year and the next, and they were sometimes at ours. The last one that I remember was in our house shortly before Steve and Clara left for Europe in the summer of 1966. I remember telling Steve that I thought a teach-in should be organized at the International Congress of Mathematicians in Moscow that year. Given what actually happened on the steps of Moscow University as Steve recounts it, a teach-in in 1966 was quite far-fetched. By the spring of 1966 I had already become Steve's graduate student. The initial problem, which was quickly done, was to prove the Kupka–Smale theorem for endomorphisms. I used to stop by at Steve's office almost daily to say hello, and tell him anything new or ask questions. He was always happy to see me and to hear anything new. But he wasn't too interested in technical details or vague ideas. When he started biting his lower lip, I knew it was time to go. On one of these visits I told Steve that I thought that the squaring map on the circle was structurally stable. Thus, the more extensive part of my thesis research began. Steve was always helpful and encouraging, and good about the big picture. I remember some advice which took place in strange circumstances. Once we encountered each other running in different directions as the police were breaking up a demonstration on Telegraph Avenue. Steve stopped for a moment and said that he understood why expanding maps were stable; they were contracting. I never found out what he meant as we had to start running again almost immediately. Another time, we were in the Greek Theater where some sort of Vietnam War protest was taking place; in the midst of watching events on the stage, Steve asked me if I could prove the expanding map conjecture if I knew the fundamental group had a nilpotent subgroup of finite index. I already knew how to do it if the groups was nilpotent.

Those years in Berkeley were heady days not only for politics but for dynamical systems. Steve returned from infinite dumensions to dynamical systems theory. After the nondensity of structurally stable systems, he proved the omega stability theorem. He was writing his 1967 *Bulletin* paper which was a distillation and amplification of his previous work. The paper is a major restructuring of ordinary differential equations from the point of view of one of the leading topologists of the time. Steve's enthusiasm and the scope of his vision created a large group working on dynamics. Charles Pugh joined the faculty at Berkeley in 1964. Moe Hirsch got involved in dynamics, partly he has claimed because Steve went on leave and Steve's students came around to talk to Moe. Jacob Palis, Nancy Kopell, and I were the first bunch. There was enough interesting work for all of us and plenty more. Some of it was important for Steve's own work as well. In 1969, I was sharing an office with Steve at Warwick during the dynamical system year. I had been puzzled by a certain aspect of the stable manifold theorem for hyperbolic sets on and off for two years. Finally, I could put my finger on my objection. A technical point in the general theorem was not correct and the omega stability theorem depended on it. Steve tried to fill the gap for a while without success. A few days later, we were in our office and Steve was calmly sitting and working on something else. I asked him how come. He said he had been asking us guys to prove the theorem for some years and that we kept saying that we could. A little later that summer, Charlie, Jacob, Moe, and I indeed did prove a version which was correct and all that Steve needed.

Steve is always conscious of what he is doing and evaluating its position within science. He is willing to undertake enormous projects over the long term on subjects he finds important. He starts out full of energy and conviction that he will do something important and perhaps a bit naively, but he is extremely flexible and learns rapidly along the way. Partly, learning proceeds from talking to people a lot and taking what they say very seriously. Partly, it comes from going to the library a lot. His time in the library and confidence that he can learn what he needs to there remind me of the story I remember (I hope correctly) about Steve's education in a one-room schoolhouse where he looked up how to solve linear equations in the encyclopedia. By the time Steve's papers are finally written, they tend to be so clear and well-organized that it is difficult to detect the enormous effort that went into them. I have been partly involved in Steve's project on the theory of computation, and Steve and I have written a few joint papers by now. In 1981 while we were working on polynomial root-finding, we got a bit competitive as is both our wonts. Steve always works very hard. But I was lucky to be on sabbatical while he had to teach, so I could (barely) hold my own.

Over the years, my friendship with Steve has deepened. Recently Steve, Clara, Beate, and I were on our terrace in New York having a drink. Beate and I have been married for two years now. Steve described his plans for his photography. During the three days Steve was in New York to give a lecture, he also scoured the city looking for the perfect photography paper. "You see, Beate," I said, "Steve is a man of no small ambitions" and, I should have added, successes. I think that is true and marvellous. Yet Steve is gentle, direct, unpretentious, and honest. His views are frequently novel and refreshing from mathematics to movies and politics. His reactions personally and politically have always been sympathetic and on the side of basic human rights and decency, as long as I've known him from the Free Speech Movement and Vietnam protests until now. I have great admiration, respect, and affection for Steve and feel very lucky to have been his student and to be his colleague and friend.

References

Aho, A., Hopcraft, J., and Ullman, J. [1979], *The Design and Analysis of Computer Algorithms*, Addison-Wesley, Reading, MA.

Blum, L. and Shub, M. [1986], Evaluating rational functions: Infinite precision is finite cost and tractable on average, *SIAM J. Comput.* **15**, 384–398.

Blum, L., Shub, M., and Smale, S. [1989], On a theory of computation and complexity over the real numbers: NP-completeness, Recursive functions and Universal Machines, *Bull. Amer. Math. Soc.* **21**, 1–46.

Blum, L. and Smale, S. [1993], The Gödel incompleteness theorem and decidability over a ring, this volume.

Branin, Jr., F.H. [1972], Widely convergent methods for finding multiple solutions of simultaneous non-linear equations, *IBM J. Res. Develop.* **16**, (5) 504–522.

Canny, J. [1988], Some algebraic and geometric computations is *P*-space, Proceedings of the 20th Annual ACM Symposium on the Theory of Computing, pp. 460–467.

Canny, J. [1990], Generalized characteristic polynomials, *J. Symbol. Comput.* **9**, 241–250.

Demmel, J. [1987a], On condition numbers and the distance to the nearest ill-posed problem, *Numerische Math.* **51**, 251–289.

Demmel, J. [1987b], The geometry of ill conditioning, *J. Complexity* **3**, 201–229.

Doyle, P. and McMullen, C. [1989], Solving the Quintic by iteration, *Acta Math.* **163**, 152–180.

Edelman, A. [1988], Eigenvalues and condition numbers of random matrices, *SIAM J. Matrix Anal. Appl.* **9**, 543–560.

Friedman, J. [1990] Random polynomials and the density of approximate zeros *SIAM J. Comput.*, **19**, 1068–1099.

Garey, M. and Johnson, D. [1979], *Computers and Intractability*, Freeman, New York.

Hirsch, M.W. and Smale, S. [1979], On algorithms for solving $f(x) = 0$, *Commun. Pure Appl. Math.* **32**, 281–312.

Kim, M.-H. [1988a], On approximate zeros and root finding algorithms, *Math. Comput.* **51**, 707–719.

Kim, M.-H. [1988b], Error analysis and bit complexity: Polynomial root finding problem, Part I, preprint, Bellcore, Morristown, NJ.

Levine, H. [1989], A lower bound for the topological complexity of Poly(D, n), *J. Complexity* **5**, 34–44.

McMullen, C. [1988], Braiding of the attractor and the failure of iterative algorithms, *Invent. Math.* **91**, 259–272.

Megiddo, N. [1993], A general NP-completeness theorem, this volume.

Neff, C.A. [1990], Specified Precision Polynomial Root Isolation Is in NC, IBM Research Report, RC 15653, and to appear *J. Computer Syst. Sci.*

Renegar, J. [1987a], On the efficiency of Newton's method in approximating all zeros of a system of complex polynomials, *Math. Oper. Res.* **12**, 121–148.

Renegar, J. [1987b], On the worst-case arithmetic complexity of approximating zeros of polynomials, *J. Complexity* **3**, 90–113.

Renegar, J. [1989], On the worst case arithmetic complexity of approximating zeros of systems of polynomials, *SIAM J. Computing* **18**, 350–370.

Renegar, J. and Shub, M. [1992], Unified complexity analysis for Newton LP methods, *Math. Programming*, **53**, 1–16.

Schönhage, A. The fundamental theorem of algebra in terms of computational complexity, preliminary report, University of Tubingen.

Shub, M. and Smale, S. [1985], Computational complexity, on the geometry of polynomials and a theory of cost: Part I, *Ann. Scient. Ecole Norm. Sup.* **18**, 107–142.

Shub, M. and Smale, S. [1986a], Computational complexity, on the geometry of polynomials and a theory of cost: Part II, *SIAM J. Computing* **15**, 145–161.

Shub, M. and Smale, S. [1986b], On the existence of generally convergent algorithms, *J. Complexity* **2**, 2–11.

Shub, M. [1993], Some remarks on Bezout's theorem and complexity theory, this volume.

Smale, S. [1976], A convergent process of price adjustment and global Newton methods, *J. Math. Econ.*, **3**, 107–120.

Smale, S. [1981], The fundamental theorem of algebra and complexity theory, *Bull. Amer. Math. Soc.* **4**, 1–36.

Smale, S. [1983a], On the average number of steps in the simplex method of linear programming, *Math. Programming* **27**, 241–262.

Smale, S. [1983b], The problem of the average speed of the simplex method, *Proceedings of the XI International Symposium on Mathematical Programming, June 1983*, Bachem et al., eds., Springer-Verlag, Berlin, pp. 530–539.

Smale, S. [1985a], On the efficiency of algorithms of analysis, *Bull. Amer. Math. Soc.* **13**, 87–121.

Smale, S. [1985b], Newton's method estimates from data at one point, *The Merging of Disciplines: New Directions in Pure, Applied and Computational Methematics*, R.E. Ewing, et al. eds., Springer-Verlag, New York.

Smale, S. [1986], Algorithms for solving equations, Proceedings of the International Congress of Mathematicians, Berkeley, CA, American Mathematical Society, Providence, R.I., pp. 172–195.

Smale, S. [1987], On the topology of algorithms I, *J. Complexity* **3**, 81–89.

Smale, S. [1988], The Newtonian contribution to our understanding of the computer, *Queen's Quarterly* **95/1**, 90–95.

Smale, S. [1989], Newton's Contribution and the Computer Revolution, *Math. Medley* **2**, pp. 51–57.

Smale, S. [1990], Some remarks on the foundations of numerical analysis, *SIAM Rev.* **32**, 211–220.

Sutherland, S. [1989], Finding roots of complex polynomials with Newton's method, preprint, Institute for Math Science, SUNY, Stoney Brook.

Thom, R. [1966], Sur les Travaux de Stephen Smale, Proceedings of International Congress of Mathematicians, Moscow, pp. 25–28.

Tischler, D. [1989], Critical points and values of complex polynomials, *J. Complexity* **5**, 438–456.

Vasiliev, V.A. [1988], Braid group cohomologies and algorithm complexity, *Funct. Anal. Appl.* **22** (3), 182–190 (English translation).

29
Smale's Work on the Dynamics of Numerical Analysis

STEVE BATTERSON*

A major theme of this conference is the interdisciplinary nature of Smale's work. I would like to comment on the areas of dynamical systems and numerical analysis. For many years it has been clear that numerical analysis could contribute to the study of dynamical systems. Smale appreciated that it was natural to apply the theory of dynamical systems to the study of certain problems in numerical analysis. Let me illustrate by describing a general problem in each area.

Assume that M is a space and G is a continuous map of the space into itself. A fundamental problem in dynamical systems is to understand the asymptotic behavior of the orbit of a point under iteration by the map [i.e., $\lim_{n\to\infty} G^n(y)$]. A large variety of techniques (e.g., stable manifold theory, symbolic dynamics, bifurcation theory, structural stability, chain recurrence) have been useful in addressing problems of this nature.

Now consider an iterative numerical algorithm. For any approximation, the algorithm provides a rule for computing a (hopefully) more accurate approximation. For each initial value x_0, the algorithm yields a sequence of approximations $\{x_0, x_1, \ldots\}$. More formally, we may view the possible initial values as a space of candidates and the solutions as targets. The rule for producing subsequent approximations is a function, F, on the space of candidates. Each iteration sequence is then the orbit of a point under iteration by the map. A good example of an iterative algorithm is Newton's method for finding roots of a function h. The space of candidates is the set of complex numbers and the iteration function is defined by $F(x) = x - h(x)/h'(x)$.

A general numerical problem with Newton's method or any iterative algorithm is to determine the likelihood that $\lim_{n\to\infty} F^n(x)$ is a solution to the problem. Ideally, this property would be satisfied for any starting point, but this is probably too much to expect. Smale used the term *generally convergent* to describe an algorithm for which there is convergence to a solution for almost any initial point. Smale saw that when an iterative algorithm is

* The author was partially supported by the NSF.

viewed as a dynamical system, the numerical problem may also be couched as a dynamical systems problem. For example, to show that an algorithm is not generally convergent, it suffices to establish the existence of a periodic sink which is not a solution.

One of the fortunate consequences of Steve's reformulation of numerical analysis is the beautiful work which Curt McMullen will discuss this afternoon. I will close by mentioning a problem in the area of eigenvalue computation which is a natural candidate for the Smale approach. For 25 years, a variant of the Francis-shifted QR algorithm has been employed to approximate the eigenvalues of real matrices. In practice, the algorithm has been remarkably reliable and fast. Whereas it is known that there exist matrices for which the algorithm will not converge to a solution, it is unknown whether the algorithm is generally convergent. More details about this problem may be found elsewhere in this volume.

Personal Reflections

I began graduate school at Northwestern University in 1971. Whereas I had no idea what area of mathematics I wanted to pursue, I was certain that it would not be algebra. At that time, Northwestern (with Bob Williams, Clark Robinson, and John Franks) had what might be described as the midwest campus of the Smale Berkeley school of dynamical systems. During my first year, I took a geometry course from John Franks. I was struck by the clarity with which he explained mathematics and decided that was the quality I needed most in a thesis advisor.

Ziggy Nitecki's book had just come out and Clark Robinson suggested that I read it to obtain the background to enter dynamical systems. Prior to the publication of Ziggy's book, Smale's 1967 *Bulletin* "survey article" had been the rite of passage. As Jacob remarked, this article is much more than a survey. I would often refer to the *Bulletin* article and marvel (as I still do) at Smale's vision and his capacity to identify so much of the essence of the subject at such an early stage. On the other hand, Ziggy's book had a lot more details and I was very grateful that it was to be the basis for my orals. I survived the exam and began to work on a problem of Joel Robbin to classify linearly induced maps of projective space.

Northwestern was a good place to be a graduate student in dynamical systems. Not only were John, Clark, and Bob all very helpful, but most of the leading figures of the field passed through to give a talk at one time or another. I was invited to attend the parties and dinners for the visitors.

I especially remember Smale's visit. I was very excited to have an opportunity to meet the person who had had the greatest mathematical influence on my work. I recall telling my friends that it would be like meeting a movie star. Bob invited me to go with a group of 10 or so to dinner. The dinner was in Chicago at a Peruvian restaurant named Piqueo. When we arrived we found that wine was not available at the restaurant and we were referred to a shop down the street. A group of us went to purchase wine and as luck would have it I was paired with Smale on the walk back to the restaurant on Clark Street. On the one hand, I was thrilled to be with the celebrity, but I also felt the need to say something profound (or at least not stupid). Fortunately, Steve asked me about my work which was at a very early stage. Steve listened and then told me that Kuiper was working on something similar.

I informed Franks of my conversation. Although I did not realize it at the time, John wrote to Kuiper to follow up Smale's remark. Kuiper sent a preprint essentially solving my thesis problem. Everyone around me seemed to think that this was a disaster, but my first thought was that at least I could stop banging my head against the impasses that I had encountered in the problem. The story has a happy ending in that I was able to use the work that I had already done in writing a thesis on linearly induced maps of Grassmann manifolds.

The theme of Smale connecting me with other mathematicians and approaches to mathematics is the one that I would like to stress. Five years ago, I read Steve's 1985 *Bulletin* article and was struck by his ideas for applying dynamical systems to the study of numerical analysis. Following the lead of Shub, I began to study eigenvalue computation from this point of view. Having grown up among dynamicists, I was virtually unacquainted with people working on computation. I went to a conference at MSRI and Steve made certain to introduce me to Eric Kostlan.

At the conference, I learned from Parlett about the problem of determining whether the Rayleigh quotient iteration algorithm was generally convergent for nonsymmetric matrices. John Smillie and I subsequently produced a negative solution. Our construction was a very technical and involved calculus argument. When we described the work to Smale he suggested that there should be a much simpler bifurcation approach. Although we were dubious at first, Smale, of course, was correct and the bifurcation argument is what we published.

As this is being written I am completing a sabbatical at Berkeley under the sponsorship of Steve Smale. The semester at Berkeley has given me the opportunity for stimulating interaction with scholars in dynamics, complexity, and numerical analysis. For this and the generosity with which he has shared his unique perspective on the "big picture," I am very grateful.

30
Steve Smale and the Geometry of Ill-Conditioning

James Demmel

1. Introduction

The work of Steve Smale and his colleagues on average case analysis of algorithms [30, 31] and modeling real computations [3, 32] introduced methods and models not previously used in numerical analysis and complexity theory. In particular, the use of integral geometry to bound the sizes of sets of problems where algorithms "go bad" and the introduction of a model of real computation to clearly formulate complexity questions have inspired a great deal of other work. In this chapter, we will survey past results and open problems in three areas: the probability that a random numerical problem is difficult, the complexity of condition estimation, and the use of regularization to solve ill-posed or ill-conditioned problems; we will define all these terms below.

All this work is based on the following three-step approach. Let X denote the space of problems, Y the class of solutions, and f the solution map carrying a problem $x \in X$ to its solution $f(x) \in Y$.

1. Within a given space of problems X, one identifies the set IP of "ill-posed" problems, i.e., those x for which a solution $f(x)$ does not exist or is not a smooth function of x. The sensitivity of $f(x)$ to changes in x is measured by the gradient $Df(x)$ if it exists. $\|Df(x)\|$, or sometimes a multiple of it, such as $\|Df(x)\|/\|f(x)\|$, is called the *condition number* of f at x (or just of x if f is understood). We will use the notation $\kappa(f, x)$ [or $\kappa(x)$] to denote the condition number. If $\kappa(x)$ is large, x is called *ill-conditioned.*
2. As x approaches IP, the condition number $\kappa(x)$ generally approaches infinity. One shows that $\kappa(x)$ is approximated by a function of the distance from x to IP: $\mathrm{dist}(x, \mathrm{IP}) = \min\{\|x - y\|: y \in \mathrm{IP}\}$. This lets one approximate the set of ill-conditioned problems $\mathrm{IC}(\bar{\kappa}) \equiv \{x: \kappa(x) \geq \bar{\kappa}\}$ by the set of all points within a given distance of IP. We will use the notation $T(\varepsilon, \mathrm{IP}) = \{x: \mathrm{dist}(x, \mathrm{IP}) \leq \varepsilon\}$ to denote this latter set.
3. One estimates the volume of the set $T(\varepsilon, \mathrm{IP})$. If one lets ε depend on $\bar{\kappa}$ appropriately, and normalizes properly, this approximates the fraction of problems whose condition number exceeds $\bar{\kappa}$. Said another way, if a prob-

lem x is chosen "at random," this yields the probability that its condition number will exceed $\bar{\kappa}$.

It turns out that for problems where X is a finite-dimensional Euclidean space and f is an algebraic function, all these steps can be carried out in essentially the same way. The result is a classification theorem for the probability distributions than can arise in step 3. For X complex, it turns out that to first order the probability distribution depends only on three integers: the dimension of X, the codimension of IP within X, and the degree of IP as an algebraic variety. For X real, an upper bound for the probability can be given by these three integers, but a first-order correct expression depends on the dim(IP)-dimensional volume of IP. This work will be surveyed in Section 2, and open problems involved in computing these volumes discussed.

The second area is the complexity of condition estimation, or of approximating $\kappa(x)$. This is of immediate interest in practical computations, where approximate error bounds for approximations to $f(x)$ are routinely computed. The goal is to approximate $\kappa(x)$ with as little work as possible, while minimizing the likelihood of approximating it badly. We discuss several methods for condition estimation and discuss a conjecture that computing $\kappa(x)$ with a guaranteed error bound is no less expensive than computing $f(x)$ itself. In other words, computing a guaranteed error bound seems to be as expensive as solving the problem in the first place.

The final area is regularization of ill-posed or very ill-conditioned problems. If one has a problem x very close to IP and so with a very large condition number $\kappa(x)$, how does one find an acceptable solution? Regularization involves changing the problem in a systematic way to reduce the condition number without changing the essential nature of the problem. There are numerous methods used in practice; here we restrict ourselves to "projection onto IP," or replacing the original problem x with a nearby problem lying exactly within IP. This seems like a bad idea, but if we now restrict allowable perturbations of x to make it lie within IP, then the corresponding restricted condition number is generally small again. For example, this is precisely how the Moore–Penrose pseudo-inverse is used in practice to solve rank-deficient linear least squares problems [14]. In the case of the pseudo-inverse, the projection is perfect in the sense that the condition number of the reduced problem is as small as possible. However, there are many other problems where the projection method can fail to find the "correct" problem on IP, for example, staircase algorithms [6, 7, 9, 10] for computing the Jordan form of a matrix and its generalizations. We argue that that difference in behavior lies in the structure of the singularities of IP and propose a way to measure the likelihood of failure of these regularization-by-projection methods.

In addition to Steve Smale, much of this work was motivated by work of Kahan [21] while the author was a graduate student at U.C. Berkeley. Indeed, the combination of the approaches of Smale and Kahan has proven very fruitful in understanding the behavior of numerical methods.

2. The Geometry of Ill-Conditioning

We begin by explaining how the condition number describes the behavior of a numerical algorithm. Our running example will be matrix inversion. Thus, the spaces X and Y of problems and solutions, respectively, can both be naturally identified with $\mathbf{R}^{n^2}$, and $f(x) = x^{-1}$. It is easy to verify that the condition number $\|Df(x)\|/\|f(x)\| = \|x^{-1}\|$ where we take $\|\cdot\|$ to be the operator norm induced by the Euclidean vector norm.

In general, roundoff and other errors prevent us from computing the true solution $f(x)$ exactly. Instead one gets alg(x), the output of some algorithm. Our goal is to bound the error alg$(x) - f(x)$. The property of the algorithm used to do this is called *backward stability*; this means that alg$(x) = f(x + \delta x)$, where $\|\delta x\|$ is small compared to $\|x\|$. $\|\delta x\|$ is called the *backward error*, to contrast it to the *forward error* $\|\text{alg}(x) - f(x)\|/\|f(x)\|$. Backward stability is a property shared by many well-designed algorithms and is discussed at length in textbooks on numerical analysis [14, 36]. If f is sufficiently smooth, this leads to the obvious error bound

$$\frac{\|\text{alg}(x) - f(x)\|}{\|f(x)\|} \approx \frac{\|Df(x)\delta x\|}{\|f(x)\|} \leq \frac{\|Df(x)\|}{\|f(x)\|}\|\delta x\| = \kappa(x)\,\|\delta x\|.$$

So the condition number times the backward error provides a bound on the forward error. If the algorithm is backward stable ($\|\delta x\|$ is small) and the condition number is moderate, the forward error will be small.

Another use of condition number is to analyze the rate of convergence of iterative algorithms for computing $f(x)$. For example, consider an approximate Newton's method for computing the matrix inverse x^{-1}. If one has an approximate inverse $\hat{x}^{-1}$, then one can use the iteration (I denotes the identity matrix):

$$z_{i+1} = z_i - \hat{x}^{-1}(xz_i - I).$$

Denoting the error by $e_i = z_i - x^{-1}$, one gets that

$$e_{i+1} = (I - \hat{x}^{-1}x)e_i$$

so the norm of the error decreases by at least $\|I - \hat{x}^{-1}x\|$ at each iteration. If $\hat{x} \approx x$, then $\|I - \hat{x}^{-1}x\|$ is bounded by $\kappa(x)\,\|x - \hat{x}\|/(1 - \kappa(x)\,\|x - \hat{x}\|)$, so the larger the condition number, the slower the convergence.

The first step of the three-step approach of Section 1 is to describe the problems within X which are "ill-posed"; this means those problems for which the condition number is infinite or undefined. In the case of matrix inversion, this set is obviously the singular matrices. In the case of finding zeros of polynomials, for example, this set is the set of polynomials with multiple roots. These are "ill-posed" in the sense that the multiple roots are not differentiable functions of the coefficients, even though they are continuous. Similarly, for the matrix eigenvalue problem, the ill-posed matrices are those with Jordan blocks of size two or more. In all these cases, note that IP is an algebraic variety within X. In the case of matrix inversion, this variety is defined by setting the determinant to zero; for polynomial zero finding, it is defined by setting the discriminant to zero; and for the matrix eigenvalue problem, it is defined by setting the discriminant of the characteristic polynomial to zero.

In all these cases, one expects just by continuity that the condition number becomes large in a neighborhood of IP. To proceed with the second step of our plan, we need a better characterization of the set $\mathrm{IC}(\bar{\kappa})$ of problems whose condition number exceeds $\bar{\kappa}$. It turns out that one can establish simple bounds between the distance $\mathrm{dist}(x, \mathrm{IP})$ from x to IP and the condition number of x. The canonical example is matrix inversion, where the condition number $\|x^{-1}\|$ is precisely the reciprocal of $\mathrm{dist}(x, \mathrm{IP})$ [4, 21]. This means that $\mathrm{IC}(\bar{\kappa})$ is precisely a tubular neighborhood $T(1/\bar{\kappa}, \mathrm{IP})$ of IP, consisting of all points within distance $1/\bar{\kappa}$ of IP. For other problems, such as polynomial zero finding and the matrix eigenvalue problem, one can prove weaker one-sided inclusions relating $\mathrm{IC}(\bar{\kappa})$ and $T(1/\bar{\kappa}, \mathrm{IP})$.

The relation that connects condition numbers to distance to IP is the following. For f an algebraic function, one can prove that one or both of the differential inequalities

$$m\kappa(x) \leq \frac{\|D\kappa(x)\|}{\kappa(x)} \leq M\kappa(x)$$

holds, where $0 \leq m \leq M$. For example, in the case of matrix inversion, $m = M = 1$, so we actually have equalities. One can show that the left-hand inequality implies $\mathrm{dist}(x, \mathrm{IP}) \leq 1/(m\kappa(x))$, and the right-hand inequality implies $\mathrm{dist}(x, \mathrm{IP}) \geq 1/(M\kappa(x))$ [4]. Note that the quantity $\|D\kappa(x)\|/\kappa(x)$ is just the condition number of $\kappa(x)$. Thus, these inequalities are equivalent to the statement that "the condition number of the condition is nearly itself," or that estimating the condition number is as sensitive a problem as solving the original problem. We return to this notion in the next section, where we conjecture that the *complexity* of $\kappa(f, x)$ is also at least as big the complexity of $f(x)$.

The third step of our plan involves estimating the volume of the tubular neighborhood $T(1/\kappa, \mathrm{IP})$. This is a problem to which many contributions have been made, dating back to Weyl and Hotelling in 1939 [5, 12, 13, 15–17, 20, 22, 24–26, 28–31, 33, 35]. Some of this work considers volumes of tubes around specific varieties (like singular matrices) and gets quite detailed information [12].

Because varieties and tubes are infinite objects, we need to explain what we mean by their volumes. We will always compute

$$v(\varepsilon, \mathrm{IP}) = \frac{\mathrm{vol}(T(\varepsilon, \mathrm{IP}) \cap B(1))}{\mathrm{vol}(B(1))},$$

where $\mathrm{vol}(\cdot)$ denotes Lebesgue measure and $B(1)$ is the unit ball centered at the origin (in $\mathbf{C}^N$ or $\mathbf{R}^N$). Thus, one can interpret $v(1/\bar{\kappa}, \mathrm{IP})$ as approximating the fraction of problems in $B(1)$ whose condition number exceeds $\bar{\kappa}$. This is quite reasonable because most condition numbers are homogeneous functions of x, so there is no loss to restricting to x of norm at most one, or even to $\|x\| = 1$.

The simplest case, the one covering most applications to condition numbers, is for IP a complex purely $2d$ (real)-dimensional homogeneous complex variety in $\mathbf{C}^N$. Then we can show [5]

$$v(\varepsilon, \mathrm{IP}) = \binom{N}{d} \deg(\mathrm{IP})\varepsilon^{2(N-d)} + o(\varepsilon^{2(N-d)}),$$

where deg(IP) is the degree of IP as an algebraic variety. Thus, for asymptotically small ε, $v(\varepsilon, \mathrm{IP})$ is the same as though IP were flat with the same area and $v(\varepsilon, \mathrm{IP})$ were a rectangular parallelepiped. This much is predicted by Weyl's results [35] for closed compact manifolds and sufficiently small ε; the challenge was to extend this to surfaces with singularities and to arbitrary ε. There is also an upper bound proportional to $\deg(\mathrm{IP})\varepsilon^{2(N-d)}(1 + N\varepsilon)^{2d}$, and a lower bound proportional to $\varepsilon^{2(N-d)}(1-\varepsilon)^{2d}$. The upper bound continues to hold if IP is nonhomogeneous, and the lower bound if IP is nonhomogeneous but passes through the origin. Thus, the details of the actual problem whose ill-posed problems are IP are not important; only the three integers N, d and deg(IP) matter.

It is instructive to compare the results with Edelman's [11], who derives the exact distribution for IP = {singular matrices}. His results are expressed as the fraction of the unit sphere in $\mathbf{C}^N$ within ε of IP. Letting $\hat{v}(\varepsilon, \mathrm{IP})$ denote this quantity, our bounds yield the following estimates for $n \times n$ matrices:

$$\hat{v}(\varepsilon, \mathrm{IP}) = n(n^2 -)\varepsilon^2 + o(\varepsilon^2),$$

$$\frac{\varepsilon^2(1-\varepsilon)^{2n^2-2}}{2n^4} \leq \hat{v}(\varepsilon, \mathrm{IP}) \leq e^2 n^5 \varepsilon^2 (1 + n^2\varepsilon)^{2n^2-2}$$

[where $e = \exp(1)$], whereas Edelman's result is the simple expression

$$\hat{v}(\varepsilon, \mathrm{IP}) = 1 - (1 - n\varepsilon^2)^{n^2-1}.$$

Thus, one can see how much one loses by dealing with a general variety.

For real varieties, the results are more complicated because the volume depends on the coefficients of the polynomial(s) defining IP, not just on the degree and dimension of IP. Still, one can derive an upper bound on $v(\varepsilon, \mathrm{IP})$ just from these quantities. Suppose IP is a real purely d-dimensional variety in $\mathbf{R}^N$, which is the complete intersection of $N - d$ polynomials of maximum degree D. Then Ocneanu [25] shows

$$v(\varepsilon, \mathrm{IP}) \leq 2(N-d) \sum_{k=N-d}^{N} \binom{N}{k} (2D\varepsilon)^k.$$

The dominant term in this is the $k = N - d$ term. We can drop the assumption of complete intersection and still prove a lower bound proportional to $\mathrm{vol}(\mathrm{IP}[1-\varepsilon])\varepsilon^{N-d}/\deg(\mathrm{IP})$, where $\mathrm{IP}[1-\varepsilon] = \mathrm{IP} \cap B(1-\varepsilon)$. For IP =

{singular matrices}, Edelman again derives an exact answer, this time in the form $\hat{v}(\varepsilon, \mathrm{IP}) = \int_\varepsilon^\infty p(s)\,ds$, where $p(s)$ is given in terms of the Gauss hypergeometric function.

Several open questions remain regarding the volume estimates. Can the upper and lower bounds for complex varieties be brought closer together without any more assumptions about the varieties? The upper bound is much too large for practical sizes of N and ε [5]. Can an upper bound for real varieties be derived that does not depend on complete intersection? A number of interesting IP sets that arise in numerical analysis are not complete intersections, such as matrices of rank deficiency 2 or greater. Are there simple upper and lower bounds which are asymptotically correct for small ε, perhaps just in the case of homogeneous varieties? Such a formula for real varieties would have to be proportional to vol(IP). For further discussion, see [5].

In Section 4, we return to volume estimates, describing a problem not handled by the general theory outlined here.

3. Condition Estimation

Our main example, as in Section 2, is matrix inversion, or just solving a single system of linear equations $Az = b$ with coefficient matrix A. As discussed in Section 2, the condition number for this problem is $\|A^{-1}\|$, where the solution is, of course, just A^{-1}, or $A^{-1}b$. In the case of matrix inversion, once A^{-1} is in hand, it is a simple matter to compute its norm (at least approximately) and so get an error bound. However, it is possible to solve $Az = b$ for z much less expensively without computing A^{-1} first. Even for dense matrices, solving $Az = b$ with Gaussian elimination is three times as fast as first computing A^{-1}, and if A is large and sparse or has other special structure, it may be orders of magnitude cheaper to solve $Az = b$ than to compute A^{-1}.

Therefore, it is of great practical interest to estimate $\|A^{-1}\|$ without first computing A^{-1}. Algorithms for this are called *condition estimators*, and there is a large literature on the subject ([8, 18, 23]; see [19] for a survey). Most of the methods produce a lower bound on $\|A^{-1}\|$ by finding a unit vector b such that $\|A^{-1}b\|$ is as large as possible. There are two approaches generally used. The first approach assumes one has performed Gaussian elimination and so factored the matrix A as $A = PLU$ where P is a permutation matrix, L is a unit lower triangular matric with entries bounded by 1 in absolute value, and U is upper triangular. L usually contributes little to A's ill-conditioning, so one considers only U. A vector u is then constructed by examining all the entries of U and choosing u to make the entries of $U^{-1}u$ grow rapidly. The second approach assumes only that it is possible to solve $Az = b$ (and possibly $A^{\mathrm{T}}z = b$) for arbitrary b, and uses various search procedures to find a b making $\|z\|$ large. Both approaches take advantage of the fact that performing Gaussian elimination, or solving $Az = b$ by other means, is much cheaper than computing A^{-1}.

Results in [8, 23] show that it is possible to estimate $\|A^{-1}\|$ with a low probability of large error, using only the ability to solve $Az = b$ and $A^{T}z = b$. We consider the question of whether it is possible to compute $\|A^{-1}\|$ with *guaranteed* accuracy with lower complexity than computing A^{-1}. We conjecture that this is not possible, i.e., that getting a guaranteed error bound is as expensive as solving the problem in the first place, and describe two ways to approach this problem.

Suppose first that our only way to extract information about A is via a "black box" subroutine that takes any vector b as input and returns $A^{-1}b$. This is a reasonable model of some practical computations. It is easy to prove that if one uses this subroutine to compute $A^{-1}b_i$ for fewer than $n = \dim(A)$ linearly independent b_i, then any estimate of $\|A^{-1}\|$ can be arbitrarily wrong. This is because given any n independent vectors forming the columns of B, and any nonsingular C, the matrix $A^{-1} = CB^{-1}$ maps the columns of B to the columns of C, and so if fewer than n columns of B and C are specified, the remaining columns can be chosen to make A as ill-conditioned as desired. But if we have to compute $A^{-1}b_i$ for n independent b_i, we can compute the inverse by choosing the b_i as consecutive columns of the identity.

Now let us take a different approach. To relate the complexity of computing $\|A^{-1}\|$ to computing A^{-1}, we consider the simpler problem of computing the determinant $\det(A)$. There are well-known results which say under a mild assumption the complexity of matrix multiplication and matrix inversion are identical, and that computing the determinant is no harder [1]. More concretely, let $M(n)$ be the complexity of multiplying two square $n \times n$ matrices, and assume $8M(n) \geq M(2n) \geq 2^{2+\varepsilon}M(n)$ for some $\varepsilon > 0$. This essentially means that $M(n)$ grows at least as fast as $n^{2+\varepsilon}$ and no faster than n^3. (For example, Strassen's method [1] guarantees that $2 + \varepsilon \leq \log_2 7 \approx 2.81$, and faster methods are known.) Then $\det(A)$, A^{-1}, and $z = A^{-1}b$ can all be computed in time $O(M(n))$.

It is apparently not known whether computing the determinant is at least as hard as inversion or solving $Ax = b$. However, it is easy to see that it is at least as hard as computing one entry of A^{-1} or of $A^{-1}b$ because of Cramer's rule.

Therefore, if we could show that computing $\|A^{-1}\|$ with some guaranteed accuracy were as complex as computing the determinant, we would have a (perhaps weak) lower bound on the complexity of condition estimation, namely, the cost of computing one component of A^{-1} or $A^{-1}b$. To do this, we consider a variation of the model in [3], that of "straight-line code." This corresponds to the model in [3] in which no branches are allowed until the end. Consider the simple inequalities

$$\frac{\|A\|_2}{\|A^{-1}\|_2^{n-1}} \leq \det(A) \leq \frac{\|A\|_2^{n-1}}{\|A^{-1}\|_2}.$$

These imply that guaranteed upper and lower bounds on $\|A^{-1}\|$ (in any reasonable norm) yield upper and lower bounds on $\det(A)$. In particular, a guaranteed upper bound on $\|A^{-1}\|$ implies that $\det(A) \neq 0$, so we consider the problem of deciding if $\det(A) = 0$. In the straightline code model, we would compute a rational function $h(A)$ of the matrix entries, and then have a single test "$h(A) = 0$?" at the end to decide if $\det(A) = 0$. To make a correct decision, we must have that $h(A) = 0$ if $\det(A) = 0$, so that $h(A) = \det(A)g(A)$ for some rational function $g(A)$ which must not be zero when $\det(A)$ is nonzero.

If A is complex and complex rational operations are performed, then $g(A)$ being a rational function of A without poles means that it is polynomial. Because it has no zeros other than those of det, and det is an irreducible polynomial, $g(A)$ must be a constant multiple of a power of $\det(A)$, so $h(A)$ is a constant multiple of a power of $\det(A)$. Thus, our straight-line program would have to do at least as much work as needed to compute a power of $\det(A)$. From this, one could extract $\det(A)$ to as much precision as desired with an amount of work independent of the dimension of A. Thus, a straight-line program doing complex arithmetic to decide if $\det(A) = 0$ must in fact do as much work as required to compute $\det(A)$.

If A is real and real rational operations are performed, one can no longer conclude that $g(A)$ is a polynomial because it could potentially have factors like $x^2 + 1$ in its denominator. However, if one can show the algorithm has the property that it would work correctly on complex data, then the above argument would apply.

More generally, it is easy to see that computing a guaranteed upper bound on the condition number $\kappa(x)$ is at least as hard as deciding whether problem $x \in \mathrm{IP}$. If IP is defined by a single irreducible polynomial $p(x)$ like $\det(A) = 0$, then the above argument shows that a straight-line program has to do at least as much work as needed to compute a power of $p(x)$.

Of course, the full model in [3] includes arbitrarily many branch statements. It is an open problem to extend this argument to that situation. Asked another way, is it any less expensive to decide whether an irreducible polynomial is zero than to evaluate it (or a power of it)?

4. Regularization of Ill-Posed Problems

Finally, we consider regularization. This means taking a problem whose condition number is very large and modifying it in a certain way to decrease the condition number. There are various methods in use, but we will consider just the "projection onto IP" method mentioned in Section 1. We consider first a case where there is a very satisfactory solution: solving rank-deficient linear least squares problems. Here, one tries to minimize $\|Ax - b\|_2$, where A is a nearly rank-deficient matrix with at least as many rows as columns. One is also given $\varepsilon > 0$ bounding the uncertainty in A; ε can be a bound for measurement error, round-off error, or truncation error, and is a problem-dependent quantity supplied by the user. An explicit solution for $\varepsilon = 0$ (no data uncertainty) is given in terms of the *singular value decomposition* $A = U\Sigma V^T$, where U and V are orthogonal matrices and $\Sigma = \text{diag}(\sigma_1, \ldots, \sigma_n)$ with the *singular values* σ_i satisfying $\sigma_1 \geq \cdots \geq \sigma_n \geq 0$. The explicit solution is $x = A^+ b$, where the *pseudo-inverse* A^+ of A is defined by $A^+ = V\Sigma^+ U^T$, $\Sigma^+ = \text{diag}(\sigma_1^+, \ldots, \sigma_n^+)$, and $\sigma_i^+ = 1/\sigma_i$ if $\sigma_i \neq 0$ and $0^+ = 0$. The corresponding condition number is σ_1/σ_r, where σ_r is the smallest nonzero singular value [14].

Regularization consists of replacing A with the matrix A_1 of lowest rank within distance ε (measured using the matrix norm $\|\cdot\|_2$) of A and using A_1^+ instead of A^+. This is easy to describe in terms of the singular value decomposition: Let $\Sigma_1 = \Sigma$ with its singular values less than or equal to ε set to zero, and let $A_1 = U\Sigma_1 V^T$. This is an orthogonal projection onto the nearest matrix within ε of A with as low rank as possible, and lowers the condition number from σ_1/σ_r to at most σ_1/ε. Thus, we have a nested sequence of sets of ill-posed problems: the matrices of rank r. Problems within each set have moderate condition numbers σ_1/σ_r until σ_r approaches 0 and the matrix is close to a matrix of rank $r - 1$.

This regularization process is well-understood and efficiently implementable. In particular, there is no ambiguity about how to regularize: The regularized matrix is unambiguously defined as the nearest matrix within ε of as low rank as possible. In particular, consider the intersection S of the set of singular matrices and the closed ε ball around A. Then S is connected, so there is no question as to which component of S on which to project (see Fig. 1).

The situation becomes more interesting when dealing with polynomial zeros or matrix eigenvalues. Here, the nested sets of ill-posed problems are the polynomials (matrices) with multiple zeros (eigenvalues). These form a hierarchy of ill-conditioned sets because zeros of multiplicity k can vary like the kth root of the perturbation to the coefficients, and a kth root is arbitrarily larger than a $(k - 1)$st root for asymptotically small perturbations. Hierarchies of this sort appear very generally [27].

Unfortunately, it is not so easy to perform regularization in these cases. In fact, one can quantitatively show that it is a much more ambiguous venture than in the simple case of least squares. We illustrate with the case of real

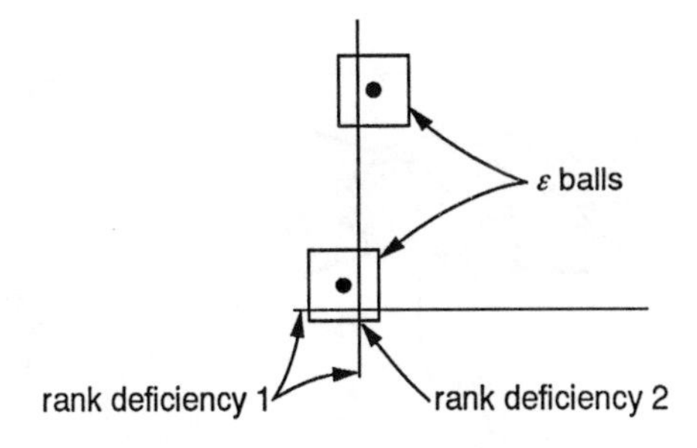

FIGURE 1: Ill-posed problems for rank-deficient least squares.

monic cubic polynomials, a real space of dimension 3 (with the polynomial coefficients as coordinates). In this case there are just two kinds of ill-posed problems: those with one double (real) root and those with one triple (real) root. It is elementary to show that the polynomials with triple roots form a one-dimensional curve which is a cusp of the two-dimensional set of polynomials with a double root. Regularization consists of finding a polynomial q with a multiple root within ε (in the Euclidean space of coefficients) of the initial polynomial p. Geometrically, one takes the intersection S of the set IP of polynomials with a multiple root and the ε ball around p. Now, because IP has a cusp, sometimes S will consist of two components, one corresponding to making roots r_1 and r_2, say, merge, and the other corresponding to making r_2 and r_3 merge. In the least squares case, the corresponding ambiguity (making either of two different singular values zero) is resolved by making both zero, and so projecting onto a subvariety of ill-posed problems. The cusp in the case of polynomials makes this possibility quite remote: it much more likely to be within ε of two different polynomials with different double roots than to be anywhere close to a polynomial with triple roots.

These two situations are illustrated schematically in Figs. 1 and 2. In Fig. 1, corresponding to rank deficient matrices IP, the two branches of IP shown (corresponding to rank deficiency 1) intersect orthogonally, and so if the ε ball surrounding A intersects both, then it also intersects the matrices of rank deficiency 2. In Fig. 2, corresponding to real monic cubic polynomials, many balls intersect both branches of IP with the nearest triply multiple root being much farther away.

In fact, in this simple case we can quantify the likelihood of three algorithmic possibilities. It is elementary to show that the cusp (a cross section of which is in Fig. 2) is essentially given by $y = \pm x^{3/2}$. Then given a random real monic cubic polynomial A, one can show that the probability that S

1. contains a single component of IP is $\Theta(\varepsilon)$,
2. contains two components of IP is $\Theta(\varepsilon^{5/3})$, and
3. contains a polynomial with a triple zero is $\Theta(\varepsilon^2)$,

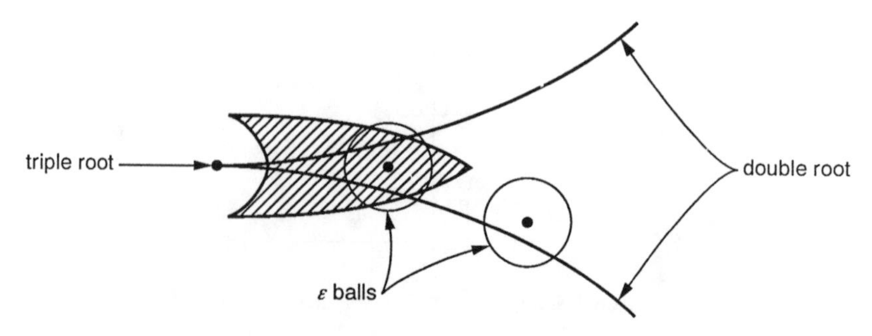

FIGURE 2: Ill-posed problems for roots of a cubic polynomial.

The $\varepsilon^{5/3}$ term is the volume of the region "doubly covered" by tubes around both branches of IP, but farther than ε away from a triple root (the shaded region in Fig. 2; there is no corresponding region in Fig. 1). Thus, the "ambiguous" case 2 is asymptotically much more likely than being able to resolve it by being in case 3. This means that we cannot unambiguously cluster the roots of polynomials into disjoint sets, each of which can be replaced by a single multiple root.

This volume problem cannot be solved for general varieties using the methods of the last section, and so estimating the volumes of regions "multiply covered" by tubes is an open problem.

A more practical open problem would be to extend this analysis to cover "staircase algorithms" [7, 9, 10] for the Jordan canonical form and its generalizations. These algorithms are of great interest in control and systems theory, but can occasionally fail to recover the correct Jordan form. This approach holds the possibility of being able to analyze these failures geometrically and probabilistically. The singularities are high-dimensional swallowtails [2], and the presence of cusps in swallowtails indicates that the ambiguous situation shaded in Fig. 2 is common in practice.

References

[1] A. Aho, J. Hopcroft, and J Ullman, *The Design and Analysis of Computer Algorithms*, Addison-Wesley, Reading, MA, 1974.

[2] V.I. Arnol'd, On matrices depending on parameters, *Russ. Math. Surveys* 26: 29–43, 1971.

[3] Lenore Blum, Mike Shub, and Steve Smale, On a theory of computation and complexity over the real numbers: NP-completeness, recursive functions and universal machines. *AMS Bull.* (*New Series*) 21(1): 1–46, 1989.

[4] James Demmel, On condition numbers and the distance to the nearest ill-posed problem, *Numer. Math.* 51(3): 251–289, 1987.

[5] James Demmel, The probability that a numerical analysis problem is difficult, *Math. Comput.* 50(182): 449–480, 1988.

[6] James Demmel and Bo Kågström, Computing stable eigendecompositions of matrix pencils, *Linear Alg. Appl.* 88/89: 139–186, 1987.

[7] James Demmel and Bo Kågström, Accurate solutions of ill-posed problems in control theory, *SIAM J. Math. Anal. Appl.* 9(1): 126–145, 1988.

[8] J.D. Dixon, Estimating extremal eigenalues and condition numbers of matrices. *SIAM J. Numer. Anal.* 20: 812–814, 1983.

[9] P. Van Dooren, The computation of Kronecker's canonical form of a singular pencil, *Linear Alg. Appl.* 27: 103–141, 1979.

[10] P. Van Dooren, The generalized eigenstructure problem in linear system theory, *IEEE Trans. Auto. Control.* AC-26: 111–128, 1981.

[11] A. Edelman, On the distribution of a scaled condition number, *Math. Comput.*, 58(197): 185–190, 1992.

[12] A. Edelman, Eigenvalues and condition numbers of random matrices, *SIAM J. Math. Anal. Appl.* 9(4): 543–560, 1988.

[13] H. Federer, *Geometric Measure Theory*. Springer-Verlag, Berlin, 1969.

[14] Gene Golub and Charles Van Loan, *Matrix Computations*, Johns Hopkins University Press, Baltimore, MD, 1983.
[15] A. Gray, Comparison theorems for the volumes of tubes as generalizations of the Weyl tube formula, *Topology* 21(2): 201–228, 1982.
[16] A. Gray, An estimate for the volume of a tube about a complex hypersurface, *Tensor* (*N. S.*) 39: 303–305, 1982.
[17] P.A. Griffiths, Complex differential and integral geometry and curvature integrals associated to singularities of complex analytic varieties, *Duke Math. J.* 45(3): 427–512, 1978.
[18] W.W. Hager, Condition estimators, *SIAM J. Sci. Stat. Comput.* 5: 311–316, 1984.
[19] N. Higham, A survey of condition number estimation for triangular matrices, *SIAM Rev.* 29: 575–596, 1987.
[20] H. Hotelling, Tubes and spheres in n-spaces, and a class of statistical problems, *Amer. J. Math.* 61: 440–460, 1939.
[21] W. Kahan, Conserving confluence curbs ill-condition, Computer Science Dept. Technical Report 6, University of California, Berkeley, CA, August 1972.
[22] E. Kostlan, Statistical complexity of numerical linear algebra, Ph. D. thesis, University of California, Berkeley, CA, 1985.
[23] J. Kuczyński and H. Woźniakowski, Estimating the largest eigenvalue by the power and Lanczos algorithms with a random start, Computer Science Department, Columbia University, Argonne, IL, March 1989.
[24] P. Lelong, *Fonctions plurisousharmoniques et formes differentielles positiv*, Gordon and Breach, New York, 1968.
[25] A. Ocneanu, On the volume of tubes about a real variety, *J. Complexity*, to be published.
[26] A. Ocneanu, On the loss of precision in solving large linear systems, Technical report, Mathematical Sciences Research Institute, Berkeley, CA, 1985.
[27] J. Renegar, Ill-posed problem instances, this volume.
[28] J. Renegar, On the efficiency of Newton's method in approximating all zeros of systems of complex polynomials, *Math. Oper. Res.* 12(1): 121–148, 1987.
[29] L.A. Santaló, *Integral Geometry and Geometric Probability*, Vol. 1 of *Encyclopedia of Mathematics and Its Applications*, Addison-Wesley, Reading, MA, 1985.
[30] Steve Smale, The fundamental theorem of algebra and complexity theory, *AMS Bull.* (*New Series*), 4(1): 1–35, 1981.
[31] Steve Smale, On the efficiency of algorithms of analysis, *AMS Bull.* (*New Series*) 13(2): 87–121, 1985.
[32] Steve Smale, Some remarks on the foundations of numerical analysis, *SIAM Rev.* 32(2): 211–220, 1990.
[33] G. Stolzenberg. *Volumes, Limits, and Extensions of Analytic Varieties*, Springer-Verlag, Berlin, 1966.
[34] P. Thie, The Lelong number of points of a complex analytic set, *Math. Ann.* 172: 269–312, 1967.
[35] H. Weyl, On the volume of tubes, *Amer. J. Math.* 61: 461–472, 1939.
[36] J. H. Wilkinson, *The Algebraic Eigenvalue Problem*, Oxford University Press, Oxford, 1965.

31
On Smale's Work in the Theory of Computation: From Polynomial Zeros to Continuous Complexity

J.F. TRAUB

I am delighted to have this opportunity to comment on the work of Steve Smale. Mike Shub was kind enough to give me an advance copy of his excellent paper (Shub [93]), which was very useful for it served as a reminder of the many areas in the theory of computation to which Steve has made major or often seminal contributions.

I am in a particularly good position to provide an appreciation of Steve's research because we overlap in about half a dozen areas. I can, therefore, testify from particular first-hand knowledge about the originality and importance of his contributions.

Today I will confine myself to just three areas. I will comment briefly on Smale's work on polynomial zeros and on average case analysis and then discuss continuous computational complexity.

I will begin with the approximate computation of polynomial zeros and Smale's influential 1981 paper (see Smale [81]) in which he introduced us to the **probabilistic** analysis of an algorithm for solving this problem. More precisely, Smale placed a Lebesgue measure on the space of coefficients of a polynomial of degree d and excluded sets of measure μ. He showed that a relaxed Newton iteration enjoys algorithm complexity which is a polynomial in the degree d and in $1/\mu$.

Perhaps even more important was that Steve indentified the complexity of the approximate computation of polynomial zeros as a central problem of computational complexity and stimulated a flood of important research which is referenced in Mike's paper. Here, I will mention only the work of Shub and Smale who established better algorithmic complexity (Shub and Smale [85, 86]) and the work of Renegar on systems of polynomials (Renegar [89])

Next, I will comment on Steve's work on average case analysis. The simplex algorithm is widely used for the solution of linear programming. It was known that in the worst case this algorithm could take an exponential number of steps, but that, in practice, the algorithm was very fast.

It was desirable to reconcile theory and practice which Smale achieved in his 1983 paper. He put a Gaussian distribution on the inputs and proved that, on the average, the number of steps in the simplex algorithm is sub-linear; see Smale [83].

The idea of trying to show that a problem which is worst-case intractable is tractable on the average is most important in computational complexity. Examples in continuous complexity include multivariate integration and multivariate approximation.

Finally, I will discuss the work of Smale and his colleagues in continuous computational complexity. Computational complexity studies the intrinsic difficulty of mathematically posed problems and seeks optimal algorithms for their solution. The intrinsic difficulty is called the problem complexity or, for brevity, the complexity. By continuous computational complexity I mean the study of continuous problems such as matrix multiplication, systems of polynomial equations, partial differential equations, and multivariate integration.

Continuous computational complexity may be contrasted with discrete computational complexity which studies, for example, discrete optimization, routing, and graph-theoretic problems. Computational complexity is an enormously rich subject and this division into continuous and discrete is an oversimplification which will serve today's purpose.

In the early seventies, and using a Turing machine model, Steve Cook and Richard Karp developed an elegant and deep theory of NP-complete problems. If, as is widely believed $P \neq NP$, then the NP-complete problems are all intractable. This theory has been used to investigate the intractability of many discrete problems.

Complexity is always relative to a model of computation. A model of computation is an abstraction of a digital computer, although it must be added that the Turing machine preceded the modern digital computer. As I mentioned above, Turing machines are used in the Cook–Karp theory of NP-complete problems.

A second model of computation is also of interest, the real number model. Recall that a model of computation is an abstraction of a digital computer. There are at least two motivations for studying the real-number model. The first is that one can then use very powerful mathematical tools to attack complexity issues. The second is that the arithmetic commonly used to solve scientific problems on digital computers is fixed precision floating point. One important property of this arithmetic is that the cost of an operation is independent of the size of an operand. In the real-number model, the cost of an operation is also *independent* of operand size. In contrast, on a Turing machine, the cost of an operation inputs *depends* on the size of the inputs.

The model of computation is crucial. A problem which is tractable relative to one model may be intractable in another.

Blum, Shub, and Smale [89] use an abstract machine over the reals. (They have also studied machines over the complex field and over the ring of integers, but I will confine myself here to the reals.) In this model of computation, they ask if $P = NP$ and they show that the problem of determining whether a system of n real polynomials of degree at most four has a real zero is

	Discrete Computational Complexity	Continuous Computational Complexity	
		Blum-Shub-Smale Complexity	Information-Based Complexity
Type of Problem	Discrete	Finite Dimensional	Infinite Dimensional
Model of Computation	Turing Machine	Real	Real

FIGURE 1. Relations among three computational complexity theories.

NP-complete. I will refer to this theory as Blum–Shub–Smale complexity. This is a very rich new area of computational complexity.

Continuous computational complexity may be divided into two branches, depending on whether the problems are finite or infinite dimensional. Blum–Shub–Smale complexity studies the complexity of finite-dimensional problems such as systems of polynomials. Information-based complexity studies the complexity of infinite-dimensional problems such as partial differential equations; see Traub, Wasilkowski, and Woźniakowski [88], Traub and Woźniakowski [91], and Werschulz [91].

Figure 1 exhibits the relation among the three computational complexity theories discussed above.

I will close by suggesting a line of future research. Information-based complexity proves certain infinite-dimensional problems are intractable by using adversary arguments on infinite-dimensional mathematical inputs. Blum–Shub–Smale complexity proves certain finite-dimensional problems are intractable, if $P \neq NP$, by showing they are NP-complete.

I suggest in the future both techniques will be combined to study the complexity of continuous problems.

References

BLUM, L, SHUB, M., AND SMALE, S.

[89] On a Theory of Computation and Complexity Over the Real Numbers: *NP*-completeness, Recursive Functions and Universal Machines, *Bull. Amer. Math. Soc.* 21 (1989), 1–46.

RENEGAR, J.

[89] On the Worst Case Arithmetic Complexity of Approximating Zeros of Systems of Polynomials, *SIAM J. Comput.* 18 (1989), 350–370.

SHUB, M.

[93] On the Work of Steve Smale on the Theory of Computation. This volume.

SHUB, M. AND SMALE, S.

[85] Computational Complexity, On the Geometry of Polynomials and a Theory of Cost: Part I, *Ann. Scient. Ecole. Norm. Sup.* 18 (1985), 107–142.

[86] Computational Complexity, On the Geometry of Polynomials and a Theory of Cost: Part II, *SIAM J. Comput.* 15 (1986), 145–161.

SMALE, S.

[81] The Fundamental Theorem of Algebra and Complexity Theory, *Bull. Amer. Math. Soc.* 4 (1981), 1–36.

[83] On the Average Number of Steps in the Simplex Method of Linear Programming, *Math. Programming* 27 (1983), 242–262.

TRAUB, J.F., WASILKOWSKI, G.W., AND WOŹNIAKOWSKI, H.

[88] *Information-Based Complexity*, Academic Press, Boston, MA, 1988.

TRAUB, J.F. AND WOŹNIAKOWSKI, H.

[91] Information-Based Complexity: New Questions for Mathematicians, *Math. Intelligencer* 13 (1991), 34–43.

WERSCHULZ, A.G.

[91] *Computational Complexity of Differential and Integral Equations*, Oxford University Press, Oxford, 1991.

32
The Gödel Incompleteness Theorem and Decidability over a Ring*

LENORE BLUM AND STEVE SMALE

Here we give an exposition of Gödel's result in an algebraic setting and also a formulation (and essentially an answer) to Penrose's problem. The notions of *computability* and *decidability over a ring R* underly our point of view. Gödel's Theorem follows from the Main Theorem: There is a definable undecidable set over **Z**. By way of contrast, Tarski's Theorem asserts that every definable set over the reals or any real closed field R is decidable over R. We show a converse to this result: Any sufficiently infinite ordered field with this latter property is necessarily real closed.

1. Introduction

Gödel showed [Gödel, 1931] that given any reasonable (consistent and effective) theory of arithmetic, there are true assertions about the natural numbers that are not theorems in that theory.[1] This "incompleteness theorem" ended Hilbert's program of formalizing mathematics and is rightfully regarded as the most important result in the foundations of mathematics in this century.

Now the concept of *undecidability* of a set plays an important role in understanding Gödel's work. On the other hand, the question of the undecidability of the Mandelbrot set has been raised by Roger Penrose [Penrose, 1989]. Penrose acknowledges the difficulty of formulating his question because "decidability" has customarily only dealt with countable sets, not sets of real or complex numbers.

Here we give an exposition of Gödel's result for mathematicians without background in logic and also a formulation (and essentially an answer) to

* Partially supported by NSF grants and (the first author) by the Letts-Villard Chair at Mills College.

[1] This formulation of Gödel's Theorem, with *consistency* in place of *ω-consistency*, is due to [Rosser, 1936].

Penrose's problem. The notions of *computability* and *decidability over a ring* R as developed in [BSS][2] underly our point of view.

A generalization of Gödel's theorem and various intermediate assertions may be formulated over an arbitrary ring or field R. If $R = \mathbf{Z}$, the integers, the specialization is essentially the original theorem. Our proof of it is valid in the case R is an algebraic number ring (i.e., a finite algebraic extension ring of $\mathbf{Z}$) or number field (a finite algebraic extension field of $\mathbf{Q}$). Using $R = \mathbf{R}$, the real numbers, the undecidability of the Mandelbrot set is dealt with.

Suppose that R is a commutative ring or field (perhaps ordered), which contains $\mathbf{Z}$, and R^k is the cartesian product, viewed as a k-dimensional vector space (or module) over R. A set $S \subset R^k$ is *decidable over* R if its *characteristic function*

$$\chi\colon R^k \to \{0, 1\}, \qquad \chi(\mathbf{x}) = 1 \quad \text{if and only if } \mathbf{x} \in S$$

is *computable over* R in the sense of [BSS].

A set $S \subset R^k$ is *definable over* R if it is of the form

$$S = \{(\mathbf{y}_1, \ldots, \mathbf{y}_k) \in R^k \mid \exists x_1 \forall x_2 \exists x_3 \ldots \forall x_n \text{ such that } (\mathbf{y}_1, \ldots, \mathbf{y}_k, x_1, \ldots, x_n) \in Y\}$$

for some *constructible* (or *semi-algebraic*) set Y in R^{k+n}.[3]

A natural question is:

[D] Is a definable set over R necessarily decidable over R?

If $R = \mathbf{Z}$, the answer is no, and this is the backbone of the Gödel Incompleteness Theorem. This may be interpreted (Section 2) as asserting there is a "polynomially defined" set of assertions over $\mathbf{Z}$,[4] and that there is no way of deciding which are the true ones. The proof is in Sections 3 and 4; the Gödel Theorem is proved in Section 5.

If $R = \mathbf{R}$, the real numbers the answer is yes. Tarski's fundamental decidability result [Tarski, 1951] for the case of the reals may be interpreted as this assertion.

These results extend to certain other rings and fields as well.

Julia Robinson extended Gödel's result. Robinson did not have the notion of decidability over a ring and put her theorems in the context of "decidable rings" (or fields). A ring R is *decidable* if the set of "first-order" sentences true in R is decidable in the traditional sense (of Tuling et al., or decidable over $\mathbf{Z}$ in our sense). Her way of showing a certain ring R is undecidable is to reduce the problem to Gödel's Theorem by defining $\mathbf{Z}$ in R. We use her algebraic results [Robinson, 1959, 1962, 1965] on the definability of $\mathbf{Z}$ to answer **[D]** negatively for finite extensions of $\mathbf{Z}$ and $\mathbf{Q}$.

---

[2] We use [BSS] to denote our main reference [Blum, Shub, Smale, 1989].

[3] That is, Y is a finite union of finite intersections of sets defined by polynomial equations and inequations (and inequalities, if R has an order) over R. (See Section 2.)

[4] By this, we mean a set of sentences of the form $\{\exists x_1 \forall x_2 \exists x_3 \ldots \forall x_n P(\mathbf{z}, x_1, \ldots, x_n) = 0\}_{\mathbf{z} \in \mathbf{Z}^k}$ where P is some polynomial over $\mathbf{Z}$.

On the positive side of **[D]**, Tarski's work was done in the generality of real closed fields. His work also implies that **[D]** has an affirmative answer for algebraically closed fields. Tarski's arguments are a developed form of elimination theory, which in a sharp complexity-theoretic form can be seen in [Renegar, 1989].

What about the answer to **[D]** for the remaining rings and field? Let us say that R has *Property D*, just in case every definable set over R is decidable over R.

We give some results (Section 6) in the direction of showing Tarski's examples exhausts all fields (sufficiently infinite) satisfying Property D. Here we follow [MacIntyre, 1971], and [MacIntyre, McKenna, van den Dries, 1983]. Indeed, our results could be considered an infinitary version of theirs. Note that Property D, as stated, is a nonuniform property. That is, each decidable set might have a different decision procedure. However, an immediate corollary to these results (for sufficiently infinite fields) is the uniformity of Property D once it is satisfied.

We would like to acknowledge helpful insights gleamed from conversations with Michel Herman, Adrien Douady and others concerning the mathematics underlying the undecidability of the Mandelbrot set. The first author would like to acknowledge helpful discussions with George Bergman, Lou van den Dries, Leo Harrington, and Simon Kochen; the personal influence of Julia Robinson is deeply felt.

[Friedman, Mansfield, 1988] and [Michaux, 1990] have results related to what we are doing here.

2. Background

We give here some background on decidability and definability over a ring R. Then we state our Main Theorem.

Let R be a commutative ring or field, which contains $\mathbf{Z}$, and for the moment is ordered. Let R^k be the direct sum of R with itself k times. If $k = \infty$, then $x \in R^k$ is a vector $(\mathbf{x}_1, \mathbf{x}_2, \ldots, \mathbf{x}_n, \ldots)$ with $\mathbf{x}_n = 0$ for sufficiently large n. In this section, we may take k finite.

The notion of a *computable function over R*,

$$\varphi_M \colon \Omega_M \to R^l,$$

is taken from [BSS] where $\Omega_M \subset R^k$, the domain of φ_M, is the *halting set* of a machine M over R.

A set $S \subset R^k$ is called *decidable over R* if its characteristic function is computable over R. We note that a set is decidable over R if and only if both it and its complement are halting sets (over R). Halting sets may be naturally thought of as "semidecidable" sets.

If $Y \subset R^k$ and $S \subset Y$, say that S is *decidable relative to Y over R* if the restriction of χ to Y, $\chi|_Y$, equals $\varphi|_Y$ for some φ computable over R. (We might *say that the set of admissible inputs* of the corresponding machine is Y.) In that case, if $Y' \subset Y$, then $S \cap Y'$ is decidable relative to Y' over R.

Next a very brief review of definability over R is given. First suppose $R = \mathbf{Z}$.

A subset S of $\mathbf{Z}^k$ is called *definable* over $\mathbf{Z}$ if there is a polynomial P in $n + k$ variables with integer coefficients such that

$$S = \{\mathbf{y} = (\mathbf{y}_1, \ldots, \mathbf{y}_k) \in \mathbf{Z}^k | \exists x_1 \forall x_2 \exists x_3 \ldots \forall x_n P(\mathbf{y}_1, \ldots, \mathbf{y}_k, x_1, \ldots, x_n) = 0\}.$$

The "defining" *formula* $\exists x_1 \forall x_2 \exists x_3 \ldots \forall x_n P(y_1, \ldots, y_k, x_1, \ldots, x_n) = 0$ contains an alternating sequence of quantifiers which could begin or end with either $\exists$ or $\forall$. In this expression, $y_1, \ldots, y_k$ are called *free variables* (and $x_1, \ldots, x_n$ *bound variables*). If we replace each free variable y_i by an integer $\mathbf{y}_i$, the expression becomes a *sentence* perhaps true, perhaps false in $\mathbf{Z}$. (If there are no free variables, the formula is already a sentence.)

Note, we have been using bold letters to denote elements (constants) or vectors and nonbold letters to denote variables. Henceforth, in the sequel we shall use nonbold letters for both; the intended meaning should be clear from context.

Quantifiers $\exists z$ or $\forall z$ with a new variable z may be added at any place in a formula without changing the set S. Thus the assumption of an alternating set of quantifiers places no restriction on the notion of definability. Moreover, "*not* $\forall$" is *logically equivalent* to "$\exists$ *not*." Similarly, "*not* $\exists$" is equivalent to "$\forall$ *not*." Thus, negations of formulas could be incorporated using

$$P(y, x_1, \ldots, x_n) \neq 0.$$

But over $\mathbf{Z}$, we have $P \neq 0$ if and only if

$$\exists z_1 \exists z_2 \exists z_3 \exists z_4 ((P - (1 + z_1^2 + z_2^2 + z_3^2 + z_4^2)) \times (P + (1 + z_1^2 + z_2^2 + z_3^2 + z_4^2)) = 0).$$

This assertion follows from the next lemma. Suppose R is an ordered ring or field. Then:

Lemma. *If $P(x)$ and $Q(x)$ are polynomials in n variables over R, and $x \in R^n$, then:*

(a) $P(x) \neq 0$ *if and only if* $-P(x) > 0$ *or* $P(x) > 0$.
(b) *If* $R = \mathbf{Z}$:

$$P(x) > 0 \text{ if and only if } \exists z_1, \ldots, z_4 \in \mathbf{Z} \text{ such that } P(x) = \sum_{i=1}^{4} z_i^2 + 1 \qquad (\textit{Lagrange}).$$

(c) $P(x) = 0$ **or** $Q(x) = 0$ *if and only if* $P(x)Q(x) = 0$.

So over $\mathbf{Z}$, negations of formulas are equivalent to formulas of the specified type.[5] In the sequel, the following equivalences will also be useful:

(d) $P(x) = 0$ **and** $Q(x) = 0$ *if and only if* $P^2(x) + Q^2(x) = 0$.
(e) $Q(x) \ngtr 0$ *if and only if* $-Q(x) > 0$ *or* $Q(x) = 0$.

Now for definability over a general ordered ring or field R, we must modify our definition to incorporate semi-algebraic sets:

A *basic semi-algebraic* set $X \subset R^n$ (*over* R) is defined as the set of $x = (x_1, \ldots, x_n)$ satisfying *basic conditions* of the type

$$P(x_1, \ldots, x_n) = 0,$$

$$Q_i(x_1, \ldots, x_n) > 0, \quad i = 1, \ldots, m,$$

where P *and* Q_i are polynomials over R. A set $X \subset R^n$ is *semi-algebraic* (*over* R) if it is generated by basic semi-algebraic sets using a finite process taking unions (i.e., "*or*'s" of basic conditions), intersections ("*and*'s"), and complements ("*not*'s").

From the above lemma, and by adding and substituting new variables, and by de Morgan's laws, it easily follows that:

Proposition 1. *Every semi-algebraic set $X \subset R^n$ can be expressed as finite union of basic semi-algebraic sets (and conversely).*

Intersections are eliminated using (d) and complements using (a) and (e).

Definition. A set $S \subset R^k$ is *definable over* R if there exists a semi-algebraic set $X \subset R^k \times R^n$ such that

$$S = \{(y_1, \ldots, y_k) | \exists x_1 \forall x_2 \exists x_3 \ldots \forall x_n \text{ such that } (y_1, \ldots, y_k, x_1, \ldots, x_n) \in X\}.$$

Note that the image of a definable set in $R^k \times R^l$ under the projection $R^k \times R^l \to R^k$ is a definable set. Indeed, definable sets over R are precisely those sets that are "derivable" from semi-algebraic sets by means of a finite sequence of projections and complements.

A (*defining*) *formula over* R for the above S is $\exists x_1 \forall x_2 \exists x_3 \ldots \forall x_k \Phi(y_1, \ldots, y_k, x_1, \ldots, x_n)$, where $\Phi(y_1, \ldots, y_k, x_1, \ldots, x_n)$ is a finite *disjunction* of *basic semi-algebraic formulas*, i.e., Φ is

$$(P^1 = 0 \,\&\, Q_1^1 > 0 \,\&\, \cdots \,\&\, Q_{m_1}^1 > 0) \quad \text{or} \quad \ldots \quad \text{or}$$

$$(P^l = 0 \,\&\, Q_1^l > 0 \,\&\, \cdots \,\&\, Q_{m_1}^l > 0),$$

the P's and Q's being polynomials over R (in the variables $y_1, \ldots, y_k, x_1, \ldots,$

[5] Note here, and in the sequel, we are implicitly moving quantifiers to the front of formulas, changing variables as necessary to avoid clashes and to maintain logical equivalence.

x_n) that describe the basic semi-algebraic pieces of X in a union as given by Proposition 1. Definitions for *free/bound variables* and *sentences over R* can be given as above for **Z**.

We remark that for ordered rings, our notion of definability over R is *equivalent* to the classical notion of definability given in "first-order" logic over the language containing the mathematical primitives $+$, $\times$, $=$ as well as $>$ *and* constants from R. That is, the sets defined are the same.

For consistency we need:

Proposition 2. *If $R = \mathbf{Z}$, our two notions of definable coincide.*

But this follows from Proposition 1 and the lemma because we can eliminate occurrences of $>$ using (b), "and's" using (d), and "or's" using (c).

Both the concepts of decidability and definability over a ring can be developed without $>$, retaining $\neq 0$. For decidability, one uses machines with branch nodes of which divide according to $h(x) = 0$ versus $h(x) \neq 0$. For definability, one omits all constructions requiring $>$. Semi-algebraic sets are replaced by what are usually called *constructible* sets, finite unions of sets satisfying *basic* conditions of the type

$$P_i(x_1, \ldots, x_n) = 0, \quad i = 1, \ldots, k,$$

$$Q(x_1, \ldots, x_n) \neq 0.$$

Similarly, a *basic constructible formula* is of the type $P_1 = 0 \,\&\, \cdots \,\&\, P_k = 0 \,\&\, Q \neq 0$ and a formula over R is a finite disjunction of basic ones.

The following results apply to rings and field with an order, or without.

Main Theorem. *Supposes R is a ring (of "algebraic integers") which is a finite extension of* **Z**, *or a field which is a finite extension of the rationals* **Q**. *Suppose $k < \infty$. Then there is a set $S \subset R^k$ which is definable over R, yet not decidable over R.*

The Main Theorem may be interpreted as saying that there is a reasonable family of sentences over R, but there is no way of deciding which are true in R. It is an immediate consequence of Propositions A and B below.

Suppose R is as in the Main Theorem. Then:

Proposition A. *For $k \leq \infty$, there is a halting set $S \subset R^k$ (of a machine) over R which is not decidable over R (i.e., S is a "semidecidable" undecidable set over R).*

Proposition A will be proved in Section 3.

Proposition B. *For $k < \infty$, any halting set $S \subset R^k$ over R is definable over R.*

Proposition B will be proved in Section 4.

Remark. We remark that over the reals $\mathbf{R}$, Proposition A is true (see [BSS]). But by [Tarski, 1951], the Main Theorem must fail over $\mathbf{R}$ for each $k < \infty$, and, thus, so must Proposition B. See Section 6 for more discussion and results along these lines.

3. Decidability: Proof of Proposition A

Proposition A is proved in part in [BSS] (see Proposition 2 below). Friedman and Mansfield have a complete proof [Friedman, Mansfield, 1988]. But to keep our paper accessible to nonlogicians, we indicate in this section a proof of Proposition A.

Say that subsets $S_1 \subset R^k$, $S_2 \supset R^l$ are *computably isomorphic* over R if there is a bijection $f: S_1 \to S_2$ such that f, f^{-1} are computable over R.

Proposition 1. *If R is as in the Main Theorem, there is a computable isomorphism*

$$f: R \to \mathbf{Z}^n \subset R^n$$

over R where n is the degree of the extension.

PROOF. Let $R = \mathbf{Z}[w_1, \ldots, w_n]$ and for $x = \sum_{i=1}^n x_i w_i$ let $f(x) = (x_1, \ldots, x_n)$. Clearly, f^{-1} is computable over R. Now list the elements of $\mathbf{Z}^n$ with norm nondecreasing. A comparison machine using this list shows that f is computable.

Similarly, if R is a finite extension of $\mathbf{Q}$ of degree n, then R is computably isomorphic to $\mathbf{Q}^n$ over R. To finish the proof, note that if R is any extension field of $\mathbf{Q}$, tnen there is a bijection $g: \mathbf{Z} \to \mathbf{Q}$ with g, g^{-1} computable over R.

The next step in our development (see [BSS]) is to associate to each machine M over a ring R, a point $\varphi(M) \in R^\infty$, its "code." This substitutes for the usual Gödel code which uses prime number factorization of positive integers. One can then construct a universal machine U which on input $(\varphi(M), x)$, $x \in U_M$, outputs $\varphi_M(x)$ and does not halt on other inputs. Then:

Proposition 2. *For any ring or field R, the halting set $\Omega_U \subset R^\infty$ is not decidable over R.*

The proof is an adaptation of the Cantor diagonal argument.

Proposition 3. *Suppose R is as in the Main Theorem. Then for any $k \leq \infty$, R and R^k are computably isomorphic over R.*

PROOF. The case $R = \mathbf{Z}$ is done using the Cantor pairing function. (See, e.g., [Davis, 1982].) Next, use Proposition 1 to obtain a computable isomorphism as the composition $R \to \mathbf{Z}^n \to (\mathbf{Z}^n)^k \to R^k$.

Propositions 2 and 3 now combine to yield Proposition A.

As remarked earlier, it is shown in [BSS] that Proposition A holds for $R = \mathbf{R}$ (although Proposition 3 does not).

Problem. Find k, R such that Proposition A fails, i.e., such that every halting set $S \subset R^k$ is decidable over R. (See also [Friedman, Mansfield, 1988].)

4. Definability: Proof of Proposition B

Suppose $S \subset R^k$ and $f: S \to R^l$, $k, l < \infty$. We say f is *definable over* R if the *graph of* $f = \{(x, f(x)) | x \in S\}$ is definable over R. Note that if f is definable over R, then both S and $f(S)$ are definable over R, and if, in addition, f is a bijection, then f^{-1} is definable over R.

Remark. For the proof of Proposition B, it suffices to consider finite dimensional machines. (See footnote, p. 28, [BSS].)

So suppose M is a finite-dimensional machine with *input space* $\bar{I}$, *output space* $\bar{O}$, and *state space* $\bar{S}$, each of the form R^k, for some $k < \infty$ (although not necessarily the same k).

Recall [BSS] that the *computing endomorphism* of M is a map $H: \bar{N} \times \bar{S} \to \bar{N} \times \bar{S}$ defined by the pair (*next node, next state*). Here $\bar{N} = \{1, \ldots, N\}$ is the set of *node* labels (1 being the label of the *input node*, N of the *output node*), and M is assumed to be in *normal form*. From [BSS], it is clear that H is definable over R.

The *halting register equations* associated to M are given by (1), (2), and (3) as follows:

(1) $z_0 = (1, I(y))$, where $I: \bar{I} \to \bar{S}$ is the *input map*, a polynomial map.
(2) $H(z_{i-1}) = z_i$, $i = 1, 2, \ldots, T$, $z_i \in \bar{N} \times \bar{S}$.
(3) $z_T = (N, x)$, some $x \in \bar{S}$.

The *halting set* $\Omega_M \subset \bar{I}$ may be described by

$\Omega_M = \{y \in \bar{I} | \exists T \in \mathbf{Z}^+, z_0, \ldots, z_T \in R^{k+1}, x \in R^k$ such that (1), (2), and (3) are satisfied$\}$.

However, this expression for Ω_M does not make Ω_M definable over R because the number of equations and variables is not a fixed finite nurnber. It depends on T. So we need more.

Generalized Gödel Sequencing Lemma. *Let R be as in the Main Theorem and $k \leq \infty$. There is a map σ:* N $\times$ *$R^k \to R^k$ such that*

(a) *given $a_0, \ldots, a_m \in R^k$, $\exists u \in R^k$ such that $\sigma(i, u) = a_i$, $i = 0, \ldots, m$,* and
(b) *if $k < \infty$, σ is definable over R.*

We now combine the register equations of a given finite-dimensional machine M as above with the function σ of the Gödel lemma to show that Ω_M is definable over R.

Let $F(y)$ denote the following "formula":

$$\exists T \in \mathbf{Z}^+ \exists u \, \exists x \, \forall i \in \mathbf{Z}$$

$$[\textbf{If } 0 < i \leq T, \textbf{ then } \sigma(0,u) = (1, I(y)), H(\sigma(i-1,u)) = \sigma(i,u), \text{ and } \sigma(T,u) = (N,x)].$$

Here R^k of the Gödel Sequencing Lemma is identified with $R \times \bar{S}$.

Now on the one hand, $\Omega_M = \{y | F(y) \text{ is true in } R\}$. This follows from the properties of the register equations and of $\sigma(i,u)$ of the Gödel lemma. On the other hand, the above expression for $F(y)$, and the definability of H and σ, and $\mathbf{Z}$ (also $\mathbf{Z}^+$ and $\mathbf{N}$) in R [Robinson, 1959, 1962, 1965], gives us the definability required in Proposition B.[6]

It remains to consider the Gödel lemma.

First observe that if the Gödel lemma is true for $k = 1$, then it is true for general k. This can be seen as follows. Suppose σ is the map of the Gödel lemma for $k = 1$. Then let σ_k: $\mathbf{N} \times R^k \to R^k$ be defined by

$$\sigma_k(i,u) = (\sigma(i,u_1), \sigma(i,u^2), \ldots),$$

where $u = (u_1, u_2, \ldots)$.

Now let $k = 1$. For $R = \mathbf{Z}$, we have the original Gödel lemma [Gödel, 1931]] which is derived from the Chinese Remainder Theorem. The general case now follows by noting the isomorphisms given in the proofs of Propositions 1 and 3 in Section 3 with $k < \infty$ are definable over R.

Problem. For which commutative rings R is the following true? If $\mathbf{Z}$ is definable in R, then for each $k < \infty$, any halting set $S \subset R^k$ over R is definable over R (and conversely).

5. Incompleteness

In this section, the Gödel Incompleteness Theorem is formulated and proved using the Main Theorem.

First fix a ring or field R, with or without order. Let Σ_R be the set of all *first-order sentences over* R, i.e., the set of sentences in the first-order language with mathematical primitives $+$, $\times$, $=$ (possibly $>$) and constants from R. (See,

[6] Here we are also using the logical equivalence of the expressions "*if P then Q*" and "*not P or Q*."

for example, [Cohen, 1960].)[7] The sentences over R described in Section 2 are first-order sentences and can be considered in "normal form" because each first-order sentence is equivalent to one of these.

One may consider Σ_R as a subset of R^∞ by an appropriate natural coding. For example, one can use pairs $(y_1, y_1'), (y_2, y_2'), \ldots$ where either y_i or y_i' is zero. If $y_i \neq 0$, then it stands for a logical symbol, not in R normally, but thus coded by an element of R. If $y_i = 0$, then $y_i' \in R$ plays its role as an element of R, as, for example, the coefficient of a polynomial used in the sentence.

One may use the codings of polynomials and rational functions of [BSS], for example.

It makes sense to talk about the subset of sentences $T_R \subset \Sigma_R$ that are *true* in R. For each sentence $\sigma \in \Sigma_R$, either $\sigma \in T_R$ or *not* σ (the negation of σ) $\in T_R$ (*completeness*), but not both (*consistency*). A set of *axioms* Y over R is simply a subset $Y \subset T_R$. Of course, Y could be empty.

Given a set of axioms Y, we will define the *derived set* Y_D, $Y \subset Y_D \subset T_R$, via (finite application of) a set of *rules of inference*. Then Y_D is the body of *theorems*, or *theory*, generated by the axioms in Y.

The rules of inference, which are convenient for our purposes are Rules A–G (pp. 9–11 of [Cohen, 1960]).

As an example we state:

Rule B. If A and $A \to B$ are sentences, then so is B.

Thus, if A and $A \to B$ both belong to Y, Y_D must include B. A similar situation prevails with the other rules.

Recall [BSS] that $S \subset R^k$, $k \leq \infty$, is an *output set over* R if there is a computable function $\varphi\colon \Omega_M \to R^k$ over R with $\varphi(\Omega_M) = S$.

We have the following:

Proposition 1. *If $Y \subset \Sigma_R \subset R^\infty$ is an output set over R, then so is the derived set Y_D. For the proof we use the following lemma.*

Lemma. *For each finite $Y' \subset \Sigma_R$, Y_D' is a halting set over R. Indeed, there is a machine M' such that $\Omega_{M'} = \bigcup_{Y' = \{y_1, \ldots, y_n\} \subset \Sigma_R} \{(y_1, \ldots, y_n)\} \times Y_D'$.*

Now suppose $\varphi(\Omega_M) = Y$. Define $F\colon \Omega_M^\infty \times \Sigma_R \to \Sigma_R$ via

[7] One can think of a "first-order sentence over R" as an ordinary mathematical sentence that is made up of variables $(x, y, z, \ldots)$, quantifiers ($\forall$ meaning "for all," $\exists$ "there exists"), connectives ($\neg$ meaning "not," & "and," $\vee$ "or," $\to$ "implies"), parentheses and mathematical symbols ($=$, $+$, $\times$, perhaps $>$ and constants from R). A sentence (as opposed to a "formula") has no free variables. i.e., all variables must be quantified. "First order" means that quantification is over elements (e.g., "there exists an element x in R such that for all elements y in R"), not over subsets.

$$F((x_1,\ldots,x_n),\sigma) = \begin{cases} \sigma & \text{if } ((\varphi(x_1),\ldots,\varphi(x_n)),\sigma) \in \Omega_{M'} \\ \text{underfined}, & \text{otherwise.} \end{cases}$$

Any element in Y_D is in some Y'_D for some finite $Y' \subset Y$. So by the above lemma, the range of F is Y_D and F is computable if φ is.

Proposition 2. *If R is as in the Main Theorem, a finite extension of* $\mathbf{Z}$ *or* $\mathbf{Q}$, *then the output sets and halting sets over R coincide.*

One directly checks $\mathbf{Z}$ and $\mathbf{Q}$. (This is quite classic essentially.) The general case follows from Proposition 1 of Section 3.

Gödel Incompleteness Theorem. *Fix a ring R, a finite extension of* $\mathbf{Z}$, *or a finite field extension of* $\mathbf{Q}$. *Let $Y \subset T_R$ be any set of axioms which is a halting set over R (i.e., an "effective" set of axioms). Then Y_D, the body of theorems derived from Y, is not T_R, the true sentences of R.*

If $R = \mathbf{Z}$, this is the usual Gödel theorem.

The proof goes as follows. On the one hand, the Main Theorem asserts there is an undecidable definable set over R. This implies T_R is not *decidable* (i.e., not a decidable subset of Σ_R) over R. For if it were, then a decision procedure would restrict to decide *any* definable set $X \subset R^k$ because $x \in X$ if and only if the sentence $\varphi(x)$ is in T_R (where φ is a given defining formula for X).

On the other hand, by the above, Y_D is a halting set over R. But any complete and consistent theory T which is a halting set is necessarily decidable: Let T be the halting set of machine M. Given $\sigma \in \Sigma_R$, to decide whether or not $\sigma \in T$ input both σ and *not* σ into M. Because one and only one of these inputs is in T, M will halt on one and only one of them, thus deciding whether or not $\sigma \in T$. Thus, $Y_D \neq T_R$. Q.E.D.

6. Decidability over R and Property D

In the following, R will denote $R_=$ (a commutative ring or field of characteristic 0) or $R_<$ (an ordered ring or field). L will denote the corresponding (first-order language $L_=$ with mathematical primitives $\{=, +, \times, 0, 1\}$ or $L_<$ with the additional primitive $<$. In case R is a field, we will also assume the primitive $\div$ is included. Without loss of generality, we may assume L has constants for each integer or each rational as appropriate. In general, if $C \subset R$, then L_C will denote the corresponding first-order language allowing additional constants from C.

Now let Σ_C denote the set of *sentences* in L_C, and T_C the subset *true* in R.

Recall that in the classical setting, a ring or field R is said to be *decidable* iff $T_\varnothing$ (viewed as a subset of $\mathbf{Z}^\infty$, as in Section 5) is decidable relative to $\Sigma_\varnothing$

over **Z**. This is what is meant for example when one says that the reals are decidable. But often one really has more. Thus, in the case of the reals, we see that $T_{R_<}$ ($\subset \mathbf{R}^\infty$) is decidable over $\mathbf{R}_<$, by a machine with parameters from **Z**. This follows immediately from Tarski's theorem that **R** admits uniform elimination of quantifiers.[8]

Thus, we are motivated to *extend* the notion of decidability over a ring.

Definition. R is *(strongly) decidable over* R iff T_R is decidable relative to Σ_R over R (by a machine with parameters from **Z**).

If R is decidable over R, then clearly the Main Theorem fails for each $k < \infty$, i.e., R has *Property D*:

Definition. R has *Property D* iff for each $k < \infty$, any $S \subset R^k$ definable over R is also decidable over R.

Property D is a weak form of elimination of quantifiers:

Proposition 1. *If R has Property D, then for each L_R-formula $\varphi_R(x_1,\ldots,x_k)$, there is a finite set $C \subset R$ and an (effectively) countable set of quantifier-free L_C-formulas $\psi_C^j(x_1,\ldots,x_k)$ such that $S = \bigcup S_j$, where S, S_j are the sets defined by φ_R, ψ_C^j, respectively. Furthermore, we can assume without loss of generality the ψ_C^j to be basic semi-algebraic (or basic constructible) formulas.*

Problem. Can we assume that C is just the set of constants occurring in φ_R? If so, we shall say R has *strong* Property D.

What is the relationship between the above notions of decidability?

Theorem. *Suppose R is a field of infinite transcendence over the rationals. Then the following are equivalent*:

1. *R admits uniform elimination of quantifiers.*
2. *R is strongly decidable over R.*
3. *R has strongly Property D.*
4. a. *R is an algebraically closed field in case $R = R_=$.*
   b. *R is a real closed field in case $R = R_<$.*

And these imply:

5. *R is decidable.*

---

[8] **Definition** (Classical). R admits elimination of quantifiers iff for each L-formula $\varphi(x_1,\ldots,x_k)$ there is a *quantifier-free* L-formula $\psi(x_1,\ldots,x_k)$ such that φ and ψ define the same subset of R^k. The elimination is *uniform* iff the map from φ to ψ is computable over **Z**.

Furthermore, if we add the condition "R is dense in its real closure in case $R = R_<$,"[9] *we can add to the above list of equivalences*:

2.′ *R is decidable over R.*
3.′ *R has Property D.*

Remark 1. The stipulations of uniformity in 1 and effectiveness (of the countable decomposition of definable sets into semi-algebraic sets) as implied by 3 (or 3′) are not necessary. These will follow from a simple analysis of the proof of the theorem.

Remark 2. We note that, in general, the notion of decidability is *weaker* than decidability over R (or Property D). For example, $\mathbf{R}_=$ is decidable, *but* because $\mathbf{R}_=$ is not algebraically closed, the theorem implies it *cannot* be decidable over $\mathbf{R}_=$.

PROOF OF THEOREM. It is easy to see that 1 implies 2 which implies 3; 4 implies 1 by Tarski's theorems [Tarski, 1951]. Clearly, 2 implies 5, and 2 implies 2′ implies 3′. Thus, we are left to show that under the appropriate hypotheses, strong Property D or Property D imply 4, i.e., the closedness conditions. Here we are inspired by, and indeed closely follow, the proof in [MacIntyre, McKenna, van den Dries, 1983] of the converse of Tarski's theorems.

We start with some preliminaries. In the following, R will be a field (of characteristic 0), and $n > 1$.

Fix n. Let $\mathrm{Poly}_R(n)$ denote the *space of* ***monic*** *degree n polynomials in one variable over R*. There is a natural *identification*, $\mathrm{Poly}_R(n) \simeq R^n$, where

$$r_0 + \cdots + r_{n-1}z^{n-1} + z^n \leftrightarrow r = (r_0, \ldots, r_{n-1}) \in R^n.$$

Let $\mathbf{S} \subset R^n$ correspond to the set of polynomials in $\mathrm{Poly}_R(n)$ *not solvable* in R. Note that

$$\mathbf{S} = \{r \in R^n | \mathbf{f}(r, z) \text{ has no root in } R\},$$

where $\mathbf{f}(r, z) = r_0 + \cdots + r_{n-1}z^{n-1} + z^n$. Thus, $\mathbf{S}$ is *defined* by the L-formula $\forall z\mathbf{f}(x_0, \ldots, x_{n-1}, z) \neq 0$. So if R has Property D, then, by the Proposition 1, $\mathbf{S} = \bigcup_{j \in J} S_j$, where J is countable and

$$S_j = \{r \in R^n | P_i^j(r) = 0, i = 1, \ldots, m_j, Q^j(r) \neq 0\} \text{ in case } R = R_=,$$

or

$$S_j = \{r \in R^n | P^j(r) = 0, Q_i^j(r) > 0, i = 1, \ldots, m_j\} \text{ in case } R = R_<.$$

Here the P's and Q's are polynomials over $\mathbf{Q}(C)$, where C is a finite subset of R.

---

[9] R *is dense in its real closure* $\hat{R}$ means: For each r and $\varepsilon > 0$ in $\hat{R}$, there is an r' in R such that $|r - r'| < \varepsilon$.

Proposition 2. *Suppose R is of infinite transcendence. If R has an algebraic extension of degree n, then for each finitely generated subfield $K \subset R$, $\mathbf{S}$ cannot be covered by a countable union of proper algebraic varieties V_j in R^n defined over K (i.e., with $V_j = \{r \in R^n | P_i^j(r) = 0,\ i = 1, \ldots, k_j\}$, the P's being nonzero polynomials with coefficients from K).*

Modulo Proposition 2, we proceed with our proof.

We first consider the case $R = R_=$ and suppose R has property D. Suppose R is not algebraically closed, i.e., for some n, R has an algebraic extension of degree n. So by Proposition 2 (and noting $\mathbf{S} \neq R^n$) we may assume for at least one j, $S_j = \{r \in R^n | Q^j(r) \neq 0\} \neq \varnothing$, and so $Q^j \not\equiv 0$. Also $S_j \subset S$. So for $r \in R^n$, *whenever $Q^j(r) \neq 0$, then $\mathbf{f}(r, z) = 0$ has no solution in R.* This contradicts

Lemma 1. *For each nonzero polynomial $Q\colon R^n \to R$, there is an $r \in R^n$ such that $Q(r) \neq 0$ and $\mathbf{f}(r, z)$ has* all *its solutions in R.*

To see this let $\sigma\colon R^n \to R^n$ be the polynomial map given by the elementary symmetric functions $\sigma_1, \sigma_2, \ldots, \sigma_n$. σ is an algebraically independent map. So for $Q \not\equiv 0$, and because R is infinite, there is $\alpha = (\alpha_1, \ldots, \alpha_n) \in R^n$ such that $Q(\sigma(\alpha)) \neq 0$. Letting $r = \sigma(\alpha)$ we have $Q(r) \neq 0$ and $\mathbf{f}(r, z) = \sigma_1(\alpha) + \sigma_2 z \cdots + \sigma_n(\alpha) z^{n-1} + z^n = \prod_{i=1}^n (z - \alpha_i)$. Thus, r has the required properties.

Now we consider the case $R = R_<$ and suppose R has Property D.

Suppose first that some odd degree polynomial has no solution in R. This implies, as above, that for some odd $n > 1$ and j in J, we may assume

$$S_j = \{r \in R^n | Q_i^j(r) > 0,\ i = 1, \ldots, m_j\} \neq \varnothing.$$

Let $\psi(c, x)$ be the formula $Q_1^j(x) > 0 \,\&\cdots\&\, Q_{m_j}^j(x) > 0$ where $x = (x_0, \ldots, x_{n-1})$ and $c \in R^m$ is a vector of all the coefficients occurring in the Q's. So, there is an $r \in R^n$ such that $\psi(c, r)$ is true in R *and* for all $r \in R^n$, *whenever $\psi(c, r)$ is true in R, then $\mathbf{f}(r, z) = 0$ has no solution in R.* Note if R has strong Property D, then without loss of generality we may assume $c \in \mathbf{Q}^m$. Then, given the hypotheses of our theorem, the following lemma will yield a contradiction.

Lemma 2. *Suppose R is dense in $\hat{R}$, its real closure, or that $c \in \mathbf{Q}^m$. Suppose $\psi(c, r)$ is true in R for some $r \in R^n$. Then there is an $r' \in R^n$ such that $\psi(c, r')$ is true in R* and *$\mathbf{f}(r', z) = 0$ is solvable in R.*

Proof. Let

$$F = \begin{cases} R & \text{if } R \text{ is dense in } \hat{R} \\ \mathbf{Q}, & \text{in case } c \in \mathbf{Q}^m. \end{cases}$$

If $\psi(c, r)$ is true in R, then $\psi(c, r)$ is true in $\hat{R}$, and so either trivially in the first case, or by the transfer property for real closed fields in the second case, $\psi(c, r^*)$ is true in $\hat{F}$, some $r^* \in \hat{F}^n$. Now, $\mathbf{f}(r^*, z) = (z - \xi)(a_0 + a_1 z + \cdots +$

$a_{n-1}z^{n-1})$, some $\xi, a_0, \ldots, a_{n-1} \in \hat{F}$. For $\xi', a'_0, \ldots, a'_{n-1} \in F$, define $r' \in F^n$ by $\mathbf{f}(r', z) = (z - \xi')(a'_0 + a'_1 z + \cdots + a'_{n-1}z^{n-1})$. Because F is dense in $\hat{F}$, we can choose $\xi', a'_0, \ldots, a'_{n-1}$ close enough to $\xi, a_0, \ldots, a_{n-1}$ to guarantee $\psi(c, r')$ is true in F, and hence in R.

Now suppose some positive element in R has no square root in R, i.e., $T = \{t \in R | t > 0 \text{ and } \sqrt{t} \notin R\} \neq \emptyset$. But then, $\{tr^2 | t \in T, r \in R, r \neq 0\} \subset T$, and so if $T \neq \emptyset$, then T must have an infinite number of algebraically independent elements. (Recall the degree of transcendence of R over $\mathbf{Q}$ is infinite.) Now T is definable and, hence, decidable over R. Thus, T can be decomposed as in the proposition, and so there is a $c \in R^m$ such that either all elements of T are algebraic over $\mathbf{Q}(c)$, or else T contains a nonempty interval $T' = (t_1, t_2) \cap R$, *some* $t_1, t_2 \in \widehat{\mathbf{Q}(c)}$. The former is not possible as we have just seen, but neither is the latter:

Let F be as in Lemma 2; then, $T'' = (t_1, t_2) \cap F \neq \emptyset$. But because F is dense in $\hat{F}$, for each $t \in T''$, we can choose $s \in F$ close enough to $\sqrt{t}$ so that s^2 will be close enough to t to be in the interval (t_1, t_2), a contradiction to the definition of T.

We have thus shown modulo Proposition 2 that R is real closed.

We now proceed as follows. Let $X \subset R^n$.

Definition. We shall say X has *property A* iff for each finitely generated subfield $K \subset R$, X cannot be covered by a countable union of proper algebraic varieties V_j in R^n defined over K.

Lemma 3. *Suppose R is of infinite transcendence. Then R^n has property A.*

The proof is by induction on n and is similar to the proof of a related result of [Amitsur, 1956][10]: Suppose R is an uncountable field. Then R^n cannot be covered by a countable union of proper algebraic varieties defined over R.

Remark 3. Here we cannot weaken the hypothesis of infinite transcendence. For example, let $R = \widetilde{Q(c)}$, where $c = (c_1, \ldots, c_m)$ and let $\{P_j(x)\}_{j \in \mathbf{Z}^+}$ be a listing of all nonzero polynomials in one variable over $\mathbf{Q}(c)$. Then, $R = \bigcup_{j \in \mathbf{Z}^+} V_j$, where $V_j = \{r \in R | P_j(r) = 0\}$.

Remark 4. If R^n has property A, then $\mathbf{X} = \{r = (r_0, \ldots, r_{n-1}) \in R^n | r_i \neq 0 \text{ some } i > 0\}$ has property A since $R^n = \mathbf{X} \cup \{r \in R^n | \prod_{i=1}^{n-1} r_i = 0\}$. More generally, if $X \subset R^n$ has property A, then so does $\{x \in X | x_i \neq 0 \text{ some } i \in I\}$, where $I \subset \{0, \ldots, n-1\}$.

Remark 5. If X has property A, and $X \subset S \subset R^n$, then S has property A.

---

[10] We are grateful to George Bergman for pointing us to this paper.

Lemma 4. *Suppose $F: R^n \to R^n$ is an algebraically independent polynomial map (i.e., for all polynomials $g: R^n \to R$, if $gF \equiv 0$ then $g \equiv 0$). Then if $X \subset R^n$ has property A, then so does $F(X)$.*

PROOF. Let $Y = F(X)$ and consider the following diagram:

$$\begin{array}{ccc} X \subset R^n & & \\ \downarrow F & \searrow gF & \\ Y = F(X) \subset R^n & \xrightarrow[g]{} & R \end{array}$$

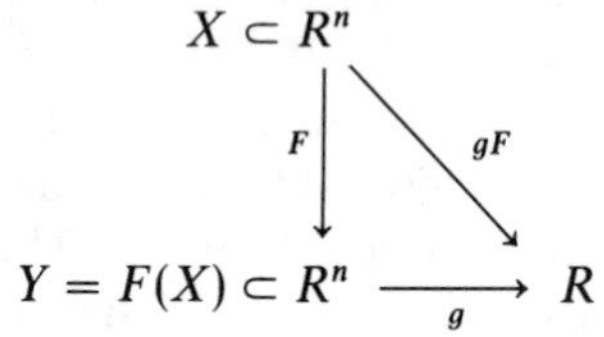

First suppose $\{Y_j\}_{j \in J}$ covers Y. Let $X_j = F^{-1}(Y_j)$. Then $\{X_j\}_{j \in J}$ covers X. Now suppose Y_j is a variety in R^n defined by a (finite) set G_j of polynomials over $K \subset R$. Then

$$X_j = \{r \in R^n | F(r) \in Y_j\} = \{r \in R^n | gF(r) = 0, \text{ all } g \in G_j\}$$

is a variety in R^n defined by the (finite) set of polynomials $\{gF | g \in G_j\}$ over $K(a_1, \ldots, a_m)$. Here $a_1, \ldots, a_m$ are the constants occurring in F.

So if $\{Y_j\}_{j \in J}$ is a countable cover of Y by varieties over a finitely generated subfield K, then $\{X_j\}_{j \in J}$ is a countable cover of X by varieties over $K(a_1, \ldots, a_m)$.

If X has property A, then $X_j = R^n$, for some $j \in J$. So for all $r \in R^n$, $gF(r) = 0$, all $g \in G_j$. Thus, because R is infinite, $gF \equiv 0$, all $g \in G_j$. But F is algebraically independent. Therefore, $g \equiv 0$, all $g \in G_j$. Therefore, $Y_j = R^n$ and so Y has property A. This proves Lemma 4.

Now let $\tilde{R}$ be the algebraic closure of R, and so $\tilde{R}^n$ can be considered the *space of* (*sequences of all*) *roots* of elements of $\text{Poly}_{\tilde{R}}(n)$.

The *Viete map* π: $\tilde{R}^n$ (roots) $\to \tilde{R}^n$ ($\simeq \text{Poly}_{\tilde{R}}(n)$) is given by

$$\pi(\xi) \leftrightarrow \prod_i (z - \xi_i) \quad \textit{for } \xi = (\xi_1, \ldots, \xi_n) \in \tilde{R}^n.$$

We consider the following diagram:

$$\begin{array}{ccccc} \widehat{\text{Poly}}_R(n-1) \times \tilde{R}^n \simeq R^n \times \tilde{R}^n & \xrightarrow{\text{Eval}} & \tilde{R}^n & \textbf{(Space of roots)} \\ \downarrow \text{Proj} & \searrow F & \downarrow \pi & \textbf{(Viete map)} \\ \widehat{\text{Poly}}_R(n-1) \simeq R^n & \xrightarrow{E_\alpha} & \tilde{R}^n \simeq \text{Poly}_{\tilde{R}}(n) & \end{array}$$

(Space of degree $n-1$ polys/R) **(Space of monic degree n polys/$\tilde{R}$)**

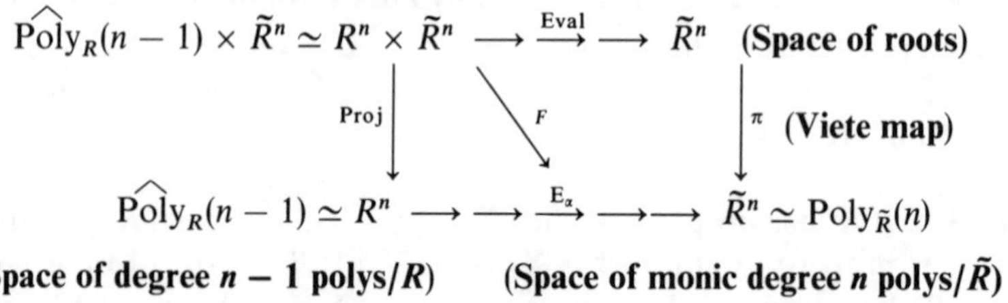

Here $\widehat{\text{Poly}}_R(n-1)$ is the *space of univariate degree* $(n-1)$ *polynomials over* R which is naturally associated with R^n via

$$r_0 + \cdots + r_{n-1} y^{n-1} \leftrightarrow r = (r_0, \ldots, r_{n-1}) \in R^n.$$

And for $r \in R^n$, $\alpha = (\alpha_1, \ldots, \alpha_n) \in \tilde{R}^n$, Eval is given by $\text{Eval}(r, \alpha)_i = r_0 + \cdots + r_{n-1}\alpha_i^{n-1}$, $F = \pi \cdot \text{Eval}$ and $F_\alpha(r) = F(r, \alpha)$. Thus,

Remark 6. $\{\mathrm{Eval}(r, \alpha)_{i=1,\ldots,n}\}$ is the set of all roots of the monic polynomial associated with $F_\alpha(r)$.

Let $R^n_* = \{\alpha \in \tilde{R}^n | \{\alpha_i\}_{i=1,\ldots,n}$ are distinct conjugates of degree n over $R\}$. So, $R^n_* \neq \varnothing$ just in case R has an algebraic extension of degree n. If $\xi \in R^n_*$, then $\pi(\xi)$ is the *minimal polynomial* of ξ_i over R.

Lemma 5. *Suppose $\alpha \in R^n_*$. Then*

(a) $F_\alpha\colon R^n \to R^n$ and
(b) F_α is algebraically independent.

Proof. See [MacIntyre, McKenna, van den Dries, 1983].

Recall $\mathbf{X} = \{r \in R^n | r_i \neq 0$, some $i > 0\}$ and $\mathbf{S} = \{r \in R^n | \mathbf{f}(r, z)$ has no roots in $R\}$.

Corollary 1. *If $\alpha \in R^n_*$, then $F_\alpha(\mathbf{X}) \subset \mathbf{S}$.*

Proof. Suppose $\alpha \in R^n_*$ and $r \in R^n$. Then by Lemma 5(a), $F_\alpha(r) \in R^n$. If, in addition, $r \in \mathbf{X}$, then $\mathrm{Eval}(r, \alpha)_i \notin R$, $i = 1, \ldots, n$. So by Remark 6, $F_\alpha(r) \in \mathbf{S}$.

Corollary 2. *Suppose the degree of transcendence of R is infinite. If $R^n_* \neq \varnothing$, then* **S** *has property A.*

Proof. By Lemma 3 and Remark 4, **X** has property A. Let $\alpha \in R^n_*$. Then, by Lemmas 5(b) and 4, $F_\alpha(\mathbf{X})$ has property A. So by Corollary 1 and Remark 5, **S** has property A.

Thus, we have proved Proposition 2 and hence the theorem.

Remark 7. Using a lemma of [Michaux, 1990], we need only assume as hypothesis in the theorem, in case $R = R_=$ (likewise, in case $R = R_<$), that R be a commutative ring without zero divisors (an ordered commutative ring) of infinite transcendence over $\mathbf{Q}$.

7. On the Undecidability of the Mandelbrot Set over R

For a heuristic discussion of the problem of the decidability of the Mandelbrot set M, one can see [Penrose, 1989]. For the mathematics of M, see [Douady, Hubbard 1984–1985].

A well-known and presumably difficult conjecture is: The boundary of M has Hausdorff dimension 2. On the other hand, it seems much easier from the work of Douady, Hubbard, Misiurewicz, Tan Lei, and others that the following holds.

Weak Conjecture. *The boundary of M has Hausdorff dimension greater than* 1.

Proposition. *If the Weak Conjecture is true, then the Mandelbrot set is undecidable over* **R**.

PROOF. Suppose the contrary. Then M is the halting set of some machine over **R** and, hence, is the countable union of basic semi-algebraic sets (see [BSS]). M is closed, and the closure of a basic semi-algebraic set is a semi-algebraic set. Thus, we may suppose

$$M = \bigcup_{i=1}^{\infty} S_i,$$

where each $S_i \subset \mathbf{C} = \mathbf{R}^2$ is a closed semi-algebraic set. For each i, we claim: $\dim(\partial M \cap S_i) \leq 1$, where ∂M is the boundary of M and dim is the Hausdorff dimension.

If $\dim S_i \leq 1$, this is immediate. However, if $\dim S_i > 1$, then $\dim S_i = 2$ because S_i is a closed semi-algebraic set. So $\dim \partial S_i \leq 1$. Next note that the interior of S_i must be contained in the interior of M. Therefore,

$$\partial M \cap S_i \text{ is contained in } \partial M \cap S_i \text{ and has dimension } \leq 1.$$

So now we have

$$\partial M = \bigcup_{i=1}^{\infty} (\partial M \cap S_i) \text{ has dimension } \leq 1,$$

contradicting the Weak Conjecture.

Remark. It is easy to show that the complement of M is a halting set, so once more we have (provisionally) an example of an undecidable "semi-decidable" set.

Added in Proof. At the Smalefest in Berkeley (August 1990), Dennis Sullivan has shown us what appears to be a direct proof that the Mandelbrot set is not the countable union of semi-algebraic sets.

References

Amitsur, S.A., "Algebras over Infinite Fields," *Proc. AMS American Math.* **7** (1956), 35–48.

Blum, L., Shub, M., and S. Smale, "On a Theory of Computation and Complexity over the Real Numbers: NP-Completeless, Recursive Functions and Universal Machines," *Bull. AMS* **21** (1989), 1–46.

Cohen, P., *Set Theory and the Continuum Hypothesis*, Benjamin, New York, 1960.

Davis, M., *Computability and Unsolvability*, Dover, New York, 1982.

Douady, A. and J. Hubbard, "Etude Dynamique des Polynomes Complex, I, 84–20, 1984 and II, 85–40. 1985, Publ. Math. d'Orsay, Univ. de Paris-Sud, Dept. de Math. Orsay, France.

Friedman, H., and R. Mansfield, "Algorithmic Procedures," preprint, Penn State, 1988.

Gödel, K., "Uber formal Unentscheidbare Satze der Principia Mathematica und Verwandter Systeme, I," *Monatsh. Math. Phys.* **38** (1931), 173–198.

MacIntyre, A., "On ω_1- Categorical Fields," *Fund. Math.* **7** (1971), 1–25.

MacIntyre, A., McKenna, K., and L. van den Dries, "Elimination of Quantifiers in Algebraic Structures," *Adv. Math.* **47**, (1983), 74–87.

Michaux, C., "Ordered Rings over which Output Sets Are Recursively Enumerable Sets," preprint, Universite de Mons, Belgium, 1990.

Penrose, R., *The Emperor's New Mind*, Oxford University Press, Oxford, 1989.

Renegar, J., "On the Computational Complexity and Geometry of the First-Order Theory of the Reals," Part I, II, III, preprint, Cornell University, 1989.

Robinson, J., "The Undecidability of Algebraic Rings and Fields," *Proc. AMS* **10** (1959), 950–957.

Robinson, J., "On the Decision Problem for Algebraic Rings," *Studies in Mathematical Analysis and Related Topics*, Gilberson et al., ed., University Press, Stanford, 1962, pp. 297–304.

Robinson, J., "The Decision Problem for Fields," *Symposium on the Theory of Models*, North-Holland, Amsterdam, 1965, pp. 299–311.

Rosser, B., "Extensions of Some Theorems of Gödel and Church," *J. Symb. Logic* **1** (1936), 87–91.

Tarski, A., *A Decision Method for Elementary Algebra and Geometry*, University of California Press, San Francisco, 1951.

33
Ill-Posed Problem Instances

JAMES RENEGAR

1. Introduction

1.0. This chapter consists of those sections of a longer work [5] which formed the basis for much of the author's talk at the Smalefest. The longer work contains complete proofs and develops much additional material.

1.1. This chapter is concerned with problems for which approximations to solutions are to be computed from approximate data. This is of interest, for example, if the actual data can only be approximated through experimentation, or if computations can only be performed with approximate data obtained by rounding the actual data. This chapter and [5] form a step toward developing a general theory of solving algebraic problems using approximate data.

An extremely general set of problems is considered, including: linear equation solving, all of the standard matrix factorization problems, eigenvalue problems, nonlinear programming problems with polynomial objective and constraint functions (e.g., linear programming, quadratic programming), various sensitivity analysis problems, and many, many other problems.

We assume data for a problem instance is specified by a vector of real numbers, $y \in \mathbb{R}^m$ (e.g., for solving a system of linear equations $Ax = b$ the data would consist of the coefficients of A and b). For formalism, we assume approximate data is available through an oracle. Input to the oracle is $\delta > 0$ and output is approximate data strictly within error δ of the data for the actual instance. The oracle might represent experimentation, or procedures for approximating the actual data (e.g., rounding). We allow very general error measurement functions in the full development presented in [5], but in the present chapter we assume errors are measured by a norm.

Using the approximate data $\bar{y}$ and strict upper bound δ on its error, the goals are to decide if the actual instance has a solution $x \in \mathbb{R}^n$ and if so, to compute a point guaranteed to be within prespecified error $\varepsilon > 0$ of a solution for the actual instance.

If the only knowledge about the actual instance used in attempting to achieve the goals is the approximate data $\bar{y}$ and error bound δ, then all problem instances with data vector $\tilde{y} \in \mathbb{R}^m$ strictly within error δ of $\bar{y}$ are candidates for the actual instance. Achieving the goals amounts to achieving the goals for all candidates; a point can be guaranteed, using only $\bar{y}$ and δ, to be within error ε of a solution for the actual instance only if it is within error ε of a solution for all instances with data vector $\tilde{y} \in \mathbb{R}^m$ strictly within error δ of $\bar{y}$.

Practitioners are willing to use everything they know about a problem instance to solve it efficiently. For example, because of the context the actual instance arises in, it might be known that if the actual instance has a solution, then it has infinitely many solutions. Such knowledge can narrow the set of candidates for the actual instance; the candidates would not simply be those instances whose data is strictly within error δ of the approximate data $\bar{y}$, but rather, candidates would also be required to satisfy the property of having either infinitely many solutions or no solutions. Knowledge that narrows the set of candidates can have the effect that less data accuracy is required to determine if the actual instance has a solution and, if so, to compute a good approximation to a solution.

A theory of solving problem instances using approximate data should thus incorporate what we call *instance knowledge*. In [5], we lay some foundations for such a theory, including some foundations for a general theory of condition numbers.

To develop a rigorous mathematical theory, it is necessary to impose conditions on the form of instance knowledge. In [5], we do that. However, in this chapter we focus solely on an intriguing question regarding instance knowledge.

1.2. The question is this: "Is it ever possible for one to know that all of the knowledge one possesses about the actual instance is insufficient to either design an algorithm to determine if the actual instance has a solution, or to design an algorithm to compute an approximation to a solution for the actual instance guaranteed to be of prespecified accuracy, assuming the algorithms would only have access to arbitrarily accurate *approximate* data?"

The question is intriguing from a logical point of view, but an answer to it is also of practical interest. A "Yes" answer for a particular problem implies that one can sometimes know it is pointless to attempt achieving the goals, given the state of knowledge regarding the actual instance. Consequently, one can then avoid wasting time and money designing algorithms and collecting accurate data. A "No" answer implies that one cannot know it is pointless when it is pointless, a somewhat disturbing situation.

We consider knowledge about instance knowledge to be instance knowledge. In particular, if one does indeed know that all of the knowledge one possesses about the actual instance is insufficient to design algorithms which require only sufficiently accurate approximate data to achieve the goals, then the knowledge of the insufficiency is part of the knowledge one possesses about the actual instance.

We assume the instance knowledge always contains a precise statement of the problem for which the instance is indeed an instance (e.g., solving a system of linear equations) including the number of coordinates in the data vector $y \in \mathbb{R}^m$ and in the (potential) solution vectors $x \in \mathbb{R}^n$. This assumption is formalized in [5].

We argue that, for the broad class of problems mentioned earlier, the answer to the above question is "No." Regardless of what one knows about the actual problem instance, if that knowledge cannot be used to design algorithms which require only sufficiently accurate approximate data to acheive the goals, one cannot know it. We also provide a simple example of a problem not in the class of problems considered for which the answer is "Yes."

Our argument does not assume the instance knowledge other than the problem statement to be of any particular mathematical form. Consequently, the argument is not completely rigorous. So we never refer to the argument as a "proof." However, the author finds the argument very convincing. It seems the argument forms an outline for a proof whenever acceptable mathematical form is imposed on the instance knowledge. The argument proceeds by showing that if one knew that one's knowledge regarding the actual instance was insufficient to design algorithms that achieve the goals, then one could use that knowledge to show that a particular algorithm in [5] does achieve the goals, a contradiction. In other words, a "Yes" answer to the question forms the basis for showing the answer is "No." So the answer must be "No."

In certain sections of [5], where we do require instance knowledge to be of a particular but very general form, we arrive at a "No" answer as a simple corollary to the main theorem of that paper. The general arguments, with no form imposed on the instance knowledge, proceed from those arguments. Although we do not present the complete arguments in this chapter, we present a sufficiently detailed outline to provide the reader with an understanding of the gist of the arguments.

In "reality," as time progresses, new knowledge may become available about the actual instance, far beyond simply new data approximations. The new instance knowledge may be superfluous, providing no real new insight that is relevant to the design of algorithms, or it may be radically different than previously available information. Besides arguing that "one cannot know if the instance knowlege is insufficient to design algorithms which require only sufficiently accurate approximate data to achieve the goals," our results shed light on the questions, "What new instance knowledge is 'radically different'?" and "How many times can radically different instance knowledge be obtained without the body of knowledge becoming sufficient to design algorithms which require only sufficiently accurate approximate data to achieve the goals?"

1.3. We now introduce several definitions. These will allow us to make more precise the above discussion.

We say a problem instance is *indistinguishable* from the actual instance if the instance knowledge does not exclude the possibility that it is the actual instance. For example, if one wishes to solve a system of linear equations and one knows nothing more about the actual instance than that it has 20 equations, 30 variables, and a solution set of dimension 13, then the systems of 20 equations in 30 variables with solution set of dimension 13 are the instances which are indistinguishable from the actual instance. Anything that can be deduced from the instance knowledge about the actual instance is also true for the instances which are indistinguishable from the actual instance.

It may be impossible to design an algorithm from the instance knowledge to determine if the actual instance has a solution even if the algorithm would have access to arbitrarily accurate approximate data. For example, this is so if the following condition holds: For each $\delta > 0$ and each $\bar{y} \in \mathbb{R}^m$ strictly within error δ of the actual data, there exist $y_1, y_2 \in \mathbb{R}^m$ which (i) are indistinguishable from the actual instance, (ii) are strictly within error δ of $\bar{y}$, and (iii) are such that y_1 has a solution but y_2 does not. If this condition holds, then we say that the actual instance is *definitely ill-posed for the decision problem.* For such an instance, exact data is required to determine if the instance has a solution.

If the actual instance has a solution, then it may be impossible to design an algorithm from the instance knowledge which computes a point guaranteed to be within error ε of a solution for the actual instance even if the algorithm would have access to arbitrarily accurate approximate data. For example, this is so if the following condition holds: For each $\delta > 0$ and each $\bar{y}$ strictly within error δ of the actual data, there does not exist $\bar{x} \in \mathbb{R}^n$ which is within error ε of a solution for each instance which (i) has a solution, (ii) is indistinguishable from the actual instance, and (iii) has data strictly within error δ of $\bar{y}$. If the actual instance has a solution and if this condition holds for each sufficiently small $\varepsilon > 0$, then we say that the actual instance is *definitely ill-posed for the approximation problem.* For such an instance, exact data is required to compute a point guaranteed to be within error ε of a solution, assuming $\varepsilon > 0$ is small.

We say the actual instance is *definitely ill-posed* if it is definitely ill-posed for either the decision problem or the approximation problem. It is the author's opinion that any reasonable definition of an ill-posed instance for the problems considered should be satisfied by instances which are definitely ill-posed.

It is very important to understand that the definition of a definitely ill-posed instance is a relative definition, relative to the instance knowledge. If new instance knowledge is obtained, from whatever source, the instance may not be definitely ill-posed relative to the broader set of instance knowledge because the set of instances indistinguishable from the actual instance may shrink.

We will have occasion to speak of an instance $\tilde{y}$ indistinguishable from the actual instance as being definitely ill-posed. By this, we mean that the definition of what is required for the actual instance to be definitely ill-posed is satisfied if one replaces the actual data with $\tilde{y}$. In other words, $\tilde{y}$ would be ill-posed if it was the actual instance and the instance knowledge remained unchanged.

At the opposite extreme are problem instances which can most certainly be solved using approximate data, i.e., well-posed instances. To formalize this notion somewhat, we begin with the following definition.

We say that an algorithm[1] for the problem is *acceptable* if the algorithm:

(1) Accepts as input any tuple $(\bar{y}, \delta, \varepsilon)$, where $\delta > 0$, $\varepsilon > 0$, and $\bar{y}$ might be provided by the oracle upon input δ.[2]
(2) Replies one of the following three statements:
  (a) "All instances which
    (i) are indistinguishable from the actual instance and
    (ii) are strictly within error δ of $\bar{y}$
    have solutions, and $\bar{x}$ is strictly within error ε of a solution for all such instances," (where $\bar{x}$ is computed by the algorithm).
  (b) "All instances which
    (i) are indistinguishable from the actual instance and
    (ii) are strictly within error δ of $\bar{y}$
    do not have solutions."
  (c) "Please provide approximate data of better accuracy."
(3) Can be proven correct, i.e., correct in the sense that whenever it replies with statement (2a) or (2b) and the input tuple satisfies the condition that $\bar{y}$ is strictly within error δ of data for an instance which is indistinguishable from the actual instance, then the statement replied is indeed true.

An acceptable algorithm corresponds to a willingness to use everything known about the actual instance to solve the instance, and an insistence that the answer arrived at be verifiably correct. An acceptable algorithm is a reusable algorithm in the following sense: If the actual instance of interest changes but the instance knowledge remains unchanged (so the original actual instance changes to one that had been indistinguishable from it, and

[1] We are very lax in what we mean by an "algorithm." If one wants to restrict consideration to rational number approximate data and finite precision arithmetic, we only require that Turing machine algorithms be considered legitimate. If one allows irrational approximate data and infinite precision arithmetic, we only require real number machine algorithms (Blum, Shub, and Smale [1]) be considered as legitimate. Beyond these requirements the reader is free to decide the boundaries as to what qualifies as an algorithm.

[2] If one wants to restrict attention to rational number approximate data, then, of course, one only requires the algorithm to accept rational coordinate tuples. It is then natural to assume the oracle provides only rational coordinate approximate data.

the set of instances indistinguishable from the actual instance remains unchanged), then the algorithm correctly computes an approximate solution or determines a solution does not exist if it replies either (2a) or (2b) upon input for which $\bar{y}$ is strictly within error δ of the data for the (new) actual instance.

The motivation for requiring that the algorithm can be proven correct in the sense of (3) is that if the algorithm does, say, reply with statement (2a) upon some input for which $\bar{y}$ is strictly within error δ of the actual instance, then one can prove, in terms of one's knowledge regarding the actual instance, that the point $\bar{x}$ is indeed within error ε of a solution for the actual instance. In other words, one can be certain in terms of what one knows about the actual instance that the algorithm will not erroneously claim a certain point to be within error ε of a solution for the actual instance when in fact it is not. *Indeed, the reader should regard the definition of an acceptable algorithm to simply formalize the requirement that one be able to trust the algorithm not to reply with an incorrect answer for the actual instance.*[3]

We say the actual instance is *definitely well-posed* if there is an acceptable algorithm available[4] which replies with statement (2a) or (2b) upon any input $(\bar{y}, \delta, \varepsilon)$ for which $\delta > 0$ is sufficiently small (where what constitutes "sufficiently small" may be dependent on ε) and $\bar{y}$ is strictly within error δ of the data for the actual instance. It is the author's opinion that for the problems considered, a definitely well-posed instance would never be ill-posed under any entirely acceptable definition of the term ill-posed.

We will also have occasion to speak of an instance $\tilde{y}$ indistinguishable from the actual instance as being definitely well-posed. By this we mean that there is an acceptable algorithm available (relative to the same instance knowledge, so the set of indistinguishable instances remains unchanged) which replies (2a) or (2b) upon any input $(\bar{y}, \delta, \varepsilon)$ for which $\delta > 0$ is sufficiently small (possibly dependent on $\tilde{y}$ and ε) and $\bar{y}$ is strictly within error δ of $\tilde{y}$. In other words, $\tilde{y}$ would be definitely well-posed if it was the actual instance and the instance knowledge remained unchanged.

1.4. In certain sections of [5] where we restrict instance knowledge to be of a particular but very general form, we prove that among the set of instances

[3] One can trust an algorithm not to reply with an incorrect answer for the actual instance if and only if one can trust it not to reply with an incorrect answer for any instance indistinguishable from the actual instance since the trust must be based only on one's knowledge. This is why we chose to phrase the reply statements made by an acceptable algorithm in terms of the instances indistinguishable from the actual instance rather than in terms of just the actual instance. If the reply statements were phrased just in terms of the actual instance, then such an algorithm might be provably correct for the particular actual instance, but the proof might make use of properties regarding the particular actual instance that are not part of one's knowledge, i.e., one would not consider the proof to be a proof.

[4] By "available" we mean "available in the literature" or, even more narrowly, "available in [5]." This comment will make more sense as the reader forges ahead.

indistinguishable from the actual instance, the set of definitely ill-posed instances is the complement of the set of definitely well-posed instances. So, in that context, it is natural to define "ill-posed" just as we defined definitely ill-posed, and "well-posed" just as we defined definitely well-posed.

In this chapter, we are not imposing mathematical form on the instance knowledge so there is no evident reason why the two sets should be complements. Still the two sets can be easily argued to be disjoint. The argument is instructive. It is similar in spirit to, but much simpler than, the argument for showing "one cannot know if the instance knowledge is insufficient to design algorithms which require only sufficiently accurate approximate data to achieve the goals." In particular, it displays why we do not call the arguments "proof."

Here is the argument showing the two sets to be disjoint. Assume instance $\tilde{y}$ is indistinguishable from the actual instance and is definitely well-posed. Then, by definition, there is an acceptable algorithm which replies (2a) or (2b) upon any input $(\bar{y}, \delta, \varepsilon)$, $\delta > 0$, $\varepsilon > 0$, for which δ is sufficiently small and $\bar{y}$ is strictly within error δ of $\tilde{y}$. Let $(\bar{y}, \delta, \varepsilon)$ denote such an input for which δ is sufficiently small and assume there is an acceptable algorithm which replies, say, (2b) upon input $(\bar{y}, \delta, \varepsilon)$. Since an acceptable algorithm can be proven correct the replied statement is indeed true. Since $\tilde{y}$ is strictly within error δ of $\bar{y}$ and since, for now, errors are measured in terms of some norm (so $\| \tilde{y} - \bar{y} \| < \delta$), the correctness of the replied statement thus implies there is an open neighborhood of $\tilde{y}$ for which all instances in the neighborhood which are indistinguishable from the actual instance have no solution. It now easily follows that $\tilde{y}$ is not definitely ill-posed since otherwise we would obtain a contradiction. An exactly analogous argument can be constructed if it is assumed an acceptable algorithm replies (2a) rather than (2b). So the two sets are disjoint.

Two questionable key points to the argument just given are: (1) an acceptable algorithm "can be proven correct" and (2) an implicit assumption that a contradiction cannot be proven. These are also the questionable key points in the harder argument for showing "one cannot know if the instance knowledge is insufficient to design algorithms which require only sufficiently accurate approximate data to achieve the goals." If there is no mathematical form imposed on the instance knowledge, to say that an acceptable algorithm "can be proven correct" is certainly ambiguous since such a proof would likely depend on the instance knowledge. Moreover, the implicit assumption that a contradiction cannot be proven does, in a sense, require something mathematical of the instance knowledge, i.e., it must be consistent. Clearly, it would be unwise to call the above argument a "proof." However, the argument is convincing to the author and seems to form a general outline of a proof for showing the two sets are disjoint.

One big difference between the above argument and other arguments in this chapter is that the above argument assumes no particular problem structure. The other arguments rely deeply on particular problem structure, but the questionable key points in those arguments are the same as in the one above and do not arise in the parts of the arguments pertaining to problem structure.

1.5. Again consider the question which started this discussion, "Is it ever possible for one to know that all of the knowledge one possesses about the actual instance is insufficient to either design an algorithm to determine if the actual instance has a solution, or to design an algorithm to compute an approximation to a solution for the actual instance guaranteed to be of pre-specified accuracy, assuming the algorithms would only have access to arbitrarily accurate approximate data?" We argue something stronger than that the answer to this question is "No" for the class of problems considered. We argue it is impossible to deduce from the instance knowledge that the actual instance is not definitely well-posed; in particular, one cannot know the actual instance is not definitely well-posed.

The argument proceeds by showing that if one can deduce from the instance knowledge the actual instance is not definitely well-posed, then one can deduce that one of the algorithms in [5] is acceptable and achieves the goals for the actual instance using only approximate data. Hence, the actual instance is definitely well-posed, a contradiction. So it is impossible to deduce the actual instance is not definitely well-posed.

An outline of the argument can be found at the beginning of Section 3. The outline was written assuming the reader has not read beyond this point but a complete understanding of the argument requires much additional material developed in [5].

1.6. Now we turn to discussing the acquisition of new instance knowledge, in particular, instance knowledge that provides radically new insight.

We speak of time periods. Whenever new instance knowledge is obtained, a new time period commences. For example, whenever the oracle is called upon to provide new approximate data, a new time period commences.

We present a simple example. The problem in the example is in the class of problems to which our arguments apply, and the instance knowledge that is postulated is of the restricted form focused on in [5], although no form is assumed in the arguments to be discussed shortly. As time progresses, new instance knowledge is obtained. We are not concerned with how the instance knowledge is obtained. Instead, we are interested in understanding something about limits on how much the body of instance knowledge can grow before a problem instance that was originally definitely ill-posed becomes definitely well-posed.

Consider the problem of deciding if a real square matrix has a real eigenvalue and, if so, of approximating the largest real eigenvalue. Assume the number of rows is even and fairly large, say, 20.

Assume that in the initial time period, period one, there is no instance knowledge other than the problem statement, including the size of the matrix.

Assume that in time period two the instance knowledge consists of the problem statement, approximate data obtained from the oracle, and the facts that the actual instance has at least one real eigenvalue and all real eigenvalues are of even multiplicity; we assume all of the new knowledge became known simultaneously with the commencement of period two. (Again, we do not care how this knowledge was obtained.)

Assume that in time period three the instance knowledge has expanded only to additionally include new approximate data obtained from the oracle and the fact that the actual instance has at least two distinct real eigenvalues.

Finally, assume that in time period four the instance knowledge has expanded only to additionally include the fact that the actual instance is a symmetric matrix.

We claim that in time periods one and two the actual instance is definitely ill-posed, and in period four, it is definitely well-posed. Without additional information we can say nothing about period three. Before establishing these claims, we make a few observations regarding them.

Note the argument that the actual instance is definitely ill-posed in period one cannot be based solely on the instance knowledge available in period one but must rely on instance knowledge not available in that period. Otherwise, thinking of the present time as being the first period, one would be able to deduce from the instance knowledge that the actual instance is definitely ill-posed, and hence not definitely well-posed, contradicting the arguments discussed previously.

The argument that the actual instance is definitely ill-posed in period one relies on the instance knowledge available in period two. Thus, thinking of the present time as being the second period, one is able to deduce from the instance knowledge that the actual instance had been definitely ill-posed in the previous period (relative to the instance knowledge available then), but, again, it is impossible to deduce that the actual instance is now definitely ill-posed (relative to the present instance knowledge). In a sense, the instance knowledge available in period two is radically different from that in period one because it allows one to deduce something very strong about period one that could not be deduced in period one. In the same sense, the instance knowledge available in period three is radically different from that in period two. Finally, the instance knowledge available in period four is certainly radically different from that in period two, not only because it contains the instance knowledge from period three (which is radically different from the instance knowledge of period two) but also because the actual instance goes from being definitely ill-posed to being definitely well-posed between periods two and four.

Now we establish the claims.

The actual instance is definitely ill-posed for the decision problem in period one because, as is known in period two, it has real eigenvalues but all real eigenvalues have even multiplicity. Consequently, by an arbitrarily slight perturbation of the actual data, one can obtain a matrix with no real eigenvalues. The actual instance is indistinguishable in period one from these perturbed matrices.

In time period two, the actual instance is definitely ill-posed for the approximation problem because all of its eigenvalues are of even multiplicity and, as becomes known in the next period, it has at least two real eigenvalues. Because of these facts there exists $\bar{\varepsilon} > 0$ such that an arbitrary slight perturbation of the data for the actual instance can result in an instance with real eigenvalues, all of even multiplicity, but whose largest real eigenvalue is at least $\bar{\varepsilon}$ less than the largest eigenvalue of the actual instance. (The perturbation is chosen to cause the largest real eigenvalue of the actual instance to split into complex numbers with imaginary parts.) The actual instance is indistinguishable in period two from these perturbed matrices.

In time period four, the actual instance is definitely well-posed because all eigenvalues of a symmetric matrix are real and because there exist computational procedures for approximating all real roots of a polynomial to specified precision using sufficiently accurate approximations to the polynomial coefficients if the polynomial is known to only have real roots.

For convenience, we say that the instance knowledge in a sequence of time periods $t_1 < \cdots < t_M$ *changes radically* if for each $i > 1$, the instance knowledge in period t_i can be used to deduce the actual instance had not been definitely well-posed in period t_{i-1}.

For the broad class of problems mentioned at the beginning of the Introduction we argue that regardless of the form of instance knowledge, if the instance knowledge in a sequence of time periods $t_1 < \cdots < t_M$ changes radically then $M \leq m + 1$, where m is the number of data vector coordinates. We argue, in fact, that if $M = m + 1$, then the actual instance is definitely well-posed in period M and forever thereafter. (If the actual instance is definitely well-posed in some period, then it is defnitely well-posed forever thereafter because, as we assume, knowledge about the actual instance is never lost once it is obtained; consequently, an acceptable algorithm remains acceptable.) We refrain from calling the argument a proof for the same reasons cited earlier. However, for instance knowledge of the particular form focused on in [5], we obtain a rigorous proof as a simple corollary of the main theorem of that paper. The general argument uses that proof as a basis. The argument is outlined in Section 3.

Here is a very simple example showing there do indeed exist sequences $t_1 < \cdots < t_M$ for which the instance knowledge changes radically and for which $M = m + 1$, i.e., the upper bound can be attained. Define a vector $y \in \mathbb{R}^m$ to have as solution the number 1 if all of its coordinates are nonnegative, and no solution otherwise. That is the statement of the problem, considering m to be fixed. Assume that at the beginning of the initial period only the problem statement is known. At the beginning of period i, where $i = 2, \ldots, m + 1$, assume that the instance knowledge consists only of the problem statement and the fact that the jth coordinate of the actual instance equals zero for all $j < i$. We leave the details to the reader.

1.7. We also establish a result regarding what can be known about the data accuracy needed to compute an approximate solution guaranteed to be within specified error of a solution for the actual instance.

We say the actual instance *requires only linear data accuracy* if there is an acceptable algorithm fulfilling the requirements of the definition, making the actual instance a definitely well-posed instance, for which the following is true:

There exists $K > 0$ such that the algorithm replies (2a) or (2b) for each input $(\bar{y}, \delta, \varepsilon)$ for which $0 < \delta < K\varepsilon$, ε is sufficiently small, and $\bar{y}$ is strictly within error δ of the data for the actual instance.

In Section 3, we argue that it is impossible to deduce from the instance knowledge that the actual instance does not require only linear data accuracy, assuming errors are indeed measured by norms. Similarly, we argue that if the instance knowledge in time periods $t_1 < \cdots < t_M$ is such that the instance knowledge in period t_i can be used to deduce the actual instance did not require only linear data accuracy in period t_{i-1}, then $M \leq m + 1$ and, in fact, if $M = m + 1$, then the actual instance requires only linear data accuracy in period t_M and forever thereafter. Again the results are a simple consequence of the main theorem of [5] if instance knowledge is restricted to be of the particular form focused on in that paper, but the general arguments do not assume instance knowledge to be of that form.

1.8. Here we present examples of problems for which it is possible to know that all of the knowledge one possesses about the actual instance is insufficient either to design an algorithm to determine if the actual instance has a solution, or to design an algorithm to compute a point guaranteed to be within prespecified accuracy of a solution for the actual instance, assuming the algorithms would only have access to arbitrarily accurate approximate data. So our "No" answer to the question that started this discussion is not correct for all problems; our arguments are for a very general, but restricted class of problems introduced in Section 2.

Consider the problem of deciding if a single equation in two variables is satisfied by a rational coordinate point and, if so, of approximating such a point. An instance of the problem corresponds to specifying the equation coefficients which may be irrational. Assume that the instance knowledge consists only of the problem statement and the fact that the set of points which satisfy the equation form a line, i.e., the actual equation is "generic." Using this instance knowledge, it is easily seen that no algorithm could be designed to determine if the actual instance has a solution (i.e., a rational coordinate point satisfying the equation), assuming the algorithm would only have access to arbitrarily accurate approximate data. With an arbitrarily slight perturbation of the data, one can obtain an equation which is indistinguishable from the actual equation and which is satisfied by a rational coordinate point, and one can obtain an equation which is indistinguishable from the actual equation and which is not satisfied by a rational coordinate point. In other words, we can deduce from the instance knowledge that the actual instance is definitely ill-posed for the decision problem.

A slight extension of the above example provides a situation where it is possible to deduce from the instance knowledge that the actual instance is definitely ill-posed for the approximation problem. Now assume the instance knowledge consists of the problem statement, the fact that the actual equation has a solution (i.e., a rational coordinate point satisfying the equation) and the fact that the set of points in $\mathbb{R}^2$ which satisfy the equation form a line of irrational slope. The actual instance is then definitely ill-posed for the approximation problem. It is easily deduced from the instance knowledge that the actual equation has exactly one (rational coordinate) solution and it is easily deduced that the actual equation can be perturbed by an arbitrarily small amount to obtain an equation indistinguishable from the actual equation whose unique solution is arbitrarily far away from that for the actual equation.

2. What Is a Problem?

In this section, we restrict the structure of problems to be considered. We work in an extremely general algebraic context defined in terms of the classical decision problem for the first-order theory of the reals. Anyone familiar with this setting already knows its great generality; anyone unfamiliar with it will become convinced. The context subsumes many of the standard algebraic problems considered in numerical analysis and mathematical programming, e.g., matrix inversion, various matrix factorization problems, eigenvalue problems, linear and quadratic programming, nonlinear programming with algebraic constraints, sensitivity analysis problems, etc., etc., etc.

The basic object in this context is a *sentence* and is composed of four ingredients. By way of example, the sentence

$$(\exists v_1 \in \mathbb{R}^{n_1})(\forall v_2 \in \mathbb{R}^{n_2})[(g_1(v_1,v_2) > 0) \vee (g_2(v_1,v_2) = 0)] \wedge (g_3(v_1,v_2) \neq 0) \tag{2.1}$$

is composed of

1) blocks of variables (i.e., v_1, v_2);
2) a quantifier for each block (i.e., $\exists$, $\forall$);
3) *atomic predicates* which are (multivariate) polynomial equalities and inequalities [e.g. $g_1(v_1,v_2) > 0$];
4) a Boolean expression into which the atomic predicates are inserted (i.e., $[\mathrm{B}_1 \vee \mathrm{B}_2] \wedge \mathrm{B}_3$).

A sentence is an assertion. The above sentence asserts that there exists $v_1 \in \mathbb{R}^{n_1}$ such that for all $v_2 \in \mathbb{R}^{n_2}$ both (i) $g_1(v_1, v_2) > 0$ or $g_2(v_1, v_2) = 0$ and (ii) $g_3(v_1, v_2) \neq 0$. Depending on the specific polynomials g_1, g_2, and g_3, the sentence is either true or it is false.

The *first-order theory of the reals* is the set of all true sentences. A *decision method* is an algorithm which, given any sentence, determines if the sentence is true. Tarski [6] first proved that decision methods exist. He proved this by constructing a decision method.

A sentence is a special case of a more general expression, a *formula*. For example,

$$(\exists v_1 \in \mathbb{R}^{n_1})(\forall v_2 \in \mathbb{R}^{n_2})[(g_1(w, v_1, v_2) > 0) \vee (g_2(w, v_1, v_2) = 0)]$$
$$\wedge (g_3(w, v_1, v_2) \neq 0)$$

is a formula. Not all variables in a formula are required to be quantified. The block of variables w are unquantified here. They are the *free variables*.

More generally, when formulas are inserted into a Boolean expression, just as the atomic predicates were inserted into the Boolean expression to build the sentence (2.1), the resulting expression is again a formula. If one quantifies some of the free variables in a formula, one again obtains a formula but with fewer free variables. If all variables are quantified, then the formula is a sentence. In these ways, one obtains all formulas and sentences.

When a specific vector is substituted for the free variables in a formula, the formula becomes a sentence. Depending on the vector, the resulting sentence is true or false. A vector which results in a true sentence is said to be a *solution* for the formula.

An overview of computational complexity results regarding decision methods and methods for constructing approximate solutions for formulas can be found in the work of Renegar [3, 4]; the best upperbounds known are proven there.

For us, a *problem* will be a formula in which the free variables are divided into two vectors, x and y. We use $P(x, y)$, or simply P, to denote a problem. We view y as representing data. As in Section 1, a problem instance corresponds to specifying particular values for y. Given $y \in \mathbb{R}^m$, we say that $x \in \mathbb{R}^n$ is a *solution for instance* y if x and y together form a solution for P, i.e., if when the values are substituted for the free variables, the resulting sentence is true.[5]

We provide a few examples. First, consider the problem of inverting $n \times n$ matrices $y := [y_{ij}]$, viewing n as fixed. The corresponding formula is simply

$$P_1(x, y) := \left[\bigwedge_i \left(1 - \sum_k x_{ik} y_{ki} = 0\right]\wedge\left[\bigwedge_{i \neq j}\left(\sum_k x_{ik} y_{kj} = 0\right)\right].$$

[5] Eaves and Rothblum [2] also considered problems as defined in this manner but with a very different motivation.

The problem of computing a real eigenvalue x for a real $n \times n$ matrix $y := [y_{ij}]$ corresponds to the formula

$$P_2(x, y) := [\det(xI - [y_{ij}]) = 0].$$

The problem of computing the largest real eigenvalue corresponds to the formula

$$P_3(x, y) := P_2(x, y) \wedge [(\forall z \in \mathbb{R})(P_2(z, y) \Rightarrow (x - z \geq 0)].$$

Here, $P_2(z, y) \Rightarrow (x - z > 0)$ means that if (z, y) is a solution for P_2, then $x - z \geq 0$. (Of course, $A \Rightarrow B$ is equivalent to $B \vee \sim A$). Similarly, it is easily seen that there are formulas corresponding to the problems of computing L–U, singular value, and Q–R factorizations of matrices, etc.

If we are concerned only with matrices possessing a given sparsity structure, then appropriate formulas are obtained simply by substituting 0 for the corresponding y_{ij}. The additional structure reflects additional knowledge regarding the instance of interest.

Here are a few more examples. The problem of computing feasible points for systems of linear inequalities of fixed dimensions

$$Ax \geq b,$$
$$x \geq 0,$$

corresponds to the formula

$$P_4(x, A, b) := \left[\bigwedge_i \left(b_i - \sum_j a_{ij}x_j \leq 0\right)\right] \wedge \left[\bigwedge_j (x_j \geq 0)\right],$$

where $A = [a_{ij}]$ and $y := (A, b)$ is the data vector. The problem of computing optimal solutions for linear programming problems of fixed dimensions,

$$\min c^{\mathrm{T}}x$$
$$\text{s.t. } Ax \geq b$$
$$x \geq 0,$$

corresponds to the formula

$$P_5(x, A, b, c) := P_4(x, A, b) \wedge [(\forall z) P_4(z, A, b) \Rightarrow (c^{\mathrm{T}}z - c^{\mathrm{T}}x \geq 0)],$$

where $y := (A, b, c)$ is the data vector. It is easily seen that sensitivity analysis problems for linear programming can be coded as formulas. Similarly, we can do all of these things for nonlinear programming problems if the objective and constraint functions are multivariate polynomials, assuming we restrict attention to problems of fixed dimensions involving polynomials of bounded degree. The data will then consist of the coefficients of the polynomials and, again, may reflect known sparsity structure.

Etc., etc., etc.

For the Turing machine context, it is important to note that most interesting problems correspond to formulas involving only integer coefficients (in fact, only 0 and 1), although the data y of the instance of interest may be irrational. It is important to note this because we will assume descriptions of problems are available for algorithms, the descriptions simply being the formulas.

3. Outlines of the Arguments

In this section, we outline the arguments discussed in the introduction. The arguments are completed in Section 12 of [5]. Their completion relies on much additional material developed there.

Following is an outline of the argument which shows that if the actual instance can be deduced from the instance knowledge not to be definitely well-posed, then it can be deduced to be definitely well-posed.

The argument is based on $\overline{m}$ algorithms, $\overline{m} \leq m + 1$, generated by a single universal algorithm, m being the number of data coordinates. The algorithms are described in Section 12 of [5]. The $\overline{m}$ algorithms accept as input any tuple $(\overline{y}, \delta, \varepsilon)$, where $\overline{y} \in \mathbb{R}^m$, $\delta > 0$, and $\varepsilon > 0$. The algorithms require only finite precision if the input consists only of rational numbers. The algorithms reply with one of the statements (2a), (2b), or (2c) in the definition of an acceptable algorithm. This is not to say the algorithms are necessarily correct when they reply (2a) or (2b).

We prove in Section 12 of [5] that the first of the $\overline{m}$ algorithms is an acceptable algorithm regardless of the instance knowledge, that is, it is indeed correct whenever it replies (2a) or (2b). We also prove that if the first algorithm replies (2a) or (2b) upon some input $(\overline{y}, \delta, \varepsilon)$, then it replies the same upon any input $(\overline{y}', \delta', \varepsilon)$ for which the set of instances strictly within error δ' of $\overline{y}'$ is a subset of the set of instances strictly within error δ of $\overline{y}$. [So, if the first algorithm replies (2a) or (2b) for some approximate data for an instance, then it replies the same for all sufficiently accurate approximate data for the instance.]

We also prove the following in Section 12 of [5] for each k, $1 < k \leq \overline{m}$: If

(I) All algorithms $1, \ldots, k - 1$ always reply (2c) for each input $(\overline{y}, \delta, \varepsilon)$ where $\overline{y}$ is strictly within error δ of an instance indistinguishable from the actual instance,

then:

(II) Algorithm k is an acceptable algorithm.

(III) If algorithm k replies (2a) or (2b) upon some input $(\overline{y}, \delta, \varepsilon)$ where $\overline{y}$ is strictly within error δ of an instance $\tilde{y}$ which is indistinguishable from the actual instance, then it replies the same for any input $(\overline{y}', \delta', \varepsilon)$, where $\overline{y}'$ is strictly within error δ' of $\tilde{y}$ and δ' is sufficiently small ("sufficiently small" possibly being dependent on $\tilde{y}$ and ε).

(IV) If $k = \overline{m}$, then algorithm k replies either (2a) or (2b) for each input $(\overline{y}, \delta, \varepsilon)$, where $\overline{y}$ is strictly within error δ of an instance $\tilde{y}$ indistinguishable from the actual instance and δ is sufficiently small ("sufficiently small" possibly depending on $\tilde{y}$ and ε).

In short, if (I), then (II), (III), and (IV). Also, the previously stated facts regarding algorithm 1 imply (II) and (III) are true for $k = 1$.

Now assume it can be deduced from the instance knowledge that the actual instance is not definitely well-posed. Then, by definition of indistinguishability, all instances which are indistinguishable from the actual instance are not definitely well-posed. Consequently, since (II) and (III) are true for algorithm 1, the definition of a definitely well-posed instance implies that (I) is true for $k = 2$.

Since whenever (I) is true then so are (II), (III), and (IV), it, thus, follows that (II) and (III) are true for $k = 2$. So we have shown that since (II) and (III) are true for $k = 1$, they are true for $k = 2$.

Proceeding inductively, we deduce that (II) and (IV) are true for $k = \overline{m}$. But (II) and (IV) for $k = \overline{m}$ are precisely what it means for the actual instance to be definitely well-posed. In other words, we have shown that if one can deduce from the instance knowledge that the actual instance is not definitely well-posed, then one can deduce that it is definitely well-posed, a contradiction.

This concludes the outline of the argument.

Next we outline the argument showing that if the instance knowledge for a sequence of time periods $t_1 < \cdots < t_M$ changes radically, then $M \leq m + 1$ and, in fact, if $M = m + 1$, then the actual instance is definitely well-posed in period t_M (and forever thereafter). We argue something slightly stronger. Using the same $\overline{m}$ algorithms as in the previous argument, and recalling that $\overline{m} \leq m + 1$, we show that if the instance knowledge in a sequence of time periods $t_1 < \cdots < t_{\overline{m}}$ changes radically, then the actual instance is definitely well-posed in period $t_{\overline{m}}$.

The argument is very similar to the previous one but we must be careful when we invoke the "if (I), then (II), (III), and (IV)" assertion since the set of instances which are "indistinguishable" from the actual instance is dependent on the time period. To expedite the argument, we rephrase the assertion as follows:

If

(I′) All algorithms $1, \ldots, k - 1$ always reply (2c) for each input $(\overline{y}, \delta, \varepsilon)$, where $\overline{y}$ is strictly within error δ of an instance indistinguishable from the actual instance relative to the instance knowledge in period t_k,

then:

(II′) Algorithm k is an acceptable algorithm relative to the instance knowledge in t_k.

(III′) If algorithm k replies (2a) or (2b) upon some input $(\overline{y}, \delta, \varepsilon)$, where $\overline{y}$ is strictly within error δ of an instance $\tilde{y}$ which is indistinguishable from the actual instance relative to the instance knowledge in period t_k, then it replies the same for any input $(\bar{y}', \delta', \varepsilon)$, where $\bar{y}'$ is strictly within error δ' of $\tilde{y}$ and δ' is "sufficiently small" ("sufficiently small" possibly being dependent on $\tilde{y}$ and ε).

(IV′) If $k = \overline{m}$, then algorithm k replies either (2a) or (2b) for each input $(\bar{y}, \delta, \varepsilon)$, where $\bar{y}$ is strictly within error δ of an instance $\tilde{y}$ indistinguishable from the actual instance relative to the instance knowledge in period t_k and δ is sufficiently small ("sufficiently small" possibly depending on $\tilde{y}$ and ε).

In short, if (I′), then (II′), (III′), and (IV′). Also, the previously stated facts regarding algorithm 1 imply (II′) and (III′) are true for $k = 1$.

Now assume the instance knowledge in the sequence of time periods $t_1 < \cdots < t_{\overline{m}}$ changes radically. Then it can be deduced from the instance knowledge in period t_2 that the actual instance, and hence all instances indistinguishable from the actual instance in period t_2, had not been definitely well-posed in period t_1 relative to the instance knowledge available in period t_1. Consequently, since (II′) and (III′) are true for $k = 1$, and since the set of instances indistinguishable from the actual instance in period t_2 is a subset of that from period t_1 (knowledge about the actual instance is not lost), it follows from the definition of a definitely well-posed instance that (I′) is true for $k = 2$.

Since whenever (I′) is true then so are (II′), (III′) and (IV′), it, thus, follows that (II′) and (III′) are true for $k = 2$. So, we have shown that since (II′) and (III′) are true for $k = 1$, they are true for $k = 2$. Proceeding inductively, we find that (II′) and (IV′) are true for $k = \overline{m}$. In other words, the actual instance is definitely well-posed in period $t_{\overline{m}}$.

This concludes the outline of the argument.

Next we discuss the argument showing that it is impossible to deduce from the instance knowledge that the actual instance does not require only linear data accuracy, assuming errors are measured by norms. The argument is exactly analogous to the first argument outlined in this section although the algorithms are somewhat different.

The argument is based on $\hat{m}$ algorithms (described in Section 12 of [5]), $\hat{m} \leq m + 1$, generated by a single universal algorithm. The $\hat{m}$ algorithms accept as input any tuple $(\bar{y}, \delta, \varepsilon)$, where $\bar{y} \in \mathbb{R}^m$, $\delta > 0$, $\varepsilon > 0$. The algorithms require only finite precision if the input consists only of rational numbers. The algorithms reply with one of the statements (2a), (2b), or (2c) in the definition of an acceptable algorithm.

We prove in Section 12 of [5] that the first of the $\overline{m}$ algorithms is an acceptable algorithm regardless of the instance knowledge.

We also prove the following in Section 12 of [5] for each k, $1 < k \leq \hat{m}$: If

(I″) All algorithms $1, \ldots, k - 1$ are acceptable, and for each instance indistinguishable from the actual instance none of algorithms $1, \ldots, k-1$ require only linear data accuracy

then:

(II″) Algorithm k is an acceptable algorithm.

(III″) If $k = \overline{m}$, then algorithm k requires only linear data accuracy for any instance indistinguishable from the actual instance.

The inductive argument for showing it is impossible to deduce from the instance knowledge that the actual instance does not require only linear data accuracy exactly mimics the earlier inductive arguments and is left to the reader.

Finally, we discuss the argument showing that if the instance knowledge in periods $t_1 < \cdots < t_M$ is such that the knowledge in period t_i can be used to deduce the actual instance did not require only linear data accuracy in period t_{i-1}, then $M \leq m + 1$ and, in fact, if $M = m + 1$, then the actual instance requires only linear data accuracy in period t_M and forever thereafter. Referring to the same $\hat{m}$ as in the previous argument, it suffices to show that if $M = \overline{m}$, then the actual instance requires only linear data accuracy in period t_M.

The argument is exactly analogous to the second argument in this section and relies on the "if (I″), then (II″) and (III″)" assertion conveniently rephrased as follows:

If

(I‴) All algorithms $1, \ldots, k-1$ are acceptable relative to the instance knowledge in period t_k, and for each instance indistinguishable from the actual instance relative to the instance knowledge in period t_k, none of algorithms $1, \ldots, k-1$ require only linear data accuracy

then

(II‴) Algorithm k is an acceptable algorithm relative to the instance knowledge in period t_k.

(III‴) If $k = \overline{m}$, then algorithm k requires only linear data accuracy for any instance indistinguishable from the actual instance relative to the instance knowledge in period t_k.

We leave the outline of the inductive argument to the reader.

References

[1] Blum, L., Shub, M., Smale, S.: On a theory of computation and complexity over the real numbers: NP-completeness, recursive functions and universal machines, *Bull. Amer. Math. Soc.* 21 (1989), 1–46.

[2] Eaves, B.C., Rothblum, U.G.: A theory on extending algorithms for parametric problems. *Math. Oper. Res.* 14 (1989), 502–533.

[3] Renegar, J.: On the computational complexity and geometry of the first order theory of the reals: Parts I, II and III. *Journal of Symbolic Computation*, Vol. 13, pp. 255–352, 1992.

[4] Renegar, J.: On the computational complexity of approximating solutions for real algebraic formulae. *SIAM J. Computing*, Vol. 21, No. 6, pp. 1008–1025, 1992.

[5] Renegar, J.: Is it possible to know a problem instance is ill-posed?; Some foundations for a general theory of condition numbers. To appear in *Journal of Complexity*.

[6] Tarski, A.: *A Decision Method for Elementary Algebra and Geometry*, University of California Press, San Francisco, 1951.

34
Cohomology of Braid Groups and Complexity

V.A. VASSIL'EV

The main subject of this talk is the following: The Topological Complexity of Algorithms Finding the Roots of Polynomials and Polynomial Systems.

First, let us define all these words.

The *topological complexity* of an algorithm is the number of its branchings (operators IF); the topological complexity of a problem is the minimal topological complexity of algorithms which solve this problem. This characteristic was introduced in [Smale], where also a topological approach to the investigation of this complexity was discovered.

A *polynomial* is an expression of the form

$$X^d + a_1 X^{d-1} + \cdots + a_d, \tag{1}$$

where all the coefficients a_i are complex numbers in the circle $|a_i| \leq 1$. The set of all such polynomials is a polydisk in $\mathbb{C}^d$, we denote them by B^d.

To Find the Roots. Let there be given a positive number ε, then the problem $P(\varepsilon, d)$ is: For any polynomial p from B^d to find d complex numbers z_i such that the roots ξ_i of p can be ordered in such a way that $|z_i - \xi_i| \leq \varepsilon$ for any i. A similar problem "to find only one root up to ε" is denoted by $P_1(\varepsilon, d)$.

Algorithm. *The notion of algorithm we use is taken from* [*Smale*]; *namely, an algorithm is a directed graph having nodes of the following four types* (*see Figure* 1): *one input node* (*in our case, the input data are* 2*d numbers—real and imaginary parts of the coefficients* a_i); *the calculation nodes, where some arithmetic operations over the input data are executed*; *branching nodes, where one of the functions calculated earlier in the algorithm is compared with zero and depending on the result of comparision we go to the right or to the left*; *and the output nodes, in any of which some collection of numbers calculated earlier is proclaimed to be the answer.*

The first example of Smale's method of estimating the topological complexity! Consider the family of polynomials of the form $X^2 - a$, $|a| \leq 1$. The topological complexity of the problem "to find one root of any such polyno-

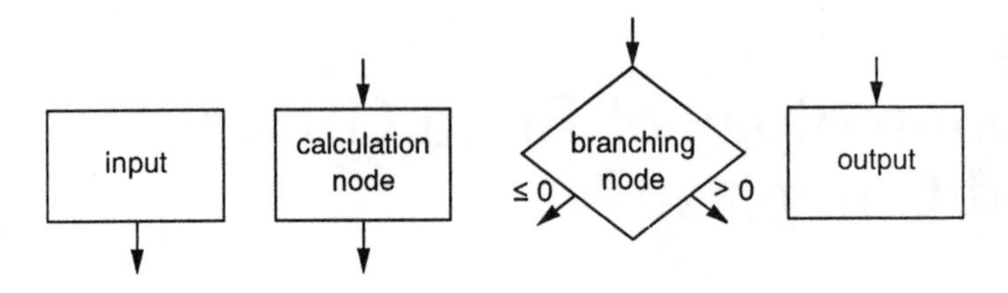

FIGURE 1

mial up to ε, $\varepsilon < 1$," is at least 1. Indeed, suppose that it equals 0, i.e., there exists an algorithm without branchings, and hence a continuous function from the unit circle to the plane $\mathbb{C}^1$ which for any a from the unit circle presents one of its square roots up to ε. So, its value at the point 1 is in the ε-neighborhood of $+1$ or -1, say of $+1$. But when a moves along the circumference and comes back, then, by continuity, this value comes into the ε-nbhd of -1; a contradiction.

It is easy to see that this proof is closely related to the fact that the 2-fold covering of the circumference has no continuous sections. The generalization of this proof to more complicated polynomial equations is based on the following notion.

Schwarz Genus. Consider a continuous map $\pi: X \to Y$ where X, Y are normal topological spaces (say, manifolds) and π is a map onto Y, i.e., $Y = \pi(X)$. The Schwarz genus $g(\pi)$ is defined as the minimal possible number of open subsets in Y such that they cover Y and over any of them there exists a continuous section of π. This characteristic of maps was introduced in [Schwarz] and rediscovered in [Smale].

Smale Principle. Let Σ be the subset in B^d consisting of polynomials having multiple roots. Then over the difference $B^d - \Sigma$, there are defined two coverings: the d-fold covering $\varphi_d: N^d \to (B^d - \Sigma)$ and $d!$-fold covering $f_d: M^d \to (B^d - \Sigma)$. The fiber of f_d over a polynomial p consists of all ordered collections of its d roots, the fiber of φ_d over p consists of pairs of the form (p, one root of p).

Theorem 1 [Smale Principle for the Problem $P(\varepsilon, d)$]. *The topological complexity of $P(\varepsilon, d)$ for ε small enough is no less than the genus of the covering f_d minus one*:

$$\tau(\varepsilon, d) \geq g(f_d) - 1.$$

The proof follows essentially from the definitions (like the proofs of many other principal statements in mathematics). First, note that the topological complexity of an algorithm plus one is exactly the number of output nodes. Suppose we have an algorithm with r output nodes. For any output node, consider a subset in B^d—the set of such input data, that if we give them to the input, then we go along the algorithm and finish exactly in the our output node. Such r subsets corresponding to the output nodes cover B^d. The desired sections of the covering f_d over the intersections of these subsets with $B^d - \Sigma$ are essentially presented by the algorithm (which to any polynomial assigns a set of its nearly roots). Here appears two troubles: that these r subsets in $B^d - \Sigma$ may be not open, and that algorithms present only approximate sections "up to ε"; but these troubles may be easily overcome and we get the theorem.

A similar inequality relates the topological complexity of the problem $P_1(\varepsilon, d)$ and the genus of the d-fold covering φ_d

$$\tau_1(\varepsilon, d) \geq g(\varphi_d) - 1.$$

Of course, the Smale principle is very general: To any calculation problem from a wide class of problems, we can assign a fiber bundle and estimate the complexity of the problem by means of the genus of this bundle. I understand my purpose in this lecture is to describe various methods of calculating and estimating the Schwarz genera and to demonstrate them on some examples related to the root-finding algorithms.

In [Smale], the first estimate for the genus of the covering f_d is given:

$$g(f_d) > (\log_2 d)^{2/3};$$

and, hence, also an estimate of the topological complexity:

$$\tau(\varepsilon, d) \geq (\log_2 d)^{2/3}$$

for ε small enough. This estimate is based on the study of the cohomology ring of the space $B^d - \Sigma$, due to Fuchs [Fuchs]. This space is a classifying space of the braid group with d strings,

$$B^d - \Sigma = K(\mathrm{Br}(d), 1)$$

and, hence, its cohomology is called the cohomology of this braid group.

In [V_1], the following estimate is proved:

Theorem 2. *$d - D_p(d) \leq \tau(\varepsilon, d) \leq d - 1$ for ε small enough; here p is an arbitrary prime, while $D_p(d)$ is the sum of digits in the p-adic expansion of d.*

The upper estimate here is straightforward: I present an algorithm of the needed complexity. (Another algorithm, which seems to be a better on other parameters like the number of arithmetical nodes, was constructed in [Kim].)

Corollaries.

1. *Put $p = 2$, then we get $\tau(\varepsilon, d) \geq d - \log_2 d$.*
2. *If d is a power of a prime p, then $D_p(d) = 1$, and we get $\tau(\varepsilon, d) = d - 1$ for small ε.*

In fact, in the last case (when $d = p^k$) a more general statement holds:

Theorem 3. *For $d =$ (a power of a prime), the topological complexity of the problem $P_1(\varepsilon, d)$ of finding one root is equal to $d - 1$ for small ε: $\tau_1(\varepsilon, d) = d - 1$.*

Unfortunately, I cannot prove that the last characteristic $\tau_1(\varepsilon, d)$ is monotonic in d (that is one of the main problems here), so, for arbitrary d, only the following is proved:

$$\tau_1(\varepsilon, d) \geq (\text{the largest factor of } d \text{ which is a power of a prime}) - 1$$
$$\text{for } \varepsilon \text{ small enough;}$$

note that the last number is asymptotically not lower than $\ln(d)$.

Both lower estimates in Theorems 2 and 3 based on the Smale principle and on the study of the genera of the corresponding coverings. The estimate in Theorem 2 is based on the following property of genera:

Suppose we have a principal covering $\pi: X \to Y$, i.e., a principal fiber bundle with a discrete group G for a fiber. Then there exists a classifying map of the base Y in the classifying space of G, cl: $Y \to K(G, 1)$, such that our covering π is isomorphic to the one induced from the canonical covering over $K(G, 1)$ by this map. For any G-module A, the map cl induces a homomorphism of cohomology groups with coefficients in A, $\mathrm{cl}^i: H^i(K(G, 1), A) \to H^i(Y, \mathrm{cl}^* A)$.

Definition. The *homological A-genus* of the covering $\pi: X \to Y$ is the smallest integer j such that the homomorphism cl^i is trivial for all $i \geq j$.

The homological A-genus is denoted by $h_A(\pi)$.

Theorem 4 [Schwarz]. *For any G-module A, and any principal G-covering π, $g(\pi) \geq h_A(\pi)$.*

For instance, the covering f_d is a principal G-covering, where G is the permutation group $S(d)$, and we can use Theorem 4. I guessed, that here it is very convenient to use the $S(d)$-module $\pm Z$, i.e., the module which is isomorphic to the integers as a group, and the odd permutations in $S(d)$ act on them as the multiplication by -1.

Theorem 5. (A) *the classifying map*

$$\mathrm{cl}^*: H^*(S(d), \pm Z) \to H^*(B^d - \Sigma, \pm Z)$$

is an epimorphism, and hence the homological genus $h_{\pm Z}(f_d)$ equals $1 + \max\{i | H^i(B^d - \Sigma), \pm Z) \neq 0\}$.

(B)
$$H^{d-1}(B^d - \Sigma, \pm Z) = \begin{cases} Z/pZ & \text{if } d = p^k, p \text{ a prime} \\ 0 & \text{if not.} \end{cases}$$

The lower estimate in Theorem 2 follows immediately from this theorem in the case d = a power of a prime and is easily deduced from it for arbitrary d; see [V_1].

In the proof of Theorem 3, where the d-fold covering φ_d is considered, we use one other estimate of the genera, also due to Schwarz.

For any fiber bundle $\pi: X \to Y$ and for any natural number k, let π^{*k} be the fiber bundle over Y obtained from π by $(k-1)$-fold join operation of its fibers with themselves, in particular $\pi^{*1} = \pi$.

Theorem 6 [Schwarz]. *The genus of arbitrary fiber bundle π is less then $k+1$ if and only if the bundle π^{*k} admits a continuous section.*

For our covering φ_d, I calculated the (only) topological obstruction to the section of the bundle $\varphi_d^{*(d-1)}$ and proved that at least for d = (a power of a prime), this obstruction is nontrivial; this also proves Theorem 3.

The Multidimensional Case

Here we study the topological complexities of the algorithms solving polynomial systems. There are many possible generalizations of the previous problems $P(\varepsilon, d)$ and $P_1(\varepsilon, d)$ to the multidimensional case. Again, in almost all of these cases our estimates are asymptotically proportional to the dimensions of the spaces of polynomial systems to be solved.

Let $d_1, \ldots, d_n$ be n natural numbers, and $Q = (q_1, \ldots, q_n)$ a system of homogeneous polynomials in $\mathbb{C}^n$ of degrees $d_1, \ldots, d_n$, having only one common root at the origin in $\mathbb{C}^n$.

Denote by [Q] the space of polynomial systems $V = (v_1, \ldots, v_n)$ in $\mathbb{C}^n$, where any v_i is a polynomial of degree d_i with leading homogeneous part equal to q_i and all the coefficients of degree $<d_i$ of this polynomial are in the disk $|\cdot| \leq 1$.

Let $P(\varepsilon, Q)$ be the problem of finding all $D = d_1 \cdots d_n$ roots of arbitrary polynomial systems in [Q] up to ε: If $n = 1$ and $Q = q_1 = X^d$, this is exactly the problem $P(\varepsilon, d)$.

In this case, the estimates are formulated in the following terms. Let $W(d)$ be the number of points $T = (t_1, \ldots, t_n)$ in the integer lattice Z^n, satisfying the following conditions: $0 \leq t_1 < [d_1/2]$, $0 \leq t_2 < d_2, \ldots, 0 \leq t_n < d_n$, $t_1 + \cdots + t_n < d_1$.

Theorem 7 [V_2]. (A) *For arbitrary $\varepsilon > 0$ and Q as before, the topological complexity of the problem $P(\varepsilon, Q)$ satisfies the inequality*

$$\tau(\varepsilon, Q) \leq \dim_{\mathbb{R}}[Q].$$

(B) *For almost any system Q of homogeneous polynomials of degrees $d_1, \ldots, d_n$, $\tau(\varepsilon, Q) \geq W(d)$ for ε small enough; in particular, it is true for the system $Q = (X_1^{d_1}, \ldots, X_n^{d_n})$.*

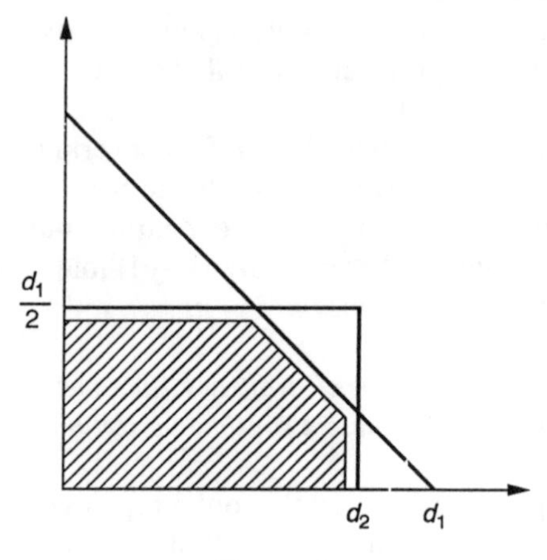

FIGURE 2

Here "almost any" means "any but an algebraic subvariety in the space of homogeneous systems", I think that, in fact, this subvariety is empty.

Corollary. *Let all the degrees d_i be equal to each other and grows to the infinity: $d_1 = \cdots = d_n \to \infty$. Then the leading terms of asymptotics of the lower and upper estimates from Theorem 7 differ by the constant multiple $2n/(1 - 2^{-n})$.*

In the case $n = 1$, each of these two estimates is twice worse than the realistic asymptotic (equal to d).

The proof of the upper estimate in this theorem is again straightforward; the lower one is again based on the Smale principle. Namely, let Σ be the subvariety in $[Q]$ consisting of the systems having multiple roots; consider the $D!$-fold covering $f(Q)$ over $[Q] - \Sigma$ whose fiber over a system consists of all ordered collections of the roots of this system.

I constructed a $W(d)$-dimensional torus T^W in the space $[Q] - \Sigma$, over which the natural D-fold covering $\varphi(Q)$ splits into a set of disjoint coverings, and some $W(d)$ of them are the twofold coverings being in the one-to-one correspondence with the multiplicative one-dimensional generators of the ring $H^*(T^W, Z_2)$ in such a way that any such generator is the obstruction to the section of the corresponding twofold covering. It is easy to check that the genus of the restriction of our covering $f(Q)$ on this torus is equal to $W(Q) + 1$; but, obviously, the genus of a fiber bundle is not less than that of its restriction and our estimate follows.

Note. The first result on the lower estimates in this problem (with a different space of systems) is due to Levine [Levine], but the estimate from Theorem 7 is stronger.

Now, consider the problem "to find only one root."

Let d be a natural number; let $[d, n]$ be the set of all systems of n polynomials $v_1, \ldots, v_n$, v_i: $\mathbb{C}^n \to \mathbb{C}$, of degrees not exceeding d. Let U be an arbitrary subset in $\mathbb{C}^n$.

The problem $P_1(\varepsilon, [d, n], U)$ is "to find one root in U up to ε of any system from $[d, n]$"; an algorithm *solves* this problem if for any system $V \in [d, n]$ it either finds (up to ε) one of its roots in the ε-neighborhood of U, or gives a true statement on the absence of roots in U.

Theorem 8 [V$_2$]. *For any $U \subset \mathbb{C}^n$ containing a nonempty open set, and for any ε small enough, the topological complexity of the problem $P_1(\varepsilon, [d, n], U)$ satisfies the inequality*

$$\tau_1(\varepsilon, [d, n], U) > (d/n)^n \left(1 + \underset{d \to \infty}{\mathrm{O}} (1/d)\right).$$

In particular, the leading term of this estimate is again proportional to the dimension of the space of systems $[d, n]$.

The proof of Theorem 8 is elementary modulo Theorem 3. Namely, without loss of generality, we can assume that U contains a neighborhood of the origin in $\mathbb{C}^n$. Consider the subset in the space $[d, n]$ consisting of systems $V = (v_1, \ldots, v_n)$ such that all the polynomials $v_2, \ldots, v_n$ are fixed and have the form $v_i = (X_i - X_{i-1}^{a_i})$ for an appropriate set of natural numbers $a_i \leq d$, and only the polynomial v_1 is varied.

These $n - 1$ polynomials $v_2, \ldots, v_n$ determine a smooth curve in $\mathbb{C}^n$; for the coordinate on this curve, we can take the coordinate X_1 and the restriction of any monomial $X_1^{b_1} \cdots X_n^{b_n}$ on this curve equals to X_1 in the degree

$$b_1 + b_2 a_2 + b_3 a_2 a_3 + \cdots + b_n a_2 \cdots a_n.$$

Such restrictions of monomials of degrees $\leq d$ in $\mathbb{C}^n$ gives all the degrees of X_1 from 0 to a very large number $N(d, n)$ without gaps. Hence, using the restrictions of v_1 on our curve, we can simulate any polynomial of very large degree in one variable X_1. To finish the proof, just use Theorem 3.

Unfortunately, I do not know any upper estimate in this case. The only nontrivial fact about them is that if the domain U is semi-algebraic, then there exists a common finite estimate not depending on ε: there exists a number $T = T(U, d, n)$ such that $\tau_1(\varepsilon, [d, n], U) \leq T$ for all $\varepsilon > 0$.

A much better situation is in the problem of finding not a root up to ε but an ε-root, that is, a point z in $\mathbb{C}^n$ such that $|v_i(z)| \leq \varepsilon$ for all i.

Let B be a subset in the space $[d, n]$ and U a domain in $\mathbb{C}^n$. By the definition, an algorithm solves the problem: "to find an ε-root in U of any polynomial system in B," if for any such system it either finds an ε-root in U or gives the true statement that there are no $\varepsilon/2$-roots in U.

Theorem 9. (1) *Let B be a subset in $[d, n]$ containing a neighborhood of the zero system. Then, for any ε small enough, the topological complexity of the problem "to find an ε-root in U of any system from B" is not less than a number $Y(d, n)$*

not depending on B and U, and the asymptotics of these numbers $Y(d, n)$ for any fixed n satisfies the same inequality as in Theorem 8:

$$Y(d, n) \geq (d/n)^n \left(1 + \underset{d \to \infty}{O}(1/d)\right).$$

(2) *If U belongs to a compact set in $\mathbb{C}^n$ and B to a compact set in $[d, n]$, then for any $\varepsilon > 0$ the topological complexity of the last problem is not greater than* $1 + \dim_{\mathbb{R}}[d, n]$.

About some other problems which concern the topological complexity, see [BSS] and [Hirsch].

I would like to finish with a statement on the cohomology of braid groups which has no immediate relation to complexity theory, but is closely related to some other topics of our conference.

THE MAY–SEGAL FORMULA FOR THE COHOMOLOGY OF THE STABLE BRAID GROUP IS A SPECIAL CASE OF THE HOMOLOGICAL SMALE–HIRSCH PRINCIPLE.

The May–Segal formula affirms that the cohomology ring of the braid group with infinitely many strings, $H^*(\mathrm{Br}(\infty))$, is isomorphic to the cohomology ring of the second loop space of a three-dimensional sphere, $H^*(\Omega^2 S^3)$. (Recall that $\Omega^r S^m$ is the space of continuous maps $(S^r, *) \to (S^m, *)$.)

The Smale–Hirsch principles are statements of the following form. Consider the space of smooth maps $M \to M$ (M and N manifolds) having no singularities of a certain kind (that is, the jet extensions of these maps must be the sections of the natural jet bundle $J(M, N) \to M$ which do not meet a certain subset in the jet space). The jet extensions embed this space of maps into the space of all continuous sections of the jet bundle, not meeting this "singular" subset. The Smale–Hirsch principle is a statement that these two spaces are similar to each other and this "similarity" is induced by the natural embedding; in particular, a homological Smale–Hirsch principle is the statement that they have similar homologies.

Now, the space $\Omega^2 S^3$ is homotopy equivalent to the space of all continuous maps $\mathbb{C}^1 \to (\mathbb{C}^2 - 0)$ with some fixed behavior at infinity. Any polynomial from the space B^d presents a map $\mathbb{C}^1 \to \mathbb{C}^2$: its complex 1-jet extension. This map does not meet the origin in $\mathbb{C}^2$ iff our polynomial is in $B^d - \Sigma$, and we get an embedding of the space $B^d - \Sigma = K(\mathrm{Br}(d), 1)$ into a space homotopy equivalent to $\Omega^2 S^3$. This is exactly an embedding from the Smale–Hirsch principle.

Theorem 10. *This embedding induces a cohomology isomorphism in dimensions not exceeding* $[d/2] + 1$.

When the number of strings d tends to the infinity, we get the May–Segal formula.

Theorem 10 has many generalizations concerning the spaces of holomorphic Morse functions in several complex variables, of real functions without complicated singularities and so on; see [V_3, V_4].

References

[BSS] L. Blum, M. Shub, S. Smale, On a theory of computations over the real numbers; *NP*-completeness, recursive functions and universal machines, *Bull. AMS* 21:1 (1989), 1–46.

[Fuchs] D.B. Fuchs, Cohomology of braid groups mod 2, *Funct. Anal. Appl.* 4:2 (1970), 46–59.

[Hirsch] Michael D. Hirsch, Applications of topology to lower bound estimates in computer science, preprint, International Computer Science Institute, 1990.

[Kim] M.-H. Kim, Topological complexity of root-finding algorithms, *J. Complexity* 5:3 (1989), 331–344.

[Levine] H. Levine, A lower bound for the topological complexity of Poly(D, n), *J. Complexity* 5 (1989) 34–44.

[Schwarz] A.S. Schwarz, The genus of a fibre bundle, *Trans. Moscow Math. Soc.* 10 (1961) 217–272.

[Segal] G.B. Segal, Configuration-spaces and iterates loop-spaces, *Invent. Math.* 21:3 (1973) 213–221.

[Smale] S. Smale, On the topology of algorithms. I, *J. Complexity* 3:2 (1987) 81–90.

[V_1] V.A. Vasiliev, Cohomology of braid groups and complexity of algorithms, *Funct. Anal. Appl.* 22:3 (1988), 15–24.

[V_2] ———, Topological complexity of algorithms of approximate solving the systems of polynomial equations, *Algebra Anal.* 1:6 (1989) 198–213.

[V_3] ———, Topology of complements to discriminants and the loop-spaces, in *Theory of Singularities and its Applications* (V.I. Arnold, ed.), spaces, Advances in Soviet Mathematics, Vol. I, Publ. AMS, 1990, 9–21.

[V_4] ———, Topology of the spaces of functions without complicated singularities, *Funct. Anal. Appl.* 23:4 (1989), 24–36.

35
The Dynamics of Eigenvalue Computation

Steve Batterson*

In recent years, there has been a great deal of interdisciplinary work between numerical analysis and dynamical systems. The theme of this chapter is that techniques from dynamical systems can be applied to the study of certain problems in numerical analysis. We will focus on the particular numerical analysis problem of approximating the eigenvalues of a real matrix. The discussion is from the point of view of dynamical systems and assumes a basic knowledge of dynamical systems [D2] and general topology [M]. A substantial portion of the chapter consists of motivating and defining the relevant numerical algorithms and no background in numerical analysis is required.

Assume that M is a space and G is a continuous map of the space into itself. A fundamental problem in dynamical systems is to understand the asymptotic behavior of the orbit of a point under iteration by the map [i.e., $\lim_{n\to\infty} G^n(y)$]. That numerical analysis was an important tool for the study of this problem has been appreciated for some time.

We will now indicate the framework for moving in the opposite direction. Consider an iterative numerical algorithm such as Newton's Method for finding roots of a function h. Associated to the algorithm is a space of candidates and a function, F, which maps an approximate solution to a (hopefully) more accurate approximate solution. In the case of Newton's Method, the candidate space consists of numbers and the function is defined by $F(x) = x - h(x)/h'(x)$. A general numerical problem with Newton's Method or any iterative algorithm is to determine the likelihood that $\lim_{n\to\infty} F^n(x)$ is a solution to the problem. In [S4], Steve Smale demonstrated that it was natural to apply the theory of dynamical systems to such problems.

Michael Shub adapted this approach for the eigenvector problem [S1]. If A is an $n \times n$ real matrix, then a vector is an eigenvector if and only if any nonzero scalar multiple is also an eigenvector. The eigenvector candidate space may be considered to be either the nontrivial elements of $\mathbf{R}^n$ or the

* The author was partially supported by the NSF.

compact manifold, $\mathbf{P}^{n-1}$. The iteration function or dynamical system is determined by the specific algorithm. Two algorithms which will be discussed in this chapter are the power method and Rayleigh quotient iteration.

EISPACK is a widely used subroutine package for dealing with eigenvalue problems. The principal EISPACK subroutine for approximating the eigenvalues of a real matrix, HQR2, involves an application of the shifted QR algorithm. We will show how convergence questions for these algorithms can be couched in terms of dynamical systems. In particular, it will be shown that there is a semiconjugacy between a map of projective space and the iteration map of shifted QR.

Whereas certain numerical questions will be central to the development of this chapter, other important numerical issues will be omitted. The reader who is interested in more information and details on numerical linear algebra is referred to the excellent references [GV, P3, S5, W4].

1. The Power Method

The power method (PM) is the simplest algorithm for approximating an eigenvector of a matrix. Suppose A is an $n \times n$ real matrix.

PM Algorithm.

1. $x_0 \neq 0$,
2. $x_k = Ax_{k-1}$.

Note that $x_k = A^k x_0$. In actual practice, it is customary to include a step to normalize x_k. This is intended to prevent overflow. For the moment, we will omit the normalization.

To appreciate the dynamics of PM, assume A has n eigenvalues of distinct moduli $|\lambda_1| < |\lambda_2| < \cdots < |\lambda_n|$. From the hypothesis on the eigenvalues, there exists a basis of corresponding unit eigenvectors v_i. Since the eigenvectors are a spanning set, x_0 may be expressed as a linear combination of the eigenvectors:

$$x_0 = \alpha_1 v_1 + \cdots + \alpha_n v_n,$$

$$x_k = A^k(x_0) = \alpha_1 \lambda_1^k v_1 + \cdots + \alpha_{n-1}\lambda_{n-1}^k v_{n-1} + \alpha_n \lambda_n^k v_n.$$

If $\alpha_n \neq 0$, then

$$x_k = \alpha_n \lambda_n^k \left[\frac{\alpha_1}{\alpha_n}\left(\frac{\lambda_1}{\lambda_n}\right)^k v_1 + \cdots + \frac{\alpha_{n-1}}{\alpha_n}\left(\frac{\lambda_{n-1}}{\lambda_n}\right)^k v_{n-1} + v_n \right].$$

Since each of the λ quotients is going to zero, x_k is approaching the one-dimensional subspace spanned by v_n, the dominant eigenvector. If $\alpha_n = 0$, then the iterates approach the one-dimensional subspace of the eigenvector for the largest i such that $\alpha_i \neq 0$. Since the convergence is to one-dimensional subspaces and the candidates are one-dimensional subspaces, it is natural to consider the iteration and the map it induces on $\mathbf{P}^{n-1}$. The following theorem is proved in [B1].

Theorem. *Let A be an $n \times n$ nonsingular matrix with n eigenvalues of distinct moduli and suppose that G is the map on $\mathbf{P}^{n-1}$ induced by the power method. G is a Morse–Smale diffeomorphism whose periodic set consists of the eigenvectors, each of which is fixed. The eigenspace corresponding to the eigenvalue of largest modulus is the only sink and attracts the orbit from almost any initial point.*

We remark that desirable global convergence properties can be obtained with weaker restrictions on the eigenvalues, but we now will shift our attention to the rate of convergence. From the expression for x_k, it can be seen that this rate is determined by λ_{n-1}/λ_n. If this quantity is close to 1, then the convergence will be very slow; but even if the quantity is small, then the rate of convergence is slow by the standards of numerical analysis (for example, if it is 0.5, then over three iterations are required for each significant digit increase).

In summary, PM has the desirable features of simplicity and nice global convergence. The drawbacks of PM are its slow rate of convergence and that it is ineffective in computing nondominant eigenvectors. The algorithms of the subsequent sections will address these issues.

2. Rayleigh Quotient Iteration

To accelerate convergence, it is useful to employ a *shift* by a scalar μ. The shifted matrix is $A - \mu I$. Note that the shift preserves the eigenvectors of the matrix, whereas the effect on the eigenvalues is to translate by the shift:

$$Ax = \lambda x \quad \Leftrightarrow \quad (A - \mu I)x = (\lambda - \mu)x.$$

If the matrix is shifted by an approximate eigenvalue, then one of the eigenvalues of the shifted matrix will be very small. The Rayleigh quotient iteration algorithm (RQI) is motivated by two observations concerning the power method. If one iterates by the inverse of the matrix, most orbits will converge to the eigenvector corresponding to the eigenvalue of smallest modulus at a rate determined by the ratio of the two smallest eigenvalues. If this iteration is performed on a matrix which is shifted by an approximate eigenvalue, then the rate of convergence should be rapid.

It remains to define the choice of an approximate eigenvalue. If A is a matrix and x is a nontrivial vector, then the *Rayleigh quotient* is defined by the quotient of inner products $\rho_A(x) = x^{\mathrm{T}}Ax/x^{\mathrm{T}}x$. The Rayleigh quotient is an approximate eigenvalue in the sense that it is the scalar, μ, which minimizes the norm of $Ax - \mu x$ (for a nice geometric argument, see [P3]).

RQI Algorithm.

1. $x_0 \neq 0$.
2. *If $A - \rho_A(x_{k-1})I$ is singular, solve for an eigenvector and stop. Otherwise,* $[A - \rho_A(x_{k-1})I]x_k = x_{k-1}$.

The iteration map is the inverse of the Rayleigh shifted matrix. In [S2], it is shown that this induces a well-defined map on an open subset of $\mathbf{P}^{n-1}$. For symmetric matrices, RQI converges to an eigenspace for almost any choice of initial vector and the rate of convergence is rapid. The following statement summarizes results proved in [O, P2, B2]. Define ϕ_k to be the angle in radians between x_k and the closest eigenvector.

Theorem. *If A is a symmetric $n \times n$ matrix, then for almost any choice of x_0 the RQI sequence $\{x_k\}$ approaches the one-dimensional subspace spanned by an eigenvector and* $\lim_{k\to\infty} (\phi_{k+1}/\phi_k^3) \leq 1$.

The rate of convergence is said to be cubic. Asymptotically, each additional iteration triples the number of significant digits. Contrast this with the power method in which several iterations are required to improve a single digit.

How effective is RQI with nonsymmetric real matrices? This question is addressed in [B3]. If the matrix is not symmetric, then the characteristic polynomial may have complex roots. Associated to each complex root (and its conjugate) is an invariant subspace. To consider RQI as an effective algorithm on nonsymmetric matrices, it seems reasonable to demand that the RQI sequence at least converge to an invariant subspace for almost every initial vector. This is the case for 2×2 matrices (see [A, B3, S2]), but not when $n \geq 3$.

Theorem. *For each $n \geq 3$, there is a nonempty open set of matrices each of which possesses an open set of initial vectors for which RQI does not converge to an invariant subspace.*

The above theorem is proved in [B3] by a bifurcation argument on the iteration maps of the two-parameter family of matrices

$$\begin{bmatrix} 0 & -c & 0 \\ \frac{1}{c} & 0 & 0 \\ 0 & 0 & d \end{bmatrix}.$$

It is shown that with appropriate choices of c and d, one can produce period doubling and pitchfork bifurcations which create a period four sink outside the invariant subspaces.

The RQI, $[A - \rho_A(u)I]^{-1}u$, induces a map on the open subset of $\mathbf{P}^{n-1}$ for which $\rho_A(u)$ is not an eigenvalue. In [B2], it is shown that this does not extend to a continuous map of $\mathbf{P}^{n-1}$. To overcome the continuity difficulty, we consider the quotient space of $\mathbf{P}^{n-1}$ obtained by identifying all the proper invariant subspaces of A to a single point, $\#_A$. The resulting space is denoted Φ_A. We now define three maps of Φ_A beginning with one induced by RQI.

Definition. $R_A\colon \Phi_A \to \Phi_A$ by

$R_A(\#_A) = \#_A$,
$R_A(u) = \#_A$ if $\rho_A(u)$ is an eigenvalue,
$R_A(u) = [A - \rho_A(u)I]^{-1}u$, otherwise.

Lemma. *W is an A-invariant subspace if and only if $W^\perp$ is an A^{T}-invariant subspace.*

PROOF. Suppose that W is an A-invariant subspace and $u \in W^\perp$. Let $\{v_1, \dots, v_k\}$ be a basis for W.

$$u \in \langle v_1, \dots, v_k\rangle^\perp \subset [A\langle v_1, \dots, v_k\rangle]^\perp,$$
$$0 = v_1^{\mathrm{T}} A^{\mathrm{T}} u, \dots, v_k^{\mathrm{T}} A^{\mathrm{T}} u,$$
$$A^{\mathrm{T}} u \in \langle v_1, \dots, v_k\rangle^\perp = W^\perp.$$

Definition. $H_A\colon \Phi_A \to \Phi_{A^{\mathrm{T}}}$ by

$H_A(\#_A) = \#_{A^{\mathrm{T}}}$,
$H_A(u) = \langle u, Au, \dots, A^{n-2}u\rangle^\perp$ otherwise.

Definition. $F_A\colon \Phi_A \to \Phi_A$ by

$F_A(\#_A) = \#_A$,
$F_A(u) = \#_A$ if $\rho_A(H_A(u))$ is an eigenvalue,
$F_A(u) = [A - \rho_A(H_A(u))I]u$, otherwise.

Theorem. *The maps F_A and $R_{A^{\mathrm{T}}}$ are topologically conjugate by H_A.*

Before proving the theorem, observe that F_A is the iteration map of an algorithm involving forward iteration of a shifted matrix. The theorem implies that this algorithm is as successful as RQI in finding an element of an invariant subspace.

PROOF. First we show that $H_A \circ F_A = R_{A^{\mathrm{T}}} \circ H_A$. This is immediate for $\#_A$ and when $\rho_A(H_A(u))$ is an eigenvalue. For other values of u, we have

$$
\begin{aligned}
&H_A \circ F_A \\
&= \langle [A - \rho_A(H_A(u))I]u, A[A - \rho_A(H_A(u))I]u, \ldots, A^{n-2}[A - \rho_A(H_A(u))I]u \rangle^{\perp} \\
&= \langle [A - \rho_{A^{\mathrm{T}}}(H_A(u))I]u, [A - \rho_{A^{\mathrm{T}}}(H_A(u))I]Au, \ldots, [A - \rho_{A^{\mathrm{T}}}(H_A(u))I]A^{n-2}u \rangle^{\perp} \\
&= [A^{\mathrm{T}} - \rho_{A^{\mathrm{T}}}(H_A(u))I]^{-1} \langle u, Au, \ldots, A^{n-2}u \rangle^{\perp} = R_{A^{\mathrm{T}}} \circ H_A.
\end{aligned}
$$

To show that H_A is a homeomorphism, it suffices to show that it is continuous at $\#_A$ and has a continuous inverse. The inverse of H_A is $H_{A^{\mathrm{T}}}$. If u is contained in an invariant subspace W and v is close to u, then $H_A(v)$ is close to $W^{\perp}$ which is in $\#_{A^{\mathrm{T}}}$.

3. *The Basic (Unshifted) QR Algorithm*

Recall that one of the drawbacks of the power method was its ineffectiveness in approximating nondominant eigenvectors. The QR algorithm addresses this problem. For the remainder of this paper our primary concern will be with eigenvalues rather than eigenvectors.

Definition. An $n \times n$ matrix, M, is (*upper*) *Hessenberg* provided $M_{ij} = 0$ whenever $i > j + 1$. *A Hessenberg matrix is* reduced *provided* $M_{ij} = 0$ *for some* $i = j + 1$.

EXAMPLE. Arbitrary entries are indicated by x:

$$
\begin{bmatrix}
x & x & x & x & x & x \\
x & x & x & x & x & x \\
0 & x & x & x & x & x \\
0 & 0 & x & x & x & x \\
0 & 0 & 0 & x & x & x \\
0 & 0 & 0 & 0 & x & x
\end{bmatrix}
$$

The first step in approximating the eigenvalues of a matrix is to perform an orthogonal similarity transformation to obtain a Hessenberg matrix M^1. The existence of an algorithm to accomplish this transformation is well-known in numerical analysis and involves applying a sequence of Householder transformations (see [GV]). Thus, to approximate eigenvalues it suffices to consider the problem on Hessenberg matrices. In choosing an iterative algorithm, we seek to achieve the following aim.

Decoupling Goal. An iteration that preserves Hessenbergness and the orthogonal similarity class and which is forward asymptotic to the reduced set.

Equipped with such an algorithm, one could perform a sufficient number of iterations that a subdiagonal element would become negligible. At that point, the problem would decouple into approximating the eigenvalues of two smaller matrices. Continue this process until all the matrices are 1×1 or 2×2 and then obtain the eigenvalues of these matrices.

Any square matrix can be factored as the product of an orthogonal matrix and an upper triangular matrix. This is clear if the matrix is nonsingular since its columns are a basis and the columns of the orthogonal matrix can be obtained from the Gram–Schmidt process. The factorization is known as the QR factorization. Details and algorithms on this and many other relevant topics can be found in [GV].

Basic (unshifted) QR Algorithm.

M^1 Hessenberg
If M^i is reduced stop, otherwise
$M^i = Q_i R_i \qquad Q_i$-orthogonal, R_i-upper triangular
$M^{i+1} = Q_i^{\mathrm{T}} M^i Q_i$.

The basic QR algorithm is a simpler and slower version of the QR algorithm which is used in eigenvalue computation. In [P1], it is shown that under fairly general Hessenberg hypotheses the basic QR algorithm converges to reduced form. [S3] and [W1] give insight into the dynamics of the basic QR algorithm as a fancy variant of the power method. [D1] employs a flow approach to analyze QR. Our approach will be to show that there is a semiconjugacy from the power method to basic QR.

The dynamical system induced by the QR algorithm maps an unreduced Hessenberg matrix to an orthogonally similar Hessenberg matrix. Thus, we may restrict our attention to an orthogonal similarity class. Let C denote a canonical form matrix in the class. To relate QR to the power method, we need to associate elements of projective space with unreduced Hessenberg matrices. The association is motivated by the following example. It demonstrates that an unreduced Hessenberg matrix in the orthogonal similarity class of C is essentially determined by the first column of the orthogonal matrix.

EXAMPLE.

$$\begin{bmatrix} x & x & x \\ x & x & x \\ 0 & x & x \end{bmatrix} = S^{\mathrm{T}} C S = \begin{bmatrix} u^{\mathrm{T}} \\ v^{\mathrm{T}} \\ w^{\mathrm{T}} \end{bmatrix} C[u \ v \ w] = \begin{bmatrix} u^{\mathrm{T}}Cu & u^{\mathrm{T}}Cv & u^{\mathrm{T}}Cw \\ v^{\mathrm{T}}Cu & v^{\mathrm{T}}Cv & v^{\mathrm{T}}Cw \\ w^{\mathrm{T}}Cu & w^{\mathrm{T}}Cv & w^{\mathrm{T}}Cw \end{bmatrix}.$$

Equating the lower left entries Cu is orthogonal to w and, thus, is in the space spanned by u and v. Since $v^{\mathrm{T}}Cu \neq 0$, v is in the span of u and Cu. Hence, v may be obtained from Gram–Schmidt on u and Cu. More generally, the columns of the orthogonal similarity transformation are determined from the first column by applying the Gram–Schmidt process to the orbit of the first column under C. The following theorem is well-known in numerical analysis.

Implicit Q Theorem. *Assume that P, Q, C are $n \times n$ matrices with P and Q orthogonal. Define $H = Q^{T}CQ$ and $G = P^{T}CP$ and suppose that both H and G are Hessenberg with H unreduced. If the first columns $q_1 = \pm p_1$, then $q_i = \pm p_i$ for $i = 2, \ldots, n$ with these columns resulting from Gram–Schmidt on the orbit of q_1 under C. The matrices H and G are equal up to sign and there exists $D = \operatorname{diag}(\pm 1, \ldots, \pm 1)$ such that $G = DHD$.*

With the following two relations define a quotient space, Ω_C, of the Hessenberg matrices orthogonally similar to C. Identify all reduced Hessenberg matrices to a single point $*_C$. Two unreduced Hessenberg matrices are in the same equivalence class provided they are orthogonally similar by a diagonal matrix for which each diagonal entry is ± 1. The map, Q_C, induced on Ω_C by the QR algorithm is well-defined by the implicit Q theorem. A careful examination of the algorithm for constructing QR iterates of Hessenberg matrices shows that Q_C is continuous. Let P_C be the map on Φ_C from the power method. The continuity of this map follows from the continuity of C.

Definition. $\Gamma_C\colon \Phi_C \to \Omega_C$ by

$\Gamma_C(\#_C) = *_C$;
$\Gamma_C(u) = S^{T}CS$, otherwise, where the columns of S result from Gram–Schmidt on $u, Cu, \ldots, C^{n-1}u$.

Theorem. *Γ_C is a continuous surjective quotient map and $\Gamma_C^{-1}(\{*_C\}) = \{\#_C\}$.*

PROOF. First we must show that Γ_C is well-defined. Suppose that u does not lie in an invariant subspace. The definition of S determines each column up to multiplication by ± 1 and, thus, $S^{T}CS$ is determined up to the equivalence relation defining Ω_C. To see that $S^{T}CS$ is Hessenberg, consider the inner product of a column s_i with Cs_j, where $i > j + 1$. Then $Cs_j \in \langle s_1, \ldots, s_{j+1} \rangle \perp s_i$. It is unreduced because s_i is the orthonormal completion of $s_1, \ldots, s_{i-1}$ in $\langle s_1, \ldots, s_{i-1}, Cs_{i-1} \rangle$.

For surjectivity, pick an unreduced Hessenberg matrix M for which there exists an orthogonal matrix P with $M = P^{T}CP$. Let u be the first column of P. By the implicit Q theorem, $\Gamma_C(u) = M$.

If u is close to an invariant subspace and $\Gamma_C(u) = S^{T}CS$, then $C\langle u, \ldots, C^{k-1}u \rangle = C\langle s_1, \ldots, s_k \rangle$ is close to $\langle s_1, \ldots, s_k \rangle$ for some $k < n$. Consequently, s_{k+1} is nearly orthogonal to Cs_k and, hence, $(\Gamma_C(u))_{k+1,k}$ is small.

Since Φ_C is compact, Ω_C is Hausdorff, and Γ_C is a continuous surjection, the map is also closed and, thus, a quotient map.

Theorem. *The power method map P_C is topologically semiconjugate to the basic QR algorithm map Q_C by Γ_C.*

PROOF. Suppose that $\Gamma_C(u) = S^T CS$. There exist orthogonal matrices W and Z such that $\Gamma_C P_C(u) = W^T CW$ and $Q_C \Gamma_C(u) = Z^T S^T CSZ$. By the implicit Q theorem, it suffices to show that SZ and W have the same first columns. Since $S^T CS = ZR$, the first column of Z is a scalar multiple of

$$S^T CSe_1 = S^T Cu = S^T P_C(u).$$

Thus, the first column of SZ is $P_C(u)$ which is also the first column of W.

Basic QR Decoupling Problem. *Which unreduced Hessenberg matrices are forward asymptotic to the reduced set under the basic QR algorithm?*

If follows immediately from the theorem that if $\lim_{m\to\infty} P_C^m(u) = \#_C$ for any $u \in \Phi_C$, then every unreduced Hessenberg matrix orthogonally similar to C is asymptotic to the reduced set under the basic QR algorithm. Since the first property depends only on the (not necessarily orthogonal) similarity class of the matrix, we have the following corollary.

Corollary. *If B is an $n \times n$ matrix with the property that $\lim_{m\to\infty} P_B^m(u) = \#_B$ for any $u \in \Phi_B$, then every Hessenberg matrix in the similarity class of B is asymptotic to the reduced set under the basic QR algorithm. If B has the property that $P_B^m(u)$ is bounded away from $\#_B$ whenever $u \neq \#_B$, then no unreduced Hessenberg matrix in the similarity class of B will decouple.*

Thus, it suffices to consider the power method on the following real canonical form from [HS].

Real Canonical Form Theorem. *Any $n \times n$ matrix is similar to a matrix of the block-diagonal form*

$$B = \begin{bmatrix} B_1 & & & 0 \\ & \cdot & & \\ & & \cdot & \\ & & & \cdot \\ 0 & & & B_i \end{bmatrix},$$

where each diagonal block has one of the following two forms:

1. $$a\ j \times j \text{ matrix } R_j(\lambda) = \begin{bmatrix} \lambda & 1 & & 0 \\ & \cdot & \cdot & \\ & & \cdot & \cdot \\ & & & \cdot\ 1 \\ 0 & & & \lambda \end{bmatrix};$$

2. $$a\ 2j \times 2j\ matrix\ C_j(a,b) = \begin{bmatrix} A & I & & 0 \\ & \cdot & \cdot & \\ & & \cdot & \cdot \\ & & \cdot & I \\ 0 & & & A \end{bmatrix}$$

$$\text{where } A = \begin{bmatrix} a & b \\ -b & a \end{bmatrix}, \quad I = \begin{bmatrix} 1 & 0 \\ 0 & 1 \end{bmatrix}.$$

Lemma 1. *If $j > 1$ and either $B = R_j(\lambda)$ or $B = C_j(a,b)$, then $\lim_{m\to\infty} P_B^m(u) = \#_B$ for any $u \in \Phi_B$.*

Proof. If $B = R_j(\lambda)$, then $\#_B$ is spanned by e_1. Consider the ratio of the j coordinate to the $j-1$ coordinate of an orbit. It can be shown that this ratio is asymptotic to zero. Proceeding inductively on the ratio of a coordinate to its predecessor completes the argument. If $B = C_j(a,b)$, then $\#_B$ is $\langle e_1, e_2 \rangle$.

From Lemma 1, it suffices to consider matrices in which for each block $j = 1$.

Lemma 2. *If any of the blocks $B_1, \ldots, B_i$ correspond to eigenvalues of distinct moduli, then $\lim_{m\to\infty} P_B^m(u) = \#_B$ for any $u \in \Phi_B$.*

From Lemmas 1 and 2, it suffices to consider matrices in which all blocks have the same modulus and $j = 1$.

Lemma 3. *Suppose every eigenvalue of B has the same modulus and*

$$B = \begin{bmatrix} B_1 & & & 0 \\ & \cdot & & \\ & & \cdot & \\ & & & \cdot \\ 0 & & & B_i \end{bmatrix}$$

with each block of the form $R_1(\lambda)$ or $C_1(\alpha, \beta)$. If there is a repeated eigenvalue, then $\Phi_B = \#_B$. If the eigenvalues are distinct, then $P_B^m(u)$ is bounded away from $\#_B$ under the nonvacuous condition that $u \neq \#_B$.

Proof. First we characterize the invariant subspaces. Consider a vector u consisting of 1×1 and 2×2 blocks corresponding to the B_k. If any of these is the trivial vector, then u is certainly in a proper invariant subspace.

Suppose that u has no trivial blocks. Then u is in a proper invariant subspace if and only if $\{u, Bu, \ldots, B^{n-1}u\}$ is a linearly dependent set. It suffices to consider the problem over the complex numbers. Over the complex field $\begin{bmatrix} a & b \\ -b & a \end{bmatrix}$ is similar to the diagonal matrix

$$\begin{bmatrix} a+bi & 0 \\ 0 & a-bi \end{bmatrix}$$

by $\begin{bmatrix} i & 1 \\ 1 & i \end{bmatrix}$. Thus, B is similar to a diagonal matrix D by a similarity transformation T. It remains to determine when $\{Tu, DTu, \ldots, D^{n-1}Tu\}$ is a linearly dependent set. Note that no coordinate of Tu vanishes.

Consider the matrix whose rows are Tu, DTu, ..., $D^{n-1}Tu$. Dividing each column by its first entry yields a Vandermonde matrix. This matrix is singular if and only if there is a repeated eigenvalue. It follows that $\Phi_B = \#_B$ when there is a repeated eigenvalue. When the eigenvalues are distinct, u will be close to $\#_B$ provided one of its blocks is negligible. Since the norm of the block is preserved by B, $P_B^m(v)$ is bounded away from $\#_B$ whenever $v \neq \#_B$.

Proposition. *If the characteristic polynomial of B has roots that are all distinct, but of the same modulus, then* $P_B^m(u)$ *is bounded away from* $\#_B$ *whenever* $u \neq \#_B$. *If B does not satisfy this condition, then* $\lim_{m\to\infty} P_B^m(u) = \#_B$ *for any* $u \in \Phi_B$.

Theorem. *If C is an* $n \times n$ *unreduced Hessenberg matrix with* $n \geq 3$, *then C will fail to decouple under the basic QR algorithm if and only if the roots of the characteristic polynomial are all distinct, but of the same modulus.*

Whereas the theorem completely solves the basic QR decoupling problem as stated, there remains the deeper question as to which matrices will continue to decouple until only 1×1 and 2×2 matrices remain. There are two reasons that our techniques fail to resolve this problem. Since all reduced Hessenberg matrices are identified to a point, the number and precise form of the decoupled matrices is not certain (this issue is addressed in [P1]). Second decoupling introduces an error, even in exact arithmetic.

4. Shifted QR Algorithms

Given the intimate connection between basic QR and the power method it is not surprising that QR's rate of convergence to reduced form depends on the spacing of the eigenvalues. To accelerate decoupling, it is advantageous to combine a shift strategy with QR.

As with RQI, the choice of shift is motivated by a coarse eigenvalue approximation. For an unreduced Hessenberg matrix, a diagonal entry (say the bottom one) is the simplest estimate.

Rayleigh Shifted QR Algorithm.

M^1 Hessenberg
If M^i is reduced stop, otherwise

$$M^i - M^i_{nn}I = Q_i R_i \qquad Q_i\text{-orthogonal, } R_i\text{-upper triangular}$$
$$M^{i+1} = Q_i^{\mathrm{T}} M^i Q_i.$$

The Rayleigh shifted QR algorithm preserves the orthogonal similarity class and Hessenberg form of the matrix. In [P2], it was shown that RQI is related to the Rayleigh shifted QR algorithm. The dynamical connection between these algorithms will now be established. In Section 3, it was shown that there is a topological conjugacy between F_C and the RQI map on the transpose, $R_{C^{\mathrm{T}}}$.

Theorem. *F_C is topologically semiconjugate to the Rayleigh shifted QR iteration map by Γ_C.*

Corollary. *RQI is topologically semiconjugate to the Rayleigh shifted QR iteration map.*

Corollary. *For each $n \geq 3$, there is a nonempty open set of unreduced Hessenberg matrices which do not decouple under the Rayleigh shifted QR algorithm.*

Proof of Theorem. The proof is similar to that of the basic QR semiconjugacy with F_C and the Rayleigh shifted QR iteration maps in place of P_C and Q_C. The first column of Z is a scalar multiple of

$$\begin{aligned}[S^{\mathrm{T}}CS - (S^{\mathrm{T}}CS)_{nn}I]e_1 = [S^{\mathrm{T}}CS - \rho_C(s_n)I]e_1 &= S^{\mathrm{T}}[C - \rho_C(H_C(u))I]Se_1 \\ &= S^{\mathrm{T}}[C - \rho_C(H_C(u))I]u = S^{\mathrm{T}}F_C(u).\end{aligned}$$

Why does the Rayleigh shifted QR algorithm fail? Recall that the shift is intended to provide a coarse approximation to an eigenvalue. The Rayleigh shift is a real number, whereas the roots of the characteristic polynomial may be complex. Thus, the Rayleigh shift might be a hopelessly coarse approximation.

We will now discuss the ingenious shift strategy of Francis [F] (see also [GV]). The basic idea is to utilize the eigenvalues of the lower right 2×2 submatrix as shifts. These provide eigenvalue approximations which are not restricted to the real field. This strategy raises two obvious questions. Which of the two eigenvalues is selected for the shift? If a real matrix is shifted by a complex number, doesn't a QR iteration of real matrices become an iteration of complex matrices?

Both of these issues are resolved by the use of a double-shift strategy. Ostensibly, each iteration is itself two QR steps, first QRing the matrix shifted by one of the eigenvalues and then QRing the result shifted by the other eigenvalue. As to the second question above, note that if one of the eigenvalues of the submatrix is complex, then the other eigenvalue is its conjugate. It can be shown that, in exact arithmetic, the result of the double step is a real matrix. Furthermore, this matrix can be computed by a single

QR step in which the factorization is performed on the matrix obtained by evaluating the submatrix characteristic polynomial at the original matrix.

Francis Shifted QR Algorithm

M^1 Hessenberg
If M^i is reduced stop, otherwise
$p_i(t) = t^2 - (M^i_{nn} + M^i_{n-1n-1})t + (M^i_{nn}M^i_{n-1n-1} - M^i_{nn-1}M^i_{n-1n})$
$p_i(M^i) = Q_iR_i$
$M^{i+1} = Q_i^T M^i Q_i$.

Note. Whereas the algorithm as stated above yields the iteration sequence of the Francis shifted QR algorithm, in actual practice a more efficient method is employed to obtain M^{i+1}. The above definition is sufficient for our purpose since we are primarily interested in the dynamics of the iteration sequence.

Francis Shifted QR Decoupling Problem. *Which unreduced Hessenberg matrices are forward asymptotic to the reduced set under the Francis shifted QR algorithm and do they comprise a set of full measure*?

As with the basic and Rayleigh shifted QR iterations, we will show that Γ_C is a semiconjugacy between an appropriate map of Φ_C and the Francis shifted QR iteration on Ω_C. First we will place shifted QR algorithms in a more general context (see also [P2, W2, W3]). A *polynomial shift iteration* is a function, p, that associates to each Hessenberg class $M \in \Omega_C$, a polynomial, $p(M)(t)$, whose coefficients are continuous in M. If p is a polynomial shift iteration, then there is a p-shifted QR algorithm.

***p*-Shifted QR Algorithm**

M^1 Hessenberg
If M^i is reduced stop, otherwise
$p(M^i)(M^i) = Q_iR_i$
$M^{i+1} = Q_i^T M^i Q_i$.

EXAMPLES.

1. $p(M)(t) = t$ is the basic QR algorithm.
2. $p(M)(t) = t - M_{nn}$ is the Rayleigh shifted QR algorithm.
3. $p(M)(t) = t^2 - (M_{nn} + M_{n-1n-1})t + (M_{nn}M_{n-1n-1} - M_{nn-1}M_{n-1n})$ is the Francis shifted QR algorithm.

A p-shifted QR algorithm induces a map on each Ω_C. There is a related map on Φ_C.

Definition. $G_p\colon \Phi_C \to \Phi_C$

$G_p(\#_C) = \#_C$.
$G_p(u) = \#_C$ if $p(\Gamma_C(u))(C)$ is singular.
$G_p(u) = p(\Gamma_C(u))(C)(u)$ otherwise.

Theorem. *If p is a polynomial shift iteration and C is an $n \times n$ matrix, then G_p is topologically semiconjugate to the p-shifted QR iteration map by Γ_C.*

PROOF. Again we follow the previous semiconjugacy arguments:

$$p(S^{\mathrm{T}}CS)(S^{\mathrm{T}}CS)e_1 = S^{\mathrm{T}}p(S^{\mathrm{T}}CS)(C)Se_1 = S^{\mathrm{T}}p(\Gamma_C(u))(C)u = S^{\mathrm{T}}G_p(u).$$

Corollary. *If C is an $n \times n$ matrix and p is the polynomial shift iteration for Francis, then $G_p\colon \Phi_C \to \Phi_C$ is semiconjugate to the Francis shifted QR iteration map on Ω_C. The map G_p is defined (in the otherwise case) by*

$$G_p(u) = [C^2 - (s_n^{\mathrm{T}}Cs_n + s_{n-1}^{\mathrm{T}}Cs_{n-1})C + (s_n^{\mathrm{T}}Cs_n \times s_{n-1}^{\mathrm{T}}Cs_{n-1} - s_n^{\mathrm{T}}Cs_{n-1} \times s_{n-1}^{\mathrm{T}}Cs_n)I](u),$$

where s_{n-1} and s_n are the last two vectors obtained from Gram–Schmidt on $\{u, Cu, \ldots, C^{n-1}u\}$.

To study the Francis shifted QR decoupling problem, it suffices to analyze the dynamics of G_p on canonical representatives of the orthogonal similarity class. An $n \times n$ matrix A is *normal* provided $AA^{\mathrm{T}} = A^{\mathrm{T}}A$. Normal matrices have nice canonical forms.

Theorem [HJ]. *An $n \times n$ matrix is normal if and only if it is orthogonally similar to a matrix of the form*

$$\begin{bmatrix} B_1 & & & 0 \\ & \cdot & & \\ & & \cdot & \\ 0 & & & B_i \end{bmatrix},$$

where the first k blocks are 2×2 matrices of the form $C_1(a_j, b_j)$ and the remaining blocks are 1×1 $R_1(\lambda_j)$'s.

It follows that there are two types of 3×3 normal matrices. There are the symmetric classes in which the canonical form is diagonal. For each nonsymmetic class, the canonical form contains both a $C_1(a, b)$ block and an $R_1(\lambda)$ block. The dynamics of G_p on these forms is examined in detail in [B4]. We summarize the results below. For the remainder of the section, G_p is defined from the Francis shift.

For the nonsymmetric normal 3×3 case, we may assume

$$C = \begin{bmatrix} a & b & 0 \\ -b & a & 0 \\ 0 & 0 & d \end{bmatrix}.$$

To obtain coordinates for Φ_C, consider the upper hemisphere of points (u_1, u_2, u_3). The target class $\#_C$ consists of $u_3 = 0$ and $u_3 = 1$. In obtaining a workable formula for G_p in [B4], we exploit the fact that $u = s_1$ is the cross-product of s_2 and s_3. It turns out that G_p preserves the latitudes $u_3 =$ constant. Hence, it suffices to consider what happens to the u_3 coordinate under G_p. This is a map of the interval which is given below.

$$L(u_3) = \left[\frac{b^2}{\tau^2}(u_3^{-2} - 1)^3 + 1\right]^{-1/2},$$

$$\tau^2 = b^2 + (a - d)^2.$$

The fixed point set of L is $\{0, \sqrt{b/(b + \tau)}, 1\}$. The derivatives at the fixed points are 0, 3, 0, respectively. It follows that 0 and 1 are superattracting sinks and all other points are repelled from the source to the sinks. The superattraction is actually cubic in the sense that

$$\lim_{z \to 0} \frac{L(z)}{z^3} = \frac{\tau}{b}, \qquad \lim_{z \to 1} \frac{L(z) - 1}{(z - 1)^3} = \frac{4b^2}{\tau^2}.$$

Proposition. *Suppose C is a normal, nonsymmetric 3×3 matrix and G_p: $\Phi_C \to \Phi_C$. Then all points are attracted to $\#_C$ at an asymptotically cubic rate except for an invariant repelling circle.*

For the 3×3 symmetric case, we have the diagonal canonical form

$$C = \begin{bmatrix} a & 0 & 0 \\ 0 & b & 0 \\ 0 & 0 & d \end{bmatrix}.$$

The target class $\#_C$ is comprised of the spaces $u_1 = 0$, $u_2 = 0$, $u_3 = 0$. In the portion of Φ_C from the positive octant, all points are asymptotic to $\#_C$ with the exception of a single global repeller. The global repeller is a period two source in which the image changes the sign of the second coordinate.

Proposition. *Suppose C is a symmetric 3×3 matrix and G_p: $\Phi_C \to \Phi_C$. Then all points are attracted to $\#_C$ except for two source orbits of period two.*

Before considering the ramifications for the Francis shifted QR decoupling problem, compare the situation to the basic QR decoupling problem. With basic QR, it suffices to consider the real canonical forms rather than orthogonal canonical forms. Depending on the characteristic polynomial of B, either every point was asymptotic to $\#_B$ under P_B or else all the orbits were bounded away from $\#_B$. For G_p from a 3×3 normal matrix, there are points which are asymptotic to $\#_C$ and there is an invariant repelling set. To solve the Francis shifted decoupling problem, an understanding of the quotient map Γ_C is required.

Proposition. (a) *If M is an unreduced Hessenberg class which is orthogonally similar to a 3×3 normal nonsymmetric canonical form C, then $\Gamma_C^{-1}(M)$ is a u_3 = constant latitude of Φ_C.*

(b) *If M is an unreduced Hessenberg class which is orthogonally similar to a 3×3 diagonal matrix, C, with distinct diagonal entries, then $\Gamma_C^{-1}(M)$ consists of four elements having the form $\langle(u_1, u_2, u_3)\rangle$, $\langle(-u_1, u_2, u_3)\rangle$, $\langle(u_1, -u_2, u_3)\rangle$, $\langle(-u_1, -u_2, u_3)\rangle$.*

Theorem. *For every orthogonal similarity class of 3×3 normal matrices with distinct roots of the characteristic polynomial, the Francis shifted QR map is topologically conjugate to a north-pole south-pole map of a sphere (i.e., there is a fixed source and a fixed sink which are the backward and forward limits of all other orbits) with the reduced class as the sink. In the nonsymmetric and symmetric cases, the sphere dimensions are one and two, respectively.*

Hence, for 3×3 normal matrices, the Francis shifted QR algorithm has excellent global convergence behavior. In each orthogonal similarity class, there is precisely one matrix (up to signs of the entries) for which the algorithm is not asymptotic to reduced form. For $n > 3$ or nonnormality, the Francis shifted QR decoupling problem is open. The above solution for 3×3 normal matrices might suggest the following generalization to the $n \times n$ normal case.

Problem. *In an orthogonal similarity class of $n \times n$ normal matrices in which the roots of the characteristic polynomial are distinct and include k conjugate pairs, is the Francis shifted QR map topologically conjugate to a north-pole–south-pole map of S^{n-k-1}?*

5. HQR2 and Other Shifts

The EISPACK subroutine for approximating the eigenvalues of a real Hessenberg matrix is HQR2 [MP, PW]. The primary weapon of HQR2 is the Francis shifted QR algorithm. To deal with the contingency of matrices that do not decouple under Francis, HQR2 is equipped with an *exceptional shift* strategy. The exceptional shift strategy is a polynomial shift iteration applied to the Rayleigh shifted matrix.

Exceptional Shift Step.

M^i unreduced Hessenberg
$e(M^i)(t) = t^2 - 1.5(|M^i_{nn-1}| + |M^i_{n-1n-2}|)t + (|M^i_{nn-1}| + |M^i_{n-1n-2}|)^2$
$e(M^i)(M^i - M^i_{nn}I) = Q_i R_i$
$M^{i+1} = Q_i^{\mathrm{T}}(M^i - M^i_{nn}I)Q_i.$

The goal of HQR2 is to perform a sequence of iterations and obtain decouplings until all the matrices are 1×1 and 2×2. The variable *its* keeps track of the number of iterations within a decoupling effort. The exceptional shift is invoked at iterations 11 and 21 in the event that Francis has failed to effect a decoupling at these junctures. After a decoupling, *its* is reset and work begins on a submatrix. The variable *itn* represents 30*n minus the total number of iterations from all the decouplings. If *itn* reaches 0, HQR2 halts and returns an error parameter.

HQR2 performs extremely well in practice. A significant measure of the effectiveness of the algorithm is the average number of iterations required for a decoupling. For a particular Hessenberg matrix, this quantity is the total number of iterations required by HQR2 divided by the number of decouplings. Empirical evidence suggests that this average is typically on the order of 2 [GV].

We will now examine the dynamics of HQR2. Note that the exceptional shift step does not preserve the orthogonal similarity class of the matrix. If C represents the original class, then the exceptional iteration induces a function E which maps an element of Ω_C to $\Omega_{C'}$, the Hessenberg classes of $C' = C - M^i_{nn}I$. As has become the custom, we will produce a related map from Φ_C to $\Phi_{C'} = \Phi_C$.

Theorem. *If C is an $n \times n$ matrix and $G_e\colon \Phi_C \to \Phi_C$ is defined by*

$$G_e(u) = [(C')^2 - 1.5(|s_n^{\mathrm{T}} C s_{n-1}| + |s_{n-1}^{\mathrm{T}} C s_{n-2}|)C' + (|s_n^{\mathrm{T}} C s_{n-1}| + |s_{n-1}^{\mathrm{T}} C s_{n-2}|)^2 I]u,$$

where *$C' = C - s_n^{\mathrm{T}} C s_n I$, then $E(\Gamma_C(u)) = \Gamma_{C'}(G_e(u))$.*

The HQR2 decoupling problem is complicated by the use of a different dynamical system on iterations 11 and 21. A Hessenberg matrix is said to be *hopeless* for HQR2 provided its orbit is bounded away from the reduced set under any sequence of Francis and exceptional iterations. Being hopeless is a sufficient condition for the algorithm to fail to approximate a solution. To find the hopeless matrices, it suffices to consider the dynamics of sequences of Francis G_p's and exceptional G_e's on Φ_C.

In [B4] there is a characterization of the 3×3 normal matrices which are hopeless. We will give a brief description of the results. First note that we need only consider the fixced sources from Francis since all other matrices are asymptotic to the reduced set under a sequence of Francis iterations.

Now consider the classes in Φ_C corresponding to the source. In the symmetric case, a direct computation shows that this source is mapped by G_e to a point which is asymptotic to $\#_C$ under G_p.

In the nonsymmetric case we may assume

$$C = \begin{bmatrix} a & b & 0 \\ -b & a & 0 \\ 0 & 0 & d \end{bmatrix}.$$

The repelling latitude for G_p is also preserved by G_e if and only if a certain equation involving a, b, and d is satisfied. The solution set of this equation is nonempty. For each orthogonal similarity class in which the canonical form satisfies the equation, the hopeless set consists of a single equivalence class of matrices, all of whose members are identical up to the signs of their entries. In all other orthogonal similarity classes of 3×3 normal matrices, there are no hopeless matrices.

It is important to remember that the theory for hopeless matrices assumes exact arithmetic. As an experiment, HQR2 was run on a Sun 4/280 with a double precision approximation of a hopeless matrix. Early iterations are nearly fixed. However, round-off and the large derivative of the source eventually come to the rescue and produce an approximate solution after 40 iterations.

In closing, we remark that a new linear algebra subroutine library, LAPACK, is currently under development. One of the motivations for LAPACK is to produce algorithms which exploit parallel architectures. Bai and Demmel [BD] have designed an eigenvalue algorithm which employs a polynomial shift iteration involving a polynomial of higher degree.

References

[A] J. Akao, The dynamics of Rayleigh quotient iteration on 2×2 matrices, honors thesis, Emory University, 1989.

[B1] S. Batterson, Stuctually stable Grassmann transformations, *Trans. Amer. Math. Soc.* **231** (1977), 385–405.

[B2] S. Batterson and J. Smille, The dynamics of Rayleigh quotient iteration, *SIAM J. Numer. Anal.* **26** (1989), 624–636.

[B3] S. Batterson and J. Smillie, Rayleigh quotient iteration for nonsymmetric matrices, *Math. Comput.* **55** (1990), 169–178.

[B4] S. Batterson, Convergence of the shifted QR algorithm on 3×3 normal matrices, *Numer. Math.*, **58** (1990), 341–352.

[BD] Z. Bai and J. Demmel, On a block implementation of Hessenberg multishift QR iteration, *Int. J. of High Speed Comput.* (1989), 97–112.

[D1] P. Deift, L. Li, T. Nanda, and C. Tomei, The Toda flow on a generic orbit is integrable, *Commun. Pure Appl. Math.* **39** (1986), 183–232.

[D2] R. Devaney, *An Introduction to Chaotic Dynamical Systems*, Addison-Wesley, Redwood City, CA, 1989.

[F] J. Francis, The QR transformation: A unitary analogue to the LR transformation, Parts I and II, *Comput. J.* **4** (1961), 265–272, 332–345.

[GV] G. Golub and C. Van Loan, *Matrix Computations*, Johns Hopkins University Press, Baltimore, MD, 1983.

[HJ] R. Horn and C. Johnson, *Matrix Analysis*, Cambridge University Press, Cambridge, 1985.

[HS] M. Hirsch and S. Smale, *Differential Equations, Dynamical Systems, and Linear Algebra*, Academic Press, New York, 1974.

[M] J. Munkres, *Topology*, Prentice-Hall, Englewood Cliffs, NJ, 1975.

[MP] R. Martin, G. Peters, and J. Wilkinson, The QR algorithm for Real Hessenberg matrices, *Numer. Math.* **14** (1970), 219–231.

[O] A. Ostrowski, On the convergence of the Rayleigh quotient iteration for the computation of characteristic roots and vectors, I, II, III, IV, V, VI, *Arch. Rat. Mech. Anal.* **1, 2, 3, 4** (1957–60).

[P1] B. Parlett, Global convergence of the basic QR algorithm on Hessenberg matrices, *Math. Comput.* **22** (1968), 803–817.

[P2] B. Parlett and W. Kahan, On the convergence of a practical QR algorithm, *Mathematical Software*, Information Processing 68, sect 1, North-Holland, Amsterdam, 1969, pp. 114–118.

[P3] B. Parlett, *The Symmetric Eigenvalue Problem*, Prentice-Hall, Englewood Cliffs, NJ, 1980.

[PW] G. Peters and J. Wilkinson, Eigenvectors of real and complex matrices by LR and QR triangularizations, *Numer. Math.* **16** (1970), 181–204.

[S1] M. Shub, The geometry and topology of dynamical systems and algorithms for numerical problems, preprint, 1983.

[S2] M. Shub, Some remarks on dynamical systems and numerical analysis, *Dynamical Systems and Numerical Analysis: Proceedings of the VII ELAM, Equinoccio*, Universidad Simon Bolivar, Caracas, 1986, pp. 69–92.

[S3] M. Shub and A. Vasquez, Some linearly induced Morse-Smale systems, the QR algorithm and the Toda lattice, preprint.

[S4] S. Smale, On the efficiency of algorithms of analysis, *Bull. Amer. Math. Soc.* **13** (1985), 87–121.

[S5] G. Stewart, *Introduction to Matrix Computations*, Academic Press, New York, 1973.

[W1] D. Watkins, Understanding the QR algorithm, *SIAM Rev.* **24** (1982), 427–440.

[W2] D. Watkins, On GR algorithms for the eigenvalue problem, preprint, 1988.

[W3] D. Watkins and L. Elsner, Convergence of algorithms of decomposition type for the eigenvalue problem, preprint, 1990.

[W4] J. Wilkinson, *The Algebraic Eigenvalue Problem*, Clarendon Press, Oxford, 1965.

36
On the Role of Computable Error Estimates in the Analysis of Numerical Approximation Algorithms

FENG GAO

Truncation error estimation for methods of numerical approximation, i.e., estimation of their errors of approximation, is an important constituent of numerical analysis. The error estimates obtained can be dichotomized according to whether an error estimate is computable from information that can be used for computing the approximations. For example, the standard error estimate for the bisection method for finding a zero of a function $f(x)$ in an interval $[a, b]$, under the assumption that f is continuous on $[a, b]$ with $f(a)f(b) < 0$, is computable. The method proceeds as follows. Let $[a_0, b_0] = [a, b]$. At Step k ($k \geq 0$), compute $f(x_k)$, where $x_k = (a_k + b_k)/2$. If $f(x_k) = 0$, an exact zero is found. Otherwise, let $[a_{k+1}, b_{k+1}] = [a_k, x_k]$ if $f(a_0)f(x_k) < 0$, or $[a_{k+1}, b_{k+1}] = [x_k, b_k]$ if $f(x_k), f(b_k) < 0$, and proceed to Step $k + 1$. Now let x_k be the kth approximation to an zero of f in $[a, b]$. Then there exists $\bar{x} \in [a, b]$ such that $f(\bar{x}) = 0$, and

$$|\bar{x} - x_k| \leq \frac{b - a}{2^k}. \tag{1}$$

The Right-hand-side of (1) is a computable and guaranteed (deterministic) bound on the error of approximation.

An example of a noncomputable error estimate is the standard one for the trapezoidal rule for approximating the integral of a function $f(x)$ over $[a, b]$,

$$\int_a^b f(x)\,dx \approx \frac{f(a) + f(b)}{2}.$$

For any $f \in C^2[a, b]$,

$$\left| \int_a^b f(x)\,dx - \frac{f(a) + f(b)}{2} \right| \leq \frac{(b - a)^3}{12} \sup_{a \leq x \leq b} |f^{(2)}(x)|. \tag{2}$$

The term $\sup_{a \leq x \leq b} |f^{(2)}(x)|$ on the right-hand side of (2) is not computable for an arbitrary C^2-function, i.e., it cannot be computed using a finite amount of data from the set $\{x, f(x)\}_{x \in [a,b]}$ as does a quadrature approximation.

This simple dichotomy of truncation error estimates according to whether an estimate is computable had not seemed to attract much attention in traditional numerical analysis. Its potential importance was perceived by Steve Smale and became the basis of his 1986 paper "Newton's method estimates from data at one point" [Smale (1986)]. Smale commented at the beginning of this paper:

> The work of Kantorovich has been seminal in extending and codifying Newton's method. Kantorovich's approach, which dominates the literature in this area, has these features: (a) weak differentiability hypotheses are made on the system, e.g., the map is C^2 on some domain in a Banach space; (b) derivative bounds are supposed to exist over the whole of this domain. In contrast, here strong hypotheses on differentiability are made; analyticity is assumed. On the other hand, we deduce consequences from data at a single point. This point of view has valuable features for computation and its theory.

Smale then presented a criterion for testing whether an approximation z_0 to a zero of a complex polynomial or complex analytic function f is a so-called approximate zero. More precisely, the criterion involves estimation of the quantity $\alpha(z, f)$,

$$\alpha(z, f) = \left|\frac{f(z)}{f'(z)}\right| \sup_{k>1} \left|\frac{f^{(k)}(z)}{k!f'(z)}\right|^{1/(k-1)} \tag{3}$$

at $z = z_0$ to see whether it is less than $\alpha_0 \approx 0.130707$. If the criterion is satisfied, then z_0 is an approximate zero, which is defined as satisfying the condition that Newton's iteration $z_k = z_{k-1} - f(z_{k-1})/f'(z_{k-1})$ starting at z_0 has the property

$$|z_k - z_{k-1}| \leq \tfrac{1}{2}^{2^{k-1}-1}|z_1 - z_0|, \quad k = 1, 2, \ldots. \tag{4}$$

It was then shown that if z_0 is an approximate zero, then there exist η, $f(\eta) = 0$, such that

$$|z_n - \eta| \leq \tfrac{1}{2}^{2^{n-1}}|z_1 - z_0|K, \quad n = 1, 2, \ldots, \tag{5}$$

where $K \leq 1\frac{3}{4}$. In other words, starting at an approximate zero, the Newton iteration approximation is superconvergent. Here, criterion (3) is computable from the coefficients of the analytic series of f at z_0 (when f is not a polynomial, I use the term "computable" just to mean dependence on data only at the single point z_0 since the supremum is taken over a countable set of numbers), and estimates (4) and (5) do not involve any noncomputable function-dependent constant. This is in contrast to the traditional Kantorovich theory [Kantorovich and Akilov (1964)], where, similar to the case of the trapezoidal rule for integral approximation, the convergence estimate involves a bound on the second derivative of f over a whole domain which is not computable locally for an arbitrary C^2-function. The new, computable estimate was used by Smale to construct new zero-finding algorithms based on Newton's method.

Even though this so-called α-theory for zero-finding of analytic functions can also be derived from the Kantorovich theory [e.g., Rheinbolt (1988)], the significance of Smale's observation on computable error estimates goes far beyond zero-finding. This is because, whereas a noncomputable error estimate provides information regarding the speed of convergence of a method of approximation, a computable error estimate can also be used as an error criterion in an algorithm. This is a very important point that I shall return to later in this chapter.

In fall 1985 at Berkeley, when Steve Smale was disseminating this α-theory in his graduate course and also in one Mathematics Department colloquium talk, I was at the preliminary stage of my Ph.D. research under his supervision. Smale's observation on computable error estimates struck me as pointing out a potentially important direction for the analysis of numerical approximation algorithms. More thinking brought me to the following conclusion: Most numerical approximation methods simply do no have guaranteed and yet computable error bounds under a weak differentiability assumption; whereas a strong differentiability assumption is one way to obtain computable error estimates, estimates applicable under a weak differentiability assumption are also important, for this assumption usually captures the generality of a method and is the starting point of many algorithms in practice; most practical algorithms, thus, use computable but not guaranteed error estimates; they are heuristics that may be incorrect some of the time but prove to be generally useful in practice.

Around this time, I was studying Smale's 1985 paper "On the efficiency of algorithms of analysis" [Smale (1985)] and, in particular, the section "On efficient approximation of integrals." I was also reading a preprint of the paper "Approximation of linear functionals on a Banach space with a Gaussian measure" by David Lee and Greg Wasilkowski. Both concern the average analysis of integral approximation using Gaussian measures on function spaces. The former focused on the average errors of some classical quadratures, whereas the latter focused more on the average approximative power of quadratures for a given number of allocation points. Upon Smale's suggestion, I was also exploring the problem of adaption for numerical integration in this average analysis setting, which resulted in an unpublished manuscript Gao [1986]. It then occurred to me that my thoughts motivated by Smale's observation on computable error estimates could perhaps be formalized in this average analysis setting. Namely, the assumption of a probability measure on a function space with weak differentiability may lead to computable conditional average error estimates—bounds on the average error given the computed data—for an approximation method, even though computable and guaranteed error bounds may not exist. And this formalism may serve as the theoretical basis for the heuristic error estimates in practical algorithms.

I decided that the key was to first show the relevance of such a formalism by establishing a close connection between the conditional average error estimates produced in such a model and the practical heuristic estimates. The case of the trapezoidal rule mentioned earlier was the right canonical example for me since I was looking at numerical integration. More specifically, I considered the following problem:

The integral $\int_a^b f(t)\,dt$ is to be evaluated for $f \in C_0^k[a,b]$, where

$$C_0^k[a,b] = \{g \in C^k[a,b]: g^{(i)}(a) = 0, i = 0, 1, \ldots, k\}.$$

A quadrature $Q_n(f) = \sum_{i=1}^n a_i f(t_i)$ is used to approximate the integral, where $a = t_0 < t_1 < \cdots < t_n = b$ and the choices of the quadrature points $\{t_i\}_{i=1}^n$ and the coefficients $\{a_i\}_{i=1}^n$ depend on n and satisfy the following two conditions:

Condition 1. There exists a constant $c > 0$, independent of n, such that

$$\frac{\max_{1 \le i \le n} |t_i - t_{i-1}|}{\min_{1 \le i \le n} |t_i - t_{i-1}|} \le c. \tag{6}$$

Condition 2. There exists a constant $\bar{c} > 0$, independent of n, such that

$$\left| \int_a^b f(t)\,dt - \sum_{i=1}^n a_i f(t_i) \right| \le \frac{\bar{c}}{n^k} \sup_{a \le t \le b} |f^{(k)}(t)| \tag{7}$$

for any $f \in C_0^k[a,b]$.

Since $\sup_{a \le t \le b} |f(t)|$ is not finitely computable from $\{x, f(x)\}_{x \in [a,b]}$, we suppose an algorithm employs the following heuristic error criterion obtained by replacing the kth derivative with the kth divided differences:

$$\left| \int_a^b f(t)\,dt - \sum_{i=1}^n a_i f(t_i) \right| \lessapprox \frac{\bar{c}}{n^k} \max_{1 \le i \le n} |[t_{i-k}, \ldots, t_i]f|. \tag{8}$$

Here $[t_{i-k}, \ldots, t_i]f$ denotes the kth divided difference of f at $t_{i-k}, \ldots, t_i$ and by convention $t_{-i} = a - t_i$, $f(t_{-i}) = 0$, $i = 1, \ldots, n$. Also assume $n > k$. The (recursive) definition of the divided difference is given by

$$[t_{i-l}, \ldots, t_i]f = \frac{[t_{i-l+1}, \ldots, t_i]f - [t_{i-l}, \ldots, t_{i-1}]f}{t_i - t_{i-l}},$$

$$i = 1, \ldots, n, \quad 1 \le l \le n.$$

The criterion (8) is not a guaranteed error bound since the actual error of approximation can be arbitrary large for a C^k-function whose kth divided differences at $\{t_i\}$ are small. It is a variant of some of the heuristic error criteria used in practical quadrature algorithms.

To analyze (8), I then assumed that the likelihood of functions in $C_0^k[a,b]$ as integrands are distributed according to the Wiener measure in $C_0^k[a,b]$. Intuitively, this measure is a Gaussian measure on $C_0^k[a,b]$ with mean $f \equiv 0$ and variance 1 and can be viewed as concentrating in

$$H_0^{k+1}[a,b] = \{f \in H^{k+1}[a,b]: f^{(i)}(a) = 0, i = 0, 1, \ldots, k\}$$

with the normal probability density distributed with respect to the norm

$$\|f\|_{H_0^{k+1}} = \left(\int_a^b (f^{(k+1)}(t))^2\,dt \right)^{1/2},$$

where

$$H_0^{k+1}[a,b] = \{f \in C_0^k[a,b]: f^{(k+1)} \in L_2[a,b]\}$$

is a Hilbert space with inner product $\langle f,g \rangle = \int_a^b f^{(k+1)}(t)g^{(k+1)}(t)\,dt$.

The following theorem became the main result of my Ph.D. thesis.

Theorem [Gao (1989)]. *There exist constants $c_1 \geq 0$, and $c_2 > 0$, such that for any quadrature $\sum_{i=1}^n a_i f(t_i)$ which satisfies* (6) *and* (7),

$$\underset{\substack{g \in C_0^k[a,b] \\ g(t_i)=f(t_i) \\ i=1,\ldots,n}}{\text{Average}} \left| \int_a^b g(t)\,dt - \sum_{i=1}^n a_i g(t_i) \right|^2 \leq \frac{c_1}{n^{2k+2}} + c_2 n \left(\frac{\bar{c}}{n^k} \max_{1 \leq i \leq n} |[t_{i-k}, \ldots, t_i]f| \right)^2 \tag{9}$$

for any $(f(t_1), \ldots, f(t_n))^{\mathrm{T}} \in R^n$. Here Average $g \in C_0^k[a,b]$, $g(t_i) = f(t_i)$, $i = 1, \ldots, n$ *denotes the average with respect to the conditional Wiener measure under the constraints $g(t_i) = f(t_i)$, $i = 1, 2, \ldots, n$.*

Condition 1 on a uniform partitioning of $[a,b]$ is too stringent for practice. In a later paper [Gao (1990b)], it was relaxed and Condition 2 was replaced with a sum of local error bounds, allowing the theory to cover adaptive quadratures.

This theorem demonstrated the close connection between the probabilistic model and the heuristic error estimates in practice. The technical results needed to prove it also gave rise to new numerical quadrature algorithms. I initially did not pay enough attention to this until it was emphasized to me by Beresford Parlett [1987)]. To use the divided-difference criterion (8) to bound the conditional average error [the left-hand side of (9)], (8) has to be used to bound the difference between the approximation by this quadrature and another approximation given by the integral of the natural spline interpolation which is the mean of the conditional measure given the constraints $g(t_i) = f(t_i)$, $i = 1, 2, \ldots, n$. Namely, to prove the theorem, the following lemma was needed.

Lemma [Gao (1989)].

$$\left| \int_a^b s_f(t)\,dt - \sum_{i=1}^n a_i f(t_i) \right| \leq \sqrt{c_2 n}\, \frac{\bar{c}}{n^k} \max_{1 \leq i \leq n} |[t_{i-k}, \ldots, t_i]f| \tag{10}$$

for any $(f(t_1), \ldots, f(t_n))^{\mathrm{T}} \in R^n$ with the convention $f(t_i) = 0$, $i \leq 0$. Here *s_f is the natural spline in $H_0^{k+1}[a,b]$ that interpolates f at $t_1, t_2, \ldots, t_n$.*

Therefore, one can view an algorithm equipped with the error criterion (8) as one that uses the natural spline interpolants as the sample integrands and terminates successfully when the difference between the two solutions are small enough. This is in the spirit of the algorithms in practice where one often compares the approximation locally (in a subinterval) with a different local approximation. The new algorithm does something stronger: It compares the approximate solution to a global one that has a minimal-norm property [see Gao (1989) for details], and yet the comparison is done implicitly, i.e., without having to compute the spline solution explicitly, and using only local error checks (divided differences, each depending on a small number of function values locally).

The main drawback again was the uniform partition condition (Condition 1) that excludes adaptive quadratures. In a later paper, Gao [1990a], this point of view was further pursued, with the uniform partition condition relaxed, to derive new adaptive quadrature algorithms.

So, these results serve as an example to illustrate the value of studying computable error estimates in understanding practical numerical approximation algorithms. They are also in the spirit of Smale's general view on studying algorithms of numerical analysis from a computational complexity viewpoint expressed in, e.g., Smale (1985):

> Experience in the use of algorithms, especially with the computer in recent decades, has given rise to certain practices and beliefs. To give a deeper understanding of this culture, we try to give reasonable underlying mathematical formulations, and eventually to prove theorems, usually confirming the experience of the practitioners. Idealizations and simplifications are made, but we try to keep the essence of the observed phenomena. There is a kind of parallel in this approach to that of theoretical physics. Our primary goal is not the design of new algorithms, but we hope that this deeper understanding will eventually be constructive in that domain too.

Compared to five years ago, today I am only more convinced of the practical and theoretical importance of studying computable error estimates to the design and analysis of numerical approximation algorithms, even though I have taken on other topics of interests in much of my research time. And I believe that I can now make better arguments for it based on the further reflections I have had.

A guaranteed error bound is important to evaluating a method of numerical approximation. As an a priori estimate, it provides information regarding the speed of convergence of the method and helps one decide whether the method is useful and where and when it should be used. A computable error estimate, on the other hand, is important as an error criterion in an algorithm of numerical approximation. As an a posteriori estimate, it allows the algorithm to decide whether a particular approximation is good enough, and if not, what the next, more refined approximation should be. Therefore, the need to distinguish between the two types of error estimates is underlied by the need to distinguish between a method of numerical approximation—an asymptotic way to approximate the exact solution of a problem—and an algorithm of numerical approximation—a finite process to construct one satisfactory approximation. Whereas an algorithm is usually based on a method of approximation, it has other, more algorithmic aspects. A numerical approximation algorithm generates a sequence of approximations to obtain a satisfactory one. To do so, it has to choose an initial approximation, and after generating each approximation it has to determine whether the approximation is satisfactory, and if not, how to generate the next approximation. These more algorithmic aspects of a numerical approximation algorithm, namely, the choice of initial approximation, the criterion for termination, and the rules for adaption, often do not come with the method of approximation the algorithm uses. This latter distinction is important because it reveals the merits as well as limitations of traditional numerical analysis. Traditional numerical analysis has focused its study on guaranteed a priori error bounds and, in doing so, has produced a good understanding of many methods of numerical approximation. However, it has not given much study to computable a posteriori error estimates that can be used in numerical approximation algorithms. This is not too surprising in hindsight since under the weak differentiability assumption most computable error estimates cannot be guaranteed error bounds; it would require more sophisticated mathematical models than the worst-case one traditional numerical analysis is accustomed to, to conduct a theoretical study. From a practical perspective, since the design of most of the error criteria in algorithms of numerical approximation have been heuristical, a theoretical study of computable error estimates can gain valuable insights to help improve existing algorithms and discover new algorithrns. From a theoretical perspective, this is a terrain, relatively unexplored, by traditional numerical analysis, where the mathematical analysis of numerical algorithms, with fresh ideas and novel tools, has the potential to establish itself and gain wider acceptance as a scientific field.

Acknowledgment. I wish to thank Jim Curry for the pointer to the paper by W.C. Rheinbolt.

References

Gao, F. (1986), Nonasymptotic error of numerical integration—an average analysis, unpublished manuscript.

Gao, F. (1989), A probabilistic theory for error estimation in automatic integration, *Numer. Math.* 56, 309–329.

Gao, F. (1990a), On the power of *a posteriori* error estimation for numerical integration and function approximation, in *Proc. AFA International Conference "Curves and Surfaces"* held in Chamonix, France, to appear.

Gao, F. (1990b), Probabilistic analysis of numerical integration algorithms, *J. Complexity*, to appear.

Lee, D. and Wasilkowski, G.W. (1986), Approximation of linear functionals on a Banach space with a Gaussian measure, *J. Complexity* 2, 12–43.

Parlett, B.N. (1987), private communication.

Rheinbolt, W.C. (1988), On a theorem of S. Smale about Newton's method for analytic mappings, *Appl. Math. Lett.* 1 (1), 69–72.

Smale, S. (1985), On the efficiency of algorithms of analysis, *Bull. Amer. Math. Soc.* 13, 87–121.

Smale, S. (1986), Newton's method estimates from data at one point, *The Merging of Disciplines: New Directions in Pure, Applied and Computational Mathematics* (R.E. Ewing, K.I. Gross, and C.F. Martin, eds.), Springer-Verlag, New York, 1986, pp. 185–196.

37
Applications of Topology to Lower Bound Estimates in Computer Science

MICHAEL D. HIRSCH*

1. Introduction

In recent years, there has been an attempt to bring techniques and tools from pure mathematics to bear on problems in computer science. In particular, the paper of Blum–Shub–Smale [2] developed in detail the concept of computation over the real numbers, showing, for the first time, that real number machines are as rich and robust as classical Turing machines. In this formal framework, one can speak of NP-completeness, problem reduction, and many of the classical problems of (discrete) computer science, on real number computers in a rigorous fashion.

Perhaps the most important consequence of this is that it puts the problems of computer science right in the middle of classical and contemporary mathematics. In the past, when the only rigorous models were discrete (bit model) computers, virtually every result was proved by combinatorial techniques—mostly with ad hoc arguments. In the new model, the input spaces are $\mathbf{R}^n$ (or $\mathbf{C}^n$) or semi-algebraic subsets of $\mathbf{R}^n$ (or $\mathbf{C}^n$) and the legal operations are rational (or sometimes analytic) functions. Such spaces are ideally suited to apply the classical theories of real and complex analysis, algebra, algebraic geometry, and topology.

Analysis and algebraic geometry have been the fields most commonly applied to the theory of computers. The recent work of Renegar [12–14] in particular shows the depth and power of classical analysis and algebra in understanding algorithms to find solutions of polynomial equations. The entire field of algebraic complexity theory (see [4] for an excellent survey) is deeply rooted in analysis and algebraic geometry.

There have been fewer examples of the use of topology in computer science, perhaps because it is not as clear how it is related to the fundamental questions. In this chapter we are especially interested in applying the techniques of topology to lower bounds. In recent years, there have been some

* Supported by a Department of Education National Need Fellowship at U.C. Berkeley.

interesting results of this type. For instance, Steele and Yao [19], Ben-Or [1], Lubiw and Racs [9], and Yao [25] used Milnor's theorem [10] about the topology of real semi-algebraic sets to find nonlinear lower bounds on the complexity of many problems—including the knapsack problem and the membership problem.

The problem of the minimal number of conditional statements in an algorithm to solve a particular problem seems particularly well-suited for the topological approach. The question can be phrased in a completely topological fashion, and there have been several successes with this route. In [17], Smale defined *topological complexity* to be this minimal number of conditional statements, then he applied topological techniques to the problem of finding the roots of a univariate complex polynomial. Vasiliev [24] has extended Smale's results and seems quite close to getting matching upper and lower bounds on the topological complexity of this problem. Levine [7] has had some success in generalizing these results to the problem of finding the zeros of a multivariate polynomial. In their paper, Blum, Shub, and Smale [2] have easy calculations of the topological complexity of the traveling salesman problem and the greatest integer function.

The results mentioned above are either very simple and use almost no topology at all, or are applications of advanced topological techniques to rather specific problems which are not so universally applicable. The goal of this research is to present simple problems in which the role played by topology is crucial, and yet can be approached using techniques that are not too esoteric. The idea is to develop a set of topological tools which can then be applied to other problems in complexity theory.

To this end we introduce a new problem, which we call the New Point Problem (NPP). The problem is as follows:

Given a list of points in some topological space X, find a new point of X not on the list.

In Section 2, we discuss several versions of this problem, each of which has a different topological complexity.

The most realistic and applicable version of the new point problem is the following. Given a list of *n distinct points* $x_1, x_2, \ldots, x_n \in X$ such that $d(x_i, x_j) > \delta$ for $i \neq j$, what is the topological complexity of finding a new point y such that $d(x_i, y) > \delta$ for all $1 \leq i \leq n$?

This version of the new point problem is sometime known as the "Burger King Problem." You own all the Burger Kings on Manhattan Island and they are all more than 5 blocks apart. You want to open another franchise and you want it to also be more than 5 blocks away from all the others. Where can you put it?

In Subsection 5.3.3 we prove:

Theorem. *The topological complexity of the above problem on the interval* $[0, 1]$, *with* $n \leq 1/2\delta$, *is* n.

This is proved by a chain of reductions starting with a less realistic version of the new point problem which is more tractable analytically, and ending with this version.

This chapter studies the topological complexity of this problem, and several of it's variants. The topological complexity depends strongly on the variant and the topology of X. All of this is discussed in Section 5.

The earlier sections set the stage for Section 5. In section 2, we define what we mean by "a problem" and define the model of computing and complexity we use. In Section 3, we prove a generalization of Schwarz's theorem which is used later to compute topological complexity and define the important concept of "problem reduction."

In Section 4, we prove an approximate extension theorem for continuous functions on arbitrary subsets which we later use to show the results of Section 5 are true in any continuous model of computing.

2. Definitions

EXAMPLE. We start with an example problem to motivate our definition of what a problem is. Consider the problem of finding all the roots of a degree d complex monic polynomial of one variable. The space of such polynomials is the input space and the space of roots (with multiplicity) of such polynomials is the output space for an algorithm which solves the problem. Each of these spaces is just $\mathbf{C}^d$ since a monic polynomial can be specified by its coefficients, and each polynomial has d roots, each a complex number. There is a natural map π, called the Vieta map, from the space of roots to the space of polynomials given by $\pi(\zeta_1, \zeta_2, \ldots, \zeta_d)$ is the polynomial $f(z) = (z - \zeta_1)(z - \zeta_2)\cdots(z - \zeta_d)$

A solution to the problem of finding the roots of a polynomial is an algorithm which, given any polynomial f, outputs a list of roots $(\zeta_1, \zeta_2, \ldots, \zeta_d)$ which satisfies $f = \pi(\zeta_1, \zeta_2, \ldots, \zeta_d)$. In other words, a solution to the problem provides a partial inverse to the map from outputs (roots) to inputs (polynomials).

This leads to the definition of a *problem* that we use here, which was first used by Smale [17].

Definition 2.1. A *problem* is a triple $(\mathscr{I}, \mathscr{O}, \pi)$, where $\mathscr{I}$ and $\mathscr{O}$ are topological spaces, and $\pi: \mathscr{O} \to \mathscr{I}$ is a continuous map. $\mathscr{O}$ is called the *solution* (or *output*) *space* and $\mathscr{I}$ the problem (or *input*) *space*.

A *subproblem* $(\mathscr{I}', \mathscr{O}', \pi')$ of $(\mathscr{I}, \mathscr{O}, \pi)$ is a problem such that $\mathscr{I}' \subset \mathscr{I}$, $\mathscr{O}' = \pi^{-1}(\mathscr{I}') \subset \mathscr{O}$, and $\pi' = \pi|\mathscr{O}'$.

When no confusion will arise, the map π will often be used to refer to the entire problem.

In the example of root finding, above, the output space was the space of all roots, which agrees well with the usual, intuitive notion of the output space of a problem. Most intuitively defined problems take inputs from an input space $\mathscr{I}$ with outputs in a space $\mathscr{S}$ of solutions. Such a situation translates into the triple $(\mathscr{I}, \mathscr{O}, \pi)$, where $\mathscr{I} = \mathscr{I}$, $\mathscr{O} = \{(x, s) \in \mathscr{I} \times \mathscr{S} | s$ is an (intuitive) solution to $x\}$ and π is the projection onto the first factor, $\pi: \mathscr{I} \times \mathscr{S} \to \mathscr{I}$ restricted to $\mathscr{O}$.

EXAMPLE. Let S_n, the symmetric group on n elements, act on $\mathbf{R}^n$ by permuting coordinates. The problem of finding a permutation which will sort n numbers has $\mathscr{I} = \mathbf{R}^n$, $\mathscr{O} = \{(x, \sigma) \in \mathbf{R}^n \times S_n |$ the coordinates of $\sigma(x)$ are increasing$\}$ and π projects onto $\mathbf{R}^n$. Alternatively, if one wishes the answer to be a sorted list, then $\mathscr{O} = \{(x, y) \in \mathbf{R}^n \times \mathbf{R}^n |$ the coordinates of y are in increasing order and there exists $\sigma \in S_n$ with $y = \sigma(x)\}$.

Sometimes we will use the intuitive description of a problem and leave it to the reader to make the translation to the rigorous definition if so desired.

An algorithm which solves the problem provides a *section* of the map π, i.e., a map $\sigma: \mathscr{I} \to \mathscr{O}$ such that the composition $\pi \circ \sigma$ is the identity map of $\mathscr{I}$. It is important to reiterate here that the problem is a map *from outputs to inputs* and not the other way around. An algorithm computes a function in the traditional direction, from inputs to outputs. To be more rigorous, we must specify our model of computation.

The model we use is essentially the frequently used *algebraic computation tree* model, except that, as in Smale [17], we charge only for branching, thus making the model equivalent to the algebraic decision tree model. It is the Blum–Shub–Smale [2] finite, nonuniform, real number machine.

An algorithm is a binary tree T with root r associated to a particular embedding $\mathscr{I} \subset \mathbf{R}^m$, $\mathscr{O} \subset \mathbf{R}^n$ of a problem $\pi: \mathscr{O} \to \mathscr{I}$. The inputs are any $(x_1, x_2, \ldots, x_m) \in \mathscr{I} \subset \mathbf{R}^m$. The outgoing edges of a node are labeled with either a computation or a comparison as follows:

(0) A node with no children is a *leaf*. It has no outgoing edges and no associated computation or comparison.

(1) A node with one child is a *computation node*. The edge exiting the node is labeled with a computation of the form $x \circ y$, where x, $y \in \mathbf{R}$ are inputs, constants, or the results of computations on edges between the current node and the root (henceforth called *earlier computations*) and $\circ$ is one of $\{+, -, *, /\}$.

(2) A node v with 2 children is a branch node. Each outgoing edge of v is labeled with a comparison of an input or previous computation to zero so that the two labels are negations of each other. For example, if one edge is labeled $x > 0$, the other must be labeled $x \leq 0$. The standard comparisons $\{=, \neq, <, \leq, >, \geq\}$ are allowed.

An algorithm T solves the problem $\pi\colon \mathcal{O} \to \mathcal{I}$ if, for each $x \in \mathcal{I}$, there is a (necessarily unique) path in T from the root to a leaf such that:

(i) no computation on the path contains a division by zero,
(ii) the comparisons on the path are all satisfied (whence uniqueness of the path), and
(iii) the last n computations, $y_1, y_2, \ldots, y_n$ define a point in $\mathcal{O} \subset \mathbf{R}^n$ and $\pi(y_1, y_2, \ldots, y_n) = x$, the original input.

Note that the restrictions on the allowed computations could easily be generalized to allow functions other than the elementary rational functions $\{+, -, *, /\}$ that we do allow. In the interest of simplicity, we will not do this, except in Section 4 where we show that allowing the machine to compute arbitrary continuous functions would not charge the results of Section 5.

On occasion, we will speak informally of the "category of computation," meaning the class of computable functions. Thus, to paraphrase the previous paragraph, we usually are dealing with the "rational category," though in Section 4 we show that our results are unchanged even in the "continuous category." In Section 3, we indicate how this can be made rigorous in the mathematical context of Category Theory.

As mentioned, associated to an algorithm (tree) is the input (problem) space $\mathcal{I}$, and the output (solution) space $\mathcal{O}$. One thinks of the algorithm as starting at the root node of the tree and being presented with the input on which it then performs the computation associated with that node's outgoing edge if it is a computation node, or branches along the edge labeled with the true inequality if that node is a branch node, and then proceeds on to the next node. The algorithm continues on down the tree in this manner, performing computations or branching at each node, until it arrives at a leaf, at which point it stops. If $\mathcal{O}$ is represented as $\mathcal{O} \subset \mathbf{R}^n$, then the last n computations are required to be the coordinates of the solution.

Thus, an algorithm T defines a map $\sigma^T\colon \mathcal{I} \to \mathcal{O}$. If $\pi \circ \sigma^T =$ identity on $\mathcal{I}$, then we say T *solves* the problem $(\mathcal{I}, \mathcal{O}, \pi)$. A map σ satisfying $\pi \circ \sigma =$ identity is called a *section* of π.

Clearly, for a problem π to have a solution, π must be surjective, i.e., $\pi^{-1}(x) \neq \emptyset$, $\forall x \in \mathcal{I}$. $\pi^{-1}(x)$ is called the *solution fiber* of x.

In this chapter we are concerned with what is called the "topological complexity" of a problem. This is quite different from the standard measure of complexity, usually just called "complexity" or sometimes "total complexity." For completeness, we define both terms.

Definition 2.2. The *complexity of an algorithm* T, $C(T)$, is the depth of T.

The *complexity of a problem* $(\mathcal{I}, \mathcal{O}, \pi)$, $C(\pi)$, is the minimum, over all algorithms which solve the problem, of the complexity of each algorithm.

Definition 2.3. The *topological complexity of an algorithm* T, $TC(T)$, is the number of branch nodes in T. Note that this is the number of leaves -1.

The *topological complexity of a problem* $(\mathscr{I}, \mathcal{O}, \pi)$, $TC(\pi)$, is the minimum, over all algorithms which solve the problem, of the topological complexity of each algorithm.

Note that for many problems, the topological complexity is trivial. For instance, finding the determinant of an $n \times n$ matrix takes no branching though this method might have unnecessarily high total complexity. The decision version of the knapsack problem can be solved with only one branch node, though this many not be the most efficient way to do it. In fact, most decision problems (i.e., problems which require only a "yes" or "no" answer) have topological complexity 1. Topological complexity is a more interesting concept for search problems.

The set of points of $\mathscr{I}$ for which the algorithm ends in a particular leaf l is the set of all points satisfying the equalities and inequalities of the branch nodes on the path to l. Such a set is called a *semi-algebraic* set. Since every point in the input space ends up at some leaf, an algorithm yields a decomposition of $\mathscr{I}$ into semi-algebraic sets V_i and rational maps σ_i^T: $V_i \to \mathscr{I}$ such that $\pi \circ \sigma_i^T(x) = x$ for $x \in V_i$. Thus, the answer to the question "What is the topological complexity of a problem?" can be bounded below by the answering the question "What is the minimal number of semi-algebraic sets V_i that $\mathscr{I}$ can be decomposed into, so that there is a rational map σ_i on each V_i, as above?"

Definition 2.4. A problem $(\mathscr{I}, \mathcal{O}, \pi)$ is *extendable* if, for all semi-algebraic $V \subset \mathscr{I}$ and all sections σ: $V \to \mathcal{O}$ of π, σ can be extended to an open neighborhood of V.

For extendable problems, the answer to the above question is closely related to the *Schwarz genus* of π, a concept due to Schwarz [16] and reinvented by Smale [17].

Definition 2.5. The *Schwarz genus* of a continuous map π: $X \to Y$, Genus(π), is the minimal number of open sets U_i needed to cover Y such that there are continuous maps σ_i: $U_i \to X$ with $\pi \circ \sigma_i$ the identity on U_i.

Proposition 2.6. *Let* $(\mathscr{I}, \mathcal{O}, \pi)$ *be an extendable problem. The its topological complexity satisfies*

$$TC(\pi) \geq \text{Schwarz genus}(\pi) - 1.$$

PROOF. This is immediate from the preceding paragraphs. The -1 occurs because there is one more leaf on a tree than there are branch nodes. □

The main problem in this paper is the new point problem (Section 5) which is extendable. The main technique used is to calculate the Schwarz genus of the problem and apply Proposition 2.6.

3. Estimating the Schwarz Genus

Not surprisingly, by reducing the problem of estimating topological complexity to a purely topological question, it becomes possible to prove theorems about it.

In this section we do two things. In the first subsection, we prove a theorem that estimates the Schwarz genus of a problem from the cohomology of the spaces involved. In the second, we define what is meant by a problem reduction and use this as a framework for using the Schwarz genus of one problem to give bounds on the Schwarz genus of another.

3.1. Schwarz's Theorem

The following is a slight generalization of a theorem originally due to Schwarz [16] and rediscovered by Smale [17] relating the Schwarz genus to cohomology. It was used by Smale to bound the topological complexity of root finding, and will be used in Subsection 5.3.2 and Section 5.4.

The standard reference for algebraic topology is Spanier [18], but for a kinder, gentler introduction, see Munkres [11].

Let $\pi\colon X \to Y$ be a continuous map and let $\{A_\alpha\}$ be a collection of subsets of Y such that

1. $\{A_\alpha\}$ is closed under finite unions, and
2. each pair (Y, A_α) or $(X, \pi^{-1}A_\alpha)$ is an excisive pair.

Under these two conditions $\bigoplus_\alpha H^*(Y, A_\alpha)$ and $\bigoplus_\alpha H^*(X, \pi^{-1}A_\alpha)$ have ring structures with cup product giving the multiplication. Let

$$K_\alpha = \ker(\pi^*\colon H^*(Y, A_\alpha) \to H^*(X, \pi^{-1}A_\alpha)).$$

Definition 3.1. Let R be a ring with product denoted by $\cup$. Then the *cup-length* of R is the maximal k such that $\gamma_1 \cup \gamma_2 \cup \cdots \cup \gamma_k \neq 0$ *for some* $\gamma_i \in R$.

Theorem 3.2 (Schwarz's Theorem). *With* $\pi\colon X \to Y$ *and* K_α *as above, let* $K = \bigoplus_\alpha K_\alpha$. *Then* K *has a ring structure and the Schwarz genus of* $\pi \geq 1 +$ *cup-length of* K.

PROOF. The proof is quite similar to that given in Smale [17].

The ring structure on K follows from assumption 2 about the collection $\{A_\alpha\}$. See Spanier [18] for a discussion of relative cup products.

Suppose Y has an open cover $\{V_i, 1 \leq i \leq k\}$ with continuous sections $\sigma_i\colon V_i \to X$ of π such that $\pi \circ \sigma_i(v) = v$ for all $v \in V$. We need to prove that

$$\gamma_1 \cup \gamma_2 \cup \cdots \cup \gamma_k = 0$$

for any $\gamma_1, \gamma_2, \ldots, \gamma_k \in K$. Since $K = \bigoplus_\alpha K_\alpha$, it is enough to prove this when $\gamma_i \in K_{\alpha_i}$ for all $1 \leq i \leq k$.

Consider the commutative diagram

$$\begin{array}{ccccc} H^*(Y, V_i \cup A_{\alpha_i}) & \xrightarrow{I_i} & H^*(Y, A_{\alpha_i}) & \xrightarrow{J_i} & H^*(V_i \cup A_{\alpha_i}, A_{\alpha_i}) \\ & & \downarrow{\scriptstyle \pi^*} & & \downarrow{\scriptstyle \phi} \\ & & H^*(X, \pi^{-1}A_{\alpha_i}) & \xrightarrow{\sigma_i^*} & H^*(V_i, V_i \cap A_{\alpha_i}) \end{array}$$

in which the top row is the long exact sequence of the triple $(Y, V_i \cup A_{\alpha_i}, A_{\alpha_i})$ and the vertical map ϕ is an excision isomorphism.

We have

$$J_i(\gamma_i) = \phi^{-1}\sigma_i^*\pi^*(\gamma_i) = \phi^{-1}\sigma_i^*(0) = 0$$

by the definition of K_{α_i}. Thus, by exactness, there is a $v_i \in H^*(Y, V_i \cup A_{\alpha_i})$ such that $I_i(v_i) = \gamma_i$.

Each V_{α_i} is open, so $(Y, V_i \cup A_\alpha)$ is an excisive pair because of the assumptions on $\{A_\alpha\}$, and we can form the cup product $v_1 \cup v_2 \cup \cdots \cup v_k \in H^*(Y, \bigcup_{i=1}^k V_i \cup A_{\alpha_i}) = H^*(Y, Y) = 0$. By naturality of the cup product $v_1 \cup v_2 \cup \cdots \cup v_k$ maps to $\gamma_1 \cup \gamma_2 \cup \cdots \cup \gamma_k$, which proves the theorem. □

3.2. *Reductions of Problems*

In mathematics and computer science, one often speaks loosely of "reducing one problem to another." Here, since we have a definition of what a problem is, it seems reasonable to ask for a definition of problem reduction as well.

Recall the following definition.

Definition 3.3. Given continuous maps $f\colon Y' \to Y$ and $\pi'\colon X \to Y$, the *pullback* by f of X over Y', written f^*X is

$$f^*X = \{(y', x) \in Y' \times X \mid f(y') = \pi(x)\}.$$

There are natural projections $\pi^*\colon f^*X \to Y'$ and $f^*\colon f^*X \to X$ such that the following diagram commutes.

$$\begin{array}{ccc} f^*X & \xrightarrow{f^*} & X \\ \downarrow{\scriptstyle \pi^*} & & \downarrow{\scriptstyle \pi} \\ Y' & \xrightarrow{f} & Y \end{array}$$

We now define problem reduction.

Definition 3.4. A problem $(\mathcal{O}', \mathcal{I}', \pi')$ *reduces* to another problem $(\mathcal{O}, \mathcal{I}, \pi)$ if there is a map $f\colon \mathcal{I}' \to \mathcal{I}$ and a map $F\colon f^*\mathcal{O} \to \mathcal{O}'$ such that the following diagram commutes:

$$\begin{array}{ccccc}
\mathcal{O}' & \xleftarrow{F} & f^*\mathcal{O} & \xrightarrow{f^*} & \mathcal{O} \\
\downarrow{\scriptstyle \pi'} & & \downarrow{\scriptstyle \pi'^*} & & \downarrow{\scriptstyle \pi} \\
\mathcal{I}' & \cong & \mathcal{I}' & \xrightarrow{f} & \mathcal{I}
\end{array}$$

Intuitively, the map f transforms the data of the first problem into the data for the second. A solution gives a map up to $\mathcal{O}$, the output space, and F retranslates this solution into a solution of the original problem. The resulting section of π' is $\phi'(y) = F(y, \phi \circ (y))$ where ϕ is the solution of the second problem. This is the way almost every problem reduction works.

A trivial example of problem reduction is a subproblem. Every subproblem of a problem reduces to the full problem with f the inclusion map and F the identity. Nontrivial examples occur in Subsections 5.3.2 and 5.3.3.

The definition can be refined to apply to any category of computation. In the study of algebraic computation trees, f and F are required to be computable by an algebraic computation tree, which would imply f and F are piecewise rational (though not necessarily continuous). More generally, one could allow more general decision trees, allowing f and F to be, for instance, piecewise analytic, or Lipschitz, or even just continuous. Since we are concerned mostly with the Schwarz genus, we only need f and F to be continuous.

Any reasonable definition of problem reduction must make the following theorem true.

Theorem 3.5. *Suppose* (f, F) *reduces* $(\mathcal{O}', \mathcal{I}', \pi')$ *to* $(\mathcal{O}, \mathcal{I}, \pi)$ *and* (g, G) *reduces* $(\mathcal{O}, \mathcal{I}, \pi)$ *to* $(\mathcal{O}'', \mathcal{I}'', \pi'')$. *Then* $(\mathcal{O}', \mathcal{I}', \pi')$ *reduces to* $(\mathcal{O}'', \mathcal{I}'', \pi'')$.

PROOF. The composition $g \circ f$ maps $\mathcal{I}'$ to $\mathcal{I}''$, so $(g \circ f)^*\mathcal{O}'' \to \mathcal{I}'$ and $(g \circ f)^*\mathcal{O}'' \to \mathcal{O}''$ are naturally defined. To prove the theorem, we need a map $(g \circ f)^*\mathcal{O}'' \to \mathcal{O}'$ so that the following diagram commutes:

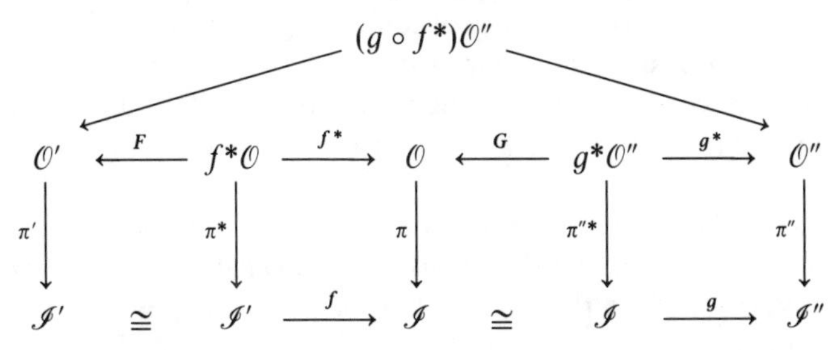

where the map $(g \circ f)^*\mathcal{O}'' \to \mathcal{I}'$ is left out of the diagram.

$(g \circ f)^*\mathcal{O}'' = f^*g^*\mathcal{O}''$ so there is a natural map to $g^*\mathcal{O}''$. G maps $g^*\mathcal{O}''$ to $\mathcal{O}$ so there is a map of $f^*g^*\mathcal{O}''$ to $\mathcal{O}$. $f^*(g^*\mathcal{O}'')$ also maps naturally to $\mathcal{I}'$, and any time a space maps to $\mathcal{O}$ and $\mathcal{I}'$, there is a natural map to $f^*\mathcal{O}$ and from there to $\mathcal{O}'$, which completes the proof. □

We can now justify the phrase "category of computation" used in Section 2. One can define *categories* (in the sense of Category Theory) in which problems are the objects and reductions are the morphisms. The previous theorem shows that reductions can be composed to get reductions, and clearly every problem reduces to itself, so reductions satisfy the morphism axiom of a category.

Thus, problems with continuous reductions can be considered the "Schwarz category" (see the next proposition for a justification of this name), problems with piecewise rational reductions, the "category of algebraic computation trees," etc.

This is only tangential to the subject of this paper, but others, such as Strassen [20–22], Hermann [6], and Lickteig [8], believe that category theory is the proper setting for unifying classical mathematics, numerical analysis, and theoretical computer science.

Reductions allow the transfer of upper and lower bounds on the Schwarz genus.

Proposition 3.6 (The Reduction Principle). *Given two problems, $(\mathcal{O}', \mathcal{I}', \pi')$ and $(\mathcal{O}, \mathcal{I}, \pi)$, suppose (f, F) reduces the first to the second. Then* $\text{Genus}(\pi') \leq \text{Genus}(\pi)$.

Proof. Suppose $\{O_i, 1 \leq i \leq k\}$ is a minimal open cover of $\mathcal{I}$ and $\phi_i\colon O_i \to \mathcal{O}$ are sections of π over O_i. Then $\{f^{-1}O_i\}$ is an open cover of $\mathcal{I}'$ since f is continuous, and $y \mapsto F(y, \sigma(f(y)))$ is a section of π' on $f^{-1}O_i$. Therefore, $\text{Genus}(\pi') \leq k$ and, by minimality of k, $\text{Genus}(\pi') \leq \text{Genus}(\pi)$. □

Versions of this theorem are true in other categories of problems as well. For instance, the same analysis will show that if f and F are piecewise rational, then any semi-algebraic decomposition of $\mathcal{I}$ with rational sections of π pulls back to a semi-algebraic decomposition of $\mathcal{I}'$ with rational sections, thus relating the topological complexity of π' to the topological complexity of π and the topological complexity of computing f. Since we will not use this fact, we refrain from stating it rigorously.

It is, however, worth stating that these reductions can also be used to relate a bound for the total complexity of one problem to a bound for another.

Proposition 3.7. *Suppose (f, F) reduces $(\mathcal{O}', \mathcal{I}', \pi')$ to $(\mathcal{O}, \mathcal{I}, \pi)$ and suppose T_f and T_F are computation trees (algorithms) which compute f and F with depth d_f and d_F, respectively. Then the complexity of π', $C(\pi') \leq d_f + d_F + C(\pi)$.*

Sketch of Proof. Let T be a computation tree solving π realizing $\text{depth}(T) = C(\pi)$. Let $\phi\colon \mathcal{I} \to \mathcal{O}$ be the section of π computed by T. We can construct a tree T' which computes $F(\cdot, \sigma \circ f(\cdot))\colon \mathcal{I}' \to \mathcal{O}'$ by attaching T to each leaf of T_f, the putting T_F at each leaf of the resulting tree. Clearly, $C(\pi') \leq \text{depth}(T') \leq d_f + d_F + C(\pi)$. □

4. Computing in the Continuous Category

One natural question is to consider a broader class of computer—one which can work with more general functions and more general sets. Often one supposes the ability to extract roots, compute analytic functions, or even on occasion to model arbitrary continuous functions. A priori, one might expect this to decrease the topological complexity of the problem, but the fact is that this is often not the case.

In most cases in which the topological complexity has been worked out, it is the topology and not the algebraic structure that is paramount. In this section, we prove an approximate extension theorem on a large class of topological spaces which says that any continuous function defined on any subset of one of these spaces can be extended to an open neighborhood of that set if you are allowed to alter the values it takes on the original set by an arbitrarily small amount. In many cases, in particular in the New Point Problem, this theorem can be used to show that the topological complexity is dependent more on the topology of the situation than on the algebra.

4.1. An Approximate Extension Theorem

Theorem 4.1. *Let X be paracompact and Y be an ANR with metric* $\mathrm{d}(\cdot, \cdot)$. *Let $A \subset X$ be arbitrary. Given continuous maps $\phi_A\colon A \to Y$ and $\varepsilon\colon Y \to (0, \infty)$, there exists an open set $U \supset A$ and a continuous function $\phi_U\colon U \to Y$ such that if $x \in A$, we have* $\mathrm{d}(\phi_U(x), \phi_A(x)) < \varepsilon(\phi_A(x))$.

Proof. First note that since any two metrics on Y are equivalent, we may assume $Y \subset \mathbf{R}^n$ with the subspace metric. Let r be a retraction to Y.

Next, it is easy to see that there is a continuous function $\tilde{\varepsilon} < \varepsilon$ such that if $\mathrm{d}(x, y) < \tilde{\varepsilon}(x)$, then $\mathrm{d}(x, y) < \varepsilon(y)$, so without loss of generality, we can prove instead that $\mathrm{d}(\phi_U(x), \phi_A(x)) < \varepsilon(\phi_U(x))$.

Fix a function $\delta\colon Y \to (0, \infty)$ small enough that

(1) if $y \in N_\delta(Y) = \{y \in \mathbf{R}^n | \exists y' \in Y, \mathrm{d}(y', y) < \delta(y')\}$, then $r(y)$ is defined and $\mathrm{d}(r(y), y) < \varepsilon(y)/2$; and
(2) if for some $y \in Y$ and $\gamma \in \mathbf{R}^n$ we have $B_{\delta(\gamma)}(\gamma) \cap B_{\varepsilon(y)/2}(y) \neq \varnothing$, then $B_{\delta(\gamma)}(\gamma) \subset B_{\varepsilon(y)}(y)$, where $B_r(x)$ means the ball of radius r centered at at x.

Let $\{W_\gamma\}$ be an open cover of $N_\delta(Y)$ by δ-balls centered at $\gamma \in Y$. $Y \subset \mathbf{R}^n$, so we can refine $\{W_\gamma\}$ to be locally finite. Let $\{V_\beta\}$ be an open cover of M by small, open sets such that for each β, $V_\beta \cap A \subset \phi_A^{-1}(W_\gamma)$ for some γ. Now, subdivide $\{V_\beta\}$ to get $\{U_\alpha\}$ such that if $U_{\alpha_1} \cap \cdots \cap U_{\alpha_k} \neq \varnothing$, then $(U_{\alpha_1} \cup \cdots \cup U_{\alpha_k}) \cap A \subset \phi_a^{-1}(W_\gamma)$ for some γ. Finally, refine $\{U_\alpha\}$ to be a locally finite cover.

Let $U = \bigcup_{\alpha'} U_{\alpha'}$ where α' ranges over $\{\alpha | U_\alpha \cap A \neq \varnothing\}$. Let $\lambda_{\alpha'}$ be a partition of unity to U subordinate to $U_{\alpha'}$. Pick $x_{\alpha'} \in U_{\alpha'} \cap A$ and define $\phi\colon U \to \mathbf{R}^N$ by $\phi(x) = \sum_{\alpha'} \lambda_{\alpha'}(x_{\alpha'})\phi_A(x_{\alpha'})$.

Claim 1. ϕ is continuous.

PROOF OF CLAIM 1. This is true because ϕ is a locally finite sum of continuous maps. □

Claim 2. Given $x \in X$, there is a $\gamma \in Y$ such that $d(\phi(x), y) < \delta(\gamma)$ and, if $x \in A$, then $d(\gamma, \phi_A(x)) < \delta(\gamma)$ as well.

PROOF OF CLAIM 2. By the construction of the U_α, given x, the corresponding $x_{\alpha'}$ all have images $\phi_A(x_{\alpha'})$ inside the same W_γ, so they all are within $\delta(\gamma)$ of γ. $\phi(x)$ is a weighted average of these points, and hence within $\delta(\gamma)$ of γ.

If $x \in A$, then $x \in U_{\alpha'}$ for some α' and $\phi(U_{\alpha'}) \subset W_\gamma = B_{\delta(\gamma)}(\gamma)$, so $d(\phi(x), \gamma) < \delta(\gamma)$. □

From this, we see $\phi(U) \subset N_\delta(Y)$ by condition (1) on δ, and we may define $\phi_U: U \to Y$ by $\phi_U = r \circ \phi$.

For $x \in A$, we have $\phi(x) \in N_\delta(Y)$ by Claim 2, so $d(\phi(x), r\phi(x)) < \varepsilon(r\phi(x))/2$ by condition (1) on δ. But both $\phi(x)$ and $\phi_A(x) \in B_{\delta(\gamma)}(\gamma)$ so, by condition (2) on δ, $B_{\delta(\gamma)}(\gamma) \subset B_{\varepsilon(r\phi(x))}(r\phi(x))$ and $\phi_A(x) \in B_{\delta(\gamma)}(\gamma)$, so

$$d(\phi_U(x), \phi_A(x)) = d(\phi(r\phi(x)), \phi_A(x)) < \varepsilon(r\phi(x)) = \varepsilon(\pi_U(x)).$$ □

4.2. Solving Problems with Continuous Decision Trees

Theorem 4.1 can be used in many cases to show that even allowing a general decomposition will not affect the number of sets required to cover a space with continuous sections on each set. In particular, the following theorem will show this is true in the case of the new point problem in Subsection 5.3.

Corollary 4.2. *Let Y be paracompact, M be a manifold with or without boundary, $X = Y \times M \backslash \Delta$ where Δ is some closed subset of $Y \times M$, and let π_1: $X \to Y$ be projection onto the first factor and π_2: $X \to M$ be projection onto the second factor. Suppose we have an arbitrary decomposition of Y, $Y = \bigcup_{i=1}^k A_i$ and continuous sections ϕ_i: $A_i \to X$ of π. Then there is an open covering U_i of Y by k open sets and a section of π_1 on each U_i.*

PROOF. On any set $U \subset Y$, {sections of π_1} is equivalent to $\{\text{maps } \phi: U_i \to M \mid (y, \phi(y)) \notin \Delta\}$. Pick a metric on M and use the induced product metric on X and Y. Define the function $\varepsilon: X \to (0, \infty)$ by $\varepsilon(y) = d(y, \Delta)$, the distance from y to Δ. Then $\varepsilon(y) \leq$ distance from y to Δ within the fiber $\pi^{-1}(\pi(y))$.

Now apply Theorem 4.1 to the sets A_i and the sections ϕ_i of the map $\pi: X \to Y$. Let $\tilde{U}_i$ and σ_i be the resulting open sets and maps. Note that the σ_i are sections, but that

$$(\text{Identity}, \pi_2 \circ \sigma_i): \tilde{U}_i \to \tilde{U}_i \times M$$

is a section of $Y \times M \to Y$. Furthermore, since $\varepsilon(y) \leq$ fiberwise distance from

Δ for all $y \in \phi_i(A_i)$, the section corresponding to $\pi_n \circ \sigma_i$ misses Δ on A_i, and, by continuity, misses on an open neighborhood, U_i, of A_i. □

Corollary 4.3. *Lower bounds for the Schwarz genus are lower bounds for the topological complexity of problems of the type in Corollary 4.2 under any model of computing in which the computable functions are continuous.*

5. The New Point Problem

In this section, we study the New Point Problem (NPP) in detail, getting nearly tight upper and lower bound in a variety of cases. We start with definitions and examples, then prove matching upper and lower bounds on a closed ball using two different techniques. The second method uses Schwarz's theorem, which is also used to get nearly matching upper and lower bounds on closed manifolds.

5.1. *Definitions*

First we recall the intuitive definition of the New Point Problem. Let X be a topological space modeled on a computer and fix a positive integer n. Given a list of n points in X, the problem is to try to find a new list.

Note. The word "list" means the same thing as "ordered list" or "n-tuple." In every model of computation, a computer can only handle items sequentially. It makes no sense to talk about sets or unordered lists. A second point is that a list can be redundant—can have some point(s) entered more than once.

We now define the New Point Problem using the definition of a problem given in Section 2. Let X be a topological space, presumably one which can be modeled on a computer. Fix a positive integer n. Then the problem (or input) space for NPP is $\mathscr{I} = X \times X \times \cdots \times X$ (n times). This is the space of lists of points in X of length n. The solution (or output) space is $\mathscr{O} = \mathscr{I} \times X \setminus \Delta$, where $\Delta = \{(x_1, x_2, \ldots, x_{n+1}) \in \mathscr{I} \times X | x_i = x_{n+1} \text{ for some } 1 \le i \le n\}$ and $\setminus$ denotes set minus.

There is the obvious projection map $\pi\colon \mathscr{O} \to \mathscr{I}$ onto the first factor. The problem facing any algorithm to solve NPP is to invert the map π.

Let $\pi_2\colon \mathscr{O} \to X$ be projection onto the second factor. Then Δ is the set of points y for which some component of $\pi(y)$ equals $\pi_2(y)$. Note that Δ has an obvious decomposition $\Delta = \bigcup_{i=1}^n \Delta_i$ where $\Delta_i = \{y \in Y | \pi_2(y) = i\text{th coordinate of } \pi(y)\}$. This will come in useful later.

5.2. *Examples*

We start with an algorithm which has at most n branch points. Here we need to assume the existence of some function f on pairs of points in X, such that

$f(x, y) = 0$ if and only if $x = y$. For instance, any space with a computable distance function will do. Fix $n + 1$ different points $x_1, x_2, \ldots, x_{n+1}$ in the space X. The algorithm, given a list $y_1, y_2, \ldots, y_n$, first tests $\prod_{i=1}^{n} f(x_1, y_i) \neq 0$. If true, it halts and returns x_1 as the answer. Otherwise, it continues testing each of the x_i until it has tested all but x_{n+1}. Since each of the first n x_i are contained in the y_i, x_{n+1} must be a new point. Thus, it need not be tested and the algorithm can just return x_{n+1}.

This simple algorithm proves that the topological complexity of NPP in a metric space is $\leq n$. Notice that this algorithm does not use the input in its choice of test points. It seems reasonable to suspect that one can do better by a cleverer choice of test points. By and large, this turns out not to be the case.

Let us consider several examples.

1. Suppose $X = [0, 1]$. It follows immediately from the intermediate value theorem that every continuous map of the interval to itself has a fixed point. This is just the one-dimensional Brouwer fixed-point theorem. Therefore, the topological complexity is bounded below by 1. On the other hand, from the comments above, it is clearly no higher than 1, so it must equal 1.
2. Suppose instead that X is the unit circle, S^1, and let $n = 1$. NPP on this space has topological complexity 0. That is, there is an rational map of the circle to the circle with no fixed points: $x \to -x$.
3. Suppose we are given n points $x_1, \ldots, x_n \in S^1$. Let $y_i = x_1$ rotated by $i/(n + 1)$ of a complete rotation for $1 \leq i \leq n$. Then $y_i \neq x_1$ for $1 \leq i \leq n$, so we only need to do $n - 1$ tests of the form above. This turns out to be optimal (Subsection 5.4).
4. Let $X = S^1$, and suppose we have the additional information that the n input points $x_1, \ldots, x_n$ are distinct. In this case, there is a 0-branching algorithm. Define $\varepsilon = \prod_{i=2}^{n} (1/\pi) \mathrm{d}(x_1, x_i)$. Then $\varepsilon < \mathrm{d}(x_1, x_i)$ for all $i = 2, \ldots, n$. Notice this can all be computed without need for branching. The map which returns the point ε to the clockwise of x_1 provides a new point. This is a purely computational solution, so it has topological complexity 0. Similarly, on $[0, 1]$ $x_1 + (x_1 - x_2)\varepsilon$, with ε as above, is a new point for the case $n \geq 2$. The case $n = 1$ still has topological complexity 1 as in the first example.
5. Suppose we are in the same situation as 4, but with the additional information that all the points differ by more than some fixed amount, δ, and with the additional requirement that the new point also differ by that amount. This is probably the version of NPP most likely to occur in an application, but intuitively it seems no easier that the original NPP because the extra information and requirement tend to cancel each other out. Suppose $\mathrm{d}(x_1, x_2) < 2\delta$. Then the δ neighborhoods of the two points overlap, you cannot fit a new point in between that pair of points, and algorithms like that in Example 4 will not work. This problem seems harder to analyze than NPP, but there appears to be no way around this difficulty. In Sub-

section 5.3.3, we show that on the interval lower bounds for NPP can be used to get lower bounds for this problem.

6. The topological complexity for NPP is trivial on the line. Let $X = \mathbf{R}$. Then given points $x_1, x_2, \ldots, x_n$ in $\mathbf{R}$, let $y = 1 + \sum_{i=1}^{n} (x_i^2 + 1)$. Then $y > x_i$ for each i, and y is a new point. Since an open interval can be mapped homeomorphically to $\mathbf{R}$ by a rational map, this shows that NPP is topologically trivial for an open interval, as well.

These examples should give some feel for the way the topology for X can affect the topological complexity of NPP. The closed unit interval behaves quite differently from the open interval, and both are different than the circle.

5.3. *The New Point Problem on a Closed Ball*

Let B be a closed ball of some finite dimension d and consider NPP on B. Our input space $\mathcal{I} = B^n$ and the output space is $\mathcal{O} = \mathcal{I} \times B \setminus \Delta$ with projection map $\pi\colon \mathcal{O} \to \mathcal{I}$. An algorithm to solve the problem is a computation tree which computes a map $\sigma\colon \mathcal{I} \to \mathcal{O}$ such that $\pi \circ \sigma(x) = x$ for $x \in \mathcal{I}$. For the first proof of the following theorem, however, it is better to ignore σ and consider, instead, $\phi = \pi_2 \circ \sigma$ where $\pi_2\colon \mathcal{O} \to B$ is the projection onto the last coordinate. An algorithm, then, computes a map $\phi\colon \mathcal{I} \to [0, 1]$ where ϕ has the property that $\phi(x_1, \ldots, x_n) \neq x_i$ for $1 \leq i \leq n$, and restricted to each leaf space V_i, $\phi_i = \phi | V_i$ is a rational map to B.

Theorem 5.1. *The New Point Problem on the ball B has Schwarz genus $\geq n$.*

We present two proofs of this theorem. The first is an application of the Brouwer fixed-point theorem and does not really need the rigorous definition of problem from Section 2. The second uses the rigorous definition of problem and Schwarz's theorem from Section 3.

5.3.1. NPP on the Ball: Fixed-Point Proof

Fixed Point Proof of Theorem 5.1. The proof uses some very useful theorems from point set topology. A good reference is Munkres [15].

It suffices to prove the following: Suppose we are given an open cover $\{O_i\colon 1 \leq i \leq n\}$ of B^n $(= \mathcal{I})$ with continuous maps $\phi_i\colon O_i \to B$ (some of the O_i may be empty). Then there exist i, j, and $x = (x_1, \ldots, x_n)$ such that $\phi_i(x) = x_j$. In fact, we will prove the stronger statement that there exist i and x such that $\phi_i(x) = x_i$.

First we choose new open sets. By the shrinking lemma for open covers, we can find smaller open sets $U_i \subset \overline{U}_i \subset O_i$ which still cover B^n. Now use the Tietze extension theorem [23] to extend each ϕ_i from $\overline{U}_i$ to all of B^n. We will use these extended ϕ_i to show there is some i and some point $x = (x_1, x_2, \ldots, x_n) \in \overline{U}_i$ such that $\phi_i(x) = x_i$.

Consider the map $\phi = (\phi_1, \phi_2, \ldots, \phi_n)\colon B^n \to B^n$. By the Brouwer fixed-point theorem [3], ϕ has a fixed point x such that $\phi(x) = (\phi_1(x), \ldots, \phi_n(x)) = x = (x_1, \ldots, x_n)$, i.e., $\phi_i(x) = x_i$, $\forall 1 \le i \le n$. But then $x \in \overline{U}_i$ for some i, and hence $\phi_i(x) = x_i$ on $\overline{U}_i$.

This proves that the Schwarz genus is bounded below by $n + 1$. □

5.3.2. NPP on the Ball: Cohomology Proof

The cohomology proof requires some notation.

Let B be a closed ball of dimension d. Let $\partial_i B^n = B \times B \times \cdots \times B \times \partial B \times B \times \cdots \times B$ where ∂B occurs in the ith position. Let $\bar{\partial}_i B^n = \partial B^n - \partial_i B^n = \bigcup_{j \neq i} \partial_j B^n$.

Clearly, the pairs $(B^n, \partial_i B^n)$ are excisive. Let $\pi\colon B^{n+1} \backslash \Delta \to B^n$ be projection onto the first n coordinates. Then the pairs $(B^{n+1} \backslash \Delta, \pi^{-1}(\partial_i B^n))$ are also clearly excisive.

Lemma 5.2. $\pi^*\colon H^d(B^n, \partial_i B^n) \to H^d(B^{n+1} \backslash \Delta, \pi^{-1}(\partial_i B^n))$ *is the zero homomorphism.*

PROOF. $\pi^{-1}(\partial_i B^n)$ is a deformation retract of $B^{n+1} \backslash \Delta$ given by a homotopy which moves the ith coordinate to the boundary of B by moving the ith coordinate radially from the point corresponding to the $(n + 1)$st coordinate.

Therefore, $H^d(B^{n+1} \backslash \Delta^i, \pi^{-1}(\partial_i B^n)) = 0$ and the lemma follows from the commutative diagram

$$\begin{array}{ccc} H^d(B^n, \partial_i B^n) & \xrightarrow{\;0\;} & H^d(B^{n+1} \backslash \Delta^i, \pi^{-1}(\partial_i B^n)) = 0 \\ & \searrow^{\pi^*} & \downarrow{i^*} \\ & & H^d(B^{n+1} \backslash \Delta, \pi^{-1}(\partial_i B^n)) \end{array}$$

□

Lemma 5.3. $H^d(B^n, \partial_i B^n) = \mathbf{Z}$. *Let* α_i *generate* $H^d(B^n, \partial_i B^n)$. *Then* $\alpha_1 \cup \alpha_2 \cup \cdots \cup \alpha_n \neq 0$.

PROOF. By Poincaré duality

$$H^d(B^n, \partial_i B^n) \cong H_{nd-d}(B^n, \bar{\partial}_i B^n).$$

$(B^{n-1}, \partial B^{n-1})$ lies in $(B^n, \bar{\partial}_i B^n)$ by the inclusion

$$(x_1, x_2, \ldots, x_{n-1}) \mapsto (x_1, x_2, \ldots, x_{i-1}, 0, x_i, \ldots, x_{n-1})$$

and is obviously a deformation retract of $(B^n, \partial_i B^n)$. Thus,

$$H^d(B^n, \partial_i B^n) \cong H_{d(n-1)}(B^{n-1}, \partial B^{n-1}) \cong H_{d(n-1)}(S^{d(n-1)}, \mathrm{pt}) \cong \mathbf{Z}.$$

The n different inclusions of B^{n-1} all meet transversely at the point $(0, 0, \ldots, 0)$, so by the equivalence between cup product and intersection product, $\alpha_1 \cup \alpha_2 \cup \cdots \cup \alpha_n \neq 0$ □

Cohomology Proof of Theorem 5.1. By the above lemmas, the kernel in cohomology of the projection $\pi: B^{n+1} \backslash \Delta \to B^n$ has cup-length $= n$. Thus, by Theorem 3.2, π has Schwarz genus $\geq n + 1$. □

Corollary 5.4. *On B^n, TC(NPP) $= n$ in any computation tree model with continuous computable functions.*

Proof. Since the new point problem is of the form $X \times M \backslash \Delta \to X$ with X and M manifolds, by Corollary 4.3 the Schwarz genus is a lower bound for the topological complexity. Hence, TC(NPP) $\geq n$.

On the other hand, in Subsection 5.2, we gave an explicit example showing TC(NPP) $\leq n$. □

Corollary 5.5. *Call the problem of finding a new point when given n points in an interval in ascending order the sorted new point problem* (*sorted* NPP). *This problem has topological complexity n.*

Proof. The genus has an upper bound for n since this is a subproblem of the general new point problem on an interval. We show it has n as a lower bound as well.

Since sorted NPP is also of the form $X \times M \backslash \Delta \to X$, it is enough to show Genus(sorted NPP) $\geq n$. To do this, we show that NPP reduces to sorted NPP.

Let $\mathscr{I} = [0, 1]^n$ be the input space for NPP and

$$\tilde{\mathscr{I}} = \{(x_1, x_2, \ldots, x_n) \in [0, 1]^n | x_1 \leq x_2 \leq \cdots \leq x_n\}$$

be the input space for sorted NPP.

Let the symmetric group on n elements S_n act on $[0, 1]^n$ by permuting coordinates. Then $\tilde{\mathscr{I}}$ is the quotient by this action.

The solution fiber of an n-tuple of points is clearly the same as the solution fiber of a permutation of the points, so the output space of NPP is the pullback of the output space of sorted NPP. Therefore, by the Reduction Principle of Subsection 3.2,

$$\text{Genus(sorted NPP)} \geq \text{Genus(NPP)} \geq n. \qquad \square$$

5.3.3. The Nonredundant δ-NPP

The most intuitively reasonable variant of the new point problem is the one given as Example 5 of Subsection 5.2. It is called the *nonredundant δ-new point problem.* The n input points $x_1, x_2, \ldots, x_n \in X$ are guaranteed to be distinct and differ by at least δ, i.e., $\mathrm{d}(x_i - x_j) > \delta$, $\forall i, j$. The output point y must also satisfy $\mathrm{d}(y - x_i) > \delta$.

An obvious condition must be satisfied to make this problem interesting: $n\,\mathrm{Vol}(\delta\text{-ball}) < \mathrm{Vol}(X)$. Without this, there is no guarantee that there is a solution.

In many case, there is still an obvious algorithm with only n branch nodes. Fix $n+1$ points, $y_1, y_2, \ldots, y_{n+1}$, with disjoint δ-neighborhoods (this does not exit for large n and small δ and X). Check each y_i in turn, halting when $y_i \neq x_i$, $\forall j$, until y_{n+1} is reached (as before, each y_i can be checked with one branch by taking the product of all the distances and comparing to 0). y_{n+1} is the new point.

We now present a matching lower bound for the case $X = [0, 1]$.

Theorem 5.6. *The topological complexity of the nonredundant δ-NPP on the interval* $[0, 1]$ *with n inputs and $n \leq 1/2\delta$ is n.*

Proof. The input space for this problem is

$$\mathscr{I}' = \{x_1, x_2, \ldots, x_n \in [0, 1] \mid |x_i - x_j| > \delta\}$$

with output space

$$\mathscr{O}' = \{(x_1, x_2, \ldots, x_{n+1}) \in \mathscr{I}' \times [0, 1] \mid |x_i - x_{n+1}| > \delta, \forall 1 \leq i \leq n\}.$$

This is clearly of the form $X \times M \backslash \Delta$, so by Corollary 4.3 we need only prove a lower bound on the Schwarz genus. We will show sorted NPP on $[0, 1 - 2n\delta]$ reduces to this problem.

Let $\tilde{\mathscr{I}}$ be the input space for sorted NPP on $[0, 1 - 2n\delta]$. Define $f: \tilde{\mathscr{I}} \to \mathscr{I}'$ by $f(x_1, x_2, \ldots, x_n) = (x_1 + \delta, x_2 + 3\delta, \ldots, x_n + (2n - 1)\delta)$. This is clearly an embedding of $\tilde{\mathscr{I}}$ in $\mathscr{I}'$.

The solution fiber of a point in the pullback is the collection of intervals $[0, x_1 - \delta) \cup (x_1 + 2\delta, x_2 - 2\delta) \cup \cdots \cup (x_n + 2n\delta]$ where some of these internals may be empty. This homeomorphic to $[0, x_1) \cup (x_1, x_2) \cup \cdots \cup (x_n, 1 - 2n\delta]$ by a map F which is easily shown to be continuous as a function of $(x_1, x_2, \ldots, x_n)$. (f, F) are then a reduction of sorted NPP on an interval to the nonredundant NPP on an interval, so Genus(nonredundant NPP) $\geq$ Genus(sorted NPP) $\geq n$. □

5.4. NPP on a Closed Manifold

In Subsection 5.2, we constructed an algorithm for NPP on the circle, S^1, which had topological complexity of only $n - 1$. In the next subsection of this section, we indicate how to show that this is optimal. To do this, we consider the new point problem on a closed manifold. It turns out that S^1 manifests all the behavior of the most general manifold, so it is just as easy (and just as hard) to discuss the general case. For a good introduction to manifolds, see Guillemin and Pollack [5]. In the final subsection, we give many examples of manifolds on which NPP has topological complexity of exactly $n - 1$.

5.4.1. Lower Bounds for NPP on Closed Manifolds

Let M be a smooth, connected, closed, m-dimensional manifold represented in some way on a machine. We are not concerned with the details of the representation. Some manifolds can be easily represented on a computer and some cannot. We assume that the machine has *some* way of representing points and some way of telling whether two points are the same and each such decision is counted as one branch in the algorithm.

Such representations are not unusual; most naturally occurring manifolds can be embedded in $\mathbf{R}^n$ as the zero set of some polynomial $p: \mathbf{R}^n \to \mathbf{R}$. For example, the n-sphere S^n can be presented as $n+1$-tuples of real numbers such that $x_1^2 + \cdots + x_{n+1}^2 = 1$. Two points $(x_1, \ldots, x_{n+1})$ and $(y_1, \ldots, y_{n+1})$ are the same if and only if

$$\sum_{i=1}^{n+1} (x_i - y_i)^2 = 0.$$

This equality clearly takes only one comparison to check. Other manifolds or other representations could take more than one simple comparison to check, but we count it as only one branch node.

The problem space $\mathscr{I}$ is $M \times M \times \cdots \times M$ (n times). The solution space $\mathscr{O} = \mathscr{I} \times M \setminus \Delta$, where $\Delta = \{(y, x) \in \mathscr{I} \times M \mid y_i = x \text{ for some } 1 \le i \le n\}$. Let $\pi: \mathscr{O} \to \mathscr{I}$ be projection onto the first factor. NPP can be stated: Given a point $y \in \mathscr{I}$, find a point $x \in \mathscr{O}$ such that $\pi(x) = y$. Let $\pi_2: \mathscr{I} \times M \to M$ be projection onto the second factor. Then Δ is the critical set—those points y for which some coordinate of $\pi(x)$ equals $\pi_2(x)$. As mentioned in Subsection 5.1, Δ has a decomposition, $\Delta = \bigcup_{i=1}^n \Delta_i$, where $\Delta_i = \{(y, x) \in \mathscr{I} \times M \mid x = i\text{th coordinate of } y\}$.

Theorem 5.7. *Let M be defined as above. Then the topological complexity for the New Point Problem on M is at least $n - 1$.*

Since the topological complexity of NPP has an upper bound of n, this theorem provides nearly tight bounds.

Proof of Theorem 5.7. As usual, by Corollary 4.3 it suffices to show Genus(NPP on M) $\ge n$, i.e., we need to compute the Schwarz genus of the projection map $\pi: \mathscr{O} \to \mathscr{I}$. Throughout the computation, all homology and cohomology classes and groups will be with $\mathbf{Z}/2\mathbf{Z}$ coefficients.

First, we need some notational conventions. Let $[m]$ (resp. μ) be the generating class of $H_m(M)$ [resp. $H^m(M)$] with $\mathbf{Z}/2\mathbf{Z}$ coefficients. Then $[m]$ and μ are dual in the sense that $\langle \mu, [m] \rangle = 1$. In fact, the Universal Coefficient Theorem shows that this property defines one in terms of the other.

By the Kunneth formula

$$H^m(\mathscr{I}) = \bigoplus_{i_1 + \cdots + i_j = m} H^{i_1}(M) \otimes \cdots \otimes H^{i_j}(M)$$

and

$$H_m(\mathscr{I}) = \bigoplus_{i_1 + \cdots + i_j = m} H_{i_1}(M) \otimes \cdots \otimes H_{i_j}(M).$$

So, in particular, $H_m(\mathscr{I})$ contains the elements $[m] \otimes 1 \otimes \cdots \otimes 1$, $1 \otimes [m] \otimes 1 \otimes \cdots \otimes 1, \ldots, 1 \otimes \cdots \otimes 1 \otimes [m]$. Call these elements $m_1, m_1, \ldots, m_n$. Similarly, $H^m(\mathscr{I})$ has the elements $\mu \otimes 1 \otimes \cdots \otimes 1$, $1 \otimes \mu \otimes 1 \otimes \cdots \otimes 1, \ldots, 1 \otimes \cdots \otimes 1 \otimes \mu$. Call these elements $\mu_1, \mu_1, \ldots, \mu_n$.

As before, we have $\langle \mu_i, m_i \rangle = 1$, $\forall i$. For each i, choose a set of generators for $H_i(M)$ and use these generators and the Kunneth formula to get a set of generators for $H_*(\mathscr{I})$. Then μ_i is the unique element of $H^*(\mathscr{I})$ such that $\langle \mu_i, m_i \rangle = 1$ and μ_i evaluated on any other generator is 0.

Recall that $\mathcal{O} = \mathscr{I} \times M \backslash \Delta$ and $\Delta = \bigcup_{i=1}^n \Delta_i$. Letting $\mathcal{O}_i = \mathscr{I} \times M \backslash \Delta_i \cup \Delta_n$, we get the following commuting diagram for all i:

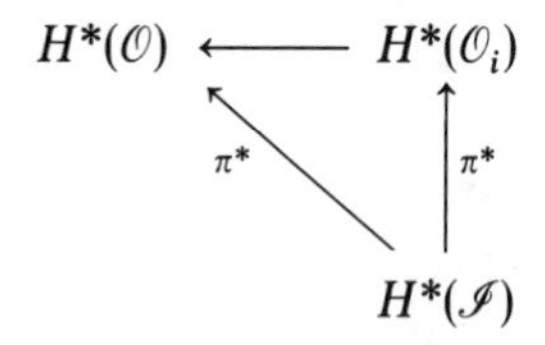

Thus, the kernel K of $H^*(\mathscr{I}) \to H^*(\mathcal{O})$ contains the kernel K_i of $H^*(\mathscr{I}) \to H^*(\mathcal{O}_i)$.

Lemma 5.8. $\mu_i + \mu_n \in K_i$.

PROOF OF LEMMA 5.8. Suppose $\pi_* \sigma \in \operatorname{Im} \pi_*$ and $\pi_* \sigma = \sum_{j=1}^n a_j m_j +$ terms involving other generators. Then $\langle \mu_i + \mu_n, \pi_* \sigma \rangle = a_i + a_n$. If $a_i = a_n$, we get

$$\langle \pi^*(\mu_i + \mu_n), \sigma \rangle = \langle \mu_i + \mu_n, \pi_* \sigma \rangle = a_i + a_n = 2a_i = 0 \ (\text{mod } 2).$$

Hence, it suffices to show $a_i = a_n$.

We have the following commuting diagram.

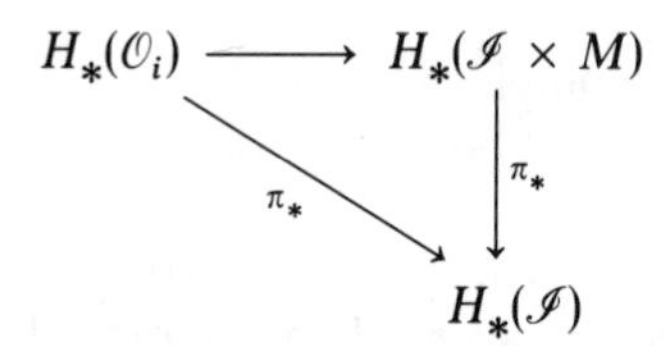

Let $\tilde{m}_1, \ldots, \tilde{m}_{n+1}$ be the distinguished generators for $H_m(M^{n+1})$ analogous to $m_1, \ldots, m_n \in H_m(\mathscr{I})$. Given $\sigma \in H_m(M^{n+1} \backslash \Delta_1 \cup \Delta_n)$, we have $i_* \sigma = \sum_{j=1}^{n+1} a_j \tilde{m}_j +$ other terms, and $\pi_* i_*(\sigma) = \sum_{j=1}^n a_j m_j +$ other terms.

Suppose $\tau \in H_*(M^{n+1})$ is $i_* \sigma$. Then the class τ has a representative which misses Δ_i and, hence, must have intersection number 0 with Δ_i (written $\tau \cdot \Delta_i = 0$). $m_j \cdot \Delta_i$ is 0 unless j is i or $n+1$ in which case $m_j \cdot \Delta_i$ is 1, so $a_i = a_{n+1}$. Similarly, $\tau \cdot \Delta_n = 0$, so $a_n = a_{n+1} = a_1$. □

It is now clear how to finish the proof of the main theorem. The lemma shows that $\mu_i + \mu_n \in K$, $\forall i$, and so

$$\begin{aligned}
&(\mu_1 + \mu_n) \cup (\mu_2 + \mu_n) \cup \cdots \cup (\mu_{n-1} + \mu_n) \\
&\quad = (\mu \otimes \mu \otimes \cdots \otimes \mu \otimes 1) + (\mu \otimes \cdots \otimes \mu \otimes 1 \otimes \mu) + \cdots + (1 \otimes \mu \otimes \cdots \otimes \mu) \\
&\quad \neq 0,
\end{aligned}$$

so the cup-length of K is at least $n - 1$. Thus, by Schwarz's theorem, the Schwarz genus is at least n, and, hence, the topological complexity is at least $n - 1$. □

Let K' be the space generated by the terms indicated by the lemma. It is of some interest to note that K is contained in the subspace of $H^*(\mathscr{I})$ generated by $\mu_1, \mu_2, \ldots, \mu_n$. The cup-length of this larger space is n since $\mu_1 \cup \mu_2 \cup \cdots \cup \mu_n \neq 0$. In fact, this space is just a truncated polynomial algebra on these generators with the relations $\mu_i \cup \mu_i = 0$, so the cup-length is just the number of generators. A subspace like K' will have cup-length equal to the number of its generators as well. Thus, K' has cup-length $= n - 1$. This does not show that the cup-length of K is $n - 1$, but it is indicative. Furthermore, we will show in the next section that there are manifolds for which the complexity of the new point problem is actually $n - 1$, thus proving that the cup-length of K for such a space is exactly $n - 1$.

5.4.2. Manifolds with Topological Complexity Exactly $n - 1$

In this subsection, we describe an algorithm for use on certain manifolds that has topological complexity of only $n - 1$, thus calculating the exact topological complexity of such manifolds. Though the manifolds are of a very particular type, they are by no means rare. The circle S^1 is the most basic example, but a list of others includes, in roughly increasing order of generality:

- Any odd dimensional sphere.
- Any Lie group.
- Any circle bundle over a manifold. This includes many common spaces, including the standard torus and the Klein bottle.
- Any manifold of Euler characteristic 0. This includes all odd-dimensional manifolds.
- Any nonclosed manifold.
- Any product of the above spaces.

Actually, what we shall do here is show the Schwarz genus of the corresponding problem and solution spaces of such manifolds is $n - 1$. The actual topological complexity of the problem depends on how well a machine can model the spaces. In the particular case of a circle, the computer can actually do it with topological complexity of $n - 1$.

Any particular instance, M, of these examples has the property that there exists a continuous map $\phi = (\phi_1, \phi_2, \ldots, \phi_n)$: $M \to M \times M \times \cdots \times M$ (n times) such that $\phi_i(x) \neq \phi_j(x)$, $\forall i \neq j$, and $\phi_i(x) \neq x$, $\forall i$. One should think of this as a function that, given an initial point, gives a sequence of points in M all different and all differing from the initial point.

Given the existence of such a ϕ we can now show that the Schwarz genus is $n - 1$. We need to construct an open cover $\{U_i\}$, $I = 1 \cdots n$ of $\mathscr{I} = M \times \cdots \times M$ (n times) and functions ϕ_i: $U_i \to M$ such that $\phi_i(x_1, \ldots, x_n) \neq x_j$,

$\forall j$. Let $U_i = \{(x_1, \ldots, x_n) | \phi_i(x_1) \neq x_j, \forall j\}$. Then the following two claims finish to computation of the Schwarz genus for this manifold.

Claim 1. *The U_i form an open cover of $\mathcal{I}$.*

PROOF. Openness is obvious. They cover because there are n different $\phi_i(x_1)$, so any given $(x_2, \ldots, x_n)$ can contain only $n - 1$ of them, leaving out, say, $\phi_k(x_1)$, and, for each i, $\phi_i(x_1) \neq x_1$ by construction. So the point $(x_1, \ldots, x_n) \in U_k$. □

Claim 2. *If $(x_1, \ldots, x_n) \in U_i$, then $\phi_i(x_1) \neq x_j$, $\forall j$.*

PROOF. This is obvious from the construction of the ϕ_i and U_i. □

This shows the Schwarz genus is $n - 1$, but what about the topological complexity? We would like to first check if $\phi_1(x_1)$ is a new point, and if not, check $\phi_2(x_1)$, etc. This works, but only if all the needed terms are defined. We must assume that M is representable on our machine and that the ϕ_i are defined as well. Slightly more subtly, we also need to be able to compare two points with the machine. We need to assume a sort of distance function which is zero when two points are the same, and nonzero otherwise. Given these assumptions, then the topological complexity is $n - 1$.

Proposition 5.9. *The examples listed above all have such a map ϕ.*

PROOF. All of these examples have a nonvanishing vector field. Such a field can be integrated to yield a flow. The fact that the field does not vanish implies that every point is moved by the flow. It is a simple matter to show that you can construct such a flow $\phi(t, \cdot)$ so that there is a time $T > 0$ with the property that there are no periodic points of period $\leq T$.

Define $\phi_i(x) = \phi((i/n)T, x_1)$. Then by the condition imposed on the periods, the $\phi_i(x)$ are all distinct and distinct from x for $1 \leq i \leq n$. □

As an example of how this can be done with an actual manifold, consider the circle S^1. S^1 can be modeled, as mentioned in Section 1, by two real numbers, x, y, such that $x^2 + y^2 = 1$, and this model requires only one comparison to check for equality.

The required map ϕ_i is just rotation by $i/(n + 1)$ of a complete rotation. One way to do this is to let $\tilde{\phi}_1(x, y) = (x, y) + \varepsilon(-y, x)$, and let

$$\phi_1(x, y) = \frac{\tilde{\phi}_1(x, y)}{\|\tilde{\phi}_1\|} \tilde{\phi}_1(x, y)$$

for some sufficiently small ε with $\|\cdot\|$ the Euclidean norm in $\mathbf{R}^2$. Now let

$\phi_i(x, y) = \phi_1^i(x, y)$ for sufficiently small ε. Thus, the new point problem on S^1 has topological complexity exactly $n - 1$.

References

[1] M. Ben-Or, Lower bounds for algebraic computation trees, *Proceedings 15th ACM STOC*, pp. 80–86, 1983.
[2] L. Blum, M. Shub, and S. Smale, On a theory of computation and complexity over the real numbers: *NP*-completeness, recursive functions and universal machines. *Bull. Amer. Math. Soc.* 21: 1–46, 1989.
[3] L.E.J. Brouwer, Über abbildung von mannigfalltigkeiten, *Math. Ann.* 71: 97–115, 1912.
[4] J. von zur Gathen, Algebraic complexity theory, Technical report 207/88, University of Toronto, Department of Computer Science, January 1988.
[5] V. Guillemin and A. Pollack, *Differential Topology*, Prentice-Hall, Englewood Cliffs, NJ, 1974.
[6] A "Geometric" View of the Dynamics of Trajectories of Computer Programs, *Acta Applicandae Mathematicae*, 18: 145–182, 1990.
[7] H. Levine. A lower bound for the topological complexity of poly(d, n), *J. Complexity*, 5: 34–44, 1989.
[8] T. Lickteig, On semialgebraic decision complexity, I.C.S.I. Tech. Report TR-90092, 1990.
[9] A. Lubiw and A. Racs, A lower bound for the integer distinctness problem, *Inform. and Comput.*, 94: No. 1, 83–92, 1991.
[10] J. Milnor, On the Betti-numbers of real varieties, *Proc. Amer. Math. Soc.* 15: 275–280, 1964.
[11] J.R. Munkres, *Elements of Algebraic Topology*, Benjamin/Cummins, Menlo Park, 1984.
[12] J. Renegar, On the computational complexity and geometry of the first-order theory of the reals, part I, Technical Report 853, School of Operations Research and Industrial Engineering, Cornell University, 1989.
[13] J. Renegar, On the computational complexity and geometry of the first-order theory of the reals, part II, Technical Report 854, School of Operations Research and Industrial Engineering, Cornell University, 1989.
[14] J. Renegar, On the computational complexity and geometry of the first-order theory of the reals, part III. Technical Report 855, School of Operations Research and Industrial Engineering, Cornell University, 1989.
[15] J.R. Munkres, *Topology: A First Course*. Prentice-Hall, Engelwood Cliffs, NJ, 1975.
[16] A.S. Schwarz, The genus of a fibre bundle, *Proc. Moscow Math. Soc.* 10: 217–272, 1961.
[17] S. Smale, On the topology of algorithms, I, *J. Complexity* 3: 81–89, 1987.
[18] E. Spanier, *Algebraic Topology*, Springer-Verlag, New York, 1966.
[19] M. Steele and A. Yao, Lower bounds for algebraic decision trees, *J. Algorithms* 3: 1–8, 1982.
[20] V. Strassen, Berechnung und program i. *Acta Inform.* 320–335, 1972.
[21] V. Strassen, Berechnung und program ii. *Acta Inform.* 64–79, 1973.

[22] V. Strassen, Relative bilinear complexity and matrix multiplication, *Crelles Zh. Reine Angew. Math.* 375/376: 406–443, 1987.

[23] H. Teitze, Über funcktionen die auf einer abgeschlossenen menge stetig sind, *Zh. Reine Angew. Math.* 145: 9–14, 1915.

[24] V. Vasiliev, Cohomology of the braid group and the complexity of algorithms, preprint, 1988.

[25] A.C.-C. Yao, Lower bounds for algebraic computation trees with integer inputs, *Proceedings 49th Annual Symposium on Fundamentals of Computer Science*, pp. 308–313, 1989.

38
On the Distribution of Roots of Random Polynomials*

E. KOSTLAN

Introduction

In the study of algebraic and numerical properties of polynomials one occasionally introduces the notion of a random polynomial. For example, this chapter was originally motivated by investigations, including [Ki, Re, SS, Sm1, Sm2], into the complexity of root-finding algorithms for polynomials. We restrict our attention to central normal measures on spaces of polynomials, i.e., we assume that the coefficients of the polynomials have a multivariate normal distribution with mean zero and known covariance.

In Section 2, we study the induced measures on the spaces of roots. For a complex polynomial of degree d in one variable, this calculation is easy (Theorem 2.1), but we have an intrinsic problem with generalizing the result to systems of polynomials. Consider the following special case of the Cayley–Bacharach Theorem: *Given two plane cubics that intersect at exactly nine points, any eight of those points determine the remaining point.*

> This fact was known in 1748 to Euler, who remarked that as a consequence polynomial functions in two or more variables would necessarily be much more complicated than in one variable, since then it is not generally the case that a set of *mn* points in plane is the common zero locus of a pair of polynomials. [GH, p. 673]

Thus, the induced probability measure on the space of collections of points is, in general, singular with respect to the Lesbegue measure. The only higher-dimensional nonlinear case for which the induced measure is nonsingular is two plane conics, for which we calculate the distribution of the roots explicitly (Theorem 2.3).

In discussions of random polynomials with real coefficients, the focus has been on the problem of calculating the expected number of real roots.

> Waring in 1782 ... stated, "... regarding cubic equations, the ratio of the probability of finding impossible [i.e. complex] roots to the probability of failing to find them is seen to be not greater than two to one [translated from the latin by [BS, p11]]."

* Work supported in part by MSRI, Berkeley, CA.

As we shall see, for a particularly natural choice of measure (the *standard normal measure* defined in Section 4), this ratio is $\sqrt{3}$. The question about the expected number of real roots is generally traced back to [BP], and other major developers of the theory include [LO] and [EO]. The problem also appears in the Soviet literature, as is evidenced by [Ma] and [Sh]. In the literature, it is usually assumed that the covariance matrix of the coefficients is the identity matrix. For this case, an exact integral formula for the expected number or real roots of a single polynomial can be found in [Kc]. We refer to such formulas as *Kac formulas.* See [BS, Ke or Ham] for a discussion. Theorem 3.3 generalizes this integral formula to *systems of polynomials* with *any* central normal measure.

Among the normal measures, one with particular properties suggests itself. We define this measure in Section 4 and discuss its characterizations. We call this measure the *standard normal measure* because it generalizes the real-valued standard normal random variable, being characterized by a combination of invariance and independence. This measure differs from the definitions of random polynomials generally found in the literature in that the variances of the coefficients are not equal. In contrast, the coefficients of a standard normal polynomial are independent central normal variables with variances related by the multinomial coefficients. This conforms with [Ka]: In practice, the coefficients of the middle terms of polynomials, coming from a variety of sources, once rescaled by a geometric series, tend to have greater variance than the coefficients of the extreme terms.

Increasing the variance of the middle coefficients increases the expected number of real roots. To be precise, Corollary 3.4 states the following.

The expected number of real roots of a system of n independent standard normal polynomials of degree d in n variables is exactly $d^{n/2}$.

This result stands in contrast to [Kc] which states the following.

The expected number of real roots of a polynomial where the coefficients are i.i.d. central normal random variables is asymptotic to $(2/\pi)\log(d)$, as $d \to \infty$.

The contrast between these two results is reminiscent of the following results on the number of real roots and the number of roots in sectors of the complex plane discussed in [Ob].

For any polynomial $f(z) = \sum_{i=0}^{d} a_i z^i$, let

$$P = \sum_{i=0}^{d} |a_i| / \sqrt{|a_0 a_d|},$$

and let r denote the number of real roots of f. Then [Sc]

$$r^2 - 2r < 4d \log P,$$

and [Sz]

$$r^2 + r < 4(d+1) \log P.$$

Furthermore, if $N(\alpha, \beta)$ denotes the number of roots in the sector $0 \leq \alpha \leq \arg(z) \leq \beta \leq 2\pi$, then [ET]

$$|N(\alpha, \beta) - (\beta - \alpha)d/2\pi| < 16\sqrt{d \log P}.$$

All these results suggest that increasing the magnitude of middle coefficients of polynomials tends to increase the number of real roots and the clustering of roots in sectors.

The number of roots of a polynomial is a special case of the volume of a projective variety. Over **C**, this volume is determined by the degree and dimension of the variety. To be precise, the volume of a variety of degree D and dimension k is given by

$$D\pi^k/k!$$

Over **R**, the volume is not determined by the degree and dimension. However, we can discuss the *expected* volume of a random algebraic variety. We assume that we have an underdetermined system of standard normal real polynomials. Then the expected volume of the real variety of dimension k in the real projective space of dimension n determined by $(n - k)$ polynomials of degree d is given by (Theorem 5.1)

$$\frac{d^{(n-k)/2}\pi^{(k+1)/2}}{\Gamma[(k + 1)/2]}.$$

1. Normal Measures on Spaces of Polynomials

We restrict our attention to homogeneous polynomials. This is not a serious restriction, as one can pass from homogeneous to inhomogeneous polynomials by replacing a variable with a constant. This may scale the coefficients by a geometric series, but will not affect the distribution of the roots.

Let $(X, \langle\ ,\ \rangle)$ by a finite-dimensional Hilbert space over **R** or **C**. We restrict our discussion to *central* normal measures, that is, we always assume that the mean is zero.

Definition. The central normal measure on $(X, \langle\ ,\ \rangle)$ is given by the volume form

$$(2\pi)^{-\dim_{\mathbf{R}}(X)/2} e^{-\langle x,x \rangle/2}\, d\mu,$$

where $d\mu$ is the standard volume form on $(X, \langle\ ,\ \rangle)$, and $\dim_{\mathbf{R}}(X)$ the dimension of X over **R**.

These measures are generally referred to as multivariate normal measures. For their elementary properties, see [BD, pp. 9–28]. We emphasize that choosing a central normal measure on a vector space is equivalent to choosing an inner product on the vector space. These measures are usually defined be means of a *covariance matrix*.

Definition. Relative to any basis of $(X, \langle\ ,\ \rangle)$, the covariance matrix is defined to be the matrix of the inner produce of the dual space $(X^*, \langle\ ,\ \rangle^*)$ with respect to the dual basis.

It is straightforward to verify that this definition of covariance matrix is equivalent to the usual definition. We may wish to consider normal measures concentrated on a proper subspace of X. This may be done by allowing the covariance matrix to be singular or, equivalently, by letting the inner product on X^* be positive semidefinite. However, although many of the results in this chapter are valid for singular covariance matrices, we assume, for simplicity, that the covariance matrices are nonsingular.

If the dimension of the Hilbert space is greater than one, we may consider the induced probability measure on the corresponding projective space. This measure is invariant under the induced action of the orthogonal or unitary groups and is, therefore, the measure induced by the Fubini–Study metric (see [GH, pp. 30–31]).

Notation. Given any Hilbert space X, the corresponding projective space will be denoted by $\mathbf{P}(X)$.

Notation. The collection of homogeneous polynomials of degree d from V to W will be denoted by $P_d(V, W)$. If W is identified with the base field, we will simply write $P_d(V)$.

We wish to define central normal measures on $P_d(V, W)$. Equivalently, we can give $P_d(V, W)$ the structure of a Hilbert space. Although $P_d(V, W)$ and $P_d(V) \otimes W$ are isomorphic as vector spaces, they may not be isomorphic as Hilbert spaces. This can present difficulties. Therefore, except in Section 4, where we discuss characterizations of inner products, we will always make the following:

Key Assumption. We assume that, after possibly changing the inner product on W, $P_d(V, W)$ and $P_d(V) \otimes W$ are isomorphic as Hilbert spaces.

Thus, when we speak of the *covariance matrix of the coefficients*, we will mean the inner product on $P_d(V)^*$, *not* the inner product on $P_d(V, W)^*$. We allow the change of inner product on W because this will not effect the induced distribution of the roots. If we are given a basis (not necessarily orthogonal) for W, we can write elements of $P_d(V, W)$ as a system of polynomials in $P_d(V)$. Thus, we have another version of our

Key Assumption. With respect to some basis in W, the polynomials in the system are i.i.d.

Thus, although the individual polynomials can have any central normal measure, they must be i.i.d. This is a genuine restriction. If the polynomials of the system are not i.i.d., then the generalized Kac formula of Section 3 is not applicable.

2. The Distribution of the Roots of a Normal Polynomial over $\mathbf{C}$

As noted in the Introduction, it is not true that almost every d^n-uplet of projective points can be realized as the roots of n homogeneous polynomials of degree d in $(n+1)$ variables. Therefore, in general, the measure on the space of d^n-uplets of points induced by a normal measure on the space of systems of polynomials is singular with respect to the Lesbegue measure. More precisely, if $\dim(V) = n + 1$, then $\dim(P_d(V)) = \binom{n+d}{n}$, and the dimension of the Grassmann manifold of k planes in $P_d(V)$ is given by $\dim(G_k(P_d(V))) = k\binom{n+d}{n} - k^2$. Thus, if almost every d^n-uplet of points in $\mathbf{P}(V)$ can be realized as the roots of n homogeneous polynomials of degree d in $(n+1)$ variables, we must have $nd^n = n\binom{n+d}{n} - n^2$. This equation has the solution $n = 1$ or $d = 1$ or $n = d = 2$.

We first consider the case $n = 1$. We are concerned with $P_d(V, W)$, where $\dim(V) = 2$ and $\dim(W) = 1$. Let v_0, v_1 be a system of coordinates for V. We introduce an inhomogeneous coordinate $z = v_1/v_0$. We consider *ordered* sets of roots, which may be identified with points in $\mathbf{C}^d$. To define the induced measure on $\mathbf{C}^d$ unambiguously, we require that the measure be invariant under permutation of the roots.

Theorem 2.1. *Let $(z_1, \ldots, z_d)$ be the ordered set of roots of a central normal polynomial of degree d. The induced probability measure is given by the volume form*

$$\det(g) \prod_{i<j} |z_i - z_j|^2 (c^\dagger g c)^{-d-1} \pi^{-d}\, d\mu,$$

where $c_k = s_k(z_1, \ldots, z_d)$, $k = 0, \ldots, d$, are the elementary symmetric functions in d variables—we take s_0 to be identically one, g is the inverse of the covariance matrix of the coefficients, and $d\mu$ is the standard volume form on $\mathbf{C}^d$.

PROOF. We first reduce to a monic polynomial by dividing by the lead coefficient. We then have an equidimensional mapping from roots to coefficients given by the symmetric functions. The Jacobian of this mapping is the Vandermonde determinant $\prod_{i<j} |z_i - z_j|^2$ [BS, pp. 150–152]. Multiplying this by the volume form induced by the Fubini–Study metric [GH, pp. 30–31] and dividing by $d!$, gives the induced volume form.

The case $d = 1$ is trivial, but we include it for completeness. Let $v_0, \ldots, v_n$ be a system coordinates for V. Introduce inhomogeneous coordinates $z_i = v_i/v_0$, $i = 0, \ldots, n$. Note that $z_0 = 1$.

Theorem 2.2. *The space of roots of a central normal linear mapping can be identified with the projective space with the measure induced by the Fubini–Study metric. In terms of inhomogeneous coordinates, the probability measure is given by*

$$\det(g^*)(z^\dagger g^* z)^{-n-1} n! \pi^{-n} \, d\mu,$$

where g^* is the covariance matrix of the coefficients and $d\mu$ is the standard volume form on $\mathbf{C}^n$.

We now consider the case of two plane conics. We are concerned with $P_d(V, W)$, where $\dim(V) = 3$, $\dim(W) = 2$, and $d = 2$. Let v_0, v_1, v_2 be a system of coordinates for V. We introduce inhomogeneous coordinates $(x, y) = (v_1/v_0, v_2/v_0)$. We consider ordered quadruplets of roots. To define the induced measure unambiguously, we require that the measure be invariant under permutation of the roots.

Theorem 2.3. *The probability measure induced from a central normal measure on the space of pairs of plane conics onto the space of ordered quadruplets*

$$((x_1, y_1), (x_2, y_2), (x_3, y_3), (x_4, y_4))$$

of plane points is given by the volume form

$$\begin{gathered} 120\pi^{-8} \det(g^*)^4 \det(Mg^*M^\dagger)^{-6} \\ \{|(x_2y_3 - x_1y_3 - x_3y_2 + x_1y_2 + x_3y_1 - x_2y_1) \\ (x_2y_4 - x_1y_4 - x_4y_2 + x_1y_2 + x_4y_1 - x_2y_1) \\ (x_3y_4 - x_1y_4 - x_4y_3 + x_1y_3 + x_4y_1 - x_3y_1) \\ (x_3y_4 - x_2y_4 - x_4y_3 + x_2y_3 + x_4y_2 - x_3y_2)|\}^6 \, d\mu, \end{gathered}$$

where

$$M = \begin{bmatrix} 1 & x_1^2 & y_1^2 & x_1y_1 & y_1 & x_1 \\ 1 & x_2^2 & y_2^2 & x_2y_2 & y_2 & x_2 \\ 1 & x_3^2 & y_3^2 & x_3y_3 & y_3 & x_3 \\ 1 & x_4^2 & y_4^2 & x_4y_4 & y_4 & x_4 \end{bmatrix},$$

g^* the covariance matrix of the coefficients, and $d\mu$ is the standard volume form on $\mathbf{C}^8$.

The four quadratic factors appearing in Theorem 2.3 are exactly the colinearity conditions on each triplet of points. The quantity $\det(Mg^*M^\dagger)$ arises from the Plücker embedding of $G_4(\mathbf{C}^6)$ into $\mathbf{P}(\mathbf{C}^{15})$ (see [GH, pp. 209–211]).

PROOF. Define θ as the function that assigns to $((x_1, y_1), (x_2, y_2), (x_3, y_3), (x_4, y_4))$ the matrix M. We can consider the rows of M as elements of $P_2(\mathbf{C}^3)^*$.

Assume that no three of the four points are colinear. Therefore, θ is a mapping from a Zariski open set in $\mathbf{C}^8$ to $G_4(P_2(\mathbf{C}^3)^*)$. There is a canonical isometry

$$\iota\colon G_4(P_2(\mathbf{C}^3)^*) \to G_2(P_2(\mathbf{C}^3)),$$

and $\iota\theta$ maps each quadruplet of points to the collection of polynomials that vanish at those points (thus, θ is essentially the evaluation mapping used in Section 3). Thus, the desired volume form on quadruplets of points is exactly the pullback of the volume form on $G_4(P_2(\mathbf{C}^3)^*)$ by θ. We now define a system of coordinates on a Zariski open subset of $G_4(P_2(\mathbf{C}^3)^*)$. We write the 4×6 matrix M as

$$[M_0 M_1],$$

where M_0 is a 4×4 matrix and M_1 is a 4×2 matrix, and set $T = M_0^{-1} M_1$. We define κ by

$$T = \kappa(M),$$

whenever $\det(M_0) \neq 0$. We also define $\tilde{T} = M_0^{-1} M$. The Fubini–Study metric on this subset of $G_4(P_2(\mathbf{C}^3)^*)$ is determined by the exterior $(1, 1)$-form

$$\omega = \frac{\iota}{2\pi}\, \partial\bar{\partial} \log \det(\tilde{T} g^* \tilde{T}^\dagger)$$

(see [KN, pp. 160–161]). $G_4(P_2(\mathbf{C}^3)^*)$ is a variety in $\mathbf{P}(\mathbf{C}^{15})$ of dimension 8 and degree 14, and, therefore,

$$8! \int_{G_4(P_2(\mathbf{C}^3)^*)} \omega^8 = 14$$

(see [St, pp. 11–14]). Thus the corresponding probability measure on $G_4(P_2(\mathbf{C}^3)^*)$ is given by the volume form

$$\frac{8!}{14}\omega^8 = \det(g^*)^4 [\det(\tilde{T} g^* \tilde{T}^\dagger)]^{-6} \left(\frac{\iota}{2\pi}\right)^8 dT_{11} \wedge d\bar{T}_{11} \wedge \cdots \wedge dT_{42} \wedge d\bar{T}_{42}.$$

As $\kappa\theta$ is a 4!-to-one mapping onto a Zariski open set, the induced volume form on the ordered quadruplets of plane points is $(1/4!)(8!/14)\theta^*\kappa^*\omega^8$. The Jacobian of $\theta^*\kappa^*$ may be calculated directly and is given by

$$\begin{aligned}
&[(x_2y_3 - x_1y_3 - x_3y_2 + x_1y_2 + x_3y_1 - x_2y_1)\\
&(x_2y_4 - x_1y_4 - x_4y_2 + x_1y_2 + x_4y_1 - x_2y_1)\\
&(x_3y_4 - x_1y_4 - x_4y_3 + x_1y_3 + x_4y_1 - x_3y_1)\\
&(x_3y_4 - x_2y_4 - x_4y_3 + x_2y_3 + x_4y_2 - x_3y_2)\det(M_0)^{-2}]^3.
\end{aligned}$$

We also have the identity

$$\det(\tilde{T} g^* \tilde{T}^\dagger) = \det(M g^* M^\dagger)|\det(M_0)|^{-2}.$$

From these two formulas the induced volume form may be calculated.

3. The Generalized Kac Formula

We now give an integral formula for the expected number of real roots of a system of random polynomials with real coefficients that have any central normal distribution. The proofs of the propositions are omitted. For a detailed discussion of these results, see [Ko1].

Lemma 3.1. *For any inner product* $\langle\ ,\ \rangle$ *on the space* $P_d(\mathbf{R}^{n+1})$,

$$\langle \mathrm{ev}(\mathbf{x}), \mathrm{ev}(\mathbf{y})\rangle^* = \sum_{I,J} g^*_{IJ} x^I y^J,$$

where ev: $\mathbf{R}^{n+1} \to P_d(\mathbf{R}^{n+1})^*$ *is the evaluation mapping,* $\mathrm{ev}(\mathbf{x})(f) = f(\mathbf{x})$, I *and* J *are multi-indices, and* g^* *is the covariance matrix of the coefficients of* $P_d(\mathbf{R}^{n+1})$.

We may thus think of $\langle \mathrm{ev}(\mathbf{x}), \mathrm{ev}(\mathbf{y})\rangle^*$ as a generating function for the covariance matrix g^*.

A collection of hyperplanes in $P_d(\mathbf{R}^{n+1})^*$ corresponds to a collection of polynomials, and the common intersection of these hyperplane with the image of ev corresponds to the common zero locus of these polynomials. A standard application of integral geometry, therefore, yields the following:

Lemma 3.2. *Let* U *be a measurable subset of* $\mathbf{P}(\mathbf{R}^{n+1})$. *The volume of the image of* U *under the mapping*

$$\mathbf{P}(\mathrm{ev})\colon \mathbf{P}(\mathbf{R}^{n+1}) \to \mathbf{P}(P_d(\mathbf{R}^{n+1})^*),$$

divided by the volume of $\mathbf{P}(\mathbf{R}^{n+1})$, *is equal to the expected number of roots in the region* U.

The map $\mathbf{P}(\mathrm{ev})$ is often referred to as the d-uple embedding [Har].

A direct calculation using Lemma 3.2 yields the following:

Theorem 3.3. *Let* $\langle\ ,\ \rangle$ *be any inner product on the space* $P_d(\mathbf{R}^{n+1})$. *Dehomogenize the standard coordinates* $\{x_i\}$ *on* $\mathbf{R}^{n+1}$ *by setting* $x_0 = 1$. *Then the expected number of roots in the region* U *is given by*

$$\frac{\Gamma[(n+1)/2]}{\pi^{(n+1)/2}} \int_U \sqrt{\det\left[\frac{\partial^2}{\partial x_i \partial y_j} \log\langle \mathrm{ev}(\mathbf{x}), \mathrm{ev}(\mathbf{y})\rangle^*\right]_{\mathbf{y}=\mathbf{x}}} \prod_{i=1}^{n} dx_i.$$

Given the covariance matrix of the coefficients, $\langle \mathrm{ev}(\mathbf{x}), \mathrm{ev}(\mathbf{y})\rangle^*$ may be calculated using Lemma 3.1. For the standard normal polynomials discussed in the following section, $\langle \mathrm{ev}(\mathbf{x}), \mathrm{ev}(\mathbf{y})\rangle^* = (1 + \sum_{i=1}^{n} x_i y_i)^d$. For the normal measure generally discussed in the literature, the monomials form an orthonormal basis for $P_d(\mathbf{R}^2)$, so we see that $\langle \mathrm{ev}(\mathbf{x}), \mathrm{ev}(\mathbf{y})\rangle^* = [1 - (x_1 y_1)^{d+1}]/(1 - x_1 y_1)$. Thus, we have at once the following two corollaries:

Corollary 3.4. *The expected number of roots in the region U of a system of n homogeneous standard normal polynomials with n + 1 unknowns is given by*

$$\frac{d^{n/2}\Gamma[(n+1)/2]}{\pi^{(n+1)/2}}\int_U \frac{\prod_{i=1}^n dx_i}{(1+\sum_{i=1}^n x_i^2)^{(n+1)/2}}.$$

In particular, the expected number of real roots for the system is $d^{n/2}$.

Corollary 3.5 (Kac Formula as Restated in [Ham]). *Assume that the coefficients of a polynomial in one variable are i.i.d. central normal random variables. Then the expected number of roots in the interval* $[a, b]$ *is given by*

$$\frac{1}{2\pi}\int_a^b \sqrt{x^{-1}\frac{d}{dx}x\frac{d}{dx}\log\left[\frac{1-x^{2d+2}}{1-x^2}\right]}\,dx.$$

4. An Invariant Normal Measure

Relative to some orthonormal basis $\{v_i\}$ for V^* and some orthonormal basis $\{W_i\}$ for W, we write a polynomial p as $\sum_{i,I} p_{iI} v^I \otimes W_i$, where I is a multi-index.

Definition. Relative to the above pair of bases, the inner product $\langle\cdot,\cdot\rangle_d$ on $P_d(V, W)$ is defined as

$$\langle p, q\rangle_d = \sum_{i,I} p_{iI}\bar{q}_{iI}\Big/\binom{d}{I}.$$

Definition. The standard normal measure on $P_d(V, W)$ is the central normal measure corresponding to the inner product $\langle\cdot,\cdot\rangle_d$. Equivalently, the standard normal measure may be defined by requiring that, relative to some pair of orthonormal bases for V^* and W, the coefficients of the monomials are independent central normal variables, where the iI^{th} coefficient has variance $\binom{d}{I}$.

The inner product can be obtained by identifying $P_d(V, W)$ with the subspace of symmetric tensors in $V^{*\otimes d} \otimes W$. The tensor product can be made into a Hilbert space by choosing as an orthonormal basis the tensor product of orthonormal bases of V^* and W. Restricting the inner product to $P_d(V, W)$ gives $\langle\cdot,\cdot\rangle_d$. This explains the appearance of the multinomial coefficients in the above definition. Thus, $\langle\cdot,\cdot\rangle_d$ depends on the choice of the inner products $\langle\cdot,\cdot\rangle_V$ and $\langle\cdot,\cdot\rangle_W$, but not on the choice of orthonormal bases. It is clear that this inner product is invariant under left and right action of the orthogonal or unitary group. As was pointed out to the author by Joseph Wolf, it is well-known that the representation of the unitary group given by polynomials of a fixed degree is irreducible. Consequently, we have the following well-known

Theorem 4.1. *If V and W are finite-dimensional complex Hilbert spaces, a Hermitian inner product $\langle \cdot, \cdot \rangle_h$ on $P_d(V, W)$ is invariant under the right action of $U(\langle \cdot, \cdot \rangle_V)$ and the left action of $U(\langle \cdot, \cdot \rangle_W)$ if and only if it is a scalar multiple of $\langle \cdot, \cdot \rangle_d$. Therefore, up to scalar multiplication, $\langle \cdot, \cdot \rangle_d$ is characterized by unitary invariance.*

As the corresponding representation for the orthogonal group is reducible, no analogue of Theorem 4.1 is possible; that is, there are other invariant inner products for spaces of real polynomials. These measures are of independent interest, but they will not be considered in this chapter. Nonetheless, over **R**, the standard normal measure may be characterized in a number of ways.

Theorem 4.2. *If V and W are finite-dimensional real Hilbert spaces, a symmetric inner product $\langle \cdot, \cdot \rangle_h$ on $P_d(V, W)$ is invariant under the right action of $O(\langle \cdot, \cdot \rangle_V)$ and the left action of $O(\langle \cdot, \cdot \rangle_W)$ and has the property that monomials are orthogonal if and only if it is a scalar multiple of $\langle \cdot, \cdot \rangle_d$. Therefore, up to scalar multiplication, $\langle \cdot, \cdot \rangle_d$ is characterized by orthogonal invariance and orthogonality of monomials.*

PROOF. Since the polynomial $\langle \mathrm{ev}(\mathbf{x}), \mathrm{ev}(\mathbf{y}) \rangle_h^*$ is orthogonally invariant, classical invariant theory implies that it must be a polynomial in $\langle \mathbf{x}, \mathbf{x} \rangle$, $\langle \mathbf{y}, \mathbf{y} \rangle$, and $\langle \mathbf{x}, \mathbf{y} \rangle$. (See [Sp, Theorem 35, p. 481]). But orthogonality of monomials implies that $\langle \mathrm{ev}(\mathbf{x}), \mathrm{ev}(\mathbf{y}) \rangle_h^*$ is a polynomial of the form $\sum a_I \mathbf{x}^I \mathbf{y}^I$ and therefore must be of the form $r\langle \mathbf{x}, \mathbf{y} \rangle^d$ for some $r \in \mathbf{R}$.

So far, we have discussed the characterization of $\langle \cdot, \cdot \rangle_d$ rather than the characterization of the standard normal measure itself. The following characterization, which holds over **R** and **C**, has the advantage that normality of the measure is not assumed. For simplicity, we only discuss the real case. The complex case is similar. We begin with

Lemma 4.4. *Let $\{X_i\}_{i=1}^n$ be a set of independent real-valued random variables, and let $\{a_i\}_{i=1}^n$ and $\{b_i\}_{i=1}^n$ be sets of nonzero real constants. Assume that $n \geq 2$. If $\sum_{i=1}^n a_i X_i$ and $\sum_{i=1}^n b_i X_i$ are independent, then the X_i are normal.*

PROOF. This is the first theorem of [Sk]. See also [Fe, Theorem XV.8.2, p. 526], for $n = 2$.

Theorem 4.5. *If $\dim(V) > 1$ or $\dim(W) > 1$, then the standard normal measure on $P_d(V, W)$ is characterized, up to scalar multiplication, by statistical independence of the coefficients of the monomials relative to some pair of ortho-*

normal bases for V^* and W, and invariance of the measure under the right action of $O(\langle\cdot,\cdot\rangle_V)$ and the left action of $O(\langle\cdot,\cdot\rangle_W)$.

PROOF. First we establish normality of the coefficients. By orthogonal invariance, independence of the coefficients of the monomials relative to some orthonormal basis is equivalent to independence of the coefficients of the monomials relative to all orthonormal bases. If $\dim(W) > 1$, choose an orthonormal basis for W and an element of $O(\langle\cdot,\cdot\rangle_W)$ that has all nonzero entries with respect to this basis. Then apply Lemma 4.4 to each monomial separately to establish normality. If $\dim(W) = 1$ and $\dim(V) > 1$, let $\{u_i\}$ and $\{v_i\}$ be two orthonormal bases for V^* such that, for all i, $\langle u_1, v_i\rangle^*$ and $\langle u_2, v_i\rangle^*$ are nonzero real constants. Relative to $\{v_i\}$, the u_1^d and u_2^d coefficients are linear forms with all nonzero components. Thus, Lemma 4.4 establishes normality.

Once normality is established, using invariance under the antipodal map in W, we establish that the distribution is central. We then apply Theorem 4.2 to deduce the covariance of the distribution.

For $d = 2$, a proof of Theorem 4.5 appears in [Me, Theorem 2.6.3, p. 52].

We conclude this section with another characterization of $\langle\cdot,\cdot\rangle_d$ over **C**. Following [La, pp. 285–286], we define an inner product on $L^2((V, d\rho), (W, \langle\cdot,\cdot\rangle_W))$ as

$$\langle f, g\rangle = \int_V \langle f, g\rangle_W \, d\rho,$$

where $d\rho$ is any $U(\langle\cdot,\cdot\rangle_V)$-invariant measure on V with at least $2d$ finite moments. The space $P_d(V, W)$ is a subspace of this Hilbert space, so we may restrict the inner product to $P_d(V, W)$. Over **C** we then have by a trivial computation (or by application of Theorem 4.1) the following:

Theorem 4.6. *The L^2 inner products on $P_d(V, W)$ are scalar multiples of $\langle\cdot,\cdot\rangle_d$.*

Note that over **R** the monomials are not L^2 orthogonal, and, thus, the L^2 inner product and $\langle\cdot,\cdot\rangle_d$ are not proportional. This reflects the fact that there are many invariant normal measures on spaces of polynomials over **R**.

5. The Expected Volume of a Real Algebraic Variety

We would like to generalize Theorem 3.3 to algebraic varieties of arbitrary dimension. We, therefore, consider a system of $(n - k)$ elements of $P_d(\mathbf{R}^{n+1})$. We may consider the expected volume of such a variety. In general, this presents difficulties, but for invariant measures integral geometry may be applied. In particular, we have the following:

Theorem 5.1. *For the standard normal measure on* $P_d(\mathbf{R}^{n+1}, \mathbf{R}^{n-k})$, *the expected volume of the real algebraic variety* (*of dimension* k) *is*

$$\frac{d^{(n-k)/2}\pi^{(k+1)/2}}{\Gamma[(k+1)/2]}.$$

For a detailed discussion of this result, see [Ko2].

Acknowledgments. Discussions with George Csordas concerning the analytic properties of polynomials and with George Wilkens and Joel Wiener Concerning integral geometry were of great help during this investigation. A conversation with Joseph Wolf concerning representation theory was also of great value. Investigation of random polynomials was suggested by research of Steve Smale, to whom this chapter is dedicated on the occasion of his sixtieth birthday.

References

[BS] Bharucha-Reid, A. and Sambandham, M., *Random Polynomials*, Academic Press, New York, 1986.

[BD] Bickel, P. and Doksum, K., *Mathematical Statistics: Basic Ideas and Selected Topics*, Holden-Day, San Francisco, 1977.

[BP] Bloch, A. and Pólya, G., On the roots of a certain algebraic equations, *Proc. London Math. Soc.* **33** (1932), 102–114.

[EO] Erdös, P. and Offord, A., On the number of real roots of a random algebraic equation, *Proc. London Math. Soc. (3)* **6** (1956), 139–160.

[ET] Erdös, P. and Turán, P., On the distribution of roots of polynomials, *Ann. Math (2)* **51** (1950), 105–119.

[Fe] Feller, W., *An Introduction to Probability Theory and Its Applications*, Vol. 2, Wiley, New York, 1971.

[GH] Griffiths, P. and Harris, J., *Principles of Algebraic Geometry*, Wiley, New York, 1978.

[Har] Hartshorne, R., *Algebraic Geometry*, Springer-Verlag, New York, 1977.

[Ham] Hammersley, J., The zeros of a random polynomial, *Proc. Third Berkeley Symp. Math. Statist. Prob.* **2** (1956), 89–111.

[Kc] Kac, M., On the average number of roots of a random algebraic equation, *Bull. Amer. Math. Soc.* **49** (1943), 314–320.

[Ka] Kahan, R., personal communication, MSRI, January conference, 1986.

[Ke] Kesten, H., The influence of Mark Kac on probability theory, *Ann. Probab.* **14** (1986), 1103–1128.

[Ki] Kim, M., Computational complexity of the Euler type algorithms for the roots of complex polynomials, Ph.D. thesis, Graduate School, CUNY, 1985.

[KN] Kobayashi, S. and Nomizu, K., *Foundations of Differential Geometry*, Vol. 2, Wiley, New York, 1969.

[Ko1] Kostlan, E., On the expected number of real roots of a system of random polynomials, preprint.

[Ko2] Kostlan, E., On the expected volume of a real algebraic variety, preprint.

[La] Lang, S., *Real Analysis*, Addison-Weseley, Reading, MA, 1969.

[LO] Littlewood, J. and Offord, A., On the number of real roots of a random algebraic equation, *J. London Math. Soc.* **13** (1938), 288–295.

[Ma] Maslova, N., On the distribution of the number of real roots of random polynomials (in Russian), *Teor. Veroyatnost. Primenen* **19** (1974), 488–500.

[Me] Mehta, M., *Random Matrices, Revised and Enlarged (2nd Ed.)*, Academic Press, San Diego, 1991.

[Ob] Obreschkoff, N., *Verteilung und Berechnung der Nullstellen Reeler Polynome*, Veb Deutscher Verlag der Wissenschaften, Berlin, 1963.

[Re] Renegar, J., On the efficiency of Newton's method in approximating all zeroes of a system of complex polynomials, preprint, Colorado State Univ., 1984.

[Sc] Schur, I., Untersuchungen über algebraische Gleichungen. I: Bemerkungen zu einem Satz von E. Schmidt, *Sitzungsber. preuss. Akad. Wiss., Phys.-Math.* (1933), 403–428.

[Sh] Shparo, D. and Shur, M., On the distribution of the roots of random polynomials (in Russian), *Vestnik Moskov. Univ. Ser. I Mat. Mekh.* **3** (1962), 40–43.

[SS] Shub, M. and Smale S., Computational complexity: on the geometry of polynomials and a theory of cost: II, *SIAM J. Comput.* **15** (1986), 145–161.

[Sk] Skitovich, V., Linear forms of independent random variables and the normal distribution (in Russian), *Izv. Akad. Nauk SSSR Ser. Mat.* **18** (1954), 185–200.

[Sm1] Smale, S., The fundamental theorem of algebra and complexity theory, *Bull. Amer. Math. Soc.* **4** (1981), 1–36.

[Sm2] Smale, S., On the efficiency of algorithms of analysis, *Bull. Amer. Math. Soc.* **13** (1985), 87–121.

[Sp] Spivak, M., *A Comprehensive Introduction to Differential Geometry (2nd Ed.)*, Vol. 5, Publish or Perish, Berkeley, 1979.

[St] Stoll, W., Invariant forms on Grassmann manifolds, *Ann. Math. Stud.* **89** (1977).

[Sz] Szegö, G., Bemerkungen zu einem Satz von E. Schmidt über algebraische Gleichungen, *Sitzungsber. preuss. Akad. Wiss., Phys.-Math.* (1934), 86–98.

39
A General NP-Completeness Theorem*

NIMROD MEGIDDO

1. Introduction

Blum, Shub, and Smale [2] formalized a model of computation over a general ring. They proved an analogue of Cook's theorem [3] over the reals. Smale [4] has recently raised the question of existence of NP-complete problems relative to "linear" machines, i.e., machines which add, subtract, multiply by constants, and branch, depending on the sign of a number.

My motivation for writing this chapter is the encouragement I received from Steve Smale. In a number of conversations we had on the subject at IMPA, I said I could prove a general theorem on the existence of NP-complete problems, but I was not particularly excited about dealing with all the details. Steve convinced me to a certain extent that there were some subtleties involved, and that I should indeed work under a precise model.

In this chapter, I indeed state a theorem about the existence of NP-complete problems in general. The model of computation unifies the results of Cook and of Blum, Shub, and Smale. I prefer to present the model in terms more familiar to computer scientists, namely, using RAMs instead of flowcharts. In Section 2, I define an abstract domain of computation. The model is described in Section 3. The problem of satisfiability over the abstract domain is discussed in Section 4 and its NP-completeness is proven.

2. An Abstract Domain

Let D be any set such that $|D| \geq 2$, and consider any *finite* set of binary operations over D. A typical operation will be denoted by $\circ$, so that for any $\xi, \eta \in D$, $\zeta = \xi \circ \eta \in D$ is well-defined. The finiteness assumption is crucial.

* Parts of this work were done during a visit to IMPA, Rio de Janeiro.

However, a single binary operation induces a possibly infinite set of unary operations corresponding to operating on constants. For example, even if the only operation is that of multiplying two reals, we may still consider the infinite set of multiplications of a variable by some real constant. Denote by $\mathcal{O}_1$ the finite set of operations which may be applied to any two variable values, and let $\mathcal{O}_2$ denote the set of operations which induce the unary operations as explained above.

Let T be a nonempty proper subset of D. Members of D are assigned with truth-values, so that the members of T are "true" and the others are "false." For $\xi \in T$, we write $\tau(\xi) = 1$, and otherwise $\tau(\xi) = 0$. For the convenience of presentation, let us fix two elements: **true** $\in T$ and **false** $\in D \setminus T$. We denote by $\mathcal{D} = \langle D, \mathcal{O}_1, \mathcal{O}_2, T \rangle$ the underlying domain of computation. If $D =$ {**true**, **false**} and $\mathcal{O}_1 = \mathcal{O}_2 = \varnothing$, our model gives Cook's original theorem. A linear machine over the reals can be handled using this formalism by choosing D to be the set of real numbers, $\mathcal{O}_1 = \{+, -\}$, $\mathcal{O}_2 = \{\times\}$ (where $\times$ is multiplication over the reals) and T is the set of positive reals. The "real number" model corresponds to $\mathcal{O}_1 = \mathcal{O}_2 = \{+, -, \times, \div\}$.

3. Random Access Machines over Abstract Domains

To distinguish the model of this chapter from that of Blum, Shub, and Smale, I talk about *programs*, rather than *machines*.

Two models of computation are considered equivalent if they induce the same set of computable functions, and the time complexities of the computable functions are related polynomially. There are many equivalent models of computation. I have chosen to generalize the model of a RAM as described by Aho, Hopcroft, and Ullman [1], so I try to stay as close as possible to the notation used by those authors.

A RAM has a memory and two tapes: a read-only input tape and a write-only output tape. The input consists of a finite sequence of members of D and a finite sequence of positive integers. The combined length of these sequences is called the *size* of the input. The output consists of a sequence of members of D. I would like to deal with computation over an abstract domain, and also use indirect addressing. Indirect addressing is important since one finite program can explicitly use only a fixed number of addresses, whereas the input problems may be of arbitrary size and need more addresses than the ones that appear in the program explicitly. The addresses are given by integers. Thus, I have to distinguish two types of memory registers. The registers of the first type are denoted $R_0, R_1, \ldots$ and each of them is capable of holding one member of D or a blank. The registers of the second type are denoted I_0, $I_1, \ldots$ and each of them is capable of holding an integer of arbitrary magnitude. These integers are used for addresseing only and, as we shall see later, the computational capability associated with them is very limited.

3.1. *Operands*

All computation over $\mathscr{D}$ takes place in the register R_0, whereas all computation over the integers takes place in I_0. Each *instruction* of the program consists of an *operation code* and an *address*. The possible operands are:

(i) $=i$, indicating the integer i.
(ii) A non-negative integer i, indicating the contents of either R_i or I_i, depending on the instruction.
(iii) $*i$, the contents of either R_j or I_j, where $j \geq 0$ is the integer currently stored in register I_i.
(iv) $=\xi$ (where $\xi \in D$), indicating the element ξ.

We denote the integer held in I_i by $j(i)$ and the element (or the blank) held in R_i by $c(i)$. The output tape is always assumed to be blank initially. By *deterministic computation*, we mean that initially $j(i) = 0$ and $c(i)$ is a blank for all $i \geq 0$. By *nondeterministic computation*, we mean that the memory may initially hold a finite number of members of D and integers in the respective registers.[1]

The *value* $v(a)$ *of an operand* a depends on the instruction and is defined as follows:

(i) $v(=i) = i$ and $v(=\xi) = \xi$.
(ii) $v(i) = j(i)$ if the instruction operates on integers, and $v(i) = c(i)$ if it operates on members of D.
(iii) $v(*i) = j(j(i))$ if the instruction operates on integers, and $v(*i) = c(j(i))$ if it operates on members of D.

3.2. *Instructions*

The RAM instructions are:

(i) HALT: Stop the execution.
(ii) ILOAD a: This instruction operates on integers. It assigns $v(a)$ to $j(0)$.
(iii) DLOAD a: This instruction operates on members of D (or blanks). If $v(a)$ is not a blank, then it assigns it to $c(0)$; otherwise, HALT.
(iv) ISTORE i: Assign $j(0)$ to $j(i)$.
(v) ISTORE $*i$: If $j(i) \geq 0$, then assign $j(0)$ to $j(j(i))$; otherwise, HALT.
(vi) DSTORE i: Assign $c(0)$ to $c(i)$.
(vii) DSTORE $*i$: If $j(i) \geq 0$, then assign $c(0)$ to $c(j(i))$; otherwise, HALT.
(viii) $\circ\, a$ (for any $\circ \in \mathcal{O}_1$): This instruction operates on members of D (or blanks). If $v(a)$ is not a blank, then assign $c(0) \circ v(a)$ to $c(0)$; otherwise, HALT.
(ix) $\circ_\xi$ (for any $\circ \in \mathcal{O}_2$ and any $\xi \in D$). Assign $\xi \circ c(0)$ to $c(0)$.

---

[1] The class NP will be defined more precisely later.

(x) INC: Assign $j(0) + 1$ to $j(0)$.
(xi) DEC: Assign $j(0) - 1$ to $j(0)$.
(xii) READ i: Assign the current input symbol to $c(i)$ (and move the input head one step ahead).
(xiii) READ $*i$: If $j(i) \geq 0$, then assign the current input symbol to $c(j(i))$; otherwise, HALT.
(xiv) WRITE a: This instruction operates on members of D (or blanks). If $v(a)$ is not a blank, then print it on the output tape (and move the output head one step ahead); otherwise, HALT.
(xv) JUMP ℓ: Go to the instruction labeled ℓ.
(xvi) JUMPT ℓ: If $c(0) \in T$, then go to the instruction labeled ℓ; otherwise, go to the next instruction.
(xvii) JUMP0 ℓ: If $j(0) = 0$, then go to the instruction labeled ℓ; otherwise, go to the next instruction.

3.3. *The Classes* P *and* NP

We assume the uniform cost model, i.e., the running time of the program equals the number of instructions executed. A program is said to run in polynomial time if its running time is bounded by some polynomial in terms of the input size. Denote by D^* the set of all finite sequences of members of D. Similarly, denote by Z^* the set of all finite sequences of integers.

A subset of $D^* \times Z^*$ is called a *language*. A language L is said to be in the class P if there exists a (deterministic) RAM which runs in polynomial time and outputs **true** if the input $(\sigma, v) \in D^* \times Z^*$ is in L and **false** otherwise.

A language $L \subseteq D^* \times Z^*$ is in the class NP if there exists a RAM with the properties as follows. There exists a polynomial n^k such that if $(\sigma, v) \in L$, then with some initialization of the registers $R_0, R_1, \ldots$ with either members of D or blanks, and with some initialization of the registers $I_0, I_1, \ldots$ with integers, the RAM runs in time n^k and outputs **true**; if $(\sigma, v) \in D^* \setminus L$, then there is no initialization of the memory with which the machine outputs **true** and halts.

NP-*Completeness*

As usual, a language L_0 is said to be NP-*complete* if for every L in NP, there exists a polynomial-time RAM which converts members of L into members of L_0 and nonmembers of L into nonmembers of L_0. We prove below the existence of an NP-complete language in the case $\mathcal{O}_1 = \mathcal{O}_2$.

4. Satisfiability over $\mathcal{D}$

In this section, we introduce a certain computational problem and prove its completeness for NP. More specifically, we show that if L is in NP, then the problem of recognizing whether or not a given input is in L can be reduced in polynomial time to the problem of recognizing the existence of members of D and integers, which together satisfy a certain system of constraints. A precise definition of this satisfiability problem is given in Subsection 4.1.

4.1. *Definition of Satisfiability over $\mathscr{D}$*

The problem of satisfiability over D is stated as a conjunction of constraints as follows. We use three types of *variables*:

(i) x_j—attains any value in D,
(ii) y_j—attains a value in $\{\mathbf{true}, \mathbf{false}\} \subseteq D$,
(iii) z_j—attains any integer value.

Using the above variables, we build *elementary propositions* of the following types:

(i) $x_j = \xi$ (for any $\xi \in D$),
(ii) $x_j = x_k$,
(iii) $x_j = x_k \circ x_\ell$ (for any $\circ \in \mathcal{O}_1$),
(iv) $x_j = x_k \circ \xi$ (for any $\circ \in \mathcal{O}_2$ and $\xi \in D$),
(v) $x_j \in T$,
(vi) $x_j \notin T$,
(vii) $y_j = \mathbf{true}$,
(viii) $y_j = \mathbf{false}$,
(ix) $z_j = i$ (for any integer i),
(x) $z_j \neq i$ (for any integer i),
(xi) $z_j = z_k$,
(xii) $z_j = z_k + 1$.

A *constraint* is either an elementary proposition or an implication of the form:

$$\phi_1 \wedge \phi_2 \wedge \phi_3 \quad \Rightarrow \quad \phi_4,$$

where ϕ_1, ϕ_2, ϕ_3, and ϕ_4 are elementary propositions.

4.2. *Satisfiability over $\mathscr{D}$ is in* NP

In this subsection, we prove that there exists a nondeterministic RAM which solves the satisfiability problem over $\mathscr{D}$ in polynomial time.

4.2.1. Encoding

We first discuss how an instance of the satisfiability problem over $\mathscr{D}$ is encoded as an element of $D^* \times Z^*$, i.e, a pair consisting of a finite string of elements of D and a finite sequence of integers. As explained above, an instance is a conjunction of constraints. We assume we have an alphabet where each element $\xi \in D$ is represented by a symbol $\underline{\xi}$ and each integer n is represented by a symbol $\underline{n}$. The integers used here are either subscripts of names of variables or integer constants that appear in the constraints. For simplicity, let us omit the underbars. Besides elements of D and integer constants, an instance of satisfiability consists of objects from a finite set for which we have to design an encoding scheme: (i) the start of a new conjunct, (ii) members of $\mathcal{O}_1$ and $\mathcal{O}_2$, and (iii) the symbols x, y, z, $=$, $\neq$, $*$, $\in$, $\notin$, T, and $\Rightarrow$. Obviously, one can easily design a binary encoding scheme using the symbols 0 and 1, say, to encode all the necessary information.

4.2.2. Testing a Solution

We claim that the problem of satisfiability over $\mathcal{D}$ can be solved in nondeterministic polynomial time. To prove this claim, we have to exhibit a RAM which receives a code of a set of constraints as input. If there are values of the variables that satisfy the constraints, the RAM confirms this fact. We have to confirm a conjunction of constraints, each of which is either an elementary proposition or an implication involving at most four elementary propositions. Thus, it suffices to show how each type of an elementary proposition can be confirmed. Let us interpret the initial contents of the registers R_i ($i = 1, \ldots, n$) as the "guesses" of values for the variables x_i, respectively. Similarly, the contents of R_{n+i} will be the guess for the y_i's and that of I_i will the guess for z_i. For each type of a constraint, there will be a segment in the program where that constraint is tested. If a violated constraint is detected, the program halts with no output. If the validity of all the constraints has been confirmed, the program outputs **true** and halts. It is straightforward to see how each type of elementary proposition is confirmed.

4.3. *Reducing Problems in* NP *to Satisfiability over* $\mathcal{D}$

We assume henceforth that $\mathcal{O}_1 = \mathcal{O}_2$. Suppose L is in NP and let n^k be a polynomial time bound for a corresponding nondeterministic RAM which we denote by Π. Without loss of generality, we also assume that n^k is a bound on the index i of any register R_i or I_i which is used throughout the execution of the algorithm. Moreover, we assume without loss of generality that the integers which are held in any register I_i are non-negative and not greater than n^k.

Given an input (σ, v) of length n, let $m = n^k$. We construct a system of constraints on the values (in D) of variables $x_{00}, x_{01}, \ldots, x_{0m}$, representing the initial contents of registers $R_0, R_1, \ldots, R_m$, respectively, and a system of constraints on the (integer) values of the variables $z_{00}, z_{01}, \ldots, z_{0m}$, representing the initial contents of registers $I_0, I_1, \ldots, I_m$. These variables may be viewed as the independent variables, whereas the values of the auxiliary variables introduced below are determined by the values of the x_{0j}'s and the z_{0j}'s. In particular, let x_{tj} denote the contents of R_j after t time units. Analogously, let z_{tj} denote the contents of I_j after t time units.

We assume the instructions of Π are labeled by consecutive integers $\ell = 1$,

..., s ($s \geq 2$). Without loss of generality, assume the instruction labeled s is the only HALT in the program.

For each instruction labeled ℓ, and for $t = 1, \ldots, m$, let $y_{t\ell}$ be a variable such that $y_{t\ell} = \mathbf{true} \in D$ if the instruction that is executed during the tth time unit is the one labeled ℓ, and $y_{t\ell} = \mathbf{false} \in D$ otherwise. Similarly, for every i ($i = 1, \ldots, n$), let u_{ti} be a variable such that $u_{ti} = \mathbf{true}$ if during the tth time unit the input head is positioned over the ith square of the input tape, and $u_{ti} = \mathbf{false}$ otherwise. Also, for every i ($i = 1, \ldots, m$), let v_{ti} be a variable such that $v_{ti} = \mathbf{true}$ if during the tth time unit the output head is positioned over the ith square of the output tape, and $v_{ti} = \mathbf{false}$ otherwise. Finally, denote by τ_j ($j = 1, \ldots, m$) contents of the jth square of the output tape at the end of the run.

Below we state constraints ensuring the correct interpretation of the variables defined above.

4.4. Constraints

We first introduce constraints to ensure that at most one instruction is executed during each time unit, and the input head and the output head are each positioned over at most one square.

4.4.1. General Constraints

For $t = 0, 1, \ldots, m$, we impose the following constraints:

(i) For $\ell = 1, \ldots, s$ and $q \neq \ell$, $q = 1, \ldots, s$,

$$(y_{t\ell} = \mathbf{true}) \quad \Rightarrow \quad (y_{tq} = \mathbf{false}),$$

and also

$$y_{01} = \mathbf{true}.$$

(ii) For $i = 1, \ldots, n$ and $j \neq i$, $j = 1, \ldots, n$,

$$(u_{ti} = \mathbf{true}) \quad \Rightarrow \quad (u_{tj} = \mathbf{false}),$$

and also

$$u_{01} = \mathbf{true}.$$

(iii) For $i = 1, \ldots, m$ and $j \neq i$, $j = 1, \ldots, m$,

$$(v_{ti} = \mathbf{true}) \quad \Rightarrow \quad (v_{tj} = \mathbf{false}),$$

and also

$$v_{01} = \mathbf{true}.$$

The RAM recognizes an instance of L by printing **true** in the first square of the output tape. Thus, we impose the constraint:

$$\tau_1 = \mathbf{true}.$$

The rest of the constraints are derived directly from the definitions of the various instructions.

Consider any instruction whose label is ℓ. We impose a set of constraints, as a function of t, for $t = 1, \ldots, m$, depending on the nature of the instruction.

(i) If the instruction is one of the following: ILOAD, DLOAD, ISTORE, DSTORE, $\circ$, $\circ_\xi$, INC, DEC, READ, and WRITE, then we impose the following constraint:

$$(y_{t\ell} = \mathbf{true}) \quad \Rightarrow \quad (y_{t+1,\ell+1} = \mathbf{true}).$$

(ii) If the instruction is one of the following: ILOAD, ISTORE, WRITE, JUMP, JUMPT, and JUMP0, then we impose the following constraint:

$$(y_{t\ell} = \mathbf{true}) \quad \Rightarrow \quad (x_{ti} = x_{t-1,i}) \quad (i = 0, \ldots, m).$$

(iii) If the instruction is one of the following: DLOAD, DSTORE, $\circ$, $\circ_\xi$, READ, WRITE, JUMP, JUMPT, JUMP0, then we impose the following constraint:

$$(y_{t\ell} = \mathbf{true}) \quad \Rightarrow \quad (z_{ti} = z_{t-1,i}) \quad (i = 0, \ldots, m).$$

(iv) For any instruction other than READ,

$$(y_{t\ell} = \mathbf{true}) \quad \Rightarrow \quad (u_{t+1,i} = u_{t,i}) \quad (i = 1, \ldots, n).$$

(v) For any instruction other than WRITE,

$$(y_{t\ell} = \mathbf{true}) \quad \Rightarrow \quad (v_{t+1,i} = v_{t,i}) \quad (i = 1, \ldots, m).$$

More instruction related constraints are as follows.

HALT:

Recall that we have assumed the only HALT in the program is the last instruction. We impose the following constraint:

$$(y_{ts} = \mathbf{true}) \quad \Rightarrow \quad (y_{t+1,s} = \mathbf{true}).$$

ILOAD *a*

Here a may be either $=i$, i, or $*i$. First, we impose

$$(y_{t\ell} = \mathbf{true}) \quad \Rightarrow \quad (z_{tj} = z_{t-1,j}) \quad (j = 1, \ldots, m).$$

Next, if a is $=i$, we write:

$$(y_{t\ell} = \mathbf{true}) \quad \Rightarrow \quad (z_{t0} = i),$$

and if a is i, we write:

$$(y_{t\ell} = \mathbf{true}) \quad \Rightarrow \quad (z_{t0} = z_{t-1,i}).$$

If a is $*i$, we write for every j, $j = 0, \ldots, m$,

$$((y_{t\ell} = \mathbf{true}) \wedge (z_{t-1,i} = j)) \quad \Rightarrow \quad (z_{t0} = z_{t-1,j}).$$

DLOAD a

Here a is either $=\xi$, i, or $*i$. First,

$$(y_{t\ell} = \textbf{true}) \quad \Rightarrow \quad (x_{tj} = x_{t-1,j}) \quad (j = 1, \ldots, m).$$

Without loss of generality, we assume that $v(a)$ is a member of D (rather than a blank). Next, if a is $=\xi$, we write:

$$(y_{t\ell} = \textbf{true}) \quad \Rightarrow \quad (x_{t0} = \xi),$$

and if a is i, we write:

$$(y_{t\ell} = \textbf{true}) \quad \Rightarrow \quad (x_{t0} = x_{t-1,i}).$$

If a is $*i$, we write for every j, $j = 0, \ldots, m$,

$$((y_{t\ell} = \textbf{true}) \wedge (z_{t-1,i} = j)) \quad \Rightarrow \quad (x_{t0} = x_{t-1,j}).$$

ISTORE i

(i) $(y_{t\ell} = \textbf{true}) \Rightarrow (z_{tj} = z_{t-1,j}) \quad (j = 0, \ldots, m, j \neq i)$.
(ii) $(y_{t\ell} = \textbf{true}) \Rightarrow (z_{ti} = z_{t-1,0})$.

ISTORE $*i$

(i) $((y_{t\ell} = \textbf{true}) \wedge (z_{t-1,i} \neq j)) \Rightarrow (z_{tj} = z_{t-1,j}) \quad (j = 0, \ldots, m)$.
(ii) $((y_{t\ell} = \textbf{true}) \wedge (z_{t-1,i} = j)) \Rightarrow (z_{tj} = z_{t-1,0}) \quad (j = 0, \ldots, m)$.

DSTORE i

(i) $(y_{t\ell} = \textbf{true}) \Rightarrow (x_{tj} = x_{t-1,j}) \quad (j = 0, \ldots, m, j \neq i)$.
(ii) $(y_{t\ell} = \textbf{true}) \Rightarrow (x_{ti} = x_{t-1,0})$.

DSTORE $*i$

(i) $((y_{t\ell} = \textbf{true}) \wedge (z_{t-1,i} \neq j)) \Rightarrow (x_{tj} = x_{t-1,j}) \quad (j = 0, \ldots, m)$.
(ii) $((y_{t\ell} = \textbf{true}) \wedge (z_{t-1,i} = j)) \Rightarrow (x_{tj} = x_{t-1,0}) \quad (j = 0, \ldots, m)$.

$\circ\, a$ **and** $\circ_\xi$

First,

$$(y_{t\ell} = \textbf{true}) \quad \Rightarrow \quad (x_{tj} = x_{t-1,j}) \quad (j = 1, \ldots, m).$$

In the case of $\circ_\xi$,

$$(y_{t\ell} = \textbf{true}) \quad \Rightarrow \quad (x_{t0} = x_{t-1,0} \circ \xi).$$

In the case of $\circ\, a$, if a is i, then

$$(y_{t\ell} = \textbf{true}) \quad \Rightarrow \quad (x_{t0} = x_{t-1,0} \circ x_{t-1,i}),$$

and if a is $*i$, then we write for every j ($j = 0, \ldots, m$)

$$((y_{t\ell} = \textbf{true}) \wedge (z_{t-1,i} = j)) \quad \Rightarrow \quad (x_{t0} = x_{t-1,0} \circ x_{t-1,j}).$$

INC and DEC

(i) $(y_{t\ell} = \textbf{true}) \Rightarrow (z_{tj} = z_{t-1,j})$ $(j = 1, \ldots, m)$.
(ii) $(y_{t\ell} = \textbf{true}) \Rightarrow (z_{t0} = z_{t-1,0} \pm 1)$ (+ in the case of INC, − in the case of DEC).

READ i

(i) $((y_{t\ell} = \textbf{true}) \wedge (u_{tj} = \textbf{true})) \Rightarrow (x_{ti} = \sigma_j)$ $(j = 1, \ldots, n)$.
(ii) $((y_{t\ell} = \textbf{true}) \wedge (u_{tj} = \textbf{true})) \Rightarrow (u_{t+1,j+1} = \textbf{true})$ $(j = 1, \ldots, n)$.

READ $*i$

(i) $((y_{t\ell} = \textbf{true}) \wedge (u_{tj} = \textbf{true}) \wedge (z_{t-1,i} = q)) \Rightarrow (x_{tq} = \sigma_j)$ $(j = 1, \ldots, n,\ q = 1, \ldots, m)$.
(ii) $((y_{t\ell} = \textbf{true}) \wedge (u_{tj} = \textbf{true})) \Rightarrow (u_{t+1,j+1} = \textbf{true})$ $(j = 1, \ldots, n)$.

WRITE i

(i) $((y_{t\ell} = \textbf{true}) \wedge (v_{tj} = \textbf{true})) \Rightarrow (\tau_j = x_{ti})$ (for $j = 1, \ldots, m$).
(ii) $((y_{t\ell} = \textbf{true}) \wedge (v_{tj} = \textbf{true})) \Rightarrow (v_{t+1,j+1} = \textbf{true})$ (for $j = 1, \ldots, m$).

WRITE $*i$

(i) $((y_{t\ell} = \textbf{true}) \wedge (v_{tj} = \textbf{true}) \wedge (z_{t-1,i} = q)) \Rightarrow (\tau_j = x_{tq})$ (for $j = 1, \ldots, m$ and $q = 1, \ldots, m$).
(ii) $((y_{t\ell} = \textbf{true}) \wedge (v_{tj} = \textbf{true}) \Rightarrow (v_{t+1,j+1} = \textbf{true})$ (for $j = 1, \ldots, m$).

JUMP ℓ'

$(y_{t\ell} = \textbf{true}) \Rightarrow (y_{t+1,\ell'} = \textbf{true})$.

JUMPT ℓ'

(i) $((y_{t\ell} = \textbf{true}) \wedge (x_{t-1,0} \in T)) \Rightarrow (y_{t+1,\ell'} = \textbf{true})$.
(ii) $((y_{t\ell} = \textbf{true}) \wedge (x_{t-1,0} \notin T)) \Rightarrow (y_{t+1,\ell+1} = \textbf{true})$.

JUMP0 ℓ'

(i) $((y_{t\ell} = \textbf{true}) \wedge (z_{t-1,0} = 0)) \Rightarrow (y_{t+1,\ell'} = \textbf{true})$.
(ii) $((y_{t\ell} = \textbf{true}) \wedge (x_{t-1,0} \neq 0)) \Rightarrow (y_{t+1,\ell+1} = \textbf{true})$.

Proposition 4.1. *For any language L in* NP *and any nondeterministic polynomial-time RAM for recognizing instances in L, given an input (σ, v), the system of constraints defined in Subsection 4.4 can be constructed in polynomial time. Moreover, the string (σ, v) is in L if and only if there exists an assignment of appropriate values to the variables x_{ti}, y_{ti}, z_{ti}, u_{ti}, v_{ti}, and τ_j, so that all the constraints are satisfied.*

Proof. The proof follows from the fact if the interpretation of the variables is correct (and this is guaranteed if the constraints are satisfied), then the state of the machine at every stage of the execution is described precisely by these variables. ■

To summarize the reduction, suppose a description of a RAM Π in NP is given. Recall that the constraints are imposed for each time unit t. It is easy to see from the above that we can write a program Π^* (i.e., a RAM) that does the following. It receives as input any pair (σ, v) which is presented to Π. Recognizing the length the input, Π^* develops the set of constraints defined above and creates an instance of the satisfiability problem.

Acknowledgment. Helpful conversations with Steve Smale are gratefully acknowledged.

References

[1] A.V. Aho, J.E. Hopcroft and J.D. Ullman, *The Design and Analysis of Computer Algorithms*, Addison-Wesley, Reading, MA, 1976.

[2] L. Blum, M. Shub, and S. Smale, "On a theory of computation and complexity over the real numbers: NP-completeness, recursive functions and universal machines," *Bull. Amer. Math. Soc.* **21** (1989), 1–46.

[3] S.A. Cook, "The complexity of theorem proving procedures," *Proceedings of the 3rd Annual ACM Symposium on Theory of Computing* (1971), pp. 151–158.

[4] S. Smale, talk at the Workshop on Computational Complexity, IMPA, Rio de Janeiro, January, 1990.

40
Some Remarks on Bezout's Theorem and Complexity Theory

MICHAEL SHUB*

We begin by establishing the smoothness and irreducibility of certain algebraic varieties. Whereas these facts must be standard to algebraic geometers, they do not seem readily available.

For d and n positive integers, let $\mathscr{F}_{d,n}$ and $\mathscr{H}_{d,n}$ denote the spaces of polynomial mappings and homogeneous polynomial mappings $f: \mathbb{C}^n \to \mathbb{C}$ of degree less than or equal to d in the case of $F_{d,n}$ and equal to d in the case of $H_{d,n}$. For a multi-index $D = (d_1, \ldots, d_k)$, let $\mathscr{F}_{D,n}$ and $\mathscr{H}_{D,n}$ be the products

$$\prod_{i=1}^{k} \mathscr{F}_{d_i,n} \quad \text{and} \quad \prod_{i=1}^{k} \mathscr{H}_{d_i,n}.$$

So an element $F \in \mathscr{F}_{D,n}$ or $\mathscr{H}_{D,n}$ is a polynomial mapping $F: \mathbb{C}^n \to \mathbb{C}^k$. The evaluation map ev: $\mathscr{F}_{D,n} \times \mathbb{C}^n \to \mathbb{C}^k$ and ev: $\mathscr{H}_{D,n} \times \mathbb{C}^n \to \mathbb{C}^k$ is just the map $(F, x) \to F(x)$. For fixed F, $F^{-1}(0) \subset \mathbb{C}^n$ is the algebraic set determined by the simultaneous vanishing of the $f_{d_i,n}$.

Lemma 1. *Any $y \in \mathbb{C}^k$ is a regular value for*

$$\text{ev}: H_{D,n} \times (\mathbb{C}^n - \{0\}) \to \mathbb{C}^k$$

and

$$\text{ev}: F_{D,n} \times \mathbb{C}^n \to \mathbb{C}^k.$$

PROOF. $D\text{ev}_{(F,x)}(h, v) = h(x) + DF_x(v)$. The values of $h(x)$ alone are sufficient to make $D_{(F,x)}\text{ev}$ surjective.

Thus, by the implicit function theorem the union of the algebraic sets determined by the F's is smooth in the product. For $F \in \mathscr{F}_{D,n}$, let $Z_F = \{x | F(x) = 0\}$, $Z_{F_{D,n}} = Z_{\mathscr{F}} = \{(F, x) | F(x) = 0\}$. Similarly for $F \in H_{D,n}$, let $Z_F = \{x \in \mathbb{C}^n - \{0\} | F(x) = 0\}$, $Z_{\mathscr{H}_{D,n}} = Z_{\mathscr{H}} = \{(F, x) \in \mathscr{H}_{D,n} \times (\mathbb{C}^n - \{0\}) | F(x) = 0\}$, and $Z_{\hat{\mathscr{H}}_{D,n}} = Z_{\hat{\mathscr{H}}} = \{(F, x) \in Z_{\mathscr{H}} | F \not\equiv 0\}$; $Z_{\mathscr{F}}$ and $Z_{\mathscr{H}}$ are $\text{ev}^{-1}(0)$, so we have:

Proposition 1. (a) *$Z_{\mathscr{F}}$ is a connected smooth variety in $\mathscr{F}_{D,n} \times \mathbb{C}^n$ of codimension k;*

* Partially supported by an NSF grant.

(b) *$Z_{\mathscr{H}}$ and $Z_{\hat{\mathscr{H}}}$ are connected smooth varieties in $\mathscr{H}_{D,n} \times (\mathbb{C}^n - \{0\})$ of codimension k.*

PROOF. (b) It remains to prove the connectedness. The group of linear isomorphisms acts transitively on $\mathbb{C}^n - \{0\}$, and on $\mathscr{H}_{D,n} \times (\mathbb{C}^n - \{0\})$ by $(F, x) \to (F \circ L^{-1}, Lx)$, this action preserves $Z_{\mathscr{H}}$ and $Z_{\hat{\mathscr{H}}}$. Thus, the maps $Z_{\mathscr{H}} \to \mathbb{C}^n - \{0\}$, $Z_{\hat{\mathscr{H}}} \to \mathbb{C}^n - \{0\}$ are surjective locally trivial fibrations with connected base and connected fiber so they are connected, as follow: Given (f_1, x_1) and (f_2, x_2) in $Z_{\hat{\mathscr{H}}}$, choose a path x_t from x_1 to x_2, for $1 \le t \le 2$. Now lift x_t to $(\hat{f}_t, x_t)$ such that $\hat{f}_1 = f_1$; the endpoint of this path is in the fiber over x_2. This is a complex linear space minus 0 and, hence, is connected, so we continue the path in the fiber to (f_2, x_2). The same argument holds for $Z_{\mathscr{H}}$; for $Z_{\mathscr{F}}$, simply replace the linear group by the affine group.

We use $P(V)$ to denote the projective space of the vector space V, i.e., $V - 0$ mod the action of the nonzero scalars $\mathbb{C}^* = \mathbb{C} - 0$ and $P\mathbb{C}(n-1)$ for the projective space of $\mathbb{C}^n$.

$$\mathbb{C}^* \times \mathbb{C}^* \text{ acts freely on } (\mathscr{H}_{D,n} - \{0\}) \times (\mathbb{C}^n - \{0\})$$

by coordinatewise multiplication. Let N denote the dimension of $\mathscr{H}_{D,n}$:

$$N = \sum_{i=1}^{k} \binom{n + d_i - 1}{d_i}.$$

The $\mathbb{C}^* \times \mathbb{C}^*$ action leaves $Z_{\hat{\mathscr{H}}} \subset \mathscr{H}_{D,n} \times \mathbb{C}^n - \{0\}$ invariant. As the action is transversal to $S^{2N-1} \times S^{2n-1}$, $Z_{\hat{\mathscr{H}}} \cap S^{2N-1} \times S^{2n-1}$ is a smooth manifold, and, therefore, the quotient of $Z_{\hat{\mathscr{H}}}$ by the $\mathbb{C}^* \times \mathbb{C}^*$ action is the same as $Z_{\hat{\mathscr{H}}} \cap S^{2N-1} \times S^{2n-1}$ by the unit complexes $S^1 \times S^1$. This later group is compact. So the quotient by the free action is a smooth subvariety $\mathscr{Z}_{D,n} = \mathscr{Z}$ of $P(\mathscr{H}_{D,n}) \times P\mathbb{C}(n-1)$. As $Z_{\hat{\mathscr{H}}}$ is connected, so is the quotient manifold $\mathscr{Z}$. A connected, smooth projective variety is irreducible.

Theorem 1. *$\mathscr{Z}_{D,n}$ is a connected, smooth irreducible projective subvariety of $P(\mathscr{H}_{D,n}) \times P\mathbb{C}(n-1)$ of codimension k.*

Let $C_{D,n} = \{F \in \mathscr{H}_{D,n} - \{0\} | \exists x \in \mathbb{C}^n - \{0\} \text{ with } F(x) = 0\}$, i.e., $C_{D,n}$ is the set of those systems with a common root. Let $\mathscr{C}_{D,n}$ be the image of $C_{D,n}$ in $P(\mathscr{H}_{D,n})$.

Corollary 1. *$\mathscr{C}_{D,n}$ is an irreducible subvariety of $P(\mathscr{H}_{D,n})$.*

PROOF. It is the projection of $\mathscr{Z}_{D,n}$ on $P(\mathscr{H}_{D,n})$; as $\mathscr{Z}_{D,n}$ is irreducible, its image must be.

The case of $(n-1)$ homogeneous polynomials in n variables is the case of Bezout's theorem; there are generically

$$\prod_{i=1}^{n-1} d_i$$

roots and $\mathscr{C}_{D,n} = P(\mathscr{H}_{D,n})$ has the same dimension as $\mathscr{Z}_{D,n}$, i.e., the map $\mathscr{Z}_{D,n} \subset P(\mathscr{H}_D) \times \mathbb{C}P(n-1)$ induced by projection on the first factor is a surjection, almost every point is a regular point of the projection, and the fiber has πd_i points. We give a proof here.

Bezout's Theorem

Let $F = (f_1, \ldots, f_{n-1}) \in P(\mathscr{H}_{D,n})$, where f_i is homogeneous of degree $d_i > 0$, not all identically zero. Let $Z_j = Z_j(F)$, $j = 1, \ldots, k$, be the connected components of $\mathscr{Z}_F \subset \mathbb{C}P(n-1)$, where $\mathscr{Z}_F$ is the projection into $\mathbb{C}P(n-1)$ of $Z_F - \{0\} = \{x \in \mathbb{C}^{n-1} - \{0\} \mid F(x) = 0\}$. Then we may assign an index $i(Z_j)$ to each Z_j which is

(a) positive,
(b) $\sum i(Z_j) = \prod_{i=1}^{n} d_i$,
(c) $i(Z_j) = 1$ for a nondegenerate isolated zero, and
(d) there are neighborhoods U_j of Z_j in $\mathbb{C}P(n-1)$ and $\mathscr{N}$ of F in $\mathscr{H}_{D,n}$ such that if $G = (g_1, \ldots, g_{n-1}) \in \mathscr{N}$, then $Z(G) \subset \bigcup_j U_j$ and $\sum i(Z_\ell(G)) = i(Z_j)$, where the sum is taken over all components of $Z(G)$ contained in U_j.

Proof. First consider closed disjoint neighborhoods U_j of Z_j in $\mathbb{C}P(n-1)$ and a ball $\mathscr{N}$ around $F = (f_1, \ldots, f_{n-1}) \in P(\mathscr{H}_{D,n})$ such that $Z(G) \subset \bigcup \mathring{U}_j$ for all $G \in \mathscr{N}$. The critical values of the projection $\pi\colon \mathscr{Z}_{D,n} \to P(\mathscr{H}_{D,n})$ lie in a subvariety $\mathscr{D}$ of $P(\mathscr{H}_{D,n})$ (the "discriminant variety"). A fairly standard calculation shows that the regular points of the projection π are precisely those F with nondegenerate zeros. If we consider the polynomial system

$$F = (f_1, \ldots, f_{n-1}), \qquad f_i(x_1, \ldots, x_n) = x_i^{d_i} - x_n^{d_i},$$

then we see that F has $\pi\, d_i$ nondegenerate zeros. Thus, the regular values of π are open and $\mathscr{D}$ has codimension at least one. Therefore, $P(\mathscr{H}_{D,n}) - \mathscr{D}$ is arc connected. Continuing the roots along paths in $P(\mathscr{H}_{D,n}) - \mathscr{D}$, we see the number of roots is constant, off $\mathscr{D}$. Moreover, $\mathscr{N} - \mathscr{D}$ is also arc connected, and we define $i(Z_j)$ to be the number of roots of any G in $\mathscr{N} - \mathscr{D}$ which lie in U_j. This establishes (b), (c), and (d). To prove (a), we return to the projection $\pi\colon \mathscr{Z}_{D,n} \to P(\mathscr{H}_{D,n})$. Given $F \in P(\mathscr{H}_{D,n})$, the connected components Z_j, $j = 1, \ldots, k$, of $\mathscr{Z}_F$, and disjoint neighborhoods V_i of $\{F\} \times Z_j$ in $\mathscr{Z}_{D,n}$, we find points in $\mathscr{N} - \mathscr{D}$ which are in $\pi(V_i)$ for each i, as follows: Since $\mathscr{Z}_{D,n}$ is irreducible, the noncritical values of π are open and dense, the image of every open set contains interior, and since $\mathscr{N} - \mathscr{D}$ is open and dense in $\mathscr{N}$, there must be points in V_i with image in $\mathscr{N} - \mathscr{D}$. Now given neighborhoods U_j of Z_j, choose $\mathscr{N}$ small enough so that the zeros of all G in $\mathscr{N}$ are in $\bigcup U_j$ and let $V_i = (\mathscr{N} \times U_i) \cap \mathscr{Z}_{D,n}$. This finishes the proof.

Remark 1. The indices $i(Z_j)$ may be given several topological definitions. Let $F = (f_1, \dots, f_{n-1})$, where the f_i are irreducible and let $\mathscr{Z}_{f_i}$ be the projection into $\mathbb{C}P(n-1)$ of the set Z_{f_i}. Thus, $\mathscr{Z}_F = \bigcap \mathscr{Z}_{f_i}$. Let

$$\mathscr{Z} = \mathscr{Z}_{f_1} \times \mathscr{Z}_{f_2} \times \cdots \times \mathscr{Z}_{f_{n-1}} \subset \underbrace{\mathbb{C}P(n-1) \times \cdots \times \mathbb{C}P(n-1)}_{n-1}$$

$$\equiv \mathbb{C}P^{n-1}(n-1).$$

Let $\Delta = \{(z_1, \dots, z_{n-1}) \in \mathbb{C}P^{n-1}(n-1) | z_i = z_j \ \forall_{i,j}\}$ be the small diagonal. $\mathscr{Z}_F$ is homeomorphic to $\mathscr{Z} \cap \Delta$. We can now compute the homological intersection of $\mathscr{Z}$ and Δ in two ways. First, since f_i is irreducible, $Z(f_i)$ represents d_i times the generator in $H_{2_{n-4}}(\mathbb{C}P(n-1))$. Now the algebraic structure on $\mathbb{C}P^{n-1}(n-1)$ and the Kunneth formula gives πd_i for the intersection of $\mathscr{Z}$ and Δ.

Next consider

$$H_0(\Delta) \leftarrow H_{(2n-4)(n-1)}(\mathbb{C}P^{n-1}(n-1), \mathbb{C}P^{n-1}(n-1) - \Delta)$$

$$H_{(2n-4)(n-1)}(\mathscr{Z}) \to H_{(2n-4)(n-1)}(\mathbb{C}P^{n-1}(n-1)) \to H_{(2n-4)(n-1)}(\mathbb{C}P^{n-1}(n-1), \mathbb{C}P^{n-1}(n-1) - \Delta)$$

$$H_{(2n-4)(n-1)}(\mathscr{Z}) \searrow H_{(2n-4)(n-1)}(\mathscr{Z}, \mathscr{Z} - \mathscr{Z} \cap \Delta) \approx \sum_{i=1}^{k} H_{(2n-4)(n-1)}(V_i, V_i - Z_i) \nearrow H_{(2n-4)(n-1)}(\mathbb{C}P^{n-1}(n-1), \mathbb{C}P^{n-1}(n-1) - \Delta)$$

where V_i are disjoint closed neighborhoods of the connected components Z_i of $\mathscr{Z} \cap \Delta$ in $\mathscr{Z}$. The intersection number is the image of the generator of $H_{(2n-4)(n-1)}(\mathscr{Z})$ in $H_0(\Delta)$. All the maps are induced by inclusion, except for the isomorphism in the bottom row which is given by excision and the map

$$H_{(2n-4)(n-1)}(\mathbb{C}P^{n-1}(n-1), \mathbb{C}P^{n-1}(n-1) - \Delta) \to H_0(\Delta)$$

which is the Thom isomorphism (see [Dold]). This gives a topological method of computing $i(Z_j)$ when $F = (f_1, \dots, f_{n-1})$ has irreducible components. If Z_i is a nondegenerate zero, $H_{(2n-4)(n-1)}(V_i, V_i - Z_i)$ contributes *a* plus 1 to the sum.

Remark 2. We can also give a topological computation of the index as the degree of the map

$$\pi_Y\colon H_{\text{top}}(\mathscr{Z}_{D,n}, \mathscr{Z}_{D,n} - \{f\} \times Z_i) \to H_{\text{top}}(P(\mathscr{H}_{D,n}), P(\mathscr{H}_{D,n}) - \{f\})$$

(see [Dold]).

Zulehner has proposed using continuation techniques along a projective line to solve systems of equations [Zulehner]. Canny's generalized characteristic polynomial methods show some of the same geometric features, which we state in the next theorem. Let L be a projective line contained in $P(\mathscr{H}_{D,n})$ and suppose that L is not contained in $\mathscr{D}$. Then for an open dense set of points U in L, $G \in U$ has πd_i zeros and $W = \mathscr{Z}_{D,n} \cap \pi^{-1}(U)$ is a πd_i-to-one covering space of U. $\pi^{-1}(L)$ is a subvariety of $\mathscr{Z}_{D,n}$ and W is Zariski dense in $R(L) = V_1 \cup \cdots \cup V_k$, where V_i are irreducible curves. We claim $R(L) = \overline{W}$ in the usual topology and is, in fact, W union a finite number of points. Suppose U' is an open set in $R(L) - W$, then it intersects some V_i in an open set and, hence, by irreducibility of V_i, W in an open set which is a contradiction. Thus, $R(L) = \overline{W}$. Moreover, $R(L) - W$ only contains points in $\pi^{-1}(L \cap \mathscr{D})$ which are π^{-1} of a finite number of points. These fibers which intersect $R(L)$ cannot contain varieties of dimension one, for these curves would then have open sets disjoint from $U \cap V_i$ for each i. So we have proven:

Theorem 2. *Let $L \subset P(\mathscr{H}_{D,n})$ be a projective line not contained in $\mathscr{D}$ and $U = L - \mathscr{D}$. Let $\mathscr{Z}_{D,n} \subset P(\mathscr{H}_{D,n}) \times \mathbb{C}P(n-1)$ be the projection of $\{(F,x) \in \mathscr{H}_{D,n} - \{0\} \times \mathbb{C}^n - \{0\} | F(x) = 0\}$ and $\pi\colon \mathscr{Z}_{D,n} \to P(\mathscr{H}_{D,n})$ the projection. Let $W = \overline{\pi^{-1}(U)}$. Then W is $\pi^{-1}(U)$ union a finite number of points. π maps W onto L. If $G_i \in U$ and converge to F, then the roots of G_i converge to $\pi^{-1}(F) \cap W$. Moreover, they have a limit point in each connected component of $\mathscr{Z}_F$.*

Proof. The discussion immediately preceding the theorem proves all assertions but the last sentence, which follows from Bezout's Theorem.

Zulehner suggests using a variant of Newton's method to do continuation of roots on real lines contained in a projective line L. Since $L \cap \mathscr{D}$ has at most degree $\mathscr{D}$ points, most real lines in L do not meet $\mathscr{D}$. We are back to a familiar situation, the projective line L is the Riemann sphere with degree $\mathscr{D}$ points removed representing the discriminant variety in L. We may choose a polynomial, say $F = f_i = z_i^{d_i} - z_n^{d_i}$, and consider real lines connecting F to the system we wish to solve, G. For example, we may take the projective line

$$uG + vF$$

and let the ratios of u and v lie on the $2 \deg \mathscr{D}$ lines of ratio

$$\frac{u}{v} = \exp\left(i \frac{2\pi j}{2 \deg \mathscr{D}}\right) \qquad \text{for } j = 1, \ldots, 2 \deg \mathscr{D}.$$

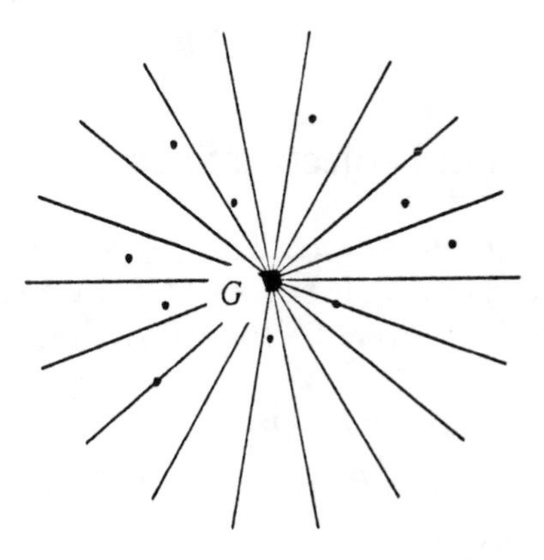

Figure 1

This is reminiscent of Smale's original approach to the fundamental theorem of algebra.

Recall that the polynomial f is thought of as taking the complex number $f: \mathbb{C} \to \mathbb{C}$. In the target space

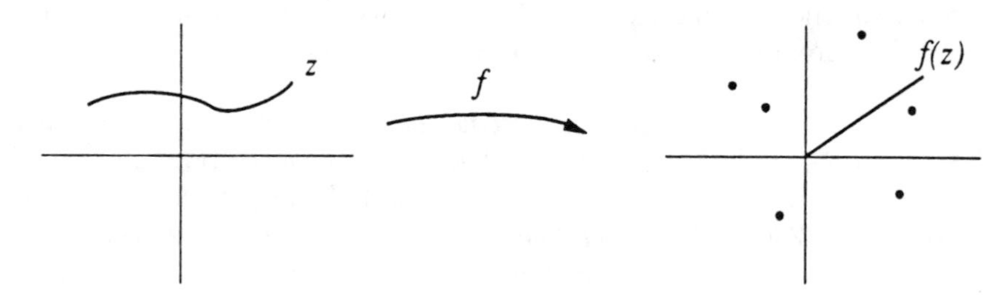

FIGURE 2

a point z is chosen in the domain and a straight line drawn from $f(z)$ to 0; as long as the critical values of f are avoided, the line can be lifted back and continued to a root ξ of f. If we think of the image plane as a plane of polynomials given by changing the constant terms, we have sloved $f - f(z)$ (with z) then drawn the straight line to f, i.e., $(1-t)(f - f(z)) + tf = G_t$, so $G_0 = f - f(z)$ and $G_1 = f$. Avoiding the critical points of f exactly says that the line of polynomials G_t never has a double root, i.e., that this line misses the discriminant variety in the space of polynomials. Now the analogy is complete. It is not so easy to find polynomials whose roots we know and which are not in $\mathscr{D}$, but $F = (f_1, \ldots, f_{n-1})$ and $f_i = z_i^{d_i} - z_n^{d_i}$, $i = 1, \ldots, n$, is a simple example.

Now we need a Newton's method. There does not seem to be a natural Newton's method on projective space. In fact, what we propose is not analytic and fails to be the classical method even for a polynomial of one complex variable which is then homogenized as a two-variable polynomial which gives us a method back on $\mathbb{C}P(1) = S^2$. But this method is well-suited for Smale's α-theory.

Newton's Method in Projective Space

Let $F \in \mathscr{H}_{D,n}$, where $D = (d_1, \ldots, d_n)$, $F: \mathbb{C}^{n+1} \to \mathbb{C}^n$, $F = (f_1, \ldots, f_n)$, and f_i is homogeneous of degree $d_i \geq 1$. For $x \in \mathbb{C}^{n+1}$, let $PT(x) = \text{Null}(\bar{x})$, i.e., if $x = (x_1, \ldots, x_{n+1})$, $\text{Null}(\bar{x}) = \{w \in \mathbb{C}^{n+1} | w = (w_1, \ldots, w_{n+1}) \text{ and } \sum \bar{x}_i w_i = 0\}$. Let $Df(x)|PT(x) = P(x)$.

The projective Newton vector field is

$$PNV(x) = PNV_F(x) = -P(x)^{-1}F(x)$$

and is defined where $P(x)^{-1}$ exists.

The projective Newton method is

$$PN(x) = PN_F(x) = x - P(x)^{-1}F(x) = x + PNV(x).$$

Proposition 2. *$PNV(x)$ and $PN(x)$ transform appropriately and, hence, are defined on $\mathbb{C}P(n)$.*

PROOF. For $\lambda \in \mathbb{C}$, let $\Delta(\lambda^{d_i})$ be the linear map which takes $(x_1, \ldots, x_n) \to (\lambda^{d_1}x_1, \ldots, \lambda^{d_n}x_n)$.

By the chain rule for nonzero $\lambda \in \mathbb{C}$, $DF(\lambda x) = \Delta(\lambda^{d_i})DF(x)\lambda^{-1}$ and $PT(\lambda x) = PT(x)$. Thus,

$$\begin{aligned} PNV(\lambda x) &= -P(\lambda x)^{-1}F(\lambda x) \\ &= -(\Delta(\lambda^{d_i})P(x)\lambda^{-1})^{-1}\Delta(\lambda^{d_i})F(x) \\ &= -\lambda P(x)^{-1}F(x) = \lambda PNV(x). \end{aligned}$$

Thus,

$$PN(\lambda x) = \lambda x + PNV(\lambda x) = \lambda(x + PNV(x)) = \lambda PN(x)$$

and PNV and PN are defined on $\mathbb{C}P(n)$. See [Zulehner] for the same differential equations.

There is also a spherical version of PN, SN

$$SN(x) = \frac{x - PNV(x)}{\|x - PNV(x)\|} \quad \text{for } x \text{ of norm one.}$$

The figure following Theorem 2 shows rays which do not intersect the discriminant variety in a projective line L. Renegar gives bounds on Newton's method in terms of the distance to the discriminant variety in $P(\mathscr{H}_{D,n})$ [Renegar]. In [Shub], I give a lower bound to the distance in $P(\mathscr{H}_{D,n})$ compared to the distance in L.

Theorem 3. *Let $H \subset \mathbb{C}P(n)$ be a hypersurface of degree m and L a projective line. Let R be the furthest distance of a point in L from H. Then, for $x \in L$,*

$$m^{1/m}\pi(\csc R)d_{\mathbb{C}P(n)}(x, H)^{1/m} \geq d_L(x, H \cap L) \geq d_{\mathbb{C}P(n)}(x, H).$$

But I do not think that this estimate is good enough to get good complexity bounds. Canny and Renegar use the u resultant and generalizations; we will return to these later. The most problematical polynomial systems are those with excess components, i.e., the algebraic set of zeros has dimension larger than zero. The simplest example is two linear equations in three variables which are dependent:

$$\begin{aligned} ax + by + cz &= 0, \\ ax + by + cz &= 0. \end{aligned}$$

The homotopy then can add t times

$$x - z = 0,$$
$$y - z = 0$$

to the system, obtaining

$$(a + t)x + by + (c - t)z = 0,$$
$$ax + (b + t)y + (c - t)z = 0.$$

This is solved by the 2×2 determinants giving

$$(-tc + t^2, -tc + t^2, t(a + b) + t^2),$$

dividing by t gives

$$(-c + t, -c + t, (a + b) + t).$$

Letting t tend to zero selects the zero $(-c, -c, a + b)$ in the line of zeros of the original system.

Moreover, the zero of the system varies nicely with t. In general, one has for $\ell_1, \ell_2 \in L$ that the distance between the sets of zeros Z_{ℓ_1}, Z_{ℓ_2} given by Theorem 3 satisfy

$$d(Z_{\ell_1}, Z_{\ell_2}) \leq C_L d(\ell_1, \ell_2)^{1/\pi_{i=1}^{n-1} d_i},$$

but the Holder constant C_L which depends on L is not bounded, as computation even on linear examples shows.

Problem. Estimate C_L in terms of the distance from L to the subvariety of problems with excess components.

Given homogeneous polynomials f_i: $\mathbb{C}^n \to \mathbb{C}$, $i = 1, \ldots, n - 1$, add one more equation $u = \sum_{i=1}^n u_i x_i = 0$, where $(u_1, \ldots, u_n) \in \mathbb{C}^n$. The resultant of $R(f_1, \ldots, f_{n-1}, u)$, called the u resultant, is then a polynomial function $R(u)$ which vanishes precisely if there is a root $(\xi_1, \ldots, \xi_n)$ of the system $f_1, \ldots, f_{n-1}$ s.t. $\sum u_i \xi_i = 0$. If there are finitely many solution rays, then $R(u)$ vanishes on a finite union of hyperplanes see [Van der Waerden]. But if the system has excess components, $R(u)$ vanishes identically. Canny gets around this problem by considering the system $\hat{f}_i = f_i - s x_i^{d_i}$. The resultant now is a polynomial function in s and u, $R(s, u)$, $R(s, u) = \sum C_k(u) s^k$. Let k be the minimum integer such that $C_k(u) \not\equiv 0$. Call $C_k(u)$ the s, u resultant of f. Canny proves that $C_k(u)$ vanishes on a union of hyperplanes containing a subset corresponding to all the isolated roots of the system [Canny].

The situation is analogous to Theorem 2. Let $g_i(x) = x_i^{d_i}$ for $i = 1, \ldots, n - 1$. Then $f_i - s g_i$ represents a projective line L contained in $P(\mathscr{H}_{D,n})$, $D = (d_1, \ldots, d_{n-1})$. When $s = \infty$, the system of equations has a unique solution ray, $(0, \ldots, 0, 1)$. This implies that the set V of s for which there are finitely many solutions is open and dense in L. Now let π: $Z_{D,n} \to P(\mathscr{H}_{D,n})$, $Z_L = \pi^{-1}(L)$, and $W = \overline{\pi^{-1}(V)}$. π maps W onto L, the map is finite-to-one, and W contains all isolated solution points of the system $f_i - sg$, for any fixed s. Z_L is W union a finite number of excess components over finitely many values of s in L; this is because those s which correspond to excess components are determined by the u resultant being identically zero, which defines an algebraic subset of L. Now since the index of any connected component of zeros is positive, there are points of W arbitrarily close to any connected component of zeros. Therefore, W has nonempty intersection with every connected component of zeros of the system $f_i - sg_i$ for all s. We have thus proven an analogue of Theorem 2 for the line L and open dense subset V. I will not bother to restate this theorem except to draw a conclusion which partially answers a question in [Canny]. The s, u resultant of the system f_i, $C_k(u)$, vanishes on precisely the hyperplanes $\sum_{i=1}^{n} w_i U_i = 0$, where $(w_1, \ldots, w_n) \in \pi^{-1}(0) \cap W$. This follows from [Canny, Theorem 3.2].

Theorem 4. *Let f_i: $\mathbb{C}^n \to \mathbb{C}$ be homogeneous polynomials of degree $d_i \geq 1$ for $i = 1, \ldots, n - 1$. Let $C_k(u)$ be the s, u resultant of the system. Then $C_k(u)$ vanishes on a finite union of hyperplanes $\sum_{i=1}^{n} w_{i,j} u_i = 0$, where $(w_{ij}, \ldots, w_{nj})$ is a solution ray for every j. Moreover, for every connected component of solutions, there is a j s.t. $(w_{ij}, \ldots, w_{nj})$ in this component.*

Problem. What continuation methods correspond to W? Are they practical?

We return to the irreducible varieties $\mathscr{C}_{D,n} \subset P(\mathscr{H}_{D,n})$, where $D = (d_1, \ldots, d_k)$, $d_i \geq 1$. $\mathscr{C}_{D,n}$ is the projectivized set of homogeneous polynomials $(f_1, \ldots, f_k)$ of degrees d_i which have a common solution ray in $\mathbb{C}^n$. We have shown that $\mathscr{C}_{D,n}$ is irreducible, and from Bezout's theorem it follows that $\mathscr{C}_{D,n} = P(\mathscr{H}_{D,n})$ for $k \leq n - 1$. Here we give the codimension of $\mathscr{C}_{D,n}$ in the other case.

Theorem 5. *$\mathscr{C}_{D,n}$ is an irreducible subvariety of $P(\mathscr{H}_{D,n})$ of codimension* $\max(0, n - k + 1)$. *For $k \geq n$, the map from $\mathscr{Z}_{D,n} \subset P(\mathscr{H}_{D,n}) \times \mathbb{C}P(n - 1)$ to $\mathscr{C}_{D,n}$ induced by projection on the first factor is generically finite-to-one. When $k = n - 1$, the generic number of points is $\prod_{i=1}^{n-1} d_i$. When $k > n - 1$, the map is generally one-to-one.*

Proof. Add to the equations $f(x) = 0$, the equations $\det(M_i Df(x)) = 0$, where $M_i Df$ runs through the $(n - 1) \times (n - 1)$ minors of the derivative of f at x. Let $W_{D,n} \subset \mathscr{Z}_{D,n}$ be the subvariety defined by the vanishing of all the $\det(M_i Df(x))$. Let $V_{D,n} \subset \mathscr{C}_{D,n}$ be the image of $W_{D,n}$. First, we claim that if $F \in \mathscr{C}_{D,n} - V_{D,n}$, then every common zero ray of F is isolated and so they are finite in number. Next, to see that $\mathscr{C}_{D,n} - V_{D,n}$ is open and dense in $\mathscr{C}_{D,n}$, it suffices to produce an open set in $\mathscr{C}_{D,n} - V_{D,n}$. For this, find $(n - 1)$ homogeneous polynomials with finitely many solution rays and with the derivative at each solution ray of rank $(n - 1)$. Complete this system to one which still has common solution rays. Now, any perturbation of the initial $(n - 1)$ polynomials still has only finitely many solutions and each has derivative of rank $(n-1)$. $f_i = x_i^{d_i} - x_n^{d_i}$, $i = 1, \ldots, n-1$, are such a system of $(n-1)$ polynomials. We can add an nth polynomial which is a power of a linear map (L^{d_n}), where L vanishes at only one of common roots of the f_i; this will establish the last statement.

Remark 3. The fact that, for overdetermined systems $k > n-1$, the map from solutions to systems is generically one-to-one implies that the solution may be given as a rational function of the coefficients on a Zariski dense set of problems.

Remark 4. For $k = n$, that is, the case of n homogeneous equations in n variables of degree $D = (d_1, \ldots, d_n)$, $\mathscr{C}_{D,n} \subset P(\mathscr{H}_{D,n})$ is an irreducible hypersurface defined by an irreducible polynomial in the coefficients of the f_i. This is the resultant polynomial [Macauley, Van der Waerden, etc.]; it has degree

$$\left(\prod_{j=1}^{n} d_j\right) \sum_{i=1}^{n} 1/d_i.$$

If we specialize to homogeneous quadratic equations in n variables, the vector space $\mathscr{H}_{D,n}$ has dimension $n^2(n+1)/2$ and the resultant polynomial, the vanishing of which is necessary and sufficient for the system to have a nonzero root, has degree $n2^{n-1}$. Even in this case one may hope to prove that the resultant cannot be computed in polynomial cost in n and, thereby, establish that $P \neq NP$ as below.

Proposition 3. *Let $H \subset \mathbb{C}^N$ be an irreducible hypersurface given by the irreducible polynomial $P\colon \mathbb{C}^N \to \mathbb{C}$. Then any machine M over $\mathbb{C}$ which solves the decision problem $(\mathbb{C}^N, H)$ must compute a rational multiple of P on a Zariski dense open subset of H.*

Proof. For each path of nodes $n_1, \ldots, n_k$ which leads an input $h \in H$ to a yes output node, we have the corresponding subset $V_{n_1,\ldots,n_k} \subset H$ which outputs on this path. There are countably many $V_{n_1,\ldots,n_k}$, each of which is either contained in a hypersurface of H and, consequently, nowhere dense in H or contains an open set of H. As H is not the union of countably many nowhere dense sets, at least one $V_{n_1,\ldots,n_k}$ must be open.

Say the path of nodes encounter ℓ branch nodes and at each branch node the composite polynomial rational function computed is f_i/g_i, $i = 1, \ldots, \ell$. If f_i is not a multiple of P, then $V_{n_1,\ldots,n_k}$ must take the $\neq 0$ branch. If f_i is never a multiple of P, then $V_{n_1,\ldots,n_k}$ has always taken the $\neq 0$ branch, but then the same is true for a neighborhood of $V_{n_1,\ldots,n_k}$ in $\mathbb{C}^n$ and M has made an error, which finishes the proof. The argument actually also shows that $V_{n_1,\ldots,n_k}$ must be open and dense in H.

Corollary 2. *Any machine over $\mathbb{C}$ which solves the decision problem $(\mathscr{H}_{D,n}, \mathscr{C}_{D,n})$ must compute a multiple of the resultant on Zariski dense open subsets of $\mathscr{C}_{D,n}$ where $D = (d_1, \ldots, d_n)$, and in particular for n-quadratic homogeneous polynomials in n unknowns.*

This corollary was independently proven by Steve Smale. He was considering the question of $P \neq NP$ in various contexts. One context was machines over the integers with input size the bit input size, cost the bit cost, and branching over $=0$ or $\neq 0$.

Let Z_+ denote the non-negative integers. I suggested that (Z, Z_+) is not in P. Here is a proof worked out with Michael Ben-Or.

Proposition 4. (Z, Z_+) *is in NP but not in P for machines over Z branching on $\neq 0$ or $=0$ and with bit input size and bit cost.*

PROOF. First, to see that (Z, Z_+) is in NP, note that any non-negative integer is the sum of four squares. Now to see that the problem is not in P, we use the following lemma:

Lemma 2. *Let $f \in \mathscr{Z}[t]$ be an integral polynomial with at least k distinct integer roots. Suppose for some $w \in \mathscr{Z}$ that $f(w) \neq 0$, then $|f(w)| \geq ((k/2)!)^2$.*

PROOF. Let $r_1, \ldots, r_k$ be distinct integral roots of f, and let $Q = \prod_{i=1}^{k} (t - r_i)$. Then $f(t) = Q(t)P(t)$, where $P(t)$ is an integral polynomial. Since $P(w) \neq 0$, $|P(w)| \geq 1$, and, hence, $|f(w)| \geq |Q(w)| \geq ((k/2)!)^2$ since it is the product of k distinct nonzero integers.

Since $(k/2)!$ has more than $k/2$ bits, we see that at a branch node if k elements of $\mathscr{Z}$ take the $=0$ branch, then the cost of the computation is at least k. Now if there is a polynomial cost machine, at each branch only polynomial many inputs of size n may take the $=0$ branch which gives a contradiction because 2^n inputs must be eliminated by polynomially many nodes.

Problem. Is this proposition still true if the cost is reduced to the number of algebraic operations?

This problem is related to Problem 5.1 of [Blum–Shub–Smale]. Given the integers Z, branching on ≥ 0 or <0, and bit input size, does the class of polynomial cost decision problems P increase if the bit cost is reduced to the number of algebraic operations? In this context, the class NP certainly gets larger because it contains undecidable problems (Hilbert's tenth). Because the outputs are just 0 and 1, we might compute mod 2 which would make the algebraic and bit cost comparable. The problem is at branching nodes where inequalities are verified. A model problem is:

For fixed k and ℓ, give $2(k+\ell)$ positive integers.
$a_i, n_i, i = 1, \ldots, k, \quad b_j, m_j, j = 1, \ldots, \ell$.
Is $\prod_{i=1}^{k} a_i^{n_i} > \prod_{j=1}^{\ell} b_j^{m_j}$?

This problem is easily seen to be in P with cost the number of algebraic operations since the powers can be taken by repeated squaring. The bit size of the numbers, however, gets exponentially large in this procedure. Yet we have the following surprising theorem.

Theorem 6. *The decision problem, a_i, n_i, b_j, m_j positive integers, $i = 1, \ldots, k$, $j = 1, \ldots, \ell$, and $\prod_{i=1}^{k} a_i^{n_i} > \prod_{j=1}^{\ell} b_j^{m_j}$ is in P over Z with branching on ≥ 0 or < 0, bit input size, and bit cost.*

PROOF. We use Baker's theorem [Baker, Theorem 3.1], and refer to [Lang, Chapter XI, Theorem 1.1]. There is a constant $B > 0$ with the following property. Let $M = \max_{i,j}(a_i, b_j)$. Then either

$$\sum_{i=1}^{k} n_i \log a_i - \sum_{j=1}^{\ell} m_j \log b_j = 0$$

or

$$\left| \sum_{i=1}^{k} n_i \log a_i - \sum_{j=1}^{\ell} m_j \log b_j \right| > B^{(k+\ell)(\log M)^{k+\ell} \log\log M}.$$

Thus, we need only compute

$$\sum_{i=1}^{k} n_i \log a_i - \sum_{j=1}^{\ell} m_j \log b$$

to accuracy 0 (input size) $^{k+\ell+1}$ to determine if it is positive, zero, or negative. This entails computing a polynomial number of bits of the logs of the a_i and b_j. This can be done rather naively by dividing by a power of two and using the Taylor series for the log near one, or by Newton's method or more sophisticatedly as in [Borwein and Borwein].

This theorem suggests the following naive problem.

Problem. Given an integral polynomial P in variables x_i and variable exponents n_{i_l}, is it always possible to determine if $P = 0$, $P > 0$, or $P < 0$ in polynomial bit cost in the bit input size of the x_i and n_{i_l}?

References

Baker, A., *Transcendental Number Theory*, Cambridge University Press, Cambridge, 1975.

Blum, L., Shub, M., and Smale, S., On a theory of computation and complexity over the real numbers: NP-completeness, recursive functions and universal machines, *Bull. Amer. Math. Soc.* **21** (1989), 1–46.

Borwein, J. and Borwein, P., *Pi and the AGM*, Wiley, New York, 1987.

Canny, J.F., Generalized characteristic polynomials, *J. Symbol. Comput.* **9** (1990), 241–250.

Dold, A., *Lectures on Algebraic Topology*, 2nd ed., Springer-Verlag, Berlin, 1980.

Lang, S., *Elliptic Curves Diophantine Analysis*, Springer-Verlag, Berlin, 1978.

Macauley, F.S., *The Algebraic Theory of Modular Systems*, Cambridge Tracts in Mathematics and Mathematical Physics, No. 10, Steckart-Hafner Service Agency, New York, 1964.

Renegar, J., On the worst case arithmetic complexity of approximating zeros of systems of polynomials, *SIAM J. Comput.* **18** (1989), 350–370.

Shub, M., On the distance to the zero set of a homogeneous polynomial, *J. Complexity* **5** (1989), 303–305.

Van der Waerden, B.L., *Modern Algebra*, Vol. II, Frederick Unger, New York, 1950.

Zulehner, W., A simple homotopy method for determining all isolated solutions to polynomial systems, *Math. Comp.* **50** (1988), 167–177.

41
Some Results Relevant to Smale's Reports*

XINGHUA WANG

The background of this paper is Smale's report to the 20th International Congress of Mathematicians in 1986 and his material [1] written for this congress. We mainly list results about iterative convergence, estimates from data at one point, and complexity of numerical integrals from [7–13].

1. The Approximate Zero and Criterion for Newton's Iteration

1.1. Definition of an Approximate Zero

Let f be an analytic map from a real or complex Banach space $\mathbb{E}$ to another $\mathbb{F}$. Consider Newton's iteration $z_{n+1} = z_n - Df(z_n)^{-1}f(z_n)$ with the initial point $z_0 = z \in \mathbb{E}$.

For the definition of an approximate zero, Smale [1, 2] has given:

Definition 1. Let $z \in \mathbb{E}$. If Newton's iteration with $z_0 = z$, $z_{n+1} = z_n - Df(z_n)^{-1}f(z_n)$, is well-defined and satisfies

$$\|z_{n+1} - z_n\| \leq (\tfrac{1}{2})^{2^n-1}\|z_1 - z_0\|,$$

we call z an approximate zero of f for Newton's iteration.

Definition 2. Let $z \in \mathbb{E}$. If Newton's iteration with $z_0 = z$, $z_{n+1} = z_n - Df(z_n)^{-1}f(z_n)$, is well-defined and for some zero ζ of f,

$$\|\zeta - z_n\| \leq (\tfrac{1}{2})^{2^n-1}\|\zeta - z\|,$$

we call z as an approximate zero of the second kind of f.

In comparison with Smale's work [4], the above definition has been reasonably modified. This modification coincides with the concept of an approx-

* The subject was supported by Zhejiang Province Natural Funds.

imate zero of second order that we introduced in [6]. Using this concept, the quadratic convergence of Newton's method can be demonstrated.

1.2. Criterion

Thus, the problem is: What criterion can decide if z is an approximate zero of f (or one of the second kind)?

For this problem, we first recall Kantorovich's theory. Provided that Df is Lipschitz continuous on some convex region $D \subset \mathbb{E}$, this theory can be applied to prove convergence and estimate the speed of convergence of Newton's method in terms of the value of criterion h, where

$$h = h(z, f, D) = \beta K \qquad \text{and} \qquad \beta = \beta(z, f) = \|Df(z)^{-1}f(z)\|.$$

$K = K(z, f, D) = \sup_{w \in D} \|Df(z)^{-1}D^2f(w)\|$, where we assume that the distance from z to the boundary of D is not less than $2\beta/(1 + \sqrt{1 - 2h})$.

Obviously, in this case the criterion is complicated because one needs information on the region D. Smale once proposed to substitute the analyticity of f at z for Lipschitz continuity of Df in D and substitute

$$\gamma = \gamma(z, f) = \sup_{n \geq 2} \left\| Df(z)^{-1} \frac{D^n f(z)}{n!} \right\|^{1/(n-1)}$$

for $K = K(z, f, D)$ such that the criterion becomes

$$\alpha = \alpha(z, f) = \beta\gamma,$$

without any additional information on the region.

Smale's article [2] proved that there exists an absolute constant α_0 of about 0.130707, such that if $\alpha(z, f) < \alpha_0$, then z is an approximate zero of f for Newton's iteration. This is called the "point estimate" in [1]. This result is one of two theorems stated in the abstract given to the participants at the International Congress of Mathematicians.

On the other hand, $\|z - \zeta\| \gamma(\zeta, f)$ constitutes a criterion for an approximate zero of the second kind of f. Smale [2] also proved that if $\|z - \zeta\| \gamma(\zeta, f) < (3 - \sqrt{7})/2$, then z is an approximate zero of the second kind for Newton's iteration.

A natural problem is: Can we estimate the value of $K(z, f, D)$ in D from the value of $\gamma(z, f)$. If possible, then the point estimate result may be reduced to Kantorovich's theory. We have the following:

Theorem 1 [9]. *Let $f: \mathbb{E} \to \mathbb{F}$ be analytic and $z \in \mathbb{E}$. Suppose $a, \kappa \in \mathbb{R}_+$ such that*

$$a\gamma(z, f) \leq 1 - \sqrt[3]{\kappa(\sqrt{1 + (8/27)\kappa} + 1)} + \sqrt[3]{\kappa(\sqrt{1 + (8/27)\kappa} - 1)},$$

then

$$\kappa a K(z, f, \overline{S(z, a)}) \leq 1,$$

where $\overline{S(z, a)}$ represents a closed ball with center z and radius a.

Now setting $a = \frac{3}{2}\beta(t, f)$, $\kappa = \frac{3}{2}$ in Theorem 1, we have from

$$\alpha(z, f) \leq \frac{3}{2}\left\{1 - \sqrt[3]{\frac{\sqrt{13}+3}{2}} + \sqrt[3]{\frac{\sqrt{13}-3}{2}}\right\} \approx 0.121522$$

that

$$h(z, f, \overline{S(z, \tfrac{3}{2}\beta(z, f))}) \leq \tfrac{4}{9}$$

follows. Hence, according to Kontorovich's theory, improved by [14, 15, 20, 21], z is an approximate zero of f for Newton's iteration.

Setting $z = \zeta$, $a = \|z - \zeta\|$, $\kappa = 3$ in Theorem 1, we have, from

$$\|z - \zeta\| \gamma(\zeta, f) \leq 1 - \sqrt[3]{\sqrt{17}+3} + \sqrt[3]{\sqrt{17}-3} \approx 0.115378$$

that

$$3\|z - \zeta\| K(\zeta, f, \overline{S(z, \|z - \zeta\|)}) \leq 1$$

follows. Thus, according to [16, 22], z is an approximate zero of the second kind of f for Newton's iteration.

The possibility of deducing the point estimate from Kotorovich's theory has been pointed out by Smale in [1, 2]. But out estimation is optimal.

1.3. A Problem of Smale

Smale [1] proved the inequality $\alpha(\gamma, \varphi_\alpha) \leq 1$ for all $\gamma \in (0, 1)$, where $\varphi_\alpha(\gamma) = \sum_{i=0}^{\alpha} \gamma^i$, $\alpha \in \mathbb{N} \cup \{\infty\}$. This inequality is very important. Using this, a generally convergent iteration is given in [1] (see Subsection 3.3). For an analytic map $f: \mathbb{E} \to \mathbb{F}$ with the form $f(z) = \sum_{i=0}^{\alpha} a_i z^i$, the inequality proves that

$$\gamma(z, f) \leq \|f\| \cdot \|Df(z)^{-1}\| \frac{\varphi_\alpha'(\|z\|)^2}{\varphi_\alpha(\|z\|)},$$

where $\|f\| = \sup_k \|a_k\|$. This shows that the criterion $\gamma(z, f)$ is easy to estimate.

Smale asked if there is a kind of L^2-version of the inequality $\alpha(\gamma, \varphi_\alpha) \leq 1$. If this is true, we can apply it to work in Hilbert space.

Problem (Smale [3]). *Is $\alpha(\gamma, \psi_\alpha) \leq 1$ true for all $\gamma \in (0, 1)$, where $\psi_\alpha(\gamma) = \sqrt{\sum_{i=0}^{\alpha} \gamma^{2i}}$.*

For this problem, we prove the following:

Theorem 2 [13]. *For all $\gamma \in (0, 1)$*

$$\alpha(\gamma, \psi_\alpha) \geq \frac{1}{2\gamma^2}.$$

Thus, if $\gamma < 1/\sqrt{2}$, then $\alpha(\gamma, \psi_\alpha) > 1$. That is, the answer to the problem is negative.

II. Exact Point-Estimate of Iterations

2.1. Newton's Method

Let $N_f(z) = z - Df(z)^{-1}f(z)$. Suppose the criteria $\alpha = \alpha(z, f)$, $\beta = \beta(z, f)$, $\gamma = \gamma(z, f)$ defined for an analytic map $f: \mathbb{E} \to \mathbb{F}$ and $z \in \mathbb{E}$. Let $h: \mathbb{R} \to \mathbb{R}$ be defined as follows:

$$h(t) = \beta - t + \frac{\gamma t^2}{1 - \gamma t}, \quad t \in \mathbb{R}.$$

It is easy to see that $\beta(0, h) = \beta$ and $\gamma(0, h) = \gamma$; so $\alpha(0, h) = \alpha$. The function h has two zeros:

$$\left.\begin{matrix} t^* \\ t^{**} \end{matrix}\right\} = \frac{1 + \alpha \mp \sqrt{(1 + \alpha)^2 - 8\alpha}}{4\gamma}$$

and we can estimate for $t_n = N_h^n(0)$ by the following expression:

$$t_{n+1} - t_n = \frac{(1 - q^{2^n})\sqrt{(1 + \alpha)^2 - 8\alpha}}{2\alpha(1 - \eta q^{2^n - 1})(1 - \eta q^{2^{n+1} - 1})} \eta q^{2^n - 1} \beta,$$

where

$$\eta = \frac{1 + \alpha - \sqrt{(1 + \alpha)^2 - 8\alpha}}{1 + \alpha + \sqrt{(1 + \alpha)^2 - 8\alpha}}, \qquad q = \frac{1 - \alpha - \sqrt{(1 + \alpha)^2 - 8\alpha}}{1 - \alpha + \sqrt{(1 + \alpha)^2 - 8\alpha}}.$$

Obviously, if $\alpha \leq 3 - 2\sqrt{2}$, t_n increases to $t^* \in \mathbb{R}$ monotonously and for Newton's sequence $z_n = N_f^n(z)$ and a zero ζ of f, we can prove that $\|z_{n+1} - z_n\|$ and $\|\zeta - z_n\|/\|\zeta - z_0\|$ are bounded by $t_{n+1} - t_n$ and $(t^* - t_n)/(t^* - t_0)$, respectively. Thus, we obtain:

Theorem 3 ([7, 10]). 1. *If $\alpha(z, f) \leq 3 - 2\sqrt{2}$ for an analytic map $f: \mathbb{E} \to \mathbb{F}$ then $z_n = N_f^n(z_0)$, $z_0 = z \in \mathbb{E}$ are well-defined and satisfy*

$$\|z_{n+1} - z_n\| \leq (\tfrac{1}{2})^n \|z_1 - z_0\|, \quad n = 0, 1, \ldots.$$

The bound $3 - 2\sqrt{2}$ is optimal with respect to this property. More generally, for $q \in (0, 1)$, we have:

2. *If*

$$\alpha(z, f) \leq \frac{1 + 4q + q^2 - (1 + q)\sqrt{1 + 6q + q^2}}{2q}$$

for any analytic map $f: \mathbb{E} \to \mathbb{F}$, then $z_n = N_f^n(z_0)$, $z_0 = z \in \mathbb{E}$ are well-defined and satisfy

$$\|z_{n+1} - z_n\| \leq q^{2^n - 1} \|z_1 - z_0\|, \quad n = 0, 1, \ldots,$$

and

$$\|\zeta - z_n\| \leq q^{2^n - 1} \|\zeta - z\|, \quad n = 0, 1, \ldots.$$

The kernel for α is optimal with respect to this property for any q.

Setting $q = \frac{1}{2}$ in theorem 3, we obtain a precise bound of criterion $\alpha(z, f)$, $(13 - 3\sqrt{17})/4 \approx 0.157671$. That is to say that z is an approximate zero of f for Newton's iteration and also is an approximate zero of the second kind.

2.2. Deformed Newton's Methods

Properly choosing a map $A_n: \mathbb{F} \to \mathbb{E}$ to substitute for $Df(z_n)^{-1}$ in Newton's iteration, the computational efficiency of the deformed Newton's iteration $z_{n+1} = z_n - A_n f(z_n)$ is often higher than Newton's. Wang, Han, and Sun [11] have studied point estimates of the following deformations:

I. $A_n = Df(z_{m[n/m]})^{-1}$ for $m \in \mathbb{N}$ given.
II. $A_{n+1} = 2A_n - A_n Df(z_{n+1}) A_n$.
III. $A_n = Df(\zeta_n)^{-1}$, $\zeta_{n+1} = \zeta_n - \frac{1}{2} A_n f(z_{n+1})$.

2.3. Halley's Family

Halley, an astronomer who discovered the Halley Comet, once used an iterative function for f:

$$H_f(z) = z - \frac{2f(z)f'(t)}{2f'(z)^2 - f(z)f''(z)}$$

with third-order convergence. In fact, $z' = H_f(z)$ is a zero of Pade's approximation at z with order [1/1]. For $k \in \mathbb{N}$, let $z' = H_{k,f}(z)$ be a zero of Pade's approximation with orders $[1/k - 1]$. Thus, we get a family of iterative functions. This family $H_{1,f} = N_f$, $H_{2,f} = H_f$, ... has increasing order of convergence. In [17], an analytic representation of $H_{k,f}(z)$ was given as

$$H_{k,f}(z) = z + \frac{kD^{k-1}f(z)^{-1}}{D^k f(z)^{-1}}.$$

But analogous to the form of Bernonlli iteration, there is no problem in Banach space to give the following expression:

$$H_{k,f}(z) = z - \left(1 + \sum_{i=2}^{k} Df(z)^{-1} \frac{D^i f(z)}{i!} \prod_{j=1}^{i-1} (H_{k-j,f}(z) - z)\right)^{-1} Df(z)^{-1} f(z).$$

Using the same majorant function as in subsection 2.1, we can prove:

Theorem 4 ([13]). *Let $f: \mathbb{E} \to \mathbb{F}$ be analytic, $z \in \mathbb{E}$. If $\alpha = \alpha(z, f) \le 3 - 2\sqrt{2}$, then for arbitrary $k \in \mathbb{N}$, Halley's iterative family is well-defined and satisfies*

$$\|\zeta - z_{k,n}\| \le \frac{1 + \alpha - \sqrt{(1+\alpha)^2 - 8\alpha}}{4\alpha} \varepsilon_{k,n} \|z_{1,1} - z\|,$$

where $z_{k,n} = H^n_{k,f}(z)$, $f(\zeta) = 0$, and

$$\varepsilon_{k,n} = \begin{cases} \left[\dfrac{1+\alpha-\sqrt{(1+\alpha)^2-8\alpha}}{1+\alpha+\sqrt{(1+\alpha)^2-8\alpha}}\left(\dfrac{3-\alpha-\sqrt{(1+\alpha)^2-8\alpha}}{3-\alpha+\sqrt{(1+\alpha)^2-8\alpha}}\right)^{1/k}\right]^{(k+1)^n-1} & \text{if } \alpha < 3-2\sqrt{2} \\ \left(\dfrac{1}{k+1}\right)^n, & \text{if } \alpha = 3-2\sqrt{2}. \end{cases}$$

III. Euler's Series, Integral, and Others

3.1. Convergent Radius of Euler's Series

Let $f\colon \mathbb{C} \to \mathbb{C}$ be analytic and $f'(z) \neq 0$ for $z \in \mathbb{C}$, f_z^{-1} represents a local inverse of f at z which maps $w = f(z)$ into z. Hence, in some neighborhood of w, $z' = f_z^{-1}(w')$ can be expanded with the powers of $(w' - w)$:

$$z' = z + \sum_{n=1}^{\infty} \frac{(f_z^{-1})^{(n)}(w)}{n!}(w' - w)^n.$$

We call this series Euler's series. Let $R(z, f)$ denote the convergence radius of this series. Smale used Loewner's result about Schlicht function theory to give an upper bound for $R(z, f)$.

To obtain a lower bound, we make a substitution $\lambda = (z' - z)/u$ and $\omega = (w - w')/w$, where $u = -f(z)/f'(z)$. Through this substitution, the Taylor series of $w' = f(z')$ at z and $z' = f_z^{-1}(w')$ at w can be reduced to

$$\omega = \lambda - \sum_{n=2}^{\infty} a_n \lambda^n \quad \text{and} \quad \lambda = \omega + \sum_{n=2}^{\infty} b_n \omega^n.$$

By means of Bell's polynomial, we can obtain

$$b_n = \sum_{m=2}^{n} v_{n,m} a_m,$$

where

$$v_{n,m} = \sum_{j_1 + 2j_2 + \cdots + (n-2)j_{n-2} = n-m} C_m^{j_1 \cdots j_{n-2}} b_2^{j_1} \cdots b_{n-1}^{j_{n-2}}$$

and

$$C_m^{j_1 \cdots j_{n-2}} = \frac{m!}{j_1! \cdots j_{n-2}!(m - j_1 - \cdots - j_{n-2})!}.$$

Hence, the series $\sum_{n=1}^{\infty} B_n \alpha^{n-1} |\omega|^n$ with positive terms is a majorant series for the reduced Euler's series, where $\lambda = \sum_{n=1}^{\infty} B_n \omega^n$ is the inverse of the series $\omega = \lambda - \sum_{n=2}^{\infty} \lambda^n$. Thus, we have:

Theorem 5 ([13]).

$$\frac{\alpha(z, f) R(z, f)}{|f(z)|} \geq 3 - 2\sqrt{2},$$

where the constant on the right is the best possible. Furthermore,

$$\left|\frac{(f_z^{-1})^{(n)}(w)}{n!}\right| \leq \frac{B_n}{|f'(z)|}\left(\frac{\alpha(z,f)}{|f(z)|}\right)^{n-1}, \quad \forall n \in \mathbb{N}$$

where constants $B_1 = 1$,

$$B_n = \sum_{n/2 \leq i \leq n} (-1)^{n-i} \frac{(2i-3)!!}{(n-i)!(2i-n)!} 3^{2i-n} 2^{i-n-2}$$

are also the best possible for all n.

Kim [23] also obtained the same lower bound using complex analysis methods. But in [13] we use the majorant series method. Moreover, there is an interesting fact that the coefficient B_n of the majorant series is just the numbers of different ways to insert parentheses in the expression $x_1 x_2 \cdots x_n$ of n letters. For example, for $x_1 x_2 x_3 x_4$, there are 11 ways:

$x_1x_2x_3x_4$, $x_1(x_2x_3x_4)$, $(x_1x_2x_3)x_4$, $x_1x_2(x_3x_4)$, $x_1(x_2x_3)x_4$,

$(x_1x_2)x_3x_4$, $x_1(x_2(x_3x_4))$, $x_1((x_2x_3)x_4)$,

$(x_1(x_2x_3))x_4$, $((x_1x_2)x_3)x_4$, $(x_1x_2)(x_3x_4)$.

3.2. Convergence of Euler's Family

With the notions above, the partial sum of series of f_z^{-1} at $w' = 0$ with k terms is $z + \sum_{n=1}^{k} b_n$, which consist of a family of iterative functions. It writes $E_{k,f}(z)$. This family is called Euler's family. The majorant function in Subsection 2.1. is also one for Euler's iterative family. The reason for this is that we can prove for $z \in \mathbb{C}$, $f'(z) \neq 0$, and $z' = E_{k,f}(z)$,

$$f(z') = \sum_{n=k+1}^{\infty} \sum_{m=2}^{n} \sum_{j_1+2j_2+\cdots+(k-1)j_{k-1}=n-m} C_m^{j_1 \cdots j_{k-1}} b_2^{j_1} \cdots b_k^{j_{k-1}} a_m f(z).$$

Here, convergence of the majorant sequence $t_{k,n} = E_{k,h}^n(0)$ cannot be given constructively as in Halley's case. But we derive it from the relation

$$DE_{k,f}(z) = (k+1) b_{k+1,f}(z), \tag{$*$}$$

where $b_{k+1,f}(z)$ denotes b_{k+1} for f at z. Hence:

Theorem 6 ([13]). *Let $f\colon \mathbb{C} \to \mathbb{C}$ be analytic and $z \in \mathbb{C}$. If $\alpha(z,f) \leq 3 - 2\sqrt{2}$, then for arbitrary $k \in \mathbb{N}$, iterations of Euler's family, $z_{k,n} = E_{k,f}^n(z)$ are well-defined and convergent to a zero of f.*

The proofs of Theorems 5 and 6 are due to a majorizing principle. Thus, its result is easy to be generalized to the case of Banach space. Perhaps further study about global Newton's method needs a form of Theorem 5 in Banach space.

Error estimation of iterative families depends on error estimation of the majorant sequence. The difference with Halley's family is that error estimation of the majorant sequence cannot be obtained explicitly. But if we apply (*) to h, we have

$$t^* - t_{k,n+1} = (k+1) \int_{t_{k,n}}^{t^*} b_{k+1,h}(t)\,dt.$$

From this we can get an estimation about the majorant sequence.

3.3. *Generally Convergent Iterative Family*

Smale once constructed an iterative method which is convergent for almost all polynomials and almost initials. We call this convergence, general convergence. Now we apply the above result to obtain a generally convergent iterative family,

$$z_{k,n+1} = z_{k,n} - \frac{f(z_{k,n})}{f'(z_{k,n})} \sum_{i=1}^{k} b_{i,f}(z_{k,n})\omega(z_{k,n})^i,$$

where

$$\omega(z) = \min\left\{1, \frac{3-2\sqrt{2}}{\alpha(z,f)}\right\}.$$

Theorem 7. *Let* **f**: $\mathbb{C} \to \mathbb{C}$ *be analytic and* $f'(z_0) \neq 0$ *for* $z_0 \in \mathbb{C}$, *then for arbitrary* $k \in \mathbb{N}$ *given,* $z_{k,n}$ *is well-defined and convergent to a zero or a critical point of* f.

This result is adopted from the manuscript of Han Danfu.

3.4. *The Efficiency of Integral Approximation*

Let $\mathscr{H}^k$ be a Sobolev space on $[0,1]$ and Q: $\mathscr{H}^k \to \mathbb{R}$ be a quadrature functional with degree at least $k-1$. Suppose J: $f \mapsto \int_0^1 f(t)\,dt$, $h = 1/n$. Define a quadrature functional QM_k by $Q(M_h(f))$ where $M_h(f)$: $C[0,1] \to C[0,1]$ is determined by

$$M_h f(t) = h \sum_{i=0}^{n-1} f(ih + th), \quad \forall f \in C[0,1], \forall t \in [0,1].$$

Under the assumption that the inner product of a unit vector e and any vector f in $\mathscr{H}^k$ obeys a standard normal distribution, the study of integral approximate efficiency can be reduced to find the average

$$\mathscr{E}^k_{Q,h} = \operatorname*{Av}_{f \in \mathscr{H}^k} |(J - QM_h)f|.$$

For this problem, we have a general result as follows:

Theorem 8 ([8])

$$\mathscr{E}_{Q,h}^{k} = \left\{(-1)^k \frac{2}{\pi}(J-Q)_t(J-Q)_s \frac{(s-t)_t^{2k-1}}{(2k-1)!}\right\}^{1/2} h^k,$$

where $\mathscr{L}_t f(t) = \mathscr{L}f$, $t_+ = \max\{t, 0\}$.

Thus, for example, if Q is given by

$$Qf = \sum_{i=0}^{n-1} p_j f(t_j), \qquad \sum_{j=0}^{m-1} p_j t_j^{r-1} = \frac{1}{r!}, \quad 1 \le r \le k,$$

then

$$\mathscr{E}_{Q,h}^{k} = \left\{(-1)^k \frac{2}{\pi}\left[\frac{1}{(2k+1)!} - \sum_{i=0}^{m-1} p_j \left(\frac{t_j^{2k} + (1-t_j)^{2k}}{(2k)!}\right.\right.\right.$$

$$\left.\left.\left. - \sum_{l=j+1}^{m-1} p_l \frac{(t_l - t_j)^{2k-1}}{(2k-1)!}\right)\right]\right\}^{1/2} h^k.$$

References

1. Smale, S., Algorithms for solving equations, *Proceedings of the International Congress of Mathematicians*, Berkeley, CA. American Mathematical Society, Providence, RI, 1986, 172-195.
2. Smale, S., Newton's method estimates from data at one point, in The Merging of Disciplines; New Directions in Pure and Applied Mathematics, Ewing, R., Cross, K., and Martin, C. (Eds), Springer-Verlag, New York, 1986.
3. Smale, S., On the efficiency of algorithms of analysis, *Bull. Amer. Math. Soc.* (*New Series*) **13** (1985), 87–121.
4. Smale, S., The fundamental theorem of algebra and complexity theory, *Bull. Amer. Math. Soc.*, **4** (1981), 1–36.
5. Shub, M. and Smale, S., On the existence of generally convergent algorithms, *J. Complexity* **2** (1986), 2–11.
6. Wang Xinghua and Xuan Xiaohua, Random polynomial space and computational complexity theory, *Scientia Sinica* (*Series A*) **7** (1987), 637–684.
7. Wang Xinghua and Han Danfu, On dominating sequence method in the point estimate and Smale theorem, *Sci. China* (*Ser. A*) **33** (2) (1990), 135–144.
8. Wang Xinghua and Han Danfu, Computational complexity in numerical integrals, preprint.
9. Wang Xinghua and Han Danfu, Domain and point estimates on Newton's iteration, *Chinese J. Numer. Math. Appl.* **12** (3) (1990), 1–8.
10. Wang Xinghua and Han Danfu, Precise judgement of approximate zero of the second kind for Newton's iteration, *A Friendly Collection of Mathematics Papers I*, Jilin University Press, Changchun, China, 19XX, pp. 22–24.
11. Wang Xinghua, Han Danfu, and Sun Fangyu, Point estimates on some eformed Newton's methods, *Chinese J. Numer. Math. Appl.* **12** (4) (1990).
12. Wang Xinghua, Zheng Shiming, and Han Danfu, The convergence of Euler's series, Euler's and Halley's families (Chinese), *Acta Math. Sinica*, **6** (1990).

13. Wang Xinghua, Shen Guangxing, and Han Danfu, Some remarks on the algorithms of solving equations, *Acta Math.* Sinica (New Series) (*Chinese J. Math.*), **8** (92), 337–348.
14. Wang Xinghua, Convergence on a iterative procedure (in Chinese), *KeXue TongBao* **20** (1975), 558–559.
15. Wang Xinghua, Error estimates on some numerical methods of equation finding (in Chinese), *Acta Math. Sinica* **5** (1979), 638–642.
16. Wang Xinghua, Convergent neighbourhood on Newton's method (in Chinese), *KeXue TongBao*, Special Issue of Mathematics, Physics and Chemistry (1980), 36–37.
17. Wang Xinghua and Zheng Shiming, A family of parallel and interval iterations for finding simultaneously all roots of a polynomial with rapid convergences, *J. Comput. Math.* **1** (1984), 70–76.
18. Xuan Xiaohua, The computational complexity of the resultant method for solving polynomial equations, *J. Comput. Math.* **3** (1985), 162–166.
19. Xuan Xiaohua, Computational complexity theory, Master's paper, Hangzhou University.
20. Gragg, G.W. and Tapia, R.A., Optimal error bounds for the Newton–Kantorovich theorem, *SIAM J. Numer. Anal.* **11** (1974), 10–13.
21. Ostrowski, A.M., *Solutions of Equations in Eulidean and Banach Spaces*, Academic Press, New York, 1973.
22. Traub, J. and Wozniakowski, H., Convergence and complexity of Newton iteration for operator equation, *J. Assoc. Comput. Mach.* **29** (1979), 250–258.
23. Kim, M.H., On approximate zeros and root finding algorithms for a complex polynomial, to appear.

42
Error Estimates of Ritz–Gelerkin Methods for Indefinite Elliptic Equations

XIAOHUA XUAN

1. Introduction

We study the error estimates of Ritz–Gelerkin methods for Dirichlet problems:

$$Lu - \lambda_0 u = f \quad \text{in } \Omega \subset \mathbb{R}^n,$$
$$u = 0 \quad \text{on } \partial\Omega.$$

In particular, we will deepen the usual error estimate

$$\|u_h - u\|_{H^1(\Omega)} \leq C(n, L, \lambda_0, \Omega)|u|_{H^2(\Omega)}h$$

by giving an explicit expression of the constant $C(n, L, \lambda_0, \Omega)$, and using the data f instead of the unknown data u. In this way, the computational cost can easily be estimated in terms of the number of linear equation solvings.

For simplicity, let Ω be a simply connected convex domain in Euclidean space $\mathbb{R}^n$ with smooth boundary $\partial\Omega$. We consider the Dirichlet problem (usual summation convention is used)

$$\begin{aligned} Lu - \lambda_0 u = -D_i(a^{ij}(x)D_j u) + b^i(x)D_i u + c(x)u - \lambda_0 u = f \quad \text{in } \Omega, \\ u = 0 \quad \text{on } \partial\Omega, \end{aligned} \tag{1}$$

where λ_0 is a real constant and the operator L has the following properties:

P1. (Uniform Ellipticity). The operator L is uniformly elliptic, i.e., there exist a constant α_0 such that, for all $x \in \Omega$, and $\xi = (\xi_1, \ldots, \xi_n) \in \mathbb{R}^n$,

$$a^{ij}(x)\xi_i\xi_j \geq \alpha_0|\xi|^2. \tag{2}$$

This is equivalent to say that the matrix $A = (a^{ij}(x))_{nn}$ is uniformly positive definite.

P2 (Boundedness of Coefficients). The coefficients functions $a^{ij}(x)$ belong to $C^1(\overline{\Omega})$ and $b^i(x)$, $c(x)$ belong to $C^0(\overline{\Omega})$. Therefore, there exist constants M_a, M_a', M_b, M_c such that, for all $x \in \Omega$ and $i, j, k = 1, \ldots, n$,

$$|a^{ij}(x)| \le M_a, \qquad \left|\frac{\partial a^{ij}(x)}{\partial x_k}\right| \le M_a', \qquad |b^i(x)| \le M_b, \qquad |c(x)| \le M_c. \quad (3)$$

We will be interested in seeking weak solutions of Eq. (1) defined as follows. (See Section 2 about the definition of Sobolev space $H^k(\Omega)$, subspace $H_0^k(\Omega)$, seminorm $|v|_{k,\Omega}$, and norm $\|v\|_{k,\Omega}$.)

Weak Solution Equation. Find $u \in H_0^1(\Omega)$ such that

$$B[u, v] - \lambda_0(u, v) = (f, v) \quad \text{for all } v \in H_0^1(\Omega), \tag{4}$$

where

$$B[u, v] = \int_\Omega a^{ij}(x)D_iuD_jv + b^i(x)D_iuv + c(x)uv,$$

$$(u, v) = \int_\Omega uv, \qquad (f, v) = \int_\Omega fv.$$

It is well-known that, except for $\lambda_0 \in \operatorname{Spec}(L)$ which is the countable set of eigenvalues of L, a unique weak solution u will exist and belong to $H^2(\Omega)$ for every $f \in L^2(\Omega)$. In these almost cases, we wish to approximate the weak solution using the Ritz–Galerkin methods. Precisely, given a family of subspaces V_h, $h > 0$, of $H_0^1(\Omega)$, we seek a element $u_h \in V_h$ from an equation similar to Eq. (4).

Ritz–Gelerkin Equation. Find $u_h \in V_h$ such that

$$B[u_h, v] - \lambda_0(u_h, v) = (f, v) \quad \text{for all } v \in V_h. \tag{5}$$

By choosing a basis of V_h, the above equation is a linear equation which can be solved by direct and iterative methods. Throughout this chapter, the approximation space V_h has the following property.

P3. (First-Order Space). There exist a constant $C_{\text{approxi}} = C_{\text{approxi}}(n, \Omega)$ such that, for every $u \in H^2(\Omega) \cap H_0^1(\Omega)$,

$$\inf_{\phi \in V_h} \|u - \phi\|_{H^1(\Omega)} \le C_{\text{approxi}} |u|_2 h.$$

Our result is to show, in the case that L is self-adjoint, that the error $\|u_h - u\|_{H^1(\Omega)}$ can be bounded by the information of Eq. (1) such as

n:	dimension
$\operatorname{diam}(\Omega)$:	diameter of the domain
α_0:	ellipticity constant
M_a, M_a', M_b, M_c:	bounds of coefficients
$\operatorname{Spec}(L) = \{\lambda_1, \lambda_2, \dots\}$:	eigenvalues of the operator L

λ_0:	constant
$\|f\|_{L^2(\Omega)}$:	norm of the right-hand side
h:	parameter
C_{approxi}:	approximation constant

Precisely, we will prove

Main Theorem. *Assume properties* P1–P3. *Suppose that the operator L is self-adjoint and $\lambda_0 \notin \operatorname{Spec}(L)$.*

1. *If $h < 2/(\sqrt{v}C_{\text{errbnd}})$, then*

$$\|u_h - u\|_{H^1(\Omega)} \le C_{\text{errbnd}}\|f\|_{L^2(\Omega)}h, \tag{6}$$

where

$$C_{\text{errbnd}} = \frac{4E^2}{\alpha_0}C_{\text{approxi}}\left(C''_{\text{regu}} + C_{\text{regu}}\max_{i>0}\frac{1}{(\lambda_0-\lambda_i)^2}\right)^{1/2},$$

$$E \le nM_a + M_b\frac{\operatorname{diam}(\Omega)}{\sqrt{n}} + (M_c + |\lambda_0|)\left(\frac{\operatorname{diam}(\Omega)}{\sqrt{n}}\right)^2,$$

$$v = \frac{nM_b^2}{2\alpha_0^2} + \frac{M_c+\lambda_0}{\alpha_0} + \frac{1}{2}, \qquad M = \max\{M_a, M'_a, M_b, M_c\},$$

$$C''_{\text{regu}} = \frac{4(n+3)}{\alpha_0^2}, \qquad C_{\text{regu}} = \frac{4(n+3)}{\alpha_0^2} + \left(\frac{2(n+3)^4M^2}{\alpha_0^2}+1\right)^4.$$

2. *Let $\phi(\lambda) = (1/\sqrt{2\pi\sigma})\exp(-\lambda^2/2\sigma)$ $(\sigma > 1)$, $\operatorname{meas}(A) = \int_A \phi(\lambda)\,d\lambda$. Given $0 < \mu < 1$, there exists a set Σ^μ such that $\operatorname{meas}(\Sigma^\mu) < \mu$, and*

$$\int_{(-\infty,\infty)-\Sigma^\mu} \|u_h - u\|_{H^1(\Omega)}\phi(\lambda_0)\,d\lambda_0 \le C(n, L, \Omega)(\sigma|\log\mu| + \sigma\log\sigma)^{(n+3)/2}. \tag{7}$$

Inequality (6) relates the error estimate to the eigenvalue distribution. This brings the result, taking $\max_{i>0}(1/|\lambda_0 - \lambda_i|)$ as a weight of a problem, close to the work on tractability estimates initiated by Smale [9]. There are many recent work in such direction. (See [10] for the motivation and references.)

Remark. When the operator L is not self-adjoint, our proof will show the following:

Assume properties P1–P3. If

$$h < \frac{\alpha_0}{2E^2\sqrt{v}C_{\text{approxi}}|(L^* - \lambda_0 I)^{-1}|},$$

then

$$\|u_h - u\|_{H^1(\Omega)} \le \frac{4E}{\alpha_0}C_{\text{approxi}}h\|f\|_{L^2(\Omega)}|(L^* - \lambda_0 I)^{-1}|, \tag{8}$$

where the operator L^* is the adjoint of L, and

$$|(L^* - \lambda_0 I)^{-1}| = \sup_{f \neq 0} \frac{|(L^* - \lambda_0 I)^{-1} f|_2}{\|f\|_{L^2(\Omega)}},$$

$$|(L - \lambda_0 I)^{-1}| = \sup_{f \neq 0} \frac{|(L - \lambda_0 I)^{-1} f|_2}{\|f\|_{L^2(\Omega)}}.$$

Section 2 contains some definitions and preliminaries. We will prove the main theorem in Section 3.

There are vast literatures on error estimates, especially for the finite-element method as a particular case. Most of them deal with the definite elliptic equations. (See Babuška and Aziz [2], Ciarlet [3].) This is in the flavor of solving the linear equation $Ax = b$ when the matrix A is positive definite. Our result is close to Schatz's work (Schatz [8]) which is in the flavor of solving the linear equation $Ax = b$ when the matrix A is nonsingular. Other related work is Werschultz [11] on the analysis of optimal error algorithms.

This work is a preparation for an analysis similar to the work of Xuan [12, 13] for nonlinear Dirichlet problems, supervised by Steve Smale. I always feel grateful for his constant encouragement on my working in this direction.

2. Definitions and Preliminaries

Let $\Omega \subset \mathbb{R}^n$ be open bounded and $L^2(\Omega)$ the set of squarely integrable functions in Ω. Throughout this chapter, we will use the following Sobolev spaces.

Definition 2.1 (Sobelev Spaces). For $k \in \mathbb{N} = \{0, 1, \dots\}$, we define

$$H^k(\Omega) = \{v \in L^2(\Omega)\colon D^\alpha v \in L^2(\Omega) \qquad \text{for all } |\alpha| \leq k\}, \tag{9}$$

and its seminorm and norm

$$|v|_k = |v|_{k,\Omega} = \left(\sum_{|\alpha|=k} \|D^\alpha v\|^2_{L^2(\Omega)} \right)^{1/2}, \qquad \|v\|_k = \|v\|_{k,\Omega} = \left(\sum_{j=0}^{k} |v|_j^2 \right)^{1/2}.$$

Moreover, we define

$$H_0^k(\Omega) = \text{Closure of } C_0^\infty(\Omega) \text{ in } H^k(\Omega), \tag{10}$$

where $C_0^\infty(\Omega)$ is the set of functions which have continuous derivatives of any order and have compact support in Ω.

Note that $H^k(\Omega)$ is a Hilbert space and its scalar inner product is given by $(u, v)_k = \sum_{|\alpha| \leq k} \int_\Omega D^\alpha u D^\alpha v$. We will also use notation $\Delta u = \sum_{i=1}^n (\partial^2/\partial x^2)$.

Lemma 2.1. 1. (*Cauchy Inequality with* ε) *For any* $a, b \in \mathbb{R}^1$, $\varepsilon > 0$,

$$|ab| \leq \frac{\varepsilon}{2}a^2 + \frac{1}{2\varepsilon}b^2. \tag{11}$$

2. *If* $u \in H^2(\Omega)$, *then* $|u|_2 = |\Delta u|_0$.
3. (*Interpolation Inequality*) *If* $u \in H^2(\Omega) \cap H_0^1(\Omega)$, $\varepsilon > 0$, *then*

$$|u|_1^2 \leq \frac{\varepsilon}{2}|u|_2^2 + \frac{1}{2\varepsilon}|u|_0^2. \tag{12}$$

PROOF. Part 1 is straightforward. Part 2 is proved in the work of Gilbarg and Trudinger [5]. To prove part 3, one uses a special case of Green's first identity (the second term on the right-hand side is zero)

$$\int_\Omega DuDu = -\int_\Omega u\Delta u + \int_{\partial\Omega} u\frac{\partial u}{\partial \tau}ds$$

and then uses part 1 and part 2.

Definition 2.2 (Weak Solution). We say that $u \in H_0^1(\Omega)$ is a weak solution of Eq. (1) if

$$B[u,v] - \lambda_0(u,v) = (f,v) \quad \text{for all } v \in H_0^1(\Omega), \tag{13}$$

where the bilinear form $B[\ ,\ ]$ and inner product $(\ ,\ )$ are

$$B[u,v] = \int_\Omega a^{ij}(x)D_iuD_jv + b^i(x)D_iuv + c(x)uv,$$

$$(u,v) = \int_\Omega uv, \qquad (f,v) = \int_\Omega fv.$$

Definition 2.3 (Eigenvalue and Eigenvector). We say that λ is an eigenvalue of the operator L if

$$Lu - \lambda u = 0 \quad \text{in } \Omega,$$

$$u = 0 \quad \text{on } \partial\Omega$$

has nontrivial weak solutions. Such nontrivial solutions are called eigenvectors. We use $\text{Spec}(L)$ or $\text{Spec}(L,\Omega)$ to denote the set of all the eigenvalues of L.

Lemma 2.2 (Poincaré Inequality). *Let* $u \in H_0^1(\Omega)$.

1. *If* λ_1 *is the smallest eigenvalue of the Laplacian operator* $-\Delta$, *then*

$$|u|_0 \leq \frac{1}{\sqrt{\lambda_1}}|u|_1. \tag{14}$$

2. *There is a following more explicit inequality*

$$|u|_0 \le \frac{\operatorname{diam}(\Omega)}{\sqrt{n}} |u|_1. \tag{15}$$

PROOF. One can prove part 1 by minimizing $J(u) = \int_\Omega |Du|^2$ with a constrain $\int_\Omega u^2 = 1, u = 0$ on $\partial\Omega$. See details in [4, 5]. The proof of part 2 can be found in [1].

Lemma 2.3 (Boundedness and Coersiveness). *Assume properties* P1 *and* P2 *and* $u, v \in H_0^1(\Omega)$.

1. (*Boundedness*) *There exists*

$$E \le nM_a + M_b \frac{\operatorname{diam}(\Omega)}{\sqrt{n}} + (M_c + |\lambda_0|)\left(\frac{\operatorname{diam}(\Omega)}{\sqrt{n}}\right)^2$$

such that

$$|B[u, v] - \lambda_0(u, v)| \le E|u|_1 |v|_1. \tag{16}$$

2. (*Coersiveness*) *There exists*

$$\nu = \frac{nM_b^2}{2\alpha_0^2} + \frac{M_c + \lambda_0}{\alpha_0} + \frac{1}{2}$$

such that

$$B[u, u] - \lambda_0(u, u) \ge \frac{\alpha_0}{2} \|u\|_1^2 - \alpha_0 \nu^2 |u|_0^2. \tag{17}$$

3. *For all* $k, j, l = 1, \ldots, n$, *and* $x \in \Omega$,

$$\left| a^{ik}(x) \frac{\partial a^{jk}(x)}{\partial x_i} \right| \le nM_a M_a'. \tag{18}$$

PROOF. First

$$\begin{aligned} |B[u, v] - \lambda_0(u, v)| &= \left| \int_\Omega a^{ij}(x) D_i u D_j v + b^i(x) D_i uv + c(x)uv - \lambda_0(u, v) \right| \\ &\le nM_a |u|_1 |v|_1 + M_b |u|_1 |v|_0 + (M_c + |\lambda_0|)|u|_0 |v|_0 \\ &\le E|u|_1 |v|_1 \qquad \text{by Poincaré Inequality (15).} \end{aligned}$$

This proves part 1. Second,

$$\begin{aligned} B[u, u] - \lambda_0(u, u) &= \int_\Omega a^{ij}(x) D_i u D_j u + b^i(x) D_i uu + c(x)uu - \lambda_0(u, u) \\ &\ge \alpha_0 |u|_1^2 - (M_b |u|_1 |u|_0 + M_c |u|_0^2) - \lambda_0 |u|_0^2. \end{aligned}$$

By using Cauchy inequality (11) with $\varepsilon = (\alpha_0/M_b)^2$ in the term $M_b|u|_1|u|_0$, the inequality becomes

$$\geq \alpha_0|u|_1^2 - \left(\frac{\alpha_0}{2}|u|_1^2 + \frac{M_b^2}{2\alpha_0}|u|_0^2 + M_c|u|_0^2\right) - \lambda_0|u|_0^2$$

$$= \frac{\alpha_0}{2}\|u\|_1^2 - \alpha_0 v|u|_0^2.$$

This proves part 2. Finally, part 3 follows directly from property P2.

The following lemma is some simple estimates with the Gaussian density function.

Lemma 2.4. *Let* $\phi(\lambda) = (1/\sqrt{2\pi\sigma})e^{(-\lambda^2/2\sigma)}$ $(\sigma > 1)$ *and* $0 < \mu < 1$.

1. *For any* $M > 0$,

$$\int_{\infty}^{-M} + \int_{M}^{\infty} \phi(\lambda)\,d\lambda \leq 2\phi(M).$$

2. *If* $M \geq 4\sigma\sqrt{|\log(1/\mu^2 2\pi\sigma)|}$, *then*

$$\text{meas}((-\infty, -M) \cup (M, \infty)) = \int_{(-\infty,-M)\cup(M,\infty)} \phi(\lambda)\,d\lambda \leq \frac{\mu}{2}.$$

3. *For any* $\lambda \in \mathbb{R}^1$,

$$\lambda^2\phi(\lambda) \leq 2\sigma^2\phi(2\sigma^2) = \sqrt{\frac{2}{\pi}}e^{-1}\sigma^{3/2}.$$

4. *Let*

$$\Sigma_i^\mu = \left[\lambda_i - \frac{\sqrt{2\pi\sigma\mu}}{4M'}, \lambda_i + \frac{\sqrt{2\pi\sigma\mu}}{4M'}\right],$$

then $\text{meas}(\Sigma_i^\mu) \leq \mu/2M'$ *and*

$$\int_{(-M,M)-\Sigma_i^\mu} \frac{\lambda^2\phi(\lambda)}{\lambda - \lambda_i|}\,d\lambda \leq C\sigma^{3/2}(\log M + \log M' \log \mu).$$

Proof. Since $\phi(x + M) \leq \phi(x)\phi(M)$ for x, $M > 0$, one has

$$\int_M^\infty \phi(\lambda)\,d\lambda = \phi(M)\int_0^\infty \phi(\lambda)\,d\lambda$$

and, similarly,

$$\int_\infty^{-M} \phi(\lambda)\,d\lambda = \phi(-M)\int_\infty^0 \phi(\lambda)\,d\lambda = \phi(M)\int_0^\infty \phi(\lambda)\,d\lambda,$$

thus

$$\int_{\infty}^{-M} + \int_{M}^{\infty} \phi(\lambda)\, d\lambda \leq 2\phi(M),$$

which proves part 1. Part 2 follows from by substituting $M = 4\sigma\sqrt{|\log(1/\mu^2 2\pi\sigma)|}$, i.e.,

$$\text{meas}((-\infty, -M) \cup (M, \infty)) \leq \frac{2}{\sqrt{2\pi\sigma e^{-(1/2)\log(16/\mu^2\sigma)}}} = \frac{2}{\sqrt{2\pi\sigma}} \frac{\mu\sqrt{2\pi\sigma}}{4} \leq \frac{\mu}{2}.$$

To prove part 3, one uses differentiation to find that the maximum happens at $\lambda = 2\sigma^2$, i.e.,

$$\max_{\lambda} \lambda^2 \phi(\lambda) = 2\sigma^2 \phi(2\sigma^2) = \sqrt{\frac{2}{\pi}} e^{-1} \sigma^{3/2},$$

which proves part 3. From this, part 4 follows easily. (c, C are some constants.)

$$\begin{aligned}
\int_{(-M,M)-E_i^\mu} \frac{\lambda^2 \phi(\lambda)}{|\lambda - \lambda_i|} d\lambda &\leq \sqrt{\frac{2}{\pi}} e^{-1} \sigma^{3/2} \int_{(-M,M)-E_i^\mu} \frac{1}{|\lambda - \lambda_i|} d\lambda \\
&\leq c\sigma^{3/2} (\log M + \log \frac{\sqrt{2\pi\sigma\mu}}{4M'} \\
&\leq C\sigma^{3/2} (\log M + \log M' + \log \mu).
\end{aligned}$$

The lemma is proved.

3. Proof of the Main Theorem

Lemma 3.1 (Existence, Uniqueness, and Regularity of the Weak Solution). *Assume properties* P1 *and* P2. *If* $\lambda_0 \notin \text{Spec}(L)$, *a at most countable set of eigenvalues of* L, *the Dirichlet problem* (1)

$$\begin{aligned}
Lu - \lambda_0 u = f &\quad \textit{in } \Omega, \\
u = 0 &\quad \textit{on } \partial\Omega
\end{aligned}$$

has a unique weak solution in $H_0^1(\Omega)$ *for each* $f \in L^2(\Omega)$. *Moreover, the weak solution* u *belongs to* $H^2(\Omega)$.

Proof. See Gilbarg and Trudinger [5].

Lemma 3.2 (Regularity Estimate of the Weak Solution). *Assume properties* P1 *and* P2. *Let* $f \in L^2(\Omega)$, *and* u *be the weak solution of Eq.* (1). *There exist constants*

$$C''_{\text{regu}} = \frac{4(n+3)}{\alpha_0^2}, \qquad C_{\text{regu}} = \frac{4(n+3)}{\alpha_0^2} + \left(\frac{2(n+3)^4 M^2}{\alpha_0^2} + 1\right)^4$$

with $M = \max\{M_a, M'_a, M_b, M_c\}$, *such that*

$$|u|_2 \leq (C''_{\text{regu}} \|f\|^2_{L^2(\Omega)} + C_{\text{regu}}|u|_0^2)^{1/2}. \tag{19}$$

Proof. First one uses two inequalities proved in the work of Grisvard [6].

1. Assume property P1, then

$$\alpha_0^2 \int_\Omega \sum_{i,j=1}^n |D_i D_j u|^2 \leq \int_\Omega a^{ik}(x) a^{jl}(x) D_j D_k u D_i D_l u. \tag{20}$$

2. If Ω is convex, then

$$\left| \int_\Omega D_i(a^{jl}(x) D_l u) D_j(a^{ik}(x) D_k u) \right| \leq \int_\Omega |D_i(a^{ij}(x) D_j u)|^2. \tag{21}$$

From the above inequalities, it follows that

$$\begin{aligned}
\alpha_0^2 |u|_2^2 = \alpha_0^2 \int_\Omega |D_i D_j u|^2 &\leq \int_\Omega a^{ik}(x) a^{jl}(x) D_j D_k u D_i D_l u \\
&= \int_\Omega D_i(a^{jl}(x) D_l u) D_j(a^{ik}(x) D_k u) - 2a^{ik} \frac{\partial a^{jl}}{\partial x_i} D_j D_k u D_l u \\
&\leq \int_\Omega |D_i(a^{ij}(x) D_j u)|^2 - 2a^{ik} \frac{\partial a^{jl}}{\partial x_i} D_j D_k u D_l u.
\end{aligned}$$

Using inequality (18), $|a^{ik}(x)[\partial a^{jk}(x)/\partial x_i]| \leq n M_a M'_a$, and then using Cauchy inequality (11) with $\varepsilon = (\alpha_0^2/2nM_aM'_a)^2$, one obtains

$$\begin{aligned}
\alpha_0^2 |u|_2^2 &\leq \int_\Omega |D_i(a^{ij}(x) D_l u)|^2 + 2nM_aM'_a |D_l u D_j D_k u| \\
&\leq \int_\Omega |D_i(a^{ij}(x) D_l u)|^2 + \frac{\alpha_0^2}{2}|u|_2^2 + \frac{(2n^4 M_a M'_a)^2}{2\alpha_0^2} |Du|_0^2.
\end{aligned}$$

Moving the $(\alpha_0^2/2)|u|_2^2$ term to left-hand side and using Eq. (1), one further obtains

$$\begin{aligned}
|u|_2^2 &\leq \int_\Omega |f - b^i(x) D_i u - (c(x) + \lambda_0) u|^2 + \frac{(2n^4 M_a M'_a)^2}{\alpha_0^2} |Du|_0^2 \\
&\leq \frac{2(n+3)\|f\|^2_{L^2(\Omega)}}{\alpha_0^2} + \frac{2(n+3)(M_c^2 + |\lambda_0|^2)|u|_0^2}{\alpha_0^2} \\
&\quad + \left(2(n+3)\frac{M_b^2}{\alpha_0^2} + \frac{(2n^4 M_a M'_a)^2}{\alpha_0^4} \right) |Du|_0^2.
\end{aligned}$$

Using Interpolation inequality (12) with

$$\varepsilon_0 = \left(\frac{2(n+3)M_b^2}{\alpha_0^2} + \frac{(2n^4 M_a M'_a)^2}{\alpha_0^4} \right),$$

and simplifying terms, one finally obtains

$$|u|_2^2 \leq \frac{4(n+3)\|f\|_{L^2(\Omega)}^2}{\alpha_0^2} + \frac{4(n+3)(M_c^2 + |\lambda_0|^2)|u|_0^2}{\alpha_0^2}|u|_0^2 + \varepsilon_0^2|u|_0^2$$

$$\leq \frac{4(n+3)}{\alpha_0^2}(\|f\|_{L^2(\Omega)}^2 + |u|_0^2 + \left(\frac{2(n+3)^4M^2}{\alpha_0^2} + 1\right)^4 |u|_0^2,$$

with $M = \max\{M_a, M'_a, M_b, M_c\}$. This proves the lemma.

The following lemma quantizes the result in [8] for Ritz–Galerkin methods with indefinite bilinear form.

Lemma 3.3. *Assume properties* P1–P3. *If*

$$h < \frac{\alpha_0}{2E^2\sqrt{v}C_{\text{approxi}}\|(L^* - \lambda_0 I)^{-1}\|},$$

then a unique approximate solution u_h defined by Eq. (5) *exists and*

$$\|u - u_h\|_{H^1(\Omega)} \leq \frac{4E^2}{\alpha_0} C_{\text{approxi}}\|f\|_{L^2(\Omega)} h\|(L^* - \lambda_0 I)^{-1}\|. \tag{22}$$

Proof. We first prove prior estimate. From the coersiveness inequality (17) and bilinearity of $B[u, v]$, one has, for $e_h = u - u_h$,

$$\frac{\alpha_0}{2}\|u - u_h\|_1^2 - \alpha_0 v^2|u - u_h|_0^2 \leq B[e, e] + \lambda_0(e, e) \leq E|u - u_h||u|.$$

Let $\Sigma^*(1) = \{v \in H^2(\Omega): \|(L^* - \lambda_0 I)v\| = 1\}$. By Nitshe's inequality (see Ciarlet [3], Nitsche [7])

$$|u - u_h|_0 \leq \frac{E^2}{\alpha_0} \sup_{v \in \Sigma^*(1)} \inf_{\phi} \|v - \phi\|_1$$

and property P3, one obtains

$$|u - u_h|_0 \leq \frac{E^2}{\alpha_0} C_{\text{approxi}} h \sup_{v \in \Sigma^*(1)} |v|_2 = \frac{E^2}{\alpha_0} C_{\text{approxi}} h|(L^* - \lambda_0 I)^{-1}|.$$

It then follows that

$$\frac{\alpha_0}{2}\|u - u_h\|_1^2 - \alpha_0 v h^2 \frac{E^4}{\alpha_0^2} C_{\text{approxi}}|(L^* - \lambda_0 I)^{-1}|^2 \leq E|e|_1|u|.$$

Therefore, if

$$h < \frac{\alpha_0}{2E^2\sqrt{v}C_{\text{approxi}}|(L^* - \lambda_0 I)^{-1}|},$$

one has

$$\|e\|_1 \leq \frac{4E}{\alpha_0}|u|_1.$$

Thus, if $u = 0$, $u_h = 0$. This implies the uniqueness and, equivalent in the finite-dimensional case, the existence of solution for Eq. (5). Replacing u by $u - \phi$, one finally proves part 1 by

$$|u_h - u|_1 \leq \frac{4E}{\alpha_0} \inf_{\phi} \|u - \phi\|_1 \leq \frac{4E}{\alpha_0} C_{\text{approxi}} h \|f\|_{L^2(\Omega)} |(L^* - \lambda_0 I)^{-1}|.$$

Lemma 3.4 (Order of Increasing of Eigenvalues). *Let $\lambda_1 \leq \lambda_2 \leq \cdots$ be the eigenvalues of L. There exist some real constants $c_1 > 0$, c_2 such that*

$$\lambda_i \geq c_1 i^{2/n} - c_2.$$

PROOF. From the proof of the coersiveness inequality (17), one has

$$B[u,u] \geq \frac{\alpha_0}{2}\|u\|_1^2 - \alpha_0 v^2 |u|_0^2$$

with

$$v = \frac{nM_b^2}{2\alpha_0^2} + \frac{M_c + \lambda_0}{\alpha_0}.$$

It then follows from the extremal property of eigenvalues (see Courant and Hilbert [4]) that $\lambda_i \geq \lambda_i'$ where $\lambda_1' \leq \lambda_2' \leq \cdots$ are the eigenvalues of a operator $-(\alpha_0/2)\Delta u + \alpha_0 v^2 u$. On the other hand, by using the proof of the asymptotic distribution of eigenvalues in [4], one knows that there, there exit real constants $c_1 > 0$, c_2 such that $\lambda_i' \geq c_1 i^{2/n} - c_2$. Therefore, one obtains the desired inequality

$$\lambda_i \geq \lambda_i' \geq c_1 i^{2/n} - c_2.$$

We now prove the Main Theorem.

PROOF OF MAIN THEOREM. With the assumption of the Main Theorem, Lemma 3.1 guarantees that a unique weak solution will exist and belong to $H^2(\Omega)$. Since L is self-adjoint and $\lambda_0 \notin \text{Spec}(L)$, one can take eigenvectors $\{u_n(x)\}$ of $L + \lambda_0 I$ as an orthonormal basis of $L^2(\Omega)$. Thus, one has the following expansion: $u(x) = \sum_{i=1}^{\infty} c_i u_i(x)$. Since the weak solution satisfies Eq. (1), it follows that

$$u(x) = \sum_{i=1}^{\infty} \frac{c_i u_i(x)}{(\lambda_0 - \lambda_i)}, \qquad c_i = (f, u_i).$$

Therefore,

$$|u|_0^2 = \sum_{i=1}^{\infty} \frac{c_i^2}{(\lambda_0 - \lambda_i)^2} \leq \max_{i>0} \frac{1}{(\lambda_0 - \lambda_i)^2} \sum_{i=1}^{\infty} c_i^2 = \max_{i>0} \frac{1}{(\lambda_0 - \lambda_i)^2} \|f\|_{L^2(\Omega)}^2.$$

Using the regularity estimate (19), one obtains

$$|(L^* - \lambda_0 I)^{-1}| = |(L^* - \lambda_0 I)^{-1}| = \sup_{f \neq 0} \frac{|u|_2}{\|f\|_{L^2(\Omega)}}$$
$$\leq \sup_{f \neq 0} \frac{(C''_{\text{regu}} \|f\|^2_{L^2(\Omega)} + C_{\text{regu}} |u|_0^2)^{1/2}}{\|f\|_{L^2(\Omega)}}$$
$$\leq \left(C''_{\text{regu}} + C_{\text{regu}} \max_{i>0} \frac{1}{(\lambda_0 - \lambda_i)^2} \right)^{1/2}.$$

Now part 1 of the theorem follows directly from Lemma 3.3. From the estimate (6) in part 1, one has

$$\|u_h - u\|_{H^1(\Omega)} \leq C(n, L, \Omega) \lambda_0^2 \left(\max_{i>0} \frac{1}{(\lambda_0 - \lambda_i)^2} \right)^{1/2} h \|f\|.$$

Let $M = 4\sigma \sqrt{|\log(1/\mu^2 2\pi\sigma)|}$ as in Lemma 2.4. If one picks the smallest integer M' larger than $[(M + c_2/c_1]^{n/2}$ with constants c_1, c_2 as in Lemma 3.4, one knows from that lemma that only at most M' number of eigenvalues of L will be belong to $(-M, M)$. Now one defines

$$\Sigma_i^\mu = \left[\lambda_i - \frac{\sqrt{2\pi\sigma\mu}}{4M'}, \lambda_i + \frac{\sqrt{2\pi\sigma\mu}}{4M'} \right], \qquad \Sigma^\mu = (-\infty, -M) \cup (M, \infty) \bigcup_{i=1}^{M'} \Sigma_i^\mu.$$

By using Lemma 2.4, one has $\text{meas}(\Sigma^\mu) \leq \mu/2 + \mu/2 = \mu$, and

$$\int_{(-\infty,\infty)-\Sigma^\mu} \|u_h - u\|_{H^1(\Omega)} \phi(\lambda_0)\, d\lambda_0$$
$$\leq c(n, L, \Omega) \int_{\bigcup_{i=1}^{M'} \Sigma_i^\mu} \lambda_0^2 \phi(\lambda_0) \max_{i>0} \frac{1}{|\lambda_0 - \lambda_i|} h \|f\|_{L^2(\Omega)}\, d\lambda_0$$
$$\leq c(n, L, \Omega) h \|f\|_{L^2(\Omega)} \sum_{i=1}^{M} \int_{(-M,M)-\Sigma_i^\mu} \lambda_0^2 \phi(\lambda_0) \frac{1}{|\lambda_0 - \lambda_i|}\, d\lambda_0$$
$$\leq c(n, L, \Omega) h \|f\|_{L^2(\Omega)} M' C \sigma^{3/2} (\log M + \log M' \log \mu)$$
$$\leq C(n, L, \Omega) \sigma^{n/2} (|\log \mu| + |\log \sigma|)^{n/4} \sigma^{3/2} (|\log \mu| + |\log \sigma|)$$
$$\leq C(n, L, \Omega)(\sigma |\log \mu| + \log \sigma)^{(n+3)/2}.$$

This finishes the proof of the Main Theorem.

References

[1] Agmon, S., *Lectures on Elliptic Boundary Value Problems*, Van Nostrand-Reinhold, Princeton, NJ, 1965.

[2] Babuška, I. and Aziz, K., "Survey lectures on the mathematical foundations of the finite element methods, *The Mathematical Foundations of the Finite Element*

Method, With Applications to Partial Differential Equations (edited by A.D. Aziz), Academic Press, New York and London, 1972.

[3] Ciarlet, P.G., *The Finite Element Method for Elliptic Problems*, North-Holland, Amsterdam; *Bifurcation Theory*, Vol. 2, Springer Verlag, New York, 1982.

[4] Courant, R. and Hilbert, D., *Methods of Mathematical Physics*, Vols. 1, 2, Wiley-Interscience, New York, 1962.

[5] Gilbarg, D. and Trudinger, N.S., *Elliptic Partial Differential Equations of Second Order*, Springer-Verlag, Berlin and New York, 1983.

[6] Grisvard, P., *Boundary Value Problems on Non-smooth Domains*, Pitman, Boston, 1985.

[7] Nitsche, J., Linear spline-funktionen und die methods von Ritz für elliptische randwertprobleme, *Arch. Rational Mech. Anal.* **36** (1970), 348–355.

[8] Schatz, A.H., An observation concerning Ritz-Galerkin methods with indefinite bilinear forms, *Math. Comput.* **28** (1974), 959–962.

[9] Smale, S., The fundamental theorem of algebra and complexity theory, *Bull. Amer. Math. Soc.* **4** (1981), 1–36.

[10] Smale, S., Some remarks on the foundation of numerical analysis, *SIAM Rev.* **32** (1990), 211–220.

[11] Werschultz, A.G., Complexity of indefinite elliptic problems, *Math. Comput.* **46** (1986), 457–477.

[12] Xuan, X., Error estimate and complexity of nonlinear Ritz-Galerkin methods, to appear.

[13] Xuan, X., Nonlinear boundary value problems: Computation, convergence and complexity, Ph.D. dissertation, University of California at Berkeley, 1990.

Part 7
Nonlinear Functional Analysis

43
Smale and Nonlinear Analysis: A Personal Perspective

ANTHONY J. TROMBA

In the fall of 1966, Smale returned from the International Congress to spend a semester at the Institute for Advanced Study. I had just passed my general exams at Princeton and was looking into a possible thesis topic. As an undergraduate at Cornell, my interest in global nonlinear analysis was ignited by the entertaining and informative lectures of Jim Eells. It was, therefore, natural for me to look for a problem in this area.

Smale's solution to the generalized Poincare conjecture had already raised him to legendary status among the graduate students who regularly met for their daily tea at Fine Hall. However, the graduate students were also aware of the excitement created by the work of Palais and Smale on an infinite-dimensional version of Morse's critical point theory.

Thus, when Smale lectured at the Institute on his infinite-dimensional version of Sard's theorem and mod(2) degree theory the room was filled to the brim. Although still a mathematical novice, the originality of Smale's viewpoint made a deep impression on me. Smale's attendance at Fine Hall teas then gave me the opportunity to approach him in the crowded and chatty atmosphere of the tea room. In particular I wanted to discuss his cryptic concluding remarks at the lecture that his theory was "related" to Leray–Schauder degree. This remark was eventually to form the core of my and David Elworthy's Ph.D. dissertations. After several conversations, Smale invited me (along with Arthur Greenspoon) to write our theses at Berkely under his direction.

Upon approaching Salomon Bochner, who was at that time in charge of graduate students, for permission to leave Princeton for Berkeley, he told me "Tromba, we don't care what you do, but if you ever write a thesis just mail it in." With this encouragement, I set out for Berkeley and on my career.

Historically the topology of, and analysis on, infinite-dimensional manifolds began at the Mexico topology conference in 1956 with the presentation by Jim Eells of his fundamental paper showing that suitable function spaces of maps between manifolds themselves form a smooth infinite-dimensional manifold. Jim Eells told the author that although he seemed to be invited everywhere to lecture on his results, no one seemed to pay any additional attention. However, the papers of Palais and Smale [5, 13] on Morse theory, the paper of Smale [14] on the infinite-dimensional Sard theorem, and the Eells–Sampson breakthrough paper [6] on the existence of harmonic mappings changed things dramatically; suddenly there was enormous interest in the field. The two papers Morse Theory and a Non-Linear Generalization of the Dirichlet Problem, which appeared in the *Annals of Mathematics* in 1964, and An Infinite Dimension Version of Sards Theorem, which appeared in the *American Journal of Mathematics* in 1965, constitute Smale's main contribution to Nonlinear Analysis.

In the first paper, Smale introduced his (and Palais' [5], who discovered it independently) *condition C*, or what is now commonly called *the Palais–Smale condition.* This work generalized the abstract critical point theory of Marston Morse to infinite-dimensional Hilbert manifolds M equiped with a complete Riemannian metric $<, >: TM \times TM \to R$. Smale considers C^2-functions $\mathscr{J}: M \to R$ satisfying the Palais–Smale condition:

Whenever a sequence $\{x_n\} \subset M$ satisfies

(i) $\mathscr{J}(x_n)$ is bounded,
(ii) $\|D\mathscr{J}(x_n)\| \to 0$

[$D\mathscr{J}(x_n)$ the derivative of $\mathscr{J}$ at x_n], then $\{x_n\}$ has a subsequence which converges (naturally to a critical point x_0 of $\mathscr{J}$; i.e., $D\mathscr{J}(x_0) = 0$].

Suppose now that x_0, $\mathscr{J}(x_0) = c$, is a nondegenerate critical point [this means that the Hessian, $D^2\mathscr{J}(x_0)$: $T_{x_0}M \times T_{x_0}M \to R$ induces an isomorphism of $T_{x_0}M$ and its dual space $T_{x_0}M^*$]. Assume further that x_0 is a critical point of finite Morse index θ; the integer θ is defined to be the dimension of the maximal subspace on which the Hessian is negative definite.[1]

Then Smale's main abstract result is that:

$\mathscr{J}^{-1}(-\infty, c+\varepsilon]$ has the homotopy type of $\mathscr{J}^{-1}(-\infty, c-\varepsilon]$ with a cell of dimension θ attached.

Condition C then implies the Morse inequalities hold as in finite dimensions, namely,

$$C_0 \geq R_0,$$

$$C_0 - C_1 \geq R_1 - R_0,$$

$$C_2 - C_1 + C_0 \geq R_2 - R_1 + R_0,$$

and the Morse equality $\sum(-1)^\theta R_\theta = \sum(-1)^\theta C_\theta$, where R_θ is the Betti number of $\mathscr{J}^{-1}[a,b]$ and C_θ is the number of critical points of index θ in $\mathscr{J}^{-1}[a,b]$ in the level set $\mathscr{J}^{-1}[a,b]$, assuming that a and b are not critical values.

Palais presents a similar theory but shows, in addition, that $\mathscr{J}^{-1}(-\infty, c+\varepsilon)$ is *diffeomorphic* to $\mathscr{J}^{-1}(-\infty, c-\varepsilon)$ with a cell of dimension θ attached. To

[1] Palais also considers critical points of infinite Morse index and shows that they are topologically irrelevant.

obtain this stronger diffeomorphism result, Palais needs a Hilbert space version of the famous Morse lemma, for which he gives an extremely elegant proof. A simple version of the Morse lemma states that if x_0 is a nondegenerate critical point, one can find a coordinate neighborhood (φ, U) about x_0 so that

$$(\mathscr{J} \circ \varphi^{-1})(x) = Q(x) + \text{constant}$$

for all $x \in \varphi(U)$, where Q is a quadratic form on the Hilbert space which models M. What is interesting and innovative in Smale's approach is that he does not use the Morse lemma, but uses instead only Taylor's theorem to analyze the change of homotopy type near the critical point. This is a fortuitous observation since the Morse lemma and condition C can be incompatible [21]. For example, Struwe's beautiful application of Palais–Smale theory to minimal surfaces [18], and surfaces of constant mean curvature [16, 17], which we later discuss, is exactly such a situation.

Smale concludes his paper by giving a concrete application to a large class of variational problems. For the purpose of exposition, let us consider a special case of Smale's theory, albeit the case that perhaps has the most interest for people in the variational calculus.

Let Ω be a domain in $\mathbb{R}^n$ with smooth boundary $\partial\Omega$. Let $J(Q)$ be the first jet bundle over Ω; i.e., $J(\Omega) = \Omega \times R \times \mathscr{L}(\mathbb{R}^n, \mathbb{R})$, where $\mathscr{L}(\mathbb{R}^n, R)$ are the linear maps from $\mathbb{R}^n$ to $\mathbb{R}$. Let $H_0^1(\Omega)$ denote the Sobolev space of functions whose derivatives are square integrable and which assume the Dirichlet boundary value 0 a.e. on $\partial\Omega$. Let $F: J^1(\Omega) \to R$ and consider the variational integral $\mathscr{J}: H_0^1(\Omega) \to R$ defined by

$$\mathscr{J}(u) = \int_\Omega F(x, u(x), Du(x))\, dx,$$

where $Du(x)$ denotes the derivative of u at x. Denote the variables of the integrand F by (x, p_0, p_1) or simply (x, p) if we denote the last two variables by one letter. Smale assumes the following conditions:

(1) $F(x, p) \le C_1 \|p\|^2 + C_2$,
(2) $F_{pp}(x, p)(\beta, \beta) \le C_3 \|\beta\|^2$,
(3) $C_4 \|p_1\|^2 - C_5 \le \int_\Omega F(x, p)\, dx$,
(4) $C_6 \|\beta\|^2 \le F_{p_1 p_1}(x, p)(\beta, \beta)$,

where subscripts denote partial differentiation. Then if $M = H_0^1(\Omega)$, Smale proves that $\mathscr{J}: M \to R$ satisfies condition C. As a consequence, one can conclude that $\mathscr{J}$ has an absolute minimum and, moreover, if the critical points are nondegenerate and finite in number, one can conclude the Morse inequalities

$$C_0 \ge 1,$$

$$C_1 - C_0 \ge -1,$$

$$C_2 - C_1 + C_0 \ge 1,$$

and so on.

Note that Smale assumed that the critical points were nondegenerate. At the time, no conditions were known when nondegeneracy held or if nondegeneracy held even in some *generic* situation.

When this paper appeared there was, as I mentioned, an initial burst of excitement. Many mathematicians worldwide looked at it to see if it applied to their problems. After the initial euphoria, people took a more sobering view of these methods. The Palais–Smale condition, in the hands of mathematicians like Paul Rabinowitz and Antonio Ambrosetti, continued to have great success in applications to variational problems with one variable.

However, in the several-variable case, it did not seem to apply to many of the most challenging variational problems of the time, as, for example, the existence of harmonic mappings between Riemannian manifolds, minimal surfaces in space or in Riemannian manifolds, or the existence of H-surfaces, to mention just a few. In fact, Smale told me in 1971 that one of his principle goals in developing the Hilbert manifold Morse theory was to apply it to Plateau's problem, something that Morse himself failed to achieve.

In addition to these criticisms, people were aware that one could obtain the existence of absolute minima under weaker conditions required for a Morse theory, and, in addition, some regularity results and, hence, the existence of classical solutions were known only for minimizers. As a guest of the Steklov Institute in 1969, Anosov and Novikov let me know that they felt there was little future in these methods, which they viewed as just a fancy way of looking at Morse's original ideas. Stefan Hildebrandt, in his interesting modern history of the "Calculus of variations today, reflected in the Oberwolfach Meetings" [10] described the feelings people had in 1968 as follows:

> We moreover decided to leave aside the Morse theory developed by Palais and Smale since there seemed to be no applications to interesting geometric problems except for those leading to one-dimensional variational problems.

By 1990, Smale's 60th birthday year, the pendulum had swung back in the other direction. By using condition C for a perturbed variational problem and then going to the limit, Sacks and Uhlenbeck were able to show the existence of minimal spheres in a large class of Riemannian manifolds. Struwe, Rheinhold Böhme, and the author were able to complete the Morse theory for minimal surfaces of disc type [2, 15, 22, 23]. Brezis, Coron, and Struwe then used these ideas in solving a conjecture on the existence of large solutions of constant mean curvature [3, 4, 16, 17]. The development of many of these ideas and methods is a long story. But we are getting ahead of ourselves. Let us turn now to Smale's second fundamental paper.

In this paper, Smale was the first to introduce the notion of a nonlinear Fredholm map, i.e., one whose derivative is linear Fredholm, and to define the Fredholm index of such mappings.

Granted, the idea of linear Fredholm maps were very much in the air at this time; Atiyah and Singer had recently proved their famous index theorem which gave a topological formula for the Fredholm index of a linear elliptic operator defined on sections of vector bundles. Nevertheless, it was Smale's vision to see that much of finite-dimensional differential topology could be carried over to infinite dimensions if one restricted one's attention to Fredholm maps and, furthermore, that these techniques could have significant applications to analysis. The first and most fundamental result in this direction is, of course, Sard's theorem.

Let M and N be second countable C^∞ smooth manifolds with M, N connected. $f: M \to N$ is said to be *Fredholm* if the sum of the dimensions of the kernd and cokernel of the derivative Df_x is finite for each $x \in M$. This implies that the range of Df_x is closed. In the case f is Fredholm, we define the Fredholm index of f (ind f) by

$$\operatorname{ind} f = \operatorname{index} f = \dim \operatorname{Ker} Df_x - \dim \operatorname{Coker} Df_x.$$

It is an exercise to show that this is independent of $x \in M$. A value $y \in N$ is said to be a regular value if either

(i) $f^{-1}(y) = \varnothing$

or

(ii) if $x \in f^{-1}(y)$, then Df_x is surjective.

The map f is said to be *proper* if the inverse image of compact sets is compact. Then Smale proves the following version of Sard's classic theorem

Theorem (Smale–Sard). *If* $f: M \to N$ *is a* C^r *smooth Fredholm map with* $r >$ $\max(\operatorname{ind} f, 0)$, *then the set of regular values is a set of second category. If* f *is proper they are open and dense.*

It was known that the theorem fails if f is not Fredholm or if $r \leq$ $\max(\operatorname{ind} f, 0)$. In addition, counterexamples for real-valued maps on open subsets of Hilbert spaces were also constructed. However, the theorem does hold for real-valued maps f if there exists a "sufficiently smooth Fredholm vector field" X transverse to f; i.e., $Df_x(X(x)) \geq 0$ where equality holds only if x is simultaneously a critical point of f and a zero of X [26].

Smale's version of Sard's theorem was the beginning of genericity results in nonlinear partial differential equations and the calculus of variations and, in particular, results describing exactly when critical points of functionals $\mathscr{J}$ could be nondegenerate; Smale pointed the way for such results with his own application:

Let $J^2(\Omega)$ be the second jet bundle over a domain $\Omega \subset \mathbb{R}^n$, again with smooth boundary $\partial\Omega$, where

$$J^2(\Omega) = \Omega \times F \times \mathscr{L}(\mathbb{R}^n, R) \times \mathscr{L}_s^2(\mathbb{R}^n, R),$$

where $\mathscr{L}_s^2(\mathbb{R}^n, R)$ are the symmetric bilinear real-valued maps on $\mathbb{R}^n$. Let $F: J^2(\Omega) \to R$ be C^∞ smooth and again denote the variables of F by

(x, p_0, p_1, p_2). F defines a nonlinear second order operator Φ by

$$\Phi(u) = F(x, u(x), Du(x), D^2u(x)).$$

Let $f\colon \partial\Omega \to R$ and define $C_f^{2,\alpha}(\Omega)$ be the Hölder-space of twice differentiable functions on Ω whose second partials satisfy a Hölder condition with exponent α, $0 < \alpha < 1$, and which agree with f on $\partial\Omega$. Then Φ defines a map from $C_f^{2,\alpha}(\Omega)$ into $C^\alpha(\Omega)$. Smale defines F (or Φ) to be *elliptic* if the derivative $F_{p_2} \in \mathscr{L}_s^2(\mathbb{R}^n, R)$ is a positive or negative definite bilinear form at an points. It then follows from linear elliptic theory that Φ defines a Fredholm map of index zero.

Application of his Sard theorem allows Smale to conclude that for almost all functions $g \in C^\alpha(\Omega)$, the set of solutions to the equation $\Phi(u) = g$ is discrete. Actually, there is a small technical error in Smale's paper, namely, that $C^{k,\alpha}(\Omega)$ is not a separable Banach space (and so cannot be second countable). Smale should have worked with the space $\Lambda^{k,\alpha}$, the closure of the C^∞ functions in $C^{k,\alpha}$, which is a separable space and for which the Hölder PDE estimates, on which Smale's theory is based, continue to hold.

Smale concludes his paper by introducing a generalized mod(2) degree for Fredholm maps of non-negative index. If $f\colon M \to N$ is a sufficiently smooth and proper (inverse image of compact sets is compact) Fredholm map of index $p > 0$, and N is connected, then the unoriented cobordism class of $f^{-1}(y)$, $[f^{-1}(y)]$, is independent of the choice of regular value y, and $[f^{-1}(y)]$ can thus be interpreted as a generalized mod(2) degree. This idea tied recent developments in topology (the cobordism of René Thom) with developments in non-linear analysis, e.g., degree theory as a tool for existence and uniqueness theorems. In fact, Smale ends his paper, as he did his talk at Princeton, with the remark that the mod(2) degree is related to Leray–Schauder degree. It is safe to say that the full potential of the ideas opened up in this paper are only beginning to be explored.

We would like to conclude this paper with a brief discussion of some results of the author and Michael Struwe which are a culmination of the ideas that Smale initiated in both of his fundamental papers.

Let $D \subset \mathbb{R}^2$ be the unit disc and $\Gamma \subset \mathbb{R}^n$ a smoothly embedded curve, say the image of an H^r smooth embedding $\alpha\colon \partial D \to \mathbb{R}^n$. A map $u\colon D \to \mathbb{R}^n$, $u \in C^\infty(D^0) \cap C^0(D)$, is said to be a *minimal surface of disc type* spanning Γ if

(1) $\Delta u = (\Delta u^1, \Delta u^2, \Delta u^n) = 0$,
(2) $u_x \cdot u_y \equiv 0$,
(3) $\|u_x\|^2 = \|u_y\|^2$,
(4) $u\colon \partial D \to \Gamma$ is a homeomorphism,

where (x, y) denote the coordinates on D and $\|\ \|$ is the R^n norm.

Conditions (2) and (3) mean that the map is a conformal parameterization of the image surface in $\mathbb{R}^n$. Since the middle of the nineteenth century, until the work of Jesse Douglas and Tibor Rado in 1931, it was unknown whether or not a map u satisfying (1)–(4) exists for any Γ. Douglas proved existence by the direct method of the calculus of variations. For this and other work, Douglas was the first (along with Lars Ahlfors) to receive a Fields medal at the International Congress in 1936.

Let $\mathcal{N}_\alpha$ be the manifold of all H^s harmonic maps $u: D \to \mathbb{R}^n$, $1 < s \ll r$, such that $u|\partial D$ is homotopic to α, and let $\mathcal{M}_\alpha$ denote those $u \in \mathcal{N}_\alpha$ which map ∂D "monotonically" onto $\alpha(\partial D) = \Gamma$ and which map three points on ∂D to three fixed points on 1?

Dirichlets energy $E_\alpha: \mathcal{N}_\alpha \to R$ is defined by

$$E_\alpha(u) = \frac{1}{2} \sum_{j=1}^{n} \int_D \nabla u^j \cdot \nabla u^j \, dx \, dy.$$

Roughly speaking, Douglas showed that E_α achieved a minimum on $\mathcal{M}_\alpha$ and that this minimum was a minimal surface.

A natural question is whether a Morse theory holds for $E_\alpha: \mathcal{M}_\alpha \to R$ as it does for the geodesics on Riemannian manifolds [5]. Before attempting to answer this question, one would like to know if it is reasonable to assume that the critical points of E_α are nondegenerate in some sense. In this case, to form the manifold $\mathcal{N}_\alpha$, we must choose $s > 1$. As a consequence, the critical points of E_α cannot possibly be nondegenerate in any classical sense:

First, Dirichlet's energy is invariant under the action of the conformal group $\mathcal{G}$ of the disc, $z \to c(z - a)/1 - \bar{a}z$, $|a| < 1$, $|c| = 1$, a noncompact three-dimensional Lie group. Therefore, the Hessian $D^2E_\alpha(u): T_n\mathcal{N}_\alpha \times T_n\mathcal{N}_\alpha \to R$ at a critical point u will always have at least a three-dimensional kernel. Second, and most importantly, if $s > 1$, even after factoring out this kernel, the Hessian cannot induce an isomorphism between a complement to the tangent space of the orbit of $\mathcal{G}$ and its dual space, which is the classical definition of nondegeneracy.

Although the theory of nondegeneracy in this context has been developed [21], for the purpose of this expository paper, we will say that a critical point u of E_α is *classically nondegenerate* if the Hessian $D^2E_\alpha(u)$ has only a three-dimensional kernel. In this case, a Morse lemma holds about such critical points.

In analogy with the theory of geodesics with fixed endpoints, one might conjecture that for almost all α, the critical points of E_α in $\mathcal{M}_\alpha$ are nondegenerate. This, in fact, turns out to be false in dimension $n = 3$. Let us try to understand why this is so, in the context of the ideas originating with Smale.

Minimal surfaces can have branch points, points where $u: D \to \mathbb{R}^n$ fails to be an immersion. Since $u_x - iu_y: D^0 \to \mathbb{C}^n$ is holomorphic, each interior branch point $z_0 \in D^0$ has an order λ defined by

$$u_x - iu_y = (z - z_0)^\lambda G(z),$$

where $G(z_0) \neq 0$.

It is less obvious, but nevertheless true, that boundary branch points also have a specific order. By a fundamental result due to Osserman, Gulliver, and Alt, minima do not have interior branch points.

Let Σ be the space of all minimal surfaces in $\bigcup_\alpha \mathcal{M}_\alpha$ and let Σ_λ^ν, $\lambda = (\lambda_1, \ldots, \lambda_p)$ and $\nu = (\nu_1, \ldots, \nu_q)$, be those minimal surfaces with p interior branch points of orders $\lambda_1, \ldots, \lambda_p$ and q boundary branch points of orders $\nu_1, \ldots, \nu_q$. Let $|\lambda| = \sum \lambda_i$ and $|\nu| = \sum \nu_i$.

The following result was proved by the author and R. Böhme [2] and strengthened by Thiel [20]. It is the two-dimensional analogue to Morse's famous theorem that on a Riemannian manifold N for fixed P and for almost all Q, the geodesics joining P and Q are nondegenerate critical points of the energy functional on paths: $\sigma \to \frac{1}{2}\int \|\sigma'(t)\|^2$. The techniques used to prove this result are, however, drastically different than those used by Morse.

Theorem (Böhme-T). *The strata Σ_λ^0 are all manifolds; if weighted Sobolev spaces are used then according to Thiel [20], the Σ_λ^ν are manifolds also. If $\pi\colon \bigcup_\alpha \mathcal{N}_\alpha \to \bigcup_\alpha \alpha = \mathcal{A}$ represents the map which sends the surface to its boundary contour and π_ν^λ the restriction of π to Σ_ν^λ, then each π_ν^λ is a nonlinear Fredholm map in the sense of Smale, with index*

$$\operatorname{index} \pi_\nu^\lambda = 2|\lambda|(2-n) + |\nu|(2-n) + 2p + q.$$

In a strong sense $\bigcup \Sigma_\lambda^\nu$ has the structure of an infinite-dimensional algebraic variety. A generalization of these results to higher genus surfaces has been proved by Tomi and Tromba [19].

For $n \geq 4$, this index is zero only on the stratum Σ_0^0 (i.e., the stratum consisting of no branch points). Let us call a surface $u \in \Sigma_0^0$ *BT nondegenerate* if $D\pi_0^0(u)\colon T_u\Sigma_0^0 \to T_\alpha\mathcal{A}$ is an isomorphism.

As a consequence of the Smale–Sard theorem and the regularity theorem of Hildebrandt [7], almost all $\alpha \in \mathcal{A}$ are regular values of $\pi|\bigcup \Sigma_\lambda^\nu$, and from the index theorem it follows that an open dense set of α in $\mathbb{R}^n$ have the property that only a finite number of BT nondegenerate minimal surfaces span the curve $\alpha(S^1)$, $S^1 = \partial D$. For $n \geq 4$, BT nondegeneracy coincides with classical nondegeneracy. Thus, for $n \geq 4$, we know there are only a finite number of minimal surfaces of disc type which span $\alpha(S^1)$ for generic α and all these minimal surfaces will be classically nondegenerate. Let α be such a generic curve, and $u_1, \ldots, u_N$ be the disc minimal surfaces spanning $\alpha(S^1)$ of Morse indices $\theta_1, \ldots, \theta_N$.[2] Then the author proved the Morse equality

$$\sum_i (-1)^{\theta_i} = 1. \qquad (**)$$

The technique of proof of the author involves a generalization of Leray–Schauder degree to manifolds (in fact, to infinite-dimensional varieties) that Smale hinted at, at the end of his Institute lecture and at the end of the Sard paper. An "analog" of formula (**) is also valid for $n = 3$.

Somewhat later, Struwe [18] was able to use the Palais–Smale condition

[2] These will always be finite.

suitably adapted to the "convex" set $\mathscr{M}_\alpha$ to show that all Morse inequalities hold; namely,

$$C_0 \geq 1,$$
$$C_1 - C_0 \geq -1,$$
$$C_2 - C_1 + C_0 \geq 1,$$

and so on, assuming all critical points are classically nondegenerate, which by the index theorem holds only for $n \geq 4$. As already mentioned, Struwe did not (and could not) use the Morse lemma to derive these inequalities, but fortunately his way (as was the way to generic nondegeneracy) had been paved by Smale.

For $n = 3$, the situation changes dramatically. Here the index is 0 on the stratum Σ_0^0 and all strata Σ_λ^0, where $\lambda = (1, \ldots, 1)$. This implies the existence of simply branched minimal surfaces u such that $D\pi_\lambda^0(u)$: $T_u\Sigma_\lambda^0 \to T_\alpha\mathscr{A}$ is an isomorphism; i.e., BT nondegenerate. Such minimal surfaces will be stable (i.e., follow along smoothly) under the perturbation of the boundary curve α. However, all such u's will be classically degenerate since the additional dimension of the kernel of the Hessian at u will be $2|\lambda|$.

Thus, surprisingly, in $\mathbb{R}^3$ *the generic contour does not admit only classically nondegenerate solutions which span it.* However again, as a consequence of Smale's generalization of the Sard theorem, the generic contour in $\mathbb{R}^3$ will admit finitely many BT-nondegenerate minimal surfaces.

The question now arises as to how such a simply branched minimal surface in $\mathbb{R}^3$ is to be "counted" in the Morse theory of Palais and Smale as adapted to Plateau's problem by Struwe.

From general theoretical principles [21], it follows that there is a nice gradient ∇E_α to E_α: $\mathscr{N}_\alpha \to R$. The theory of winding numbers of such vector fields was developed in [24]. The winding numbers of a gradient field around a classically nondegenerate minimal surface will always be ± 1. However, what about the branched minimal surfaces u in $\mathbb{R}^3$ which are BT nondegenerate? The surprising answer is that it is not ± 1, but $\pm 2^p$ where p is the number of branch points for u. Thus, remarkably, the winding number reflects the number of singularities in the surface. As a consequence, it follows that for the Morse inequalities, each such u must be counted as 2^p classically nondegenerate surfaces; and in fact, if the boundary contour for such a surface is perturbed into $\mathbb{R}^4$, there will be at least 2^p disc minimal surfaces spanning the perturbed contour.

These results complete Morse's original program of applying his theory to Plateau's problem, a natural geometric variational problem in two variables.

These techniques were further adapted to attack an unsolved problem in the theory of surfaces of constant mean curvature, bounding a contour in $\mathbb{R}^3$. For example, cut a smooth region Ω of the topological type of the disc from the sphere S_R^2 of radius R and the remaining region. $S_R^2 - \Omega$ is a surface of constant mean curvature $1/R$ spanning the curve $\partial\Omega$.

Hildebrandt [8] was able to prove a general existence theorem, namely, that if H is a real number and $\Gamma \subset B_R[0]$, the ball of radius R about the origin, and if $|H|R < 1$, then there exists a surface of the type of the disc with mean curvature H and boundary Γ. Such solutions arise as a minimum to a variational problem and are called *small* solutions. Examples indicate that these should also be a second solution, a large H-surface spanning Γ. For, instance, it is easy to see that a simple closed contour on a sphere bounds two solutions.

Struwe [16] (as well as Brezis and Coron [3]) suitably adapting the ideas of Palais and Smale was able to prove the existence of the large solution under the same conditions mentioned for the existence of the small solution.

We cannot fail to mention the wonderful results of Sacks and Uhlenbeck [11, 12] who achieved the first major advance on the existence of harmonic mappings since the work of Eells and Sampson mentioned earlier.

We state two results which follow from their techniques, namely:

Theorem 1. *Let M be a compact surface without boundary, N a compact Riemannian manifold, $g\colon M \to N$ continuous. Then there exists a harmonic map $u\colon M \to N$ such that the induced maps on the fundamental groups agree*; *i.e.*,

$$u_{\#} = g_{\#}\colon \pi_1(M) \to \pi_1(N).$$

If $\pi_2(N) = 0$, then u is homotopic to g.

Theorem 2. *Suppose the universal cover $\tilde{N}$ of the compact Riemannian manifold N is not contractible. Then there exists a nontrivial conformal harmonic map $u\colon S^2 \to N$. Moreover, conformal harmonic maps generate $\pi_2(N)$ as a $Z(\pi_1(N))$ module, and each such harmonic map is energy minimizing in its free homotopy class.*

Harmonic maps are critical points of the Dirichlet energy functional

$$E(v) = \tfrac{1}{2}\int_M |dv|^2.$$

It is known that E does not satisfy condition C on an appropriate space of manifolds of mappings. Sacks and Uhlenbeck get around this by considering a perturbed functional

$$E_\alpha(v) = \tfrac{1}{2}\int_M (1 + |dv|^2)^\alpha$$

for $\alpha > 1$. This does satisfy the Palais–Smale condition and, consequently, has a minimum v_α. They obtain their results by leting $\alpha \to 1$ and showing that v_α's have a convergent subsequence.

These results should be viewed as a triumph of the Palais–Smale theory even though it does not apply directly to the problem at hand.

Finally, we mention another beautiful application of the degree theory that

Smale foresaw, again to minimal surfaces. Brian White [27] has recently shown, using such a theory that minimal surfaces of the type of the torus exist in the three sphere with any metric of positive Ricci curvature.

Surely, the ideas stemming from these two major papers have yet to realize their full potential.

References

[1] Ambrosetti, A. and Rabinowitz, P., Dual variational methods in critical point theory and applications, *J. Funct. Anal.* **14** (1973), 349–381.

[2] Böhme, R. and Tromba, A.J., The index theorem for classical minimal surfaces, *Ann. Math.* **113** (1981), 447–499.

[3] Brézis, H. and Coron, J.M., Sur la conjecture de Rellich pour les surfaces à coubure moyenne prescrite, *C.R. Acad. Sci. Paris Sér. I* **295** (1982), 615–618.

[4] Brézis, H. and Coron, J.M., Multiple solutions of H-systems and Rellich's conjecture, *Commun. Pure Appl. Math.* **37** (1984), 149–187.

[5] Palais, R., Morse theory on Hilbert manifolds, *Topology* **2** (1963), 299–340.

[6] Eells, J. and Sampson, J.H., Harmonic mappings of Riemannian manifolds, *Amer. J. Math.* **86** (1964), 109–160.

[7] Hildebrandt, S., Boundary behavior of minimal surfaces, *Arch. Rat. Mech. Anal.* **35** (1969), 47–82.

[8] Hildebrandt, S., On the Plateau problem for surfaces of constant mean curvature, *Commun. Pure Appl. Math.* **23** (1970), 97–114.

[9] Hildebrandt, S., Über Flächen konstanter mittlerer Krümmung, *Math. Z.* **112** (1969), 107–144.

[10] Hildebrandt, S., Calculus of variations today, reflected in the Oberwolfach meetings. *Perspective in Mathematics, Anniversary of Oberwolfach 1984*, Birkhäuser, Basel.

[11] Sacks, J. and Uhlenbeck, K., The existence of minimal immersions of 2-spheres, *Ann. Math.* **113**

[12] Sacks, J. and Uhlenbeck, K., Minimal immersions of closed Riemann surfaces, *Trans. Amer. Math. Soc.* **271**, 639–652.

[13] Smale, S., Morse theory and a non-linear generalization of the Dirichlet problem, *Ann. Math.* (2) **80** (1964), 382–396.

[14] Smale, S., An infinite dimensional version of Sard's theorem, *Amer. J. Math.* **87** (1965), 861–866.

[15] Struwe, M., On a critical point theory for minimal surfaces spanning a wire in $\mathbb{R}^n$, *J. Reine Angew. Math.* **349** (1984), 1–23.

[16] Struwe, M., Large H-surfaces via the mountain-pass-lemma, *Math. Ann.* **270** (1985), 441–459.

[17] Struwe, M., Nonuniqueness in the Plateau problem for surfaces of constant mean curvature, *Arch. Rat. Mech. Anal.* **93** (1986), 135–157.

[18] Struwe, M., *Plateau's Problem and the Calculus of Variations, Mathematical Notes* No. 35, Princeton University Press, Princeton, NJ, 1988.

[19] Tomi, F. and Tromba, A.J., The index theorem for minimal surfaces of higher genus, preprint.

[20] Thiel, U., On the stratification of branched minimal surfaces, *Analysis* **5** (1985), 251–274.

[21] Tromba, A.J., A general approach to Morse theory, *J. Diff. Geom.* **32** (1977), 47–85.

[22] Tromba, A.J., On the number of minimal surfaces spanning a curve, Mem. *AMS* **194** (1977).

[23] Tromba, A.J., Degree theory on oriented infinite dimensional varieties and the Morse number of minimal surfaces spanning a curve in $\mathbb{R}^n$. Part I: $n \geq 4$, *Trans.* **290** (1985), 385–413; Part II: $n = 3$, *Amer. Math. Soc. Manuscriptà Math.* **48** (1984), 139–161.

[24] Tromba, A.J., The Euler characteristic of vector fields on Banach manifolds and a globalization of Leray–Schauder degree, *Adv. Math.* **28** (1978), 148–173.

[25] Tromba, A.J., Intrinsic third derivatives for Plateau's problem and the Morse inequalities for disc minimal surfaces in $\mathbb{R}^3$, SFB 256, preprint 58, Bonn.

[26] Tromba, A.J., The Morse–Sard–Brown theorem for functionals and the problem of Plateau, *Amer. J. Math.* **99** (1977), 1251–1256.

[27] White, B., Every three sphere of positive Ricci curvature contains a minimally embedded torus, *Bull. Amer. Math. Soc.* **21** (1) (1989).

44
Discussion

R. ABRAHAM AND F. BROWDER

Ralph Abraham: I'm very happy to be here at this 22nd reunion of the class of '68, a reunion under the banner of unity and diversity. Of course, diversity is represented by six sessions on different subjects. And about unity, I think it's appropriate to say that the unity of these subjects was not only through the personality and the genius of Steve Smale, for there was a fantasy at the time. The balloon of the fantasy, called global analysis, went up in 1956 in Mexico City and came down here in Dwinelle Plaza in 1968. And this balloon of fantasy hoped, in the early sixties, to achieve the reunification of the whole of mathematics. Maybe I'm overstating the case, but we were young and idealistic, hoping to achieve a bifurcation from the divergence of mathematics into different fields of specialty into a convergence. And we attained a feeling of community throughout the mathematical research community.

This balloon did get wounded, I think, by external events in the early sixties: the Bay of Pigs; the assassination of J.F.K.; the Free Speech Movement, the Vietnam Day Committee, the Students for a Democratic Society, and so on. And certainly this was an aspect, I think, of the inflation of this balloon because we were very much working in a state of mind, a miraculous, creative state of mind.

After 1968, when there was a conference here in Berkeley organized by Chern and Smale, there appeared a paper by Steve Smale called "What is Global Analysis?" There the definition given in just two or three sentences said more or less that global analysis was the study of dynamical systems. That was not an "in" joke so much as the inside view on our hopes for the reunification of mathematics through these techniques based on Banach manifolds, manifolds of mappings, and so on. That was, I would say, the DNA of it. The germ of this reunification fantasy was the infinite-dimensional manifolds of maps that began with Jim Eells' paper in 1956. My own work in the early 1960s and Dick Palais' work also (I think we worked independently and almost identically through 1963) was on this technical paraphernalia of manifolds and maps. So I'll just end with this question: Why is this session called "functional" analysis and not "global" analysis?

Felix Browder: I have known Steve Smale for over 30 years, especially since 1960 when we were both in Rio de Janeiro, he for six months and I for two. This was the epoch, made famous by his reference to the beaches of Rio, in which, he made a big impact on the mathematical world by his solution of the Poincaré conjecture for $n \geq 5$. It was clear to me at the time, and I am sure to most of the people who met him, that he was a very significant and very unusual personality on the mathematical scene. He had one big idea and most people who get noticed tend to have one big central idea—and that was the idea of relating many kinds of mathematical and scientific problems to the study of vector fields on manifolds and the properties of the related flows. That was his central theme and central mechanism. Whatever he has done since that time, and he has certainly varied his interests in terms of domains of application, he has never given up that central tool. The other central tool was genericity as is illustrated in Tony Tromba's lecture on functional analysis (and it is, in fact, a lecture on functional analysis, not global analysis because generally speaking, global analysis means the application of these central tools to everything).

The other day, Paul Robinowitz remarked to me that one definition of a concrete analyst is someone who doesn't believe in genericity. In that sense, I am really only a part-time concrete analyst because I believe in genericity occasionally. It is obviously the case that the people who came into the study of infinite-dimensional manifolds and its applications to functional analysis from differential topology saw the property of genericity as much more significant than those people who came in through the study of problems in partial differential equations as I did. The early sixties were a period of revival of the study of nonlinear partial differential equations. The theory of such equations had been studied for many decades, as, for example, the Plateau problem, and went back to the beginning of the twentieth century to the work of Sergei Bernstein in the first decade of the century and the work of Leray and Schauder in the early 1930s. On the other side, there was the development by Marston Morse (who has been mentioned often) and by Lyusternik and Schnirelman and their school in the Soviet Union of global methods of studying variational problems on manifolds. It is interesting to explore the question of how the work of Smale and the school which he developed relate to these things. First of all, it is clear that almost all the ideas in the study of the calculus of variations on infinite-dimensional manifolds are present to some degree in the work both of Morse and of the Lyusternik school and the latter group did work extensively on infinite-dimensional variational problems. The principal difference in Smale's work is a relentless focus on simple basic geometric conditions and an avoidance of technical complications.

Smale was undoubtedly the central promoter of the systematic study of infinite-dimensional manifolds. There were many predecessors who had put forward general definitions and structures on this terrain producing nice theories without any particularly interesting applications. One of Smale's central contributions (aside from his emphasis on nonlinear Fredholm mappings and the Sard–Smale theorem) was to point the theory to a potentially important field of applications, the study of nonlinear partial differential equations. This illustrates one of the typical and most important characteristic of his work: his refusal to see mathematical theories and disciplines as isolated from one another and from potential applications in other fields of science and of knowledge. In recent years, from the point of view of many pure mathematicians, he has become some strange kind of semipure mathematician working in fields like economics, operations research, and computation.

However, there is a fundamental unity of approach which has been typical of him which consists of ignoring the differences in these domains and letting the ideas carry him wherever they may apply and not just saying: "I must produce a nice clear-cut set of definitions and prove a beautiful, well-disciplined set of theorems, and let that suffice": Smale's style is still not fashionable in mathematics, and even with the brilliant school which he has generated, he remains very isolated in the mathematical community. Frankly, I find that very regrettable.

Let me now make some remarks to illuminate Tony Tromba's talk in terms of present day results in analysis. For partial differential equations, the crucial property which must be verified in both the variational theory and degree theory is the property of mapping being *proper*. For the Fredholm mapping corresponding to a second-order nonlinear elliptic operator acting in the appropriate function spaces (or for more general nonlinear equations), the properness property corresponds to another great mathematical paradigm, this one beloved of concrete analysts working on PDE, the paradigm of *a priori estimates.* At least for the Fredholm theory in Hölder spaces, properness requires *strong a priori estimates.* For equations much more general than a single (scalar) second-order nonlinear equation, such estimates are almost certainly not valid. (This is one of the central reasons why I, since the early 1960s, have worked on situations in which (like the monotonicity theories) you have techniques that work without strong a priori estimates). For scalar second-order strongly elliptic equations, the program of proving such estimates begun by Bernstein in 1910 has finally been completed on domains on R^n through an important breakthrough by Krylov as well as Caffarelli, Nirenberg, and their collaborators. This is the culmination of a long and very difficult program which, in the early 1960s, nobody had real reasons to believe would actually be completed.

In the case of variational problems, condition C, the Palais–Smale condition, is a refinement and localization of the notion of properness. This turns out in a fairly direct way to be what is needed to carry through the proof of the Morse theory of the Lyusternik–Schnirelman theory by proofs not substantially different from the finite-dimensional cases. At this point, we may apply a poetic analogy with the complimentarity principle of Niels Bohr: The negation of a deep truth is a deep truth. We might formulate it this way: The negation of a deep paradigm is an even deeper paradigm. If condition C is the initial paradigm, the really interesting cases are those in which condition C fails. Thus, for example in the problem of periodic orbits for Hamiltonian systems which Paul Rabinowitz began to study a number of years ago by variational methods, the question becomes how to deal with variational problems when the original problem (at lease without modifications) does not satisfy the original properness condition. In some recent investigations, you may have to modify the manifold by compactifying with respect to the given variational problem. Condition C becomes the focal point of the study because where it fails is where you have to look. It becomes the crucial point of the investigation, the touch stone or crux of the problem. It is an interesting twist that in many significant studies in geometry and mathematical physics, the major point to be settled is where does condition C fail and how to create a mechanism to analyze that failure in a precise way.

Part 8
Applications

45
Steve Smale and Geometric Mechanics

JERROLD E. MARSDEN*

1. Some Historical Comments

In the period 1960–1965, geometric mechanics was "in the air." Some key papers were available, such as Arnold's work on KAM theory and a little had made it into textbooks, such as Mackey's book on quantum mechanics and Sternberg's book on differential geometry. In this period, Steve was working on his dynamical systems program. His survey article (Smale [1967]) contained important remarks on how geometric mechanics (specifically Hamiltonian systems on symplectic manifolds) fits into the larger dynamical systems framework. In 1966 at Princeton, Abraham ran a seminar using a preprint of the survey article and it was through this paper that I first encountered Smale's work. After he visited the seminar, the importance of what he was doing was obvious; also, it became evident that there was great power in asking simple, penetrating, and sometimes even seemingly naive questions. I should add that in the mathematical physics seminar at Princeton that I also had the good fortune of attending, Eugene Wigner had a remarkably similar aura.

Smale's dynamical systems work suggested developing similar ideas like structural stability, dynamic bifurcations, and genericity in the context of mechanics. Structural stability aspects were developed by Abraham, Buchner, Robinson, Robbin, and others. The generic bifurcations of equilibria, relative equilibria, Hamiltonian–Krein–Hopf bifurcations that can occur in Hamiltonian systems has been studied by Williamson, Arnold, Meyer, van der Meer, Duistermaat, Cushman, Golubitsky, Stewart, and others. See Abraham and Marsden [1978], Marsden [1992] and Delliniz, Melbourne and Marsden [1992] for further information and references.

In 1966–1967, two important personal events occured. First, Abraham gave his lectures on mechanics at Princeton from which our book *Foundations*

* Research partially supported by NSF Grant DMS-8922704 and DOE Contract DE-FGO3-88ER25064.

of Mechanics arose. Second, Arnold's [1966] paper on rigid-body mechanics and ideal fluid mechanics appeared. The latter paper influenced me profoundly; it showed how the dynamics of these systems could be interpreted as geodesic flow on $SO(3)$ with a left-invariant metric and on $\mathrm{Diff}_{\mathrm{Vol}}(\Omega)$—the volume-preserving diffeomorphism groups of, say, a region Ω in $\mathbb{R}^3$—with the right-invariant metric defined by the kinetic energy of the fluid. This paper, together with the emerging work of Kostant and Souriau on the role of symmetry groups and the momentum map, laid important ideas latent in the traditional approach to mechanics, in clear and concise geometric terms.

Smale gave a course of lectures on mechanics in the fall of 1968 at Berkeley, the semester I arrived. This led to his two-part paper Topology and Mechanics (Smale [1970]).

In the same period, I completed a paper with Ebin (Ebin and Marsden [1970]) in which we put Arnold's work on fluid mechanics in the context of Sobolev (H^s) manifolds and showed the remarkable fact that Arnold's geodesic flow on $H^s\text{-}\mathrm{Diff}_{\mathrm{vol}}(\Omega)$ (the volume-preserving diffeomorphisms of Ω to itself of Sobolev class H^s) comes from a *smooth* geodesic spray. This fact is remarkable because it allows one to solve the initial value problem using only Picard iteration and ordinary differential equations theory on $TH^s\text{-}\mathrm{Diff}_{\mathrm{vol}}(\Omega)$ and one would not expect this since the Euler equations for fluid mechanics are rather nasty PDEs, not ODEs! This led to a number of interesting analytic and numerical developments in fluid mechanics. Around this same time, I began work with Fischer on the Hamiltonian structure of general relativity (see, for example, Fischer and Marsden [1972]). At this stage I had only a rough idea how these two topics might be related to Smale's work.

Already around 1971, with so much happening in geometric mechanics, Abraham and I started work on the second edition of *Foundations of Mechanics* with the help of Ratiu and Cushman. Doing so prompted thoughts about how all of these ingredients might fit together. Especially interesting was the question of how Smale's papers might be linked with Arnold's. Smale used symmetry ideas in the context of tangent and cotangent bundles of configuration spaces with Hamiltonians of the form kinetic plus potential energy; that is, he dealt with *simple mechanical systems*. The examples and some of the theory were concerned with *abelian symmetry* groups. In this context, Smale's work contained some of the essential ideas of what we now call *reduction theory*. It was natural to attempt to put Smale's ideas and those of Arnold in the more general and unifying context of symplectic manifolds. Doing so led to the paper with Weinstein (Marsden and Weinstein [1974]) that was completed in early 1972. Some of these ideas were found independently by Meyer [1973] whose paper appears in the proceedings of the 1971 conference organized by Peixoto that was, in effect, a large conference on Steve's work on dynamical systems as a whole.

This effort led to the now fairly well-developed area of reduction theory. There are expositions of this subject available in the 10 or so texts and monographs that are currently devoted to geometric mechanics (Abraham and Marsden [1978], Guillemin and Sternberg [1984], and Arnold [1989] are examples). We will come back to a description of one of the reduction theorems later (the cotangent bundle reduction theorem) and give an indication of why this approach made so much fall beautifully into place. Briefiy, if one starts with a cotangent bundle T^*Q, and a Lie group G acting on Q, then the quotient $(T^*Q)/G$ is a bundle over $T^*(Q/G)$ with fiber $\mathfrak{g}^*$, the dual of the Lie algebra of G One has the following structure of the Poisson reduced space $(T^*Q)/G$ (its symplectic leaves are the symplectic reduced spaces of Marsden and Weinstein [1974]).

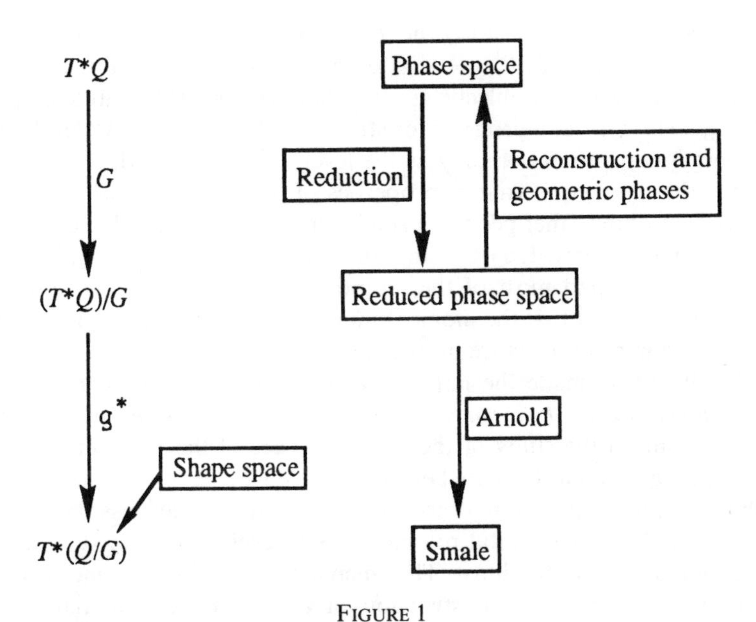

FIGURE 1

Thus, one can say—perhaps with only a slight danger of oversimplification—that reduction theory synthesises the work of Smale, Arnold (and their predecesors of course) into a bundle, with Smale as the base and Arnold as the fiber. This bundle has interesting topology and carries mechanical connections (with associated Chern classes and Hannay–Berry phases) and has interesting singularities (Arms, Marsden, and Moncrief, Guillemin and Sternberg, Atiyah, and others). We will describe some of these features later.

2. Highlights from Topology and Mechanics

One of Smale's main goals was to use topology, especially Morse theory, to estimate the number of relative equilibria in a given simple mechanical system with symmetry, such as the n-body problem, and to study the associated bifurcations as the energy and momentum are varied. This approach proved to be quite successful and has been carried on by Palmore, Cushman, Fomenko, and others to give more detailed information in the n-body problem and basic information in other problems like vortex dynamics, rigid-body mechanics, and other integrable systems.

One should also mention that this paper of Smale did a lot for the subject itself. The paper attracted worldwide attention and brought many excellent young people into the field. The idea of using topology and geometry in a classical subject to bring new insights and fresh ideas must have been quite appealing.

Smale's strategy was to study the topology of the level sets of the energy-momentum map $H \times \mathbf{J}: P \to \mathbb{R} \times \mathfrak{g}^*$ on a given phase space P with a given Hamiltonian H and a symplectic group action having a momentum mapping $\mathbf{J}: P \to \mathfrak{g}^*$. He lays out a program for studying bifurcations in the level sets of the energy-momentum mapping as the level value changes. In doing so, he sets out the basic equivariance properties of momentum maps, apparently independent of the other people normally credited with introducing the momentum map, namely, Lie, Kostant, and Souriau. (An interesting historical note is that Lie had most of the essential ideas, including—according to Weinstein—the fact that the momentum map is a Poisson map, its equivariance, the symplectic structure on the coadjoint orbits, and more, all back in 1890!) Smale also made the important observation that *a point $z \in P$ is a regular point of* **J** *iff the symmetry* (*isotropy*) *group of z is discrete.* This idea surfaces again in the study of the solution space of the Einstein or Yang–Mills equations, as we shall see below.

One of the most important objects that Smale introduced was the *amended potential* V_μ that plays a vital role in current developments in stability and bifurcation of relative equilibria. The amended potential is a geometric generalization of the classical construction of the "effective potential" in the (planar) two-body problem, which is obtained by adding to the given potential $V(r)$, the centrifugal potential at angular momentum value μ; in this simple case,

$$V_\mu = V + \frac{\mu^2}{2r^2}.$$

In this situation, reduction corresponds to the elimination of the angular variable θ (division by the group $G = S^1$) and the replacement of the potential V by the potential V_μ.

As we shall see later, in special situations, such as the abelian case, the reduction of a simple mechanical system is again a simple mechanical system, but the general situation, even for groups like the rotation group, is more complicated. The fact that the abelian reduction of a simple mechanical system is again a simple mechanical system is essentially contained in Smale's paper, but it has a long history with a surprising amount contained in the work of Routh around 1860 in his books on mechanics. See, for example,

Routh [1877]. In fact, Routh's work contains results that get rediscovered from time to time in the modern literature, but that is another story.

For problems like rigid-body mechanics and fluid mechanics, one has to deal with Lie–Poisson structures on $\mathfrak{g}^*$ (the "Arnold fiber") and magnetic terms on $T^*(Q/G)$ (the "Smale base"). The magnetic terms modify the canonical symplectic structure on $T^*(Q/G)$ with the addition of the μ-component of the curvature of a connection on $Q \to Q/G$ called the *mechanical connection.* This connection, defined below, is given implicitly in Smale's paper (it is introduced in §6 of his paper).

Smale studies relative equilibria by applying critical point theory to V_μ. The function V_μ contains much of the information of $E \times \mathbf{J}$ through the general fact proved by Smale that a point of Q is the configuration of a relative equilibrium if and only if it is a critical point of V_μ. (Some of these concepts are recalled in the next section.)

Smale's examples deal with the abelian case (when the "Arnold fiber" has trivial Poisson structure); in this situation, V_μ defines a function on Q/G, and on this quotient space, one expects the critical points to be generically nondegenerate. In the general case (such as a rotating rigid body with internal structure), one has to carefully synthesize the analysis of Arnold and Smale to get the sharpest information.

This basic theory, and some simple but very informative examples, comprise part I of "Topology and Mechanics." Part II is concerned with the planar n-body problem in which $G = S^1$ is the planar rotation group and Q is $\mathbb{R}^{2n}$, minus collision points. The study of relative equilibria is done by determining the global topology of the level sets of $E \times \mathbf{J}$ and their quotients by S^1—Theorems A and B of the paper. Theorem C relates this to critical points of V_μ and Theorem D determines the bifurcation set. Theorem E relates the topology of the rduced phase space to that of the configuration space. Corollaries give more details for $n = 2$ and $n = 3$. An interesting consequence of these results is Moulton's theorem stating that there are $n!/2$ classes of *colinear* relative equilibria. The fact that one is looking for colinear relative equilibria enables one to reduce the problem to one of finding critical points of a function on *real* projective $n - 2$ space minus collisions. In fact, a combinational argument shows that this space has $n!/2$ components and the corresponding function has a single nondegenerate maximum on each component. There have, of course, been many important contributions to the n-body problem since 1970, such as those of Palmore, McGehee, Mather, Meyer, and others. The book of Meyer and Hall [1991] can be consulted for some of the relevant literature.

3. A Glimpse at Reduction Theory

In this section we will focus on *some* of the ways Smale's paper is connected with *some* of the current research in geometric mechanics. No attempt is made at thoroughness here—the focus is on selected topics of personal interest only! In particular, we focus on some aspects of reduction theory, one of the most fruitful outgrowths of Smale's and Arnold's work. Of course, others also deserve much credit for setting the foundations, especially Lie, Kostant, Kirillov, and Souriau.

Let P be a symplectic manifold and G be a group acting symplectically on P. Let $\mathfrak{g}$ be the Lie algebra of G and $\mathfrak{g}^*$ its dual. Let G act on $\mathfrak{g}^*$ by the coadjoint action and let $\mathbf{J}: P \to \mathfrak{g}^*$ be an equivariant momentum map; that is, $\mathbf{J}$ is equivariant and $\mathbf{J}$ generates the group action in the sense that for each $\xi \in \mathfrak{g}$,

$$X_{\langle \mathbf{J}, \xi \rangle} = \xi_P,$$

where X_f is the Hamiltonian vector field determined by the function f and ξ_P is the infinitesimal generator of the action on P.

If G_μ is the isotropy subgroup of $\mu \in \mathfrak{g}^*$, the *reduced space* at μ is

$$P_\mu = \mathbf{J}^{-1}(\mu)/G_\mu.$$

Equivariance guarantees that G_μ acts on $\mathbf{J}^{-1}(\mu)$, so the quotient makes sense. If μ is a (weakly) regular value and the quotient is nonsingular, the *reduction theorem* states that P_μ is a symplectic manifold and that G-invariant Hamiltonian systems on P decend to Hamiltonian systems on P.

Given a G-invariant hamiltonian H on P, a *relative equilibrium* (in the terminology of Poincaré) is a point in P whose dynamic orbit equals a one-parameter group orbit. Relative equilibria correspond to critical points of $H \times \mathbf{J}$ and to (dynamically) fixed points on P_μ.

There are three interrelated special cases. First, if $P = T^*G$, then the reduced space at μ is the coadjoint orbit through μ and its reduced symplectic structure is that of Kirillov, Kostant, and Souriau. Second, if $\mu = 0$ and $P = T^*Q$ (with the canonical cotangent structure), then $P_0 = T^*(Q/G)$ with the canonical symplectic structure. Finally, if G is abelian (or $G = G_\mu$), then $P_\mu = T^*(Q/G)$ although the structure on $T^*(Q/G)$ need not be canonical.

We need to also recall that the coadjoint orbits $\mathcal{O} \subset \mathfrak{g}^*$ are the symplectic leaves in the *Lie–Poisson structure*

$$\{F, K\}(\mu) = \pm\left\langle \mu, \left[\frac{\delta F}{\delta \mu}, \frac{\delta K}{\delta \mu}\right]\right\rangle,$$

where one uses "$-$" for left actions and "$+$" for right actions. Here $\delta F/\delta\mu \in \mathfrak{g}$ is the *generalized functional derivative* defined by

$$\left\langle \frac{\delta F}{\delta \mu}, v \right\rangle = \mathbf{D}F(\mu) \cdot v$$

for all $v \in \mathfrak{g}^*$. As is well-known [or follows using Poisson reduction in the form $(T^*G)/G \cong \mathfrak{g}^*$], this bracket makes $\mathfrak{g}^*$ into a Poisson manifold. This term "Lie–Poisson" structure was coined by Marsden and Weinstein [1983] since this expression occurs *explicitly* in Lie's work around 1890.

If $\mathcal{O}$ is the coadjoint orbit through μ, then following Marle [1976] and

Kazhdan, Kostant, and Sternberg [1978], one finds a symplectic identification

$$P_\mu \cong \mathbf{J}^{-1}(\mathcal{O})/G \subset P/G$$

which shows that P_μ may be identified with a symplectic leaf in the Poisson reduced space P/G. An account of this, along with some additional information is given in Marsden [1981] . Reduction theory has been applied to a large number of interesting situations—the literature is too vast to survey here. We just mention a few: see Cushman and Rod [1982] for a penetrating application to resonances, Deprit [1983] for a solution of the problem of Jacobi's elimination of the node in the n-body problem, Bobenko et al. [1989] for integrable systems and a group-theoretic resolution of the integrability of the Kowalewski top, Marsden and Weinstein [1982, 1983] and Marsden, Ratiu and Weinstein [1984] for applications in fluid and plasma dynamics, and David, Holm, and Tratnik [1990] for applications to polarization lasers. Consult Guillemin and Sternberg [1984] for some applications involving representation theory.

Perhaps the most interesting case is the one considered by Smale: $P = T^*Q$ with the canonical symplectic structure. We assume G acts on Q and hence by cotangent lift on T^*Q. We also assume we are dealing with a Hamiltonian of the form kinetic plus potential energy, where the metric g on Q is G-invariant and where the potential $V: Q \to \mathbb{R}$ is G-invariant.

The *cotangent bundle reduction theorem* states that *the reduced space P_μ is a bundle over $T^*(Q/G)$ with fiber $\mathcal{O}$, the orbit through μ.* The corresponding Poisson statement is that the space $(T^*Q)/G$ is a $\mathfrak{g}^*$-bundle over $T^*(Q/G)$. The corresponding description of the symplectic or Poisson structure is a nontrivial synthesis of the Lie–Poisson structure on $\mathfrak{g}^*$ and the canonical (plus magnetic) structure on $T^*(Q/G)$. This was worked out in Montgomery, Marsden, and Ratiu [1984] motivated by the cases of the Hamiltonian structure for the interaction of a fluid or plasma with an electromagnetic field and the work of Sternberg and Weinstein on the geometry of Wong's equations that describe the motion of a particle in a Yang–Mills field (see Montgomery [1984] and references therein).

The proof of the cotangent bundle reduction theorem (see, for example, Marsden [1992] for a recent account) utilizes two crucial ideas, each of which is in Smale's paper (one of them implicitly). The first is the *mechanical connection* and the second is the associated *momentum shift.*

The *locked inertia tensor* is the map $\mathbb{I}: Q \to \mathfrak{g}^* \otimes \mathfrak{g}^* \cong L(\mathfrak{g}, \mathfrak{g}^*)$ defined by

$$\langle \mathbb{I}(q)\xi, \eta \rangle = \langle\langle \xi_Q(q), \eta_Q(q) \rangle\rangle,$$

where $\langle \cdot, \cdot \rangle$ denotes the natural pairing and $\langle\langle \cdot, \cdot \rangle\rangle$ is the metric pairing. If the action is locally free, then $\mathbb{I}(q)$ is a positive definite symmetric tensor. Define $\alpha: TQ \to \mathfrak{g}$ by

$$\alpha(v_q) = \mathbb{I}(q)^{-1}\mathbf{J}(\mathbb{F}L(v_q)),$$

where $v_q \in T_qQ$ and $\mathbb{F}L: TQ \to T^*Q$ is the Legendre transformation determined by the metric. The above formula for α is equivalent to the one given by Smale.

The first person to note and exploit the fact that $\alpha: TQ \to \mathfrak{g}$ is a G-connection on the bundle $Q \to Q/G$ seems to have been Kummer [1981]. The μ-component of the curvature of α is added to the canonical symplectic structure on $T^*(Q/G)$ in the reduction process. This was observed, without the language of connections, and the phrase "magnetic term" coined by Abraham and Marsden [1978].

The picture of $Q \to Q/G$ as a G-bundle carrying the connection α is the beginning of the story of the "gauge theory of deformable bodies" and the remarkable work of Wilczek, Shapere, and Montgomery on the link between optimal control and the motion of a colored particle moving in the Yang–Mills field α. See Shapere and Wilczek [1989], Montgomery [1990] and references therein.

The μ-component of α defines a one-form $\alpha_\mu: Q \to T^*Q$. One of the properties of a connection translates to

$$\alpha_\mu \in \mathbf{J}^{-1}(\mu),$$

which is the way Smale thought of α. The mechanical connection was explicitly used by Smale to describe the amended potential as the composition of the Hamiltonian with α_μ.

The momentum shift $T^*Q \to T^*Q$ taking a covector p_q at q to the covector $p_q - \alpha_\mu(q)$ therefore maps $\mathbf{J}(\mu)$ to $\mathbf{J}^{-1}(0)$. This, in effect, replaces reduction at μ by reduction at 0, which gives $T^*(Q/G)$. It is by this means that the cotangent bundle reduction theorem is proved.

Reduction has its counterpart, *reconstruction*, which is part of the theory. This concerns how one constructs dynamic trajectories in $\mathbf{J}^{-1}(\mu) \subset P$ given the reduced dynamic trajectory in P_μ. It turns out that α induces a connection on the G_μ-bundle $J^{-1}(\mu) \to P_\mu$, and the horizontal lift and holonomy of this connection play a basic role in reconstruction and in the interpretation of geometric phases (Hannay–Berry phases). See Marsden, Montgomery, and Ratiu [1990] for details. In particular, in this reference one will find a beautiful formula of Montgomery for the phase shift of a rigid body—when a rigid body undergoes a periodic motion in its reduced (body angular-momentum space), then the actual body does not return to its original position, but undergoes a rotation about the constant spatial angular momentum vector through an angle given by

$$\Delta\theta = -\Lambda + \frac{2ET}{\|\mu\|},$$

where Λ is the solid angle on the sphere of radius $\|\mu\|$ enclosed by the trajectory in body angular-momentum space, E is its energy, and T is the period. The first term, *the geometric phase*, is the holonomy of the canonical one form regarded as an S^1 connection on the Hopf bundle $\mathbf{J}^{-1}(\mu) \to S^2$. This type of formula proves to be useful in a number of problems, such as the control of the attitude of a rigid body with internal rotors; see Bloch, Krishnaprasad, Marsden, and Sánchez de Alvarez [1992].

Geometric phases come up in a variety of other problems as well, and the ideas of reduction and reconstuction can be useful for understanding them. Some of these are described in Marsden, Montgomery, and Ratiu [1990] and Montgomery [1990]. Many other applications involve integrable systems—one of these is the phase shift that one sees when two solitons interact. This is described in Alber and Marsden [1992]. Others that seem likely are phenomena like Stokes' drift in fluid mechanics.

4. Other Directions

Here we describe a few other recent research directions, each of which has a specific link with Smale's paper.

4.1. General Methodology

Most academics get judged on specific contributions—in mathematics, one is ideally judged on specific theorems. In many circumstances, this is a sound procedure, but what often turns out to be more valuable for science as a whole is the point of view or pedagogical approach that is developed. Smale (along with Poincaré, Arnold, Atiyah, Singer, and a few others) gave us the valuable and influential point of view of dynamical systems and, more generally, of *global or geometric analysis*. This view has profoundly influenced a whole generation of workers and has had a pervasive effect, often taken for granted. It has also indirectly influenced areas Steve never worked in. For instance, global analysis ideas have proved useful in nonlinear elasticity, even to the point of designing better numerical codes. The book of Marsden and Hughes [1983] is typical of many works showing this impact.

4.2. Stability of Relative Equilibria

The context of the cotangent bundle reduction theorem provides a setting for another synthesis of the works of Arnold [1966] and Smale [1970]. This concerns explicit (computable) criteria for the dynamic stability of relative equilibria. The main recent works on this point are Simo, Posbergh, and Marsden [1990] and Simo, Lewis, and Marsden [1991].

The Lie–Poisson reduction methods of Arnold are built around the fact, mentioned above, that $(T^*G)/G \cong \mathfrak{g}^*$. In this case, Arnold worked out explicit stability criteria and applied them to rigid bodies and fluids. Working in the context of fluids, to overcome technical difficulties with the PDEs involved, he developed what is now known as the *Arnold method* or the *energy-Casimir method* in Arnold [1969] and related references. This technique was developed by a number of authors and was applied to a variety of fluid and plasma problems; Holm et al. [1985] contains a fairly complete survey and bibliography to that date.

In the general case, Smale's work suggests that one should test for stability by looking at the second variation $\delta^2 V_\mu$ at a critical point in Q. However, should one view it on Q/G_μ or on $\mathfrak{g}^* \times Q/G$? How does the Arnold stability criterion fit in? This is an interesting point because the philosophies of the two approaches are rather different. From the point of view of Smale, things are considerably simpler in the abelian case and here the criterion is clear—test $\delta^2 V_\mu$ for positive definiteness on Q/G. A more general suggestion is to test $\delta^2 V_\mu$ for definiteness on Q/G_μ (this criterion is an exercise in *Foundations of Mechanics* but is implicit in Smale's paper). Note that if $Q = G$, then Q/G_μ is the orbit through μ, so one can expect $\delta^2 V_\mu$ to correspond to the Arnold criterion on a coadjoint orbit. However, Arnold's philosophy was different: the energy-Casimir method is more tractable if one relaxes the restriction to orbits in the spirit of the Lagrange multiplier theorem. Namely, we add to the Hamiltonian a function that Poisson commutes with every other function, that is, with a *Casimir function.* (As an aside, we note for amusement only that some would like to call such a Casimir function a "Casimirian," so it would sound just like a Hamiltonian or a Lagrangian. Unfortunately, English is neither logical nor perfect—we also do not call a "Green's function" a "Greenian," even though it probably is more correct to do that.)

At this point in the history of geometric mechanics, it was not clear whether the energy-Casimir method of Arnold or the second variation method suggested by Smale's work was the more appropriate. A motivation for looking more deeply into this problem came from nonlinear elasticity. Here, the complexity of the orbits of $\mathfrak{g}^*$ means that Casimir functions are difficult or impossible to find. Arnold already realized this for *three*-dimensional ideal flow (where the only known Casimir is the helicity) and this fact surely was a discouragement for the method. Abarbanel and Holm [1987] made some progress on this problem by working directly in material representation, before reduction. (It would be interesting to return to this and related questions in plasma physics studied by Morrison [1987] in the light of the block diagonalization work described below, and the work in progress of Bloch, Krishnaprasad, Marsden and Ratiu on *instability* criteria with the addition of dissipation obtained using Chetaev's method.)

All of these factors led to the development of the *energy-momentum method* (or block-diagonalization, or reduced energy-momentum method). The key to this method is the development of a *synthesis* of the Arnold and Smale methods. One splits the space of variations of (a concrete realization of) Q/G_μ into variations in G/G_μ and variations in Q/G. With the appropriate splitting, one gets the block-diagonal structure

$$\delta^2 V_\mu = \begin{bmatrix} \text{Arnold form} & 0 \\ 0 & \text{Smale form} \end{bmatrix},$$

where the Smale form means $\delta^2 V_\mu$ computed on Q/G. This method turns out to be an extremely powerful one when applied to specific systems such as spinning satellites with flexible appendages.

Perhaps even more interesting is the structure of the linearized *dynamics* near a relative equilibrium. That is, both the augmented Hamiltonian $H_\xi = H - \langle \mathbf{J}, \xi \rangle$ *and* the symplectic structure can be *simultaneously* brought into the following normal form:

$$\delta^2 H_\xi = \begin{bmatrix} \text{Arnold form} & 0 & 0 \\ 0 & \text{Smale form} & 0 \\ 0 & 0 & \text{Kinetic energy} > 0 \end{bmatrix}$$

and

$$\text{Symplectic Form} = \begin{bmatrix} \text{Coadjoint orbit form} & * & 0 \\ -* & \text{Magnetic (coriolis)} & I \\ 0 & -I & 0 \end{bmatrix}$$

where the columns represent the *coadjoint orbit variable* (G/G_μ), the *shape variables* (Q/G), and the *shape momenta*, respectively. The term $*$ is an *interaction term* between the group variables and the shape variables. The magnetic term is the curvature of the μ-component of the mechanical connection, as we described earlier.

For $G = SO(3)$, this form captures all the essential features in a well-organized way: centrifugal forces in V_μ, coriolis forces in the magnetic term, and the interaction between internal and rotational modes. In fact, in this case, the splitting of variables solves an important problem in mechanics: *how to efficiently separate rotational and internal modes near a relative equilibrium.*

4.3. Bifurcation and Symmetry Breaking

Smale realized, as pointed out earlier, that the symmetry group of a point in phase space determines how degenerate it is for the momentum map. Correspondingly, one expects, from the work of Golubitsky and co-workers, that these symmetry groups will play a vital role in the bifurcation theory of relative equilibria and its connections with dynamic stability theory. The beginnings of this theory has started and it will be tightly tied with the normal form methods of Subsection 4.2. Smale concentrated on the topology of the level sets of $H \times \mathbf{J}$ and their associated bifurcations as the level sets vary. However, in many problems one also wants to vary other system parameters as well.

A simple example will perhaps help here. Consider the dynamics of a particle moving without friction in a rotating circular hoop, as in Fig. 2.

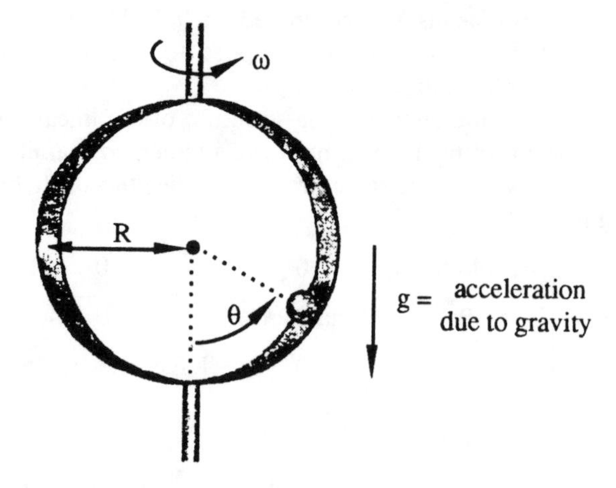

FIGURE 2. A ball in a rotating hoop.

As the angular velocity ω of the hoop increases past $\sqrt{g/R}$, a Hamiltonian pitchfork bifurcation occurs near the central equilibrium point, as in Fig. 3.

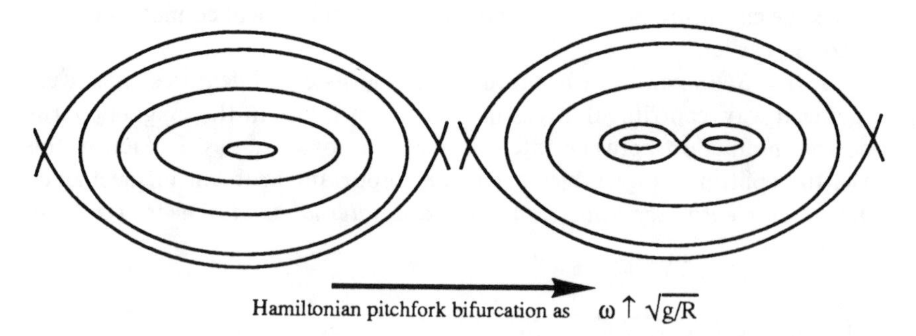

FIGURE 3. The Hamiltonian bifurcation for the ball in the rotating hoop.

The stability of the central point, which has $\mathbb{Z}_2$ symmetry, gets transferred to the bifurcating solutions, for which the $\mathbb{Z}_2$ symmetry is lost.

Related ideas appear in the work of Golubitsky and Stewart [1987] and in the study of a rotating planar liquid drop (with a free boundary held with a surface tension τ) in Lewis, Marsden, and Ratiu [1987] and Lewis [1989]. In the latter, a circular drop loses its circular symmetry to a drop with $\mathbb{Z}_2 \times \mathbb{Z}_2$ symmetry as the angular momentum of the drop is increased (although the stability analysis near the bifurcation is somewhat delicate). There are also interesting stability and bifurcation results in the dynamics of vortex patches, especially those of Wan in a series of papers starting with Wan and Pulvirente [1984].

4.4. Discrete Symmetries

In mechanics, the time-honored discrete symmetry is reversibility—the *anti-symplectic* involution $(q, p) \to (q, -p)$. However, there are many interesting discrete symmetries that are *symplectic*—the spatial $\mathbb{Z}_2$ symmetry of the ball in the hoop in Fig. 2 being a simple example. In bifurcation theory with symmetry, the Golubitsky school shows that discrete symmetries (and their corresponding fixed point sets, etc.) play an important role in the theory. A similar thing is true in the Hamiltonian case. For instance, discrete *and* continuous symmetries play a key role in the wonderful work of Bobenko, Reyman, and Semenov-Tian-Shansky [1989] that puts the integrability of the Kowalewski top into a reduction-theoretic framework. These ideas have been put into a general framework of *discrete reduction* by Harnard, Hurtubise, and Marsden [1991]. It would be of interest to go back to Smale's program with these discrete symmetry ideas to see their effect.

4.5. Singularity Structures in Solution Spaces

We already noted that Smale observed that singular points of $\mathbf{J}$ are points with symmetry. This is a simple but a profound observation with far-reaching implications. Abstractly, it turns out that level sets of $\mathbf{J}$ typically have *quadratic* singularities at its singular (=symmetric) points, as was shown by Arms, Marsden and Moncrief [1981]. In the abelian case, the images of these symmetric points are the vertices, edges, and faces of the convex polyhedron $\mathbf{J}(P)$ in the Atiyah–Guillemin–Sternberg–Kirwan convexity theory. (See Atiyah [1982] and Guillemin and Sternberg [1984].)

These ideas apply in a remarkable way to solution spaces of relativistic field theories, such as Einstein's equations of general relativity and the Yang–Mills equations. Here the theories have symmetry groups and, appropriately interpreted, corresponding momentum maps. The relativistic field equations split into two parts—Hamiltonian hyperbolic evolution equations and elliptic constraint equations. The solution space structure is determined by the elliptic constraint equations, which, in turn, say nothing other than the momentum map vanishes.

A fairly long story of both geometry and analysis is needed to really establish this, but the result is easy to understand in the terms we have given: The *solution space has a quadratic singularity precisely at those field points that have symmetry*. For further details, see Fischer, Marsden, and Moncrief [1980] and Arms, Marsden and Moncrief [1982].

Whereas these results were motivated by perturbation theory of classical solutions (gravitational waves as solutions of the linearized Einstein equations, etc.), there is some evidence that these singularities have quantum implications. For example, there appears to be evidence that, in the Yang–Mills case, wave functions tend to concentrate near singular points (see, for example, Emerich and Romer [1990]). It would be of interest to explore these ideas further using the theory developed by Sjamaar [1990] and Sjamaar and Lerman [1991].

4.6. Mechanical Integrators

With Steve's more recent interests in computation, it might be appropriate to note that there is quite a bit of activity in developing numerical codes that respect the underlying structure of a mechanical system with symmetry. For example, one can develop codes that preserve exactly the energy-momentum map $H \times \mathbf{J}$ *or* that preserve the symplectic structure and $\mathbf{J}$ (it turns out that one *cannot* do *all* of these; see Ge and Marsden [1988]). There are too many references to adequately survey here, but the one just cited, Channell and Soovel [1990], references therein, and recent works of Feng, Krishnaprasad, and Simo, will give one a start.

To obtain an integrator preserving $\mathbf{J}$ is related to finding an algorithm $F_{\Delta t}: P \to P$ that is consistent with the symmetry. To get one that preserves H, one can base the analysis on a discretization of the variational principle, and to get one preserving the symplectic structure, one can discretize the Hamilton–Jacobi equation.

One of the interesting things about these integrators is that they seem to perform better than conventional ones (such as Runga–Kutta schemes) in long-term integrations where chaotic dynamics becomes important.

4.7. Homoclinic Chaos

Of course, Smale is noted for the famous "horseshoe" that is associated to homoclinic tangles. The technical way that this is handled is via what is usually called the *Birkhoff–Smale theorem*, which associates an invariant Cantor set having a well-understood symbolic dynamics to a homoclinic tangle. This phenomena had its origins in the work of Poincaré on the three-body problem and led to the *Poincaré–Melnikov–Arnold* technique for explicitly finding homoclinic tangles in specific systems. (See Wiggins [1988] for a thorough account of these topics.) We note that the method has proved effective in establishing homoclinic chaos for PDEs by using infinite-dimensional versions of the Poincaré–Melnikov–Arnold theorem and the Birkhoff–Smale theorem; see Holmes and Marsden [1981].

It is also interesting to note that horseshoes and reduction fit nicely together and this is needed when one proves that various systems with symmetry (such as rigid bodies with attachments) have homoclinic chaos (see Holmes and Marsden [1983]). Here ones sees that Smale's work on chaos in dynamical systems and his work on symmetry and mechanics fit together in a mutually suportive way. This is, of course, just one example of the many connections running through different parts of Smale's work when it is viewed on a global scale.

Epilogue

In this paper, I have sketched only some of Smale's involvement with mechanics. He was also interested in problems such as rotating fluid masses and provided much good advice about such problems. He was also interested in elementary particles, and using topology to help classify them—see Abraham [1960]. This subject is of course now in vogue with people like Witten, who is a good example of someone who has a blend of analysis, geometry, and topology in the Smale spirit.

A curious twist in Smale's work involves his recent work and the work of others on linear programming and computational complexity described elsewhere in this volume. We are now witnessing the beginnings of deep links between this work and mechanics by people like Deift, Brockett, Bloch, Flaschka, Ratiu, and others. For example, efficient ways of diagonalizing martices can be done by following the dynamics of integrable Hamiltonian systems (for instance, of Toda type) on appropriate spaces of matrices. This is one of many nice illustrations of the conference theme "unity and diversity" that runs through Smale's work and the approach he takes to his topics. Whereas there is a broad *diversity* in the subject matter, there is a *deep unity*, not only the obvious one of using global analysis methodology throughout his work, but nonobvious ones, like the preceeding link between computational techniques and mechanics, that repeats in unexpected yet beautiful ways in each of the subjects that he treated.

References

Abarbanel, H.D.I., and D.D. Holm [1987] Nonlinear stability analysis of inviscid flows in three dimensions: incompressible fluids and barotropic fluids. *Phys. Fluids* **30**, 3369–3382.

Abraham, R. [1960] Lectures of Smale on Differential Topology. Mimeographed notes, Columbia University.

Abraham, R., and J. Marsden [1978] *Foundations of Mechanics.* Addison-Wesley Publishing Co., Reading, Ma.

Alber, M., and J.E. Marsden [1992] On geometric phases and the phase shift of solitons, *Comm. Math. Phys.* **149**, 217–240.

Arms, J.M. [1981] The structure of the solution set for the Yang–Mills equations. *Math. Proc. Camb. Phil Soc.* **90**, 361–372.

Arms, J.M., J.E. Marsden, and V. Moncrief [1981] Symmetry and bifurcations of momentum mappings. *Commun. Math. Phys.* **78**, 455–478.

Arms, J.M., J.E. Marsden, and V. Moncrief [1982] The structure of the space solutions of Einstein's equations: II. Several Killings fields and the Einstein–Yang–Mills equations. *Ann. Phys.* **144**, 81–106.

Arnold, V.I. [1966] Sur la géometrie differentielle des groupes de Lie de dimenson infinie et ses applications à l'hydrodynamique des fluids parfaits. *Ann. Inst. Fourier, Grenoble* **16**, 319–361.

Arnold, V.I. [1969] On an a priori estimate in the theory of hydrodynamical stability. *Amer. Math. Soc. Transl.* **79**, 267–269.

Arnold, V.I. [1989] *Mathematical Methods of Classical Mechanics. Second Edition.* Springer-Verlag, New York.

Atiyah, M. [1982] Convexity and commuting Hamiltonians. *Bull. Lond. Math. Soc.* **14**, 1–15.

Bloch, A.M., P.S. Krishnaprasad, J.E. Marsden, and G. Sánchez de Alvarez [1992] Stabilization of rigid body dynamics by internal and external torques, *Automatica* **28**, 745–756.

Bobenko, A.I., A.G. Reyman, and M.A. Semenov-Tian-Shansky [1989] The Kowalewski Top 99 years later: A Lax pair, generalizations and explicit solutions. *Commun. Math. Phys.* **122**, 321–354.

Channell, P., and C. Scovel [1990] Symplectic integration of Hamiltonian systems. *Nonlinearity* **3**, 231–259.

Cushman, R., and D. Rod [1982] Reduction of the semi-simple 1:1 resonance. *Physica D* **6**, 105–112.

David, D., D.D. Holm, and M. Tratnik [1990] Hamiltonian chaos in nonlinear optical polarization dynamics. *Phys. Rept.* **187**, 281–370.

Dellnitz, M., I. Melbourne and J.E. Marsden [1992] Generic bifurcation of Hamiltonian vector fields with symmetry, *Nonlinearity* **5**, 979–996.

Deprit, A. [1983] Elimination of the nodes in problems of N bodies. *Celestial Mech.* **30**, 181–195.

Ebin, D.G., and J.E. Marsden [1970] Groups of diffeomorphisms and the motion of an incompressible fluid. *Ann. Math.* **92**, 102–163.

Emmrich, C., and H. Römer [1990] Orbifolds as configuration spaces of systems with gauge symmetries. *Commun. Math. Phys.* **129**, 69–94.

Fischer, A.E., and J.E. Marsden [1972] The Einstein equations of evolution—a geometric approach. *J. Math. Phys.* **13**, 546–68.

Fischer, A.E., J.E. Marsden, and V. Moncrief [1980] The structure of the space of solutions of Einstein's equations, I: One Killing field. *Ann. Inst. H. Poincaré* **33**, 147–194.

Ge, Zhong, and J.E. Marsden [1988] Lie–Poisson integrators and Lie–Poisson Hamiltonian–Jacobi theory. *Phys. Lett. A* **133**, 134–139.

Golubitsky, M., and I. Stewart [1987] Generic bifurcation of Hamiltonian systems with symmetry. *Physica* **24D**, 391–405.

Guillemin, V., and S. Sternberg [1984] *Symplectic Techniques in Physics*, Cambridge University Press, Cambridge.

Harnad, J., J. Hurtubise, and J. Marsden [1991] Reduction of Hamiltonian systems with discrete symmetry, preprint.

Holm, D.D., J.E. Marsden, T. Ratiu, and A. Weinstein [1985] Nonlinear stability of fluid and plasma equilibria. *Phys. Rept.* **123**, 1–116.

Holmes, P.J., and J.E. Marsden [1981] A partial differential equation with infinitely many periodic orbits: chaotic oscillations of a forced beam. *Arch. Rat. Mech. Anal.* **76**, 135–166.

Holmes, P.J., and J.E. Marsden [1983] Horseshoes and Arnold diffusion for Hamiltonian systems on Lie groups. *Indiana Univ. Math. J.* **32**, 273–310.

Kazhdan, D., B. Kostant, and S. Sternberg [1978] Hamiltonian group actions and dynamical systems of Calogero type. *Commun. Pure Appl. Math.* **31**, 481–508.
Kummer, M. [1981] On the construction of the reduced phase space of a Hamiltonian system with symmetry. *Indiana Univ. Math. J.* **30**, 281–291.
Lewis, D., [1989] Nonlinear stability of a rotating planar liquid drop. *Arch. Rat. Mech. Anal.* **106**, 287–333.
Lewis, D., J.E. Marsden, and T.S. Ratiu [1987] Stability and bifurcation of a rotating liquid drop. *J. Math. Phys.* **28**, 2508–2515.
Lewis, D., J.E. Marsden, T. Ratiu, and J.C. Simo [1990] Normalizing connections and the energy-momentum method. *Proceedings of the CRM Conference on Hamiltonian systems, Transformation Groups, and Spectral Transform Methods*, edited by Harnad and Marsden. CRM Press, Montreal, pp. 207–227.
Marle, C.M. [1976] Symplectic manifolds, dynamical groups and Hamiltonian mechanics. *Differential Geometry and Relativity*, edited by M. Cahen and M. Flato. D. Reidel, Boston.
Marsden, J.E. [1981] *Lectures on Geometric Methods in Mathematical Physics.* SIAM, Philadelphia, PA.
Marsden, J.E. [1992] *Lectures on Mechanics*, London Math. Soc. Lecture Notes **174**, Cambridge Univ. Press.
Marsden, J.E., and T.J.R. Hughes [1983] *Mathematical Foundations of Elasticity.* Prentice-Hall, Redwood City, CA.
Marsden, J.E., R. Montgomery, and T. Ratiu [1990] Reduction, symmetry, and phases in mechanics. *Mem. AMS* **436**.
Marsden, J.E., and T.S. Ratiu [1986] Reduction of Poisson manifolds. *Lett. Math. Phys.* **11**, 161–170.
Marsden, J.E., T. Ratiu, and A. Weinstein [1984] Semi-direct products and reduction in mechanics. *Trans. Amer. Math. Soc.* **281**, 147–177.
Marsden, J.E., and A. Weinstein [1974] Reduction of symplectic manifolds with symmetry. *Rep. Math. Phys.* **5**, 121–130.
Marsden, J.E., and A. Weinstein [1982] The Hamiltonian structure of the Maxwell–Vlasov equations. *Physica D* **4**, 394–406.
Marsden, J.E., and A. Weinstein [1983] Coadjoint orbits, vortices and Clebsch variables for incompressible fluids. *Physica D* **7**, 305–323.
Meyer, K.R. [1973] Symmetries and integrals in mechanics. *Dynamical Systems*, edited by M. Peixoto Academic Press, New York, pp. 259–273.
Meyer, K.R., and G. Hall [1991] *Hamiltonian Mechanics and the n-body Problem.* Springer-Verlag, New York.
Montaldi, J.A., R.M. Roberts, and I.N. Stewart [1988] Periodic solutions near equilibria of symmetric Hamiltonian systems. *Phil. Trans. R. Soc. Lond. A* **325**, 237–293.
Montgomery, R. [1984] Canonical formulations of a particle in a Yang–Mills field. *Lett. Math. Phys.* **8**, 59–67.
Montgomery, R. [1990] Isoholonomic problems and some applications. *Commun. Math Phys.* **128**, 565–592.
Montgomery, R., J. Marsden, and T. Ratiu [1984] Gauged Lie–Poisson structures. *Contrib. Math. AMS* **28**, 101–114.
Morrison, P.J. [1987] Variational principle and stability of nonmonotone Vlasov–Poisson equilibria. *Z. Naturforsch.* **42a**, 1115–1123.
Oh, Y.G., N. Sreenath, P.S. Krishnaprasad and J.E. Marsden [1989] The dynamics of

coupled planar rigid bodies. Part 2: Bifurcations, periodic solutions, and chaos. *Dynam. Diff. Eqs.* **1**, 269–298.

Patrick, G., [1989] The dynamics of two coupled rigid bodies in three space. *Contrib. Math. AMS* **97**, 315–336.

Routh, E.J., [1877] Stability of a given state of motion. Reprinted in *Stability of Motion*, edited by A.T. Fuller. Halsted Press, New York, 1975.

Shapere, A., and F. Wilczeck [1989] Geometry of self-propulsion at low Reynolds number. *J. Fluid Mech.* **198**, 557–585.

Simo, J.C., D. Lewis, and J.E. Marsden [1991] Stability of relative equilibria I: The reduced energy momentum method. *Arch. Rat. Mech. Anal*, **115**, 15–59.

Simo, J.C., T.A. Posbergh, and J.E. Marsden [1990] Stability of coupled rigid body and geometrically exact rods: block diagonalization and the energy-momentum method. *Physics Rept*, **193**, 280–360.

Sjamaar, R. [1990] Singular orbit spaces in Riemannian and symplectic geometry. Thesis, Utrecht.

Sjamaar, R., and E. Lerman [1991] Stratified symplectic spaces and Reduction, *Ann. of Math.* **34**, 375–422.

Smale, S [1967] Differentiable dynamical systems. *Bull. Amer. Math. Soc.* **73**, 747–817.

Smale, S [1970] Topology and mechanics. *Invent. Math.* **10**, 305–331; **11**, 45–64.

Smale, S [1980] *The Mathematics of Time*. Springer-Verlag, New York.

Wan, Y.H., and M. Pulvirente [1984] Nonlinear stability of circular vortex patches. *Commun. Math. Phys.* **99**, 435–450.

Wiggins, S. [1988] *Global Bifurcations and Chaos*. Springer-Verlag, and Applied Mathematical Sciences, New York.

46
Smale's Topological Program in Mechanics and Convexity

TUDOR S. RATIU*

1. A Personal View

In the early seventies, the full impact of modern dynamical systems theory and the geometrization of mechanics have made themselves felt also in Romania. These ideas trickled in over a period of several years both from the West and the Soviet Union into a country which was, for all practical purposes, in complete intellectual isolation. The late sixties saw a slight respite from a very harsh rule and a few mathematicians were allowed to travel. They brought back into the country any material they could lay their hands on and made it readily available to their colleagues and their students. In July 1972, a Chinese-style cultural revolution was imposed by the Central Committee of the Communist Party with disastrous consequences, for mathematics in particular. A complete return to Stalinism occurred. By 1975, the Central Institute of Mathematics in Bucharest, the only in the country, was dissolved and mathematicians were looking for jobs. Those whose views were perceived to "infect" the Marxist indoctrination of students were fired from their university positions. This was just one of the many purges which periodically hit the higher education system, except that this time it took on an even more sinister twist: Ideological purity became again codified into various governmental regulations (before it was harshly applied but it had only a convoluted legal basis) and only those deemed worthy by the Communist Party were even allowed to register to the graduate study admission competitions at different universities and institutes. Membership in the party became a necessary, but not sufficient, condition to be admitted to the exam.

It was in this context that I did my undergraduate work at the University of Timisoara. During those years, I came into contact with Steve Smale's ideas and the course of my life was changed forever. Students were taught a rigid curriculum of standard mathematics courses, but there was a lot of informal talk about the new topological and geometric ideas in analysis,

* Research partially supported by NSF Grant DMS 9142613.

differential equations, and mechanics in the hallways of the department. In my junior year, I was lucky to have as my mechanics professor Gh. Drecin, a man who genuinely loved the subject, understood its importance at the infer-face between pure and applied mathematics, and who tried to keep in touch with the last developments. He was not a deep researcher, but had taste, a very acute judgment of what was important and of where the subject was going, and was not afraid to state his views even if they ran counter to both the official position of the Ministry of Education and to the larger part of the Romanian research community in mechanics. I distinctly remember the day when he took me and a very good friend of mine, M. Puta, to a discussion in his office and told us what we must read; it was Steve Smale's work: his 1967 paper on dynamical systems and his 1970 papers on mechanics. He also told us that we will need some solid background which we should learn from the 1968 edition of Abraham and Marsden's *Foundations of Mechanics* and he offered to help us in the understanding of this material. The problem was, of course, to even be able to get it; all copy machines were under lock and under the strict supervision of the secret police. It took us several weeks and quite a bit of money to bribe the only person in the entire university cleared to have access to the copy machine and to convince him to make us copies of the papers and the book.

Little did I know as I started reading this work that my whole life will be changed. In a few weeks, it became clear to me that I have made a lifetime commitment to work in this area of mathematics. Steve's ideas were so stunning and elegant that I dropped everything else and devoted my entire time trying to understand them. In the meantime, my ongoing troubles with the secret police worsened and I was eventually stopped from entering graduate study since I did not fulfill even the necessary condition of being admitted to the exam: I repeatedly refused to join the party. I was also barred from any job nationwide having anything to do with mathematical research. It is only due to the efforts of several mathematicians, and Steve Smale in particular, that I was finally expelled from Romania instead of being sent to a concentration camp. For all his efforts, I will remain forever grateful to him. I came to Berkeley after Steve's most active political years so I cannot comment on his successes in that field. But I am living proof of one of his successful political actions, so, after all, there is even unity in Steve's mathematical and political work!

2. The Topological Program and Convexity

Smale's [1970] work in mechanics has set the stage for the topological study of the bifurcation phenomena occurring in classical mechanical systems. His main goal was to use topological methods, especially Morse theory, to estimate the number and the nature of relative equilibria in classical mechanical systems in general and in the n-body problem in particular. One cannot overemphasize enough the influence this work has had in the past twenty years both in theoretical investigations and in the methodology employed in attacking concrete complicated mechanical systems. In the beginning there was a flurry of activity to use his methods to either extend his work on the n-body problem or to study other classical systems such as the rigid body; see the works of J. Palmore [1973, 1975, 1976], A. Iacob [1971, 1973], and S. Katok [1972]. Important as this work is, it turns out that the 1970 papers of Smale have had an even more profound influence on subsequent research in mechanics simply by the types of questions it has raised and the techniques employed. The main legacy of this work is the permanent link it has forged between the topological nature of the level sets of conserved quantities and the dynamic behavior of the mechanical system.

My short presentation is not the place to catalogue the extremely fruitful outgrowth of Smale's topological ideas in the research of the past twenty years in mechanics and symplectic geometry; this can be found in the presentations of J. Marsden and A. Fomenko. I want to concentrate on some recent developments linking convexity properties of the momentum map to the dynamic behavior of a classical integrable system, the Toda lattice. The question raised is the one put by Smale: How does the geometry of the momentum map determine the nature of the dynamic behavior of the Hamiltonian system? If the mechanical system is integrable, what comes to mind is the collection of the independent commuting integrals of motion. The surprise in the link discussed below is that the momentum map considered is not given by the integrals of motion, but by the action of another abelian group on a phase space, transversal to that of the Toda lattice. On this second phase space, there is a natural gradient system depending on several parameters, which when set to take on specific values (of Lie algebraic nature) force this gradient system to coincide with the Toda lattice equations on the intersection of the two phase spaces. As will be discussed below, the convexity of this simple momentum map dictates to a large extent the dynamic behavior of the Toda system. It may turn out that this example is the prototype of a general phenomenon (a) certain Hamiltonian systems (for example, integrable ones) are linked to gradient systems on different phase spaces and the momentum map of one system governs the dynamics of the other and (b) these two systems live in a larger Poisson manifold and fit together via some complex structure. As will be seen, the relationship in (b) is not simply that the gradient and Hamiltonian systems are the real and imaginary parts of a complex vector field; it is considerably more subtle. The presentation below is based on Bloch, Brockett, and Ratiu [1990, 1991] and on Bloch, Flaschka, and Ratiu [1990]; for a more detailed review of these ideas, see Bloch and Ratiu [1991].

Throughout this presentation the notations, sign conventions, and definitions in Marsden's contribution to this volume will be used. We begin by reviewing some of the convexity results to be referred to in the sequel. Schur [1923] and Horn [1954] showed that if $\{\mathbf{a}\}$ denotes the set of diagonals of all

Hermitian $n \times n$ matrices with given eigenvalues, $\boldsymbol{\lambda} = (\lambda_1, \ldots, \lambda_n)$, then $\{\mathbf{a}\}$ is the convex hull of $S_n \cdot \boldsymbol{\lambda}$, where the symmetric group S_n acts on $\mathbb{R}^n$ by permutation of the coordinates. This result was shown by Kostant [1973] to be the $SU(n)$ case of the following statement. Let G be a connected semisimple or a compact connected Lie group and T a maximal torus. Let $\mathfrak{g}$ and $\mathfrak{t}$ denote the Lie algebras of G and T, respectively. Fix a G-invariant metric on $\mathfrak{g}$. Then the orthogonal projection of an adjoint orbit $\mathcal{O}$ onto $\mathfrak{t}$ is the convex hull of the corresponding Weyl group orbit $\mathcal{O} \cap \mathfrak{t}$. Atiyah [1982] and Guillemin and Sternberg [1982, 1984] have generalized this result for the action of a n-dimensional torus T on a compact symplectic manifold P by establishing the famous *Convexity Theorem* of the momentum map $J: P \to \mathbb{R}^n$ of such an action. Namely, $J(P)$ is the convex hull of the finite set $J(P^T)$, where P^T is the fixed point set of the action characterized equivalently to be the image of all connected components of the set where the derivative of J vanishes. *Kostant's Convexity Theorem* quoted above follows from this result by observing that (a) the G-invariant metric identifies coadjoint and adjoint orbits and, thus, taking $P = \mathcal{O}$, the orthogonal projection is just the momentum map of the T-action on $\mathcal{O}$ and (b) the fixed point set of the T-action on $\mathcal{O}$ is the Weyl group orbit. Kirwan [1984] generalized the Atiyah–Guillemin–Sternberg convexity theorem by replacing a torus with a general compact connected Lie group G, acting in a Hamiltonian fashion on P with momentum map $J: P \to \mathfrak{g}^*$. The result states that the intersection of $J(P)$ with the positive Weyl chamber in the Lie algebra of a maximal torus in G is convex.

There is another convexity result, also due to Atiyah [1982], of a slightly different flavor which will be referred to later on. If G is a compact connected Lie group and T is a maximal torus, we shall denote by $G_{\mathbb{C}}$ and $T_{\mathbb{C}}$ their complexifications. By identifying coadjoint orbits with quotients of G by centralizers of subtori and recalling that these homogeneous spaces are diffeomorphic to quotients of $G_{\mathbb{C}}$ by the corresponding parabolic subgroups and, hence, are naturally complex manifolds, homogeneous under the $G_{\mathbb{C}}$-action, it follows that the coadjoint orbits are homogeneous complex manifolds. Fix an invariant metric on G and identify coadjoint and adjoint orbits. If A is the noncompact part of $T_{\mathbb{C}}$, i.e., $A = \exp \mathfrak{a}$, $\mathfrak{a} = i\mathfrak{t}$, $\mathfrak{t}$ the Lie algebra of T, denote by Y the closure of an A-orbit in $\mathfrak{g}$ which lies in the adjoint orbit $\mathcal{O}$ of G. If $J: \mathcal{O} \to \mathfrak{t}$ is the orthogonal projection then *Atiyah's Kähler Convexity Theorem* states that $J(Y)$ is the convex hull of the points in $\mathfrak{t}$ which are the images of the points in Y where the derivative of J vanishes. In addition, J establishes a homeomorphism of Y with this closed convex polytope. For generic Y, $J(Y) = J(\mathcal{O})$.

3. Brockett's Gradient System and the Toda Lattice

For the rest of the presentation, K will always denote a compact connected semisimple Lie group, $\mathcal{O}_{\mu_0}$ an adjoint orbit through $\mu_0 \in \mathfrak{k}$, and κ the Killing form on $\mathfrak{k}$. We shall have to deal with two different metrics on the orbit. The first is the *normal metric* which is obtained in the following way. Left translate $-\kappa(\cdot, \cdot)$ from $\mathfrak{k} = T_eK$ to T_gK for every $g \in K$. This metric on K is actually bi-invariant since κ is invariant under the adjoint action. The quotient of K by the stabilizer of μ_0 inherits in this way a Riemannian structure. Thus, $\mathcal{O}_{\mu_0}$ becomes a Riemannian manifold. To write an explicit formula for the normal metric, let $\mu \in \mathfrak{k}$, $\mathfrak{k}_\mu = \{\xi \in \mathfrak{k} \mid [\xi, \mu] = 0\}$, $\mathfrak{k}^\mu = \operatorname{im} \operatorname{ad} \mu$, and denote for $\eta \in \mathfrak{k}$ by η^μ the $\mathfrak{k}^\mu$-component of η in the direct sum decomposition $\mathfrak{k} = \mathfrak{k}_\mu \oplus \mathfrak{k}^\mu$; the two summands are orthogonal relative to $-\kappa(\cdot, \cdot)$. Then for $[\mu, \eta], [\mu, \zeta] \in T_\mu \mathcal{O}_{\mu_0}$, the normal metric has the expression

$$\langle [\mu, \eta], [\mu, \xi] \rangle_n = -\kappa(\eta^\mu, \zeta^\mu). \tag{3.1}$$

This metric is not Kähler. The second metric is one of the family of K-invariant Kähler metrics, which, by Borel's theorem, are in one-to-one correspondence with the points of the positive Weyl chamber. Among all of these, we single out the one corresponding to the intersection point of $\mathcal{O}_{\mu_0}$ with the positive Weyl chamber. If $A(\mu) = \sqrt{-(\operatorname{ad} \mu)^2}$ is the positive square root of $-(\operatorname{ad} \mu)^2$, then this Kähler metric is given by

$$\langle [\mu, \eta], [\mu, \zeta] \rangle_K = \langle A(\mu)[\mu, \eta], [\mu, \zeta] \rangle_n. \tag{3.2}$$

One can be more explicit by using the root space decomposition of the complexification $\mathfrak{g}_{\mathbb{C}}$ of $\mathfrak{k}$. The notation $\mathfrak{g}$ is reserved for the Lie algebra $\mathfrak{g}_{\mathbb{C}}$ thought of as a real Lie algebra of twice the dimension of $\mathfrak{k}$. If T is a fixed maximal torus of K and $\mathfrak{t}$ is its Lie algebra, then its complexification $\mathfrak{h} = \mathfrak{t} \oplus \mathfrak{a}$, $\mathfrak{a} = i\mathfrak{t}$ is a Cartan subalgebra of $\mathfrak{g}_{\mathbb{C}}$. Let $\{\alpha_1, \ldots, \alpha_l\}$ be the simple roots of $\mathfrak{g}_{\mathbb{C}}$. $\mathfrak{g}_\alpha$ denotes the α-root space; it is complex one-dimensional and generated by a vector e_α. Fix a Chevalley basis $\{h_i, e_\alpha \mid i = 1, \ldots, l, \alpha \text{ a root}\}$ and recall that $h_i = [e_{\alpha_i}, e_{-\alpha_i}]$. Relative to this basis, $\mathfrak{k}$ becomes the compact real form of $\mathfrak{g}_{\mathbb{C}}$. Let $\mathfrak{b}$ be the direct sum of $\mathfrak{a}$ with all the positive root spaces. We have $\mathfrak{g} = \mathfrak{k} \oplus \mathfrak{b}$ and we think of this relation over $\mathbb{R}$. Let G, K, T, A, B denote the Lie groups underlying the Lie algebras $\mathfrak{g}$, $\mathfrak{k}$, $\mathfrak{t}$, $\mathfrak{a}$, $\mathfrak{b}$, respectively. For the Iwasawa decomposition $G = KAN = KB$, we will write the factorization of $g \in G$ as $g = \mathbf{k}(g)\mathbf{b}(g)$ for $\mathbf{k}(g) \in K$, $\mathbf{b}(g) \in B$. We shall denote by $\mathfrak{g}_n$ the normal real form of $\mathfrak{g}$ determined by this Chevalley basis.

Returning to the Kähler metric on the generic orbit $\mathcal{O}_{\mu_0}$ (which is diffeomorphic to K/T), note that it is uniquely determined by a T-invariant inner product on $\mathfrak{g}/\mathfrak{t}$. Since $\mathfrak{g}/\mathfrak{t} = \bigoplus_{\alpha > 0} V_\alpha$, where V_α are the two-dimensional root spaces, the negative of the Killing form defines on each V_α an inner product. Any other inner product is obtained by multiplying this one on each V_α by a positive scalar. If $\mu \in \mathcal{O}_{\mu_0}$ is the unique element in the interior of the positive Weyl chamber, the Kähler metric is obtained by choosing these scalars to be $\alpha(\mu)$. The inner product on $\mathfrak{g}/\mathfrak{t}$ is chosen so that V_α and V_β are orthogonal for $\alpha \neq \beta$, α, β positive roots.

Brockett's [1988, 1989] gradient system on $\mathcal{O}_{\mu_0}$ is obtained in the following way (see Bloch, Brockett, and Ratiu [1990, 1991]). Fix two elements $Q, N \in \mathfrak{k}$ and define the function $F: K \to \mathbb{R}$ by

$$F(\theta) = \kappa(Q, \operatorname{Ad}_\theta N). \tag{3.3}$$

The gradient flow of F on K is given by

$$\dot{\theta} = \theta \cdot [\mathrm{Ad}_{\theta^{-1}} Q, N], \tag{3.4}$$

where $\theta \cdot P$ denotes the left translate of $P \in \mathfrak{k}$ by $\theta \in K$ to $T_\theta K$. Let $\mathcal{O}$ be the adjoint orbit containing Q. The projection of K to $\mathcal{O}$ is given by

$$\theta \mapsto \mathrm{Ad}_{\theta^{-1}} Q. \tag{3.5}$$

Via this map, a direct computation using the explicit form of the normal metric (3.1) shows that the gradient flow (3.4) transforms to

$$\dot{L} = [L, [L, N]] \tag{3.6}$$

which is again a gradient flow on the orbit $\mathcal{O}$ through Q relative to the linear function

$$H(L) = \kappa(L, N). \tag{3.7}$$

The gradient system (3.6) is called *Brockett's double bracket equation* or *Brockett's gradient system.* The behavior of its flow turns out to be determined by the momentum map on $\mathcal{O}$ which equals the orthogonal projection from $\mathcal{O}$ to $\mathfrak{t}$ given by the normal metric. Using Kostant's convexity theorem reviewed in Section 2, Bloch, Brockett, and Ratiu [1990, 1991] show that if F_t denotes the flow of (3.6), the set of equilibria equals the Weyl group orbit $\mathcal{O} \cap \mathfrak{t}$ whose convex hull is the image of $\mathcal{O}$ under the orthogonal projection from $\mathcal{O}$ to $\mathfrak{t}$. The projection of $F_t(\mathcal{O})$ lies entirely in this convex set. Moreover, the only sink of (3.6) is the element of the Weyl group orbit that lies in the Weyl chamber of $-N$; the only source is the eqilibrium in the Weyl chamber of N. The (real) dimension of the stable manifold at any of the equilibria equals twice the length of the Weyl group element which maps the Weyl chamber of N to the Weyl chamber of the equilibrium under consideration.

The orbits which are the phase spaces of Brockett's gradient system are the symplectic leaves of the Lie–Poisson structure on $\mathfrak{k}$. There is, however, a second Poisson structure on $\mathfrak{k}$, namely, the Lie–Poisson structure on $\mathfrak{b}^*$; the nondegenerate bilinear pairing between $\mathfrak{k}$ and $\mathfrak{b}$ is given by $\mathrm{Im}\,\kappa$. Explicitly, in this Poisson structure on $\mathfrak{k}$, the Lie–Poisson bracket of any two functions f, $h\colon \mathfrak{k} \to \mathbb{R}$ is given by

$$\{f, h\}(\xi) = -\mathrm{Im}\,\kappa(\xi, [\pi_{\mathfrak{b}} \nabla \tilde{f}(\xi), \pi_{\mathfrak{b}} \nabla \tilde{h}(\xi)]), \tag{3.8}$$

where $\tilde{f}$, $\tilde{h}$ are arbitrary extensions to $\mathfrak{g}$ of $f, h\colon \mathfrak{k} \to \mathbb{R}$, ∇ is the gradient relative to $\mathrm{Im}\,\kappa$, i.e.,

$$d\tilde{f}(\xi) \cdot \delta\xi = \mathrm{Im}\,\kappa(\nabla \tilde{f}(\xi), \delta\xi), \tag{3.9}$$

$\pi_{\mathfrak{b}}$ is the projection onto $\mathfrak{b}$ corresponding to the decomposition $\mathfrak{g} = \mathfrak{k} \oplus \mathfrak{b}$, and ξ, $\delta\xi \in \mathfrak{k}$. Moreover, if f is an invariant function on $\mathfrak{g}$, i.e., $[\nabla f(\zeta), \zeta] = 0$ for all $\zeta \in \mathfrak{g}$, then $f|\mathfrak{k}$ defines, relative to the Poisson structure (3.8), the following Hamiltonian system:

$$X_{f|\mathfrak{k}}(\xi) = [\pi_{\mathfrak{k}} \nabla f(\xi), \xi], \quad \xi \in \mathfrak{k}, \tag{3.10}$$

where $\pi_{\mathfrak{k}}: \mathfrak{g} \to \mathfrak{k}$ is the projection defined by the direct sum $\mathfrak{g} = \mathfrak{k} \oplus \mathfrak{b}$. The Adler–Kostant–Symes theorem (see, e.g., Kostant [1979] or Symes [1980]) guarantees that all Hamiltonian vector fields on $\mathfrak{k}$ defined by restrictions of invariant functions on $\mathfrak{g}$ commute.

Among all symplectic leaves of the Poisson structure (3.8), the minimal dimensional one is given by

$$\mathrm{Jac} = \left\{ L = i \sum_{j=1}^{l} [b_j h_j + a_j(e_{\alpha_j} + e_{-\alpha_j})] \mid a_j, b_j \in \mathbb{R} \right\}. \tag{3.11}$$

The elements of Jac are called Jacobi elements. Every $L \in \mathrm{Jac}$ is, up to a factor of i, a typical element of a "Toda orbit" in $\mathfrak{g}_n$ as defined in Kostant [1979] or Symes [1980]. Note that $\dim_{\mathbb{R}} \mathrm{Jac} = 2l$. Let $I_1, \ldots, I_l$ be a set of homogeneous generators of the ring of invariant polynomials on $\mathfrak{g}_{\mathbb{C}}$ chosen such that their restrictions to $\mathfrak{k}$ are real and generate the invariant polynomials on $\mathfrak{k}$. Hamilton's equations

$$\dot{L} = [\pi_{\mathfrak{k}} \nabla I_j(L), L] \tag{3.12}$$

on Jac are called the *Toda hierarchy*. Fix an element Λ in the interior of the positive Weyl chamber and let $\mathcal{O}_\Lambda$ be the K-adjoint orbit through Λ. Then the Toda hierarchy leaves $\mathcal{O}_\Lambda \cap \mathrm{Jac}$ invariant. The Toda hierarchy is a completely integrable system on Jac; for a proof see Kostant [1979] and Symes [1980, 1982a]. The l independent integrals in involution are $I_1, \ldots, I_l$ restricted to Jac. For $\mathfrak{g}_{\mathbb{C}} = A_l$, $\mathcal{O}_\Lambda$ consists of $(l+1) \times (l+1)$ matrices with fixed spectrum and $\mathcal{O}_\Lambda \cap \mathrm{Jac}$ is the isospectral manifold. After multiplying by i, the first vector field in the Toda hierarchy for A_l [i.e., with Hamiltonian $I_1(L) = \frac{1}{2}\kappa(L, L)$] takes the familiar form $\dot{L} = [B, L]$, where

$$L = \begin{bmatrix} b_1 & a_1 & & & \\ a_1 & b_2 - b_1 & a_2 & & \\ & \ddots & \ddots & \ddots & \\ & & & b_l - b_{l-1} & a_l \\ & & & a_l & -b_l \end{bmatrix},$$

$$B = \begin{bmatrix} 0 & a_1 & & & \\ -a_1 & 0 & a_2 & & \\ & \ddots & \ddots & \ddots & \\ & & & 0 & a_l \\ & & & -a_l & 0 \end{bmatrix}. \tag{3.13}$$

For an arbitrary Lie algebra, L has the form in (3.11) and

$$B = \sum_{j=1}^{l} a_j(e_{\alpha_j} - e_{-\alpha_j}). \tag{3.14}$$

The link between the Toda lattice and Brockett's gradient system is given by the following result of Bloch, Brockett, and Ratiu [1990, 1992]. If N is minus i times the sum of the simple coweights of $\mathfrak{g}_{\mathbb{C}}$, then for $L \in \text{Jac}$, the equation $\dot{L} = [L, [L, N]]$ gives the generalized Toda flow on the isospectral set $\text{Jac} \cap \mathcal{O}_\Lambda$. Explicitly, $N = \sum_{j=1}^{l} i x_j h_j$, where $(x_1, \ldots, x_l)$ is the unique solution of the system $\sum_{j=1}^{l} x_j \alpha_k(h_j) = -1$, $k = 1, \ldots, l$. Thus, the generalized Toda lattice is a gradient flow on $\text{Jac} \cap \mathcal{O}_\Lambda$, where $\mathcal{O}_\Lambda$ is equipped with the normal metric. In particular, the scattering behavior of its solution is governed by the behavior of Brockett's gradient system and, therefore, ultimately by the convexity properties of the momentum map of $\mathcal{O}_\Lambda$, thereby obtaining a result typical of Smale's topological program. We list below the coefficients x_j in N for all simple Lie algebras.

$\mathfrak{g}$	x
A_l	$x_i = -\frac{1}{2}i(l-i+1), i = 1, 2, \ldots, l$
B_l	$x_i = -\frac{1}{2}i(2l-i+1), i = 1, 2, \ldots, l-1$
	$x_l = -\frac{1}{4}l(l+1)$
C_l	$x_i = -\frac{1}{2}i(2l-i), i = 1, 2, \ldots, l$
D_l	$x_i = -\frac{1}{2}i(2l-1-i), i = 1, 2, \ldots, l-2$
	$x_{l-1} = x_l = -\frac{1}{4}l(l-1)$
G_2	$x_1 = -3, x_2 = -5$
F_4	$x = -11, x_2 = -21, x_3 = -15, x_4 = -8$
E_6	$x_1 = -8, x_2 = -11, x_3 = -15,$
	$x_4 = -21, x_5 = -15, x_6 = -8$
E_7	$x_1 = -17, x_2 = -49/2, x_3 = -33, x_4 = -48,$
	$x_5 = -75/2, x_6 = -26, x_7 = -27/2$
E_8	$x_1 = -46, x_2 = -68, x_3 = -91, x_4 = -135,$
	$x_5 = -110, x_6 = -84, x_7 = -57, x_8 = -29$

By the properties of Brockett's gradient system, the projection of the Toda flows to $\mathfrak{t}$ lie in the interior of the polytope which is the convex hull of $\mathcal{O}_\Lambda \cap \mathfrak{t}$. However, the polytope is *not* filled by all flows of the Toda hierarchy. In the next section we shall address precisely this issue and will discuss a result due to Bloch, Flaschka, and Ratiu [1990] which shows that by reimbedding the noncompact toral orbit $\text{Jac} \cap \mathcal{O}_\Lambda$ in $\mathcal{O}_\Lambda$, this polytope can be completely covered by all Toda flows in the hierarchy.

In view of these results one sees that Brockett's gradient system provides a generalization of the QR algorithms based on the scattering behavior of the Toda lattice; see, e.g., Deift, Nanda, and Tomei [1983], Lagarias [1988], and Symes [1982b].

4. The Momentum Level Set and Convexity

In Smale's topological program, one studies the topology of the energy-momentum map. In the case of the Toda lattice, this map is formed by the collection of all independent involutive constants of the motion $I_1, \ldots, I_l$

restricted to $\mathfrak{k}$, I_1 being the energy function. In the previous section, we saw that this level set equals $\text{Jac} \cap \mathcal{O}_\Lambda$ for a fixed Λ in the interior of the positive Weyl chamber, Jac being a B-orbit in $\mathfrak{b}^* \cong \mathfrak{k}$ and $\mathcal{O}_\Lambda$ a K-adjoint orbit in $\mathfrak{k}$. Since the Toda system is completely integrable,

$$\mathcal{J}_\Lambda^0 = \{L \in \mathcal{O}_\Lambda \cap \text{Jac} | a_j > 0 \text{ for all } j = 1, \ldots, l\} \tag{4.1}$$

is an orbit of the noncompact torus given by the flows of the commuting Hamiltonian vector fields defined by $I_1, \ldots, I_l$. The closure of $\mathcal{J}_\Lambda^0$ is

$$\mathcal{J}_\Lambda = \{L \in \mathcal{O}_\Lambda \cap \text{Jac} | a_j \geq 0 \text{ for all } j = 1, \ldots, l\} \tag{4.2}$$

and its boundary consists of $2^l - 1$ strata, each one being defined by the vanishing of some of the a_j's. We shall refer to $\text{Jac} \cap \mathcal{O}_\Lambda$ as the *isospectral manifold* and to $\mathcal{J}_\Lambda$ as the *closed isospectral set*. This terminology comes from the case of $K = SU(l + 1)$. Then a typical Jacobi matrix is of the form

$$L = i \begin{bmatrix} b_1 & a_1 & & & \\ a_1 & b_2 - b_1 & a_2 & & \\ & \ddots & \ddots & \ddots & \\ & & & b_l - b_{l-1} & a_l \\ & & & a_l & -b_l \end{bmatrix}, \tag{4.3}$$

and

$$\mathcal{J}_\Lambda^0 = \{L | L \text{ is conjugate to } \Lambda \text{ and } a_j > 0 \text{ for all } j\},$$

$$\mathcal{J}_\Lambda^0 = \{L | L \text{ is conjugate to } \Lambda \text{ and } a_j \geq 0 \text{ for all } j\}.$$

In view of Atiyah's Kähler convexity theorem discussed in Section 2, the map $(I_1, \ldots, I_l)$ to $\mathbb{R}^l$ has convex image. But we just remarked at the end of Section 3 that the orthogonal projection does not have convex image. Therefore, the question naturally arises: Can one embed $\mathcal{J}_\Lambda$ in a different way in $\mathcal{O}_\Lambda$ such that its projection is convex? The answer to this question is positive and solved in Bloch, Flaschka, and Ratiu [1990]; we will describe below the main ideas.

Before outlining the results, a very simple example is in order. Let $\mathfrak{k} = su(2)$ and identify $su(2)^*$ with $\mathbb{R}^3$. The Lie–Poisson structure of $su(2)^*$ has the concentric spheres and the origin as symplectic leaves. Represent $\Lambda = \text{diag}(i\lambda, -i\lambda)$, $\lambda > 0$, by the point $(0, 0, \lambda)$ so that $\mathcal{O}_\Lambda$ is a sphere of radius λ centered at the origin. The symplectic form is, up to a factor of $-1/\lambda$, the area form, and the complex structure, making $\mathcal{O}_\Lambda$ into a Kähler manifold, is essentially that of the Riemann sphere. The orbit $\mathcal{O}_\Lambda$ consists in this simple example of Jacobi elemens. The other Poisson structure of $\mathbb{R}^3 \cong \mathfrak{b}^*$ has leaves which are open half-planes containing the vertical axis and the points of the vertical axis if one lets B be the set of upper triangular matrices in $SU(2)$. The compact torus $T = \{\text{diag}(e^{i\theta}, e^{-i\theta}) | \theta \in \mathbb{R}\}$ acts by rotating the sphere $\mathcal{O}_\Lambda$ about the vertical axis; namely,

$$L \in \begin{bmatrix} ib & a \\ -\bar{a} & -ib \end{bmatrix} \in \mathcal{O}_\Lambda \subset su(2),$$

$b \in \mathbb{R}$, $a \in \mathbb{C}$, is sent to

$$\begin{bmatrix} ib & ae^{2i\theta} \\ -\bar{a}e^{-2i\theta} & -ib \end{bmatrix}.$$

The momentum map J of this action maps $L \in \mathcal{O}_\Lambda$ to b, i.e., it projects the point on the sphere $\mathcal{O}_\Lambda$ parallel to the horizontal plane onto the vertical axis. The image of the sphere $\mathcal{O}_\Lambda$ and of the "meridian," which is the intersection of $\mathcal{O}_\Lambda$ with a B-orbit (a "page" of the "book" $\mathfrak{b}^*$) coincide and equal the convex interval $[-\lambda, \lambda]$.

This example generalizes almost completely. What does not go through is the projection part, which, due to the low dimensionality of this example, is in this case misleading. In fact, it is precisely this part which turns out to be technically the most difficult. What is true is that $\mathcal{J}_\Lambda$ (here the "meridan") can be reimbedded in $\mathcal{O}_\Lambda$ in a very special way so that its projection is a convex polytope.

This convexity result is built on two main ingredients. The first one is the explicit solution of the Toda lattice hierarchy in terms of the Iwasawa decomposition; this can be found in Kostant [1979], Reyman and Semenov-Tijan-Shanskii [1979], Symes [1980], or Goodman and Wallach [1984]. The solution of

$$\dot{L}(t) = [\pi_{\mathfrak{t}} \nabla I_j(L(t)), L(t)], \quad L(0) = L \tag{4.4}$$

is

$$L(t) = \mathrm{Ad}_{\mathbf{k}(\exp(t\nabla I_j(L)))^{-1}} L = \mathrm{Ad}_{\mathbf{b}(\exp(t\nabla I_j(L)))} L, \tag{4.5}$$

where $g = \mathbf{k}(g)\mathbf{b}(g)$ is the factorization of $g \in G$ defined by $G = KB$. Since, generically in L, $\nabla I_j(L)$, $j = 1, \ldots, l$, generate $i\mathfrak{t}$ and since $L \in \mathcal{O}_\Lambda$, i.e., $L = \mathrm{Ad}_k \Lambda$ for some $k \in K$, it follows that $\mathrm{Ad}_{k^{-1}} \nabla I_j(L) \in \mathfrak{a} = i\mathfrak{t}$. Consequently, if $\mu \in \mathfrak{a}$, $X := \mathrm{Ad}_k \mu$, $L := \mathrm{Ad}_k \Lambda$, the curve $L(t) = \mathrm{Ad}_{\mathbf{k}(\exp(tX)^{-1})} L$ is a solution of a linear combination of Eqs. (4.4). Rewriting $L(t)$ as

$$L = \mathrm{Ad}_k \Lambda \mapsto L(t) = \mathrm{Ad}^{-1}_{\mathbf{k}(\exp(t\mu)k^{-1})} \Lambda, \quad \exp(t\mu) \in A, \tag{4.6}$$

we see that this defines a left action of A on $\mathcal{O}_\Lambda$; this is the *left dressing action* as in Lu and Weinstein [1990] and Lu [1990]. Unfortunately, the A-action (4.6) is not the diagonal action

$$L = \mathrm{Ad}_k \Lambda \mapsto \mathrm{Ad}_{k(hk)} \Lambda, \quad h \in T, \tag{4.7}$$

and Atiyah's Kähler convexity theorem does not apply. Let J denote the momentum map of the diagonal T-action, i.e., the orthogonal projection onto $\mathfrak{t}$ relative to the normal metric. The image $J(\mathcal{J}_\Lambda)$ is neither convex, nor a polytope. One can turn, however, (4.6) into (4.7) by "inversion." Namely, define $\iota\colon \mathcal{J}_\Lambda \to \mathcal{O}_\Lambda$ by $\iota(k\Lambda k^{-1}) = k^{-1}\Lambda k$ and the Toda flow (4.5) becomes

$$\iota(L) = \iota(\mathrm{Ad}_k \Lambda) \mapsto \mathrm{Ad}_{\mathbf{k}(\exp(t\mu)k^{-1})} \Lambda \tag{4.8}$$

which is exactly the diagonal action. Therefore, $J(\iota(\mathcal{J}_\Lambda))$ will be a convex polytope by Atiyah's theorem. The trouble is that the "inversion" $\mathrm{Ad}_k \Lambda \mapsto$

$\mathrm{Ad}_{k^{-1}}\Lambda$ makes no sense on $\mathcal{O}_\Lambda$: When one replaces k by kh, $h \in T$, the image under $\imath$ *will* depend on h.

More precisely, the element $L \in \mathcal{O}_\Lambda \cap \mathrm{Jac}$, $L = \mathrm{Ad}_k\Lambda$, determines k only up to the right multiplication of an element of T; remember, Λ is chosen once and for all in the interior of the positive Weyl chamber of $\mathfrak{t}$. This leads to the second main ingredient in the proof of the convexity result, namely, k must be chosen in a smooth way and uniquely for each $L \in \mathcal{J}_\Lambda$. That this is possible is ultimately based on the Bruhat decomposition of K into cells and constitutes the main technical part of the proof. It was carried out by Bloch, Flaschka, and Ratiu [1990] and was inspired by ideas that show up in Flaschka and Haine [1991a, b]. The bottom line is that the "inversion" $\imath$ can be extended continuously to the boundary $\mathcal{J}_\Lambda \backslash \mathcal{J}_\Lambda^0$ and that it is a diffeomorphism on $\mathcal{J}_\Lambda^0$ and a homeomorphism on $\mathcal{J}_\Lambda$. All of these considerations can be made very explicit in the case of $SU(l+1)$. Let $L = k\Lambda k^{-1}$. The columns of k are orthonormalized eigenvectors of L. Let now $L \in \mathcal{J}_\Lambda^0$; then the first row entries r_j of k do not vanish. Fix k by the requirement that $r_j/r_1 \in \mathbb{R}$. This is the smooth unique choice of k for a given L and, hence, defines the "inversion" $\imath$ as discussed above. The Toda hierarchy acts on $\imath(\mathcal{J}_\Lambda^0)$ as the noncompact torus A as follows. Consider the element

$$a = \exp(p_1 c_1 \Lambda t_1 + \cdots + p_{l+1} c_{l+1} \Lambda^{l+1} t_{l+1}) \in A,$$

where $c_k = i$ for k odd, $c_k = 1$ for k even, and $p_1, \ldots, p_{l+1}$ are arbitrary real constants. We have the mapping

$$a\colon \imath(L) \mapsto \mathbf{k}(ak^{-1})\Lambda\mathbf{k}(ak^{-1})^{-1}.$$

It is a straightforward check to see that this action preserves the normalization of the r_j's described above.

Returning to the general case, we can conclude that the image of $\mathcal{J}_\Lambda$ under the map $J \circ \imath\colon \mathcal{J}_\Lambda \to \mathfrak{t}$ is the convex hull of the Weyl group orbit through Λ. Duistermaat, Kolk, and Varadarajan [1983] show that the action of the one-parameter group $\exp(t\mu)$ for $\mu \in \mathfrak{a}$, on $\mathcal{O}_\Lambda$, is a gradient flow generated by $-\kappa(\mu, \cdot)$ relative to the Kähler metric. Combining this with the result above and those of Section 3, it follows that the Toda flows in $\mathcal{J}_\Lambda$ are gradient flows in the metric induced by pulling back the Kähler metric of $\imath(\mathcal{J}_\Lambda)$ to $\mathcal{J}_\Lambda$. It should be emphasized that the functions which generate the gradient flows are *not* the Toda Hamiltonians. This result is complementary to that of Section 3, where the Toda lattice was seen to be a gradient relative to the normal metric. Note also that the Toda lattice flow given by the equation $\dot{L} = [L, [L, N]]$ on $\mathrm{Jac} \cap \mathcal{O}_\Lambda$ can be projected to K/P, for P a parabolic subgroup containing T, where $\mathcal{O}_\Lambda$ is diffeomorphic to K/T. This gives rise to interesting gradient systems and generalizes Moser's [1975] gradient flows associated with the Toda lattice.

For example, let $K = SU(l+1)$, $P_0 = \mathrm{diag}(1, 0, \ldots, 0) - I/(l+1)$, and consider the K-orbit $\mathcal{O}_{iP_0} = \{k(iP_0)k^{-1} \mid k \in SU(l+1)\}$. This orbit consists of rank-one projection matrices minus a multiple of the identity and so can be identified with $\mathbb{CP}^l$. A direct computation shows that for iP in this orbit, $-(\mathrm{ad}(iP))^2([iP,\eta]) = [iP,\eta]$ and hence the normal and Kähler metric coincide on $\mathcal{O}_{iP_0}$. This is also true for general Grassmannians and the considerations below have a direct generalization to an arbitrary compact Lie group K; see Bloch, Flaschka, and Ratiu [1990]. Let $\varphi_k(iP) = -cTr(iP\Lambda^k)$ for $\Lambda = \mathrm{diag}(\lambda_1,\ldots,\lambda_{l+1})$ and c is 1 or i as needed to make $\varphi_k\colon \mathbb{CP}^l \to \mathbb{R}$. The gradient flow is

$$i\dot{P} = [iP,[iP,c\Lambda^k]],$$

so that taking $P = r\otimes\bar{r}$, $r = (r_1,\ldots,r_{l+1})$ with $\|r\|^2 = 1$, we get

$$\dot{r}_j = -r_j\left(\lambda_j^k - \sum_{i=1}^{l+1}\lambda_i^k|r_i|^2\right), \quad j = 1,\ldots,l+1,$$

which is Moser's form of the Toda lattice. By applying the complex structure $[iP,\cdot]$, we see that this flow is conjugate to the Hamiltonian flow $i\dot{P} = [iP,c\Lambda^k]$.

Atiyah's theorem applies also to K/P. Ignoring the constant shift in P_0, the momentum map $K/P \to \mathfrak{t}$ is given by $k^{-1}(\mathrm{mod}\,P) \mapsto i\,\mathrm{diag}(r\otimes\bar{r})$. The Jacobi matrix L is thus sent to $(i|r_1|^2,\ldots,i|r_{l+1}|^2)$ and the image of $\mathfrak{t} = \mathbb{R}^l$ is the standard simplex $\sum_{i=1}^{l+1}\xi_i = 1$.

References

Atiyah, M.F. [1982] Convexity and commuting Hamiltonians, *Bull. Lond. Math. Soc.* **14**, 1–15.

Bloch, A.M. [1990] The Kähler structure of the total least squares problem, Brockett's steepest descent equations, and constrained flows, *Realization anl Modelling in Systems Theory*, edited by M. Kaashoek, J.H. van Schuppen, and A.C.M. Ran. Birkhäuser, Boston.

Bloch, A.M., R.W. Brockett, T.S. Ratiu [1990] A new formulation of the generalized Toda lattice equations and their fixed point analysis via the momentum map, *Bull. Amer. Math. Soc.* **23**, 477–485.

Bloch, A.M., R.W. Brockett, T.S. Ratiu [1992] Completeley integrable gradient systems, *Commun. Math. Phys.*, **147**, 57–74.

Bloch, A.M., H. Flaschka, T.S. Ratiu [1990] A convexity theorem for isospectral manifolds of Jacobi matrices in a compact Lie algebra, *Duke Math. J.* **61**(1), 41–66.

Bloch, A.M., T.S. Ratiu [1991] Convexity and integrability, *Symplectic Geometry and Mathematical Physics, Actes du Colloque en l'honneur de Jean-Marie Souriau*, edited by P. Donato, C. Duval, J. Elhadad, and G. Tuynman. Birkhäuser, Boston.

Brockett, R.W. [1988] Dynamical systems that sort lists and solve linear programming problems, *Proc. 27th IEEE Conf. on Decision and Control*, pp. 799–803.

Brockett, R.W. [1989] Least squares matching problems, *Linear. Alg. Appl.* **122/123/124**, 761–777.

Deift, P., T. Nanda. C. Tomei [1983] Differential equations for the symmetric eigenvalue problem, *SIAM J. Numer. Anal.* **20**, 1–22.

Duistermaat, J.J., J.A. Kolk, V.S. Varadarajan [1983] Functions, flows, and oscillatory integrals on flag manifolds and conjugacy classes in real semisimple Lie groups, *Comp. Math.* **49**, 309–398.

Flaschka, H., L. Haine [1991a] Torus orbits in *G/P*, *Pacific. J. Math.* **149**, 251–292.

Flaschka, H., L. Haine [1991b] Variétés des drapeaux et réseaux de Toda, *C.R. Acad. Sci. Paris Ser. I* **312**, 255–258.

Goodman, R., N. Wallach [1984] Classical and quantum mechanical systems of Toda lattice type, II, *Commun. Math. Phys.* **94**, 177–217.

Guillemin, V., S. Sternberg [1982] Convexity properties of the moment mapping, *Invent. Math.* **67**, 491–513.

Guillemin, V., S. Sternberg [1984] Convexity properties of the moment mapping II, *Invent. Math.* **77**, 533–546.

Horn, A. [1954] Doubly stochastic matrices and the diagonal of a rotation matrix, *Amer. J. Math.* **76**, 620–630.

Iacob, A. [1971] Invariant manifolds in the motion of a rigid body with a fixed point, *Rev. Roum. Math. Pures Appl.* **16**, 1497–1521.

Iacob, A. [1973] *Topological Methods in Mechanics* (in Romanian), Ed. Acad. Rep. Soc. Romania, Bucuresti.

Katok, S. [1912] Bifurcation sets and integral manifolds in the problem of motion of a heavy rigid body (in Russian), *Usp. Akad. Nauk* **27** (2).

Kirwan, F. [1984] Convexity properties of the moment mapping III, *Invent. Math.* **77**, 547–552.

Kostant, B. [1973] On convexity, the Weyl group, and the Iwasawa decomposition, *Ann. Sci. Ec. Norm. Sup.* **6**, 413–455.

Kostant, B. [1979] The solution to a generalized Toda lattice and representation theory, *Adv. Math.* **34**, 195–338.

Lagarias, J. [1988] Monotonicity properties of the generalized Toda flow and *QR* flow, *SIAM J. Matrix Anal. Appl.*, to appear.

Lu, J.-H. [1990] Multiplicative and affine Poisson structures on Lie groups, Ph.D. Thesis, U. C. Berkeley.

Lu, J.-H., A. Weinstein [1990] Poisson Lie groups, dressing transformations and Bruhat decomposition, *J. Diff. Geom.* **31**, 501–526.

Moser, J. [1975] Finitely many mass points on the line under the influence of an exponential potential—an integrable system, Springer Lecture Notes in Physics No. 38, Springer-Verlag, Berlin.

Palmore, J. [1973] Classifying relative equilibria I, *Bull. Amer. Math. Soc.* **79**, 904–908.

Palmore, J. [1975] Classifying relative equilibria II, III, *Bull. Amer. Math. Soc.* **81**, 489–491; *Lett. Math. Phys.* **1**, 71–73.

Palmore, J. [1976] Measure of degenerate equilibria I, *Ann. Math.* **104**, 421–429.

Reyman, A.G., M.A. Semenov-Tijan-Shanskii [1979] Reduction of Hamiltonian systems, affine Lie algebras, and Lax equations; I, *Invent. Math.* **54**, 81–100; II, **63** [1981], 423–432.

Schur, I. [1923] Über eine Klasse von Mittelbildungen mit Anwendungen auf der Determinantentheorie, *Sitzungsber. Berlin. Math. Gesellschaft* **22**, 9–20.

Smale, S. [1970] Topology and Mechanics; I, *Invent. Math.* **10**, 305–311; II, **11**, 45–64.

Symes, W.W. [1980] Hamiltonian group actions and integrable systems, *Physica D* **1**, 339–374.

Symes, W.W. [1982a] Systems of Toda type, inverse spectral problems and representation theory, *Invent. Math.* **59**, 13–51.

Symes, W.W. [1982b] The *QR* algorithm and scattering for the nonperiodic Toda lattice, *Physica D* **4**, 275–280.

47
Discussion

A. WEINSTEIN AND AUDIENCE MEMBERS

Alan Weinstein: Perhaps I should also begin with a personal reminiscence, although Steve's influence on my work is not as dramatic as on Tudor's. Around 1970—Jerry and I can't quite seem to agree on the day this occurred—but it was during a period of time when universities were in a state of revolution perhaps, chaos at least, and many of us wanted to turn our attention from so-called pure mathematics to things that seemed to have more to do with the real world. And at that time, even astronomy seemed closer to the real world than some of the things I was doing. So I was very excited about Steve's giving a series of lectures on celestial mechanics. And I think it was listening to those lectures that sort of brought me back actively to symplectic geometry, where I've been ever since. It's a little bit like the grand tour of the planets, where you're in outer space. Every once in awhile you come close to a planet, a real solid mass, and you take a quick trip around the planet, and then you're back in outer space again.

On the mathematical side, what I wanted to do today was mention a couple of relations between different parts of mathematics that Steve was interested in. First, a remark about Morse theory, which doesn't immediately impinge on symplectic geometry or mechanics except in one application. I think it was in Karen's talk that she mentioned this play, or tension, between two ways of using Morse theory, namely, to use analysis via Morse theory to derive results in topology, and then using topology via Morse theory to get results in analysis on the existence of solutions. I think I first learned from Klingenberg the idea that you really want to play both of these games at once. For example, in studying closed geodesics on certain manifolds like spheres or symmetric spaces, a very important idea is the following: You start with a nice, perfectly symmetric metric which you understand completely and in which you understand all the geodesics, and therefore understand all the critical structure of the energy functional. You use that nice function to compute the topology of the loop space, and then having computed the topology of the loop space you use that with Morse theory in the other direction to get information about geodesics for arbitrary metrics on those spaces. So you go from one function, to the topology of a space, to information about other functions, and that's a very productive idea.

An even more productive idea is that of Floer. I'm trying to remember, somebody gave an explanation of radio versus the telephone, and the idea is this: Someone's child was asking him to explain radio. He said, "Well, let's imagine you have a dog here, and you twist the dog's tail and the dog barks. That's an example of a telephone. Well, radio is the same thing, except without the dog. Well, what Floer has done in his version of Morse theory is to eliminate the topology of the space; namely, you simply start with one function which you understand very well, you look at its critical points, and then you deform this function to another one, and you deduce things about the critical points of the second function by a kind of continuation procedure, just knowing the critical structure of the first one. And, in fact, the topology of the space is completely absent from this discussion. Actually, Floer himself credits someone else, namely, Charles Conley, with some of the basic ideas behind this continuation method. I really should mention this.

Another thing I wanted to mention was a connection between reduction of Hamiltonian systems and surgery. Recently, Guillemin, Sternberg, and MacDuff have been studying circle actions on manifolds. Perhaps I'll use a corner of the board here. The simplest example, perhaps, to which these reduction ideas apply is the case where one has the symplectic manifold P and an action of the circle, in which case the momentum map is just the real-value function which is the Hamiltonian generating that circle action:

$$P \xrightarrow{J} \mathbb{R}.$$

And if you look at the Morse theory of the function J, it's natural to look not just at the level sets of J, but at the level sets of J divided by the group action. So instead of simply looking at J inverse of A, you look at it divided by S^1, and then ask how the topology of this space changes when you pass a critical point. Of course, if it weren't for the dividing by S^1, one is simply adding handles here.

Recently it's been observed by, as I mentioned, Guillemin, Sternberg, and MacDuff that what goes on in this quotient space is a combination of operations, basically from algebraic geometry, namely, the blowing down and blowing up of points: replacing points by complex projective spaces. Now it's conceivable that the "Hamiltonian" for a circle action is multiple-valued (the primitive of a closed form which is not exact). In this case, the levels would look something like the picture here, with a tube connecting the top and bottom:

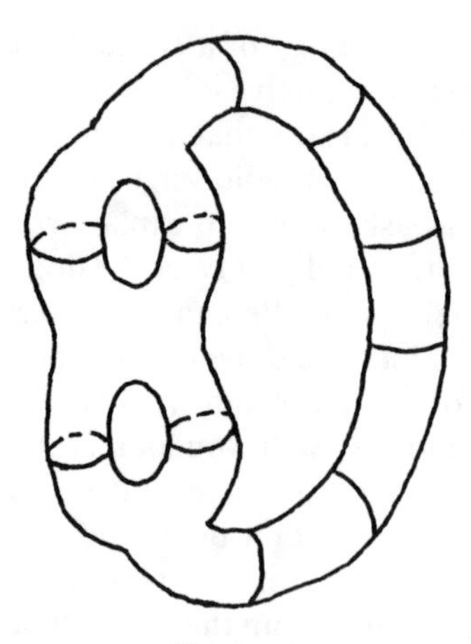

FIGURE 1

In fact, when P is four-dimensional, MacDuff proved that if there exists a fixed point, then in fact you can't have a circular picture iike that. In fact, there must be a top and bottom. And the argument is a very beautiful one; namely, by using these ideas of blowing up and blowing down, she looks at the reduced manifold as a function of A. It's going to be a two-dimensional manifold, a two-dimensional compact symplectic manifold, so in fact just a compact Riemann surface. You find that every time you pass a critical point, the Euler characteristic of this surface increases, strictly. Well, if you go around the circle and it keeps increasing, you've got a problem. So, in fact, there is no connecting tube around here, or there are no critical points. And so you've proved the existence of a momentum map for a circle action with fixed points on a four-manifold.

Well, I think that I'll close with this example which was very much in the spirit of a whole range of Steve's work involving Hamiltonian actions as well as topology.

Ralph Abraham: Well, we do have about 15 minutes left for questions on Jerry's survey or on the discussants or answers.

Jerry Marsden: The formula for the reduced potential doesn't need the dimension of the reduced space; it is completely explicit, and easy to compute in examples.

Karen Uhlenbeck: I was really startled with the fact that the idea of momentum mapping was that recent. Is that correct that the idea for momentum mapping actually dates from the 1970s?

Jerry Marsden: It's right around that era. It's something that I would like to know the history of.

Karen Uhlenbeck: Jerry, your idea certainly is absolutely essential. I'm sorry I left that part out of my talk before, because what I was doing was very much connected with momentum structures. But, it really dates to the 1970s?

Jerry Marsden: Knowing how to compute conserved quantities via Noether's theorem certainly has been around for a long time. Assembling this together as an equivariant map from P to $\mathfrak{g}^*$ is what the question is, and I think probably Kirillov and Kostant would assert that it's in their work in the mid-1960s. Certainly it's in Souriau's book in 1970, more or less in this form. Steve's paper in 1970 has it for the case of cotangent lifts on the cotangent bundle. So certainly by 1970 it was clear what the general situation was. But I'd say before 1965 or so it wasn't around. At least, that's my view of it.

Alan Weinstein: I just want to put in my usual word for Sophus Lie. I don't think one finds the momentum map, but certainly in Lie's big volume on Lie groups, one finds the dual of the Lie algebra and the coadjoint representation, and the fact that the orbits in the coadjoint representation carry symplectic structures. All that and the basic ideas of Poisson maps and so on are in the work of Lie. Souriau is the person who coined the words "momentum map"—he just called it "momentum"—in his book, or perhaps in a paper slightly earlier than that, in the general context of Lie groups acting on symplectic manifolds. In Steve's paper, he referred, I think, only to the energy momentum map, the combination of the two together, and only in the case of simple mechanical systems. There's a footnote that you find in his paper, I think added in proof, that he learned about Souriau's work just about the time that this paper went to press, so it would be an independent discovery.

Finally, there is something that one does find in the literature occasionally, the term "moment map," which is roughly synonymous with "momentum map," but I just want to make a plug for the idea that this is just basically a mistranslation of Souriau's *moment*, which is is the word that he, in fact, used for "momentum map." Jerry and I used the wrong word, in fact, in the paper we wrote in 1974 on reduction. We were temporarily blinded to the truth and called this thing the "moment map." Richard Cushman, shortly thereafter, convinced us that "momentum" was really the word that was being conveyed by Souriau, and which was correct in this place, too.

48
On Paradigm and Method*

PHILIP HOLMES

There is no philosophy that is not founded upon knowledge of the phenomena, but to get any profit from this knowlege it is absolutely necessary to be a mathematician.

—*Daniel Bernoulli*, from a letter to his nephew John III Bernoulli, 1763 (quoted by Truesdell [1984]).

1. Introduction

This is not a technical paper. Rather than turning in a brief summary of a recent research enthusiasm, I have taken the occasion of this celebration of Stephen Smale's many contributions to mathematics to reflect a little on the role of mathematics in general and dynamical systems theory in particular, the conduct of mathematicians, and the relations of all these to turbulence studies in engineering physics. The technical preoccupations which lie behind and have prompted these somewhat loose musings may be found in the works of Aubry [1990], Aubry et. al. [1988, 1990], Aubry and Sanghi [1989, 1990], Armbruster et al. [1988, 1989], Berkooz et al. [1991a, 1991b], Stone and Holmes [1989, 1990, 1991], and Holmes [1990a]. The last of these contains an introductory survey of the uses (and some misuses) of dynamical systems theory and other tricks in turbulence studies. Other papers in that same volume describe yet more tricks. These tricks are important: It seems to me that no one style or speciality of mathematics is adequate when faced with problems of this complexity. This is one of the main themes of the present essay.

* The research that occasioned these reflections was variously supported by AFOSR, ARO, NSF, and ONR.

2. From the Particular to the General and Back

Poincaré's interest in the stability of the solar system and, more particularly still, a restricted three-body problem, started it all (Poincaré [1890, 1899]; cf. Holmes [1990b].) Seventy years later, when some fruits of his ideas, cultivated by Smale and his students, were ripening, dynamical systems theory had largely become the preserve of a few "pure" mathematicians. In the West at least, it was a part of global analysis, rather than mechanics, although Soviet mathematicians remained somewhat more broad-minded. This retreat into the empyrean was probably essential, for with it concepts such as structural stability, generic properties, and the crucial role of transversality theory might never have emerged. (In fact, many of these ideas arose in the Soviet school and were subsequently used to generalize Smale's work on gradient systems.) The mechanician, "applied" mathematician or physicist, steeped in his specific problem and blinkered by technique, only occasionally takes the larger view, even when he is capable of it. Nor are "pure" mathematicians free from this myopia: Smale's work is unusual and powerful largely because he has crossed so many boundaries *within* mathematics.

In what follows I shall argue that the "big picture" of dynamical systems and their bifurcations, developed from an standpoint independent of particular equations and mappings, has been and remains of great importance for the study of specific problems. But let us not forget that, even in its development, specific problems still played a crucial part: One must pay attention to detail. Directed by Levinson [1949] to Cartwright and Littlewood's [1945] study of the Van der Pol oscillator, Smale [1963, 1967] created his famous horseshoe map; cf. Smale [1980]. Nonetheless, his major paper on this topic appeared in a volume on differential and combinatorial topology; hardly the normal reading of mechanicians or applied mathematicians in 1963, or even, alas, today. To further complicate the moral, we should recall that Littlewood's major contributions were to pure mathematics: For him the Van der Pol paper was something of an excursion.

3. The Infusion of Generic Properties into Turbulence Studies

Ruelle and Takens [1971], in bringing structural stability and generic properties to the bifurcation "scenarios" of Landau [1944] and Hopf [1948], provided a nice resolution of the apparent paradox that a deterministic dynamical system—the Navier–Stokes equation—could generate an essentially random solution. Their hypothesis did not require that solutions break down in a finite time, presumably thereafter to be mysteriously resuscitated, since turbulence clearly does not end, neither with a bang nor a whimper. It was indeed a strangely attractive idea. Alas, it has also encouraged hundreds of papers in which the Navier–Stokes equations, which we have no reason to doubt provide an excellent model for hydrodynamics, are cast aside and replaced by a mapping (often one-dimensional), a cellular automaton, or a coupled map lattice whose chief advantage seems to be the ease with which it may be numerically simulated, since it often appears as resistant to rigorous analysis as the Navier–Stokes equation itself.

Here I must say that I have nothing against numerical simulation per se: Massive computations have already become a valuable complement to experimental work in fluid mechanics, allowing one to probe flows in hitherto impossible spatial detail. But like experimental observation and measurement, simulations do not by themselves *explain* anything: They must be *interpreted* and for this some form of mathematical modeling and analysis, preferably based on Navier–Stokes, seems essential. It is the place of dynamical systems theory in this which interested Ruelle and Takens and which interests me.

Even though remarkably little mathematical evidence has been provided for the hypothesis that the Navier–Stokes equations, in interesting flow geometries, two or three dimensional, open or closed, possess a strange attractor, it seems to me a duty of the serious mathematical student of turbulence to continue to strive for such. Before an attractor can be strange, it must attract, and in this respect recent developments in (approximate) inertial manifolds and Hausdorff dimension estimates for "universal attractors" of Navier–Stokes are encouraging. (cf. Témam [1988] and Constantin et al. [1989]). Whereas these advances sprang from nonlinear PDE and employed its analytical tools, their geometric setting was certainly guided by a dynamical perspective. Inertial manifolds are intimately related to and contain global (center-) unstable manifolds, although they also contain pieces of stable manifolds, and the notion of normal hyperbolicity is more than merely similar to the spectral blocking property used in many existence proofs. Let us hope that the rich theory of bifurcations of (finite-dimensional) dynamical systems due to Smale, Arnold, Thom, and their students and colleagues will be able to contribute as much to the conjectured strangeness of Navier–Stokes as invariant manifold theory has to their attractivity. I will return to this in Section 6.

4. Paradigms Are Productive—Preconceptions Are Perilous

Here is a personal story with a moral. In our own work on the near wall region of turbulent boundary layers (Aubry et al. [1988], cf. Holmes [1990a]), after a long process involving Galerkin projection of Navier–Stokes into a (low-dimensional) subspace spanned by empirical eigenfunctions (Lumley [1967, 1970]), coaxed from experimental measurements (Herzog [1986]), we arrived at a set of ten coupled nonlinear ordinary differential equations (ODEs). The dependent variables were the amplitudes of modal coefficients which we expected to interact, producing "turbulence." Since "chaos" is "generic" for nonlinear systems of three or more equations, we eagerly set out to find it and, of course, we did (in numerical simulations of the ODEs). But we also found something rather more interesting, which we could understand, even prove theorems about and which, moreover, seems to have a close connection to the intermittent character of turbulence production in the boundary layer. These are *heteroclinic cycles*: objects which, in a generic system, should not exist at all since they are structurally unstable. We quickly realized that the physical symmetries of translations and reflections spanwise to the mean flow, inherited as $O(2)$-equivariance in our amplitude equations, could and had stabilized these cycles (Armbruster et al. [1988, 1989]). Heteroclinic cycles rapidly began appearing in other physical contexts; cf. Kevrekidis et al. [1990] and Nicolaenko and She [1990a, 1990b]. They had, of course, been there all along.

In the same work, small, pseudo-random perturbations to the ODEs played an important role, destroying the attractivity of the "simple" heteroclinic cycle and creating a characteristic probability distribution of passage times during which the solution remains near a saddle point. Thus, to fully understand these cycles, we needed to appeal to stochastic differential equations (Ornstein–Uhlenbeck processes), as well as dynamical systems theory.

The physical interpretation is as follows: A solution attracted to such a cycle spends along periods near unstable equilibria or other limit sets, punctuated by rapid transits. The former correspond to the relatively quiescent flow which obtains over much of the wall "most" of the time; the latter to the violent ejections of low speed fluid up into the outer region, referred to by experimentalists as the bursting phenomenon (Kline [1978], Blackwelder [1989]). The random pertubations mentioned above derive from pressure fluctuations in this outer region passing over the "boundary" of the inner layer, which is all that our model encompasses. They represent coupling to the outside world. See the references cited above and Stone and Holmes [1989, 1990, 1991] for more detail.

Rand's [1982] work on $SO(2)$ equivariance in the Taylor–Couette problem, and the resulting lack of "generic" resonances and frequency locking, provides another nice example of the crucial influence of symmetries. One could cite many more.

This is not to say that strange attractors have no role in the description of turbulent boundary layers; only that a different and "simpler" attractor also plays an important part. In fact, at parameter values near those at which the heteroclinic cycles lose their asymptotic stability, we observe chaotically modulated traveling waves—apparently a strange attractor with $SO(2)$-symmetry—which drive intermittent bursts of heteroclinic activity in the otherwise quiescent, higher wave-number modes (Aubry and Sanghi [1990]).

The moral is simple and clear. Each particular problem selects its own "generic properties"; the mathematician cannot impose them. The excesses of "applied catastrophe theory" should have been enough to caution us. One must not be too literal in one's interpretation of the paradigms, nor should one neglect the details of one's particular problem. There is no philosophy which is not founded upon knowledge of the phenomena.

5. On Paradigm and Method

By "paradigm" I do not intend the grand concept of scientific revolution due to Kuhn [1970]. I do not believe that "chaos theory," even if it existed, would be a "new science" (Gleick [1987]). However, Gleick is correct in pointing out that something *has* changed in recent years. For a start, mathematicians and scientists who might formerly have remained blinkered within their subdisciplines are talking to each other, reading each others' papers, attending each others' conferences, even listening to each other. This can only be healthy (provided the "applied" scientists among them keep their feet firmly on the ground of experience). It is rather difficult, unfortunately, to extract a specific approach to research problems from this vague enthusiasm. In my attempt to do so, a more modest kind of paradigm is central.

In what follows, the reader may wish to broaden "dynamical systems" to the vaguer term "nonlinear dynamics," but my message should be clear. The central concepts and theorems of dynamical systems provide *paradigms* of phenomena: behavior that one can *expect* to find in the particular model problems of interest. (Indeed, as Smale [1990] has pointed out in passing, dynamical systems theory *has* provided a unifying structure for the study of differential equations.) The word *model* is important: One must remember that dynamical systems theory concerns only *mathematical* objects such as differential equations and iterated maps; it has no *direct* contact with or implications for the physical, or any other, world. Model building is an important aspect of theoretical mechanics and physics, much underrated by a certain kind of mathematician. It is with this in mind that I argue that "chaos theory," as the phrase seems generally intended, does not and cannot exist as a coherent scientific account of phenomena comparable, say, to quantum mechanics or relativity theory.

But these paradigms, examples, counterexamples, and even models are not enough. To unravel the secrets of a particular model, one also needs appropriate *methods*. If the abstract theory provides a list of known species, it is no guarantee that they will all be sighted on a given day; and none of them will be identified correctly without the proper equipment.

It is important to remark here that the notions of generic properties and structural stability and the (finite) classification and unfolding theorems of elementary catastrophe theory and bifurcation theory have pointed out a way in which mathematics can contribute directly to model building. The lists of species can suggest what one might put into a model just as well as they do behaviors which might emerge from a model obtained otherwise. As

I have said, this viewpoint has been responsible for some excesses, but it has also helped considerably in the formulation of models in relatively "unmathetized" fields, such as neurobiology (cf. the papers of Kopell and Rand et al. in Cohen et al. [1988]).

When I first became interested in dynamical systems, I was dazzled by the abstract theory, by the rich array of phenomena, generic, hyperbolic, stable, and otherwise, and simultaneously depressed by the apparent inability of classical tools such as linearization, weakly nonlinear Taylor series expansions, and regular perturbation methods to reveal their presence or absence in specific systems. It was also already becoming clear that the (uniformly) hyperbolic theory would be inadequate to deal with many physically interesting examples. Since 1975, matters have improved somewhat, although as far as methods are concerned I feel that we still have a bag of a tricks rather than the elegant system of, say, linear algebra. We have learned to use Taylor series expansions to approximate center manifolds and perform normal form transformations and so bring local bifurcation analysis within the range of phase plane methods. Melnikov's [1963] ideas have been extended and generalized (cf. Wiggins [1988]), although they still remain about the only broadly applicable rigorous way to prove that transverse homoclinic orbits (and hence chaos and even a kind of strange attractor) occur in specific differential equations. Classification theorems for degenerate singularities and their unfoldings in the spirit of Thom [1975] and Arnold [1983] have been enormously extended and beautifully adapted to systems equivariant under symmetry groups (cf. Golubitsky et al. [1985, 1988]). Many of the tedious computations have been (semi-) automated by computer algebra (Rand and Armbruster [1987]). The development of these tools has been guided by the abstract theory, as must be their application. Applied mathematicians and mechnicians are, hopefully, rather *less* obessed by method and technique, now that they have more to choose from. A sweet paradox indeed!

If our range of methods is growing, it is still small. Faced with a strongly nonlinear third-order ODE having no obvious integrable limit or symmetry, we are still almost helpless, although we can always simulate (compute) and try to interpret the results "geometrically" in the light of our paradigms. Guckenheimer and Williams' [1979] and Williams' [1977, 1979] work on the geometrical Lorenz [1963] systems provides a nice example, which incidentally stimulated much more general work on kneading theory for piecewise continuous maps which might never have been done but for this motivation. Rigorous computer-assisted proofs, numerical and symbolic, will probably play an important part in the verification of such geometrical constructions, as they have already in the renormalization theories of bifurcation cascades for maps, (Feigenbaum [1978], Lanford [1982]). Whereas there is still a lot of room, and need, for developments in "classical" perturbation and phase space methods (I count inertial manifolds and their approximations among such), we need some new ideas. It is absolutely necessary to be a mathematician.

6. Return to Turbulence

In this closing section, I will risk some modest predictions on the role which dynamical systems might play in turbulence studies of open flows such as jets, wakes, and boundary layers. There has already been a substantial amount of work on closed, spatially constrained systems such as Rayleigh–Bénard convection and Taylor–Couette flow; cf. Swinney and Gollub [1981]. The classical division of the wave-number spectrum into low modes, in which turbulent energy is largely produced, an intermediate or inertial range, through which it cascades, and a high wave-number range, in which it is dissipated, seems to me very useful. Even if an adequate global existence theory for 3D Navier–Stokes is obtained and inertial manifolds (or approximate manifolds, or inertial sets) are proven to exist, the associated attractor dimensions for fully developed flows will be enormous, certainly of $O(1000–10{,}000)$. Even in a channel flow at a modest Reynolds number, a Lyapunov dimension of $O(750)$ has been estimated from numerical simulation (Keefe [1989]). It is doubtful that detailed analysis of phase spaces of such dimensions will be possible, or even if possible in a world of artificial intelligences, useful to human intelligences. Science advances largely by the creation of rather simple models and governing laws. For this, it seems to me we must capitalize on the physically motivated split of spatial scales mentioned above to obtain a vastly reduced system from the Navier–Stokes equations.

As Majda [1990] has suggested, an homogenization or averaging principle might be applied to the inertial and dissipative ranges to produce a set of PDEs governing the larger spatial scales. Various formal procedures, such as the generalized Lagrangian mean, are already in use (Andrews and McIntyre [1978]; cf. Phillips [1990]). Concepts such as eddy viscosity are common in the engineering literature (cf. Tennekes and Lumley [1972]). The resulting equations would presumably be "milder" than Navier–Stokes and it may be possible to prove the existence of inertial manifolds and attractors for such an averaged system, and even to construct approximate finite-dimensional subsystems, the attracting sets of which could be analyzed in detail. Here, as in the proper orthogonal decomposition used by Aubry et al. [1988], a subtle mixture of probabilistic and deterministic methods will be required. I do not believe that the paradigms of (deterministic) dynamical systems alone will be sufficient.

Since they are so mathematically intractible, some may object to my clinging to the Navier–Stokes equations as an important basis for attempts to model and understand turbulence. They may observe, correctly, that period doubling and Ruelle–Takens "routes to turbulence" are seen in, say, the Rayleigh–Bénard experiment (Swinney and Gollub [1981]). These are "universal" features of dynamical systems and one does not require the Navier–Stokes or, indeed, *any* specific equations to "explain" them. The problem with this view is that many other routes are *also* observed in differing geometries and flow conditions. Without an underlying model, we have no way of unifying these scattered observations and even less hope of predicting anything. This is not to say that we cannot profit from the bifurcation and other scenarios of universal or generic theory, only that they should be applied to the analysis and interpretation of Navier–Stokes dynamics and not *in vacuo*. Hence, my wish to *simplify*, but not abandon, these fundamental equations.

If such finite (even low-) dimensional models can be produced, they and their behavior will clearly differ greatly depending on the flows and physical geometries from which they derive. Whereas the turbulence transport and dissipation mechanisms and models probably should have a certain universality, production mechanisms vary widely, as experiments and simulations of shear layers, boundary layers, wakes, and jets show. With this in mind, I feel that at least in turbulence studies, the search for universal scaling laws and models, beloved of certain physicists and generically minded mathematicians, is somewhat misguided. The paradigms and methods of dynamical system theory, however, would be fairly directly applicable to such reduced (finite-dimensional) models of specific fluid flows. The work of Aubry et al. [1988], referred to earlier, in which a rational hierarchy of models of increasing dimension is constructed by Galerkin projection onto subspaces spanned by empirical eigenfunctions, provides a modest step in this direction. If it could be complemented by a rationally averaged model of higher wavenumber transport and dissipation, then we may have a second step. However, the staircase is long enough to keep us all climbing for some time.

7. Afterword

Several persons have remarked that Daniel Bernoulli meant by "mathematician" something quite different from what is generally meant today. This may be so: His algebra and calculus, his methods have been extended and transformed utterly; indeed the mathematician whose contributions we are celebrating has played a significant part in this. Nonetheless, I believe that the spirit of Bernoulli's advice remains as true today as 230 years ago.

Acknowledgment. I would like to thank John Guckenheimer for his comments on an earlier version of these remarks.

References

D.G. Andrews and M.E. McIntyre [1978] An exact theory of non-linear waves on a Lagrangian-mean flow. *J. Fluid Mech.* **89**, 609–646.

D. Armbruster, J. Guckenheimer, and P. Holmes [1988] Heteroclinic cycles and modulated travelling waves in system with $O(2)$ symmetry. *Physica* **29D**, 257–282.

D. Armbruster, J. Guckenheimer, and P. Holmes [1989] Kuramoto–Sivashinsky dynamics on the center-unstable manifold. *SIAM J. Appl. Math* **49**, 676–691.

V.I. Arnold [1983]. *Geometrical Methods in the Theory of Ordinary Differential Equations*, Springer-Verlag, New York, Heidelberg, Berlin.

N. Aubry [1990] Use of experimental data from an efficient description of turbulent flows. *Appl. Mech. Rev.* **43**, 5(2), S240–S245.

N. Aubry and S. Sanghi [1989] Streamwise and spanwise dynamics of the turbulent wall layer. *Forum on Chaotic Flow* (ed. K.N. Ghia), ASME, New York.

N. Aubry and S. Sanghi [1990] Bifurcation and bursting of streaks in the turbulent wall layer. *Turbulence 89: Organized Structures and Turbulence in Fluid Mechanics* (ed. M. Lesieur and O. Métais), Kluwer Academic Publishers, Amsterdam.

N. Aubry, P. Holmes, J. Lumley, and E. Stone [1988] The dynamics of coherent structures in the wall region of a turbulent boundary layer. *J. Fluid Mech.* **192**, 115–173.

N. Aubry, J.L. Lumley, and P.J. Holmes [1990] The effect of modeled drag reduction on the wall region. *Theor. Comput. Fluid Dynam.* **1**, 229–248.

G. Berkooz, P.J. Holmes, and J.L. Lumley [1991a] *J. Fluid Mech.* **230**, 75–95. Intermittent dynamics in simple models of the turbulent wall layer.

G. Berkooz, P.J. Holmes, and J.L. Lumley [1991b]. Turbulence, dynamical systems and the unreasonable effectiveness of empirical eigenfunctions *Proceedings of ICM-90*, pp. 1607–1617, *Kyoto*, Springer-Verlag, Tokyo.

R.F. Blackwelder [1989]. Some ideas on the control of near wall eddies. AIAA 2nd Shear Flow Conference, AIAA paper, 89-1009.

M.L. Cartwright and J.E. Littlewood [1945] On nonlinear differential equations of the second order, I: The equation $\ddot{y} - k(1 - y^2)\dot{y} + y = b\lambda k \cos(\lambda t + a)$, k large. *J. Lond. Math. Soc.* **20**, 180–189.

P. Constantin, C. Foias, R. Teman, and B. Nicolaenko [1989] *Integral Manifolds and Inertial Manifolds for Dissipative Partial Differential Equations*, Springer Verlag, New York.

A.H. Cohen, S. Rossignol, and S. Grillner (eds.) [1988] *Neural Control of Rhythmic Movements in Vertebrates*, Wiley, New York.

M.J. Feigenbaum [1978]. Quantitative universality for a class of nonlinear transformations. *J. Statist. Phys.* **19**, 25–52.

J. Gleick [1987] *Chaos: Making a New Science*, Viking, New York.

M. Golubitsky and D.G. Schaeffer [1985] *Singularities and Groups in Bifurcation Theory, I.* Springer-Verlag, New York.

M. Golubitsky, I. Stewart, and D.G. Schaeffer [1988] *Singularities and Groups in Bifurcation Theory, II.* Springer-Verlag, New York.

J. Guckenheimer and P. Holmes, [1983] *Nonlinear Oscillations, Dynamical Systems and Bifurcations of Vector Fields*, Springer-Verlag, New York. [Corrected (third printing, 1990)].

J. Guckenheimer and R.F. Williams [1979] Structural Stability of Lorenz attractors. *Inst. Hautes Etudes Sci. Publ. Math.* **50**, 59–72.

S. Herzog [1986] The large scale structure in the near-wall region of turbulent pipe flow. Ph.D. thesis, Cornell University, Ithaca, NY.

P.J. Holmes [1990a] Can dynamical systems approach turbulence? *Whither Turbulence? Turbulence at the Crossroads* (ed. J.L. Lumley), Springer-Verlag, New York, pp. 195–249, 306–309.

P.J. Holmes [1990b] Poincaré, celestial mechanics, dynamical systems theory and "chaos." *Phys. Rept.* **193** (3), 137–163.

E. Hopf [1948] A mathematical example displaying the features of turbulence. *Commun. Pure Applied Math* **1**, 303–322.

L. Keefe [1989] Comparison of calculated and predicted forms for Lyapunov spectra of Navier–Stokes equations. *Bull. Amer. Phys. Soc.* **34**, 2296–2301.

I.G. Kevrekidis, B. Nicolaenko, and J.C. Scovel [1990] Back in the saddle again: A computer assisted study of the Kuramoto-Sivashinksy equation. *SIAM J. Appl. Math.* **50**, 760–790.

S.J. Kline [1978]. The role of visualization in the study of the turbulent boundary layer. *Coherent Structure of Turbulent Boundary Layers*, Proc. AFOSR/Lehigh Workshop (ed. C.R. Smith and D.E. Abbotts). pp. 1–26.

T.S. Kuhn [1970]. The structure of scientific revolutions. *International Encyclopedia of Unified Science*, Vol. 2, No. 2, University of Chicago Press, Chicago.

L. Landau [1944]. On the problem of turbulence. *Dokl. Akad. Nauk. SSSR.* **44**, 339–342.

O.E. Lanford, III, [1982]. A computer assisted proof of the Feigenbaum conjecture. *Bull. Amer. Math. Soc.* **6**, 427–434.

N. Levinson [1949] A second-order differential equation with singular solutions. *Ann. of Math.* **50**, 127–153.

E.N. Lorenz [1963]. Deterministic nonperiodic flow. *J. Atmos. Sci.* **20**, 130–141.

J.L. Lumley [1967]. The structure of inhomogeneous turbulent flows. *Atmospheric Turbulence and Radio Wave Propagation* (eds. A.M. Yaglom and V.I. Tatarski), Nauka, Moscow, pp. 166–178.

J.L. Lumley [1970]. *Stochastic Tools in Turbulence.* Academic Press, New York.

A. Majda [1990]. Comments made at IMA workshop, Dynamical Theories of Turbulence in Fluid Flows, Minnesota, June 2–6, 1990.

V.K. Melnikov [1963]. On the stability of the center for time periodic perturbations. *Trans. Moscow Math. Soc.* **12**, 1–57.

B. Nicolaenko and Z.S. She [1990a]. Temporal intermittency and turbulence production in the Kolmogorov flow. *Topological Dynamics of Turbulence*, Cambridge University Press.

B. Nicolaenko and Z.S. She [1990b]. Symmetry-breaking homoclinic chaos in Kolmogorov flows, Arizona State University, preprint.

W.R.C. Phillips [1990]. Coherent structures and the generalized Lagrangian mean equation. *Appl. Mech. Rev.* **43**, 5(2), S227–S231.

H. Poincaré [1890]. Sur les équations de la dynamique et le problème des trois corps. *Acta. Math.* **13**, 1–270.

H. Poincaré [1899] *Les Methodes Nouvelles de la Mécanique Celeste* (3 Vols.), Gauthier-Villars, Paris.

D.A. Rand [1982]. Dynamics and symmetry: predictions for modulated waves is rotating fluids. *Arch. Rat. Mech. Anal.* **79**, 1–37.

R.H. Rand and D. Armbruster [1987]. *Perturbation Methods, Bifurcation Theory and Computer Algebra.* Springer-Verlag, New York.

D. Ruelle and F. Takens [1971]. On the nature of turbulence. *Commun. Math. Phys.* **20**, 167–192; **23**, 343–344.

S. Smale [1963]. Diffeomorphisms with many periodic points. *Differential and Combinatorial Topology* (ed. S.S. Cairns), Princeton University Press, Princeton, NJ, pp. 63–80.

S. Smale [1967]. Differentiable dynamical systems. *Bull. Amer. Math. Soc.* **73**, 747–817.

S. Smale [1980]. *The Mathematics of Time: Essays on Dynamical Systems, Economic Processes and Related Topics*, Springer-Verlag, New York, Heidelberg, Berlin.

S. Smale [1990] Some remarks on the foundations of numerical analysis *SIAM Rev.* **32**, 211–220.

E. Stone and P. Holmes [1989]. Noise induced intermittency in a model of a turbulent boundary layer. *Physica* **D37**, 20–32.

E. Stone and P. Holmes [1990]. Random perturbations of heteroclinic attractors. *SIAM J. Appl. Math.* **50**, 726–743.

E. Stone and P. Holmes [1991]. Heteroclinic cycles, exponential tails and intermittency in turbulence production. *Proceedings of the Symposium in Honor of J.L. Lumley*, eds. T.B. Gatski, S. Sarkar, and G.E. Speziale, pp. 179–189, Springer-Verlag, New York.

H.L. Swinney and J.P. Gollub (eds) [1981] *Hydrodynamic Instabilities and the Transition to Turbulence*, Springer-Verlag, New York. (Second edition, 1985.)

R. Témam [1988]. *Infinite-Dimensional Dynamical Systems in Mechanics and Physics*, Springer-Verlag, New York.

H. Tennekes and J.L. Lumley [1972] *A First Course in Turbulence*. MIT Press, Boston, MA.

R. Thom [1975]. *Structural Stability and Morphogenesis* (trans. D. Fowler), W.A. Benjamin, Reading, MA. (Original edition, Paris, 1972.)

C. Truesdell [1984]. *An Idiot's Fugitive Essays on Science*, Springer-Verlag, New York.

S.R. Wiggins [1988]. *Global Bifurcations and Chaos: Analytical Methods*, Springer-Verlag, New York.

R.F. Williams [1977]. The structure of Lorenz attractors. *Turbulence Seminar, Berkeley 1977/77*, eds. A Chorin, J.E. Marsden, and S. Smale, Springer Lecture Notes in Math. No. 615, Springer-Verlag, New York. pp. 94–116.

R.F. Williams [1979]. The structure of Lorenz attractors. *Inst. Hautes Etudes Sci. Publ. Math.* **50**, 73–99.

49
Dynamical Systems and the Geometry of Singularly Perturbed Differential Equations

N. Kopell

I was a student of Smale's during the heady and wonderful years of the mid-sixties described so well by Jacob Palis. At that time, I thought of dynamical systems as very ambitious and very pure. Along with the ambition and the purity came abstractness: I cannot recall in my four years at Berkeley having seen many actual differential equations.

At some point, my interests veered off the broad path laid down by Steve, and I got drawn from questions about generic properties into questions that come partially from outside of mathematics. Though at first I felt uncomfortably like an intellectual expatriate, on looking back it now seems to me that my evolution was a very natural one; indeed, at least one branch of dynamical systems has itself evolved in a similar way, and now the subject lives on the almost fractal boundary between pure and applied mathematics.

The subject of this chapter is not a survey, but rather a case study that illustrates a few aspects of the dialogue between pure and applied mathematics within dynamical systems. In particular, it points out the usefulness of working with specific examples: In a concrete equation, one may be forced to confront difficulties one might otherwise shove under a theoretical rug. This can lead to the creation of new machinery and new theory, which pushes along the development of the subject. It also shows that the concepts developed and/or fostered by Steve and the group around him continue to be an expressive language for yielding new insights into brand-name as well as generic equations.

Most people familiar with Steve's work in dynamical systems know about his horseshoe example and of the many applications it has found. I will discuss instead some phenomena that are less familiar to those in the more pure branch of dynamical systems, but which are widespread in applied mathematics. The questions have to do with so-called singularly perturbed equations, also known as fast–slow systems. These are equations with at least two time scales, e.g., those that can be written in one of the following two forms:

$$\begin{aligned} \varepsilon\dot{X} &= F(X, Y, \varepsilon), \\ \dot{Y} &= G(X, Y, \varepsilon), \end{aligned} \tag{1}$$

or

$$\begin{aligned} X' &= F(X, Y, \varepsilon), \\ Y' &= \varepsilon G(X, Y, \varepsilon). \end{aligned} \tag{2}$$

Here $X \in R^m$, $Y \in R^q$, and $\varepsilon \ll 1$. The singular nature of the equation shows up in two different ways in Eqs. (1) and (2), which are equivalent up to a change in time scale. In Eq. (1), at $\varepsilon = 0$, the equations drops dimension and the X equation reduces to the constraint

$$0 = F(X, Y, 0). \tag{3}$$

In Eq. (2), at $\varepsilon = 0$, there is no drop in dimension, but there is a manifold of equilibrium points that is given by the above constraint.

Singularly perturbed equations are important in applied mathematics because they allow the calculation of (what are alleged to be) approximate solutions. These are called "singular solutions." The basic idea is to separate the equations into the X system and the Y system, solve each separately and "match" together some carefully selected solutions from each. For the fast (X) system, one treats the slow variables Y as constant. For the slow (Y) system, one treats the fast variables as if they are "equilibrated," so that X and Y lie on the constraint manifold, [Eq. (3)]. For a portion of this manifold on which Eq. (3) can be solved for $X = \hat{X}(Y)$, the slow equations are $Y' = \varepsilon G(\hat{X}(Y), Y, 0)$.

The mysterious part of this technique, which is expounded in many methods books on applied mathematics [4, 27], is how to do the matching. (This is the first stage of what is known more generally as the method of matched asymptotic expansions.) In the hands of those with good intuition, the method is very powerful, computing solutions whose accuracy can usually be supported by numerical simulation. The range of the technique goes well beyond what is so far understood by rigorous mathematical ideas. This has both strengths and drawbacks; on the one hand, it allows one to get answers which, if sometimes miraculous, are at least consistant. On the other hand, there are simple examples in which the very methods that have worked so well before, fail spectacularly. That is, one can compute singular solutions that are nowhere near any actual solution. (One of the most renowned of these examples is the so-called resonance problem. See [6, 21] for some of the literature it has generated.) This state of affairs has justifiably piqued the interest of quite a number of mathematicians, and a variety of approaches have been used to try to put some order into the subject. These approaches include differential inequalities [3], functional analytic [10, 11, 13, 26], nonstandard analysis [7], topological/fixed-point methods [1, 5, 12, 14], and dynamical systems [9, 20, 21]. The general question is: when does the existence of a singular solution imply the existence of an actual one for which the singular solution is an approximation?

The above question is very broad, and though I believe that all the problems within it are amenable to a dynamical systems approach, there does not yet exist a unified treatment. Hence, I will specialize to a subclass to discuss some essential difficulties and why dynamical systems concepts are especially appropriate for handling them. The equations in this subclass include ones that are obtained from parabolic partial differential equations when one looks for solutions of a special type. For example, consider the Fitzhugh–Nagumo equations

$$\begin{aligned} u_t &= u_{xx} + f(u) - w, \\ w_t &= \varepsilon(u - \gamma w), \end{aligned} \tag{4}$$

where u, w are scalar functions, $f(u)$ is the cubic function $u(1-u)(u-\kappa)$ for some $0 < \kappa < 1$, and γ is a positive constant. The particular kind of solution desired is a "traveling pulse"; this is a solution such that u, w depend only on $\zeta \equiv x + \theta t$ for some wave speed θ, and $u, w \to 0$ as $\zeta \to \pm\infty$. Such a solution satisfies the ODE

$$\begin{aligned} u' &= v, \\ v' &= \theta v - f(u) + w, \\ w' &= \frac{\varepsilon}{\theta}(u - \gamma w) \end{aligned} \tag{5}$$

with boundary conditions $(u, v, w) \to (0, 0, 0)$ as $\zeta \to \pm\infty$, i.e., it is a homoclinic solution to Eqs. (5). Note that Eqs. (5) is singularly perturbed, with a manifold of equilibrium points $v = 0$, $w = f(u)$ at $\varepsilon = 0$; for $\varepsilon \neq 0$, the origin is the unique equilibrium point. The θ is not known a priori, and it is part of the solution to be found. There are many similar partial differential equations of various dimensions for which traveling wave solutions are of interest [19, 25].

For Eqs. (5), the fast pieces of the singular solutions satisfy

$$\begin{aligned} u' &= v, \\ v' &= \theta v - f(u) + w, \end{aligned} \tag{6}$$

where w is taken to be a constant. The slow pieces satisfy

$$w' = \frac{\varepsilon}{\theta}(u_{\pm}(w) - \gamma w), \tag{7}$$

where $u_-(w)$ and $u_+(w)$ correspond respectively to the right and left branches of $w = f(u)$ solved for u. (See Fig. 1). θ is determined by choosing that value θ_* for which there is a solution to Eqs. (6) with $w = 0$ (called a "front") with the property that $(u, v) \to (0, 0)$ as $\zeta \to -\infty$ and $(u, v) \to (1, 0)$ as $\zeta \to +\infty$. The other fast piece, known as the "back", is a solution to Eqs. (6) with $\theta = \theta_*$ and w chosen so that there is a trajectory joining $u_+(w)$ to $u_-(w)$ as ζ goes from $-\infty$ to $+\infty$. The two slow pieces then track the cubic and join the endpoints of the front and back. See Fig. 1. The problem is to show that, for $\varepsilon \neq 0$ but small, there is an actual homoclinic solution to Eqs. (5) near the singular orbit for some value of θ near θ_*.

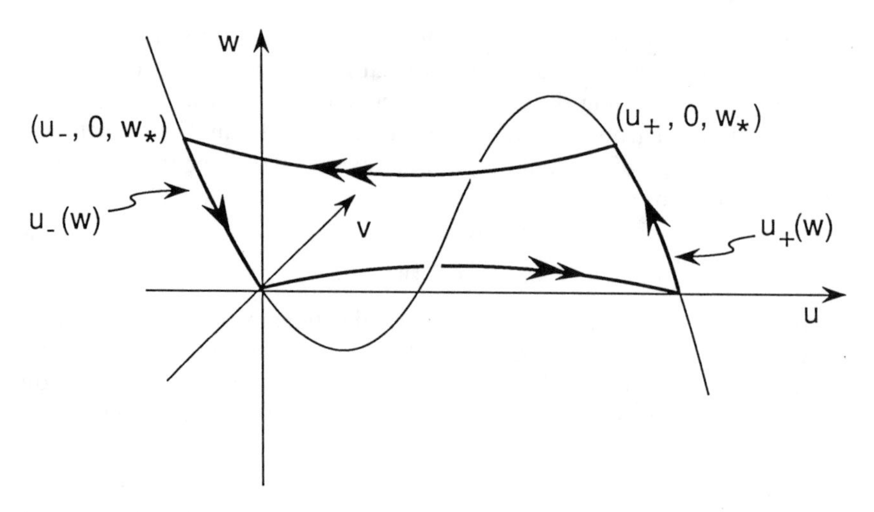

FIGURE 1. The cubic curve of equilibria at $\varepsilon = 0$ and the singular homoclinic solution.

The first really insightful approach to these equations was taken by Gail Carpenter under the directions of Charles Conley, using index methods to capture the desired orbit [1]. However, index methods do not make full use of the structure that is in the equation and do not give maximally strong results. For example, one cannot see from such methods whether the solution is locally unique. Since a homoclinic orbit is, by definition, the intersection of the stable and unstable manifolds of the equilibrium point, it is natural to try to construct such a solution as the above intersection. Adding the equation $\theta' = 0$, one can hope to show a *transverse* (and, therefore, locally unique) intersection of the center stable and center unstable manifolds of the origin.

Shortly after Carpenter wrote her thesis, I undertook such a project with Robin Langer [23]. We succeeded, at least for Eqs. (5), but with many pages of unconceptual estimates that convinced us and others that this dynamical systems approach was too cumbersome to be used on still harder equations. (It also pointed up rather dramatically the difference in ease between drawing pictures of intersections of stable and unstable manifolds and verifying such an intersection in any given case.) Ten years later, in work with Chris Jones [16, 18], we see with the help of ideas from differential geometry that the dynamical systems approach can be made clean, conceptual and general. An added advantage is that the information obtained from examining the geometry of the stable and unstable manifolds is useful in cletermining if the trajectory is stable as a solution to the original PDE [17].

Like the formal classical approach, the dynamical systems approach exploits the separation into two time scales. The underlying idea is that the "fast flow," defined as the equations at $\varepsilon = 0$, contains all the information necessary to decide whether a singular solution does indeed approximate an actual one for $\varepsilon \neq 0$. That is, there are certain submanifolds of solutions when

$\varepsilon = 0$ whose transversality is sufficient to ensure the existence of the relevant solution for $\varepsilon \neq 0$. This must be checked for any equation to which one wishes to apply the method. However, this check is far easier than working with the full equations at $\varepsilon \neq 0$, mainly because the check is done in lower dimensions. (To see how to do this for Eqs. (5), see [16].) The harder part of the method is a theory that turns the transversality at $\varepsilon = 0$ into existence for $\varepsilon > 0$. However, unlike the check. this part is done by a "machine" that is problem-independent and can merely be quoted. I shall illustrate the necessary transversality at $\varepsilon = 0$ using Eqs. (5) and discuss the machine in more generality.

Consider Eqs. (5) as four-dimensional, with the equation $\theta' = 0$ appended. At $\varepsilon = 0$, the manifold of equilibrium points is then two dimensional, given by $v = 0$ $w = f(u)$. Almost all of these points [with the exception of those points (u, v, w, θ) for which $f'(u) = 0$] are normally hyperbolic, i.e., there are two eigenvalues whose real parts are not zero: one negative and one positive. Consider all the points on the manifold of equilibrium points with u on the right branch of the cubic. Each has a one-dimensional unstable manifold, so the union of these manifolds over those points forms a three-dimensional manifold in (u, v, w, θ) space we shall call $W^u(R)$. Similarly, one defines the three-dimensional stable manifold $W^S(R)$ of the right branch and the three-dimensional stable and unstable manifolds $W^S(L)$ and $W^u(L)$ of the left branch. The transversality condition to be checked involves submanifolds of the above manifolds. There are two conditions, corresponding to transver-

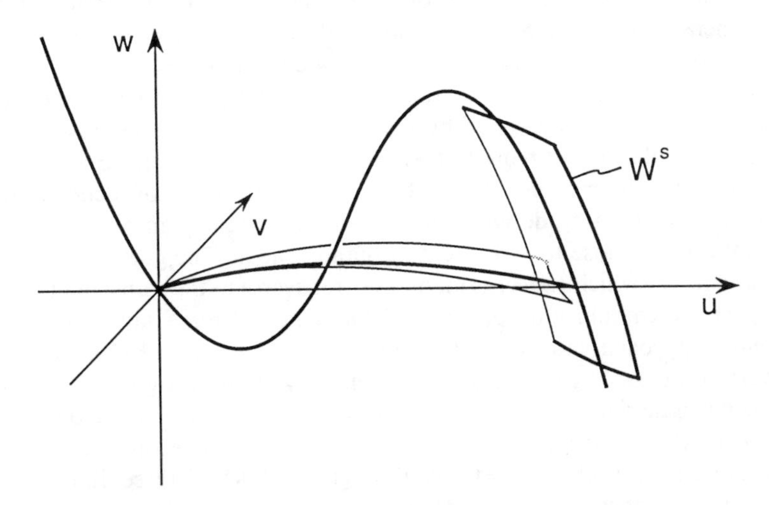

FIGURE 2. $W^u(L)$ is the union (over an interval of values of θ) of the one-dimensional unstable manifold of $(u, v, w) = (0, 0, 0)$. $W^S(R)$, the stable manifold of the right-hand branch of the equilibria points in (u, v, w, θ) space, is a three-dimensional manifold, here depicted as two dimensional, with the θ direction suppressed. These manifolds are transverse in (u, v, w, θ)-space. There is a similar picture for transversality across the back.

sality across the front and transversality across the back. For the first, we require that

$$W^u(L)|_{w=0} \quad \text{transversely intersects} \quad W^S(R). \tag{8}$$

The second condition is that

$$W^u(R)|_{\theta=\theta_*} \quad \text{transversely intersects} \quad W^S(L). \tag{9}$$

It follows from the existence of the singular solution and the choice of θ_* that there is a nonzero intersection in Eqs. (8) and (9). The transversality requires further work, done in [16]. See Fig. 2.

From "Singular Transversality" to Existence

We must now show how conditions (8) and (9) can be translated into a proof of the existence of a homoclinic solution for Eqs. (5). The aim is to use those conditions to follow the manifold of potential solutions from the origin as it passes near the singular orbit. The manifold M in question is the union of the unstable manifolds of the equilibrium point $(u, v, w) = (0, 0, 0)$ over an interval of values of θ around θ_*. For each such θ, the unstable manifold is one-dimensional, so M is two dimensional. Note that the stable manifold of $(0, 0, 0)$ for each fixed θ near θ_* is two dimensional since one of the eigenvalues that is zero at $\varepsilon = 0$ is small and positive when $\varepsilon > 0$. Thus, the union of the stable manifolds is three dimensional and has the right dimensions for a one-dimensional transverse intersection with M in (u, v, w, θ) space. Such an intersection would be the desired locally unique solution.

There are several difficulties associated with attempting to carry out such a plan, which can be illustrated by Eqs. (5). One is that, on the time scale of Eqs. (5), the solutions turn out to take a time of order $1/\varepsilon$ to move slowly past the left and right branches of the cubic, making estimation delicate. It is possible to make a crude guess at the fate of tangent planes to M as this manifold passes close to, say, the right-hand branch. The guess comes from the "λ-lemma," which concerns manifolds transverse to the stable manifold of a hyperbolic equilibrium point [24]. This famous lemma says that such a manifold approaches the unstable manifold in increasing time. (Our situation is more complex: In the case at hand, when $\varepsilon \neq 0$, there are no equilibrium points near the right-hand branch of the cubic. Even at $\varepsilon = 0$, the points are only *normally* hyperbolic.) The union of all the unstable directions of the right-hand branch of the $\varepsilon = 0$ manifold [i.e., $W^u(R)$] is three dimensional. From this, we guess that as t increases (at least for a while), M approaches some two-dimensional submanifold of $W^u(R)$. However, we do not know which submanifold. In [23], the approach was to use variational equations of Eq. (5) to follow tangent vectors to get that information. The difficulty with this is that all the relevant trajectories eventually "jump off" the cubic and head almost in the direction of the fast flow (6) (see Fig. 3.) Thus, in following

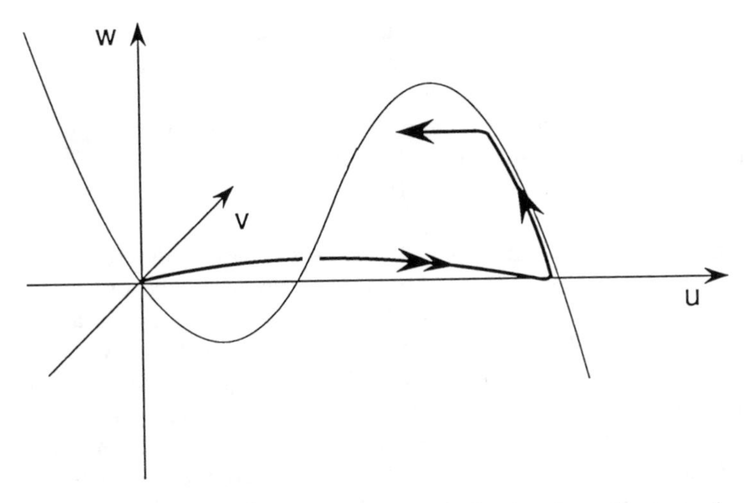

FIGURE 3. A schematic picture of a trajectory near the singular orbit. Away from the slow manifold S_ε the tangent vector points approximately along the fast flow.

individual trajectories, one loses information about the variation as θ changes, which is precisely the information that is needed. The solution to this difficulty, not thought of in 1980, lies in following not individual trajectories but the entire tangent plane to M. The key technical lemma, known as the "Exchange Lemma," gives information about the evolution of the planes tangent to an invariant manifold as that manifold passes close to a branch of $\varepsilon = 0$ equilibrium points.

Before stating the Exchange Lemma, we need a technical preliminary. We go back to the general equations (2) and assume that at $\varepsilon = 0$ there is a q-dimensional manifold S of normally hyperbolic equilibrium points, with k expanding directions and l contracting ones. To control the exponential expansion for a time of order $1/\varepsilon$, it is important to have a convenient coordinate system in which to do estimates, one adapted to the structure of the trajectories near S when $\varepsilon \neq 0$. It follows from the normal hyperbolicity that for ε small there is a nearby invariant manifold S_ε [8, 15]. S_ε is a kind of *global* center manifold that appears in singularly perturbed problems [2]; it is not unique, but the nonuniqueness is unimportant. Furthermore, coordinates can be chosen in a neighborhood of S_ε such that in those coordinates the flow has the form

$$\begin{aligned}\mathbf{a}' &= \lambda(\mathbf{a}, \mathbf{b}, \mathbf{c}, \varepsilon)\mathbf{a},\\ \mathbf{b}' &= \mu(\mathbf{a}, \mathbf{b}, \mathbf{c}, \varepsilon)\mathbf{b},\\ \mathbf{c}' &= \varepsilon h(\mathbf{a}, \mathbf{b}, \mathbf{c}, \varepsilon),\end{aligned} \tag{10}$$

where $\mathbf{a} \in \mathbf{R}^k$, $\mathbf{b} \in \mathbf{R}^l$, $\mathbf{c} \in \mathbf{R}^q$ [9]. Note that S_ε corresponds to $\mathbf{a} = 0$, $\mathbf{b} = 0$. The coordinates can be chosen so that in a box B around a compact region of S_ε

(say $|\mathbf{a}| < \Delta$, $|\mathbf{b}| < \Delta$ for some Δ), the **a** coordinates expand at a rate bounded uniformly away from 0 and the **b** coordinates decay at such a rate. In the problems we are considering, the relevant trajectories of the $\varepsilon = 0$ "slow flow" on S can be rectified, so we can choose coordinates so that $h_i(0, 0, \mathbf{c}, \varepsilon) = 0$, where h_i is the ith component of h. In the case of Eqs. (5), the slow flow is given by Eq. (7), $c_1 = w$ and $c_2 = \theta$. In general, c_1 is to be thought of as "slow time."

To track an invariant manifold M, one uses equations describing the evolution of the values of certain differential forms evaluated on the tangent planes of M. The manifolds we wish to follow, as in Eqs. (5), are $(k + 1)$ dimensional, where k is the number of expanding directions. As in Eqs. (5), the extra dimension in general comes from taking the union of unstable manifolds over some parameter interval. Thus, we use $(k + 1)$-forms, which are dual to $(k + 1)$-planes. The reason for using forms rather than planes is that it is very easy to write the equations for the evolution of the forms; they are a natural generalization of the variational equations. The latter describes the evolution (along a trajectory) of 1-forms that are the differentials of the coordinate functions. The $(k + 1)$-forms are created from the 1-forms by wedge product, and satisfy equations derived from the variational equations using the product rule. (For example, for Eqs. (5) there are six 2-forms, P_{ab}, P_{aw}, $P_{a\theta}$, P_{bw}, $P_{b\theta}$, $P_{w\theta}$, where $P_{ab} = \delta a \wedge \delta b$, etc., and δa, δb, δw, $\delta\theta$ satisfy the variational equations of Eqs. (5).) It can be seen that the forms involving δa_i could grow exponentially along a trajectory near S_ε. To control this, one works in a space of projectivized forms obtained by dividing by the fast-growing form $P_{a_1 a_2 \ldots a_k c_1}$. In this coordinate system, the plane generated by the tangents to the $a_1, \ldots, a_k, c_1$ direction is easily described as that plane for which all the projectivized $(k + 1)$-forms vanish. For this reason, this system is an especially convenient coordinate system for proving:

The Exchange Lemma (C.K.R.T. Jones and N. Kopell). *Suppose that M is a $(k + 1)$-dimensional invariant manifold such that $\Gamma_\varepsilon \equiv M \cap \{|\mathbf{b}| = \Delta\}$ is transverse to $\{\mathbf{a} = 0\}$. Let $p \in \Gamma_\varepsilon$ be a point whose trajectory hits $\{|\mathbf{a}| = \Delta\}$ at $\bar{p}$ after time $\tau = 0(1/\varepsilon)$. Then at $\bar{p}$, the tangent plane to M is close to the space spanned by the tangents to the $a_1, \ldots, a_k, c_1$ directions.* (*See Fig.* 4.)

The exchange lemma follows the tangent planes to an invariant manifold along a trajectory that stays in the box for a sufficiently long amount of time [i.e., $O(1/\varepsilon)$]. The hypothesis is that the manifold enters the box transverse to the contracting directions. Analogous to the transversality hypothesis in the λ-lemma, the normal hyperbolicity then guarantees that the evolving manifold splays out in the expanding directions. The Exchange Lemma also says that all components of the tangent plane in center directions other than that of slow time are close to zero. (This important conclusion of the lemma is absent from Fig. 4 since the other center directions are suppressed in the picture for lack of ability to draw in four dimensions.)

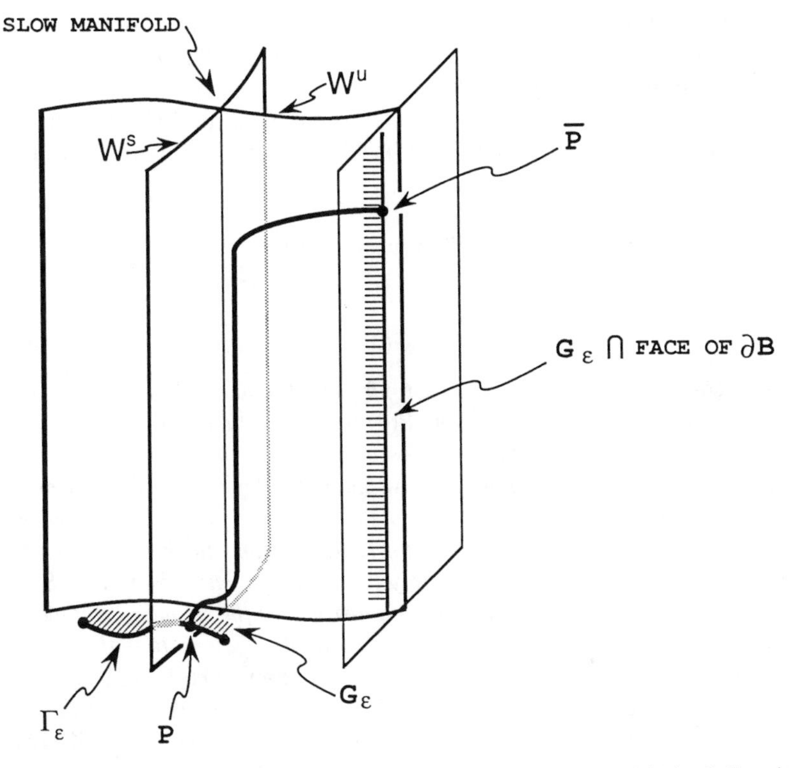

FIGURE 4. Behavior near a portion of S_ε, here depicted as a curve with the θ direction suppressed. The trajectory through p first decays quickly to near the slow manifold, then creeps along the latter, then jumps off at $\bar{p}$. The tangent plane at $\bar{p}$ is almost entirely in the direction spanned by the tangents to the a and w directions, with almost no θ component.

The name of the Exchange Lemma comes from the context in which it is used. In problems such as Eqs. (5), the manifold M is constructed to be the union of certain manifolds as a parameter is varied. Furthermore, the hypothesis that M is transverse to the contracting directions is satisfied because of the variation in that parameter. [In the case of Eqs. (5), this is condition (8).] When any such trajectory exits from the box, the tangent plane has no component in the direction of that parameter. Instead, the tangent plane is partly generated by a vector in the direction of slow time. Thus, information about transversality with respect to the original parameter at the entrance to the box is exchanged for information about transversality with respect to the direction of slow time at the exit of the box. This is exactly what one needs to continue to follow the manifold across the next fast front or back. In Eqs. (5) for example, condition (9) plus the output of the Exchange Lemma for trajectories traveling near the right branch of S_ε allows one to conclude that after flowing near the singular back, M will arrive near the left-hand branch of S_ε

with a tangent plane transverse to the contracting directions of the left-hand S_ε. Another application of the exchange lemma then allows one to follow M along trajectories flowing near S_ε.

The Exchange Lemma can be used to track invariant manifolds in situations much more general than Eqs. (5). The following theorem gives such an application.

Theorem (N. Kopell and C.K.R.T. Jones). *Consider*

$$\begin{aligned} X' &= F(X, Y, \theta, \varepsilon), \\ Y' &= \varepsilon G(X, Y, \theta, \varepsilon), \end{aligned} \tag{11}$$

where $X \in \mathbf{R}^{k+l}$, $Y \in \mathbf{R}^q$. Assume that for each θ and $\varepsilon \neq 0$, there is a locally unique hyperbolic equilibrium point $P(\theta)$ with k unstable directions and $l + q$ stable directions; of the eigenvalues associated with the latter, q tend to zero with ε. Let $\{S_i\}$ denote a family of slow manifolds for the $\varepsilon = 0$ equation (with the equilibrium point for $\varepsilon \neq 0$ in S_0) and assume that for each i, S_i is normally hyperbolic with splitting k stable, l unstable. Assume further that there is a singular homoclinic orbit, with finitely many jumps, each from S_i to S_{i+1} for some i (where the $\{S_i\}$ are not necessarily distinct, so the singular orbit may visit the same slow manifold more than once). Finally, assume that the following transversality conditions hold for the $\varepsilon = 0$ system: Let $W^S(S_i)$ and $W^u(S_i)$ denote the stable and unstable manifolds of the S_i as above and $[P(\theta), \theta]$ the graph as θ is varied of the $\varepsilon = 0$ limit of the $\varepsilon \neq 0$ equilibrium point. We require that

$$W^u(S_0)|_{[P(\theta),\theta]} \quad \text{transversely intersects } W^S(S_1) \tag{12}$$

$$W^u(S_i)|_{\text{sing. orbit}} \quad \text{transversely intersects } W^S(S_{i+1}), \quad \forall i > 0, \tag{13}$$

Then for ε sufficiently small, there is a locally unique homoclinic solution to Eqs. (11) *near the singular solution.*

Remark. Note that condition (12) reduces in Eqs. (5) to condition (8), i.e., transversality across the front as θ is varied and the critical point is held fixed. The other conditions (13) reduce in Eqs. (5) to condition (9), in which θ and all slow variables except the one corresponding to slow time are fixed, and transversality is with respect to the latter.

Epilogue and Conjecture

This class of singularly perturbed equations is one of many I have stumbled across while thinking about some problem in chemistry, biology, or physics (e.g., [22]). In each case, what would classically be construed as "matching conditions" for some matched asymptotic expansion turns out (with some work) to be formulated as transversality conditions. Furthermore, the classical conditions can fail when there is a lack of transversality, and the correct methods can be deduced by thinking geometrically. (See, e.g., [21] and references in that article.) In general, I conjecture that the framework of dynamical systems, suitably extended, can provide a widely applicable theory for so far unjustified asymptotic methods, and that the concepts of invariant manifolds and transversality will play a central role in such a theory.

References

[1] G. Carpenter, A geometric approach to singular perturbation problems with applications to nerve impulse equations. *J. Diff. Eqs.* 23:335–367, 1977.

[2] J. Carr. *Applications of Center Manifold Theory*. Springer-Verlag, NY, 1980.

[3] K.W. Chang and F.A. Howes, *Nonlinear Singular Perturbation Phenomena: Theory and Applications*. Springer-Verlag, NY, 1984.

[4] J. Cole. *Perturbation Methods in Applied Mathematics*. Ginn-Blaisdell, Waltham, MA, 1968.

[5] C. Conley, On travelling wave solutions of nonlinear diffusion equations in dynamical systems, theory and applications, Springer Lecture Notes in Physics, No. 38, Springer-Verlag, Berlin, 1975, pp. 498–510.

[6] P. deGroen, The singularly perturbed turning point problem: a spectral approach, *Singular Perturbations and Asymptotics*. Academic Press, New York, 1980, pp. 149–172; The nature of resonance in a singular perturbation problem of turning point type. *SIAM J. Math. Anal.* 11:1–22, 1980.

[7] F. Diener and G. Reeb, *Analyse Non Standard*. Hermann, Editeurs des Sciences et des Arts, Paris, 1989.

[8] N. Fenichel, Persistence and smoothness of invariant manifolds for flows. *Indiana Univ. Math. J.* 21:193–226, 1971.

[9] N. Fenichel, Geometric singular perturbation theory for ordinary differential equations. *J. Diff. Eqs.* 31:53–98, 1979.

[10] P.C. Fife, Boundary and interior transition layer phenomena for pairs of second order differential equations. *J. Math. Anal. Appl.* 54:497–521, 1976.

[11] H. Fujii, Y. Nishiura, and Y. Hosono, On the structure of multiple existence of stable stationary solutions in systems of reaction-diffusion equations. *Studies Math. Appl.* 18:157–219, 1986.

[12] R.H. Gardner and J. Smoller, Travelling wave solutions of predator-prey systems with singularly perturbed diffusion. *J. Diff. Eqs.* 47:133–161, 1983.

[13] J. Hale and K. Sakamoto, Existence and stability of transition layers. *Japan J. Appl. Math.* 5:367–405, 1988.

[14] S. Hastings, On travelling wave solutions of the Hodgkin-Huxley equations. *Arch. Rat. Mech. Anal.* 60:229–257, 1976.

[15] M.W. Hirsch, C.C. Pugh, and M. Shub, *Invariant Manifolds*, Lecture Notes in Mathematics No. 583, Springer-Verlag, New York, 1977.

[16] C. Jones, N. Kopell, and R. Langer, Construction of the Fitzhugh-Nagumo pulse using differential forms. *Patterns and Dynamics in Reactive Media*, H. Swinney, G. Aris and D. Aronson, eds., Springer-Verlag, New York, 101–116, 1991.

[17] C.K.R.T. Jones, Stability of the travelling wave solutions of the Fitzhugh-Nagumo system. *Trans. AMS* 286:431–469, 1984.

[18] C.K.R.T. Jones and N. Kopell, Tracking invariant manifolds of singularly perturbed systems using differential forms, *J. Diff. Eqs.*, in press 1993.

[19] J. Keener, Frequency dependent decoupling of parallel excitable fibers. *SIAM J. Appl. Math.* 49:210–230, 1989.

[20] N. Kopell, Waves, shocks and target patterns in an oscillating chemical reagent, *Nonlinear Diffusion. Research Notes in Mathematics.* W.E. Fitzgibbon and H.F. Walker, eds. Pitman, London, 1977.

[21] N. Kopell, The singularly perturbed turning-point problem: a geometric approach, *Singular Perturbations and Asymptotics.* Academic Press, New York, 1980, pp. 173–190; A geometric approach to boundary layer problems exhibiting resonance. *SIAM J. Appl. Math.* 37:436–458, 1979.

[22] N. Kopell and L.N. Howard, Target patterns and horseshoes from a perturbed central force problem: some temporally periodic solutions to reaction-diffusion equations. *Studies Appl. Math.* 64:1–56, 1981.

[23] R. Langer, Existence and uniqueness of pulse solutions to the Fitzhugh-Nagumo equations. Ph. D. thesis, Northeastern University, 1980.

[24] J. Palis and W. deMelo, *Geometric Theory of Dynamical Systems.* Springer-Verlag, New York, 1980.

[25] J. Rinzel, Models in neurobiology, *Nonlinear Phenomena in Physics and Biology.* R. Enns, B.L. Jones, R.M. Miura and S.S. Rangnekar, eds., Plenum, New York, 1981, pp. 345-367.

[26] K. Sakamoto, Invariant manifolds in singular perturbation problems for ordinary differential equations, *Proc. Roy. Soc. Ed.* 116A: 45–78, 1990.

[27] C.C. Lin and L.A. Segel, *Mathematical Applied to Deterministic Problems in the Natural Sciences*, Macmillan, New York, 1980.

50
Cellular Dynamata

R. ABRAHAM

1. Introduction

At the end of the sixties, our program for dynamics lost momentum, and some of us turned to applications for inspiration. My own attempts at modeling in the biological sciences led me to *complex dynamics*, a blend of our own style of dynamical systems theory with system dynamics. A complex dynamical system is a directed graph, with a dynamical scheme (dynamical system with parameters) at each node, and a coupling function on each edge, expressing the control parameters of one scheme as a function of the states of another.[1]

A related type of structure is a *cellular dynamaton* or CD. This is a special kind of complex dynamical system, in which the graph is embedded in a physical space. To my knowledge, the first CD was the one-dimensional array of oscillators introduced by Anderson in 1924 as a model for cat gut.[2] Chris Zeeman introduced me to the CD idea with his model for memory, which he described to me in Amsterdam in 1972.[3] As I understood him then, there would be a two-dimensional array of Duffing cells, with a spatial pattern of attractors corresponding to a memory engram. This is close to Walter Freeman's recent model for memory in the olfactory cortex.[4]

2. Definitions

A CD is specified by three sets of data,

A. *the standard cell*, a single dynamical scheme, envisioned as a response (or bifurcation) diagram,

[1] For more details, see (Abraham, 1984).

[2] For this and other early references, see (Kopell, 1983).

[3] See his own description of a single memory cell in (Zeeman, 1977).

[4] See (Freeman, 1991).

B. *the spatial substrate*, a regular lattice in a manifold, usually Euclidean space, and
C. *the connections*, a set of coupling functions, one for each node in the lattice, expressing the local control parameters as functions of the states at all other nodes. These data determine a single, massive, dynamical scheme, which has its own response diagram.[5] There are three worlds of CDs, depending on whether the standard cell is generated by a map, a diffeomorphism, or a vector field. Throughout this brief survey, we will assume the latter.[6]

3. Examples

The first and most studied CD is a lattice of oscillators, in a one-dimensional substrate. Here, the coupling is usually to nearest neighbors only, as in Anderson's model for peristalsis. Van der Pol oscillators are the usual choice for the standard cell. Freeman's olfactory bulb model belongs to this class.

Another early class of models comes from the discretization of PDEs, especially the wave equation, the heat equation, or reaction-diffusion equations.[7]

A simple neural net (given an arbitrary spatial substrate if neccesary) is also a CD, in which the usual standard cell is the simplest possible dynamical scheme (linear vector field in one dimension, plus a constant for control parameter), with global, linear coupling. These nets have simple cells, but complex coupling, and are capable of very complex behavior (that is, lots of static attractors with thin basins).[8]

Closely related to the simple neural net is the cuspoidal net, in which the standard cell is a cusp catastrophe, and the coupling is local and linear. As two coupled cusps can oscillate, as Kadyrov has shown,[9] a cuspoidal net can behave either as a simple neural net, or as a lattice of oscillators. They may be useful for spatial economic models[10] as well as for artificial intelligence.

---

[5] For more details, see (Abraham, 1986).

[6] For practical reasons, the iterated map CD is the type most commonly explored in computer simulation. For a review of the results with one-dimensional substrates, see (Crutchfield, 1987) . A color atlas of spatial patterns may be seen in (Abraham, 1991b).

[7] For a history of this class, see (Abraham, 1991c).

[8] On the existence of static attractors, see (Hirsch, 1989).

[9] See (Abraham 1991a). This is similar to Steve Smale's two-cell oscillator (Smale, 1976).

[10] See (Abraham, 1990b), (Beckmann, 1985), and (Beckmann, 1990).

4. The Future

So far, the development of this new subject, now in its early period, depends crucially on explorations. Currently, explorations of two-dimensional CDs with massively parallel supercomputers are on the frontier. Besides the computational cost, these explorations tax our cognitive strategies for the recognition of basic concepts of behavior of attractors and bifurcations. New methods of reducing space-time patterns to symbol sequences are needed.[11] A number of ingenious ideas based on classical analysis have evolved, especially in the context of oscillator CDs.[12] A surprising feature on the current frontier is the role played by applications. Many of the new ideas are coming from modelers in the biological and social sciences.[13] Morphogenesis remains the Everest of the scientific modeling art.

Acknowledgments. It is a pleasure to acknowledge the creative impetus of my contacts over these 30 years with Steve Smale, René Thom, and Chris Zeeman, and more recently with the scientists in various fields who have patiently instructed me in their work, and tempered my mathematical arrogance with their good humor. Special thanks to Jack Corliss and John Dorband for introducing me to massively parallel computing.

References

Abraham, 1984
Ralph H. Abraham, Complex dynamical systems, *Mathematical Modelling in Science and Technology*, eds. X.J.R. Avula, R.E. Kalman, A.I. Leapis, and E.Y. Rodin, Pergamon Elmsford, NY, pp. 82–86.

Abraham, 1986
Ralph H. Abraham, Cellular dynamical systems, *Mathematics and Computers in Biomedical Applications, Proc. IMACS World Congress, Oslo, 1985*, eds. J. Eisenfeld and C. DeLisi, North-Holland, Amsterdam, pp. 7–8.

Abraham, 1990a
Ralph H. Abraham, Cuspoidal nets, *Toward a Just Society for Future Generations*, eds. B.A. and B.H. Banothy, Int'l Society for the Systems Sciences, pp. 66–683.

Abraham, 1990b
Ralph H. Abraham, Visualization techniques for cellular dynamata, *Introduction of Nonlinear Physics*, ed. Lui Lam, Springer-Verlag, Berlin.

---

[11] See (Abraham, 1990a) for one such proposal, based on the response diagram of the standard cell.

[12] See, for example, (Kopell, 1983) or (Kaneko, 1990).

[13] See, for example, the double-logistic map lattices arising in economic models, in (Gaertner, 1991).

Abraham, 1991a
Ralph H. Abraham, Gottfried Mayer-Kress, Alexander Keith, and Matthew Koebbe, Double cusp models, public opinion, and international security, *Int. J. Bifurcations Chaos* **1**(2), 417–430.
Abraham, 1991b
Ralph H. Abraham, John B. Corliss, and John E. Dorband, Order and chaos in the toral logistic lattice, *Int. J. Bifurcations Chaos* **1**(1), 227–234.
Abraham, 1991c
Ralph H. Abraham, Cellular dynamata and morphogenesis: a tutorial, preprint.
Beckmann, 1985
Martin Beckmann and Tonu Puu, *Spatial Economics: Density, Potential, and Flow*, North-Holland Amsterdam.
Beckmann, 1990
Martin Beckmann and Tonu Puu, *Spatial Structures*, Springer-Verlag, Berlin.
Crutchfield, 1987
James P. Crutchfield and Kunihiko Kaneko, Phenomenology of spatio-temporal chaos, *Directions in Chaos*, ed. Hao Bai-Lin, World Scientific, Singapore.
Freeman, 1991
Walter Freeman, The physiology of perception, *Sci. Am.* **264**(2), 78–85.
Gaertner, 1991
Wulf Gaertner and Jochen Jungeilges, A model of interdependent consumer behavior: nonlinearities in R2, preprint.
Hirsch, 1989
Morris W. Hirsch, Convergent activation dynamics for continuous time networks, *Neural Networks* **2**, 331–349.
Kaneko, 1990
Kunihiko Kaneko, Simulating Science with Coupled Map Lattices, *Formation, Dynamics, and Statistics of Patterns*, ed. K. Kawasaki, A. Onuki, and M. Suzuki, World Scientific, Singapore.
Kopell, 1983
Nancy Kopell and G.B. Ermentrout, Coupled oscillators and mammalian small intestines, *Oscillations in Mathematical Biology*, ed. J.P.E. Hodgson, Springer-Verlag, Berlin, pp. 24–36.
Smale, 1976
Steve Smale, A mathematical model of two cells via Turing's equation, *The Hopf Bifurcation and its Applications*, eds. J.E. Marsden and M. McCracken, Springer-Verlag, New York.
Zeeman, 1977
E. Christopher Zeeman, Duffing's equation in brain modelling, *Catastrophe Theory*, ed. E. Christopher Zeeman, Addison-Wesley, Reading, MA, pp. 293–300.

51
Rough Classification of Integrable Hamiltonians on Four-Dimensional Symplectic Manifolds

A.T. FOMENKO

1. Introduction

This work develops, in particular, important ideas from the famous paper of S. Smale "Topology and mechanics" [1]. The theory which is the subject of the present paper gives, in particular, the answer to some questions formulated by Marsden, Ratiu, Novikov, Arnold, Weinstein, Adler, van Moerbeke, Scovel, Flaschka, Kozlov, Gray, McLaughlin, and Haine. These scientific discussions were important and the author expresses his thanks to all these specialists.

My special thanks to the Organizing Committee of SmaleFest 1990 and Chairman M. Hirsch for the invitation to take part in the work of the Conference at the University of California, Berkeley.

In references [5–9], we obtained a complete topological classification of all integrable Bottian nonresonant Hamiltonians on three-dimensional isoenergy surfaces. In this chapter, we discover a new topological invariant of integrable Hamiltonians on *four-dimensional* symplectic manifolds. It turns out that integrable Hamiltonians are roughly topologically equivalent iff their invariants coincide. This theorem allows us to recognize equivalent and nonequivalent (strongly nondegenerate) Hamiltonian systems of differential equations on 4-manifolds. In particular, the set of all classes of integrable Hamiltonians is *discrete*.

The author thanks A.V. Bolsinov and Nguen Tyen Zung for useful discussions.

2. Some Definitions

Let M^4 be a smooth four-dimensional symplectic manifold (compact or noncompact) and $v = \text{sgrad}\, H$ be a Hamiltonian system of differential equations with a smooth Hamiltonian H. Here, sgrad is the skew-symmetric gradient of the function H (corresponding to the symplectic structure ω on M). Let us suppose that the system v is nonresonant and completely integrable in the

Liouville sense on some open domain U in M with the help of some *Bottian integral* f (U can coincide with M). In this case, we call the Hamiltonian H *integrable*. This means that the smooth function f is independent of H, is in involution with H, and the restriction of f on each regular isoenergy 3-surface $Q_h^3 = \{H = h = \text{const}\}$ is a Bott function, i.e., all critical submanifolds of the function f in Q are *nondegenerate*. We will assume (for simplicity) that the domain U is given in the manifold M by the system of inequalities (for the Hamiltonian H), i.e., $U = \{x \in M: a_i \leq H(x) \leq b_i\}$, where a_i and b_i are some numbers. Then the domain U is foliated on the three-dimensional (isoenergy) level surfaces of the function H. Let us assume (again for simplicity) that all surfaces Q_h are *compact*. Almost all connected regular common level surfaces of the integrals H and f are two-dimensional tori (*Liouville tori*). The system v is called *nonresonant* in the domain U if its integral trajectories form the *irrational winding* on almost all Liouville tori in U. We shall also speak about the *nonresonant Hamiltonian H*. We will consider only nonresonance Hamiltonian equations which are integrable with the help of the Bottian integrals. Let us recall that two integrable nonresonance systems on isoenergy 3-manifolds Q and Q' are called *topologically equivalent* if there exists a diffeomorphism $\lambda: Q \to Q'$ that transforms the Liouville tori of the first system into the Liouvolle tori of the second system. Usually in real physical systems v, the Hamiltonian H is *uniquely determined* by the physical nature of a given problem. Consequently, we need to classify the integrable systems *with their Hamiltonians*. We fix the Hamiltonian H (for each integrable system v) and we consider the *pair* (v, H). Because the function H is fixed, the foliation of U (as the union of the three-dimensional level surfaces Q_h) also is fixed. Let us consider two systems: $v = \text{sgrad}\, H$ on the domain U in the manifold M and $v' = \text{sgrad}\, H'$ on the domain U' in the manifold M'.

Definition. Two integrable systems of Hamiltonian differential equations (or two integrable Hamiltonians H and H') are called *topologically equivalent* on the four-dimensional domains in symplectic manifolds, if there exists a homeomorphism $\tau: U \to U'$, which transforms the connected components of level surfaces of H into the connected components of level surfaces of H', which transforms (by diffeomorphism) the regular connected isoenergy 3-manifolds of the first system in the connected isoenergy 3-manifolds of the second system, and which transforms the Liouville tori into Liouville tori.

Let v be an integrable system on U. Let us cut U along some isoenergy 3-manifold Q_h and glue again two copies of Q_h (as the boundary components) using some diffeomorphism, which preserves the orientation and the foliation of Q_h as the union of the Liouville tori (Fig. 1). We obtain some new 4-manifold U' with some new Liouville foliation on it. We will say that the foliation (the system) on U' is obtained from the foliation (from the system) on U by *twisting along the isoenergy surface* Q_h.

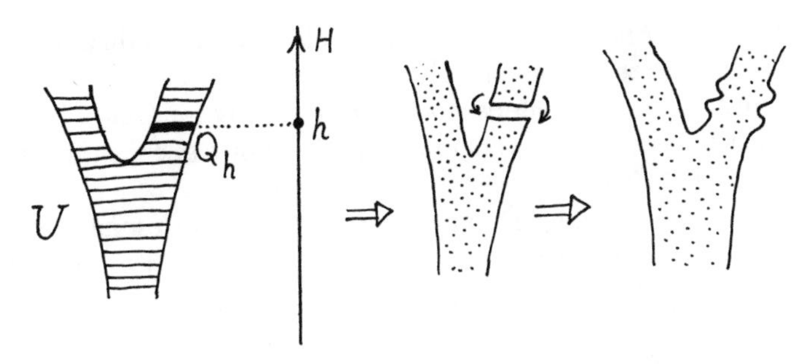

FIGURE 1

Definition. Two integrable systems are called *roughly topologically equivalent* iff the corresponding Liouville foliations on U and U' are obtained one from another by twistings along 3-surfaces Q.

If two systems are roughly topologically equivalent on 4-manifolds, then they are *topologically equivalent* on the corresponding isoenergy 3-manifolds.

If the value h is changed, then 3-manifold Q_h is smoothly deformed inside M. When h crosses the *critical value* c of the Hamiltonian H, then the system v is transformed. But the system can change its topological type at the moment h crosses some *regular* value of the function H (such examples can be constructed easily). Here Q does not change its topological type.

Definition. The value $h = c$ of the Hamiltonian H is called *bifurcation value* if the system v changes its topological type on the 3-surface Q_c^3. We will consider those Hamiltonians which have *only isolated* bifurcation values.

Definition. The integrable system v with the Hamiltonian H is called *Bottian* on the four-dimensional manifold U in M if:

(1) The function H is Bottian on U and the restriction of H to every regular level surface of the function f also is Bottian.
(2) We can choose the second integral f in the neighborhood of every regular level surface Q_h^3 of the function H in such a way that its restriction f_h to Q_h is the Bottian function, and the function f_h remains Bottian if the perturbation of h is small.
(3) The restriction of f to every saddle critical submanifold of the function H also is Bottian, and the regular level surfaces of the functions f and H intersect one to another *transversally* in the neighborhood of this critical manifold.

Thus, we consider the most strong Bottian conditions for the pair of functions f and H. It is an interesting problem (not yet solved)—how can we weaken these conditions with perserving the main results of the present theory? Remark: Condition (2) is concerned also with the values $h = c$ which are bifurcation values but noncritical ones.

The second integral f is choosen *nonuniquely* (only the fact of its *existence* is important). Our theory *does not depend* of the choice of concrete integral f.

Definition. The integrable Hamiltonian Bottian nonresonant system is called *strongly nondegenerate* if at all points of the domain U, at least one of the gradients, grad H or grad f is nonzero.

Real concrete physical integrable systems can, of course, have points x, where both gradients are zero. In these cases, we will throw out from M all level surfaces of the function H which contain such *bad points* x. As a result, we obtain the manifold with strongly nondegenerate system.

Thus, we consider *Hamiltonian, nonresonant, Bottian, strongly nondegenerate* differential equations on four-dimensional open domains U in symplectic manifolds M^4. We will speak (for brevity) simply about *strongly nondegenerate Hamiltonians H*.

3. The Classification of Integrable Hamiltonians on the Isoenergy 3-Manifolds is the Classification of Framed Molecules

According to [5–9], if we restrict the Hamiltonians H (of the type described) on the 3-manifolds Q_h, then they are classified with the help of so-called *framed molecules*. A *molecule* is a closed 2-surface P^2 which contains the graph K. This graph cuts P^2 into the union of flat rings (annulus). Each ring is marked by some rational parameter. Each connected component of the graph K is obtained from the finite collection of the circles by gluing some pairs of points (on these circles). Only two different arcs of the circles meet

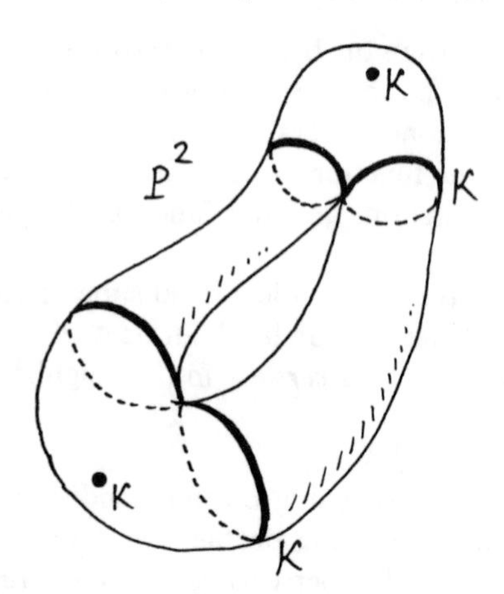

FIGURE 2

one another in each such point (the vertex of the graph K). In other words, in each such vertex precisely 4 edges of the graph meet one another. Besides these vertices, the graph K can contain the vertices of multiplicity 2 (the "stars") and isolated vertices (Fig. 2).

It turns out that the classification of the integrable systems on *four-dimensional* manifolds forced us to consider some remarkable "chemical reactions" between the molecules.

4. Classification of Critical Submanifolds and Normal Tubular Neighborhoods

Let L^k be a *connected* critical submanifold (of dimension k) of a strongly nondegenerate Hamiltonian H on U. The *separatrix diagram of the critical manifold* is the union of all segments of the integral trajectories of the vector field grad H, which go in L^k and go out from L^k.

Statement 1.

(1) *If $k = 3$, then L^3 coincides with the critical level surface Q_c of the Hamiltonian H, and H has a local minimum or local maximum on L^3. The normal tubular neighborhood of the manifold L^3 in U is diffeomorphic to the skew product $L^3 \times_\alpha D^1$ of the manifold L^3 with the segment D^1. The parameter α classifies these skew products and is determined by the subgroups G of the index 2 in the fundamental group $\pi_1(L^3)$. The boundary of the normal neighborhood is the two-sheeted covering over L^3, which can be written as $L^3 \times_\alpha S^0$ (where S^0 is a pair of points).*

(2) *If $k = 2$, then L^2 is a torus T^2 or a Klein bottle K^2.*
   (a) *If L^2 is a minimax manifold (for H), then L^2 coincides with the critical level surface Q_c of the function H. The normal neighborhood is diffeomorphic to the skew product $T^2 \times_\beta D^2$ or $K^2 \times_\eta D^2$, where β and η are the elements of corresponding cohomology groups and they classify the fibrations with the fiber disk D^2 over the bases T^2 and K^2, respectively. The boundary of the normal neighborhood is $T^2 \times_\beta S^4$ or $K^2 \times_\eta S^1$.*
   (b) *If L^2 is a saddle critical manifold (for H), then the normal neighborhood is diffeomorphic to the skew product $T^2 \times_\varepsilon D^2$ or $K^2 \times_\mu D^2$, where ε and μ classify the fibrations with the fiber D^1 over T^2 and K^2, respectively. The separatrix diagram is the skew product $T^2 \times_\varepsilon D^1$ or $K^2 \times_\mu D^1$. Its boundary is $T^2 \times_\varepsilon S^1$ or $K^2 \times_\mu S^1$.*

(3) *If $k = 1$, then L^1 is diffeomorphic to the circle S^1.*
   (a) *If S^1 is a minmax manifold, then S^1 coincides with the critical surface Q_n of the function H. The normal neighborhood is diffeomorphic to the skew product $S^1 \times_\delta D^3$ and the boundary of the tubular neighborhood is the skew product $S^1 \times_\delta S^2$. Here δ classifies the fibrations over S^1 with fiber D^3.*
   (b) *If S^1 is a saddle critical manifold of the index 1, then the tubular normal neighborhood is the skew product $S^1 \times_\nu D^3$ and its boundary is $S^1 \times_\nu S^2$.*

Here v classifies the fibrations over S^1 with fiber D^1. The separatrix diagram is a skew product $S^1 \times_v D^1$.

(c) *If S^1 is a saddle critical manifold of the index 2, then the normal tubular neighborhood is the skew product $S^1 \times_\xi D^3$ and the boundary of the neighborhood is $S^1 \times_\xi S^2$, where ξ classifies the fibrations over S^1 with fiber D^2. The separatrix diagram is $S^1 \times_\xi D^2$.*

5. The Formulas of the Reactions between the Molecules

Let W^* be some marked (framed) molecule, i.e., a pair (P, K) with numerical marks (parameters). According to [5–9], this molecule determines (in a unique way, up to diffeomorphism) the 3-manifold Q and a Liouville foliation on Q. The change of the energy value (i.e., the value of H) transforms the molecule. We can, generally speaking, transform the set of molecules in a different way. It is natural to call all such transformations *reactions*. Then we call the reactions, which really appear in Hamiltonian geometry, *chemical reactions*.

Let us assume that the connected compact 4-manifold Y is fibrated over the base L^{4-k} with fiber a k-dimensional disk D^k, where $k = 1$, 2, or 3, and L^{4-k} is some compact-oriented manifold (Fig. 3). The boundary of Y is a connected compact 3-manifold Q, which is fibrated over the base L^{4-k} with fiber the sphere S^{k-1}. The disk D^k is fibrated into the union of the spheres S_t^{k-1}, where t is the radius of the sphere. Let $0 \leq t \leq 1$, and when $t = 0$, then we obtain the point belonging to the submanifold L^{4-k}, which is embedded in Y as the zero cross section. Thus, Y is fibrated into the union of 3-manifolds $L^{4-k} \times_\varphi S_t^{k-1} = Q_t$, which are the skew products of L^{4-k} on the sphere S^{k-1}. If $t = 0$, we obtain L^{4-k}, which is called the *axis* of the manifold Y. Let, on Y, the symplectic structure ω be given and the boundary of Y (i.e., Q) is the

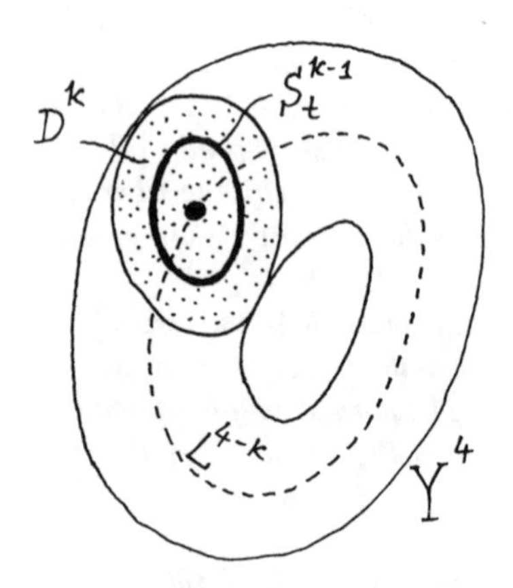

FIGURE 3

isoenergy 3-surface of some integrable Bottian system. This system is uniquely determined (up to topological equivalence) by some molecule W^*.

Definition. We will say that the molecule W^* on Q is *bordant to zero* if there exists some integrable strongly nondegenerate Hamiltonian H on Y such that (see Fig. 3):

(1) Its regular level surfaces coincide with the manifolds Q_t (for all positive t) and for $t = 0$ we obtain the critical level surface L, which coincides with the set of all critical points of H on Y, and H has a local minimum or maximum on L.
(2) The Hamiltonian system corresponding to H on each level Q_t is topologically equivalent to the initial system v on Q.
(3) We can choose the second integral f in such a way that grad $f \neq 0$ everywhere on Y.

The description of some interesting classes of the molecules, which are bordant to zero, recently was obtained by Bolsinov.

The Definition of the General Class L^3 of the Minmax Reactions of Index $\lambda = 0$ or $\lambda = 1$

Here L^3 is some critical submanifold for H on M^4.

The Minmax Reaction of Type $(L^3, \alpha = 0, W^*)$— Direct Product

The maximal reaction (*of index* 1): Two identical framed molecules W^* interact and disappear. Here W^* is an *arbitrary* molecule on the 3-manifold Q_h that is homeomorphic to L^3 [Fig. 4(1)].

The minimal reaction (*of index* 0): Two identical framed molecules W^* "are born from the vacuum." It is convenient to represent the reactions by the disk (or circle) with the notation of the reaction placed inside the disk. Each molecule that interacts in this reaction corresponds to some edge (segment). If the molecule enters in the reaction, then the edge goes in the disk. If the molecule is the result of the reaction, then the edge goes out of the disk [Fig. 4(1)]. We write the notation of the reaction near the end of the corresponding edge.

Remark. All other *minmax* reactions (listed below) contain only the reactions that are bordant to zero.

The Minmax Reaction of Type $(L^3, \alpha \neq 0, W^*)$— Two-Sheeted Covering

Here the parameter α classifies all nonequivalent coverings over L^3, which are different from the direct product.

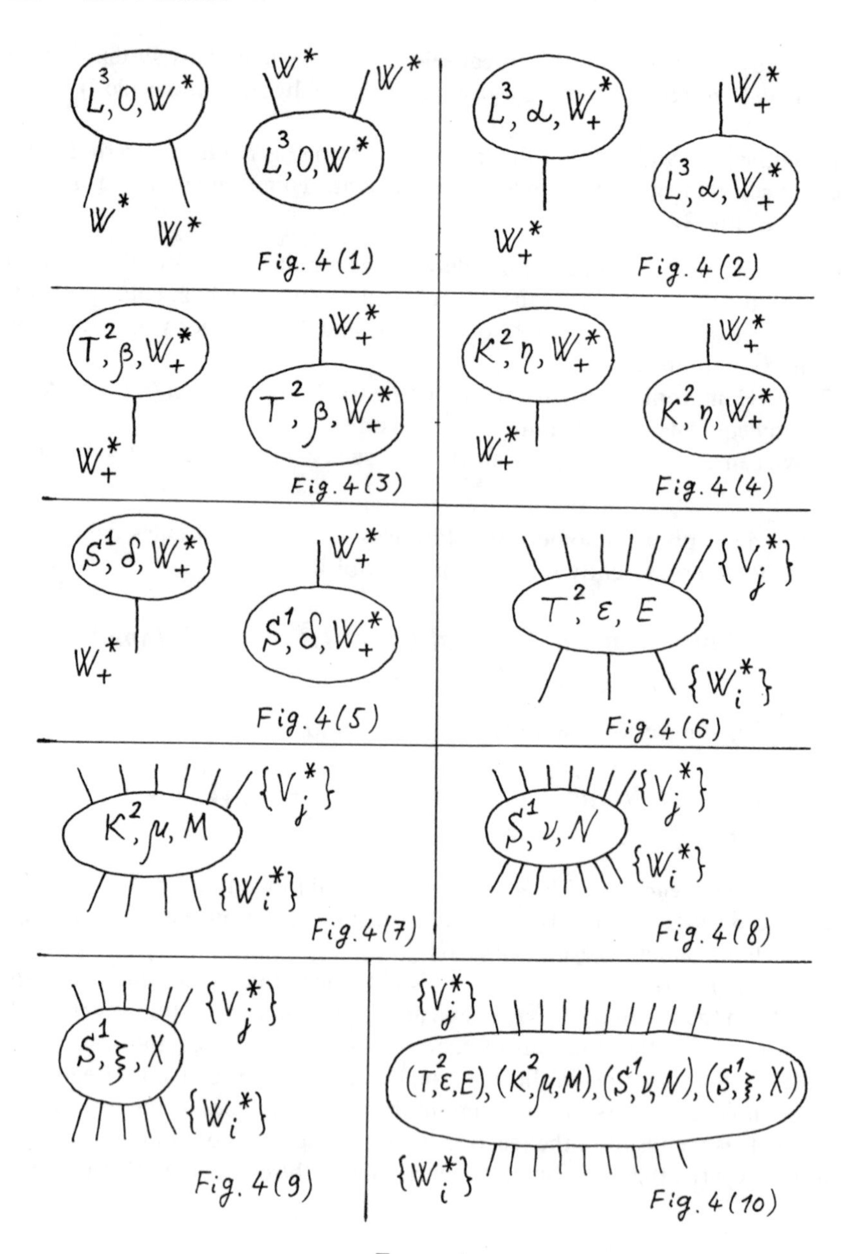

FIGURE 4

Let us consider a 3-manifold Q corresponding to the molecule W^* and let $\pi_1(Q)$ be its fundamental group. These groups were calculated in [10]. Let G be an arbitrary subgroup *of index two* in $\pi_1(Q)$ (the subgroup is considered up to conjugacy). We consider a 4-manifold $L^3 \times_\alpha D^1$ with the boundary $Q =$

$L^3 \times_\alpha S^0$. This manifold is foliated into the union of 3-manifolds, diffeomorphic to Q and contracting onto the exceptional fiber $Q/\mathbb{Z}_z$. Then we consider the molecule W^*, which is bordant to zero and corresponds to some system on Q. We will denote such molecules by W^*_+ and call them *admissible for a given reaction.*

The maximal reaction (*of index* 1): The admissible molecule W^*_+ is given on a 3-manifold Q, where $L = Q/\mathbb{Z}_2$, and is determined by the subgroup G in $\pi_1(Q)$. This molecule *disappears* (when the value of H grows) [Fig. 4(2)].

The minimal reaction (*of index* 0): The admissible molecule W^*_+ is given on the 3-manifold $Q = L^3 \times_\alpha S^0$ and "is born from the vacuum" [Fig. 4(2)].

The Definition of the General Class L^2 of Minmax Reactions of Index $\lambda = 0$ or $\lambda = 2$

Here L^2 is the critical manifold for H on M^4.

The Minmax Reaction of Type (T^2, β, W^*_+)

The parameter β classifies all nonequivalent fibrations over T^2 with fiber D^2. Here $Q = T^2 \times_\beta S^1$.

The maximal reaction (*of index* 2): The admissible molecule disappears [Fig. 4(3)].

The minimal reaction (*of index* 0): The admissible molecule W^*_+ "is born from the vacuum" [Fig. 4(3)].

The Minmax Reaction of Type (K^2, η, W^*_+)

The parameter η classifies all nonequivalent fibrations over the Klein bottle K^2 with the fiber D^2. Here $Q = K^2 \times_\eta S^1$.

The maximal reaction (*of index* 2): The admissible molecule W^*_+ disappears [Fig. 4(4)].

The minimal reaction (*of index* 0): the admissible molecule W^*_+ "is born from the vacuum" [Fig. 4(4)].

The Definition of the General Class L^1 of Minmax Reactions of Index $\lambda = 0$ or $\lambda = 3$. The Minmax Reaction of Type (S^1, δ, W^*_+)

The parameter δ classifies all nonequivalent fibrations over S^1 with fiber D^2. Here $Q = S^1 \times_\delta S^2$.

The maximal reaction (*of index* 3): The admissible molecule W^*_+ disappears [Fig. 4(5)].

The minimal reaction (*of index* 0): The admissible molecule W^*_+ "is born from the vacuum" [Fig. 4(5)].

The Definition of the General Class L^2 of Saddle Reactions of Index 1. The Saddle Reaction of Type (T^2, ε, E)

The parameter ε classifies all nonequivalent fibrations over T^2 with fiber D^2.

General Remark. At a saddle, reactions can take part with arbitrary molecules W^*, i.e., not necessary bordant to zero.

Let $\{W_i^*\} = \{W_1^*, \ldots, W_p^*\}$ be an arbitrary set of molecules. If $\varepsilon \neq 0$, then we draw an arbitrary smooth self-nonintersecting circle γ on some molecule W_i^*. If $\varepsilon = 0$, then we draw two nonintersecting and self-nonintersecting circles γ and $\bar{\gamma}$. They can be placed on the one molecule or on a different molecules.

The graph K determines the subdivision of the surface P into the union of rings (see the definition of K in [5–7] and Fig. 5). Each ring is foliated into the union of standard circles (Fig. 5). This foliation is defined uniquely up to isotopy. We call the connected arc of the circle γ *transversal* if the arc transversally intersects the fibers of the foliation in almost all its points and the arc is tangent to the foliation only in a finite number of its points (Fig. 5). The connected arc (segment) of the circle γ is called *fiber-like* if the arc coincides with some part of some fiber (of our foliation on P). The circle γ can be *totally transversal, totally fiber-like*, and *mixed transversal fiber-like*.

In the case of a strongly Bottian Hamiltonian H, the circle γ can be only of two types: totally transversal and totally fiber-like.

Let us cut the surface P along γ or along γ and $\bar{\gamma}$. Each circle generates *one coast* or *two coasts* of the cut. One coast appears in the case in which the circle γ is embedded in P as nonoriented circle. Two coasts appear in the case of oriented embedding of the circle. In the present case (which we consider in this section), we do not have nonoriented embeddings of γ. Consequently, we

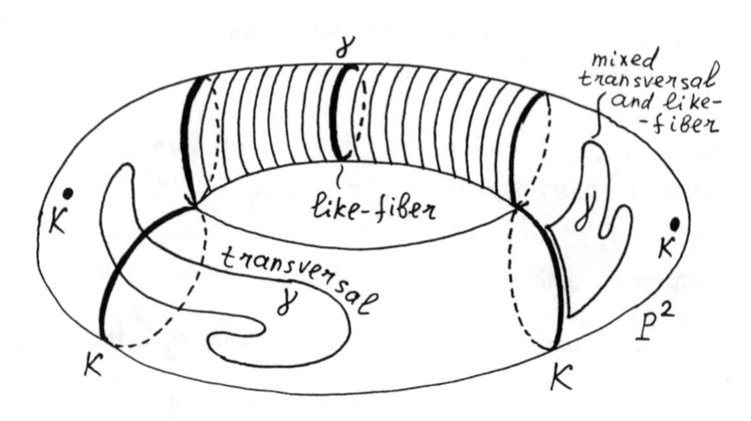

FIGURE 5

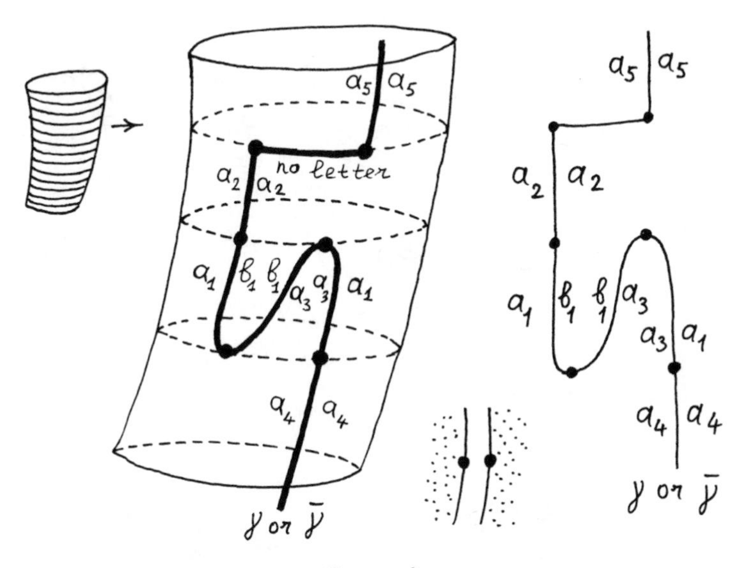

FIGURE 6

can assume that all embeddings of γ and $\bar{\gamma}$ are oriented. Let us consider, for instance, γ and let us suppose that it has the general type: mixed transversal fiber-like.

Let us mark all fiber-like segments on γ and all tangent points with the fibers of the foliation on P (Fig. 6). Each segment is duplicated after cutting and appears on each coast of the cut. Let us consider an arbitrary tangent point of the transversal segment. Each such point is duplicated after a cut (Fig. 6). This point belongs to some connected part of the fiber of the foliation on P. We call this fiber *tangent*. The marked part of the fiber meets some other coast of some cut. Let us mark all the points where the tangent fibers meet the coasts of cuts (Fig. 5). These points divide the transversal part of the cut into the union of some segments. Each such segment determines the set of "parallel" segments of the fibers of the initial foliation on P. All these segments intersect part of the coast. Now we can map the transversal segment of the coast along "parallel fibers" in some different transversal segment of the coast of some cut. We mark these two corresponding segments by identical letters. We choose different letters for different pairs of corresponding segments (Fig. 5). The fiber-like segments of the coast do not have letters.

Thus, we obtain letters from both sides of each circle on P. Each transversal segment of the circle is supplied precisely with two letters. These letters can coincide and can be different (Fig. 5).

We call this circle a *framed* circle. There are isotopies which do not change this *frame* (i.e., the sequence of the letters on γ). Of course, these isotopies are not arbitrary.

Let us demonstrate now the corresponding transformation of the mole-

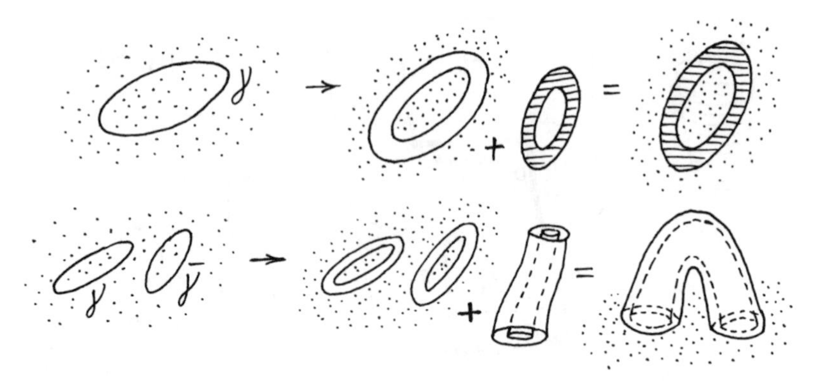

FIGURE 7

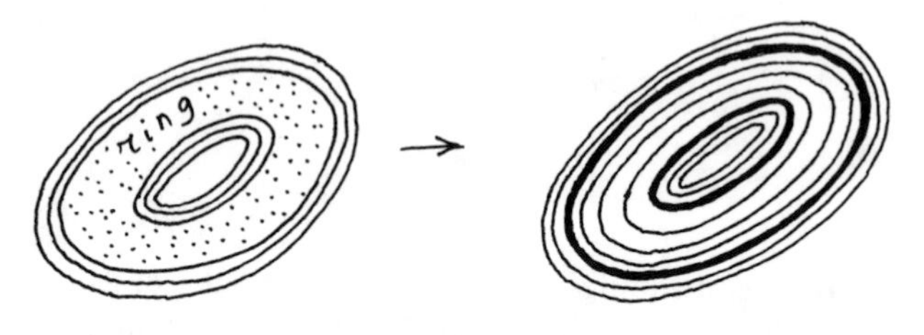

FIGURE 8

cules. The set of molecules $\{W_i^*\} = \{W_1^*, \ldots, W_p^*\}$ is transformed into a set of new molecules $\{V_j^*\} = \{V_1^*, \ldots, V_q^*\}$. In contrast to the case of minmax reactions, here the molecules do not disappear: They are transformed in some way and are "still alive" (but the total number of molecules can be changed).

If we have only *one* circle γ (in the reaction), then we correspond to it precisely *one* ring $S^1 \times D^1$. If the reaction is determined by *two* circles γ and $\bar{\gamma}$, we correspond with it *two* rings. Let us cut P along γ or γ and $\bar{\gamma}$ (Fig. 7). In the first case, we glue the ring to the two coasts of the cut. In the second case, we glue two rings to the coasts of the cuts γ and $\bar{\gamma}$. In the second case, two boundary components of one ring are glued to the coasts of *different* circles (Fig. 7).

This operation can be described in a following equivalent way. Let us consider the tubular neighborhoods of γ and $\bar{\gamma}$ and let us glue to these two rings (neighborhoods), the foot of the direct product $S^1 \times D^1 \times D^1$. Then we can transform P (Fig. 7) and finally glue two handles to P.

The molecule W^* determines a foliation on the surface P with cuts along γ and $\bar{\gamma}$. We add, after transformation of P, one or two rings to the surface P. We need to continue the initial foliation inside these rings. Note: *The new foliation must be nonsingular inside the new rings* (i.e., no singular points of the foliation in the rings). A simple example is shown in Fig. 8, where both boundary circles are fiber-like. In the general case, we need to consider the

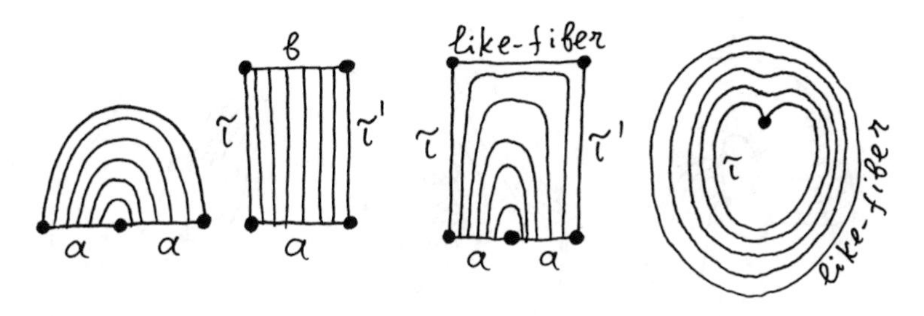

FIGURE 9

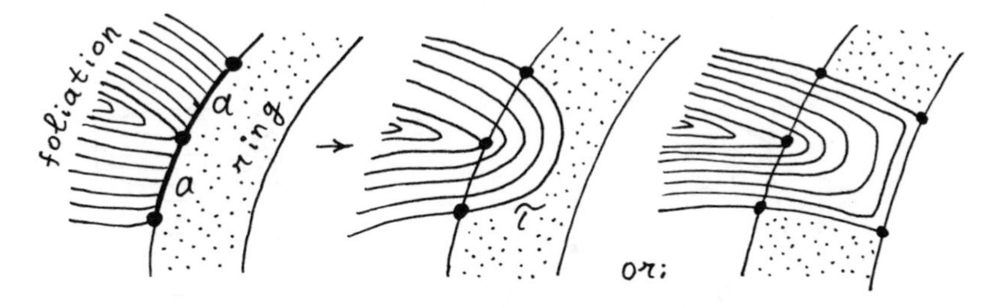

FIGURE 10

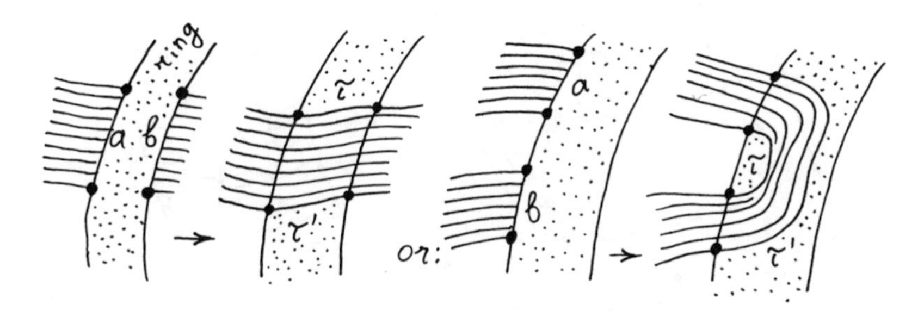

FIGURE 11

elementary bricks shown in Fig. 9. Now, if the coast of the cut contains two identical letters a, a, which are neighbors, then we continue the corresponding part of the foliation inside the ring as shown in Fig. 10. We realize this procedure for all pairs of neighbor-duplicates on the boundary of the ring. Thus, we need to consider other segment-letters, where the initial foliation is *transversal*. We divide the remaining part of the ring into the union of the curvilinear rectangles in such a way that:

(1) All their boundary segments are nondegenerate; two of them connect two transversal segment-letters (Fig. 11).
(2) The union of all rectangles is the whole ring.
(3) The rectangles either do not intersect or their intersection is some common boundary part.

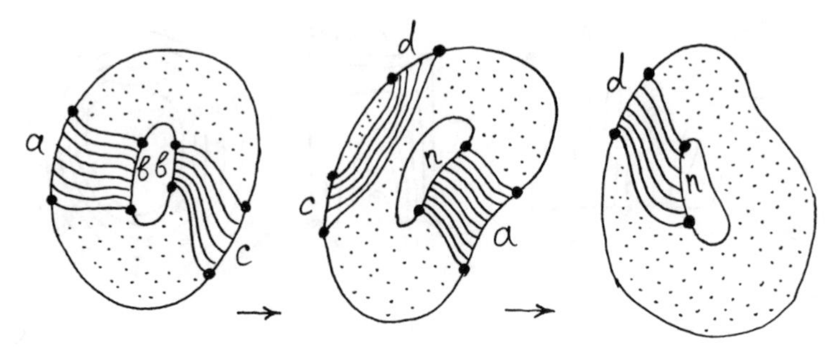

FIGURE 12

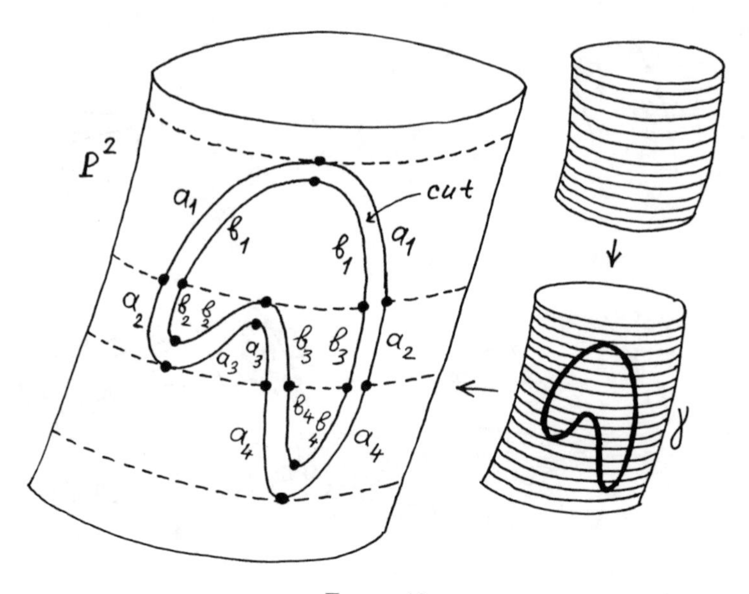

FIGURE 13

(4) If the system of rectangles is fixed, then it determines a mapping of the letters. Namely, the letter which is located on one side of the rectangle is mapped to the letter on the opposite side of the rectangle. We demand that after iterations of this mapping, the initial letter maps to its duplicate (each letter has a unique duplicate, see above). See Fig. 12.

(5) The final foliation (which we define on the new surface P' by the process described) has the following property: If we remove all singular fibers (i.e., the fibers containing some singular points of the foliation), then the surface P' transforms into the union of the rings foliated in a standard way.

In other words, the new foliation on P' determines the graph K', which is included in the definition of the molecule. An example of the transformation of the foliation is shown in Figs. 13 and 14. The transformation is uniquely

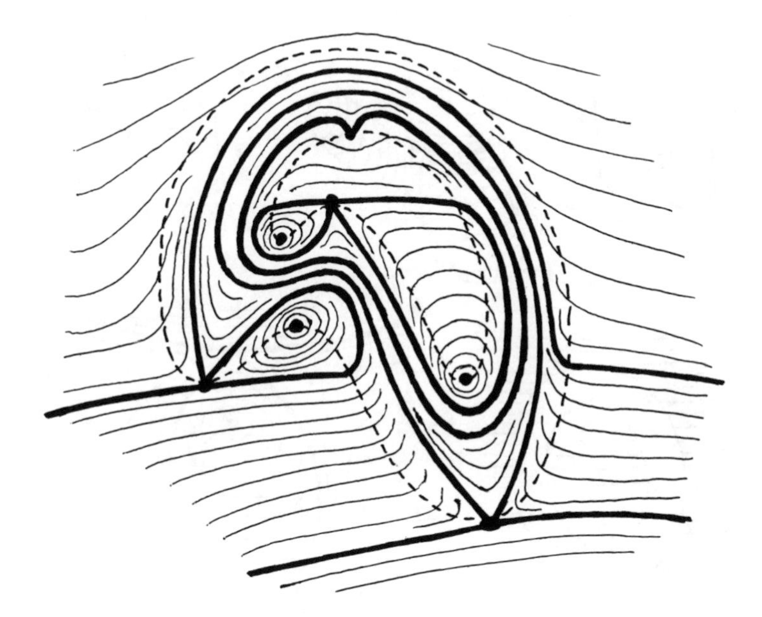

FIGURE 14

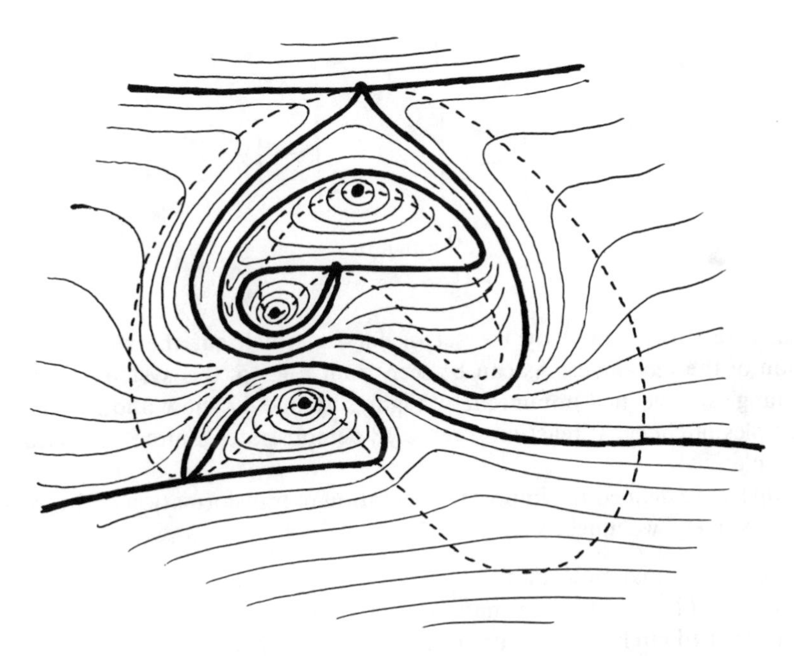

FIGURE 15

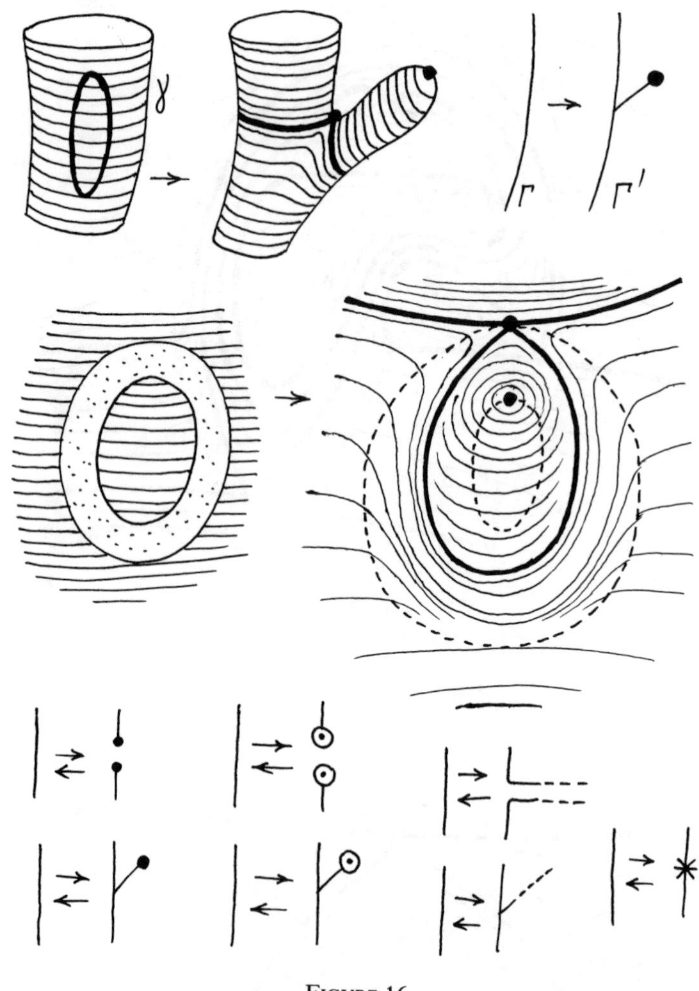

FIGURE 16

defined (up to isotopy) by the set of curves, which cut the ring into the union of the rectangles. As can be seen from Fig. 15, this system of curves (rectangles) is defined nonuniquely. Figures 16 and 17 show another simple examples of chemical reaction. Figure 18 demonstrates a more complicated example.

Finally, we defined the important operation of transformation of the molecules. Namely, we must fix:

(a) the set of initial molecules,
(b) the type (T^2, ε) of transformation,
(c) the framed circles on the surface P,

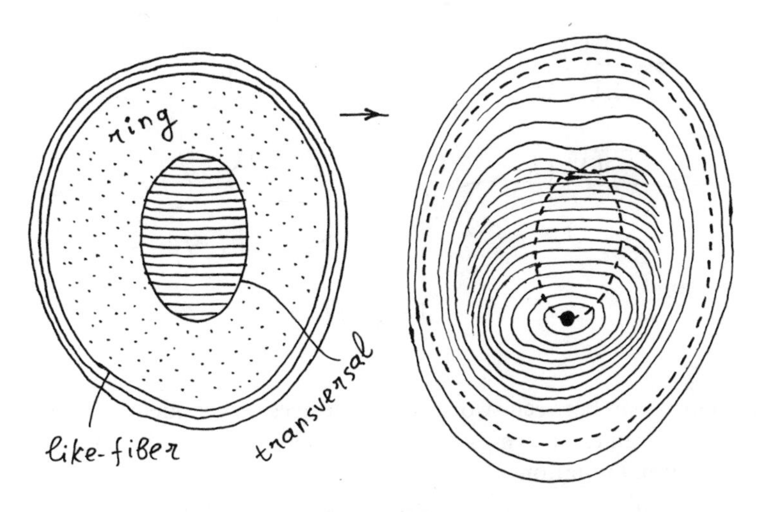

FIGURE 17

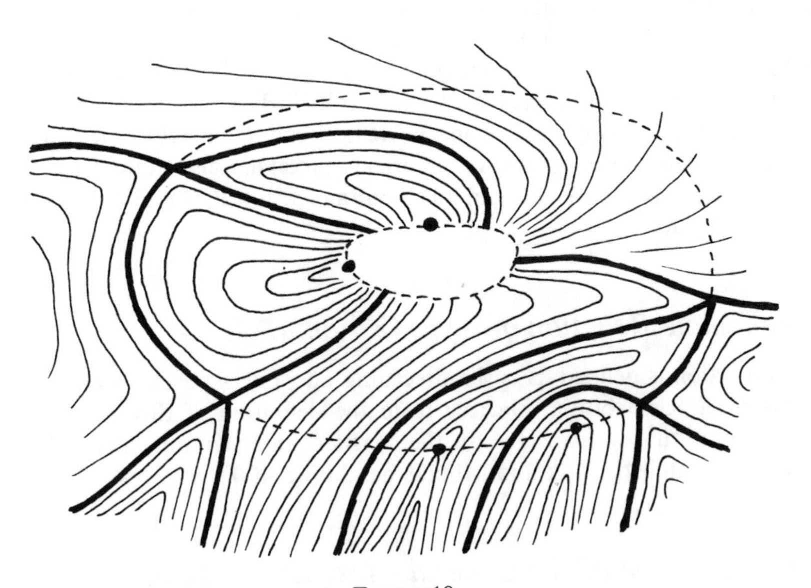

FIGURE 18

(d) the subdivision of the ring (or the rings) into the union of the rectangles and degenerate rectangles (Fig. 10) with properties (1)–(5).
(e) the numerical marks (this operation is presented below).

Let us denote this transformation of the molecules by E. Finally we can write the total transformation (described above) as (T^2, ε, E). We call this object a *saddle reaction*. Two reactions are called *equivalent* if there exists a

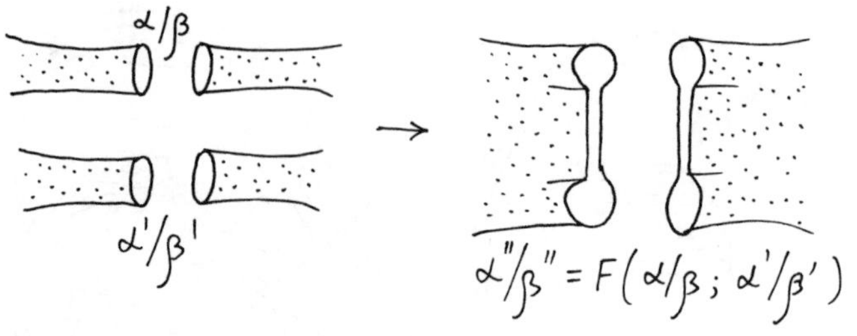

FIGURE 19

homeomorphism between two sets of molecules, which identifies the framed circles, their frames, the rings, and their subdivision into the sum of rectangles. For formal notation, see Fig. 4(6).

The set of all reactions of the type (T^2, ε, E) is *discrete.*

How can one calculate the new molecules V_j^*? Let us consider the surface P' obtained from the surface P by the procedure described above. We constructed some foliation on P' (see above). Let us consider all fibers of this foliation, which contain singular points (we call such fibers *exceptional*). Exceptional fibers (some curves) form the graph K'. The pair (P', K') is the new molecule V. It is easily to see from the definition of the reaction that the new numerical rational parameters (numerical marks) are algorithmically calculated on the base of the old numerical parameters (on P) (Fig. 19). Details are presented below.

The Saddle Reaction of Type (K^2, μ, M)

This reaction is defined as is the preceding one, but we need to replace the torus by a Klein bottle. Starting from the definition of this reaction, we should note that we can glue (to the coasts of the cuts) not only a ring or two rings, but also a disk and Möbius band [Fig. 4(7)].

The Definition of the General Class L^1 of Saddle Reactions of Index $\lambda = 1$ or $\lambda = 2$. The Saddle Reaction of Type (S^1, ν, N) of Index 1

The definition is the same as in the case (T^2, ε, E), but we need to replace the torus by the circle (Fig. 4(8)].

The Saddle Reaction of Type (S^1, ξ, X) of Index 2

The definition copies the case (T^2, ε, E) [Fig. 4(9)].

The General Type of Mixed Saddle Reaction

The general saddle reaction of the molecules is the composition of an arbitrary finite number of elementary saddle reactions described above. The general saddle reaction transforms the set of molecules $\{W_i^*\}$ in the set of new molecules $\{V_j^*\}$ [Fig. 4(10)].

6. Main Theorem

Let us consider an arbitrary collection of minmax and saddle reactions. Each reaction has some "ends," i.e., the ends of the edges which go into the circle (representing the reaction) or go out. These ends are "free" and each of them corresponds to some molecule W^*. Suppose that we have two edges with isomorphic ends W^*. Then we can glue these ends and, consequently, edges (Fig. 20). Assume that after all such identifications we obtain a *connected object* = some graph. Some of its edges can be free (i.e., with free ends). The vertices of this graph are minmax and saddle reactions. Let us *reduce* the graph, namely, we annihilate all reactions of type (L^3, O, W^*) (Fig. 21). We erase the corresponding circle and glue two edges. We denote the new reduced graph R. Let us call this object a *connected reaction of molecules.* Two

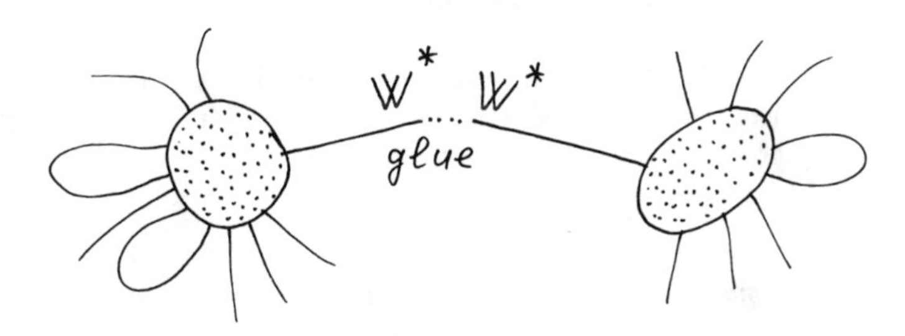

FIGURE 20

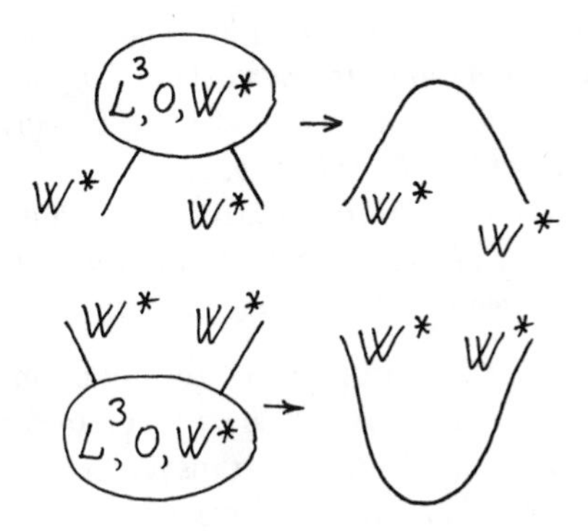

FIGURE 21

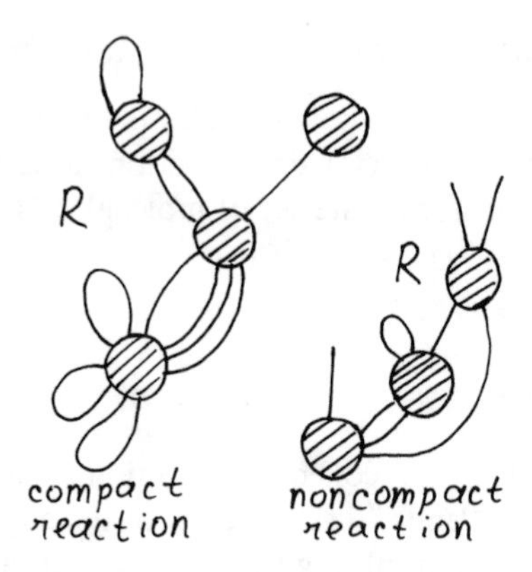

FIGURE 22

such reactions we consider as isomorphic (identical) if there exists the mapping of the first graph onto the second one, which identifies the elementary reactions. Of course, we demand that this mapping identifies the corresponding foliations (up to isotopy).

Definition. We call the reaction *chemical* if this reaction corresponds to some integrable strongly nondegenerate Hamiltonian system.

Definition. We call the connected reaction *compact* if the reaction does not have free ends (free edges), and *noncompact* in the opposite case (Fig. 22).

Theorem 1 (Main Theorem).

(1) *Each integrable strongly nondegenerate Hamiltonian H on a connected symplectic 4-manifold U determines in a unique way (up to isomorphism) some connected chemical reaction. A Hamiltonian H on a compact (resp. noncompact) 4-manifold determines a compact (resp. noncompact) chemical reaction.*
(2) *Two integrable Hamiltonians are roughly topologically equivalent if and only if their chemical reactions are isomorphic. So, the correspondence* Hamiltonian $H \to$ chemical reaction *allows us to classify the set of all integrable Hamiltonians (of the described type).*
(3) *The set of all integrable Hamiltonians (considered up to rough topological equivalence) is discrete. The reaction R is a discrete topological invariant of the corresponding system of differential equations.*

It follows from the Bottian property of the functions H and f that the circles γ (which determine the surgery, see above) are only of the following types: (a) totally fiber-like and (b) totally transversal.

There are some restrictions on the circle γ: Not every γ can appear in real chemical reactions. We should mention, for example, the following necessary condition (Nguen Tyen Zung and A.V. Bolsinov): γ is always contained inside some "family" on the molecule W^*. For the definition of the *family*, see [8].

Definition. Two integrable Hamiltonians H on U and H' on U' are called *two-sheeted equivalent* if the 4-manifold U' is the two-sheeted covering over the manifold U, and the corresponding projection $\pi\colon U' \to U$ has the following properties: $\omega' = \pi^*\omega$, where ω and ω' are the symplectic structures on U and U' (resp.) and $H' = \pi^*H$.

Theorem 2. *The classification of all integrable strongly nondegenerate Hamiltonians, which are considered up to rough topological equivalence and two-sheeted equivalence, is given by the chemical reactions not containing the elementary minmax reactions of type* (L^3, α, W^*).

All integrable Hamiltonians can be collected into a discrete table. Important problem: Describe its *physical zone* (*region*), i.e., the part which is filled by the Hamiltonians, *discovered in mathematical physics and mechanics.*

7. The Sketch of the Proof

Step 1. Let us consider all *critical* level surfaces given by the equation $H = c = \text{const}$ and let $\{L\}$ be the set of all *connected* critical manifolds of the function H on the level c.

Step 2. The complete classification of all such manifolds L and their normal tubular neighborhoods; see Statement 1.

Step 3. If L^3 is a critical manifold, then it is necessarily *minmax* (for H). Consequently, its normal tubular neighborhood is homeomorphic to either a direct product or a skew product. In the first case, the molecule W^* can be arbitrary. In the second case, W^* should be *bordant to zero.*

Step 4. If the minmax L is two-dimensional or one-dimensional, then we obtain the reaction listed above. Here the molecules W^* are bordant to zero.

Step 5. The case of saddle L^2 and L^1 is most complicated. The separatrix diagram of the manifold L consists of two parts: ingoing and outgoing. Let us denote by X the ingoing part of the diagram, and by ∂X its boundary (Fig. 23). If λ is the index of L, then $\dim X = k + \lambda$. The manifold ∂X is called *the foot of the separatrix diagram* (on the level $h = c - \varepsilon$). Let B be the tubular neighborhood of the foot ∂X in Q and ∂B its boundary (also in Q) (Fig. 23).

Lemma 1. *The Euler characteristic of* ∂B *is equal to zero, i.e., each connected component of* ∂B *is homeomorphic to either a torus* T^2 *or Klein bottle* K^2.

Step 6. According to Statement 1, we should begin the analysis with the case (T^2, ε, E). The manifold X is $X^3 = T^2 \times_\varepsilon D^1$. Then, the boundary of X is

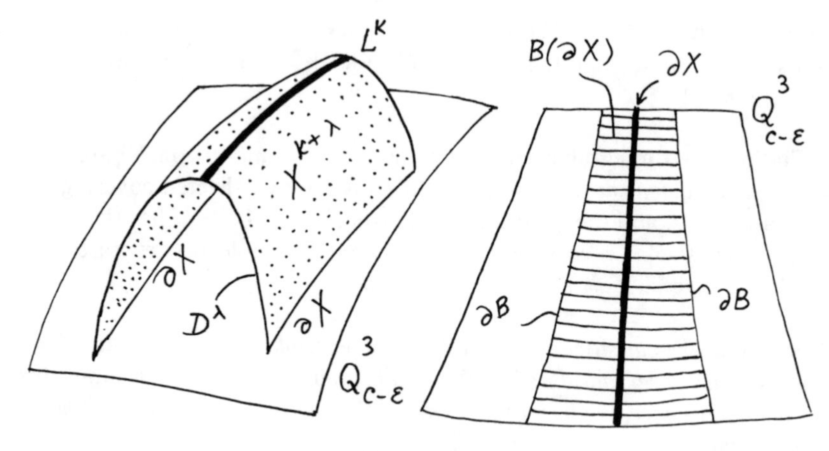

FIGURE 23

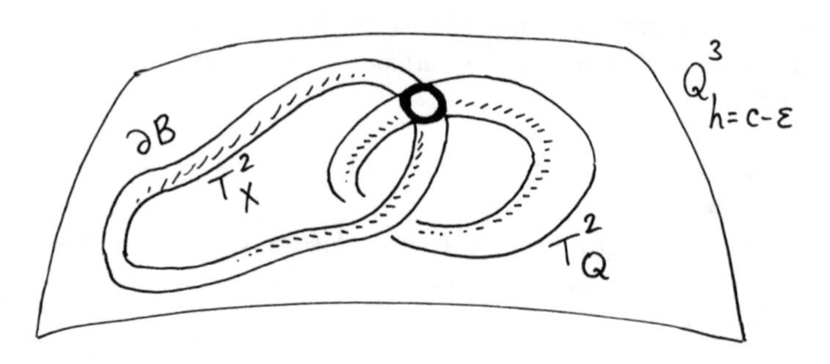

FIGURE 24

$\partial X = T^2 \times_\varepsilon S^0$. If $\varepsilon = 0$, then $\partial X = 2T^2$, the nonconnected union of two tori. If $\varepsilon \neq 0$, then the manifold ∂X is *connected* and homeomorphic to T^2, which covers (in two-sheeted way) the torus $L = T^2$.

Step 7. Let T_x^2 be the torus from the foot of separatrix diagram and T_Q^2 the arbitrary Liouville torus inside the 3-manifold Q_h (Fig. 24). Each connected component of ∂B is either a torus or a Klein bottle.

Lemma 2. *In the case of the strongly nondegenerate Bottian Hamiltonian H (in general position), we have, that the connected component of ∂B and the torus ∂X do not intersect in Q, or coincide, or their intersection consists of a finite number of smooth self-nonintersected (and nonintersected) circles. In the last case, each of the circles realize a nontrivial cycle in ∂B (or in its two-sheeted covering = the torus) and in T_Q^2. Consequently, all these circles are homologous one to another inside each of these 2-surfaces. They cut the Liouville torus T_Q^2 and ∂B in a union of open rings.*

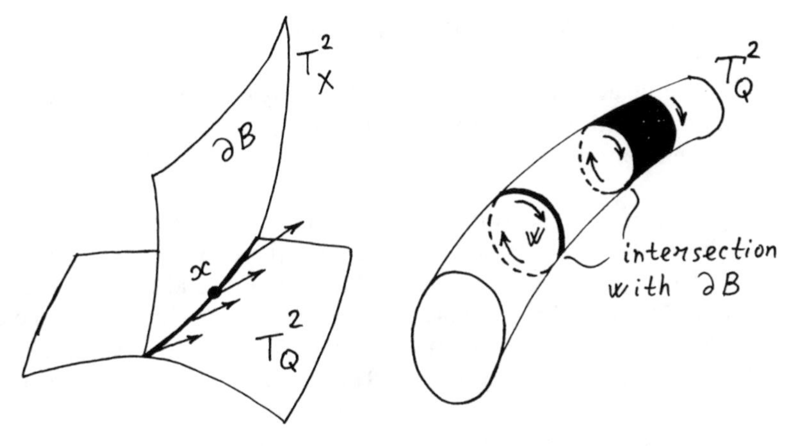

FIGURE 25

If we consider the weaker condition of mutual Bottian property for the functions H and f, then ∂B and T_Q^2 can intersect along the *rings* (Fig. 25), and each of them also realizes a nontrivial cycle. This effect corresponds to the appearance of fiber-like segments on the circle γ.

PROOF. Let us suppose for simplicity, that ∂B is the torus T_x^2 (the case of a Klein bottle is treated by the same method). Let $x \in T_x^2 \cap T_Q^2$ and $w =$ sgrad f, the vector field. This flow (field) preserves the manifold Q; consequently, w is tangent to Q and T_Q^2. We can assume that T_x^2 is close to $L^2 = T^2$ (Fig. 25). The field w is nontrivial on L^2 ($w \neq 0$); consequently, $w \neq 0$ at all points of X which are close to L. The manifold X is not invariant (generally speaking) with respect to w, but X is "almost invariant." Thus, w is "almost tangent" to $T_x^2 \cap T_Q^2$. This intersection is contained in T_Q^2 and, consequently, is "almost fibrated" into the union of integral trajectories of this field. According to the Main Lemma from [5], we can deduce that the intersection $T_x^2 \cap T_Q^2$ realizes a nontrivial cycle in the Liouville torus. On the other hand, this intersection is contained in ∂B and realizes a cycle in ∂B because the intersection is an "almost integral" submanifold of an "almost tangent" vector field $w \neq 0$.

Let us consider a cycle which is transverse to the intersection cycle inside the Liouville torus (intersected with ∂B). We can construct the factor-space of the Liouville torus with respect to this transversal cycle. Then, the torus T_x^2 transforms in the circle γ on P (!). This circle is determined uniquely (up to isotopy preserving the *frame*, i.e., the sequence of the letters along the circle; see above). Remark: Formally, in some Riemannian metric we need to consider grad and the separatrix diagram X; but really we can consider the limit when $\varepsilon \to 0$ and, consequently, the segments of separatrices also tend to points (on L).

Step 8. The cases (K^2, μ, M), (S^1, ν, N), and (S^1, ξ, X) are considered in a similar way.

Step 9. We investigated the case of precisely *one* connected critical submanifold on the critical level surface of H. In the general case, we will obtain the transformations of molecules, which are the compositions of elementary reactions.

Let us consider γ on P and the corresponding torus ∂B inside of $Q_{c-\varepsilon}$. We have a nontrivial cycle on this torus, namely, the intersection of ∂B with the Liouville torus. Let us consider all Liouville tori and all critical two-dimensional fibers of the Liouville foliation, which intersect (in Q) the 2-surface ∂B. Each critical fiber determines some critical circle, a periodic solution. Because all intersection cycles on ∂B are homologous to one another, all these critical periodic solutions are also homologous one to another inside Q (this is a remark of Nguen Tyen Zung).

To summarize, if some integrable strongly nondegenerate Hamiltonian H is given, then we can construct (calculate) its topological invariant R. The end of the proof follows the scheme of [6] and [8].

We will describe in a subsequent publication the values of the invariant R (not every reaction is *chemical reaction* !).

8. The Marked (Framed) Chemical Reaction as a Topological Invariant which Classifies Integrable Hamiltonians up to Topological Equivalence

Following the theory of the three-dimensional classification of integrable Hamiltonians, we can also construct a more delicate topological invariant R^*, which we call a *marked (framed) chemical reaction.* We will supply the graph R by some *marks (frame).* But now these marks are non-numerical and have a more complicated "noncommutative type." Let us consider the edge of the graph K connecting, for instance, two saddle reactions. This edge appears after gluing two copies of the molecule W^*. Let us consider two copies of the corresponding surface P^2. Each of them contains the graph K. Consequently, this graph determines in unique way (up to isotopy) a set of g circles $\alpha_1, \ldots, \alpha_g$, where g is the genus of the surface P^2. The circles $\alpha_1, \ldots, \alpha_g$ are nontrivial and nonhomologous cycles on the surface. We can choose these cycles as the middle circles on the rings, where the union of the rings is the surface $P \setminus K$ (when we cut P along K). We can add (to these circles) some other circles $b_1, \ldots, b_g$ conjugate to the circles $\alpha_1, \ldots, \alpha_g$. The whole set $\{\alpha_i, b_i\}$ form the basis of the cycles on P. But the second part of the basis (i.e., circles b_i) is choosen nonuniquely. Let us fix some variant of this choice. The same procedure can be realized on the second copy of P. As a result, we obtain the basis on each copy of P. The edge of the graph R defines some gluing of two copies of the 3-manifold Q in U. Consequently, we obtain some diffeomorphism identifying two copies of P (with a marked basis on each copy). Because the foliations on these copies of P should be identified, we obtain some element λ of the group $\operatorname{Aut} W^*$ where $\operatorname{Aut} W^*$ is the group of all automorphisms of the molecule W^*. This automorphism λ depends on the choice of the "second part" of the basis on P. Consequently, we can correctly define only the class $\lambda \bmod Z$, where Z is the subgroup of the group $\operatorname{Aut} W^*$ generated by the transformations, which change the second part of the basis.

We need to put this parameter λ in the center of the corresponding edge of the graph R. This object is similar to the rational parameter r_i (see [8]).

According to the idea of [8], we can also define the noncommutative version of the "family" and of the parameter n_k (or μ_k). The details will be published in a subsequent publication.

Finally, we obtain the *marked (framed) chemical reaction*. This new topological invariant classifies all integrable Hamiltonians up to *topological equivalence*.

References

1. Smale S. Topology and mechanics: I, *Invent. Math.* **10** (1970).
2. Abraham R. and Marsden J.E., *Foundations of Mechanics*, 2nd ed., Benjamin/Cummings, New York, 1978.
3. Lerman L.M. and Umansky Ya.L., Integrable Hamiltonian systems and Poisson action, *Methods of Qualitative Theory of Differential Equations*, Gorjky University Press, Gorjky, 1984, pp. 126–139.
4. Marsden J., Ratiu T., Weinstein A., Semidirect products and reduction in mechanics, *Trans. Amer. Math. Soc.* **281** (1984), 147–178.
5. Fomenko A.T., The topology of surfaces of constant energy in integrable Hamiltonian systems, and obstructions to integrability, *Math. USSR Izvestiya* **29** (1987), 629–658 (English).
6. Fomenko A.T., Topological invariants of Hamiltonian systems integrable in Liouville sense, *Funct. Anal. Appl.* **22** (4) (1988), 38–51. (See the corresponding English translation.)
7. Fomenko A.T., Symplectic topology of completely integrable Hamiltonian systems, *Uspechi Matem. Nauk.* **44** (1) (1989), 145–173. (See the corresponding English translation.)
8. Fomenko A.T. and Zieschang H., Criterion of topological equivalence of integrable Hamiltonian systems with two degree of freedom, *Izvestiya Akad. Nauk. SSSR* (1990). (See the corresponding English translation.) [See also the English paper: *Institut Hautes Etudes Sci.* **35** (1988), 1–45.]
9. Bolsinov A.V., Fomenko A.T., Matveev S.V., Topological classification and arrangement with respect to complexity of integrable Hamiltonian systems of differential equations with two degrees of freedom. The list of all integrable systems of low complexity. *Uspechi Matem. Nauk* (1990).
10. Fomenko A.T. and Zieschang H., On typical topological properties of integrable Hamiltonian systems, *USSR Izv.* **31** (2) (1989), 385–412 (English).

11. Matveev S.V. and Fomenko A.T., Constant energy surfaces of Hamiltonian systems, enumeration of three-dimensional manifolds in increasing order of complexity, and computation of volumes of closed hyperbolic manifolds, *Russian Math. Surv.* **43** (1) (1988), 3–24.
12. Fomenko A.T., Topological classification of all integrable Hamiltonian differential equations of general type with two degrees of freedom. The Geometry of Hamiltonian Systems, T. Ratiu, ed., Math. Sci. Res. Inst. Pub., v. 22, Springer, N.Y., 1989.

Part 9
Final Panel

52
Final Panel

R.F. WILLIAMS, D. KARP, D. BROWN, O. LANFORD, P. HOLMES, R. THOM, E.C. ZEEMAN, M.M. PEIXOTO, AND AUDIENCE MEMBERS

R.F. (Bob) Williams: Each of our experts will speak for ten minutes, giving their overview of their respective fields. One of the things that might be very good to get is a view of the future. It's been pointed out that maybe that should come more from the younger people here than us gray-beards. So, going in the order suggested by the program, I will start with Karp, who will speak about computations.

Dick Karp: First of all, I haven't had a chance to wish you happy birthday, Steve, so happy birthday!

The theory of computational complexity has traditionally been based on discrete, rather than continuous, mathematics. One of Steve's major achievements has been to reorient this topic so that it becomes more interlinked with fields such as topology, analysis, and geometry. And, in so doing, Steve has cast the subject in a form more compatible with the way people who actually compute think about numerical analysis and scientific computation. I'd like to trace some of Steve's contributions and indicate their impact and meaning for the field of computational complexity.

Steve's investigation of polynomial zero finding was perhaps his first venture into computational complexity. This topic was well covered by Mike Shub's talk and I won't say anything further about it.

My principal overlap with Steve has been in the the area of linear programming. One of the most fundamental and useful algorithms in existence is the simplex method of George Dantzig. The simplex method solves the problem of maximizing a linear function of many variables subject to linear inequality constraints. This is a fundamental model in many areas of operations research and economics. For reasons that are somewhat mysterious, the simplex method is wonderfully successful in practice. One can construct pathological examples on which the simplex method performs extremely poorly, and yet in practice it can be depended on to perform very rapidly. For many years it was a puzzle to explain the remarkable success of the simplex algorithm.

One natural approach is to put a probability distribution on the space of instances of the linear programming problem, and then show that in some expected sense the simplex method was good. Many people were aware of this possibility, but the two who first made significant progress were Steve and Karlheinz Borgwardt in Germany, using rather different models. Steve, with his penchant for homotopy approaches, chose a particular parametric version of the simplex algorithm and was able to prove that, under a certain type of probability distribution, it ran in polynomial expected time. Later, a number of other investigators proved stronger results under weaker hypotheses with somewhat simpler proofs, but it was Steve who first opened the door through his original analysis.

It's not very clear, of course, whether these probabilistic models really explain the success of the simplex method because the probability distributions that are assumed may have no relation to the examples that actually come up in practice. But, nevertheless, these kinds of analyses have been a landmark in the study of linear programming.

Later, Steve introduced a very interesting new concept into computation complexity theory: the topological complexity of a function. Roughly speaking, topological complexity corresponds to the number of conditional tests or branches that a program must use to evaluate a given function. Steve was able to relate this quantity to fundamental ideas in topology. He showed that topological complexity is related to the following question: Given a continuous function, into how many open sets must the domain be partitioned in order that, within each part, the function has a continuous inverse? This notion of topological complexity provides an important link between topology and the theory of computation.

Of all Steve's work in the theory of computation, his recent paper with Lenore Blum and Mike Shub has the broadest implications. The paper develops a theory of computation over the reals or, more generally, over an ordered ring. The basic point of departure of the paper, which distinguishes it from standard theories of computation, is that a real number is considered as a monolithic entity, rather than an infinite string of binary or decimal digits. Thus, an arithmetic operation such as addition or multiplication of two real numbers is considered to be a primitive step. This is not exactly what happens in a digital computer, but it corresponds pretty well to the way applied mathematicians think about computation. This viewpoint leads to a rich theory which is very different from more conventional theories of computation. For example, within this theory any discrete combinatorial problem is trivially solvable since one can encode the solutions to its countably many instances into a single real number which can be thought of as simply a constant input to the problem. To solve any instance, all you have to do is extract the appropriate bits of this constant.

So the theory is completely different from standard discrete models, and yet within it one can define natural analogues of the recursive and recursively enumerable sets, the class P of problems solvable in polynomial time, the class NP of problems for which solutions are checkable in polynomial time, the class of NP-complete problems, and other basic constructs from computability theory and complexity theory.

The theory has very interesting links with dynamical systems. It turns out, for example, that for most rational maps the basin of attraction constitutes a recursively enumerable set which is not recursive. This is a kind of connection that could not possibly arise within the standard discrete theories of computation.

Backing off from Steve's specific contributions, I'd like to mention that there are further untapped connections between the theory of dynamical systems and some important issues in theoretical computer science. For example, there is a lot of interest in the study of pseudo-random number generators—the kind that are used to generate "random" numbers for Monte Carlo experiments. A pseudo-random number generator is nothing more than a discrete dynamical system with the property that, without knowing its initial state, no polynomial-time algorithm can determine its future outputs from its past outputs. Viewed this way, pseudo-random number generators are very similar to chaotic dynamical systems, with their unpredictability due to sensitive dependence on initial conditions. This connection really cries out to be investigated in a more formal way. Also, more broadly, any computer system or computer network—the AT&T network, for example—can be regarded as a dynamical system, and it's tempting to wonder whether some of the instabilities that are observed are related to chaos. Bemardo Huberman at Xerox, Palo Alto, has conducted some investigations along these lines.

I'd like to take the liberty of concluding in a personal vein. To me, the most striking aspects of Steve's character are his courage and his confidence. His courage has been manifested in his political work, as we all know. In mathematics, Steve has always had the courage to plunge into new areas, pick up whatever background he needed, and make an attack on the central problems. In the rest of his life Steve shows the same courage. Few of us, I think, would spend a summer sailing to the most remote island group in the world, just to see if we could do it without getting shipwrecked. This is characteristic of how Steve lives life to the hilt. With respect to confidence, I've learned a lot from Steve. Many of us, as we get up into our 50s and 60s, begin to back away from research a little bit and get interested in serving on departmental committees and doing all those important things that need to be done. It's good that people do that, but it's also a retreat from facing the challenges and frustrations of research. Steve's confidence has never flagged. He remains committed to his research and we're very glad that he is.

So once again, Steve, congratulations. It's been a pleasure to be associated with you.

Bob Williams: Our next speaker is Don Brown, who will speak about economics and its relation with Steve Smale's work.

Don Brown: Well, it's hard to know what to say after the three talks we heard last Sunday. Debreu, Geanakoplos, and MasCollel talked about Steve's impact not only on my own work, but on the economics profession. I think I'm going to take a suggestion from Howard Oliver and try to talk about problems which we haven't solved, and how some of the work that Steve has done might help us resolve them. In order to do that, I want to put Steve's contribution in a different perspective than that given on Sunday and look at where mathematical economics was prior to Steve's interest in the field in the 1970s. The best reference is Debreu's *Theory of Value*, where the central mathematical framework is the theory of dual locally convex linear topological vector spaces. One vector space corresponds to the space of commodities and the other space corresponds to the space of prices. In this model, agents' characteristics are described in terms of convex subsets of the commodity space and concave functions on the relevant sets in the commodity space. The whole analysis, that is, the proof of the existence of equilibria and the proof of the welfare theorems, proceeds within this structure. The primary two theorems which are used for this analysis are the Separating Hyperplane Theorem and Brouwer's or Kakutani's Fixed-Point Theorem.

Now we look at Steve's contribution. From my perspective, and I think the perspective of many people who've studied Steve's work, his primary contribution is not in giving new proofs of existence or the welfare theorems, using different methods of analysis, but his conception of a new framework in which we ought to do general equilibrium analysis. What Steve did was to suggest that linear vector spaces was not the only appropriate framework, but that one should consider economic models as being modeled by manifolds. This conceptual breakthrough, in fact, led to the class of economic models introduced in John's talk, where the Grassmanian, an abstract manifold, is an essential ingredient of the model. What are Steve's methods of analysis? Well, he uses only two results from global analysis, as he says in his article in Volume I of the *Handbook of Mathematical Economics*. One is the Implicit Function Theorem, the other is Sard's Theorem.

If you look at the current policy and the research agendas in economics, say over the last five to six years, you see a tremendous number of commodities. Such models arise naturally when one thinks about the act of production or the act of consumption being contingent on either the state of the world, or the time of production, or the time of consumption. Let me give you a specific example. A question that economists are actively studying now, of great importance, is why it is that countries have developed at different rates. What is it that is similar or dissimilar about the Taiwan experience as opposed to the Korean experience?

Well, we have models of economic growth and these models are most naturally formulated as economies where the space of commodities is an infinite diminishing function space. One wants to know: If you write down a model with an infinite number of commodities (and say you posit a finite number of households and firms), is this general equilibrium model consistent? That is, are the basic propositions of the *Theory of Value* true in this world? Consistent in the sense that there exist prices that clear markets and the resulting allocation is Pareto optimal? The first proof of consistency in this sense was given here at Berkeley in 1972 by Truman Bewly, who has Ph.D.s in both mathematics and economics from Berkeley. In Bewley's model, the commodity space is L^∞ and the price space is L^1. They're put in duality and the topology on the commodity space is the Mackey topology.

But if you go back to the finite-dimensional case, you get more from the global analytic or smooth approach than you get from the linear vector space approach. Not only do you get existence and optimality, but you also have a way of counting the number of equilibria. The question that needs to be solved, which we have no answer to, is how do we count equilibria when we have an infinite-dimensional commodity space? In particular, the following problem is open. In a regular economy, you are able to replicate Debreu's results and Smale's results, that for almost all economies there is a finite number of equilibria.

The only result in the literature is for the special case of additively separable preferences and the commodity space is ℓ^∞. This is a very unsatisfactory state of affairs. It's a bit surprising that we've been unsuccessful—probably in part because Steve is now working on other things—because there is this wonderful extension of Sard's Theorem, due to Smale, in the infinite-dimensional setting. But if you argue by analogy that it is Sard's Theorem and the Implicit Function Theorem which suffice for the finite-dimensional case, then you would think that in infinite dimensions we should be able to prove the same results.

The problem is that in economic models, the assumptions needed to invoke those theorems aren't satisfied. Typically, in economic settings we aren't able to reduce the analysis to a separable Banach space, which is one of the hypotheses in the Smale–Sard theorems. That's one problem. A more basic problem is that these theorems are usually applied in a Banach space setting where the domains of the functions are open; this is not true in Bewley's model. We hope that when Steve finishes his work in complexity he'll come back to economics and think about the infinite-dimensional case.

Finally, in relation to Steve's work on complexity, consider the following "inverse problem" to the problem of existence of an equilibrium. Suppose you observe prices and the income distribution. When can we say that these observations are consistent with some specification of utility functions and production functions in the economy? Remarkably, the answer to this question can be expressed as a family of real semi-algebraic equations. At present, the only tool for analyzing the existence of solutions to this family of equations is the Tarski–Seidenberg Theorem. Despite their structure, arising out of the economic model, we know nothing about their complexity, nor do we have practical algorithms for solving them. Maybe Steve can also lend us a helping hand on the complexity issue. Thank you.

Bob Williams: Thank you. The next expert is Oscar Lanford, who will talk to us about physics.

Oscar Lanford: Thank you. I'm both deeply honored and really pleased to be invited to participate in this meeting honoring Steve on his 60th birthday. That said, I feel quite uncomfortable up here. It's difficult to find something useful to say after this very successful meeting.

The general theme that I want to address is the influence of Smale's work on dynamical systems on our fundamental understanding of long-time behavior of dissipative dynamical systems in nature. Now, I should first make clear that my perspective on this is that of an outsider. It's true that I was around Berkeley in the late 1960s. I participated in the seminar out of which the 1967 paper grew. Rufus Bowen and I even wrote a paper answering a question that came up in that seminar. But, fundamentally, I didn't understand the subject. I was really working on other things—statistical mechanics, in those days—and I only carne into the dynamical systems business in about 1975 through an interest in the Lorenz system. So, I came into this enterprise after the very successful period of development of the theory of uniformly hyperbolic sets that went on here in the 1960s, and from that perspective what leaps to my mind is the difficulty in making the connection between what was done in that mathematical development and applications to physics. So, I'm going to give a slightly contrarian talk here. I want to try to describe what I think the difficulties were.

In general, it seems to me that the interaction between mathematics and physics and other kinds of applications is difficult and complicated. I liked Nancy Kopell's phrase this morning about the fractal boundary. I also think that this is not the only instance of the difficulties. I've lived through another such interaction—between operator algebra theory and physics—and the qualitative similarities in these developments suggest to me that the history is not merely of anecdotal interest.

I think the biggest contribution of the Smale program to physics—it's already been said here—was in the way of looking at problems: the geometric viewpoint, the serious use of the notion of genericity, the fact that you could say something useful about differential equations without talking about any particular one, and then in somewhat more detail, the serious exploitation of the fundamentally important notion that many systems have nearby orbits which tend to separate exponentially; none of these ideas was new, but they were in a certain sense legitimated by the developments here. The quality of the output really meant that those ideas had to be taken very seriously by anybody wanting to work on the subjects.

Those are what I think the biggest direct contributions were. Now what about the difficulties? First of all, I think there was an important change in emphasis that had to be made in order to get to the kind of applications I have in mind. That change in emphasis was that for applications to dissipative systems in physics, attractors are very much more important than other kinds of hyperbolic sets. Steve's motivation in 1966 and 1967 was the search for a general structure theory, and for a general structure theory all kinds of hyperbolic sets are equally important. But if you want to understand turbulence, it seems clear that attractors are much more important than other things. Now, one of the little annoyances in this subject is that uniformly hyperbolic attractors simply haven't turned up in real examples. I still hope that that's a phenomenon having to do with the fact that we tend to look at low-dimensional examples, and that when we get to higher-dimensional ones, hyperbolic attractors may appear. But that's the way it is.

Then there was what I regard as a major red herring dragged across the path of things, and this was the precise notion that was given of genericity. I said that I thought that the use of genericity considerations was a real contribution to the subject. However, there was a precise definition of genericity, that a generic property was one which held on the complement of a set of first Baire category. Now, there are also measure-theoretic notions of genericity, and those are less attractive mathematically. In the first place, these things tend to be quantitative rather than qualitative. They're also intrinsically finite dimensional because when I say measure-theoretic I really mean with respect to a Lebesgue measure, therefore less in the spirit of global analysis. Nevertheless, in instances where topological and measure-theoretic genericity considerations are in conflict, I think we've learned, in the past 10 years, that the measure theoretic ones are probably a better guide to what you're likely to see. So that was something that needed to be sorted out.

Then there was the question of stability. Again, a big emphasis in the investigation of uniformly hyperbolic sets was the study of structural stability, and we've had to learn, over the years, to live with very unstable systems. This, to my mind, started with thinking about the Lorenz system, where the fine details of the topology change, so to say, continuously as the parameters are changed. Nevertheless, in the Lorenz system there's still a reasonable amount of stability, such that the average behavior of the system seems to change nicely as the system is changed. Rather more dramatic things happen in things like the quadratic family and, we believe also, the Hénon mapping. To take the example of the quadratic family, what we believe is that there is an open dense set of parameters for which there is an attracting cycle pulling in almost all orbits (in the measure-theoretic sense), but that, nevertheless, there is another set of parameter values with positive measure for which typical orbits behave chaotically (and so, in particular, there is no attracting cycle). That's a very bad instance of nonstability, and it's been necessary to learn to cope with that.

The next thing I'd like to say is that one of the great successes, or one of the things which is particularly suggestive for applications, in the uniformly hyperbolic enterprise is the Ruelle–Bowen Ergodic Theorem. I also think that this is an instance of the interaction between mathematics and physics at its best. As I remember the history, and I haven't had a chance to check this carefully, Bowen started by trying to study the measure of maximal entropy, and Ruelle, because of his interest in making a theory which reflected typical behavior of systems with attractors of zero measure, focused on the role of Lebesgue measure. This led to a different maximization problem which, in the end, with hindsight, seems to be more natural mathematically than the maximum entropy problem. I think also it's important to note that the work of Benedix, Carlson, and Young for the Hénon system is one of the very important new developments which shows that things like the Ruelle–Bowen Theorem work in the context of things like the Hénon system which really do seem to turn up in examples.

Then I wanted to make two very general and, therefore, nearly empty remarks about the general question of applicability of these things. The first remark is that the potential applicability of a result, at least in the short run, seems to be badly correlated with its mathematical depth. I don't mean negatively correlated, but badly correlated. The very deep stability analysis of uniformly hyperbolic sets, which was brought recently to its very satisfactory close with the work of Mañé and Palis on the stability conjecture, seems to me to be an instance of a very important result which probably doesn't have direct applications. I contrast this to the Ruelle–Bowen Theorem and the recent developments by Benedix, Carlson, and Young which seem to me also deep, 'so important, and close to applications.

There's a second observation which has been implicit already, and that is that, to paraphrase Ruelle, the confrontation of real mathematics with real physics—and I'd substitute "applications" for "physics" except that the word "real" has a special implication in physics here—this confrontation is likely to influence the mathematics at least as much as the physics. It was true both in C^*-algebras and in dynamical systems theory that this kind of confrontation between different approaches to things has proved to be profitable for both sides.

And finally I'd like to close with mildly pessimistic remarks on the difficulty of interaction between two vigorous disciplines with different objectives and different habits of thought. For that purpose I'd like to show you two of my favorite quotes on this subject, and I've put them on a transparency. These quotes are not ones that I found myself, and I think the origin is relevant. The first one many of you have probably seen; it's the motto for David Ruelle's book on mathematical statistical mechanics. It's a quote from Franz Kafka: "Richtiges Auffassen einer Sache und Missverstehen der gleichen Sache schliessen einander nich vollstandig aus." It really shouldn't be translated, but let me try to tell you approximately what it says. It says that the correct view of something and misunderstanding of that same thing do not completely exclude each other.

The second quote is even more pessimistic, but nevertheless I like it very much. It's a quote from Peter Debye, which I learned from a book of Richard Hamming, and it says that "If a problem is clearly stated, it has no more interest to the physicist." It's very tempting to add a gloss, but I think I'll end with that.

Phil Holmes, delivering remarks on "Dynamical Systems, Paradigm and Detail": A number of people whose work I learned about in the early 1970s, who really got me interested in moving in the direction of "mathematical engineering" are here, and I'd like to say thank you to them.

René Thom remarked in his lunch-time speech yesterday on the name "catastrophe theory;" maybe it was an unfortunate choice of a name. When I was just finishing my Ph.D. in experimental mechanics in 1973, I saw a small poster on the announcement board in the engineering department. It said that David Chillingworth was going to give a series of lectures on catastrophe theory, and I thought, "My God, that sounds exciting! I'll go and listen." So, that name had an influence on my life.

Shortly after that, David Rand and I started working on nonlinear oscillations together, and we wrote a paper about the Duffing equation which was sent to Christopher Zeeman to referee. He wrote a referee's report that was longer than the paper and much more interesting. It was very encouraging. And then I guess there was a conference that Rand and other people organized at Southampton in 1976 which Nancy Kopell, John Guckenheimer, and Jerry Marsden attended. Shortly after that, I came to Berkeley and met Smale and other people. This was some time after all the action had happened, but I was fascinated by all this stuff and thought, "Surely we can begin to apply it to some real problems," whatever that means. So let me, in the next few minutes, try and distill a bit of my own experience.

I think it certainly is true that applied mathematicians and engineers, perhaps engineers more than physicists, tend to be obsessed by particular problems, by particular methods, and become almost subject to the tyranny of techniques; those of us who work in institutions with institutes of technology are particularly subject to this. So a very important thing has been the larger view, the idea of paradigms, not perhaps versus methods, but paradigms *as well as* methods; of looking at generic families, of trying to come up with general properties, of going to the big picture. I think that certain pieces are very important fall-out for applied mathematicians and applied scientists in general. One of them is that to understand the problem, it's often better to embed it in a much bigger family of problems. If you want to understand a behavior in the neighborhood of a bifurcation point, it might be better to look at a more degenerate bifurcation than the one that you're actually having your problems with: a larger family. One can use universal unfoldings, look at the symmetric problems which preserve various instances of the symmetry and so on. This idea of embedding your own particular problem in a much large context is very important. However, in terms of how to do that and how to look at generating a larger structure, I think perhaps people in dynamical systems are beginning to appreciate more the role of problems in this. After all, it was a specific problem, a specific example of the Van der pol equation, which led Steve to the creation of the horseshoe. Specific problems really do tell us interesting things. They tell us important things to look for.

Now on the issue of interdisciplinary work, and the kind of tensions between the disciplines, I like Michael Fisher's remark: "There is no interdisciplinary work without the disciplines." It's very important for most of us, I think, to do sort of normal disciplinary science most of the time. And I think it's going to be increasingly important in this interdisciplinary work that collaborative groups from different disciplines are going to play a large role. But these groups must not become politicized; these groups should not be embalmed in the amber of institutes. They should be allowed to form and unform and dissolve, and whenever you start to talk to a colleague in another department, you should not hire a secretary, really.

There's a very nice constellation here at Berkeley which isn't, as far as I know, involved in this conference. There's knot theory and DNA-topology, very interesting collaborations of this kind between, in some sense, the purest mathematicians and the most applied people. There is much more to and fro than formerly. Now, Jim Gleick would like to see this as a new science—chaos theory. But one can argue, I think, maybe not too cynically, that what is scientific isn't very new. It really started with Poincaré, and Smale's paper has been around for a long time. What's scientific isn't very new, and what's new is perhaps not very scientific. However, I think all this sort of frantic activity will lead to a useful residue. You know, there will be a lot of useful things coming out of it. But I think one has to keep one's feet on the ground if one, like me, is a transient; or at least keep one's feet in a discipline. You really have to try to be honest with the terms of some particular problem if you're trying to solve the problem. You cannot afford just to make huge, broad descriptions; you make those and *then* you try to justify them in specific problems.

From that point of view, there's one more little instance of the way in which I think dynamical systems theory has had an important influence on applied mathematics and on mechanics. (It hasn't always been a good influence.) Maybe some of the other panel members might like to address this. The notions of genericity, of universal unfoldings, the idea of a classification theorem and catastrophe theory—they had an enormous influence from the point of view of coming into a problem and saying, "Aha! I recognize these sorts of universal features." And it became clear that mathematicians could perhaps play a larger role in *modeling*, in the creation of models, than they have done in the past. We've had this image of mathematicians as kind of garbage collectors who clean up after the physicists. As Oscar Lanford remarked, a consequence of that is that by the time the cleaning up has been done, the physics is "solved" and the physicists have moved on. So, from that point of view, it's very nice to see mathematical principles used in a more intimate way in building models. However, one mustn't let the principles, the paradigms, kind of run away with everything. One must pay attention to detail. I think I've said enough.

René Thom: Well, I would think that we would be very naive if we believed that everything which is objective knowledge can be reduced to strict quantitative mathematization. It is obvious that there are a lot of domains in science which, up to now, resist mathematization. And it's not enough to invoke this magical word "complexity" to indulge, to excuse, our probable inability to dominate this part of science. I think one has to try to consider the borderline between the part of science which is *really* mathematized in the strict sense, and which reduces practically to physics and perhaps a part of chemistry. I don't think biology has much more to offer in terms of strict quantification, except perhaps some genetics, some mathematical genetics, but this is not, at least to my view, fundamental. And so I believe that, on the other side, we have the whole amount of knowledge which is described in ordinary language. Ordinary language allows us, for instance, to speak about causality, to speak about cause. And the concept of cause is not part of the dynamical systems paradigm; up to now there has been no way of introducing it inside the system of classical dynamics. nor in any kinds of mathematical modelization. So, my program, as I see it, will be to consider this borderline science in which mathematization, strictly speaking, fails. Still, we can try using the flexibility offered by quantifiable dynamics in order to, in some sense, extend the intelligibility offered by ordinary language—translating the intelligibility associated to causality into some construction of quantifiable dynamics. This is the general program as I see it. And, myself, I am trying to develop this kind of thing essentially in biology, in the so-called theoretical biology. Of course, people might argue that this is not science. Well, perhaps this is not science; perhaps it is only philosophy, perhaps it is bad philosophy. And nevertheless, I believe this approach is worth trying, and I would not be ashamed to indulge in advertising my own work, "Semiophysics: A sketch" Addison-Wesley, 1989. Those among you who know about it may at least try to understand what it does, and may come to some sound judgment on this analysis.

Christopher Zeeman: Thank you very much, René. Today we have biological applications as well as physical applications, so perhaps I should say a word or two about biology. Nancy gave a beautiful talk this morning. At the end, she said it wasn't actually connected with the actual biology itself; it was more an understanding of the equations and the roots of biology. I've had a fair amount of experience over the years talking to biologists and medical people and psychologists and neurologists. The reason why I've talked is because I started out as a pure mathematician and I was released and enabled to become an applied mathematician thanks to the great, creative efforts of Steve and his school, and René Thom, in making qualitative dynamics available. That enabled me to begin to speak to the biologists. But, of course, what they say to me is, "Do you really have to learn all that junk? Is it worth investing so many hours of our lives into learning all this mathematics if it can't be of any use to predict something?" Now, if you actually go to the papers of Steve and René, they don't actually do a thing in biology. What they do is address other mathematicians in hope that they might actually go and take the gospel to the biologists. It's very difficult. Already chaos is being used in biology. For example, I was talking to Roy Anderson the other day, and he said if you count the number of HIV viruses in the bloodstream of an infected HIV person before he actually gets AIDS, then this count will be chaotic. The reason being is that the virus increases and then the antibody system gets hold of it and decreases it. Meanwhile, another one evolves in a slightly different shape and that increases, and then the antibody repeats that and so on, up and down, until one emerges which beats the antibody system. And then you have it taking hold of the antibody system, and then you're over into AIDS, of course. So, it looks like a chaotic attractor. But we're back to the big problem: Can you predict anything? All you're saying is, this chaotic attractor is a description, an after-the-event description. They link the data and look at it, and maybe they reduce it and draw some beautiful picture, that is, a description. But they still come back and say, "Predict something." And that is very educative because then you have to be really cunning if you want to use qualitative dynamics.

I'll give you an example of a little prediction I made once: Just the observation that if a fish's brain could be modeled by differential equations, then the attractors obey histeresis. Now, I predicted that if the fish was on his nest in a protective mood, and an invader came in and he flipped into the aggressive mood and chased the invader out, until he flipped back to the protective mood again, he will obey histeresis, and that would mean two boundaries to the territory: there would be an inner boundary where he attacked and an outer boundary where he stopped attacking. I've said it to you in 30 seconds. But it was a totally new idea to biologists that the territory should have two radii. Then they went off and did the experiments and they found a fish in a lake in Toronto that did have two radii. The inner radius was 13 centimeters, and the outer radius was 18 centimeters. That led to a numerical experiment, which was then confirmed with these two numbers, and so it's a very concrete biological experiment in which you actually predict. The conceptual idea was just to have the courage to think of a huge dynamical system representing the brain. So, I think one has to be cunning with these dynamical systems, and then all you've gotten are simple predictions that you can actually persuade the biologists to do.

Another lesson that I learned was if you're going to have a piece of applied mathematics, and you've got some standard model that you want to apply it to, and you want to transfer the qualitative properties of this model onto your piece of applied mathematics, then you've got to have a diffeomorphism between them. Now, that hits right at the basis of structural stability, because structural stability is based on topological equivalence as opposed to differential equivalence, and therefore I have a fundamental criticism of structural stability based on topological equivalence because it doesn't hand you those equivalence or differentiable models. So I would struggle to develop theories in which you remained within the differentiable category that you're modeling, and then even in these rather flexible biological areas, you can still make qualitative predictions.

I think, in looking to the future, that mathematical biology will be *the* great domain of the next century. I once went to a series of conferences run by C.H. Waddington, the father of Dusa McDuff. I remember, René, these were from 1967 to 1971, and they were in Italy, for four successive years. And in the first year, Waddington selected pure mathematicians, applied mathematicians, physicists, chemists, and biologists. And the first year, each one spoke to the next one. But the fourth year, the biologists were speaking directly to the pure mathematicians, to the fury of all between them. And I think that's what's going to happen in the next century. The pure mathematicians are going to speak to the biologists, and the really good biologists know exactly what's unknown and what's unsolved. And I remember Crick saying, at one of those meetings, that the three great unsolved areas of biology are evolution, development from the egg to the embryo, and intelligence. These are the three marvelous areas which just call out for geometrical modeling and dynamical systems well into the next century, so perhaps for our children and our grandchildren, that would be one of the most exciting areas of mathematics.

Mauricio Peixoto: Well, in a more pedestrian vein I would like to add a few words, related to the area of dynamical systems, about my contacts with Steve. The gist of it I have already said at that banquet the other night. In 1960, Steve spent six months in Rio de Janeiro. During that period he talked about mathematics with very few people, and I was one of them. Besides, he was still learning many classical things about differential equations, and this added some kind of charm to our conversations. One thing that impressed me very much was his instinct for the simple and for the fundamental, his knack for distinguishing what is relevant from what is not. I remember I learned from him, at that time, the meaning of the expression "red herring," which he used frequently. In the early 1960s I was very impressed, for example, by the balance and sure-footedness Steve exhibited in handling the problem of the closing lemma, avoiding a head-on confrontation with it. Also, a fact that has not been mentioned here is Steve's enormous influence on the mathematics of Brazil. (See Fig. 1.)

To close, I think that I express the sentiment of this audience by thanking Steve on his 60th birthday for the wonderful mathematics that he has produced and which, in some way, many of us were fortunate enough to share with him.

Bob Williams: We've had a certain success. We've had seven people give 10 minute talks, and it fits in an hour.

Nancy Kopell: I would like to follow up on the remarks that Christopher made about biology, almost all of which I agree with, because I think, in fact, that that is going to be a very exciting area in the future. I was talking this morning about biology, but in fact most of what I've been doing in that 10-year span I didn't talk about. It was talking with biologists and working

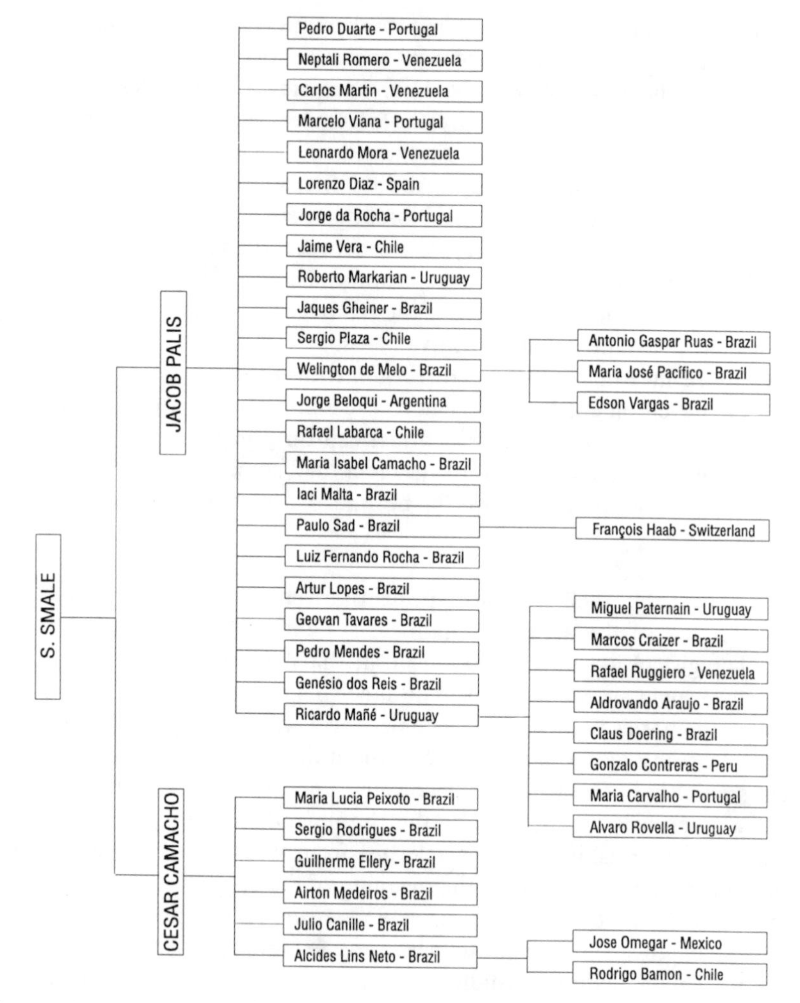

FIGURE 1. Brazilian branch of Smale's mathematical family. (Prepared by Jacob Palis.)

on mathematical models involving trying to understand particular "wet" neural networks. And in that case, dynamical systems plays a very fundamental role which I can summarize in a sentence or two. The point is that, for many large neural networks, one really does not understand the fine details of what's going on in the neural networks. One has no hope in working with networks of understanding every cell and every synapse. But one does understand a certain amount of the gross structure, and what one tries to do is reason from that, in a qualitative way, to what the consequences of what you know are, and that creates various predictions which the experimentalists can then go out and find—and they have—which also helps guide them to find more of the fine detail of what's going on in the networks. And it turns out that the particular neural networks I'm describing happen to be for production of rhythmic motion like walking and chewing and swimming, etc.; and in the gross structure there is large systems of coupled oscillators. And it turns out that if one takes this gross structure into account, one starts building a large framework within which one can reason about the experimental facts that can then be obtained and placed on that framework. This is beginning to excite a fair number of experimentalists who are joining us in this whole big project. So I see that as a very central use in biology of the whole framework of dynamics.

Eric Kostlan: Yes, I would like to speak to political work in the 1960s. I think the 1960s was perhaps the local maximum in the supply of political involvement by the public and particularly by the academic community. And it's ironic; I think you'll find that the 1990s is maybe the local maximum of the demand for political involvement by the academic community. We've talked a lot about confidence and about courage. Well, with recent world and economic events within the developed nations, as well as problems in the underdeveloped or developing nations, it's very difficult to have confidence. But at least I hope that we have courage, and I would like to see the 1990s as a period where the academic community rediscovers its global political obligations.

Mike Shub: I would like to say one word about the distinction that Oscar drew between the measure theory results and the theorem characterizing structural stability. I would say that from one perspective, the techniques and the entire structure theory developed in pursuit of the theorem on hyperbolicity and structural stability are all being carried forward today: stable manifolds, unstable manifolds, spectral decomposition, Markov partitions, and the resulting measure theory that was done by Sinai, Bowen, and Ruelle. Independent of whether or not that particular theorem turns out to be relevant, the structure is absolutely crucial to the understanding of what is coming now.

Oscar Lanford: Yeah, I entirely agree. The point I was trying to make, however, was that the interface, the match between what mathematicians find exciting to prove and what people in areas like physics find directly useable, is a complication.

Christopher Zeeman: It's true, though, isn't it still, that quantum theory is relatively incompatible with our math?

Oscar Lanford: Oh, I don't know.

Moe Hirsch: Let me say a few words about the role of mathematics in science.

Contrary to some philosophers of science, accurate description of reality is not the only role for mathematics in the natural sciences. Mathematics offers something which is not fully appreciated by people oriented towards the harder sciences, harder in all senses: namely, the possibility of achieving *insight*. From simple mathematical models, it often happens that one can obtain useful insight into a natural system situation of great inherent complexity, even though the model may be useless for predicting reality.

An interesting and perhaps the first example of this is Reverend Malthus' 1798 theory of population growth, *An Essay on the Principle of Population as it Affects the Future Improvement of Society*: While food supply grows linearly, human population will increase geometrically in the absence of war, pestilence, and famine. He gloomily concluded that there's no point in trying to improve the lot of common people because if you do, they will increase exponentially and starve to death anyway.

This theory was not predictive, as neither experiments nor observations were possible. It was neither verifiable nor falsifiable. It was all in the subjunctive mood: If you were to do this, then such-and-such would happen. The numerical examples were not realistic. An interesting question is: Why did Malthus' work have (and continue to have) such an enormous impact?

I think for two reasons. In the first place, his quantitative reasoning, although almost hopelessly simplified as a representation of reality, is nevertheless both precise and seemingly related qualitatively to reality: *something like it* seems clearly true about the real world. Second, the mathematical argument is extremely *robust*: The exact formulas for the exponential and linear growth functions are irrelevant—all you need to get his conclusion is that one grows much faster than the other. Thus, Malthus' mathematics provided a strong, unforeseen link from a biological hypothesis to a sociological conclusion. While his conclusion is surprising, the mathematics it is based on is unassailable. and his model is robust. Thus, he provided a new *insight* into a complex real world situation where accurate measurement and prediction was impossible.

This is an interesting way of using mathematics in a situation where quantitative considerations are paramount and strongly suggestive, but there is insufficient theory or observation to derive exact formulas. Instead of precise equations, a robust *class* of equations is resorted to, no one of which is accurate, but which are plausible as a class. In this way, mathematics offers insights to the natural sciences that probably cannot be obtained in any other way.

Christopher Zeeman: The reply to Malthus is just that population growth is a sigmoid curve.

Moe Hirsch: That's a different biological theory than he had. You're attacking it on biological terms.

Christopher Zeeman: Right.

Karen Uhlenbeck: This last discussion calls something to mind that I want to bring up. I just want to comment that having spent the last few years of my life conversing with theoretical physicists, I agree with Nancy and with Christopher. I think there's a point about mathematics which is actually missed by both mathematicians and people in other disciplines. I actually got this out of a conversation with Jim Glimm, to whom I was probably pointing out that physicists never admit that mathematicians do anything, in the sense that they ignored the theory of distributions because all you really need to understand is delta functions. And they didn't know gauge theory; we knew it, but they created it all. They didn't need us. But, in fact, if you observe closely, they actually do use the pedagogical models that we develop. That is, it's okay to sort of figure out groups for yourself and rediscover $SU(3)$ and $SU(4)$. But you can't afford to waste all that time with your students, so that they rethink all of group theory for themselves and so forth. And people in other disciplines don't redo the mathematics in a way that it can be taught. It's just a set of rules. And, in fact, people do take over our pedagogical methods very thankfully. They don't give us credit for it very often but, in fact, you will notice, if you watch people in other disciplines, that they will take on the logic of the presentation that mathematicians use. And I think mathematicians actually are closer to a sort of traditional model of combining research and teaching. I personally think that that's one of the real strengths of mathematics. In this conversational dialogue between my people and other disciplines in mathematics, I really felt that I wanted to bring this point forward. I really think we should be grateful for the fact that we're a little closer to students than other scientists, and I think we are valuable to the community in a way that is rarely recognized. I agree that sometimes we present things in ways that are impossible to understand, but so do other people.

Christopher Zeeman: The point is that scientists have to take 99% of their facts on trust; they have to believe other people's experiments. But we take nothing on trust.

Phil Holmes: You don't read through the proof of every theorem you're going to use each time you need one?

Christopher Zeeman: I used to when I was young.

Before we close, could I say one word? I would like to say to Hirsch, because Moe has so often been coupled in great results with Steve, and was responsible for the primary organization of this thing: Moe's a sweetie.